Lecture Notes in Physics

D0900877

Volume 837

For further volumes:
http://www.springer.com/series/5304

The Lecture Notes in Physics

The series Lecture Notes in Physics (LNP), founded in 1969, reports new developments in physics research and teaching—quickly and informally, but with a high quality and the explicit aim to summarize and communicate current knowledge in an accessible way. Books published in this series are conceived as bridging material between advanced graduate textbooks and the forefront of research and to serve three purposes:

- to be a compact and modern up-to-date source of reference on a well-defined topic
- to serve as an accessible introduction to the field to postgraduate students and nonspecialist researchers from related areas
- to be a source of advanced teaching material for specialized seminars, courses and schools

Both monographs and multi-author volumes will be considered for publication. Edited volumes should, however, consist of a very limited number of contributions only. Proceedings will not be considered for LNP.

Volumes published in LNP are disseminated both in print and in electronic formats, the electronic archive being available at springerlink.com. The series content is indexed, abstracted and referenced by many abstracting and information services, bibliographic networks, subscription agencies, library networks, and consortia.

Proposals should be sent to a member of the Editorial Board, or directly to the managing editor at Springer:

Christian Caron
Springer Heidelberg
Physics Editorial Department I
Tiergartenstrasse 17
69121 Heidelberg/Germany
christian.caron@springer.com

Miguel A. L. Marques · Neepa T. Maitra
Fernando M. S. Nogueira · Eberhard K. U. Gross
Angel Rubio
Editors

Fundamentals of Time-Dependent Density Functional Theory

 Springer

Dr. Miguel A. L. Marques
Laboratoire de Physique de la Matière
    Condensée et Nanostructures
Université Lyon 1 et CNRS
blv. du 11 novembre 1918 43
69622 Villeurbanne Cedex
France
e-mail: marques@tddft.org

Dr. Neepa T. Maitra
Department of Physics and Astronomy
Hunter College and the Graduate Center
The City University of New York
Park Avenue 695, New York
NY10065
USA
e-mail: nmaitra@hunter.cuny.edu

Prof. Dr. Fernando M. S. Nogueira
Department of de Física
Universidade de Coimbra
Rua Larga
3004-516 Coimbra
Portugal
e-mail: fnog@teor.fis.uc.pt

Prof. Dr. Eberhard K. U. Gross,
Max-Planck-Institut für
Mikrostrukturphysik
Weinberg 2
06120 Halle (Saale)
Germany
e-mail: hardy@mpi-halle.mpg.-de

Prof. Dr. Angel Rubio
Department of Materials Science
Faculty of Chemistry
University of the Basque Country
    UPV/EHU
Centro Joxe Mari Korta
Avenida de Tolosa 72
20018 Donostia-San Sebastian
Spain
e-mail: Angel.Rubio@ehu.es

ISSN 0075-8450
ISBN 978-3-642-23517-7
DOI 10.1007/978-3-642-23518-4
Springer Heidelberg Dordrecht London New York

e-ISSN 1616-6361
e-ISBN 978-3-642-23518-4

Library of Congress Control Number: 2011940844

Printed on acid-free paper

Springer is part of Springer Science+Business Media (www.springer.com)

Preface

Back in 2004 while celebrating the twentieth anniversary of the discovery of time-dependent density functional theory (TDDFT) we decided to take on the endeavor of making TDDFT really accessible to all students and researchers. Although a relatively young field, TDDFT was already beginning to rise as a method of choice in Materials Science and Quantum Chemistry, for the description of spectroscopic and non-equilibrium properties of both finite and extended systems. To this end, we set up the Benasque TDDFT series of school and workshops, a 2 week intensive meeting on TDDFT held every 2 years, nested in the beautiful Pyrenees. The Benasque TDDFT meeting is now well-established as the key training event and conference in the field. Now that we are facing the fifth edition of the event, we felt it was timely to have a pedagogical edition of our original 2006 book on TDDFT. The present volume is not a re-edition but a real new project that shares a few parts with the old book, but that has far more focus on the fundamentals and also includes new developments of the last few years of this rapidly-evolving field. Thus we give it the title of "Fundamentals of Time-Dependent Density Functional Theory".

A Little History

The first International School and Workshop on Time-Dependent Density Functional Theory was hosted by the Benasque Center for Science, Spain from August 28th to September 12th, 2004. The aim of the School was to introduce theoretical, practical, and numerical aspects of TDDFT to young graduate students, post-docs and even older scientists who were envisaging a project for which TDDFT would be the tool of choice. The School has an equal share of theoretical and practical classes. This eases the learning of the techniques and provides the students with the practical knowledge of the numerical aspects and difficulties, while at the same time introducing them to well-established open source numerical codes (e.g., OCTOPUS, http://www.tddft.org/programs/octopus). The School is followed by a

Workshop where talks are presented by leading scientists on their current research, with a schedule designed for plenty of informal discussion. This introduces the students to the forefront of TDDFT research and rounds out their training well. At the end the participants should have sufficient working knowledge to pursue their projects at their home institution. The outstanding success of the first School led to the organization of another four events, held again in Benasque, from August 27th to September 11th 2006; August 31st to September 15th 2008, January 2nd to 15th 2010, and January 4th to 17th 2012. Simultaneously, a similar sequence of events happened on the other side of the Atlantic: From June 5–10 2004, a Summer School on TDDFT was held in Santa Fe, USA. This event sparked the establishment of a series of Gordon Research Conferences on TDDFT which began in 2007.

A very tangible outcome of the two events in 2004 was the publication of a Springer Lecture Notes book, [*Time-Dependent Density-Functional Theory* edited by M. A. L. Marques, C. Ullrich, F. Nogueira, A. Rubio, K. Burke, and E.K.U. Gross, vol. 706 of *Lecture Notes in Physics* (Springer Verlag, Berlin, 2006)]. This consists of contributions from speakers at the first Benasque School and Workshop in 2004 as well as contributions from the Summer School in Santa Fe. The book was the first comprehensive review of the field. It is now on the bookshelf of many scientists and, perhaps more importantly, has been used by hundreds of students and postdocs to enter the field of TDDFT. It was also the main reference used in the later Benasque TDDFT Schools, helping over 250 students to be introduced to the theory and its applications. However, TDDFT is a fast evolving field, and much progress has been achieved in the past 5 years, which motivated us to consider compiling a revised edition. After extensive discussions among ourselves, with other scientists, and with many students of the last Benasque school, a set of recommendations emerged, both for the book as a whole, and also for the individual chapters. We concluded that we could take this opportunity not only to update the book, but also to refocus it to be more coherent, fundamental, and pedagogical (for the students), as well as to sometimes provide a different perspective on TDDFT. The result is the current volume you hold in your hands. We hope you will enjoy and learn from it as much as we have enjoyed and learned from putting it together.

A User's Guide

Time-dependent density-functional theory is a rigorous reformulation of the non-relativistic time-dependent quantum mechanics of many-body systems that places the time-dependent one-body density of a many-body interacting system at center stage. It is an extension of ground-state density functional theory, to which it is similar in philosophy, but its formulation and functionals are very different, and contain different physics. Today, the use of TDDFT is increasing in all areas where interactions are important but the direct solution of the Schrödinger equation is too demanding. Indeed, we have witnessed exponential growth of the number of articles published in this field, not unlike what occurred in ground-state density functional theory twenty years ago. High-level scientific meetings focusing on TDDFT have materialized; notably, aside from the Benasque School and Workshop, the Gordon Research Conferences in TDDFT, focused sessions and pre-meeting tutorials on TDDFT at the American Physical Society and American Chemical Society meetings, and CECAM workshops.

Despite tremendous effort focused over the years, a first-principles theoretical and practical description of the interaction of molecules with time-dependent electromagnetic fields is still a challenging problem. In fact, we are still lacking a definitive and systematic methodology, capable of bridging the different spatial and time scales that are relevant for the description of light-induced processes in nanostructures, biomolecules and extended systems with predictive power. Due to its unparalleled balance between the computational load that it requires and the accuracy that it provides, TDDFT has repeatedly shown its usefulness in the last decade when attempting this challenge. TDDFT is now a tool of choice to get quite accurate and reliable predictions for excited-state properties in solid state physics, chemistry and biophysics, both in the linear and non-linear regimes. It is routinely used for the description of photo-absorption cross section of molecules and nanostructures, electron-ion dynamics in the excited state triggered by either weak or intense laser fields, van der Waals interactions, applications to biological systems (chromophores), transport phenomena, optical spectra of solids and low-dimensional structures (such as nanotubes, polymers, surfaces, et cetera).

At the same time, however, there are important cases for which the functional approximations in use today perform poorly. One of the major on-going challenges is in the development of approximations of improved accuracy especially for phenomena important in applications of urgent interest today, such as charge-transfer processes and photo-dynamics in solar cell devices. To develop accurate and reliable approximations, a solid understanding and appreciation of the fundamental theory, as well as recent developments in theory and applications, is required. We hope this book will be useful in this regard.

In this book we focus largely on the fundamentals of the theory, but also in setting direct links with the different experimental observables and tools. We introduce all the basic concepts and build up in complexity all the way to the open problems we are facing nowadays. This book is divided into six parts. Part I presents an overview of the experimental spectroscopic techniques in use today and puts in context the need for a theoretical framework capable of describing the non-equilibrium dynamics of complex systems at different time and length scales. Part II addresses all the basic theory and fundamentals of TDDFT. More advanced concepts related to the construction of exchange-correlation functionals including dispersion forces and open quantum systems are addressed in part III. The next part addresses the realm of real-time TDDFT, namely the simulation in real-time of the combined electron-ion dynamics of real systems, from non-equilibrium excited state dynamics to molecular transport. Numerical details of the implementation of the theory discussed in the earlier sections are addressed in detail in part V, including new developments for massive parallel architectures and graphic processing units (GPUs). Part VI places TDDFT in comparison with other related theoretical frameworks developed over the years to address similar phenomena.

The Editors

Acknowledgments

Editors: The editors are very grateful to many sponsors who have helped us in the last ten years to build what is now the reference meeting for TDDFT and who have also helped us in shaping what is the book you are reading now. We are indebted to the "Centro de Ciencias de Benasque Pedro Pascual" for its continuous support, since the earliest days under its founder Prof. Pedro Pascual to present directors Prof. Jose Ignacio Latorre and Manuel Asorey, and to the invaluable staff members Tracey Paterson, Anna Gili, and David Fuentes. We also benefited greatly from the generous support of the Spanish MICINN, the European Theoretical Spectroscopy Facility (ETSF), CECAM, Psi-k, the Universities of Basque Country and Coimbra, the European Science Foundation through the SimBioMa, Intelbiomat and Ligh-Net networks, and the US National Science Foundation MCC travel award program. We most warmly thank the contributors to this volume, and the teachers and speakers at all the past four Benasque Schools and Workshops, without whom, of course, this book would not exist. Finally, we would like to thank our children, Rodrigo, Gonçalo, Samuel, Sara, Sophie, Julia, Marco, and Luca for inspiring us every day with the most natural, wonderful, and often unpredictable, demonstrations of interacting many-body systems.

Chapter 2: S. Huotari would like to thank Mikko Hakala, Arto Sakko, Tuomas Pylkkänen, and Keijo Hämäläinen for invaluable discussions. Financial support from the University of Helsinki research funds (grant 490076) is gratefully acknowledged.

Chapter 3: S. Botti would like to acknowledge financial support from EUs 7th Framework Programme under grant agreement No. 211956.

Chapter 4: N. T. Maitra gratefully acknowledges financial support from the National Science Foundation grant CHE-0647913 and from a Research Corporation Cottrell Scholar Award. The authors thank David Tempel and Lucas Wagner for their comments.

Chapter 5: L. O. Wagner and K. Burke gratefully acknowledge support of DOE grant DE-FG02-08ER46496, and thank Stephan Kümmel and Mark Casida for their input, as well as Neepa Maitra for many helpful suggestions on the manuscript.

Chapter 5 C: M. Ruggenthaler greatfully acknowledges financial support by the Erwin Schrödinger Fellowship J3016-N16 of the FWF (Austrian Science Fund).

Chapter 7: D. A. Strubbe acknowledges support from the US National Science Foundation, Grant No.DMR10-1006184 and a graduate fellowship. L. Lehtovaara and M. A. L. Marques acknowledge support from the French ANR (ANR-08-CEXC8-008-01). A. Rubio acknowledges funding by the European Research Council Advanced Grant DYNamo (ERC-2010-AdG—Proposal No. 267374) Spanish MICINN (FIS2010-21282-C02-01), ACI-promociona project (ACI2009-1036), "Grupos Consolidados UPV/EHU del Gobierno Vasco" (IT-319-07), and the European Community through e-I3 ETSF project (Contract No. 211956). S. G. Louie was supported by the Director, Office of Science, Office of Basic Energy Sciences, Materials Sciences and Engineering Division, U.S. Department of Energy under Contract No. DE-AC02-05CH11231.

Chapter 8: N. T. Maitra gratefully acknowledges financial support from the National Science Foundation grant CHE-0647913 and from a Research Corporation Cottrell Scholar Award.

Chapter 10: D. G. Tempel and A. Aspuru-Guzik gratefully acknowledge NSF PHY-0835713 for financial support. A. Aspuru-Guzik acknowledges support from the Camille and Henry Dreyfus Teacher-Scholar award, and the Sloan Research Fellowship. J. Yuen-Zhou and A. Aspuru-Guzik acknowledge support from DARPA grant FA9550-08-1-0285. We thank H. Appel, for very useful discussions concerning connections between the Master equation and Stochastic Schroedinger approaches.

Chapter 11: H. Appel would like to acknowledge useful discussions with Massimiliano Di Ventra, Matt Krems, David Tempel and Alán Aspuru-Guzik. Financial support from DOE under grant DE-FG02-05ER46204 and Lockheed Martin is gratefully acknowledged.

Chapter 6: S. Kümmel is grateful for discussions with D. Hofmann, T. Körzdörfer, and M. Mundt. Financial support by the German Science Foundation GRK 1640, the German Academic Exchange Service, and the German Israeli Foundation is gratefully acknowledged.

Chapter 23: T. Van Voorhis acknowledges financial support from an NSF CAREER grant No. CHE-0547877 and a Packard Fellowship.

Chapter 23: A. Castro acknowledges financial support by the research project FIS2009-13364-C02-01 (MICINN, Spain).

Chapter 14: B. Natarajan would like to acknowledge a scholarship from the "Fondation Nanoscience". M. E. Casida would like to acknowledge especially profitable discussions with Lorenz Cederbaum, Felipe Cordova, Imgard Frank, Todd Martinez, Mike Robb, Enrico Tapavicza, and Ivano Tavernelli. This work has been carried out in the context of the French Rhône-Alpes "Réseau thématique de recherche avancée (RTRA): Nanosciences aux limites de la nanoélectronique" and the Rhône-Alpes Associated Node of the European Theoretical Spectroscopy Facility (ETSF).

Chapter 15: J. L. Alonso, A. Castro and P. Echenique aknowledge support from the research grants E24/3 (DGA, Spain), FIS2009-13364-C02-01 (MICINN,

Spain). P. Echenique aknowledges support from the research grant 2009801064 (CSIC, Spain) and ARAID and Ibercaja grant for young researchers (Spain). A. Rubio acknowledges funding by the Spanish MEC (FIS2007-65702-C02-01), ACI-promcionaroject (ACI2009-1036), "Grupos Consolidados UPV/EHU del Gobierno Vasco" (IT-319-07), the European Research Council through the advance grant DYNamo (267374), and the European Community through projects e-I3 ETSF (Contract No. 211956) and THEMA (228539).

Chapter 17: S. Kurth acknowledges funding by the "Grupos Consolidados UPV/EHU del Gobierno Vasco" (IT-319-07) and the European Community's Seventh Framework Programme (FP7/2007-2013) under grant agreement No. 211956.

Chapter 19: S. Baroni and R. Gebauer gratefully acknowledge a fruitful collaboration with D. Rocca, O. B. Malcioğlu, Y. Saad, A. M. Saitta, and B. Walker at various stages of development of the methods exposed here.

Chapter 20: L. Lehtovaara acknowledges support from the French ANR (ANR-08-CEXC8-008-01).

Chapter 21: X. Andrade and L. Genovese acknowledge Roberto Olivares-Amaya and Andrea Marini for useful discussions. X. Andrade acknowledges support from the European Community through the e-I3 ETSF project (Contract No. 2119566) and the US National Science Foundation through SOLAR (DMR-0934480) and CDI (PHY-0835713) awards.

Chapter 25: I. V. Tokatly would like to acknowledge financial support of MICINN (FIS2010-21282-C02-01), "Grupos Consolidados UPV/EHU del Gobierno Vasco" (IT-319-07), and the European Union through e-I3 ETSF project (Contract No. 211956).

Chapter 26: K. J. H. Giesbertz, O. V. Gritsenko, and E. J. Baerends would like to acknowledge financial support of the Netherlands Foundation for Scientific Research (NWO), project no. 700.52.302, and by WCU program at Dep. of Chemistry, Pohang University of Science and Technology, funded by the Korea Science and Engineering Foundation, project no. R32-2008-000-10180-0. K. J. H. Giesbertz also acknowledges the Finnish academy for financial support.

Contents

Contributors

José L. Alonso Departamento de Física Teórica, Universidad de Zaragoza, Pedro Cerbuna 12, 50009 Zaragoza, Spain; Instituto de Biocomputación y Física de Sistemas Complejos, Universidad of Zaragoza, Mariano Esquillor s/n, 50018 Zaragoza, Spain, e-mail: alonso.buj@gmail.com

Heiko Appel Max-Planck-Gesellschaft, Fritz-Haber-Institut der, Faradayweg 4-6, 14195 Berlin-Dahlem, Germany, e-mail: appel@fhi-berlin.mpg.de

Xavier Andrade Department of Chemistry and Chemical Biology, Harvard University, 12 Oxford Street, Cambridge, MA 02138, USA; ESTF and Departamento de Física de Materiales, University of the Basque Country UPV/EHU, Av. Tolosa 72, 20018 San Sebastian, Spain, e-mail: xavier@tddft.org

Alán Aspuru-Guzik Department of Chemistry and Chemical Biology, Harvard University, 12 Oxford Street, Room M113, Cambridge, MA 02138, USA, e-mail: aspuru@chemistry.harvard.edu

Evert Jan Baerends Section Theoretical Chemistry, VU University, De Boelelaan 1083, 1081 HV, Amsterdam, The Netherlands; Pohang University of Science and Technology, San 31, Hyojadong, Namgu, Pohang, 790-784, Republic of Korea, e-mail: baerends@few.vu.nl

André D. Bandrauk Département de Chimie, Faculté de Sciences, Université de Sherbrooke, Sherbrooke, QC, J1K 2R1, Canada, e-mail: Andre.Bandrauk@USherbrooke.ca

Stefano Baroni SISSA – Scuola Internazionale Superiore di Studi Avanzati, via Bonomea 265, 34136 Trieste, Italy, e-mail: baroni@sissa.it

Silvana Botti Laboratoire des Solides Irradiés and ETSF, École Polytechnique, CNRS, CEA-DSM, 91128 Palaiseau, France; LPMNC, Université Claude Bernard Lyon I and CNRS, 69622 Villeurbanne, France, e-mail: silvana.botti@polytechnique.edu

Kieron Burke Departments of Chemistry and Physics, University of California, Irvine, CA 92697, USA, e-mail: kieron@uci.edu

Mark E. Casida Département de Chimie Molécularie, Laboratoire de Chimie Théorique, (DCM, UMR CNRS/UJF 5250), Institut de Chimie Moléculaire de Grenoble (ICMG, FR2607), Univ. Joseph Fourier (Grenoble I), 301 rue de la Chimie, BP 53, 38041 Grenoble Cedex 9, France, e-mail: Mark.Casida@UJF-Grenoble.Fr

Alberto Castro Instituto de Biocomputación y Física de Sistemas Complejos, Universidad of Zaragoza, Mariano Esquillor s/n, Edificio I+D, 50018 Zaragoza, Spain, e-mail: acastro@bifi.es

Thierry Deutsch CEA, INAC, SP2M, L_Sim, 38054 Grenoble Cedex 9, France, e-mail: thierry.deutsch@cea.fr

John F. Dobson Micro and Nano Technology Centre and School of BPS, Griffith University, Nathan, QLD, 4111 Australia, e-mail: j.dobson@griffith.edu.au

Pablo Echenique Instituto de Química Física "Rocasolano", CSIC, Serrano 119, 28006, Madrid, Spain; Instituto de Biocomputación y Física de Sistemas Complejos, Universidad of Zaragoza, Mariano Esquillor s/n, 50018 Zaragoza, Spain; Departamento de Física Teórica, Universidad de Zaragoza, Pedro Cerbuna 12, 50009 Zaragoza, Spain, e-mail: echenique.p@gmail.com

Ralph Gebauer The Abdus Salam International Centre for Theoretical Physics, Strada Costiera 11, 34151 Trieste, Italy, e-mail: rgebauer@ictp.it

Matteo Gatti Nano-Bio Spectroscopy Group and ETSF Scientific Development Centre, Departamento de Física de Materiales, Universidad del País Vasco, Centro de Física de Materiales CSIC-UPV/EHU-MPC and DIPC, Avenida Tolosa 72, 20018 San Sebastián, Spain, e-mail: matteo.gatti@ehu.es

Luigi Genovese European Synchtrotron Radiation Facility, 6 rue Horowitz, BP220, 38043 Grenoble, France; Laboratoire de Simulation Atomistique Commissariat á l'Énergie Atomique et aux Énergies Alternatives, INAC/SP2M, 17 avenue des Martyrs, 38054 Grenoble, France, e-mail: luigi.genovese@cea.fr

Klaas J. H. Giesbertz Department of Physics, Nanoscience Center, University of Jyväskylä, 40014 Jyväskylä, Finland, e-mail: klaas.giesbertz@jyu.fi

Oleg V. Gritsenko Section Theoretical Chemistry, VU University, De Boelelaan 1083, 1081 HV, Amsterdam, The Netherlands; Pohang University of Science and Technology, San 31, Hyojadong, Namgu, Pohang 790-784, Republic of Korea, e-mail: ov.gritsenko@few.vu.nl

Eberhard K. U. Gross Max-Planck-Institut für Mikrostrukturphysik, Weinberg 2, 06120 Halle (Saale), Germany, e-mail: hardy@mpi-halle.mpg.de

Simo Huotari Department of Physics, Division of Material Science, P.O. Box 64, 00014 Helsinki, Finland, e-mail: simo.huotari@helsinki.fi

Jürg Hutter Institute of Physical Chemistry, University of Zurich, Winterthurerstrasse 190, 8057 Zurich, Switzerland, e-mail: hutter@pci.uzh.ch

Stephan Kümmel Theoretical Physics IV, Department of Physics, University of Bayreuth, 95440 Bayreuth, Germany, e-mail: stephan.kuemmel@uni-bayreuth.de

Stefan Kurth Nano-Bio Spectroscopy Group, Dpto. de Física de Materiales, Universidad del País Vasco UPV/EHU, Centro Fisica de Materiales CSIC-UPV/EHU, Av. Tolosa 72, 20018 San Sebastian, Spain; IKERBASQUE Basque Foundation for Science, European Theoretical Spectroscopy Facility (ETSF), 48011 Bilbao, Spain, e-mail: stefan_kurth@ehu.es

Lauri Lehtovaara Laboratoire de Physique de la Matière Condensée et Nanostructures, Université Lyon 1 et CNRS, blv. du 11 novembre 1918 43, 69622 Villerbanne, France, e-mail: lauri.lehtovaaral@iki.fi

Franck Lépine Laboratoire de Spectrométrie Ionique et Moléculaire, University Claude Bernard Lyon 1 and CNRS, 43 boulevard du 11 novembre 1918, 69622 Villeurbanne, France, e-mail: lepine@lasim.univ-lyon1.fr

Steven G. Louie Department of Physics, University of California, 366 LeConte Hall MC 7300, Berkeley, CA 94720-7300, USA; Materials Sciences Division Lawrence Berkeley National Laboratory, 1 Cyclotron Road, Berkeley, CA 94720, USA, e-mail: sglouie@berkeley.edu

Neepa T. Maitra Department of Physics and Astronomy, Hunter College and the Graduate Center, The City University of New York, Park Avenue 695, New York, NY 10065, USA, e-mail: nmaitra@hunter.cuny.edu

Miguel A. L. Marques Laboratoire de Physique de la Matière Condensée et Nanostructures, Université Lyon 1 et CNRS, blv. du 11 novembre 1918, 43, 69622 Villeurbanne, France, e-mail: miguel.marques@tddft.org

Bhaarathi Natarajan Département de Chimie Molécularie, Laboratoire de Chimie Théorique (DCM, UMR CNRS/UJF 5250), Institut de Chimie Moléculaire de Grenoble (ICMG, FR2607), Univ. Joseph Fourier (Grenoble I), 301 rue de la Chimie, BP 53, 38041 Grenoble Cedex 9, France; CEA, INAC, SP2M, L_Sim, 38054, Grenoble Cedex 9, France, e-mail: bhaarathi.natarajan@UJF-Grenoble.FR

Dmitrij Rappoport Department of Chemistry and Chemical Biology, Harvard University, 12 Oxford Street, Cambridge, MA 02138, USA, e-mail: rappoport@chemistry.harvard.edu

Angel Rubio Nano-Bio Spectroscopy Group and ETSF Scientific Development Centre, Departamento de Física de Materiales, Universidad del País Vasco, Centro de Física de Materiales CSIC-UPV/EHU-MPC and DIPC, Avenida Tolosa 72, 20018 San Sebastián, Spain; Fritz-Haber-Institut der Max-Planck-Gesellschaft, Faradayweg 4–6, 14195 Berlin-Dahlem, Germany, e-mail: angel.rubio@ehu.es

Michael Ruggenthaler Department of Physics, Nanoscience Center, University of Jyväskylä, 40014 Jyväskylä, Finland, e-mail: michael.ruggenthaler@jyu.fi

David A. Strubbe Department of Physics, University of California, 366 LeConte Hall MC 7300, Berkeley, CA 94720-7300, USA; Materials Sciences Division Lawrence Berkeley National Laboratory, 1 Cyclotron Road, Berkeley, CA 94720, USA, e-mail: dstrubbe@berkeley.edu

David G. Tempel Department of Physics, Harvard University, 17 Oxford Street, Cambridge, MA 02138, USA, e-mail: tempel@physics.harvard.edu

Ilya V. Tokatly Departamento de Física de Materiales, Universidad del País Vasco UPV/EHU, 20018, San Sebastian, Spain; IKERBASQUE Basque Foundation for Science, 48011 Bilbao, Spain, e-mail: ilya_tokatly@ehu.es

Carsten A. Ullrich Department of Physics, University of Missouri-Columbia, Columbia, MO 65211, USA, e-mail: ullrichc@missouri.edu

Giovanni Vignale Department of Physics, University of Missouri-Columbia, Columbia, MO 65211, USA, e-mail: vignaleg@missouri.edu

Robert van Leeuwen Department of Physics, Nanoscience Center, University of Jyväskylä, 40014 Jyväskylä, Finland, e-mail: robert.vanleeuwe@jyu.fi

Troy Van Voorhis Department of Chemistry, Massachusetts Institute of Technology, Cambridge, MA 02139, USA, e-mail: tvan@mit.edu

Oleg A. Vydrov Department of Chemistry, Massachusetts Institute of Technology, Cambridge, MA 02139, USA, e-mail: vydrov@mit.edu

Lucas Wagner Department of Physics and Astronomy, University of California, Irvine, CA 92697, USA, e-mail: lwagner@uci.edu

Joel Yuen-Zhou Department of Chemistry and Chemical Biology, Harvard University, 12 Oxford Street, Room M138, Cambridge, MA 02138, USA, e-mail: joelyuen@fas.harvard.edu

Abbreviations

1D	One-dimensional
1RDM	One-body reduced density matrix
2D	Two-dimensional
3D	Three-dimensional
AA	Adiabatic approximation
AC	Alternating current
ACF	Adiabatic-connection formula
ACFD	Adiabatic-connection and fluctuation-dissipation
AIMD	Ab initio molecular dynamics
ALDA	Adiabatic local density approximation (see TDLDA)
ALL	Andersson, Langreth, and Lundqvist
ARPES	Angle-resolved photoemission spectroscopy
ATI	Above-threshold ionization
BIS	Bremsstrahlung isochromate spectroscopy (see IPES)
B3LYP	Becke's three-parameter hybrid with LYP correlation
BLYP	Becke 88 exchange with LYP correlation (a GGA)
CASSCF	Complete active space self-consistent field
c	Correlation
c.c.	Complex conjugate
CD	Circular dichroism
CIS	Configuration interaction singles
CL	Cathodoluminescence
CREI	Charge resonance enhanced ionization
CP	Car-Parrinello
CPMD	Car-Parrinello molecular dynamics
CPU	Central processing units
DC	Direct current
DDCS	Double differential cross section
DFPT	Density-functional perturbation theory
DFT	Density functional theory
DMLS	Density matrix Löwdin-Shull

DOS	Density of states
dRPA	Direct random-phase approximation
DSPA	Dressed single-pole approximation
EELS	Electron-energy loss spectroscopy
ELNES	Energy-loss near edge spectroscopy
EXAFS	Extended x-ray absorption fine structure
EXX	Exact exchange
FDT	Fluctuation dissipation theorem
GEA	Gradient expansion approximation
GGA	Generalized gradient approximation
GK	Gross and Kohn
GPU	Graphic processing units
GS	Ground state
gsBOMD	Ground-state Born-Oppenheimer MD
HAXPES	Hard x-ray photoemission spectroscopy
h.c.	Hermitian conjugate
HF	Hartree-Fock
HHG	High-harmonic generation
HOMO	Highest occupied molecular orbital
HPC	High-performance computing
HPT	Harmonic potential theorem
INS	Inelastic neutron spectroscopy
IPES	Inverse photoemission spectroscopy (see BIS)
IR	Infrared
ISB	Intersubband
ISTLS	Inhomogeneous Singwi-Tosi-Land-Sjolander
IXS	Inelastic x-ray scattering spectroscopy
LB94	Van Leeuwen and Baerends 1994
l.h.s.	Left hand side
LL	Liouville-Lanczos
LRC	Long-range correction
LYP	Lee, Yang, and Parr
KLI	Krieger, Li, and Iafrate
KS	Kohn-Sham
LDA	Local density approximation
LDOS	Local density of states
LEED	Low-energy electron diffraction
LIED	Laser-induced electron diffraction
LIERC	Laser-induced electron recollision
LIMP	Laser-induced molecular potentials
LUMO	Lowest unoccupied molecular orbital
MAE	Mean absolute error
MAPE	Mean absolute percentage error
MBPT	Many-body perturbation theory
MCSCF	Multi-configuration self-consistent field

MD	Molecular dynamics
ME	Mean error
MP2	Moeller-Plesset second-order
MPI	Message-passing interface
NACME	Nonadiabatic coupling matrix elements
NEGF	Nonequilibrium Green's function
NLSE	Nonlinear Schrödinger equation
NO	Natural orbital
occ	Occupied
OEP	Optimized effective potential
OQS	Open quantum systems
PGG	Petersilka, Gossmann, and Gross
PAW	Projector augmented-wave
PBE	Perdew, Burke, and Ernzerhof
PBE0	Hybrid functional based on PBE's GGA
PCM	Polarizable continuum model
PES	Photoemission spectroscopy
PILS	Phase including Löwdin-Shull
PT	Perturbation theory
PW	Plane-waves
PW86	Perdew and Wang 1986
QM/MM	Quantum mechanics / molecular mechanics
QOCT	Quantum optimal control theory
RDMFT	Reduced density-matrix functional theory
RG	Runge and Gross
r.h.s.	Right hand side
RIXS	Resonant inelastic x-ray scattering
ROKS	Restricted open-shell Kohn-Sham
RPA	Random-phase approximation
SAE	Single active electron
SAPT	Symmetry adapted perturbation theory
SCF	Self-consistent field
SEM	Scanning electron microscope
SF	Spin-flip
SIC	Self-interaction correction
SOSEX	Second order screened exchange
TD	Time-dependent
TDA	Tamm-Dancoff approximation
TDCDFT	Time-dependent current-density functional theory
TDLDefA	Time-dependent local deformation approximation
TDDFT	Time-dependent density functional theory
TDEHF	Time-dependent extended Hartree-Fock
TDELF	Time-dependent electron localization function
TDHF	Time-dependent Hartree-Focky
TDKLI	Time-dependent Krieger, Li, and Iafrate

TDKS	Time-dependent Kohn-Sham
TDLDA	Time-dependent local density approximation (see ALDA)
TDMCDFT	Time-dependent multicomponent DFT
TDOEP	Time-dependent optimized effective potential
TDRDMFT	Time-dependent reduced density-matrix functional theory
TDSE	Time-dependent Schrödinger equation
TEM	Transmission electron microscope
UEG	Uniform electron gas
unocc	Unoccupied
UPS	Ultraviolet photoemission spectroscopy
UV	Ultraviolet
vdW	van der Waals
VV	Vydrov and Van Voorhis
xc	Exchange-correlation
x	Exchange
XANES	X-ray absorption near-edge structure
XAS	X-ray absorption
XPS	X-ray photoemission spectroscopy
XRS	X-ray Raman spectroscopy

Notation

General

r	A point in 3-D space, (r_1, r_2, r_3)
t	An instant in time
ω	Frequency (Fourier transform of time)
x	Combined space and time coordinates $(\mathbf{r}, t)$
$f(r)$	f is a *function* of the variable $\mathbf{r}$
$f[n]$	f is a *functional* of the function n

DFT

E	Ground-state total energy
E_{xc}	Ground-state exchange-correlation energy functional
e_{xc}	Exchange-correlation energy per electron
$n(r)$	Ground-state electronic density
$\rho(r, r')$	Ground-state electronic density matrix
$\varphi i(r)$	Ground-state Kohn-Sham wave-function
ε_i	Ground-state Kohn-Sham eigenvalue
n_i	Occupation number of state i
A	Quantum-mechanical action
$n(r, t)$	TD electronic density
$j(r,t)$	TD electronic current
$v(r,t)$	TD electronic velocity
$v_{KS}(r, t)$	TD Kohn-Sham potential
$v_{ext}(r, t)$	TD external potential
$v_H(r, t)$	TD Hartree potential
$v_{xc}(r, t)$	TD exchange-correlation potential
$\varphi_i(r, t)$	TD Kohn-Sham single-particle wave-function
$\Psi(r_1, r_2, \ldots, r_N, t)$	Interacting many-body wave-function
$\Phi(r_1, r_2, \ldots, r_N, t)$	Kohn-Sham many-body Slater determinant

Operators

$\hat{H}$	Hamiltonian
$\hat{H}_{KS}$	Kohn-Sham Hamiltonian
$\hat{T}$	Kinetic energy
$\hat{V}$	Potential
$\hat{V}_{ee}$	Two-body (Coulomb) interaction
$\hat{\mathcal{T}}$	Time-ordering
$\hat{u}(t,t')$	Evolution operator
$\hat{\mathcal{L}}$	Laplace transform
$\check{A}$	$\check{A}$ is a super-operator

Many-Body and Linear Response

$\varepsilon(\boldsymbol{r},\boldsymbol{r}',\omega)$	dielectric function
$G(\boldsymbol{r},\boldsymbol{r}',\omega)$	Green's function
$\Sigma(\boldsymbol{r},\boldsymbol{r}',\omega)$	Self energy
$v_{ee}(\boldsymbol{r},\boldsymbol{r}')$	The bare Coulomb interaction $(1/|\mathbf{r}-\mathbf{r}'|)$
$W(\boldsymbol{r},\boldsymbol{r}',\omega)$	Screened Coulomb interaction
$\mathcal{X}(\boldsymbol{r},\boldsymbol{r}',\omega)$	Density-density response function
$f_{xc}(\boldsymbol{r},\boldsymbol{r}',\omega)$	Exchange-correlation kernel

Varia

v	Volume
η	Positive infinitesimal
β	Inverse temperature
μ	Chemical potential
r_s	Wigner-Seitz radius
ω_p	Plasma frequency
I	Ionization potential

Part I
Theory and Experiment:
Why We Need TDDFT

Chapter 1
Short-Pulse Physics

Franck Lépine

1.1 Introduction

The last century has seen the development of coherent light sources that have pushed our capability to probe properties of matter to a high level of sophistication. Femtosecond (fs) laser technology has paved the way to what is known as ultrafast science and led, in particular, to the mature field of femtochemistry (Hertel and Radloff 2006). Concurrently, short pulses allowed to reach unprecedented photo-excitation conditions in which the coherent absorption of a large number of photons occurs, producing highly nonlinear phenomena. With the beginning of the twenty-first century, this race has certainly not stopped and tremendous improvements on light sources have allowed crossing the sub-femtosecond barrier. Nowadays, attosecond (as) pulses are routinely produced in several laboratories over the world using table-top laser systems (Krausz and Ivanov 2009). The attosecond regime is about to be reached by large infrastructures, the free electron lasers, that are dedicated to the X-ray wavelength range with very high photon flux. From the fundamental research point of view, the motivations behind the developments of light sources are obvious:

- Increase of the light intensity in order to have access to non-linear mechanisms at any wavelength.
- Decrease of the laser pulse duration in order to investigate faster and faster dynamics.
- Improved tunability to access wider range of states in photoexcitation (valence, core excitation, etc.).

F. Lépine (✉)
Laboratoire de Spectrométrie Ionique et Moléculaire,
University of Claude Bernard Lyon 1 and CNRS,
43 boulevard du 11 novembre 1918,
69622 Villeurbanne Cedex, France
e-mail: lepine@lasim.univ-lyon1.fr

M. A. L. Marques et al. (eds.), *Fundamentals of Time-Dependent Density Functional Theory*, Lecture Notes in Physics 837, DOI: 10.1007/978-3-642-23518-4_1,
© Springer-Verlag Berlin Heidelberg 2012

It is fascinating that the current light sources have already reached a very broad range of accessible wavelength, pulse duration and photon flux. In this short introduction we wish to emphasize recent experimental results that show the new possibilities offered by these light sources. We will mainly focus our discussion on the physics of gas phase isolated species. However, some of the experiments described in this chapter have, or will have, equivalence in liquid or solid-state physics. With these new experiments often come new challenges that experimentalists can propose to the expertise of theoreticians, and TDDFT is certainly a promising tool to answer some of these questions.

1.2 Spectroscopic Tools

In order to investigate the properties of matter with light pulses, experimentalists have developed numerous spectroscopic tools that give access to meaningful observables. Historically the first spectroscopy techniques were absorption and emission spectroscopy, which consist in the measurement of the yield of photon loss or emission, respectively. Nowadays, this technique remains useful, for instance, to study nanoparticles. The optical response is of fundamental interest as it characterizes the oscillator strength of optical dipolar (or multipolar) transitions. It has also found applications in attosecond experiments for studying transient dynamics.

In contrast with "passive" spectroscopies in which the system remains intact after the photoexcitation, experiments using short pulses make use of "active" spectroscopy techniques where the system is ionized or fragmented and the emitted particles are analyzed. This strategy is naturally relevant in the case of the interaction with a short, intense pulse or with short wavelength light that usually induces ionization and/or fragmentation mechanisms. As an example, photoelectron and photoion spectroscopy are widely used in such experiments.

Due to its simple design, one of the most popular spectrometers is known as velocity map imaging spectrometer, VMIS (Eppink and Parker 1997) (see Fig. 1.1). In a VMIS experiment, the 3D momentum distribution of emitted charged particles is reconstructed from its 2D projection onto a position sensitive detector. In a typical VMIS image, the radius is proportional to the velocity of the particle while the angle represents the emission direction with respect to the laser polarization. A color scale represents the number of particles collected. The angular integration of the distribution corresponds to the kinetic energy spectrum of the emitted particles as it is measured in more traditional approaches (magnetic bottle, time-of-flight spectrometer). More sophisticated techniques using coincidence or co-variance aim at revealing the correlations between the particles and allow us to disentangle several paths from a complex reaction (see for instance the COLTRIMS experiments (Dorner et al. 2000)). At the same time, recent investigations have drawn a path to new spectroscopic tools in which, unlike traditional spectroscopy, a direct access to the shape of molecular orbitals is possible. Finally, there has been a growing interest in control strategies that permit the manipulation of molecular rotational and vibrational degrees

Fig. 1.1 Schematic of a velocity map imaging spectrometer. The velocity distribution of the particles is measured on a position sensitive detector. The radius of the image is proportional to the velocity vector of the particles (Remetter et al. 2006)

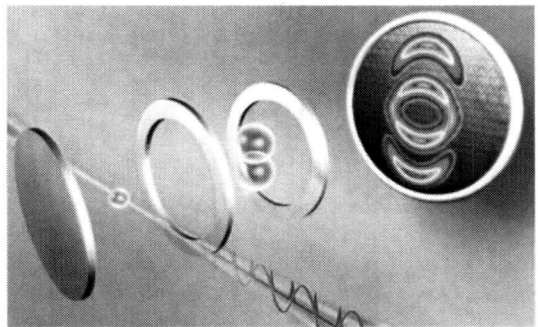

of freedom. A very useful approach known as field free molecular alignment (Rouzée et al. 2009) is based on the creation of a rotational wavepacket with a non-resonant short laser pulse via stimulated Raman transitions. The wavepacket periodically re-phases after the photoexcitation. These revivals correspond to a situation where a macroscopic number of molecules are aligned with respect to the laboratory frame, at a well-defined delay after the initial excitation pulse, determined by the rotational constant of the molecule. This offers the opportunity to perform "molecular frame" measurements, which brings detailed information on the physics of the molecule and renders the analysis of the results considerably simpler.

1.3 Physics with Intense Short Laser Pulses

When the light intensity becomes high enough, many photons can be coherently absorbed by the system up to the point where the electromagnetic force induced by the light is comparable to the force that maintains electrons and nuclei together. In general, atomic or molecular potentials can be described as "dressed" by the electric field of the light. This has a crucial impact on the ionization and fragmentation dynamics of the species (Posthumus 2004). In the following, we will mention several important electronic mechanisms that are of primary importance in intense light pulse physics. These mechanisms carry information on the electron dynamics as well as on static properties of molecules. In general, when an electron is ionized from an atom or a molecule, it remains driven by the strong light electric field. During this interaction, the removed electron is accelerated by the time-dependent electric field of incident light, leading to a variety of possible phenomena including inelastic scattering, elastic scattering or recombination. If the electron eventually reaches the continuum, a commonly occurring process is above threshold ionization (ATI) whereas its recombination leads to photon emission. The conversion of the fundamental light (typically in the near IR domain) that interacts with the atom to high energy photons is known as high harmonic generation (HHG).

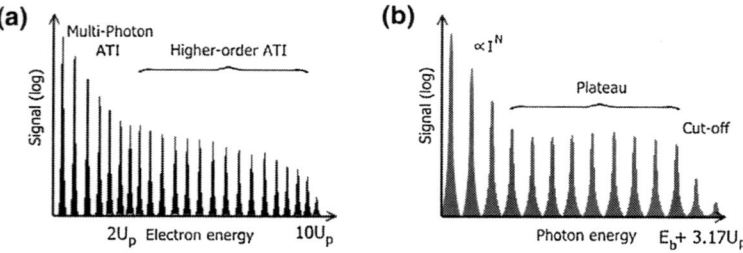

Fig. 1.2 Typical ATI (**a**) and HHG (**b**) spectra showing the cut-off and plateau regions

1.3.1 Above Threshold Ionization

In the photon picture, the ATI (Agostini et al. 1979) corresponds to a situation where an atomic system absorbs more photons than what is energetically needed to reach the ionization threshold. A signature of the ATI process is the electron kinetic energy spectrum (see Fig. 1.2), which exhibits individual peaks separated by the energy of one photon. In a quantum mechanical picture, the ionization by an oscillating electric field leads to a situation where electronic wavepackets are periodically emitted. There exists a number of possible electron paths that lead to a given final electron momentum. Therefore, the time periodicity of the ionization mechanism leads to interferences in momentum space and ATI is the result of this interference. This was beautifully illustrated in a few cycle light experiments in which either one or two light cycles allowed the ionization of an atom (Lindner et al. 2005). ATI patterns were observed in the energy spectrum when two interfering wavepackets were produced by the two light cycles. Traditionally, the ATI mechanism is characterized by the Keldysh parameter γ that measures the ratio between tunneling timescale and the timescale of the Coulomb barrier oscillations in the light electric field. A common interpretation makes a distinction between a situation where $\gamma > 1$, which corresponds to a multiphoton ionization, and a situation where $\gamma < 1$, that corresponds to a quasi-static tunneling ionization. However, this transition is not strict and a deeper interpretation was given by Ivanov et al. (2005) in terms of a non-adiabatic mechanism leading to ionization.

After ejection, the electron still interacts with the field and is accelerated. This acceleration induces a ponderomotive energy (UP). During the acceleration, the electron can re-scatter on the atomic potential and gain additional energy. In the electron kinetic energy spectrum this process appears at energies above the $2 \times$ UP cut-off and below $10 \times$ UP. In molecules, variations of the total yield of scattered electrons due to the molecular potential were observed (Cornagia 2008, 2009, 2010). In general, it has been shown that such high energy electrons carry information on the molecular structure, but the "know-how" to extract detailed information remains in its infancy.

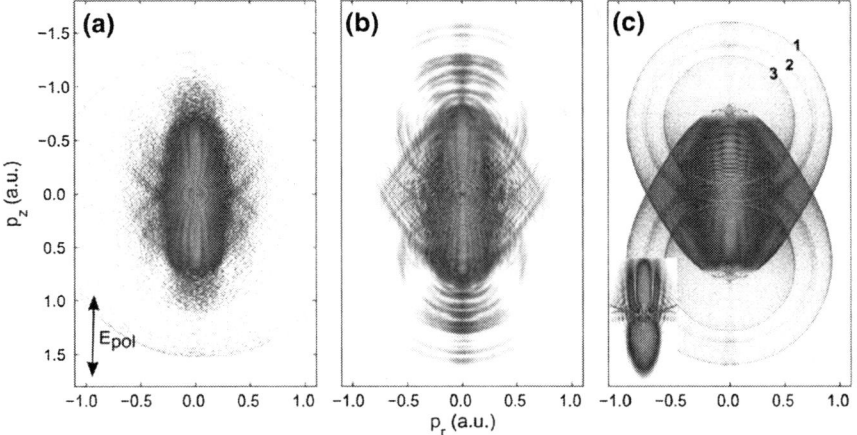

Fig. 1.3 Electron velocity distribution measured for Xe atoms interacting with a strong far IR laser pulse. The photoelectron angular distribution shows "spider-like" patterns that correspond to holographic effects (Huismans et al. 2010)

There exist other manifestations of the wavepacket interferences in strong field ionization. Whereas ATI is related to the one cycle periodicity of the laser field, holographic patterns involving half cycle dynamics have been observed in the photoelectron angular distribution (see Fig. 1.3). In a recent experiment at the far infrared intracavity free electron laser (FELICE) (Huismans et al. 2010), the electron momentum distribution of ionized Xe atoms was recorded. At high light intensity, complex electron angular distribution patterns appeared. This is understood as a holographic pattern that is created within a single half cycle of the light: the electron wavepacket is split into two parts, a first one that corresponds to direct ionization, meaning that the electron directly reaches the continuum, while the other part of the wavepacket is accelerated and scatters on the atomic potential. The two parts of the wavepacket eventually interfere leading to oscillations in the photoelectron angular distribution. In addition, scattering of the electron on the atomic potential creates off-centered rings that appear as a direct manifestation of the electron dynamics. The holographic pattern contains an "image" of the atomic potential and future theoretical work should teach us how to extract this image and how to extend this approach to molecular systems.

1.3.2 High Harmonic Generation

While in the ATI process the electron reaches the continuum, it may also recombine with the ionic core and emit high energy photons (see Fig. 1.2). The ionization, acceleration and recombination sequence is known as the 3-step model (Corkum 1993) and is our current framework to intuitively describe HHG. HHG is the key process that allowed the development of sources of short XUV pulses and the design of attosecond pulses (AP). XUV photons are typically obtained by the highly nonlinear interaction between a standard 800 nm fs laser pulse and a gas jet. The timescale at which the harmonic generation (i.e., recombination) occurs, lies in the attosecond domain. Therefore the control of the HHG process leads to the synthesis of XUV attosecond pulses and we will discuss later how these pulses are used in experiments.

On the way to control the production of APT, it was realized that XUV light emitted in HHG contains information on the sample used for the generation. Nowadays, there is a very active field of research in which information on molecular systems is extracted from HHG. One of the most fascinating aspect is the so-called molecular tomography (Itatani et al. 2004) experiment where the light properties allow for the reconstruction of molecular orbitals. The method is based on the careful analysis of the transition dipole moment of the recombination. HHG is performed with an intense IR fs pulse interacting with an aligned molecular sample. The harmonic spectrum is measured for different orientations of the molecule with respect to the polarization of the IR fs pulse from which a 3D reconstruction of molecular orbitals is obtained. Interestingly, in HHG the recombination mechanism can be considered as the reverse photoionization mechanism that is described by the same transition matrix element. However, there is a major difference since the electron + core remain coherent during the whole recombination process. As a consequence, while the photoionization from different orbitals would lead to an incoherent superposition of all the contributions, in HHG these contributions interfere (Wörner et al. 2010).

The development of the strong field tomography technique meets major challenges when increasingly complex molecules are investigated: the structure of the continuum (Higuet et al. 2011), multielectron dynamics (Shiner et al. 2011), nuclear motion and complex polyatomic molecular structures (Trallero-Herrero et al. 2009) show direct signatures in the harmonic spectrum. Therefore, there is a need for theoretical investigations that would reveal how the properties of the harmonics evolve in these cases and whether tomographic reconstruction remains feasible.

1.4 Femtosecond Science

Time-resolved experiments using short laser pulses refer to experimental protocols in which a first laser pulse prepares a non-stationary state and therefore initiates a temporal evolution of a nuclear and/or electronic wavepacket. In order to map the time-dependent mechanism of interest, the wavepacket is probed at several time delays by a second laser pulse. This scheme has been widely used since the development of ps/fs laser pulses. Nuclear dynamics in molecules have been extensively investigated leading to the broad field of femtochemistry that has shed light on transition states, isomerisation, etc.

Abundant literature can be found on time-resolved electron spectrometry (Stolow 2003). A current trend concentrates on the use of short VUV (or XUV) light pulses and very promising results have been obtained using HHG sources or free electron lasers. In contrast with multiphoton excitation induced by short intense pulses, a single photon excitation permits to perform controlled experiments where well identified states are excited and probed. An example of this trend is given by Krikunova et al. (2011) who investigated Auger decay in molecular iodine through fragmentation. Recent experiments in time-resolved photoelectron spectroscopy have shown the possibility to directly observe changes in the molecular orbitals during dissociation (Wernet et al. 2009). In these experiments a first pulse, referred to as the pump, initiates the dissociation of the molecule while a second pulse ionizes the molecule. The kinetic energy spectrum of the removed electrons is measured. When the molecule is bound, the electron spectrum characterizes the molecular orbitals. When the atoms of the molecule part, the electron spectrum shows additional peaks characteristic of the atomic orbitals. The time-dependent process shows the evolution from molecular to atomic orbitals and therefore the breaking of the chemical bond.

In a recent paper, the Br_2 molecule was studied, and single photon excitation at 395 nm (60 fs) was used to initiate the dissociation of the molecule (via the $1\Pi_u$ state). The dissociation was probed by a VUV pulse (23.5 eV, 120 fs) that ionizes the molecule, and the electron kinetic energy spectrum was measured with a standard magnetic bottle. A few 100 fs after the pump pulse, additional lines corresponding to the $3p^2$, $1d^2$, $1s^0$ atomic states arise in the kinetic energy spectrum. This experiment allows for a direct observation of chemical bond rearrangements via the time evolution of valence states. Even in this simple example, the theoretical description did not offer a fully accurate description of the results. When more complex photophysical reactions are explored in polyatomic molecules, the theoretical description will have to deal with multidimensional energy surfaces, autoionizing states, etc.

1.5 Attosecond Science

While pump-probe experiments on the fs/ps timescale are widely used in the atomic and molecular physics community, extension to the attosecond timescale is still emerging (Scrinzi et al. 2006). So far, there is no equivalent experiment where an attosecond pump is followed by an attosecond probe. However, with rapid instrumentation developments such experiment will be soon available. Until now, in all attosecond physics experiments that use actual AP, the experimental protocol is based on the utilization of an IR fs pulse that is synchronized with an attosecond pulse (or a sequence of AP:APT). The key idea relies on the synchronization of the pulses that is stable on the attosecond timescale, meaning that the time variation of the fs light electric field is fixed with respect to the attosecond pulse. Two kinds of experiments can be distinguished. The first one relies on the creation of a sequence of AP that is intrinsically created via HHG from a multicycle fs pulse. Each attosecond pulse occurs at the same periodic optimum of the IR laser field. It is experimentally possible to change the phase between the APT and the IR field. In a second design, a slightly more complex set-up makes possible to generate a single attosecond pulse

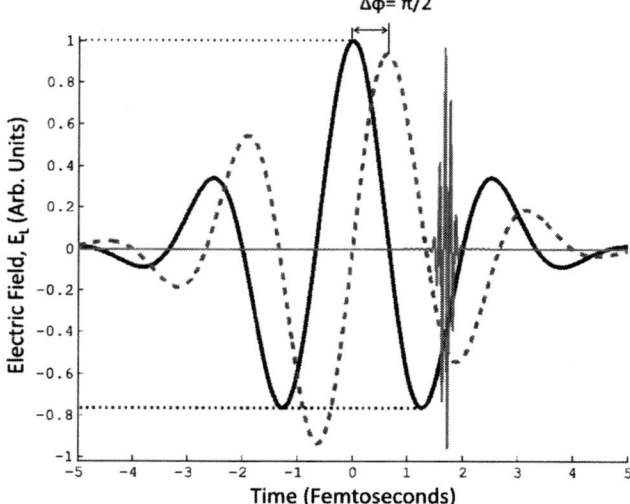

Fig. 1.4 Electric field of a few cycle pulse for two different CEP (*solid* and *dashed curve*). The pulse is synchronized with a single attosecond pulse (*thin curve*)

(SAE). It requires the use of a carrier envelope phase (CEP) stabilized pulse and the so-called polarization gating technique that allows isolating a single recombination event during the interaction with the IR pulse. Here again the attosecond pulse is synchronized on the attosecond timescale with a short few-cycle pulse (typically a 7 fs pulse) (see Fig. 1.4). Therefore the time-resolution results from the well-defined phase between the IR electric field and the attosecond pulse.

1.5.1 Electron Spectroscopy: RABBIT and Streaking

Several experimental protocols using the phase locked pulses described above have been developed in order to access attosecond dynamics. Very fruitful approaches are based on the measurement of the electron kinetic energy spectrum. In the following, we will discuss the so-called streaking and RABBIT approaches that were initially used for the characterization of the pulses themselves. In such experiments the attosecond photoionization occurs at a certain phase of the IR field. The key idea of a streaking measurement relies on the fact that the measured electron kinetic energy depends on the instant of the creation of the electronic wavepacket in the IR field. The streaking technique (Drescher et al. 2001) measures the acceleration of the electron by the electric field of the light. By changing the delay between the attosecond pulse and the short fs IR pulse, the electron acquires different possible energies. Therefore we observe variations of the maximum electron kinetic energy that maps

Fig. 1.5 RABBIT scheme.
Electron energy peaks for
single XUV photon
absorption (*long arrow*) are
shown. Additional sidebands
appear (*short arrow*) when a
"dressing" IR field is added

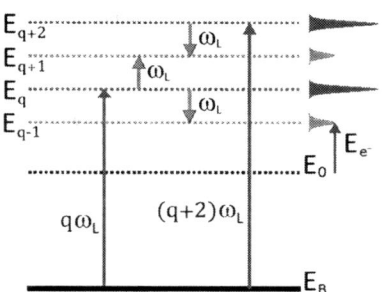

the vector potential of the IR pulse. This can be used to characterize the laser pulse and
more recently to study relaxation of core electrons in atoms, molecules and surfaces.
In these experiments the streaking is used as a probe of the time dependent processes
induced by the attosecond XUV pump pulse.

Streaking experiments are possible with an APT as well and can serve for electron wavepacket reconstruction (Remetter et al. 2006). However, it is important to
point out that APT offers the possibility to obtain both high temporal and spectral resolution, which is not directly attainable with the broad bandwidth SAE.
An APT consists of frequency combs formed by a combination of even harmonics of
the fundamental light. This is crucial for the RABBIT (reconstruction of attosecond
beating by interference of two-photon transitions) mechanism (Paul et al. 2001).
When an APT ionizes a sample, a series of peaks in the kinetic energy spectrum of
the electrons is measured. By adding an IR field, we observe additional peaks in the
spectrum (sidebands) at the energy of 1 XUV photon + 1 IR photon (see Fig. 1.5).
In fact, the oscillator strength leading to this signal is a coherent superposition of
several quantum paths where a single XUV photon is absorbed and several IR photons
are either absorbed or emitted. One has shown that the intensity of the sidebands
depends on the phase difference between the harmonics plus an additional phase
intrinsic of the ionized sample. By changing the delay between the APT and the IR
field, it is possible to recover these phases and reconstruct the train of pulses.

Phase measurements using attosecond electron spectroscopy has also led to a new
application for the determination of the time delay in photoelectron emission process
(Schultze et al. 2010). In a recent experiment, a delay of 20 as in the photoemission
from the 2s orbitals of Ne atoms was measured with respect to the emission from 2p
state. A careful study of the relation between emission time and phase has shown the
importance of the IR field, which induces an additional time delay in the measurement (Klünder et al. 2011). In general, the direct measurement of the time that an
electron takes to reach ionization continuum is certainly a very powerful observable to study electron correlation in atoms, molecules (Caillat et al. 2011), surfaces.
Again, thorough theoretical investigations are compulsory.

1.5.2 Attosecond Transient Absorption

Contrary to attosecond electron spectroscopy, in attosecond transient absorption spectroscopy, the attosecond pulse is used as a probe of the strong field mechanisms that are induced by a strong IR fs pulse. During the interaction between an atom or a molecule and a strong field, a coherent superposition of states is created. The corresponding dynamics is probed on the attosecond timescale. Theses states can be described in terms of the density matrix. In (Goulielmakis et al. 2010) the authors measured density matrix elements corresponding to the krypton ionic states produced in a strong field interaction. The population and the coherence of these states is therefore determined, giving access to the full quantum mechanical electronic motion in Kr atoms. This approach will be pushed to investigate coherence and electronic motion in molecular systems.

1.5.3 Ion Spectroscopy: Electron Localization
on the Attosecond Timescale

A pioneering experiment using an attosecond pump-probe arrangement was performed on H_2 and D_2 molecules (Sansone et al. 2010), addressing a dissociative ionization process. In this experiment a SAE was used to ionize a neutral diatomic molecule. The attosecond wavepacket created during this process evolved on the dissociative molecular energy surface. In a usual dissociation process the remaining charge has an equal probability to localize on either nucleus. In this experiment a second CEP controlled pulse is used to modify the molecular potential energy surfaces: the IR pulse creates a time-dependent superposition of bound and dissociative states that classically corresponds to an oscillatory motion of the electron from one nucleus to the other. During the dissociation, the nuclei part up to the situation where the remaining electron finally localizes on one of the two nuclei. The final localization of the charge is controlled by the delay between the two pulses. The localization of the particle is measured through the asymmetry with respect to the laser polarization of the asymptotic velocity distribution of the final ion product (see Fig. 1.6). In addition, the attosecond photoexcitation can also lead to doubly excited states that relax to dissociative states via autoionization. Therefore, this experiment addresses several crucial fundamental questions: what is the typical timescale and how do multielectronic states relax? How does non Born–Oppenheimer dynamics occur in molecules? How does the interplay between nuclear and electronic degrees of freedom determine the photophysical process? Importantly, it also shows that attosecond pulses can be used to modify the final output of a photo-induced reaction by acting directly on the electronic degrees of freedom. In this experiment, it is the charge localization that is controlled after a single or multiple electron excitation.

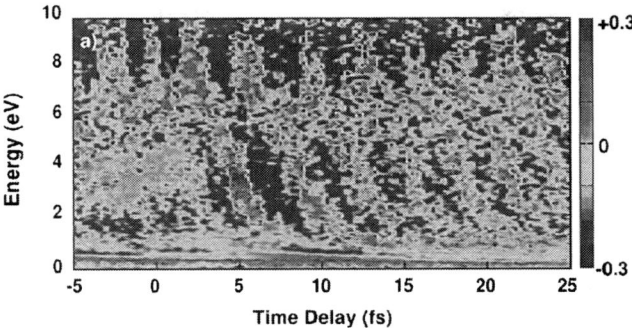

Fig. 1.6 Asymmetry of the D^+ ion signal (from -0.3 to $+0.3$) created in the attosecond pump-probe experiment. This asymmetry is monitored for all ion kinetic energies versus the delay between the pulses (Sansone et al. 2010)

1.5.4 Control of Dissociative Ionization

Another important experiment was performed on the dissociative ionization of diatomic molecules using a combination of attosecond pulse train and IR fs pulse (Kelkensberg et al. 2011). In this experiment, it was possible to control the dissociative states of the molecule using mildly strong IR light. Modulation of the fragment yield as well as the angular distribution of the ions where monitored for various ionic states as a function of the delay between the APT and the IR fs pulse. The interpretation of this result relies on a RABBIT-like excitation scheme in which electronic states are coherently coupled by XUV + IR photons. Obviously, future experiments will deal with more complex molecular systems. The excitation with a broad laser pulse of a large system with a high density of electronic states will probably make the task more challenging. However, it has been shown that such a control can already be attained for simple (H_2) or more complex multielectronic systems (O_2). In this type of experiments, future investigations will deal with the dynamics of electro-nuclear wavepackets evolving on many dissociative channels.

1.6 Conclusions

The use of short light pulses to investigate the properties of matter is driving an intense research activity. We have presented several examples of the work that is currently performed in laboratories. Short pulses can be used to concentrate a large number of photons, which induces non-linear mechanisms during the interaction with matter. Our ability to control and shape these pulses makes it possible to investigate atoms, molecules or solid state materials in extreme conditions. Current efforts are oriented towards direct measurements of orbitals or molecular structure, and strong laser fields appear as very suitable tools for these investigations. Short pulses are also crucial for time-resolved experiments, the use of short wavelengths offering new opportunities in that respect. Probably one of the most fascinating developments concerns AP that have pushed our understanding of photoinduced mechanisms to a very high level of accuracy. It is expected that these tools will be further developed to study complex systems. The most recent results obtained on molecules are also benchmark cases where thorough theoretical descriptions are needed. The TDDFT formalism has certainly an essential role to play in the development of this new area.

Chapter 2
Spectroscopy in the Frequency Domain

Simo Huotari

2.1 Introduction

In the same way as we can not visit a distant extrasolar planet to study its properties, the atomic world lies beyond our direct reach. The human being is only capable of handling objects of a certain size—both the very large and the very small are outside of our immediate reach. In these cases we can only get information by sending probes and receiving messengers. Since the atomic world is composed of elementary particles, the natural probes and messengers are precisely elementary particles—such as photons, electrons, neutrons and atoms. In this context, by a "spectrum" we refer to an object's response to a probe as a function of probe or messenger energy (or energy loss). The experimental tool for the measurement of spectra is spectroscopy. It just happens that each one of us is equipped with a pair of eyes that are excellent visible-light spectrometers with a high energy resolution, extremely good quantum efficiency, and a large dynamic range. To achieve similar characteristics with man-made instruments turns out to be challenging. However, there are several reasons to attempt this task, from the need for extending the energy range outside the narrow region of visible light, to the convenience of using other particles besides photons. This chapter describes a few experimental spectroscopic tools that may be encountered by the user of TDDFT.

Spectroscopy is not just a tool to study the structure of atomic levels. It aims at answering the question *why* a given sample behaves as it does, e.g., why it has a certain color, or what is the driving mechanism behind possible phase transitions. Unfortunately there is no universal spectroscopy for everything. Each probing technique has its own domain of application and its own unique characteristics, such as

S. Huotari (✉)
Division of Materials Science,
Department of Physics,
P.O. Box 64, 00014 Helsinki, Finland
e-mail: simo.huotari@helsinki.fi

M. A. L. Marques et al. (eds.), *Fundamentals of Time-Dependent Density Functional Theory*, Lecture Notes in Physics 837, DOI: 10.1007/978-3-642-23518-4_2, © Springer-Verlag Berlin Heidelberg 2012

sensitivity either to the bulk or the surface, element specificity, and resolving power in energy, momentum, time and space.

2.2 Probe–Electron Interaction

In a spectroscopy experiment one sends a probe, which then interacts with the system. Mathematically this can be described by adding to the Hamiltonian the interaction term $\hat{H}_{int}$. This causes a transition from the initial state $|i\rangle$ to a final state $|f\rangle$, possibly via intermediate states $|n\rangle$, where the states represent the total wave functions which include the electron states (Ψ_i, Ψ_n and Ψ_f,) and those of the probe. The transition probability w is given by Fermi's golden rule which, up to second order, reads

$$w = 2\pi \left| \langle \Psi_f | \hat{H}_{int} | \Psi_i \rangle + \sum_n \frac{\langle \Psi_f | \hat{H}_{int} | \Psi_n \rangle \langle \Psi_n | \hat{H}_{int} | \Psi_i \rangle}{E_i - E_n} \right|^2 \delta(E_f - E_i - \omega). \quad (2.1)$$

This is the starting point for practically all spectroscopies. The differences between the various experimental techniques arise from the chosen $\hat{H}_{int}$ and the set of possible Ψ_i and Ψ_f. For many applications, it is desirable that the probe wave function can be separated from the one of the target system and that the interaction $\hat{H}_{int}$ is weak, so that we are in the range of validity of the Born approximation.

Scattering experiments are usually alternatively quantified by a double differential cross section (DDCS), which gives the probability of scattering of a particle with an initial energy E_1 into the solid angle element $[\Omega, \Omega + d\Omega]$ and into the range of energies $[E_2, E_2 + dE_2]$.

2.2.1 Photon Probe

Imagine that a photon with energy ω_1, wave vector k_1, and a polarization state ϵ_1, described by the vector potential operator A, interacts with an electron with a momentum operator k and an energy E_i. In the absence of an external electromagnetic field, and in the non-relativistic case, the interaction is described by the Hamiltonian (Blume 1985)

$$\hat{H}_{int} = \hat{H}_{int}^{(1)} + \hat{H}_{int}^{(2)} + \hat{H}_{int}^{(3)} + \hat{H}_{int}^{(4)} \quad (2.2)$$

with the following terms:

$$\hat{H}_{int}^{(1)} = \frac{e^2}{2mc^2} A^2 \quad (2.3a)$$

$$\hat{H}_{int}^{(2)} = -\frac{e}{mc} A \cdot k \quad (2.3b)$$

$$\hat{H}_{\text{int}}^{(3)} = -\frac{e}{mc}\boldsymbol{\sigma} \cdot [\nabla \times \boldsymbol{A}] \tag{2.3c}$$

$$\hat{H}_{\text{int}}^{(4)} = -\frac{e}{2m^2c^2}\frac{e^2}{c^2}\boldsymbol{\sigma} \cdot [\dot{\boldsymbol{A}} \times \boldsymbol{A}], \tag{2.3d}$$

where e and m are respectively the modulus of the charge and the mass of the electron, c is the velocity of light in vacuum, and $\boldsymbol{\sigma}$ is the vector of Pauli matrices. The vector potential $\boldsymbol{A}$ is linear in the creation and annihilation operators. This means that within first order perturbation theory, $\hat{H}_{\text{int}}^{(1)}$ and $\hat{H}_{\text{int}}^{(4)}$ result in scattering (two photons are involved; one in- and one out-going). On the other hand, $\hat{H}_{\text{int}}^{(2)}$ and $\hat{H}_{\text{int}}^{(3)}$ describe absorption and emission (photon number changes by one). $\hat{H}_{\text{int}}^{(2)}$ and $\hat{H}_{\text{int}}^{(3)}$ gives also rise to scattering in second order with a resonance denominator. $\hat{H}_{\text{int}}^{(3)}$ and $\hat{H}_{\text{int}}^{(4)}$ involve the spin of the electron and thus give rise to, e.g., magnetic dichroism. For most of the purposes of the discussion below these two terms will be neglected unless otherwise stated.

Inelastic scattering of a photon (or any other elementary particle) with any energy can be used to study excitations much lower in energy than the probing particle's initial energy. In the inelastic scattering process, the probe gives away an amount of energy $\omega = \omega_1 - \omega_2$ (using here the photon formalism) and momentum $\boldsymbol{q} = \boldsymbol{k}_1 - \boldsymbol{k}_2$ to the sample; $\boldsymbol{k}_2$ and ω_2 are the wave vector and the energy of the photon after the scattering process, respectively. Note that while the maximum energy transfer is ω_1, the maximum momentum transfer magnitude $q_{\max} = 2k_1$. Most commonly these inelastic scattering spectroscopies are performed with electrons (García de Abajo 2010), visible light (Raman scattering) (Devereaux and Hackl 2007), X-rays (Schülke 2007), and neutrons (Hippert et al. 2006).

The DDCS for photon scattering from terms $\hat{H}_{\text{int}}^{(1)}$ and $\hat{H}_{\text{int}}^{(2)}$, in first and second order, is

$$\frac{\mathrm{d}^2\sigma}{\mathrm{d}\Omega\mathrm{d}\omega_2} = r_0^2 \left(\frac{\omega_2}{\omega_1}\right)\left|\langle\Psi_f|e^{i\boldsymbol{q}\cdot\boldsymbol{r}}|\Psi_i\rangle(\boldsymbol{\epsilon}_1 \cdot \boldsymbol{\epsilon}_2) + \frac{1}{m}\sum_n\right.$$
$$\times \left[\frac{\langle\Psi_f|\boldsymbol{\epsilon}_2 \cdot \boldsymbol{k}e^{-i\boldsymbol{k}_2\cdot\boldsymbol{r}}|\Psi_n\rangle\langle\Psi_n|\boldsymbol{\epsilon}_1 \cdot \boldsymbol{k}e^{i\boldsymbol{k}_1\cdot\boldsymbol{r}}|\Psi_i\rangle}{E_i - E_n + \omega_1 + i\Gamma_n}\right.$$
$$\left.\left.+ \frac{\langle\Psi_f|\boldsymbol{\epsilon}_1 \cdot \boldsymbol{k}e^{i\boldsymbol{k}_1\cdot\boldsymbol{r}}|\Psi_n\rangle\langle\Psi_n|\boldsymbol{\epsilon}_2 \cdot \boldsymbol{k}e^{-i\boldsymbol{k}_2\cdot\boldsymbol{r}}|\Psi_i\rangle}{E_i - E_n - \omega_2}\right]\right|^2 \delta(E_i - E_f + \omega). \tag{2.4}$$

This is known as the Kramers–Heisenberg formula. The second term inside the brackets depends strongly on ω_1, and it only contributes significantly in the resonance condition $\omega_1 \approx E_n - E_i$. Indeed, it gives rise to resonant inelastic scattering. The first term describes non-resonant inelastic scattering. It is often encountered in the form

$$\frac{\mathrm{d}^2\sigma}{\mathrm{d}\Omega\mathrm{d}\omega_2} = r_0^2 \left(\frac{\omega_2}{\omega_1}\right)(\boldsymbol{\epsilon}_1 \cdot \boldsymbol{\epsilon}_2)^2 S(\boldsymbol{q}, \omega) \tag{2.5a}$$

$$S(\boldsymbol{q}, \omega) = -\frac{q^2}{4\pi^2 e^2} \mathfrak{Im}\left[\epsilon^{-1}(\boldsymbol{q}, \omega)\right], \tag{2.5b}$$

defining the dynamic structure factor $S(\boldsymbol{q}, \omega)$ which is in turn related to the imaginary part of the inverse of the dielectric function $\epsilon(\boldsymbol{q}, \omega)$.

2.2.2 Electron–Electron Scattering

When an electron scatters from another electron, it is crucial that the probe can be distinguished from the target. Thus, high-energy electrons are used as probes, with wave functions that are nearly plane waves both before ($e^{i\boldsymbol{k}_1 \cdot \boldsymbol{r}}$) and after ($e^{i\boldsymbol{k}_2 \cdot \boldsymbol{r}}$) the scattering event. The scattering is mediated by the Coulomb interaction $v_{ee} = e^2/|\boldsymbol{r} - \boldsymbol{r}_i|$, where $\boldsymbol{r}$ and $\boldsymbol{r}_i$ are the positions of the probe and target electrons, respectively. The transition probability is (Platzman and Wolff 1973) (still denoting the energy transfer by ω)

$$w = 2\pi \left(\frac{4\pi e^2}{q^2}\right)^2 \left|\langle \Psi_f | e^{i\boldsymbol{q} \cdot \boldsymbol{r}} | \Psi_i \rangle\right|^2 \delta(E_i - E_f + \omega) \tag{2.6}$$

and in DDCS form

$$\frac{d^2\sigma}{d\Omega d\omega} = -\frac{1}{(\pi e q)^2} \mathfrak{Im}\left[\epsilon^{-1}(\boldsymbol{q}, \omega)\right]. \tag{2.7}$$

Note the similarity between (2.5a, b) and (2.7). The main difference between electron and non-resonant photon scattering is in the kinematic prefactor, which for electrons is relatively large and scales as q^{-2}, whereas for photons it is smaller and scales as q^2.

2.2.3 Finite Momentum Transfers

In the case of X-ray and electron scattering, an important insight concerning the transition operator encountered above is given by its expansion in a Taylor series,

$$e^{i\boldsymbol{q} \cdot \boldsymbol{r}} = 1 + i\boldsymbol{q} \cdot \boldsymbol{r} + \frac{1}{2}(i\boldsymbol{q} \cdot \boldsymbol{r})^2 + \ldots \tag{2.8}$$

As the unity operator does not induce transitions, the first important term is the dipole operator $\boldsymbol{q} \cdot \boldsymbol{r}$. For optical photons, for instance, it is the only prominent term since the corresponding photon momentum q is very small and the higher order terms become negligible. Also the photon absorption operator $\hat{H}_{\text{int}}^{(2)}$, even for X-rays, has the form of a dipole operator. Note, however, that electron and X-ray scattering do not have to respect the dipole and optical limits, and q in (2.8) may be very large. For useful applications using finite q for valence and non-dipole inner-shell excitations, (see, e.g., Sternemann et al. 2005; Weissker et al. 2006; Balasubramanian et al. 2007, 2001; Huotari et al. 2010a; Sakko et al. 2010).

2.3 Properties to Study

2.3.1 Response Functions

Often there are several methods to study the same property of a given system, or quantities measured by different techniques can be otherwise related to each other. Below we consider certain well understood material-specific properties that can be probed with more than one complementary technique.

The observable of many spectroscopies can be reduced to the dielectric function. It describes the response of a dielectric material to an alternating electric field, as explained in detail in Chap. 3. We encountered this quantity already in (2.5b) and (2.7),

$$\epsilon(\boldsymbol{q}, \omega) = \epsilon_1(\boldsymbol{q}, \omega) + \mathrm{i}\,\epsilon_2(\boldsymbol{q}, \omega). \tag{2.9}$$

The real part $\epsilon_1(\boldsymbol{q}, \omega)$ gives the polarization induced by the field, and the imaginary part $\epsilon_2(\boldsymbol{q}, \omega)$ describes absorption. Other optical functions (sometimes heretically called optical constants) can be deduced from the dielectric function. These include the complex refractive index $v + \mathrm{i}\kappa$, optical absorption coefficient α, and reflectance R. They have the relations (dependence on $\boldsymbol{q}$ and ω dropped for simplicity)

$$\epsilon_1 = v^2 - \kappa^2 \quad \epsilon_2 = 2v\kappa \quad \alpha = 4\pi\kappa\omega/c \tag{2.10a}$$

$$v = \left[\left(\sqrt{\epsilon_1^2 + \epsilon_2^2} + \epsilon_1\right)/2\right]^{1/2} \quad R = \frac{(v-1)^2 + \kappa^2}{(v+1)^2 + \kappa^2} \tag{2.10b}$$

For instance, the absorption coefficient α is used when considering the transmittance of a sample, which is described by the Beer–Lambert law,

$$I_1 = I_0 \mathrm{e}^{-\alpha d}, \tag{2.11}$$

telling us that if the intensity of light is initially I_0, it has decreased to a value I_1 after passing through a sample that has a thickness d. Inelastic scattering techniques, in turn, measure the loss function, which we in fact already quietly introduced in (2.5b) and (2.7),

$$L(\boldsymbol{q}, \omega) = -\Im\left[\frac{1}{\epsilon(\boldsymbol{q}, \omega)}\right] = -\frac{\epsilon_2(\boldsymbol{q}, \omega)}{[\epsilon_1(\boldsymbol{q}, \omega)]^2 + [\epsilon_2(\boldsymbol{q}, \omega)]^2}. \tag{2.12}$$

The real and imaginary parts of the dielectric function can be retrieved from loss-function measurements using the Kramers–Kronig relations,

$$\Re\left[\epsilon(\boldsymbol{q},\omega)^{-1}\right] = 1 + \frac{1}{\pi}\mathcal{P}\int\frac{\mathrm{d}\omega'}{\omega'-\omega}\Im\left[\epsilon(\boldsymbol{q},\omega)^{-1}\right] \tag{2.13a}$$

$$\Im\left[\epsilon(\boldsymbol{q},\omega)^{-1}\right] = -\frac{1}{\pi}\mathcal{P}\int\frac{\mathrm{d}\omega'}{\omega'-\omega}\Re\left[\epsilon(\boldsymbol{q},\omega)^{-1}-1\right], \tag{2.13b}$$

where $\mathcal{P}$ denotes the Cauchy principal value of the integral. Optical spectroscopies measure essentially the limit $\epsilon(0,\omega)$ because the optical photon momentum is practically zero. This is not a limitation for electron and X-ray spectroscopies since in those cases momenta can be very large—recall Sect. 2.2.3.

2.3.2 Typical Excitations

Excitations can be seen in the frequency-dependent linear response, and hence in the density-density response function as discussed in Chap. 4. A few examples of excitations that will be encountered in the applications of TDDFT are shown schematically as a loss function spectrum in Fig. 2.1.

- The infrared (IR) range includes vibrational excitations (phonons), and can be studied via IR absorption or high-resolution inelastic scattering techniques. For electronic excitations, the typical example in this region is free-carrier (electrons or holes) absorption. Electrical properties such as conductivity and carrier concentration can be then studied.
- Spin excitations (magnons) (100–500 meV), studied typically by neutrons (Hippert et al. 2006), due to the coupling of the magnetic moments of the neutron probe and the target atom, or resonant inelastic X-ray scattering (Ament et al. 2011).
- Valence spectra including excitations across the band gap in semiconductors (up to a few eV), excited by visible- or ultraviolet (UV) light absorption or Raman scattering, are routinely studied with TDDFT (Albrecht et al. 1998) and are treated in detail in this book.
- Collective plasmon modes [a few meV for ISB plasmons in semiconductor nanostructures (Ullrich and Vignale 2001), or several eV in bulk condensed matter (Weissker et al. 2006)]. Plasmons can also be seen in photoemission spectra as satellite peaks due to extrinsic energy losses of measured photoelectrons.
- Inner-shell-electron excitations ($\gtrsim 100$ eV) (Stöhr 1992).
- Compton recoil scattering, that yields information on the ground-state momentum density (Cooper et al. 2004).

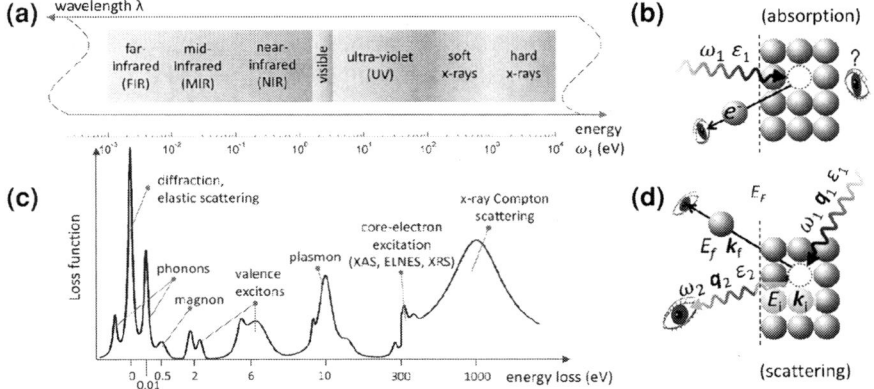

Fig. 2.1 a The spectrum of electromagnetic radiation in the regions of vibrational and electronic excitations. **b** An excitation of an electron above the Fermi level E_F in a photoabsorption process. **c** The different excitations that can be revealed by the loss function. **d** An inelastic scattering experiment, shown here for photons

2.4 Techniques

2.4.1 Ellipsometry

Ellipsometry (Fujiwara 2007; Tompkins and McGahan 1999) is a powerful tool for measuring $\epsilon(0, \omega)$ in solids, especially for surfaces and thin films. The true power of ellipsometry is that both real and imaginary parts are obtained simultaneously and independently, without the need to resort to Kramers–Kronig analysis. The readers of this book will encounter examples of ellipsometric data in many of the following chapters, e.g., the optical spectra of silicon of (Lautenschlager et al. 1987). The ellipsometric measurement analyzes the change of light polarization when it is reflected by the sample surface or transmitted through the sample. The photon energies range from IR to UV.

The name of the technique derives from the fact that the measured light is in general elliptically polarized, and it is this degree of polarization that is determined in the experiment. Namely, it measures the ratio of the reflectances of polarization components perpendicular and parallel to the sample surface, denoted R_s and R_p, respectively. This ratio is given usually as an amplitude and phase shift, denoted $\tan(\Psi)$ and Δ respectively, i.e., $R_p/R_s = \tan(\Psi)e^{i\Delta}$.

A typical example of an ellipsometry measurement is a three-phase model of an ambient—thin film—substrate ensemble. In this kind of system, ellipsometry can measure the thickness and optical properties of the thin film, assuming the substrate's optical properties are well known in advance. On the other hand, the same model is often used to study any substrate that has an oxide over-layer. The dielectric function is not directly measured by the ellipsometry experiment. Its extraction from the measured data requires the use of certain computational models and it is not possible to give the relation of the dielectric function and the measured quantity of R_p/R_s in a closed form. Thanks to the advancements in computing power and the consequent automatization and commercialization of ellipsometry instruments since the 1990s, the technique has become rather popular.

2.4.2 Photoemission Spectroscopies

Photoemission spectroscopy (PES, or XPS for X-ray photoemission spectroscopy) has a long and successful history. A good review on the subject is, e.g., Hüfner (2007). Since Hertz discovered the photoelectric effect in 1887, Einstein was awarded the Nobel Prize in 1921 for its explanation, and Kai Siegbahn in 1981 for its use as an analytical tool. A large part of our current understanding of the electronic structure of materials is due to photoemission spectroscopy. The relatively straightforward interpretation of the measured results and their direct connection to the electronic structure has provided a good basis for photoemission to become a standard research tool.

In a photoemission experiment, a photon with energy ω_1 impinges on a sample; ω_1 can belong to the UV range (produced with a discharge lamp) up to X-rays (produced by a X-ray tube or a synchrotron). The photon gets absorbed ($\hat{H}_{int}^{(2)}$) and removes a photoelectron which can be detected and its kinetic energy E_K measured by an electrostatic analyzer. From the measurement of E_K and the knowledge of the photoelectron direction, the full photoelectron momentum vector k is obtained. The original binding energy E_B of the electron before emission can be obtained from the photoelectric equation:

$$E_K = \omega_1 - \phi - |E_B|, \qquad (2.14)$$

where the work function ϕ is the energy required for electrons to escape the material surface, typically of the order of 4–5 eV for metals. Note that in the PES jargon, the values for E_B are usually taken to be positive, and $E_B = 0$ at the Fermi energy E_f.

In a solid-state sample we have to consider both the valence band and core levels. Due to higher resolving power in k and E_B, valence-band studies are usually done with UV-excitation ultra-violet photoemission spectroscopy (UPS) or very low-energy X-rays. Core levels require X-ray excitation, because to access a given E_B requires naturally $\omega_1 > E_B + \phi$.

The final state of a photoemission process involves an electron removed completely from the system, and the sample is left with a positive total charge. The hole left behind interacts strongly with the rest of the sample, and thus the final state is in fact quite complex. The description of PES can be first of all simplified by the so-called sudden approximation, in which one assumes that the final state electron does not interact anymore with the hole left behind or with the other electrons of the sample. This makes it possible to factorize the photoelectron wave function from the

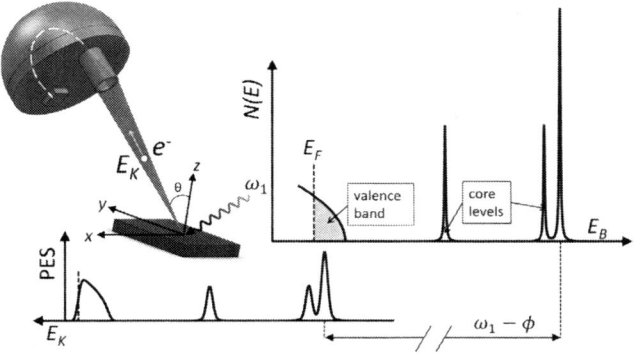

Fig. 2.2 PES experiment. The measured PES as a function of the electron kinetic energy E_K corresponds to the occupied density of states $N(E_B)$ of the electron system

wave function of the system of the other $N - 1$ electrons. Furthermore, one usually divides the photoemission process into three independent and sequential steps. This is known as the three-step model (Hüfner 2007): (i) Excitation of the photoelectron within the sample, i.e., the transition between the ground and an excited Bloch state within the solid $\Psi_i^{k_i}$ and $\Psi_f^{k_f}$. If the photon momentum is neglected, they both share the momentum $k_i = k_f = k$. This step is quantified by $M_{f,i}^{k} = \langle \Psi_f^{k} | \hat{H}_{int}^{(2)} | \Psi_i^{k} \rangle$. (ii) Electron propagation to the surface. (iii) Escape from the surface to the vacuum. The total PES intensity is then given by the product of probabilities for these individual processes. All information on the electronic structure of the ground state is contained in the step (i). Step (ii) depends on the mean free path of the electron in the solid. The probability for step (iii) depends on E_K and Ψ. The PES can be reduced to (Hüfner 2007; Damascelli et al. 2003)

$$\text{PES} \propto \sum_{f,i} \left| M_{f,i}^{k} \right|^2 A(k, E) \delta \left(E_K + E_f^{N-1} - E_i^{N} - \omega_1 \right), \qquad (2.15)$$

where $A(k, E) = \sum_m |\langle \Psi_m^{N-1} | \hat{c}_k | \Psi_i^{N} \rangle|^2$ and $\hat{c}_k$ is the destruction operator of an electron of momentum k. The function A is called the spectral function and describes the PES spectrum, since usually one considers the matrix elements $M_{f,i}^{k}$ to be constant throughout the PES measurement. In the non-interacting-particle picture $A(k, E)$ reflects the occupied density of electron states by being a series of peaks located at the single-particle orbitals with energies E_B, schematically shown in Fig. 2.2. The calculation of photoemission spectra is discussed more in detail in Chap. 3.

Angle-resolved PES (ARPES) (Damascelli et al. 2003) could as well be called band-resolved PES. With a small kinetic energy of photoexcited valence electrons it is possible to achieve a very high resolution in both their binding energy and initial momentum, thus mapping effectively the occupied band structure $E(k)$. ARPES is a very powerful and a direct tool for studying the electronic structure of the surface of solids. Since the photon momentum can be neglected in the UV-range typically used in ARPES, the electron momentum is conserved for the in-plane component $k_\parallel = \sqrt{2mE_K} \sin\theta$.

Hard-X-ray PES (HAXPES) (Horiba et al. 2004; Panaccione et al. 2006) differs from traditional UPS and XPS by using hard X-rays ($\omega_1 \gtrsim 5\,\text{keV}$). PES is usually highly surface-sensitive and as such a great tool for studying the electronic structure of the surfaces of solids. For gaining access to information on the bulk electronic states, one can utilize the fact that the electron mean free path increases with increasing kinetic energy when working in the X-ray region. Probing depth can be increased to 15–20 nm with $\omega_1 \approx 6\,\text{keV}$, compared to 0.5–1 nm in UPS. The disadvantage is a lower cross section and the requirement of a synchrotron laboratory as a light source.

Inverse photoemission (IPES) probes instead the unoccupied states. If an electron with an initial kinetic energy E_K impinges on the sample and fills an unoccupied state with an energy E_{unocc}, a photon with an energy $E_K - E_{\text{unocc}}$ may be emitted and detected. From the measurement of the IPES intensity as a function of the photon energy, the density of unoccupied states can be deduced. The technique can also be called Bremsstrahlung isochromate spectroscopy (BIS) especially if photons are detected with a fixed energy $E_K - E_{\text{unocc}}$ and the energy of incident electrons E_K is varied. A typical example of a combination of PES and BIS is that of the determination of the band gap of NiO (Sawatzky and Allen 1984).

2.4.3 Photon Absorption

Absorption spectroscopy measures how well a sample absorbs or transmits electromagnetic radiation at a given range of photon frequencies. Photoabsorption is a result of the same $\hat{H}_{\text{int}}^{(2)}$ as photoemission, but now the electron is not removed from the sample. It is merely lifted to an unoccupied state above the Fermi level. The electron stays within the vicinity of the created ion and feels the presence of the core hole; the electron-hole interaction has thus to be explicitly included in any theoretical description. When a sample absorbs a photon with energy ω_1, an excitation with that particular energy is created—the relevant energy ranges are depicted in Fig. 2.1a. For experimentalists, the hugely different energy ranges mean different practicalities and the measurement techniques vary greatly.

Circular dichroism (CD) (Berova et al. 1994) is an effect where the absorption coefficient is different for left ($-$) and right ($+$) circularly polarized light. This is typically observed in the optical response of chiral molecules, where the response to the different circular polarization depends on the handedness of the molecule. In the X-ray regime, an effect with a similar name (magnetic CD) is due to the interaction Hamiltonian $\hat{H}_{\text{int}}^{(3)}$. It is used to probe the difference in the unoccupied electron states bearing different spin, giving access to the orbital magnetic moment (Thole et al. 1992; Lovesey and Collins 1996). A CD measurement is very similar to a regular absorption measurement, and only differs in one way: the incident photon beam is circularly polarized, and the transmitted intensity is measured for two different polarizations. The CD signal is then the relative difference of the transmitted intensities for the two polarizations, $(I_+ - I_-)/(I_+ + I_-)$.

Infrared absorption probes excitations in the meV-range. Many of them are vibrational in character. IR spectroscopy is often used to identify molecules via their well-characterized vibrational excitations. They are superimposed on the spectra of low-lying electronic excitations such as magnetic excitations or electron excitations in nanostructures that have energies in the sub-eV range. The majority of IR absorption instruments are Fourier-transform spectrometers, which are able to collect spectra over a wide energy range simultaneously. Complementary probes for the same excitations in this energy range can be found within the Raman and resonant-Raman spectroscopies and inelastic neutron spectroscopy.

UV-visible absorption (UV/Vis) covers typically excitations of 3d electrons in transition metals, important for studies of strongly correlated oxides, and band gaps in semiconductors. Also many organic molecules also absorb light in the UV/Vis range. Excitations in this regime may also be studied with inelastic X-ray and electron scattering spectroscopies via the loss function.

X-ray absorption (XAS) (Stöhr 1992) is one of the most common X-ray spectroscopies and TDDFT has much to offer in this field (Brancato et al. 2008). The largest difference between X-ray and optical absorption studies is that in XAS the initial electron state is a dispersionless, narrow, and deeply bound core state. This partly facilitates the analysis of the spectra since to a rather good approximation they measure the unoccupied states directly, although always in the presence of the deep core hole and broadened by the core hole lifetime. Even more importantly, XAS is element specific due to the involvement of the core state, and can thus be used as an accurate tool even in many otherwise complex systems. XAS is a quite recent addition to the spectroscopy family since tunable X-ray sources have been recently introduced with the advent of synchrotron radiation. Depending on the sample and the range of X-ray energies, the experiments can be performed by measuring the ratio of the transmitted and incident photon intensities, or indirectly, by observing consequent processes such as secondary-particle yield (e.g., Auger processes or photoemission). The fine structure of the spectra near the edge X-ray absorption near-edge structure (XANES), yields information on the local electron density of unoccupied states at the site of the absorbing atom, in the presence of the core hole. At increasingly larger photon energies, the photoelectron has more kinetic energy and can probe also the surrounding environment by scattering from the neighboring atoms. This gives rise to extended X-ray absorption fine structure (EXAFS), which exhibits characteristic oscillations due to the interference of the outgoing and backscattered photoelectron wave functions. After expressing the oscillations as a function of the photoelectron momentum, a Fourier transformation of the signal gives information on the neighboring atoms in a form of an effective pair distribution function (Rehr and Albers 2000). Complementary probes to XAS can be found from inelastic scattering of electrons and X-rays.

2.4.4 Inelastic Scattering

Originally V. Raman noted that it was possible to measure vibrational excitations, which have energies that fall into the far-infrared regime, by observing characteristic energy losses in the spectrum of visible light. For this finding he was awarded the Nobel prize in 1930. The effect was immediately realized to be due to inelastic light scattering. For light this is possible only in the particle picture (Nobel prize of A. H. Compton 1927), and indeed inelastic scattering is a property of particles (neutrons and electrons are canonical examples). An important difference to absorption is that inelastic scattering allows investigations of a possible q-dispersion of the excitations.

Raman spectroscopy (Devereaux and Hackl 2007) is complementary to IR absorption, and is often used in the studies of vibrational and electronic excitations in the meV-energy range in molecules and solids. Its flavor called resonant-Raman spectroscopy allows to distinguish between charge-density, spin-density, and single-particle excitations. In resonant Raman experiments, the energy of the incident photon is tuned to a specific electron excitation (often to that of a band gap), and other low-energy excitations coupled to the resonance gain spectral weight considerably. This is especially useful in complex systems where certain classes of excitations can be selectively studied.

Electron-energy loss spectroscopy (EELS) and electron microscopy (García de Abajo 2010) are widely used and powerful tools to study the electronic properties, especially of thin films and surfaces with a very high spatial resolution. Due to the large electron-electron scattering cross section, the measurements are fast and high resolving power is easy to achieve in all quantities: energy, momentum transfer, and space—in particular, electrons are very easy to focus on nm-size spots on the sample. There are several different types of electron spectroscopy. The transmission electron microscope (TEM) uses high-energy electrons that are made to pass through a thin (≤ 100 nm) sample film, or a gas. Measuring the energy loss of the electrons measures essentially the loss function $L(q, \omega)$ as discussed in Sect. 2.3. Inelastic electron scattering gives rise to cathodoluminescence (CL) (Ozawa 1990) which can be measured and analyzed in the scanning electron microscope (SEM). CL was also produced in the old-fashioned cathode-ray tubes, such as television sets and computer monitors, before the flat-screen revolution. Just as PES, EELS can be used in chemical analysis by measuring the core-electron excitations (Hitchcock 2000). This flavor of EELS is sometimes called energy-loss near edge spectroscopy (ELNES), yielding similar information as XAS. EELS is a useful probe for both bulk and surface excitations which are both typically present in the measured spectra. The large scattering cross section, especially at low q values, can sometimes be a disadvantage due to the large possibility for multiple scattering. This is well known in solids (Bertoni and Verbeeck 2008; Stöger-Pollach 2008), but should also be carefully considered even in gases (Bradley et al. 2010).

Inelastic X-ray scattering spectroscopy (IXS) (Schülke 2007) can also be used to study the loss function $L(q, \omega)$. Historically, the first observation of the IXS process was the well-known Compton effect, where X-rays scattered off a sample suffer a shift in wavelength, $\Delta\lambda = \frac{h}{mc}(1 - \cos 2\theta)$; the scattering angle 2θ is the angle between the incident and scattered photon. This effect is due to recoil scattering from the electrons. The scattering electron is usually not at rest, and its ground-state momentum gives an additional Doppler shift to the recoil photon. By measuring the Compton-recoil IXS spectrum, the ground state initial momentum of the electron can be deduced (Cooper et al. 2004; Huotari et al. 2010b). The Compton spectrum can be obtained from $L(q, \omega)$ measured at large q. IXS studies of the loss function at smaller values of q have recently advanced to the level of a standard tool due to the fact that monochromatic and energy-tunable X-rays with high enough intensity can only be produced by modern synchrotron radiation facilities. The difference between IXS and EELS arises mostly from a much larger scattering probability in EELS especially at small q. This has made EELS experiments often faster, and has for a long time limited IXS to low-Z systems, where the large probing depth compensates the low cross section. However, modern high-brilliance synchrotron radiation sources have now made IXS experiments possible in all samples. IXS has the advantages of (i) being bulk-sensitive, yielding access also to samples in environments impermeable for electrons which scatter already at the surface of materials, and (ii) being able to perform measurements for an almost unlimited range of momentum transfers, only limited by $q_{\max} = 2k_1$. The ELNES counterpart for inner-shell excitations in IXS is called X-ray Raman spectroscopy (XRS)—a name which describes well the inelastic process but is not to be confused with optical Raman spectroscopy. Just as in the case of resonant-Raman spectroscopy, it is possible to perform IXS in resonant conditions in a resonant inelastic X-ray scattering (RIXS) spectroscopy experiment. In this case $\omega_1 \approx E_i - E_n$ and the second term under the square in the DDCS (2.4) dominates (Ament et al. 2011). Now the intermediate state Ψ_n corresponds to the final state in XAS—i.e., a deep core hole and an extra electron in a previously unoccupied valence state. The involvement of an intermediate state lifts off many selection rules such as the dipole rule. RIXS is element specific due to the involvement of the core hole, and thus used often in complex samples where the local electronic structure of a specific element is studied. Typical applications are strongly correlated systems, e.g., charge-transfer excitations in transition metal oxides (Schülke 2007).

Inelastic neutron spectroscopy (INS) (Hippert et al. 2006) is a widely used probe for phonons and magnetic excitations (magnons), as well as crystal field excitations within partially filled d or f electron shells. Phonons are excited in INS due to the neutron-nucleus interaction and magnetic excitations due to the magnetic-moment coupling of neutrons and atoms. The spectra are usually expressed in terms of the dynamic structure factor $S(q, \omega)$. The nuclear scattering probability is represented by the scattering length b, which is not a monotonic function of the target atomic number, and depends also on the isotope. INS is a well established tool for phonon and magnon measurements, recently also complemented by IXS (Scopigno et al. 2005) and RIXS (Ament et al. 2011).

2.4.5 Non-linear Optics

Non-linear optics is thoroughly reviewed in Chap. 18. In this case the first Born approximation does not apply, since the polarization caused by the probe is compa-rable to the field of the probe itself. This may lead to rather unexpected phenomena, such as high-harmonic generation which makes the sample to respond at multiples of the initial probe frequency (Salières et al. 1999). These techniques constitute a rapidly evolving field. Its latest additions are in the X-ray regime thanks to the advent of X-ray free electron lasers, which could very well turn out to be the light sources of the future for spectroscopists.

2.5 Summary

There is a large number of open questions in physics that spectroscopy could answer in near future. Some of the most important ones relate to many-body effects, and the lack of exact functionals to account for them. Especially strong electron-electron interaction effects for instance are manifested in low-dimensional systems—i.e., in the nanoscale. The understanding of strongly correlated systems is one of the forefront questions in the contemporary condensed matter physics and experimental results are often difficult to interpret without an ab initio theoretical counterpart. Such systems are often probed by resonant spectroscopies such as resonant inelastic X-ray scattering, where the development of rigorous ab initio theories has been especially slow. Also, the interpretation of the behavior of disordered matter such as glasses and liquids has been more difficult than those of crystalline systems due to their lack of periodicity.

Luckily spectroscopy is a large field that is developing on a fast pace. As is the case generally in science, spectroscopic experiments and theory are in a constant dialogue. Improvements in one always lead to a surge of advances in the other as a response. Sometimes experimentalists discover new phenomena with no obvious theoretical explanation, forcing the theorists to improve their techniques in the search for one. On the other hand, theoretical predictions have been made for experimentally previously unseen phenomena that have demanded a completely new level of accuracy from experiments in order to be confirmed. In the twenty first century, our understanding of physics will probably be revolutionized due to the advent of powerful new techniques and light sources allowing novel spectroscopic studies in previously inaccessible time and energy scales.

Chapter 3
The Microscopic Description of a Macroscopic Experiment

Silvana Botti and Matteo Gatti

3.1 Introduction

The interaction between electromagnetic radiation (or particles) and matter creates elementary excitations in an electronic system. On one hand, this leads to phenomena (often complicated) that can be relevant in a variety of technological fields, including electronics, energy production, chemistry and biology. On the other hand, there is a fundamental interest in perturbing a material: the perturbation can indeed reveal essential material properties that were not detectable in the ground state of the system. To visualize this idea we can imagine a bell. In its ground state we cannot know what is the sound that it will produce when it is hit by a hammer. By perturbing the system, one can hear which is the characteristic frequency of the system, i.e. measure its elementary excitations.

According to the experimental setup, one can get access to different excitation properties. For example, in an experiment when an electron is added or removed, it is possible to gain insight into elementary one-particle excitations, i.e. the quasi-particles. When the number of electrons is conserved, the experiment will instead

S. Botti (✉)
Laboratoire des Solides Irradiés and ETSF,
École Polytechnique, CNRS, CEA-DSM,
91128 Palaiseau, France
e-mail: silvana.botti@polytechnique.edu

and
LPMCN, Université Claude Bernard Lyon I and CNRS,
69622 Villeurbanne, France

M. Gatti
Nano-Bio Spectroscopy Group and ETSF Scientific Development Centre,
Departamento de Física de Materiales, Universidad del País Vasco,
Centro de Física de Materiales CSIC-UPV/EHU-MPC and DIPC,
Avenida Tolosa 72, E-20018 San Sebastián, Spain
e-mail: matteo.gatti@ehu.es

M. A. L. Marques et al. (eds.), *Fundamentals of Time-Dependent Density Functional Theory*, Lecture Notes in Physics 837, DOI: 10.1007/978-3-642-23518-4_3,
© Springer-Verlag Berlin Heidelberg 2012

provide information on neutral excitations, such as excitons (electron-hole pairs) and plasmons (coherent electron oscillations). In a similar way, thanks to the powerful combination of state-of-the-art quantum-based theories and dedicated software in continuous development, it is now possible to study electronic excitations in complex materials. Within the many-body framework, to simulate a process involving one-particle excitations one deals with quantities related to the one-particle Green's function. In the case of neutral excitations the process can be described by response functions, a particular class of two-particle Green's functions.

A review of the different theoretical approaches and experimental techniques is beyond the aim of this chapter. We will focus on two specific aspects. First, one should keep in mind that while experiments usually measure macroscopic properties, ab initio calculations yield microscopic functions that need to be processed to obtain quantitative information comparable with measured spectra. In order to bridge the gap between the microscopic and the macroscopic worlds, one needs appropriate physical models that relate microscopic and macroscopic response functions. Second, depending on the details of the physical problem under study, one has to choose which theoretical approach and which level of approximation is more suitable for calculating the microscopic response functions. Moreover, one has to be aware of the fact that the accuracy and efficiency of calculations vary necessarily with the theoretical framework employed.

3.2 Theoretical Spectroscopy

We learned from Chap. 2 that spectroscopies are ideal tools to investigate the electronic properties of extended and finite systems. What we commonly call "spectrum" is the response of a sample to a perturbation, which can be produced by an external electromagnetic field (photons) or by other particles (e.g. electrons, neutrons). This response is measured and plotted as a function of the frequency (or equivalently of the wavelength) of the incident particle.

By determining the energy (and possibly the wavevector and the spin) of the incoming and outcoming particles, it is possible to extract important information on the elementary excitations that were induced in the matter by the perturbation. Here, we restrict our interest to electronic properties, and more in particular to excitations involving valence electrons. The excitation energies will therefore be in the interval which goes from infra-red to ultraviolet radiation (up to some tens of eV), and which contains the visible portion of the electromagnetic spectrum. For such photon energies, the wavelength of the incoming radiation is always much larger than inter-atomic distances, which means that we can safely assume to be working in the long-wavelength limit. Moreover, we will consider only the first-order (linear) response, and we will not deal with strong-field interaction (e.g. intense lasers), nor with magnetic materials.

Existing spectroscopy techniques are numerous and can be classified according to different criteria. In Table 3.1. we summarize the characteristics of a selected set

Table 3.1 The spectroscopy techniques discussed in this chapter can be classified according to the probe used ('in'), the particle collected after the interaction with the sample ('out'), and the possible change in the number N of electrons in the sample

	In	Out	Number of e^-
Direct photoemission	photon	electron	$N \to N - 1$
Inverse photoemission	electron	photon	$N \to N + 1$
Photoabsorption	photon	photon	$N \to N$
Electron energy loss	electron	electron	$N \to N$

of processes. They are classified according to the probe used (electromagnetic field, electron, etc.), what is collected and measured after the interaction with the electronic system (again, electromagnetic field, electron, etc.), and whether the number of electrons in the system remains fixed, or if electrons are added or removed. The following sections are devoted to an introduction to the phenomenology of these spectroscopies, in order to specify which are the physical quantities that are measured, and that we therefore want to calculate.

In Sect. 3.3 we give examples of experiments involving one-electron excitations, i.e. in which an electron is added or removed from the system. In a photoemission (PES) process the system of interacting electrons absorbs one photon with enough energy to excite an electron (called photoelectron) above the vacuum level. The missing electron leaves a hole in the system, which generally remains in an excited state. The kinetic energy distribution of the photoelectron can be measured by the analyser, yielding the photoemission spectrum. In the simple independent-particle picture, photoemission spectra give an image of the density of the occupied electronic states of the sample. Inverse photoemission can be considered as the time-reversal process of photoemission: in this case the system absorbs an electron and a photon is emitted, whose energy distribution can be related in the independent-particle picture to the density of empty states. However, one should not forget that the sample is a many-body system. In the many-body framework, one has rather to deal with one-particle Green's functions and spectral functions, in order to extract from them physical quantities to compare with photoemission spectra.

In Sect. 3.4 we discuss two examples of processes involving neutral excitations: optical absorption and electron energy loss. In an absorption experiment the incident beam of light looses photons that are absorbed by the system. Their energy is used to excite an electron from an occupied to an empty state: an electron-hole pair is then created in the system. Instead, in electron energy-loss spectroscopy (EELS) experiments an electron undergoes an inelastic scattering with the sample. The analyser measures its energy loss and its deflection. Again, the energy lost by the incoming electron has been used to induce excitations in the system. Response functions, a particular class of two-particle Green's functions, are the key quantities to explore neutral excitations and will be presented in Sect. 3.5. From the theoretical point of view, the frequency and wavevector dependent dielectric functions $\epsilon(\boldsymbol{q}, \omega)$ and its inverse $\epsilon^{-1}(\boldsymbol{q}, \omega)$ are the most natural quantities for the description of the neutral excitations of an extended system. In a non-isotropic medium such functions must be replaced by tensors. Moreover, it should be emphasised that a electromagnetic field is a transverse field (i.e. the field is perpendicular to the direction of propagation), while impinging electrons are associated to longitudinal fields (i.e. their field is along the direction of propagation). This implies the existence of transverse (T) and longitudinal (L) dielectric functions. In Sect. 3.6 we will finally show how dielectric functions, inserted in appropriate semi-classical models, and through appropriate averaging procedures, yield the frequency-dependent spectra to be compared with experimental data. Atomic units will be used throughout the chapter.

3.3 Photoemission Spectra and Spectral Functions

At the heart of photoemission spectroscopy lays the photoelectric effect. Discovered by Hertz in 1887, its experimental study was worth a Nobel prize for Lenard and its theoretical explanation another Nobel prize for Einstein. An example of experimental valence photoemission spectrum, taken from Guzzo et al. (2011), is displayed in Fig. 3.1. In this section we will discuss how to calculate a quantity directly comparable with such a spectrum starting from first principles. For a more general description of photoemission spectroscopy we refer to Chap. 2 and to the many books and reviews available in literature [e.g. (Hüfner 2003; Schattke and Van Hove 2003; Damascelli et al. 2003; Almbladh and Hedin 1983)].

The simplest phenomenological interpretation of the photoemission process is given by the so-called three-step model (Berglund and Spicer 1964), which describes photoemission as a sequence of the actual excitation process, the transport of the photoelectron to the crystal surface, and the escape into the vacuum. A better interpretation is obtained in the one-step model (Schaich and Ashcroft 1970; Mahan 1970; Pendry 1976), where the three different steps are combined in a single coherent process, described in terms of direct optical transitions between many-body wavefunctions that obey appropriate boundary conditions at the surface of the solid. The final state of the photoelectron is a time-reversed low-energy electron-diffraction (LEED) state, which has a component consisting of a propagating plane-wave in vacuum with a finite amplitude inside the crystal.

The measured data are the energy of the incoming photon and the kinetic energy of the outcoming electron. In an angular-resolved experiment (ARPES) also the angle of emission is detected, which allows the evaluation of the wavevector of the emitted photoelectron. Moreover, using a Mott detector it is also possible to perform a spin analysis.

When an electron is removed from an electronic system, the measured photocurrent $J_k(\omega)$ is given by the probability per unit time of emitting an electron with momentum k when the sample is irradiated with photons of frequency ω. From Fermi's golden rule one obtains the relation (Almbladh and Hedin 1983; Hedin 1999):

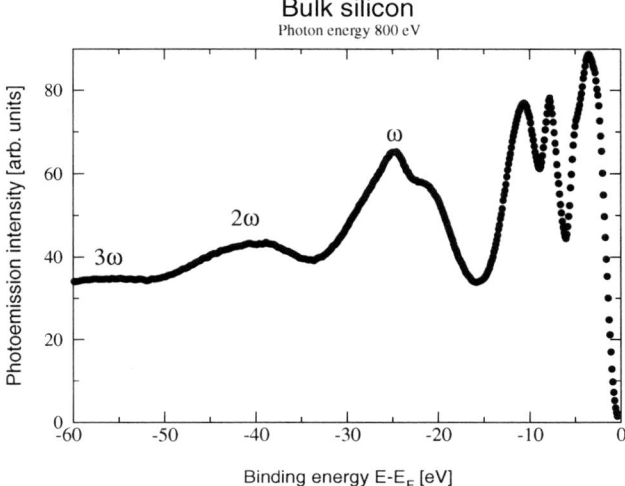

Fig. 3.1 Experimental photoemission spectrum of bulk silicon, measured at the TEMPO beamline at the Synchrotron SOLEIL with 800 eV photon energy. Besides the prominent quasiparticle peaks, corresponding to the valence bands, multiple plasmon satellites at higher binding energies are clearly visible. Figure from Guzzo (2011)

$$J_k(\omega) = \sum_m \xi_{km} \delta(E_k - E_m - \omega), \tag{3.1}$$

where

$$\xi_{km} = |\langle N - 1, m; k|A \cdot p + p \cdot A|N\rangle|^2 \tag{3.2}$$

is a matrix element of the coupling to the photon field and $|N\rangle$, $|N - 1, m; k\rangle$ are many-body states. The perturbation induces a transition from the initial N-electron ground state $|N\rangle$ to the final state $|N - 1, m; k\rangle$. The final state is represented by a system with the photoelectron with momentum k and the sample in the excited state m with $N - 1$ electrons. In order to assure the energy conservation, it is required that the photon energy ω is equal to the kinetic energy of the photoelectron E_k minus the electron removal energy, $E_m = E(N) - E(N - 1, m)$. This reads $\omega = E_k - E_m$. By knowing ω and measuring E_k, one can have access to the energy of the excited state m.

In the sudden approximation, the photoelectron is completely decoupled from the sample. One assumes that the photoelectron does not interact with the hole left behind and does not affect the state of the $(N - 1)$ electron system. Under such hypothesis, the total photocurrent $J_k(\omega)$ can be rewritten as:

$$J_k(\omega) = \sum_i |\xi_{ki}|^2 A_{ii}(E_k - \omega), \tag{3.3}$$

where we have introduced the matrix elements of the spectral function $A_{ij}(\omega)$ (for ω below the Fermi lever) and the creation and annihilation operators $\hat{c}^\dagger$ and $\hat{c}$:

$$A_{ij}(\omega) = \sum_m \left\langle N \left| \hat{c}_i^\dagger \right| N-1, m \right\rangle \langle N-1, m | \hat{c}_j | N \rangle \delta(\omega - E_m), \qquad (3.4)$$

and, moreover, we have assumed that there exists a one-particle basis in which the diagonal elements of A_{ij} are the relevant terms. For each many-body state m, the spectral function (3.4), defined rigorously as the imaginary part of the one-particle Green's function G, gives the probability to remove an electron from the ground state $|N\rangle$ and leave the system in the excited state $|N-1, m\rangle$. The total intensity of the photoemission spectrum is then the sum of the diagonal terms of spectral function A_{ii}, weighted by the photoemission matrix elements ξ_{ki}. Therefore, when the matrix elements are not zero, photoemission measurements give direct insight into the spectral function A.

The matrix elements ξ_{ki} describe the dependence of the spectra on the energy, momentum and polarization of the incoming photon. However, a common approximation is to neglect those matrix elements or, equivalently, to assume they are all equal to a constant $\bar{\xi}$, which yields:

$$J_k(\omega) = |\bar{\xi}|^2 \sum_i A_{ii}(E_k - \omega). \qquad (3.5)$$

Hence, to a first approximation, the quantity that one needs to calculate to simulate photoemission spectra is the trace of the spectral function (Gatti et al. 2007a).

In a simplified independent-particle picture, by measuring the kinetic energy of the emitted electron, one would obtain directly the energy of the one-particle level that the electron was occupying before being extracted from the sample. In this picture the many-body wavefunctions are Slater determinants and the spectral function (3.4) can be simplified to yield

$$A_{ij}(\omega) = \delta_{ij}\delta(\omega - E_i)\theta(\mu - \omega). \qquad (3.6)$$

The total photocurrent

$$J_k(\omega) = |\bar{\xi}|^2 \sum_i^{occ} \delta(E_k - \omega - E_i) \qquad (3.7)$$

turns out to be given by a series of delta peaks in correspondence to the energies E_i of the one-particle Hamiltonian. The photoemission spectrum is hence described by the density of occupied states:

$$\mathrm{DOS}(E_k - \omega) = \sum_i^{occ} \delta(E_k - \omega - E_i), \qquad (3.8)$$

evaluated at the energy $E_k - \omega$.

However, the real system is made of interacting electrons, and the distribution of kinetic energies that one actually obtains is somewhat different from this ideal picture. In fact, it is easy to understand that the emitted electron leaves a hole in the electronic system. A hole is a depletion of negative charge, hence it carries a positive charge which induces a relaxation of all the other electrons in order to screen it. Therefore, the creation of a hole promotes an excited state in the system, with a finite lifetime. This explains why in a photoemission spectrum one does not find a delta-peak in correspondence to a particular one-particle binding energy.

If one wants to perform calculations beyond the independent-particle picture, the many-body wavefunctions can then be seen as linear combinations of Slater determinants. For each excited state m, there are many non-vanishing contributions to the spectral function (3.4) that give rise to a more complex structure around the energy E_m. If in this more complex structure a main peak is still identifiable, then one can associate this peak with a quasiparticle excitation.

More complex phenomena can actually happen when the electron is extracted from the sample. In fact, the hole left in the system is itself a perturbation that can induce additional excitations. For example, a plasmon can be additionally excited in the system. In this case, the incoming photon energy is used to create more than one excitation. This additional excitations will produce a peak in the photoemission spectrum at a higher binding energy and with a smaller intensity than the corresponding quasiparticle peak, which reflects the smaller probability of a combination of events and its larger energy cost. This kind of additional structures in a photoemission spectrum is called a satellite and forms the incoherent part of the spectrum.

A clear example is given by the experimental spectrum of bulk silicon in Fig. 3.1. The three prominent peaks at lower binding energy are the quasiparticle peaks that correspond to the valence bands of silicon. Together with these peaks, we can see three additional structures with decreasing intensities, located at distances from the quasiparticle peaks equal to multiples of the plasmon energy of bulk silicon (around 16 eV). They are plasmon replicas: satellites that are the signature of the additional simultaneous excitation of one, two and three plasmons, respectively. This is a striking example that shows that photoemission spectroscopy is able to measure not only one-particle-like excitations (the quasiparticles), but also collective excitations, like plasmons (which can be directly measured in electron energy-loss spectroscopy, see Sect. 3.4).

In the extreme case of strong-correlation effects in the system, the one-particle nature of the excitation can be completely lost. In this case, in fact, it is no more possible to distinguish quasiparticle peaks, as all excitations involved have an intrinsic many-body character and the incoherent part of the spectrum dominates.

A spectral function for the metallic phase of VO_2, calculated in the GW approximation (Hedin 1965), is shown in Fig. 3.2. In order to gain a deeper insight into the nature of the structures of the spectral function, we can rewrite its matrix elements (3.4) in terms of the self-energy Σ (Hedin 1999):

$$A_{ii}(\omega) = \frac{1}{\pi} \frac{|\Im m\, \Sigma_{ii}(\omega)|}{\{\omega - \varepsilon_i - [\Re e\, \Sigma_{ii}(\omega) - v_{xci}]\}^2 + [\Im m\, \Sigma_{ii}(\omega)]^2}, \tag{3.9}$$

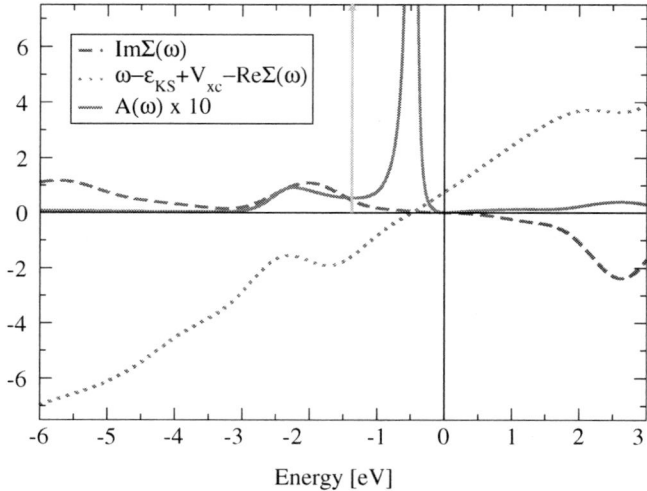

Fig. 3.2 Spectral function A_{ii} (*red solid line*) relative to the top valence at $k = \Gamma$ for the metallic phase of VO_2, calculated using the GW approximation. The corresponding Kohn–Sham eigenvalue, calculated in the local-density approximation, is represented by the *arrow*. The real and imaginary parts of the self-energy Σ are also shown (see the discussion in the text). The zero of the energy axis is set at the GW Fermi energy.

where we have assumed that the self-energy is diagonal in the one-particle basis chosen to represent the spectral function V_{xci} are the matrix elements of the xc potential and ε_i are the Kohn–Sham eigenvalues. The self-energy is an essential quantity in the Green's function formalism. It represents, in fact, the effective non-local and dynamical potential that the extra particle feels for the polarization that its propagation induces and for the exchange effects, due to the fact that it is a fermion. Analogously to the exchange-correlation potential v_{xc} of density-functional theory (DFT), the self-energy encompasses all the effects of exchange and correlation in the system.

Equation 3.9 allows us to see that the quasiparticle peak in the spectral function shown in Fig. 3.2 is determined by the zero of $\omega - \varepsilon_i - [\Re e\,\Sigma_{ii}(\omega) - v_{xc\,i}]$. The width of the peak is given by the imaginary part of the self-energy: the quasiparticle excitation has finite lifetime (which is the inverse of the width of the peak). Additional structures, the satellites, are linked to structures of $\Im m\,\Sigma_{ii}(\omega)$ or to additional zeros of $\omega - \varepsilon_i - [\Re e\,\Sigma_{ii}(\omega) - v_{xc\,i}]$, where also $\Im m\,\Sigma_{ii}(\omega)$ is not too large. Note that a necessary requirement for the presence of satellites is the dynamical nature of Σ. In particular, in the case of VO_2 shown in Fig. 3.2, we can see a satellite in the spectral function due to a structure of $\Im m\,\Sigma_{ii}(\omega)$, which corresponds to the additional excitation of a plasmon (Gatti 2007b). This situation is analogous to what we have discussed for the experimental spectrum of silicon in Fig. 3.1. In a Kohn–Sham calculation within DFT, the spectral function would reduce to a delta peak, represented by the arrow in Fig. 3.2. The same would occur in the Hartree–Fock approximation, in which the self-energy is real and static. In Hartree–Fock, thanks to Koopmans' theorem (Koopmans 1934), the eigenvalues have a physical meaning as approximate removal energies, contrary to the Kohn–Sham eigenvalues, for which Koopmans' theorem does not hold.

When one goes beyond the sudden approximation (Hedin 1998; Almbladh 2006) and considers scattering processes of the photoelectron on its way out of the sample, new features in the photoemission spectrum can appear. In this case one talks of extrinsic losses, in contrast to the structures of the spectral function discussed so far, and that are due to intrinsic losses.

Moreover, in order to restore a quantitative agreement between calculated and measured spectra, one should keep in mind that it is essential to remove the assumption that the photoemission matrix elements ξ_{ki} are constant. In fact, measured spectra generally show large variations in their profile according to the photon energy used in the experiment (Papalazarou 2009).

Since the electron escape depth is of the order of 10–50 Å for kinetic energies of 10–2000 eV (Zangwill 1988), photoemission is a surface sensitive technique. Obtaining bulk information can be achieved by using atomically flat and clean surfaces and by working at high photon energies [e.g. hard X-rays (Panaccione 2006)], thereby increasing the electron escape depth. At the same time, higher photon energies are in better agreement with the hypothesis of the sudden approximation, as the photoelectron is emitted with a high kinetic energy, and therefore its removal process can be more safely considered as instantaneous. Nevertheless, angular-resolved experiments are performed at lower photon energies, where the intrinsic momentum resolution is better (Damascelli 2003).

3.4 Microscopic Description of Neutral Excitations

The propagation of electromagnetic waves in dissipative media is described by Maxwell's equations (Jackson 1962), supplemented by appropriate constitutive equations. Optical phenomena (reflection, propagation, transmission) can be quantified by a number of parameters that determine the properties of the medium at the macroscopic level. Microscopic (semiclassical) models and averaging procedures then allow us to calculate these macroscopic parameters. In this section we focus on the microscopic models, while the averaging procedure will be the subject of Sect. 3.6.

A photon impinging on a sample can be absorbed, reflected or transmitted. When it is absorbed, its energy is used to create a neutral excitation in the system. In an independent-particle picture, this can be represented by the promotion of a valence electron from an occupied to an empty one-particle state, conserving its crystal momentum k, since a photon in the energy range of interest can transfer only a negligible momentum. In contrast with what happens in a photoemission experiment, the excited electron remains inside the sample, and it cannot therefore be considered decoupled from the other electrons.

According to the semi-classical picture provided by the macroscopic Maxwell equations (Jackson 1962), an electric field E polarized along the $\hat{e}$ direction in a dielectric medium propagates as a damped wave:

$$E(z,t) = E_0\hat{e}\,e^{i\frac{\omega}{c}(nz-ct)} = E_0\hat{e}\,e^{i\frac{\omega}{c}(n_1 z-ct)}e^{-\frac{\omega}{c}n_2 z}, \tag{3.10}$$

where n is the complex refractive index $n = n_1 + in_2$, and c is the velocity of light in vacuum. As a consequence, also the intensity of the field is exponentially decaying:

$$I(z) = |E(z)|^2 = E_0^2 e^{-2\frac{\omega}{c}n_2 z}. \tag{3.11}$$

The absorption coefficient α^{abs} is defined as the inverse of the distance where the intensity of the field is reduced by $1/e$:

$$\alpha^{abs} = \frac{2\omega n_2}{c}. \tag{3.12}$$

Equivalently, we can introduce the concept of macroscopic complex dielectric function $\epsilon_M = \epsilon_1 + i\epsilon_2 = n^2$.

The absorption coefficient can be easily rewritten in terms of the imaginary part of the macroscopic dielectric function:

$$\alpha^{abs} = \frac{\omega\epsilon_2}{cn_1}. \tag{3.13}$$

Since n_1 can be usually assumed constant in the small frequency ranges of interest, absorption spectra are usually expressed in terms of $\epsilon_2(\omega) = \Im m\epsilon_M(\omega)$.

Within the basic approximations, i.e. the independent-particle and the dipole approximation, Fermi's golden rule (Grosso 2000) gives the well-known expression for the absorption spectrum in the linear regime, as a sum of one-particle independent transitions:

$$\epsilon_2^{TT}(\omega) = \frac{8\pi^2}{\mathcal{V}\omega^2}\sum_{vc}|\langle\varphi_c|\hat{e}\cdot p|\varphi_v\rangle|^2\delta(E_j - E_i - \omega), \tag{3.14}$$

where p is the momentum operator, $\mathcal{V}$ is the unit cell volume, $|\varphi_{v,c}\rangle$ are one-particle states (e.g. Kohn–Sham states) and E_i are one-particle energies (e.g. Kohn–Sham energies). In (3.14) T stands for transverse, as the transitions from occupied to empty states $|\varphi_c\rangle$ to $|\varphi_v\rangle$ are due to the interaction with a transverse electromagnetic plane wave of frequency ω and polarization $\hat{e}$. In general one should consider different polarization directions, which means that $\hat{\epsilon}_2$ is a tensor. This tensor reduces to a scalar quantity when the system is isotropic (cubic symmetry).

Note that in the simple case of a sum of independent one-particle transitions the macroscopic dielectric function is simply the spatial average of the microscopic dielectric function $\epsilon(r, r', t - t')$. This means that classical depolarization effects (local-field effects) are not included in Eq. 3.14. We will go back to this problem, related to spatial averaging, in the following section.

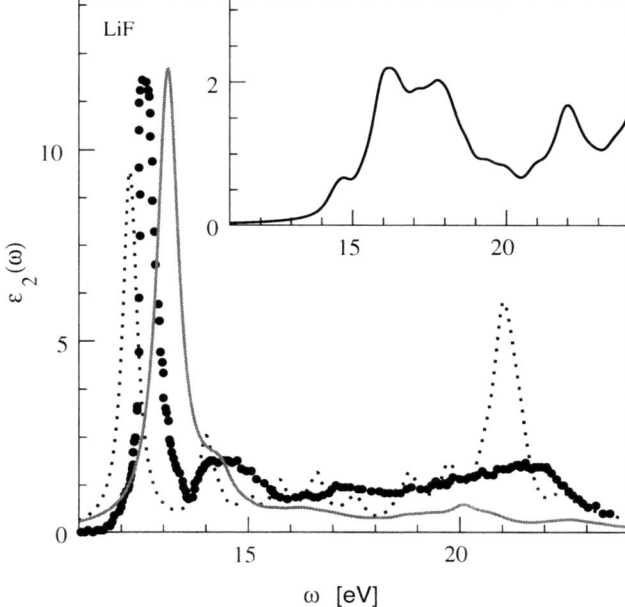

Fig. 3.3 Imaginary part of the macroscopic dielectric function for LiF. *Dots,* experiment from (Roessel 2005); *dotted line,* Bethe-Salpeter equation (BSE) calculation; *solid line,* TDDFT calculation using the dynamical model kernel derived from BSE. In the *inset,* RPA calculation using quasiparticle energies. Figure from Botti (2005)

Local-field effects are not the only missing physical ingredient. In fact, the independent-transition picture is often inadequate to capture the complex physics of the many-body problem and, even if local-field effects are properly accounted for, the quality of the calculated spectra still depends dramatically on the use of more involved approximations to include electron-electron and electron-hole interactions. Indeed, one should not forget that in an absorption experiment, contrary to photoemission, the excited electron remains inside the system. It is essential then to take into account the interaction between the electron and the hole that the excitation of the electron leaves behind, the so-called excitonic effect. This can be done in different frameworks that go beyond the independent-particle approximation, such as many-body perturbation theory, or time-dependent density-functional theory.

We consider as an example a LiF crystal (Botti 2005), whose experimental absorption spectrum (dots) is shown in Fig. 3.3. If one performs a calculation in the independent-particle approximation, even after including local-field effects within the random phase approximation (RPA) (inset), the worst disagreement concerns the absence of the large excitonic peak at about 12.5 eV. In the framework of many-body perturbation theory, it is possible to include two-quasiparticle effects by solving the Bethe-Salpeter equation (BSE) (Salpeter 1951). In this way, one can reproduce the bound exciton peak (dotted-dashed curve). Alternatively, one can use, within time-dependent density functional theory (TDDFT), model exchange-correlation kernels derived from the Bethe-Salpeter equation (Botti 2007), which improve remarkably the RPA results with a much smaller computational effort than the one required by the solution of the Bethe-Salpeter equation.

Another example of neutral excitations comes from electron energy-loss spectroscopy , where the probe is a high-energy electron which undergoes inelastic scattering. As a consequence, it looses energy and has its path deflected. Both the energy loss and the deflection angle can be measured. Inelastic interactions can involve different energy ranges, depending on their origin: phonon excitations, inter- or intra-band transitions, plasmon excitations, inner-shell ionizations, Auger processes, etc. In the energy range that we consider (up to some tens of eV), the excitations of the system that can be involved are either band transitions or plasmon excitations.

We consider the rate by which a charged fast particle transfers energy and momentum to a material. As long as the fractional changes of energy and momentum of the fast electron are small, the electron can be considered as a classical point charge moving with uniform velocity v in the medium. The probability that an impinging electron of velocity v transfers in a unit time an energy dW and a momentum q to the electronic system can be expressed by the energy-loss rate (Luth 2001):

$$\frac{dW}{dt} = \frac{1}{(2\pi)^3} \int d\omega \int d^3q \frac{\omega}{q^2} \Im \left\{ -\frac{1}{\epsilon_M(q,\omega)} \right\} \delta(\omega + q \cdot v). \quad (3.15)$$

In (3.15) one assumes that the impinging electron is a classical particle: it can be treated in a non-relativistic approximation and neglecting quantum indetermination effects on its position. The energy-loss rate is then proportional to the loss function $L(q,\omega)$:

$$L(q,\omega) = -\Im \left\{ \frac{1}{\epsilon_M(q,\omega)} \right\} = \frac{\epsilon_2(q,\omega)}{\epsilon_1^2(q,\omega) + \epsilon_2^2(q,\omega)}. \quad (3.16)$$

From Eq. 3.16 one can conclude that spectral features in EELS can be seen in the case that either ϵ_2 has a peak corresponding to some interband transition or ϵ_2 is small and $\epsilon_1 = 0$. The latter condition determines the frequencies of the collective plasmon excitations. In energy-loss experiments one can therefore excite plasmons, that are longitudinal excitations and that are not generally observed in optical spectroscopy, since photons are a transverse perturbation. Note that the loss function (3.16) is also the quantity measured in inelastic X-rays scattering (IXS) experiments.

An interesting situation can be found in the study of EELS for core states. In fact, at high energies $\epsilon_2(q,\omega) \to 0$ and $\epsilon_1(q,\omega) \to 1$. This observation allows a simplification of the loss function (3.16):

$$L(q,\omega) = -\Im \left\{ \frac{1}{\epsilon_M(q,\omega)} \right\} \to \epsilon_2(q,\omega). \quad (3.17)$$

Using again the simple independent one-particle transition picture, Fermi's golden rule allows us to evaluate the imaginary part of the longitudinal dielectric function as Grosso (2000):

$$\epsilon_2^{LL}(\boldsymbol{q}, \omega) = \frac{8\pi^2}{Vq^2} \sum_{ij} \left| |\langle \varphi_j | e^{i\boldsymbol{q} \cdot \boldsymbol{r}} | \varphi_i \rangle| \right|^2 \delta(E_j - E_i - \omega), \tag{3.18}$$

where LL means that we are evaluating the response to a longitudinal perturbation. In this picture plasmon peaks in EELS are associated to coherent oscillations of independent particles. Once again, the quality of the calculated spectra depends dramatically on the possibility to go beyond the independent-particle description of Eq. 3.18. For extended systems, EELS and IXS at small and large momentum transfer are often well reproduced simply by including local-field effects, i.e. performing RPA calculations or TDDFT calculations within the adiabatic local-density approximation (TDLDA—Botti 2007). This is in contrast with the fact that TDLDA fails in the calculation of optical ($q = 0$) spectra of non-metallic solids. To explain this failure, the wrong asymptotic limit of the exchange-correlation kernel is crucial, while the wrong behaviour of the exchange-correlation potential is less relevant (Botti 2007).

Note again that we defined two different microscopic dielectric functions, ϵ_2^{LL} and ϵ_2^{TT}, specific for longitudinal and transverse perturbations, respectively. However, since it holds (Del Sole 1993):

$$\boldsymbol{v} = \lim_{\boldsymbol{q} \to 0} \frac{1}{q} \left[\hat{H}, e^{i\boldsymbol{q} \cdot \boldsymbol{r}} \right], \tag{3.19}$$

it can be proved that in the long-wavelength limit longitudinal (3.18) and transverse (3.14) dielectric functions of an isotropic system coincide:

$$\epsilon_M^{LL}(\omega) = \epsilon_M^{TT}(\omega). \tag{3.20}$$

This is due to the invariance for $\boldsymbol{q} \to 0$ between the velocity gauge (transverse perturbation $\boldsymbol{A} \cdot \boldsymbol{v}$) and the length gauge (longitudinal perturbation $\boldsymbol{E} \cdot \boldsymbol{r}$). When $\boldsymbol{q}$ is non-vanishing and the system does not have cubic symmetry, this extreme simplification is no more valid, and one has to consider the general tensor.

3.5 Microscopic Response Beyond the Independent-Particle Picture

We will now deal with the general problem of the determination of the microscopic response of the system, beyond an independent-particle picture and Fermi's golden rules 3.14 and 3.18. In the following section we will then discuss how to relate microscopic and macroscopic dielectric functions. Linear-response theory can be applied to study the response of an electronic system to a small time-dependent perturbation $\hat{V}_{ext}(t)$, such as the incoming electromagnetic field or the high-energy electron which undergoes inelastic scattering that we discussed before. We consider here a system described by the many-body Hamiltonian $\hat{H}$, subjected to a time-dependent external perturbation $\hat{V}_{ext}(t)$. The total Hamiltonian becomes:

$$\hat{H}_{tot} = \hat{H} + \hat{V}_{ext}(t). \tag{3.21}$$

For a sufficiently small perturbation, the response of the system can be expanded into a Taylor series with respect to the perturbation. The linear coefficient linking the response to the perturbation is the response function. In this context, the density-density response function χ is defined as

$$\delta\rho(\mathbf{r}, t) = \int dt' \int d^3r' \chi(\mathbf{r}, \mathbf{r}', t - t')v_{ext}(\mathbf{r}', t'), \tag{3.22}$$

where $\delta\rho$ is the first-order variation of the electron density and v_{ext} the external perturbation. A response function is independent on the perturbation and depends only on the system. The density-density response function is also called reducible polarizability. The irreducible polarizability P can be defined in a similar way:

$$\delta\rho(\mathbf{r}, t) = \int dt' \int d^3r' P(\mathbf{r}, \mathbf{r}', t - t')v_{tot}(\mathbf{r}', t'), \tag{3.23}$$

where v_{tot} is the total classical potential. Hence, the irreducible polarizability P describes the first-oder variation of the electron density with respect to the total classical potential, which includes the polarization of the system.

In fact, as a consequence of the polarization of the system due to the applied perturbation, the total potential becomes a sum of the external potential and the induced potential: $v_{tot}(\mathbf{r}, t) = v_{ext}(\mathbf{r}, t) + v_{ind}(\mathbf{r}, t)$. The induced potential v_{ind} and the induced density $\delta\rho$ are moreover related by $v_{ind} = v_{ee}\delta\rho$ (v_{ee} is the Coulomb interaction). The basic quantity that gives information about the screening of the system in linear response is the microscopic dielectric function, which relates the total potential to the applied potential:

$$v_{tot}(\mathbf{r}, t) = \int dt' \int d^3r' \epsilon^{-1}(\mathbf{r}, \mathbf{r}', t - t')v_{ext}(\mathbf{r}', t'). \tag{3.24}$$

Remembering the definition of the density-density response function χ (3.22) we can easily deduce that the microscopic dielectric function ϵ and χ are related by

$$\epsilon^{-1}(\mathbf{r}, \mathbf{r}', t - t') = \delta(\mathbf{r} - \mathbf{r}')\delta(t - t') + \int d^3r'' v_{ee}(\mathbf{r} - \mathbf{r}'')\chi(\mathbf{r}', \mathbf{r}'', t - t'). \tag{3.25}$$

The microscopic dielectric function can also be related to the irreducible polarizability:

$$\epsilon(\mathbf{r}, \mathbf{r}', t - t') = \delta(\mathbf{r} - \mathbf{r}')\delta(t - t') - \int d^3r'' v_{ee}(\mathbf{r} - \mathbf{r}'')P(\mathbf{r}'', \mathbf{r}', t - t') \tag{3.26}$$

Moreover, the reducible and irreducible polarizabilities satisfy the Dyson equation

$$\chi(\boldsymbol{r}, \boldsymbol{r}', t - t') = P(\boldsymbol{r}, \boldsymbol{r}', t - t') + \int d^3 r'' \int d^3 r''' P(\boldsymbol{r}, \boldsymbol{r}'', t - t') v_{\text{ee}}$$
$$\times (\boldsymbol{r}'' - \boldsymbol{r}''') \chi(\boldsymbol{r}''', \boldsymbol{r}', t - t'). \tag{3.27}$$

The dielectric function is a quantity that is usually used to define the response of an extended system. Similar quantities and formulas can, however, be obtained also for finite systems. In the linear-response regime and dipole approximation, and considering that in the range of frequencies of interest (optical frequencies) the dimension of the system is much smaller than the radiation wavelength, the induced density variation in the finite system is again proportional to the perturbing field as defined by Eq. 3.22. For a dipole electric field along the direction z, the photoabsorption cross section along z, σ_{zz}, is

$$\sigma_{zz}(\omega) = -\frac{4\pi\omega}{c} \, \Im \left\{ \int d^3 r \int d^3 r' z \chi(\boldsymbol{r}, \boldsymbol{r}', \omega) z' \right\} = \frac{4\pi\omega}{c} \Im \{\alpha_{zz}(\omega)\}, \tag{3.28}$$

the above expression also serving as the definition of the dynamical polarizability $\hat{\alpha}$. Like ϵ, the dynamical polarizability is in general a tensor and the photoabsorption cross section σ can be written in the form:

$$\sigma(\omega) = \frac{4\pi\omega}{3c} \Im \left\{ \text{Tr}\hat{\alpha}(\omega) \right\}. \tag{3.29}$$

Another quantity that is often used for finite systems is the dipole strength function S: $\sigma(\omega) = (2\pi^2/c) S(\omega)$.

In linear-response theory, in general, we deal with perturbations that couple a time-dependent field $F(t)$ to an observable of the system $\hat{P}$:

$$\hat{V}_{\text{ext}}(t) = \int d^3 r_1 F(\boldsymbol{r}_1, t) \hat{P}(\boldsymbol{r}_1, t). \tag{3.30}$$

Typically the perturbation is an electromagnetic field coupled with densities or currents. In linear-response one assumes that variations δF of the field F are small. From Kubo's formula (Kubo 1957) the variation $\delta\langle \hat{O}(\boldsymbol{r}_1, t) \rangle$ of an observable $\hat{O}$ is then obtained by using first-order perturbation theory as:

$$\delta\langle \hat{O}(\boldsymbol{r}_1, t) \rangle = \int dt' \int d^3 r_2 \chi_{\text{OP}}(\boldsymbol{r}_1, \boldsymbol{r}_2, t - t') \delta F(\boldsymbol{r}_2, t'), \tag{3.31}$$

where

$$\chi_{\text{OP}}(\boldsymbol{r}_1, \boldsymbol{r}_2, t - t') = -i\theta(t - t') \langle N | [\hat{O}(\boldsymbol{r}_1, t), \hat{P}(\boldsymbol{r}_2, t')] | N \rangle$$
$$= \frac{\delta\langle \hat{O}(\boldsymbol{r}_1, t) \rangle}{\delta F(\boldsymbol{r}_2, t')} \tag{3.32}$$

is a (causal or retarded) response function and $|N\rangle$ is the many-body ground-state wavefunction. In fact, χ_{OP} is zero for $t < t'$, because there cannot be any response at a time t before a perturbation has occurred at a time t'.

By inserting the completeness relation and taking the Fourier transform, the response function can be written in the so-called Lehmann representation:

$$\chi_{OP}(r_1, r_2, \omega) = \lim_{\eta \to 0^+} \sum_m \left[\frac{O_m(r_1) P_m^*(r_2)}{\omega + E_0 - E_m + i\eta} - \frac{O_m^*(r_1) P_m(r_2)}{\omega + E_m - E_0 + i\eta} \right], \quad (3.33)$$

where

$$O_m(r_1) = \langle N | \hat{O}(r_1) | N, m \rangle, \quad (3.34)$$

(P_m is analogously defined) and η is a positive infinitesimal. The response function has poles at the excitation energies $\pm(E_0 - E_m)$, corresponding to transitions between the ground state $|N\rangle$ and the many-body excited state $|N, m\rangle$ of the unperturbed Hamiltonian $\hat{H}$. The first term in (3.22) is given by resonant transitions, the second by antiresonant transitions. The form of (3.22) is valid for finite systems with discrete eigenvalues. For extended systems, on the other hand, the spectrum is continuous, and the sum in (3.22) turns into an integral that gives rise to a branch cut along the real energy axis. The infinitely close-lying resonances thus merge into broad structures that can be identified with elementary excitations, such as plasmons or excitons. As these structures have a certain width, they are described by poles in the complex plane with a real part (the energy of the excitation) and a finite imaginary part (whose inverse is proportional to the excitation lifetime).

Since response functions represent the causal response of the system to external perturbations, they must obey several analytic properties and sum rules. The response function $\chi_{OP}(\omega)$, continued in the complex plane is analytic for all $\Im m\{\omega\} > 0$ in the upper half-plane and has poles only in the lower half-plane. Using contour integration in the complex plane, it is possible to obtain the Kramers-Kronig relations (Kramers 1927; Kronig 1926) that link the real and imaginary parts of χ_{OP}:

$$\Re e \chi_{OP}(\omega) = -\frac{1}{\pi} \int\limits_{-\infty}^{+\infty} d\omega' \, \frac{\Im m \chi_{OP}(\omega')}{\omega' - \omega}, \quad (3.35a)$$

$$\Im m \chi_{OP}(\omega) = -\frac{1}{\pi} \int\limits_{-\infty}^{+\infty} d\omega' \, \frac{\Re e \chi_{OP}(\omega')}{\omega' - \omega}. \quad (3.35b)$$

The density-density response function χ of Eq. 3.22 is an example of the more general response function of Eq. 3.31. It can be written in the Lehmann representation

$$\chi(r, r', \omega) = \sum_j \left[\frac{\rho_j(r)\rho_j^*(r')}{\omega - E_j + E_0 + i\eta} - \frac{\rho_j^*(r)\rho_j(r')}{\omega + E_j - E_0 + i\eta} \right], \quad (3.36)$$

and it obeys the Kramers–Kronig relations Eq. 3.35. It is easy to see that the imaginary part of χ has the form of a joint density of states weighted by matrix elements.

When we are interested in small perturbations induced by electromagnetic fields, which can be represented by a scalar potential v_{ext} and a vector potential A_{ext}, besides defining density-density response functions, one has also to define density-current, current-density and current-current response functions. We will see in the next section that the macroscopic response of the system to a longitudinal (and within some limitations also to a transverse) perturbation can be recast in terms of the density-density response function χ alone.

Now that we have understood how to provide a microscopic description of the (linear) response of the electronic system under perturbation, we are ready to discuss how averaging procedures allow to extract the desired macroscopic physical quantities.

3.6 Microscopic–Macroscopic Connection

For periodic systems, the most natural way to deal with spatial periodicity is to apply a Fourier transform:

$$
\begin{aligned}
\chi_{GG'}(q, \omega) &= \chi(q+G, q + G', \omega) \\
&= \int d^3r \int d^3r' e^{-i(q+G)\cdot r} \chi(r, r', \omega) e^{i(q+G')\cdot r'},
\end{aligned} \tag{3.37}
$$

where G is a vector of the reciprocal lattice, while q is a vector in the first Brillouin zone. Therefore, we rewrite the microscopic response function (3.25) in the reciprocal space

$$
\epsilon_{GG'}^{-1}(q, \omega) = \delta_{GG'} + v_{ee}G(q)\chi_{GG'}(q, \omega). \tag{3.38}
$$

We are now in principle able to calculate a microscopic dielectric function. However, what is measured in an experiment is usually a macroscopic quantity, which involves averaging over regions of space that are large in comparison with the interparticle separation, but small compared to the wavelength of the perturbation. Special care is then necessary to bridge the gap between the microscopic and the macroscopic worlds in the case of extended systems.

Moreover, we should not forget that we are dealing with different dielectric functions, that are specific for transverse and longitudinal perturbations. For a non-vanishing q and a non-cubic symmetry:

$$
D(q, \omega) = \overleftrightarrow{\epsilon_M}(q, \omega)E^{tot}(q, \omega), \tag{3.39}
$$

where

$$
D = \begin{pmatrix} D^L \\ D^T \end{pmatrix} = \begin{pmatrix} \epsilon_M^{LL} & \epsilon_M^{LT} \\ \epsilon_M^{TL} & \epsilon_M^{TT} \end{pmatrix} \begin{pmatrix} E^L \\ E^T \end{pmatrix}.
$$

This means that in general, a purely longitudinal or purely transverse perturbation will induce a longitudinal and a transverse response at the same time. Besides, transverse and longitudinal components of the dielectric tensor are no more equal, and only $\epsilon_M^{LL}(q, \omega)$ can be easily calculated, as we will show below, while $\epsilon_M^{TT}(q, \omega)$ obeys a complicated expression (Del Sole 1984). Only the high symmetry of the system can guarantee that a longitudinal (transverse) perturbation induces only a longitudinal (transverse) response. This happens when off-diagonal elements of (3.6) are zero. When the symmetry is lower, but still one is interested in the response in the long-wavelength limit, it is always possible to use the fact that the components of the dielectric tensor are analytic functions of q in the limit $q \to 0$ and rewrite the tensor along the principal axes of the crystal. Applying a longitudinal perturbation along one of the principal axis induces only a longitudinal response. Of course, in the long-wavelength limit this same expression gives also the transverse dielectric function.

One can conclude that for perturbations with vanishing momentum transfer, the longitudinal and transverse responses (measured respectively in EELS and absorption experiments) coincide (see Eq. 3.20). There is no easy way to calculate the transverse response if it does not coincide with the longitudinal one. In that case, the general expression involves not only the density–density response function, but also current-current response functions. In view of the above, we consider here only the simpler longitudinal case. For a more general discussion the reader can refer to the work of Del Sole and Fiorino (1984) and the review of Strinati (1988).

In order to compare with optical and EELS spectra, we want to determine the longitudinal component of the macroscopic dielectric tensor (or possibly the three components along the principal axis). The (longitudinal) microscopic dielectric function can be obtained by the density–density response function χ by applying Eq. 3.38. When the system is perturbed by an external scalar potential v_{ext}, the total potential (sum of the external and the induced potential) felt by a test-charge is:

$$v_{tot}(q + G, \omega) = \sum_{G'} \epsilon_{G,G'}^{-1}(q, \omega) v_{ext}(q + G', \omega). \qquad (3.40)$$

The total potential in general has different wavevector components than the perturbing potential for the presence of microscopic fluctuations induced by the inhomogeneities of the material. The difference between the microscopic potentials and their macro-scopic average are the already mentioned local-field effects.

To connect the microscopic and the macroscopic quantities one has to take a spatial average over a distance that is large compared to the lattice parameters and small compared to the wavelength of the external perturbation (Ehrenreich 1966). Since the microscopic quantities are lattice periodic, this procedure is equivalent to take the spatial average over a unit cell.

A microscopic potential $v(r, \omega)$ can be expanded in its Fourier components as:

$$v(r, \omega) = \sum_{qG} v(q + G, \omega) e^{i(q+G)\cdot r}, \qquad (3.41)$$

or:

$$v(r, \omega) = \sum_q e^{iq \cdot r} \sum_G v(q + G, \omega)e^{iG \cdot r} = \sum_q e^{iq \cdot r} v(q, r, \omega), \qquad (3.42)$$

where:

$$v(q, r, \omega) = \sum_G v(q + G, \omega)e^{iG \cdot r}. \qquad (3.43)$$

$v(q, r, \omega)$ is periodic with respect to the Bravais lattice and hence is the quantity that one has to average to get the corresponding macroscopic potential $v_M(q, \omega)$:

$$v_M(q, \omega) = \frac{1}{V} \int dr\, v(q, r, \omega). \qquad (3.44)$$

Inserting (3.43) in (3.44), one has:

$$v_M(q, \omega) = \sum_G v(q + G, \omega) \frac{1}{V} \int d^3r\, e^{iG \cdot r} = v(q + 0, \omega). \qquad (3.45)$$

Therefore the macroscopic averaged potential v_M is given by $G = 0$ component of the corresponding microscopic potential v.

In particular, in the standard spectroscopy experiments discussed in this chapter, the external perturbing potential is a macroscopic quantity. For instance it can be the electromagnetic field impinging on a sample that one can measure in an absorption experiment. For Eq. 3.45 only the $G = 0$ component of $v_{ext}(q + G, \omega)$ is different from 0. Therefore the macroscopic average $v_{tot,M}$ of the microscopic total potential v_{tot} in Eq. 3.40 is:

$$v_{tot,M}(q, \omega) = \epsilon^{-1}_{G=0, G'=0}(q, \omega)v_{ext}(q, \omega). \qquad (3.46)$$

Equation 3.46 is a relation between macroscopic potentials, so it defines the macroscopic inverse dielectric function ϵ^{-1}_M: $v_{tot,M} = \epsilon^{-1}_M v_{ext}$. In this way one obtains:

$$\epsilon^{-1}_M(q, \omega) = \epsilon^{-1}_{G=0, G'=0}(q, \omega). \qquad (3.47)$$

Therefore, the macroscopic dielectric function turns out to be defined as:

$$\epsilon_M(q, \omega) = \frac{1}{\epsilon^{-1}_{G=0, G'=0}(q, \omega)}. \qquad (3.48)$$

This corresponds to the result found by Adler (1962) and Wiser (1963): the macroscopic dielectric function ϵ_M is the reciprocal of the head (i.e. the $G = 0, G' = 0$ element) of the inverse of the microscopic dielectric function ϵ^{-1}.

It is important to note here that, since v_{tot} is in general a microscopic quantity, because it contains the microscopic fluctuations due to the polarization of the medium, one could not obtain directly the macroscopic dielectric function ϵ_M from the inverse of the relation (3.40):

$$v_{ext}(q + G, \omega) = \sum_{G'} \epsilon_{G,G'}(q, \omega) v_{tot}(q + G', \omega). \qquad (3.49)$$

Since v_{ext} is macroscopic, (3.49) is just:

$$v_{ext}(q, \omega) = \sum_{G'} \epsilon_{G=0,G'}(q, \omega) v_{tot}(q + G', \omega). \qquad (3.50)$$

Hence:

$$v_{ext}(q, \omega) = \epsilon_{G=0,G'=0}(q, \omega) v_{tot,M}(q, \omega) + \sum_{G' \neq 0} \epsilon_{G=0,G'}(q, \omega) v_{tot}(q + G', \omega). \qquad (3.51)$$

Since $v_{ext} = \epsilon_M v_{tot,M}$, one has that $\epsilon_M = \epsilon_{G=0,G'=0}$ only if the off-diagonal terms $(G' \neq 0)$ are neglected in Eq. 3.51. These off-diagonal terms correspond to the rapidly oscillating contributions to the microscopic total potential and are responsible for the crystal local-field effects. In fact, from Eq. 3.48, one has that $\epsilon_M = 1/\epsilon_{G=0,G'=0}^{-1} = \epsilon_{G=0,G'=0}$ only if $\epsilon_{G,G'}$ is diagonal, which corresponds to a dielectric function in the real space that depends only on the distance between r and r'.

It is possible to prove (Hanke 1978; Onida 2002) that ϵ_M can be directly obtained using a modified density–density response function $\bar{\chi}$:

$$\epsilon_M(q, \omega) = 1 - v_{ee\ G=0}(q) \bar{\chi}_{G=0,G'=0}(q, \omega). \qquad (3.52)$$

Whereas the density-density response function χ satisfies the Dyson equation (1.3.eq:chi-P) $\chi = P + P v_{ee} \chi$, the modified response function $\bar{\chi}$ is obtained from

$$\bar{\chi} = P + P \bar{v}_{ee} \bar{\chi}, \qquad (3.53)$$

where the macroscopic component of $\bar{v}_{ee}$ is set to zero: $\bar{v}_{ee G=0} = 0$ and for all the other components it holds $\bar{v}_{ee} = v_{ee}$. Neglecting local-field effects (NLF), the macroscopic dielectric function is (see Eq. 3.51):

$$\epsilon_M^{NLF}(q, \omega) = \epsilon_{G=0,G'=0}(q, \omega) = 1 - v_{ee\ G=0}(q) P_{G=0,G'=0}(q, \omega). \qquad (3.54)$$

Therefore, comparing (3.52) and (3.54), one can see that it is the microscopic part of the Coulomb interaction $\bar{v}$ that is responsible for the local-field effects. In fact, setting $\bar{v} = 0$ in (3.53) would imply that $\bar{\chi} = P$ and hence $\epsilon_M^{NLF} = \epsilon_M$.

Local-field effects are tightly related to spatial inhomogeneities in the system. Whenever a system is inhomogeneous, even if the external field is slowly varying,

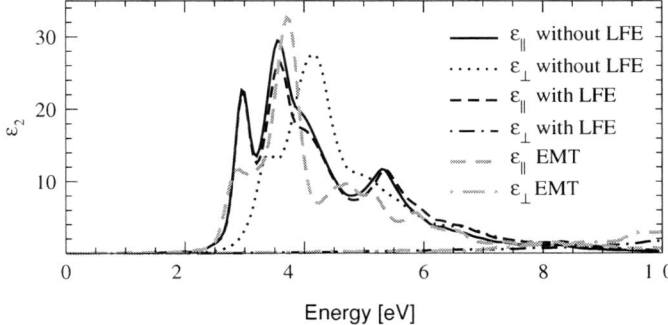

Fig. 3.4 Imaginary part of the dielectric function for a [110] Si nanowire, with a diameter of 2.2 nm. Calculations are performed for light polarized along ($\epsilon_\parallel$) and perpendicularly ($\epsilon_\perp$) to the wire axis, starting from a DFT-LDA ground state and using the RPA approximation either including local-field effects (with LFE) or neglecting them (without LFE). The result of a classical effective medium theory (Maxwell-Garnett 1904) (which accounts for classical local-field effects) is shown for comparison (*grey lines*). Figure from Bruneval (2005)

the induced charges can have rapid spatial variations. This means that the off-diagonal elements of the dielectric function are important. Typical examples where local fields play an important role (Botti 2007) are layered systems, nanotubes or nanowires, and finite systems like nanoclusters, which represent themselves an inhomogeneity in the vacuum space. Instead, bulk solids of *sp* semiconductors are often examples of homogeneous systems where local fields play a minor role. In Fig. 3.4 (Bruneval 2005) we show the optical absorption of a Si nanowire with a diameter of 2.2 nm, calculated including and neglecting local-field effects. By neglecting local-field effects one obtains a similar optical absorption for the two components of the imaginary part of the dielectric functions along and perpendicular to the wire axis (solid and dotted lines). When local-field effects are included, no significant changes in the parallel component (dashed line) are observed. In fact, the inhomogeneity of the nanowire is pronounced in the direction perpendicular to the wire axis, which explains why it turns out that $\epsilon_\perp$ (dash-dotted line) is extremely sensitive to the local-field effects: the absorption is shifted to higher energies and the nanowire becomes transparent up to 6–7 eV. This result can be better understood if we use the fact that, for large wires, the *ab initio* calculations tend to reproduce the classical limit given by the effective-medium theory (Maxwell-Garnett 1904), that accounts for the classical effects due to the gathering of charges at the polarized surface of the wire.

These criteria to evaluate the importance of local-field effects actually apply only for polarizable systems, where the initial and final states involved in a particular transition are localized in a common spatial region (Aryasetiawan 1994). On the contrary, even in strongly inhomogeneous systems, if the inhomogeneity is not very polarizable, the induced potentials are small and consequently local fields are not so important.

Absorption and electron energy-loss experiments measure $\Im m\{\epsilon_M\}$ and $-\Im m\left\{\epsilon_M^{-1}\right\}$, respectively. Once $\epsilon_{G,G'}$ has been calculated, the spectra are obtained from:

$$\text{Abs}(\omega) = -\lim_{q\to 0}\Im m\{\epsilon_M(\omega)\} = -\lim_{q\to 0}\Im m\left\{\frac{1}{\epsilon_{G=0,G'=0}^{-1}(q,\omega)}\right\}, \quad (3.55a)$$

$$\text{EELS}(\omega) = -\lim_{q\to 0}\Im m\left\{\epsilon_M^{-1}(\omega)\right\} = -\lim_{q\to 0}\Im m\left\{\epsilon_{G=0,G'=0}^{-1}(q,\omega)\right\}, \quad (3.55b)$$

where, in particular, we considered EELS at vanishing momentum transfer. Equivalently:

$$\text{Abs}(\omega) = -\lim_{q\to 0}\Im m\left\{v_{ee\,G=0}(q)\bar{\chi}_{G=0,G'=0}(q,\omega)\right\}, \quad (3.56a)$$

$$\text{EELS}(\omega) = -\lim_{q\to 0}\Im m\left\{v_{ee\,G=0}(q)\chi_{G=0,G'=0}(q,\omega)\right\}. \quad (3.56b)$$

The only difference between $\bar{\chi}$ and χ is the absence of the long-range term $v_{ee\,G=0}$ of the Coulomb interaction in the Dyson equation (3.53). Therefore $v_{ee\,G=0}$ is the responsible for the difference between absorption and EELS spectra in solids (Sottile 2005). Bulk silicon absorbs in the energy range between 3–5 eV and the plasmon resonance is at 16.8 eV. In finite systems, instead, the long-range term becomes negligible. For this reason, when the limit $q \to 0$ is assumed, EELS and absorption mathematically coincide in finite systems.

3.7 Conclusions

In this chapter we have discussed how to model processes involving one-particle and neutral excitations, such as photoemission, inverse photoemission, optical absorption and electron energy loss spectroscopies, in order to calculate spectra directly comparable with experimental data.

We analyzed which physical approximations are involved in the microscopic modeling of the interaction of radiation (or particles) and matter, and how they affect the comparison with experimental spectra. When it is necessary to obtain macroscopic response functions from microscopic ones, a special care must be taken in performing appropriate averaging procedures.

Only in the simple case of a cubic system and a vanishing wavevector the dielectric function is a scalar quantity, otherwise we must consider a tensor that contains components related to the response of the system to a longitudinal and transverse perturbation. In the long-wavelength limit the longitudinal and transverse responses coincide, which is essential to calculate optical absorption, as a simple expression in terms of the density-density response function exists for the longitudinal component of the dielectric tensor.

Part II
Basic Theory

Chapter 4
Introduction to TDDFT

Eberhard K. U. Gross and Neepa T. Maitra

4.1 Introduction

Correlated electron motion plays a significant role in the spectra described in the previous chapters. Further, placing an atom, molecule or solid in a strong laser field reveals fascinating non-perturbative phenomena, such as non-sequential multiple-ionization (see Chap. 18), whose origins lie in the subtle ways electrons interact with each other. The direct approach to treat these problems is to solve the (non-relativistic) time-dependent Schrödinger equation for the many-electron wavefunction $\Psi(t)$:

$$\hat{H}(t)\Psi(t) = i\frac{\partial \Psi(t)}{\partial t}, \quad \hat{H}(t) = \hat{T} + \hat{V}_{ee} + \hat{V}_{ext}(t) \tag{4.1}$$

for a given initial wavefunction $\Psi(0)$. Here, the kinetic energy and electron–electron repulsion, are, respectively:

$$\hat{T} = -\frac{1}{2}\sum_{i=1}^{N} \nabla_i^2, \quad \text{and} \quad \hat{V}_{ee} = \frac{1}{2}\sum_{i \neq j}^{N} \frac{1}{|\boldsymbol{r}_i - \boldsymbol{r}_j|}, \tag{4.2}$$

and the "external potential" represents the potential the electrons experience due to the nuclear attraction and due to any field applied to the system (e.g. laser):

E. K. U. Gross (✉)
Max-Planck Institut für Mikrostrukturphysik,
Weinberg 2, 06120 Halle, Germany
e-mail: hardy@mpi-halle.mpg.de

N. T. Maitra
Hunter College and the Graduate Center,
The City University of New York,
695 Park Avenue, New York,
NY 10065, USA
e-mail: nmaitra@hunter.cuny.edu

M. A. L. Marques et al. (eds.), *Fundamentals of Time-Dependent Density Functional Theory*, Lecture Notes in Physics 837, DOI: 10.1007/978-3-642-23518-4_4,
© Springer-Verlag Berlin Heidelberg 2012

$$\hat{V}_{\text{ext}}(t) = \sum_{i=1}^{N} v_{\text{ext}}(\boldsymbol{r}_i, t). \tag{4.3}$$

For example, $v_{\text{ext}}(\boldsymbol{r}_i, t)$ can represent the Coulomb interaction of the electrons with a set of nuclei, possibly moving along some classical path,

$$v_{\text{ext}}(\boldsymbol{r}, t) = -\sum_{\nu=1}^{N_n} \frac{Z_\nu}{|\boldsymbol{r} - \boldsymbol{R}_\nu(t)|}, \tag{4.4}$$

where Z_ν and $\boldsymbol{R}_\nu$ denote the charge and position of the nucleus ν, and N_n stands for the total number of nuclei in the system. This may be useful to study, e.g., scattering experiments, chemical reactions, etc. Another example is the interaction with external fields, e.g. for a system illuminated by a laser beam we can write, in the dipole approximation,

$$v_{\text{ext}}(\boldsymbol{r}, t) = E f(t) \sin(\omega t) \boldsymbol{r} \cdot \boldsymbol{\alpha} - \sum_{\nu=1}^{N_n} \frac{Z_\nu}{|\boldsymbol{r} - \boldsymbol{R}_\nu|} \tag{4.5}$$

where $\boldsymbol{\alpha}$, ω and E are the polarization, the frequency and the amplitude of the laser, respectively. The function $f(t)$ is an envelope that describes the temporal shape of the laser pulse. We use atomic units ($e^2 = \hbar = m = 1$) throughout this chapter; all distances are in Bohr, energies in Hartrees (1 H = 27.21 eV = 627.5 kcal/mol), and times in units of 2.419×10^{-17} s.

Solving Eq. 4.1 is an exceedingly difficult task. Even putting aside time-dependence, the problem of finding the ground-state scales exponentially with the number of electrons. Moreover, Ψ contains far more information than one could possibly need or even want. For example, consider storing the ground state of the oxygen atom, and for simplicity, disregard spin. Then Ψ depends on 24 coordinates, three for each of the eight electrons. Allowing ourselves a modest ten grid-points for each coordinate, means that we need 10^{24} numbers to represent the wavefunction. Assuming each number requires one byte to store, and that the capacity of a DVD is 10^{10} bytes, we see that 10^{14} DVD's are required to store just the ground-state wavefunction of the oxygen atom, even on a coarse grid. Physically, we are instead interested in integrated quantities, such as one- or two-body probability-densities, which, traditionally can be extracted from this foreboding Ψ. However, a method that could yield such quantities directly, by-passing the need to calculate Ψ, would be highly attractive. This is the idea of density-functional theories. In fact, in 1964, Hohenberg and Kohn (1964), proved that *all* observable properties of a static many-electron system can be extracted exactly, in principle, from the one-body ground-state density alone. Twenty years later, Runge and Gross extended this to time-dependent systems, showing that all observable properties of a many-electron system, beginning in a given initial state $\Psi(0)$, may be extracted from the one-body time-dependent density alone (Runge and Gross 1984). What has made (TD)DFT so incredibly successful is the Kohn–Sham (KS) system: the density of the interacting many-electron system is obtained as the density of an auxiliary system of non-interacting fermions, living in a one-body potential. The exponential scaling with system-size that the solution to Eq. 4.1 requires is replaced in TDDFT by the much gentler N^3 or N^2 scaling (depending on the implementation) (Marques 2006), opening the door to the quantum mechanical study of much larger systems, from nanoscale devices to biomolecules. (See Chaps. 19–21 for details on the numerical issues). Although the ground-state and time-dependent theories have a similar flavor, and modus operandi, their proofs and functionals are quite distinct.

Before we delve into the details of the fundamental theorems of TDDFT in the next section, we make some historical notes. As early as 1933, Bloch proposed a time-dependent Thomas–Fermi model (Bloch 1933). Ando (1977a, b), Peuckert (1978) and Zangwill and Soven (1980a, 1980b) ran the first time-dependent KS calculations, assuming a TDKS theorem exists. They treated the linear density response to a time-dependent external potential as the response of non-interacting electrons to an effective time-dependent potential. Ando calculated resonance energy and absorption lineshapes for intersubband transitions on the surface of silicon, while Peuckert and Zangwill and Soven studied rare-gas atoms. In analogy to ground-state KS theory, this effective potential was assumed to contain an exchange-correlation part, $v_{xc}(\mathbf{r}, t)$, in addition to the time-dependent external and Hartree terms. Peuckert suggested an iterative scheme for the calculation of v_{xc}, while Ando, Zangwill and Soven adopted the functional form of the static exchange-correlation potential in LDA. Significant steps towards a rigorous foundation of TDDFT were taken by Deb and Ghosh (1982), Ghosh and Deb (1982, 1983a, 1983b) for time-periodic potentials and by Bartolotti (1981, 1982, 1984, 1987) for adiabatic processes. These authors formulated and explored Hohenberg–Kohn and KS type theorems for the time-dependent density for these cases. Modern TDDFT is based on the general formulation of Runge and Gross (1984).

TDDFT is being used today in an ever-increasing range of applications to widely-varying systems in chemistry, biology, solid-state physics, and materials science. We end the introduction by classifying these into four areas. First, the vast majority of TDDFT calculations today lie in spectroscopy (Chaps. 1–3 earlier), yielding response and excitations of atoms, molecules and solids. The laser field applied to the system, initially in its ground-state, is weak and perturbation theory applies. We need to know only the exchange-correlation potential in the vicinity of the ground-state, and often (but not always), formulations directly in frequency-domain are used (Chap. 7 and later in this chapter). Overall, results for excitation energies tend to be fairly good (few tenths of an eV error, typically) but depend significantly on the system and type of excitation considered, e.g. the errors for long-range charge-transfer excitations can be ten times as large. For solids, to obtain accurate optical absorption spectra of insulators needs functionals more sophisticated than the simplest ones (ALDA), however ALDA does very well for electron-energy-loss spectra. We shall return to the general performance of TDDFT for spectra in Sect. 4.8.

The second class of applications is real-time dynamics in non-perturbative fields. The applied electric field is comparable to, or greater than, the static electric field due to the nuclei. Fascinating and subtle electron interaction effects can make the "single active electron" picture often used for these problems break down. For dynamics in strong fields, it is pushing today's computational limits for correlated wavefunction methods to go beyond one or two electrons in three-dimensions, so TDDFT is particularly promising in this regime. However, the demands on the functionals for accurate results can be challenging. Chapter 18 discusses many interesting phenomena, that reveal fundamental properties of atoms and molecules. Also under this umbrella are coupled electron-ion dynamics. For example, applying a laser pulse to a molecule or cluster, drives both the electronic and nuclear system out of equilibrium; generally their coupled motion is highly complicated, and various approximation schemes to account for electronic-nuclear "back-reaction" have been devised. Often in photochemical applications, the dynamics are treated beginning on an excited potential energy surface; that is, the dynamics leading to the initial electronic excitation is not explicitly treated but instead defines the initial state for the subsequent field-free dynamics of the full correlated electron and ion system. We refer the reader to Chaps. 14–16 for both formal and practical discussions on how to treat this challenging problem.

The third class of applications returns to the ground-state: based on the fluctuation-dissipation theorem, one can obtain an expression for the ground-state exchange-correlation energy from a TDDFT response function, as is discussed in Chap. 22. Such calculations are significantly more computationally demanding than usual ground-state calculations but provide a natural methodology for some of the most difficult challenges for ground-state approximations, in particular van der Waal's forces. We refer the reader to Chaps. 22 and 23 for a detailed discussion.

The fourth class of applications is related to viscous forces arising from electron-electron interactions in very large finite systems, or extended systems such as solids. Consider an initial non-equilibrium state in such a system, created, for example, by a laser pulse that is then turned off. For a large enough system, electron interaction subsequently relaxes the system to the ground-state, or to thermal equilibrium. Relaxation induced by electron-interaction can be in principle exactly captured in TDDFT, but for a theoretically consistent formulation, one should go beyond the most commonly used approximations. A closely related approach is time-dependent current-density functional theory (TDCDFT) (Sect. 4.4.4 and Chap. 24), where an xc vector-potential provides the viscous force (D'Agosta and Vignale 2006; Ullrich 2006b). Dissipation phenomena studied so far, using either TDDFT or TDCDFT, include energy loss in atomic collisions with metal clusters (Baer and Siam 2004), the stopping power of ions in electron liquids (Nazarov et al. 2005; Hatcher 2008; Nazarov 2007), spin-Coulomb drag (d'Amico and Ullaich 2006, 2010), and a hot electron probing a molecular resonance at a surface (Gavnholt et al. 2009). Transport through molecular devices (Chap. 17) is an important subgroup of these applications: how a system evolves to a steady-state after a bias is applied (Stefanucci Almbhdh 2004a; Kurth et al. 2005, 2010; Khosravi et al. 2009; Stefanucci 2007; Koentopp et al. 2008; Zheng 2010). In this problem, to account for coupling of electrons in the molecular wire to a "bath" of electrons in the leads, or to account for coupling to external phonon modes, one is led to the "open systems" analyses, reviewed in Chaps. 10 and 11 (Gebauer and Car 2004a; Burke 2005; Yuen-Zhou et al. 2010;

Di Ventra D'Agosta 2007; Chen 2007; Appel and di ventra 2009). In Ullrich (2002b) the linear response of weakly disordered systems was formulated, embracing both extrinsic damping (interface roughness and charged impurities) as well as intrinsic dissipation from electron interaction.

The present chapter is organized as follows. Section 4.2 presents the proof of the fundamental theorem in TDDFT. Section 4.3 then presents the time-dependent KS equations, the "performers" of TDDFT. In Sect. 4.4 we discuss several details of the theory, which are somewhat technical but important if one scratches below the surface. Section 4.5 derives the linear response formulation, and the matrix equations which run the show for most of the applications today. Section 4.6 briefly presents the equations for higher-order response, while Sect. 4.7 describes some of the approximations in use currently for the xc functional. Finally, Sect. 4.8 gives an overview of the performance and challenges for the approximations today.

4.2 One-to-One Density-Potential Mapping

The central theorem of TDDFT (the Runge–Gross theorem) proves that there is a one-to-one correspondence between the external (time-dependent) potential, $v_{\text{ext}}(r, t)$, and the electronic one-body density, $n(r, t)$, for many-body systems evolving from a fixed initial state Ψ_0 (Runge and Gross 1984). The density $n(r, t)$ is the probability (normalized to the particle number N) of finding any one electron, of any spin σ, at position r:

$$n(r, t) = N \sum_{\sigma, \sigma_2 .. \sigma_N} \int d^3 r_2 \cdots \int d^3 r_N |\Psi(r\sigma, r_2\sigma_2 \cdots r_N\sigma_N, t)|^2 \qquad (4.6)$$

The implications of this theorem are enormous: if we know only the time-dependent density of a system, evolving from a given initial state, then this identifies the external potential that produced this density. The external potential completely identifies the Hamiltonian (the other terms given by Eq. 4.2 are determined from the fact that we are dealing with electrons, with N being the integral of the density of Eq. 4.6 over r.) The time-dependent Schrödinger equation can then be solved, in principle, and all properties of the system obtained. That is, for this given initial-state, the electronic density, a function of just three spatial variables and time, determines all other properties of the interacting many-electron system.

This remarkable statement is the analogue of the Hohenberg-Kohn theorem for ground-state DFT, where the situation is somewhat simpler: the density-potential map there holds only for the ground-state, so there is no time-dependence and no dependence on the initial state. The Hohenberg-Kohn proof is based on the Rayleigh-Ritz minimum principle for the energy. A straightforward extension to the time-dependent domain is not possible since a minimum principle is not available in this case.

Instead, the proof for a 1–1 mapping between time-dependent potentials and time-dependent densities is based on considering the quantum-mechanical equation of motion for the current-density, for a Hamiltonian of the form of Eqs. 4.1–4.3. The proof requires the potentials $v_{\text{ext}}(\boldsymbol{r}, t)$ to be time-analytic around the initial time, i.e. that they equal their Taylor-series expansions in t around $t = 0$, for a finite time interval:

$$v_{\text{ext}}(\boldsymbol{r}, t) = \sum_{k=0}^{\infty} \frac{1}{k!} v_{\text{ext},k}(\boldsymbol{r}) t^k. \tag{4.7}$$

The aim is to show that two densities $n(\boldsymbol{r}, t)$ and $n'(\boldsymbol{r}, t)$ evolving from a common initial state Ψ_0 under the influence of the potentials $v_{\text{ext}}(\boldsymbol{r}, t)$ and $v'_{\text{ext}}(\boldsymbol{r}, t)$ are always different provided that the potentials differ by more than a purely time-dependent function:

$$v_{\text{ext}}(\boldsymbol{r}, t) \neq v'_{\text{ext}}(\boldsymbol{r}, t) + c(t). \tag{4.8}$$

The above condition is a physical one, representing simply a gauge-freedom. A purely time-dependent constant in the potential cannot alter the physics: if two potentials differ only by a purely time-dependent function, their resulting wavefunctions differ only by a purely time-dependent phase factor. Their resulting densities are identical. All variables that correspond to expectation values of Hermitian operators are unaffected by such a purely time-dependent phase. There is an analogous condition in the ground-state proof of Hohenberg and Kohn. Equation 4.8 is equivalent to the statement that for the expansion coefficients $v_{\text{ext},k}(\boldsymbol{r})$ and $v'_{\text{ext},k}(\boldsymbol{r})$ [where, as in Eq. 4.7, $v'_{\text{ext}}(\boldsymbol{r}, t) = \sum_{k=0}^{\infty} \frac{1}{k!} v'_{\text{ext},k}(\boldsymbol{r}) t^k$] there exists a smallest integer $k \geq 0$ such that

$$v_{\text{ext},k}(\boldsymbol{r}) - v'_{\text{ext},k}(\boldsymbol{r}) = \frac{\partial^k}{\partial t^k} \left[v_{\text{ext}}(\boldsymbol{r}, t) - v'_{\text{ext}}(\boldsymbol{r}, t) \right] \Big|_{t=0} \neq \text{const.} \tag{4.9}$$

The initial state Ψ_0 need not be the ground-state or any stationary state of the initial potential, which means that "sudden switching" is covered by the RG theorem. But potentials that turn on adiabatically from $t = -\infty$, are not, since they do not satisfy Eq. 4.7 (see also Sect. 4.5.3).

The first step of the proof demonstrates that the current-densities

$$\boldsymbol{j}(\boldsymbol{r}, t) = \langle \Psi(t) | \hat{\boldsymbol{j}}(\boldsymbol{r}) | \Psi(t) \rangle \tag{4.10}$$

and

$$\boldsymbol{j}'(\boldsymbol{r}, t) = \langle \Psi'(t) | \hat{\boldsymbol{j}}(\boldsymbol{r}) | \Psi'(t) \rangle \tag{4.11}$$

are different for different potentials v_{ext} and v'_{ext}. Here,

$$\hat{j}(r) = \frac{1}{2i} \sum_{i=1}^{N} [\nabla_i \delta(r - r_i) + \delta(r - r_i)\nabla_i] \qquad (4.12)$$

is the usual paramagnetic current-density operator. In the second step, use of the continuity equation shows that the densities n and n' are different. We now proceed with the details.

Step 1 We apply the equation of motion for the expectation value of a general operator $\hat{Q}(t)$,

$$\frac{\partial}{\partial t} \langle \Psi(t) | \hat{Q}(t) | \Psi(t) \rangle = \langle \Psi(t) | \left(\frac{\partial \hat{Q}}{\partial t} - i[\hat{Q}(t), \hat{H}(t)] \right) | \Psi(t) \rangle, \qquad (4.13)$$

to the current densities:

$$\frac{\partial}{\partial t} j(r, t) = \frac{\partial}{\partial t} \langle \Psi(t) | \hat{j}(r) | \Psi(t) \rangle = -i\langle \Psi(t) | [\hat{j}(r), \hat{H}(t)] | \Psi(t) \rangle \qquad (4.14a)$$

$$\frac{\partial}{\partial t} j'(r, t) = \frac{\partial}{\partial t} \langle \Psi'(t) | \hat{j}(r) | \Psi'(t) \rangle = -i\langle \Psi'(t) | [\hat{j}(r), \hat{H}'(t)] | \Psi'(t) \rangle, \qquad (4.14b)$$

and take their difference evaluated at the initial time. Since Ψ and Ψ' evolve from the same initial state

$$\Psi(t = 0) = \Psi'(t = 0) = \Psi_0, \qquad (4.15)$$

and the corresponding Hamiltonians differ only in their external potentials, we have

$$\frac{\partial}{\partial t} \left[j(r, t) - j'(r, t) \right] \Big|_{t=0} = -i\langle \Psi_0 | [\hat{j}(r), \hat{H}(0) - \hat{H}'(0)] | \Psi_0 \rangle$$
$$= -n_0(r)\nabla \left[v_{\text{ext}}(r, 0) - v'_{\text{ext}}(r, 0) \right] \qquad (4.16)$$

where $n_0(r)$ is the initial density. Now, if the condition (4.9) is satisfied for $k = 0$ the right-hand side of (4.16) cannot vanish identically and j and j' will become different infinitesimally later than $t = 0$. If the smallest integer k for which Eq. 4.9 holds is greater than zero, we use Eq. 4.13 $(k + 1)$ times. That is, as for $k = 0$ above where we used $\hat{Q}(t) = \hat{j}(r)$ in Eq. 4.13, for $k = 1$, we take $\hat{Q}(t) = -i[\hat{j}(r), \hat{H}(t)]$; for general k, $\hat{Q}(t) = (-i)^k [[[\hat{j}(r), \hat{H}(t)], \hat{H}(t)] \ldots \hat{H}(t)]_k$ meaning there are k nested commutators to take. After some algebra[1]:

$$\left(\frac{\partial}{\partial t} \right)^{k+1} [j(r, t) - j'(r, t)] \Big|_{t=0} = -n_0(r)\nabla w_k(r) \neq 0 \qquad (4.17)$$

---

[1] Note that Eq. 4.17 applies for all integers from 0 to this smallest k for which Eq. 4.9 holds, but not for integers larger than this smallest k.

with

$$w_k(r) = \left(\frac{\partial}{\partial t}\right)^k [v_{\text{ext}}(r, t) - v'_{\text{ext}}(r, t)]\Big|_{t=0}. \qquad (4.18)$$

Once again, we conclude that infinitesimally later than the initial time,

$$j(r, t) \neq j'(r, t). \qquad (4.19)$$

This first step thus proves that the current-densities evolving from the same initial state in two physically distinct potentials, will differ. That is, it proves a one-to-one correspondence between current-densities and potentials, for a given initial-state.

Step 2 To prove the corresponding statement for the densities we use the continuity equation

$$\frac{\partial n(r, t)}{\partial t} = -\nabla \cdot j(r, t) \qquad (4.20)$$

to calculate the $(k + 2)$nd time-derivative of the density $n(r, t)$ and likewise of the density $n'(r, t)$. Taking the difference of the two at the initial time $t = 0$ and inserting Eq. 4.18 yields

$$\left(\frac{\partial}{\partial t}\right)^{k+2} [n(r, t) - n'(r, t)]\Big|_{t=0} = \nabla \cdot [n_0(r)\nabla w_k(r)]. \qquad (4.21)$$

Now, if there was no divergence-operator on the r.h.s., our task would be complete, showing that the densities $n(r, t)$ and $n'(r, t)$ will become different infinitesimally later than $t = 0$. To show that the divergence does not render the r.h.s. zero, thus allowing an escape from this conclusion, we consider the integral

$$\int d^3r n_0(r)[\nabla w_k(r)]^2 = -\int d^3r w_k(r)\nabla \cdot [n_0(r)\nabla w_k(r)]$$
$$+ \oint dS \cdot [n_0(r)w_k(r)\nabla w_k(r)], \qquad (4.22)$$

where we have used Green's theorem. For physically reasonable potentials (i.e. potentials arising from normalizable external charge densities), the surface integral on the right vanishes (Gross and Kohn 1990) (more details are given in Sect. 4.4.1). Since the integrand on the left-hand side is strictly positive or zero, the first term on the right must be strictly positive. That is, $\nabla \cdot [n_0(r)\nabla w_k(r)]$ cannot be zero everywhere. This completes the proof of the theorem.

We have shown that densities evolving from the same initial wavefunction Ψ_0 in different potentials must be different. Schematically, the Runge-Gross theorem shows

$$\Psi_0 : v_{\text{ext}} \overset{1-1}{\longleftrightarrow} n. \tag{4.23}$$

The backward arrow, that a given time-dependent density points to a single time-dependent potential for a given initial state, has been proven above. The forward arrow follows directly from the uniqueness of solutions to the time-dependent Schrödinger equation.

Due to the one-to-one correspondence, for a given initial state, the time-dependent density determines the potential up to a purely time-dependent constant. The wavefunction is therefore determined up to a purely-time-dependent phase, as discussed at the beginning of this section, and so can be regarded as a functional of the density and initial state:

$$\Psi(t) = e^{-i\alpha(t)} \Psi[n, \Psi_0](t). \tag{4.24}$$

As a consequence, the expectation value of any quantum mechanical Hermitian operator $\hat{Q}(t)$ is a *unique* functional of the density and initial state (and, not surprisingly, the ambiguity in the phase cancels out):

$$Q[n, \Psi_0](t) = \langle \Psi[n, \Psi_0](t) | \hat{Q}(t) | \Psi[n, \Psi_0](t) \rangle. \tag{4.25}$$

We also note that the particular form of the Coulomb interaction did not enter into the proof. In fact, the proof applies not just to electrons, but to *any* system of identical particles, interacting with any (but fixed) particle-interaction, and obeying either fermionic or bosonic statistics.

In Sect. 4.4 we shall return to some details and extensions of the proof, but now we proceed with how TDDFT operates in practice: the time-dependent Kohn–Sham equations.

4.3 Time-Dependent Kohn–Sham Equations

Finding functionals directly in terms of the density can be rather difficult. In particular, it is not known how to write the kinetic energy as an explicit functional of the density. The same problem occurs in ground-state DFT, where the search for accurate kinetic-energy density-functionals is an active research area. Instead, like in the ground-state theory, we turn to a non-interacting system of fermions called the Kohn–Sham (KS) system, defined such that it exactly reproduces the density of the true interacting system. A large part of the kinetic energy of the true system is obtained directly as an orbital-functional, evaluating the usual kinetic energy operator on the KS orbitals. (The rest, along with other many-electron effects, is contained in the exchange-correlation potential.) All properties of the true system can be extracted from the density of the KS system.

Because the 1–1 correspondence between time-dependent densities and time-dependent potentials can be established for any *given* interaction $\hat{V}_{\text{ee}}$, in particular also for $\hat{V}_{ee} \equiv 0$, it applies to the KS system. Therefore the external potential $v_{KS}[n; \Phi_0](r, t)$ of a non-interacting system reproducing a given density $n(r, t)$, starting in the initial state Φ_0, is uniquely determined. The initial KS state Φ_0 is almost always chosen to be a single Slater determinant of single-particle spin-orbitals $\varphi_i(r, 0)$ (but need not be); the only condition on its choice is that it must be compatible with the given density. That is, it must reproduce the initial density and also its first time-derivative (from Eq. 4.20, see also Eqs. 4.29–4.30 shortly). However, the 1–1 correspondence only ensures the uniqueness of $v_{KS}[n; \Phi_0]$ but not its existence for an *arbitrary* $n(r, t)$. That is, the proof does not tell us whether a KS system exists or not; this is called the non-interacting v-representability problem, similar to the ground-state case. We return to this question later in Sect. 4.4.2, but for now we assume that v_{KS} exists for the time-dependent density of the *interacting* system of interest. Under this assumption, the density of the interacting system can be obtained from

$$n(r, t) = \sum_{j=1}^{N} |\varphi_j(r, t)|^2 \tag{4.26}$$

with orbitals $\varphi_j(r, t)$ satisfying the time-dependent KS equation

$$i\frac{\partial}{\partial t}\varphi_j(r, t) = \left[-\frac{\nabla^2}{2} + v_{KS}[n; \Phi_0](r, t)\right]\varphi_j(r, t). \tag{4.27}$$

Analogously to the ground-state case, v_{KS} is decomposed into three terms:

$$v_{KS}[n; \Phi_0](r, t) = v_{ext}[n; \Psi_0](r, t) + \int d^3r' \frac{n(r', t)}{|r - r'|} + v_{xc}[n; \Psi_0, \Phi_0](r, t), \tag{4.28}$$

where $v_{ext}[n; \Psi_0](r, t)$ is the external time-dependent field. The second term on the right-hand side of Eq. 4.28 is the time-dependent Hartree potential, describing the interaction of classical electronic charge distributions, while the third term is the exchange-correlation (xc) potential which, in practice, has to be approximated. Equation 4.28 *defines* the xc potential: it, added to the classical Hartree potential, is the difference between the external potential that generates density $n(r, t)$ in an interacting system starting in initial state Ψ_0 and the one-body potential that generates this same density in a non-interacting system starting in initial state Φ_0.

The functional-dependence of v_{ext} displayed in the first term on the r.h.s. of Eq. 4.28 is not important in practice, since for real calculations, the external potential is given by the physics at hand. Only the xc potential needs to be approximated in practice, as a functional of the density, the true initial state and the KS initial state. This functional is a very complex one: knowing it implies the solution of all time-dependent Coulomb interacting problems.

As in the static case, the great advantage of the time-dependent KS scheme lies in its computational simplicity compared to other methods such as time-dependent configuration interaction (Errea et al. 1985; Reading and Ford 1987; Krause 2005;

Krause 2007) or multi-configuration time-dependent Hartree–Fock (Zanghellini et al. 2003; Kato and Kono 2004; Nest 2005; Meyer 1990; Caillat 2005). A TDKS calculation proceeds as follows. An initial set of N orthonormal KS orbitals is chosen, which must reproduce the exact density of the true initial state Ψ_0 (given by the problem) and its first time-derivative:

$$n(\boldsymbol{r}, 0) = \sum_{i=1}^{N} |\varphi_i(\boldsymbol{r}, 0)|^2 = N \sum_{\sigma, \sigma_2 \ldots \sigma_N} \int d^3 r_2 \cdots \int d^3 r_N |\Psi_0(\boldsymbol{x}, \boldsymbol{x}_2 \ldots \boldsymbol{x}_N)|^2$$
$$(4.29)$$

(and we note there exist infinitely many Slater determinants that reproduce a given density (Harriman 1981; Zumbach and Maschke 1983)), and

$$\dot{n}(\boldsymbol{r}, 0) = -\nabla \cdot \mathfrak{Im} \sum_{i=1}^{N} \sum_{\sigma} \varphi_i^*(\boldsymbol{r}, 0) \nabla \varphi_i(\boldsymbol{r}, 0)$$

$$= -N \nabla \cdot \mathfrak{Im} \sum_{\sigma, \sigma_2 \ldots \sigma_N} \int d^3 r_2 \cdots \int d^3 r_N \Psi_0^*(\boldsymbol{x}, \boldsymbol{x}_2 \ldots \boldsymbol{x}_N) \nabla \Psi_0(\boldsymbol{x}, \boldsymbol{x}_2 \ldots \boldsymbol{x}_N)$$
$$(4.30)$$

using notation $\boldsymbol{x} = (\boldsymbol{r}, \sigma)$. The TDKS equations (Eq. 4.27) then propagate these initial orbitals, under the external potential given by the problem at hand, together with the Hartree potential and an approximation for the xc potential in Eq. 4.28. In Sect. 4.7 we shall discuss the approximations that are usually used here.

The choice of the KS initial state, and the fact that the KS potential depends on this choice is a completely new feature of TDDFT without a ground-state analogue. A discussion of the subtleties arising from initial-state dependence can be found in Chap. 8. In practice, the theory would be much simpler if we could deal with functionals of the density alone. For a large class of systems, namely those where both Ψ_0 and Φ_0 are non-degenerate ground states, observables are indeed functionals of the density *alone*. This is because any non-degenerate ground state is a unique functional of its density $n_0(\boldsymbol{r})$ by virtue of the traditional Hohenberg–Kohn theorem (Hohenberg and Kohn 1964). In particular, the initial KS orbitals are *uniquely* determined as well in this case. We emphasize this is often the case in practice; in particular in the linear response regime, where spectra are calculated (see Sect. 4.5). This is where the vast majority of applications of TDDFT lie today.

4.4 More Details and Extensions

We now discuss in more detail some important points that arose in the derivation of the proof above. Several of these are discussed at further length in the subsequent chapters.

4.4.1 The Surface Condition

It is essential for Step 2 of the RG proof that the surface term

$$\oint dS \cdot n_0(r) w_k(r) \nabla w_k(r) \tag{4.31}$$

appearing in Eq. 4.22 vanishes. Let us consider realistic physical potentials of the form

$$v_{\text{ext}}(r, t) = \int d^3r \frac{n_{\text{ext}}(r', t)}{|r - r'|} \tag{4.32}$$

where $n_{\text{ext}}(r, t)$ denotes normalizable charge-densities external to the electronic system. A Taylor expansion in time of this expression shows that the coefficients $v_{\text{ext},k}$, and therefore w_k fall off at least as $1/r$ asymptotically so that, for physical initial densities, the surface integral vanishes (Gross and Kohn 1990). However, if one allows more general potentials, the surface integral need not vanish: consider fixing an initial state Ψ_0 that leads to a certain asymptotic form of $n_0(r)$. Then one can always find potentials which increase sufficiently steeply in the asymptotic region such that the surface integral does not vanish. In (Xu and Rajagopal 1985; Dhara and Ghosh 1987) several examples of this are discussed, where the r.h.s. of Eq. 4.21 can be zero even while the term inside the divergence is non-zero. These cases are however largely unphysical, e.g. leading to an infinite potential energy per particle near the initial time. It would be desirable to prove the one-to-one mapping under a physical condition, such as finite energy expectation values.

An interesting case is that of extended periodic systems in a uniform electric field, such as is often used to describe the bulk of a solid in a laser field. Representing the field by a scalar linear potential is not allowed because a linear potential is not an operator on the Hilbert space of periodic functions. The periodic boundary conditions may be conveniently modelled by placing the system on a ring. Let us first consider a finite ring; for example, a system such as a nanowire with periodic boundary conditions in one direction, and finite extent in the other dimensions. Then an electric field going around the ring cannot be generated by a scalar potential: according to Faraday's law, $\nabla \times E = -dB/dt$. Such an electric field can only be produced by a time-dependent magnetic field threading the center of the ring. The situation for an infinite periodic system in a uniform field also requires a vector potential if one wants the Hamiltonian to preserve periodicity. The vector potential is purely time-dependent in this case, and leads to a uniform electric field via $E(t) = -(1/c)dA(t)/dt$. Once again, TDDFT cannot be applied, even though $B = \nabla \times A = 0$. That is, for a finite ring with an electric field around the ring, there is a real, physical magnetic field, while for the case of the infinite periodic system there is no physical magnetic field, but a vector potential is required for mathematical reasons. In either case, TDDFT does not apply (Maitra 2003). However, fortunately, the theorems of TDDFT have been generalized to include vector potentials (Ghosh and Deb 1988), leading to time-dependent *current-density* functional theory (TDCDFT) (Vignale and Kohn 1996), which will be discussed in detail in Chap. 24. Moreover, if the optical response is instead obtained via the limit $q \to 0$, the problem can be formulated as a scalar field and TDDFT does apply. In this case, the surface in the RG proof can be chosen as an integral multiple of q and the periodicity of the system, and the surface term vanishes. Finally, note that a uniform field does not usually pose any problem for any finite system, where the asymptotic decay of the initial density in physical cases is fast enough to kill the surface integral.

4.4.2 Interacting and Non-interacting v-Representability

The RG proof presented above proves uniqueness of the potential that generates a given density from a given initial state, but does not prove its existence. The question of whether a given density comes from evolution in a scalar potential is called v-representability, a subtle issue that arises also in the ground-state case (Dreizler and Gross 1990). In both the ground-state and time-dependent theories, it is still an open and difficult one. (See also Fig. 4.1.) Some discussion for the time-dependent case can be found in (Kohl and Dreizler 1986; Ghosh and Deb 1988).

Perhaps more importantly, however, is the question of whether a density, known to be generated in an interacting system, can be reproduced in a non-interacting one. That is, given a time-dependent external potential and initial state, does a KS system exist? This question is called "non-interacting v-representability" and was answered under some well-defined conditions in (van Leeuwen 1999). A feature of this proof is that it leads to the explicit construction of the KS potential. Chapter 9 covers this in detail.

One condition is that the density is assumed to be time-analytic about the initial time. The Runge-Gross proof only requires the potential to be time-analytic, but that the density is also is an additional condition required for the non-interacting v-representability proof of (van Leeuwen 1999). In fact, it is a much more restrictive condition than that on the potential, as has been recently discussed in (Maitra 2010). The entanglement of space and time in the time-dependent Schrödinger equation means that spatial singularities (such as the Coulomb one) in the potential can lead to non-analyticities in time in the wavefunction, and consequently the density, even when the potential is time-analytic. Again, we stress that non time-analytic *densities* are covered by the RG proof for the one-to-one density-potential mapping (Sect. 4.2). Two different non-time-analytic densities will still differ in their formal time-Taylor series at some order at some point in space, and this is all that is needed for the proof.

The other conditions needed for the KS-existence proof are much less severe. The choice of the initial KS wavefunction is simply required to satisfy Eqs. 4.29 and 4.30 (and be well-behaved in having finite energy expectation values).

It is interesting to point out here that much less is known about non-interacting v-representability in the ground-state theory.

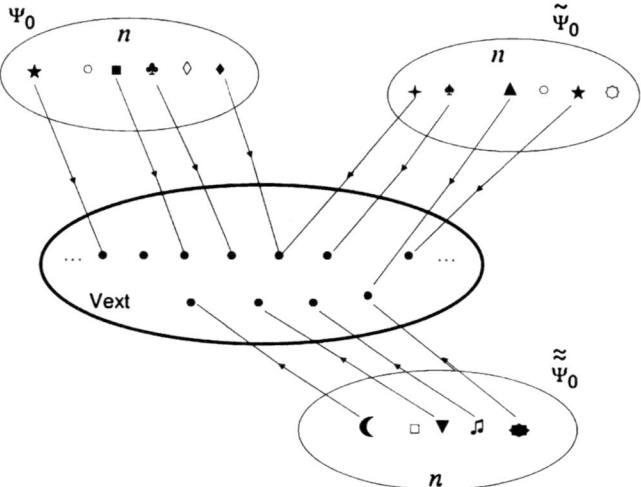

Fig. 4.1 A cartoon to illustrate the RG mapping. The three outer ellipses contain possible density evolutions, $n(r, t)$, that arise from evolving the initial state labelling the ellipse in a one-body potential v_{ext}; the potentials are different points contained in the central ellipse. Different symbols label different $n(r, t)$ and one may find the same $n(r, t)$ may live in more than one ellipse. The RG theorem says that no two lines from the same ellipse may point to the same v_{ext} in the central ellipse. Lines emanating from two different outer ellipses may point to either different or same points in the central one. If they come from identical symbols, they must point to different points (i.e. two different initial states may give rise to the same density evolution in two different potentials). The initial state labelling the lower ellipse has either a different initial density or current than the initial states labelling the upper ellipses; hence no symbols inside overlap with those in the upper. However, they may be generated from the same potential v_{ext}. Some densities, the open symbols, do not point to any v_{ext}, representing the non-v-representable densities. Finally, if an analogous cartoon was made for the KS system, whether all the symbols that are solid in the cartoon above, remain solid in the KS cartoon, represents the question of non-interacting v-representability

4.4.3 A Variational Principle

It is important to realise that the TDKS equations do *not* follow from a variational principle: as presented above, all that was needed was (i) the Runge-Gross proof that a given density evolving from a fixed Ψ_0 points to a unique potential, for interacting and non-interacting systems, and (ii) the assumption of non-interacting v-representability.

Nevertheless, it is interesting to ask whether a variational principle exists in TDDFT. In the ground-state, the minimum energy principle meant that one need only approximate the xc energy as a functional of the density, $E_{xc}[n]$, and then take its functional derivative to find the ground-state potential: $v_{xc}^{GS}[n](r) = \delta E_{xc}[n]/\delta n(r)$. Is there an analogue in the time-dependent case? Usually the action plays the role of the energy in time-dependent quantum mechanics, but here the situation is not as simple: if one *could* write $v_{xc}[n](r, t) = \delta A_{xc}[n]/\delta n(r, t)$ for some xc action functional $A_{xc}[n]$ (dropping initial-state dependence for simplicity for this argument), then we see that $\delta v_{\rm xc}[n](r, t)/\delta n(r', t') = \delta^2 A_{\rm xc}[n]/\delta n(r, t)\delta n(r', t')$ which is symmetric in t and t'. However that would imply that density-changes at later times $t' > t$ would affect the xc potential at earlier times, i.e. causality would be violated. This problem was first pointed out in (Gross et al. 1996). Chapter 9 discusses this problem, as well as its solution, at length. Indeed one can define an action consistent with causality, on a Keldysh contour (van Leeuwen 1998), using "Liouville space pathways" (Mukamel 2005) or using the usual real-time definition but including boundary terms (Vignale 2008).

4.4.4 The Time-Dependent Current

Step 1 of the RG proof proved a one-to-one mapping between the external potential and the current-density, while the second step invoked continuity, with the help of a surface condition, to prove the one-to-one density-potential mapping. In fact, a one-to-one mapping between current-densities and vector potentials, a special case of which is the class of scalar potentials, has been proven in later work (Xu and Rajagopal 1985; Ng 1989; Ghosh and Deb 1988; Gross et al. 1996). But as will be discussed in Chap. 24, even when the external potential is merely scalar, there can be advantages to the time-dependent current-density functional theory (TDCDFT) framework. Simpler functional approximations in terms of the time-dependent current-density can be more accurate than those in terms of the density: in particular, local functionals of the current-density correspond to non-local density-functionals, important in the optical response of solids for example, and for polarizabilities of long-chain polymers.

In TDCDFT, the KS system is defined to reproduce the exact current-density $j(r, t)$ of the interacting system, but in TDDFT the KS current $j_{\rm KS}(r, t)$ is *not* generally equal to the true current. As both current densities satisfy the continuity equation with the same density, i.e.

$$\dot{n}(r, t) = -\nabla \cdot j(r, t) = -\nabla \cdot j_{\rm KS}(r, t), \tag{4.33}$$

we know immediately that the longitudinal parts of j and $j_{\rm KS}$ must be identical. However, they may differ by a rotational component:

$$j(r, t) = j_{\rm KS}(r, t) + j_{\rm xc}(r, t) \tag{4.34}$$

where

$$j_{\rm xc}(r, t) = \nabla \times C(r, t) \tag{4.35}$$

with some real function $C(r, t)$. We also know for sure that

$$\int d^3r\, j_{\rm xc}(r, t) = 0 \tag{4.36}$$

because $\int d^3r\, \boldsymbol{j}(\boldsymbol{r}, t) = \int d^3r\, \boldsymbol{r}\dot{n}(\boldsymbol{r}, t)$ and the KS system exactly reproduces the true density.

The question of when the KS current equals the true current may equivalently be posed in terms of v-representability of the current: Given a current generated by a scalar potential in an interacting system, is that current non-interacting v-representable? That is, does a scalar potential exist in which a non-interacting system would reproduce this interacting, v-representable current exactly? By Step 1 of the RG proof applied to non-interacting systems, if it does exist, it is unique, and, since it also reproduces the exact density by the continuity equation, it is identical to the KS potential. There are two aspects to the question above. First, the initial KS Slater determinant must reproduce the current-density of the true initial state. Certainly in the case of initial ground-states, with zero initial current, such a Slater determinant can be found (Harriman 1981; Zumbach and Maschke 1983), but whether one can be always found for a general initial current is open. Assuming an appropriate initial Slater determinant can be found, then we come to the second aspect: can we find a scalar potential under which this Slater determinant evolves with the same current-density as that of the true system? It was shown in (D'Agosta and Vignale 2005) that the answer is, generally, no. (On the other hand, non-interacting A-representability, in terms of TDCDFT, has been proven under certain time-analyticity requirements on the current-density and the vector potentials (Vignale 2004)). In the examples of (D'Agosta and Vignale 2005), even if no external magnetic field is applied to the true interacting system, one still needs a magnetic field in a non-interacting system for it to reproduce its current.

4.4.5 Beyond the Taylor-Expansion

The RG proof was derived for potentials that are time-analytic. It does not apply to potentials that turn on, for example, like e^{-C/t^n} with $C > 0, n > 0$, or t^p with p positive non-integer. Note that the first example is infinitely differentiable, with vanishing derivatives at $t = 0$ while higher-order derivatives of the second type diverge as $t \to 0^+$. This is however only a mild restriction, as most potentials are turned on in a time-analytic fashion. Still, it begs the question of whether a proof can be formulated that does apply to these more general cases. A hope would be that such a proof would lead the way to a proof for non-interacting v-representability without the additional, more restrictive, requirement needed in the existing proof that the density is time-analytic (Sect. 4.4.2).

A trivial, but physically relevant, extension is to piecewise analytic potentials, for example, turning a shaped laser field on for some time T and then off again. These potentials are analytic in each of a finite number of intervals. The potential need not have the same Taylor expansion in one interval as it does in another, so the points where they join may be points of nonanalyticity. It is straightforward to extend the RG proof given above to this case (Maitra et al. 2002b).

There are three extensions of RG that go beyond the time-analyticity requirement on the potential. The first two are in the linear response regime. In the earliest (Ng and Singwi 1987), the short-time density response to "small" but arbitrary potentials has been shown to be unique under two assumptions: that the system starts from a stationary state (not necessarily the ground state) of the initial Hamiltonian and that the corresponding linear density-response function is t-analytic. In the second (van Leeuwen 2001), uniqueness of the linear density response, starting from the electronic ground-state, was proven for any Laplace-transformable (in time) potential. As most physical potentials have finite Laplace transforms, this represents a significant widening of the class of potentials for which a 1:1 mapping can be established in the linear-response regime, from an initial ground state. This approach is further discussed in Chap. 9, where also a third, completely new way to address this question is presented via a global fixed-point proof: the one-to-one density-potential mapping is demonstrated for the full non-linear problem, but only for the set of potentials that have finite second-order spatial derivatives.

The difficulty in generalizing the RG proof beyond time-analytic potentials, was discussed in Maitra et al. (2010), where the questions of the one-to-one density-potential mapping and v-representability were reformulated in terms of uniqueness and existence, respectively, of a particular time-dependent non-linear Schrödinger equation (NLSE). The particular structure of the NLSE is not one that has been studied before, and has, so far, not resulted in a general proof, although Chap. 9 discusses new progress in this direction (Ruggenthaler 2011b). On a lattice, the NLSE reverts to a system of nonlinear ordinary differential equations; this has been exploited in Tokatly (2011b) to prove existence and uniqueness of a TDCDFT for lattice systems. The framing of the fundamental theorems of time-dependent density-functional theories in terms of well-posedness of a type of NLSE first appeared in Tokatly (2009) where it arises naturally in Tokatly's Lagrangian formulation of TD current-DFT, known as TD-deformation functional theory (see Chap. 25). The traditional density-potential mapping question is avoided in TD-deformation functional theory, where instead this issue is hidden in the existence and uniqueness of a NLSE involving the metric tensor defining the co-moving frame. Very recently the relation between the NLSE of TDDFT and that of TD-deformation functional theory has been illuminated in Tokatly (2011a).

4.4.6 Exact TDKS Scheme and its Predictivity

The RG theorem guarantees a rigorous one-to-one correspondence between time-dependent densities and time-dependent external potentials. The one-to-one correspondence holds both for fully interacting systems and for non-interacting particles. Hence there are two unique potentials that correspond to a given time-dependent density $n(\mathbf{r}, t)$: one potential, $v_{\text{ext}}[n, \Psi_0](\mathbf{r}, t)$, that yields $n(\mathbf{r}, t)$ by propagating the interacting TDSE with it with initial state Ψ_0, and another potential, $v_{\text{KS}}[n, \Phi_0](\mathbf{r}, t)$, which yields the same density by propagating the non-interacting TDSE with the initial state Φ_0:

$$v_{\text{ext}}[n, \Psi_0](r, t) \overset{1-1}{\longleftrightarrow} n(r, t) \overset{1-1}{\longleftrightarrow} v_{\text{KS}}[n, \Phi_0](r, t). \qquad (4.37)$$

In terms of these two rigorous mappings, the exact TD xc functional is defined as:

$$v_{\text{xc}}[n; \Psi_0, \Phi_0](r, t) \equiv v_{\text{KS}}[n, \Phi_0](r, t) - v_{\text{ext}}[n, \Psi_0](r, t) - \int d^3 r' \frac{n(r', t)}{|r - r'|}. \qquad (4.38)$$

In the past years, the exact xc potential Eq. 4.38 has been evaluated for a few (simple) systems (Hessler et al. 2002; Rohringer et al. 2006; Tempel et al. 2009; Helbig et al. 2009; Thiele et al. 2008; Lein and Kummel 2005). The purpose of this exercise is to assess the quality of approximate xc functionals by comparing them with the exact xc potential (Eq. 4.38).

Normally, one has to deal with the following situation: We are given the external potential $v_{\text{ext}}^{\text{given}}(r, t)$ and the initial many-body state Ψ_0. This information specifies the system to be treated. The goal is to calculate the time-dependent density from a TDKS propagation. The question arises: can we do this, at least in principle, with the exact xc functional, i.e. can we propagate the TDKS equation

$$i\partial_t \varphi_j(r, t) = \left[\frac{-\nabla^2}{2} + v_{\text{ext}}^{\text{given}}(r, t) \right.$$
$$\left. + \int d^3 r' \frac{n(r', t)}{|r - r'|} + v_{\text{xc}}[n; \Psi_0, \Phi_0](r, t) \right] \varphi_j(r, t) \qquad (4.39)$$

with the exact xc potential given by Eq. 4.38? In particular, what is the initial potential with which to start the propagation?

To answer this, we must understand how the functional-dependences in Eq. 4.38 look at the initial time. To this end, we differentiate the continuity equation Eq. 4.20 once with respect to time, and find (van Leeuwen 1999)

$$\ddot{n}(r, t) = \nabla \cdot [n(r, t)\nabla v_{\text{ext}}(r, t)] - \nabla \cdot a(r, t), \qquad (4.40)$$

where

$$a(r, t) = -i\langle \Psi(t)|[\hat{j}(r), \hat{T} + \hat{V}_{\text{ee}}]|\Psi(t)\rangle. \qquad (4.41)$$

At the initial time, Eq. 4.40 shows that as a functional of the density and initial state, $v_{\text{ext}}[n, \Psi_0](r, t = 0)$ depends on $n(r, 0)$, $\ddot{n}(r, 0)$, and $\Psi(0)$ (through Eq. 4.41). Although $n(r, 0)$ is determined by the given $\Psi(0)$ [and so is $\dot{n}(r, 0)$], $\ddot{n}(r, 0)$ is not. Since, at the initial time, we cannot evaluate the time-derivatives to the left [i.e. via $n(t), n(t - \Delta t), n(t - 2\Delta t)$], the start of the propagation may appear problematic. But, in fact, we are *given* the external potential by the problem at hand, as in usual time-dependent quantum mechanics. Hence in Eq. 4.39 the initial external potential is known, the Hartree potential is specified since the initial wavefunction determines the initial density, and the remaining question is: what is the functional dependence of the initial xc potential? If, like v_{ext}, this depends on $\ddot{n}$, then we would have a problem starting the TDKS propagation. Fortunately, it does not.

Applying Eq. 4.40 to the KS system, we replace v_{ext} with v_{KS} in the first term on the right, $\Psi(t)$ with $\Phi(t)$ in Eq. 4.41 and put $v_{ee} = 0$ there to obtain:

$$\ddot{n}(r, t) = \nabla \cdot [n(r, t)\nabla v_{KS}(r, t)] - \nabla \cdot a_{KS}(r, t), \qquad (4.42)$$

where

$$a_{KS}(r, t) = -i\langle \Phi(t)|[\hat{j}(r), \hat{T}]|\Phi(t)\rangle. \qquad (4.43)$$

Subtracting Eq. 4.42 from 4.40 for the interacting system, we obtain

$$\nabla \cdot \left\{ n(r, t)\nabla \left[\int d^3 r' \frac{n(r', t)}{|r - r'|} + v_{xc}(r, t) \right] \right\} = \nabla \cdot [a_{KS}(r, t) - a(r, t)]. \qquad (4.44)$$

Evaluating this at $t = 0$, we see that v_{xc} depends only on the initial states, $\Psi(0)$ and $\Phi(0)$. No second-derivative information is needed, and we can therefore propagate forward.

Analysis of subsequent time-steps was done in Maitra et al. (2008) and showed that at the kth time-step, v_{xc} is determined by the initial states and by densities only at previous times as expected from the RG proof. This shows explicitly that the TDKS scheme is predictive.

4.4.7 TDDFT in Other Realms

A number of extensions of the time-dependent density functional formalism to physically different situations have been developed. In particular, for spin-polarized systems (Liu and Vosko 1989): one can establish a one-to-one mapping between scalar spin-dependent potentials and spin-densities, for an initial non-magnetic ground state. Extension to multicomponent systems, such as electrons plus nuclei can be found in Li and Tong (1986); Kreibich et al. (2004) and is further discussed in Chap. 12, to external vector potentials (Ghosh and Deb 1988; Ng 1989) further discussed in Chap. 24, and to open systems, accounting for coupling of the electronic system to its environment (Burke 2005c; Yuen-Zhou et al. 2010; Appel and Di Ventra 2009; Di Ventra and D'Agosta 2007), covered in Chaps. 10 and 11. Other extensions include time-dependent ensembles (Li and Li 1985), superconducting systems (Wacker 1994), and a relativistic two-component formulation that includes spin-orbit coupling (Wang 2005b; Peng et al. 2005; Romaniello and de Boeij 2007).

4.5 Frequency-Dependent Linear Response

In this section we derive an exact expression for the linear density response $n_1(r, \omega)$ of an N-electron system, initially in its ground-state, in terms of the Kohn–Sham density-response and an exchange-correlation kernel. This relation lies at the basis of TDDFT calculations of excitations and spectra, for which a variety of efficient methods have been developed (see Chap. 7 and also (Marques 2006b)). In fact one of these methods follows directly from the formalism already presented: simply perturb the system at $t = 0$ with a weak electric field, and propagate the TDKS equations for some time, obtaining the dipole of interest as a function of time. The Fourier transform of that function to frequency-space yields precisely the optical absorption spectrum. However, a formulation directly in frequency-space is theoretically enlightening and practically useful. After deriving the fundamental linear response equation of TDDFT in Sect. 4.5.1, we then derive a matrix formulation of this whose eigenvalues and eigenvectors yield the exact excitation energies and oscillator strengths.

4.5.1 The Density–Density Response Function

In general response theory, a system of interacting particles begins in its ground-state, and at $t = 0$ a perturbation is switched on. The total potential is given by

$$v_{\text{ext}}(r, t) = v_{\text{ext},0}(r) + \delta v_{\text{ext}}(r, t) \tag{4.45a}$$

$$\delta v_{\text{ext}}(r, t) = 0 \quad \text{for } t \leq 0. \tag{4.45b}$$

The response of any observable to δv_{ext} may be expressed as a Taylor series with respect to δv_{ext}. In particular, for the density,

$$n(r, t) = n_{\text{GS}}(r) + n_1(r, t) + n_2(r, t) + \dots \tag{4.46}$$

Linear response is concerned with the first-order term $n_1(r, t)$, while higher-order response formalism treats the second, third and higher order terms (see Sect. 4.6).

Staying with standard response theory, n_1 is computed from the density–density linear response function χ as

$$n_1(r, t) = \int_0^\infty dt' \int d^3 r' \chi(rt, r't') \delta v_{\text{ext}}(r', t'). \tag{4.47}$$

where

$$\chi(rt, r't') = \left. \frac{\delta n(r, t)}{\delta v_{\text{ext}}(r', t')} \right|_{v_{\text{ext},0}} \tag{4.48}$$

Ordinary time-dependent perturbation theory in the interaction picture defined with respect to $v_{ext,0}$ yields (Wehrum and Hermeking 1974; Fetter and Walecka 1971)

$$\chi(rt, r't') = -i\theta(t - t')\langle\Psi_0|[\hat{n}_{H_0}(r, t), \hat{n}_{H_0}(r', t')]|\Psi_0\rangle \qquad (4.49)$$

where $\hat{n}_{H_0} = e^{iH_0 t}\hat{n}(r)e^{-iH_0 t}$ and $\theta(\tau) = 0(1)$ for $\tau < (>)0$ is the step function. The density operator is $\hat{n}(r) = \sum_{i=1}^{N}\delta(r - \hat{r}_i)$. (Note that the presence of the step-function is a reflection of the fact that $v_{ext}[n](r, t)$ is a causal functional, i.e. the potential at time t only depends on the density at earlier times $t' < t$.) Inserting the identity in the form of the completeness of interacting states, $\sum_I |\Psi_I\rangle\langle\Psi_I| = \hat{1}$, and Fourier-transforming with respect to $t - t'$ yields the "spectral decomposition" (also called the "Lehmann representation"):

$$\chi(r, r', \omega) = \sum_I \left[\frac{\langle\Psi_0|\hat{n}(r)|\Psi_I\rangle\langle\Psi_I|\hat{n}(r')|\Psi_0\rangle}{\omega - \Omega_I + i0^+} - \frac{\langle\Psi_0|\hat{n}(r')|\Psi_I\rangle\langle\Psi_I|\hat{n}(r)|\Psi_0\rangle}{\omega + \Omega_I + i0^+}\right]$$
$$(4.50)$$

where the sum goes over all interacting excited states Ψ_I, of energy $E_I = E_0 + \Omega_I$, with E_0 being the exact ground-state energy of the interacting system. This interacting response function χ is clearly very hard to calculate so we now turn to TDDFT to see how it can be obtained via the noninteracting KS system.

First, the initial KS ground-state: For $t \leq 0$, the system is in its ground-state and we take the KS system also to be so. The initial density $n_{GS}(r)$ can be calculated from the self-consistent solution of the ground state KS equations

$$\left[-\frac{\nabla^2}{2} + v_{ext,0}(r) + \int d^3 r' \frac{n_{GS}(r')}{|r - r'|} + v_{xc}[n_{GS}](r)\right]\varphi_j^{(0)}(r) = \varepsilon_j\varphi_j^{(0)}(r) \quad (4.51)$$

and

$$n_{GS}(r) = \sum_{\text{lowest } N} |\varphi_j^{(0)}(r)|^2. \qquad (4.52)$$

Adopting the standard response formalism within the TDDFT framework, we notice several things. Because the system begins in its ground-state, there is no initial-state dependence (see Sect. 4.3), and we may write $n(r, t) = n[v_{ext}](r, t)$. Also, the initial potential $v_{ext,0}$ is a functional of the ground-state density n_{GS}, so the same happens to the response function $\chi = \chi[n_{GS}]$. Since the time-dependent KS Eqs. 4.26–4.28 provide a formally exact way of calculating the time-dependent density, we can compute the exact density response $n_1(r, t)$ as the response of the non-interacting KS system:

$$n_1(r, t) = \int_0^\infty dt' \int d^3 r' \chi_{KS}(rt, r't')\delta v_{KS}(r', t'), \qquad (4.53)$$

where δv_{KS} is the effective time-dependent potential evaluated to first order in the perturbing potential, and $\chi_{KS}(rt, r't')$ is the density–density response function of non-interacting particles with unperturbed density n_{GS}:

$$\chi_{KS}(rt, r't') = \left. \frac{\delta n(r, t)}{\delta v_{KS}(r', t')} \right|_{v_{KS}[n_{GS}]} \tag{4.54}$$

Substituting in the KS orbitals $\varphi_j^{(0)}$ (calculated from Eq. 4.51) into Eq. 4.50 we obtain the Lehmann representation of the KS density-response function:

$$\chi_{KS}(r, r', \omega) = \lim_{\eta \to 0^+} \sum_{k,j} (f_k - f_j) \delta_{\sigma_k \sigma_j} \frac{\varphi_k^{(0)*}(r)\varphi_j^{(0)}(r)\varphi_j^{(0)*}(r')\varphi_k^{(0)}(r')}{\omega - (\varepsilon_j - \varepsilon_k) + i\eta}, \tag{4.55}$$

where f_k, f_j are the usual Fermi occupation factors and σ_k denotes the spin orientation of the kth orbital.

The KS density–density response function Eq. 4.55 has poles at the bare KS single-particle orbital energy differences; these are *not* the poles of the true density–density response function Eq. 4.50 which are the true excitation frequencies. Likewise, the strengths of the poles (the numerators) are directly related to the optical absorption intensities (oscillator strengths); those of the KS system are not those of the true system. We now show how to obtain the true density-response from the KS system. We define a time-dependent xc kernel f_{xc} by the functional derivative of the xc potential

$$f_{xc}[n_{GS}](rt, r't') = \left. \frac{\delta v_{xc}[n](r, t)}{\delta n(r', t')} \right|_{n=n_{GS}}, \tag{4.56}$$

evaluated at the initial ground state density n_{GS}. Then, for a given δv_{ext}, the first-order change in the TDKS potential is

$$\delta v_{KS}(r, t) = \delta v_{ext}(r, t) + \int d^3 r' \frac{n_1(r', t)}{|r - r'|}$$
$$+ \int d^3 t' \int d^3 r' f_{xc}[n_{GS}](rt, r't')n_1(r', t'). \tag{4.57}$$

Equation 4.57 together with Eq. 4.53 constitute an exact representation of the linear density response (Petersilka 1996a, b). These equations were postulated (and used) in Ando (1977a, b); Zangwill and Sovea (1980a, b); Gross and Kohn (1985) prior to their rigorous derivation in Petersilka (1996a). That is, the *exact* linear density response $n_1(r, t)$ of an interacting system can be written as the linear density response of a non-interacting system to the effective perturbation $\delta v_{KS}(r, t)$. Expressing this directly in terms of the density-response functions themselves, by substituting Eq. 4.57 into Eq. 4.53 and setting it equal to Eq. 4.47, we obtain the Dyson-like equation for the interacting response function:

$$\chi[n_{GS}](rt, r't') = \chi_{KS}[n_{GS}](rt, r't')$$

$$+ \int dt_1 \int d^3r_1 \int dt_2 \int d^3r_2 \chi_{KS}[n_{GS}](rt, r_1t_1)$$

$$\times \left[\frac{\delta(t_1 - t_2)}{|r_1 - r_2|} + f_{xc}[n_{GS}](r_1t_1, r_2t_2) \right] \chi[n_{GS}](r_2t_2, r't').$$

$$(4.58)$$

This equation, although often translated into different forms (see next section), plays the central role in TDDFT linear response calculations.

4.5.2 Excitation Energies and Oscillator Strengths from a Matrix Equation

We now take time-frequency Fourier-transforms of Eqs. 4.56–4.58 to move towards a formalism directly in frequency-space. The objective is to set up a framework which directly yields excitation energies and oscillator strengths of the true system, but extracted from the KS system. We write

$$\chi(\omega) = \chi_{KS}(\omega) + \chi_{KS}(\omega) \star f_{Hxc}(\omega) \star \chi(\omega) \tag{4.59}$$

where we have introduced a few shorthands: First, we have defined the Hartree-xc kernel:

$$f_{Hxc}(r, r', \omega) = \frac{1}{|r - r'|} + f_{xc}(r, r', \omega). \tag{4.60}$$

Note that $f_{xc}[n_{GS}](r, r', \omega)$ is the Fourier-transform of Eq. 4.56; the latter depends only on the time difference $(t - t')$, like the response functions, due to time-translation invariance of the unperturbed system, allowing its frequency-domain counterpart to depend only on one frequency variable. Second, we have dropped the spatial indices and introduced the shorthand $\star$ to indicate integrals like $\chi_{KS}(\omega) \star f_{Hxc}(\omega) = \int d^3r_1 \chi_{KS}(r, r_1, \omega) f_{Hxc}(r_1, r', \omega)$ thinking of χ, χ_{KS}, f_{Hxc} etc as infinite-dimensional matrices in r, r', each element of which is a function of ω. Now, integrating both sides of Eq. 4.59 against $\delta v_{ext}(r, \omega)$, we obtain

$$\left[\hat{1} - \chi_{KS}(\omega) \star f_{Hxc}(\omega) \right] \star n_1(\omega) = \chi_{KS}(\omega) \star \delta v_{ext}(\omega). \tag{4.61}$$

The exact density-response $n_1(r, \omega)$ has poles at the true excitation energies Ω_I. However, these are not identical with KS excitation energies $\varepsilon_a - \varepsilon_i$ where the poles of χ_{KS} lie, i.e. the r.h.s. of Eq. 4.61 remains finite for $\omega \to \Omega_I$. Therefore, the integral operator acting on n_1 on the l.h.s. of Eq. 4.61 cannot be invertible for $\omega \to \Omega_I$, as it must cancel out a pole in n_1 in order to create a finite r.h.s. The true excitation energies Ω_I are therefore precisely those frequencies where the eigenvalues of the integral operator on the left of Eq. 4.61, $\left[\hat{1} - \chi_{KS}(\omega) \star f_{Hxc}(\omega)\right]$, vanish. Equivalently, the eigenvalues $\lambda(\omega)$ of

$$\chi_{KS}(\omega) \star f_{Hxc}(\omega) \star \xi(\omega) = \lambda(\omega)\xi(\omega) \tag{4.62}$$

satisfy $\lambda(\Omega_I) = 1$. This condition rigorously determines the true excitation spectrum of the interacting system.

For the remainder of this subsection, we will focus on the case of spin-saturated systems: closed-shell singlet systems and their spin-singlet excitations, to avoid carrying around too much notation. We shall return to the more general spin-decomposed version of the equations in Sect. 4.5.4.

Before we continue to cast Eq. 4.62 into a matrix form from which excitation energies and oscillator strengths may be conveniently extracted, we mention two very useful approximations that can shed light on the workings of the response equation. The idea is to expand all quantities in Eq. 4.62 around one particular KS energy difference, $\omega_q = \varepsilon_a - \varepsilon_i$ say, keeping only the lowest-order terms in the Laurent expansions (Petersilka 1996a, b). This is justified for example, in the limit that the KS excitation of interest is energetically far from the others and that the correction to the KS excitation is small (Appel et al. 2003). This yields what is known as the single-pole approximation (SPA), which, for spin-saturated systems, is:

$$\Omega = \omega_q + 2\Re\mathfrak{e} \int d^3r \int d^3r' \Phi_q^*(\boldsymbol{r}) f_{Hxc}(\boldsymbol{r}, \boldsymbol{r}', \omega_q)\Phi_q(\boldsymbol{r}') \tag{4.63}$$

where we have defined the transition density

$$\Phi_q(\boldsymbol{r}) = \varphi_i^*(\boldsymbol{r})\varphi_a(\boldsymbol{r}). \tag{4.64}$$

This approximation is equivalent to neglecting couplings with all other excitations. If also the pole at $-\omega_q$ is kept (the backward transition), i.e.

$$\chi_{KS} \approx 2 \left[\frac{\Phi_q^*(\boldsymbol{r}')\Phi_q(\boldsymbol{r})}{\omega - \omega_q + i0^+} - \frac{\Phi_q(\boldsymbol{r}')\Phi_q^*(\boldsymbol{r})}{\omega + \omega_q + i0^+} \right] \tag{4.65}$$

(where again the factor of 2 arises from assuming a spin-saturated system), then we obtain the "small-matrix approximation"(SMA):

$$\Omega^2 = \omega_q^2 + 4\omega_q \int d^3r \int d^3r' \Phi_q(\boldsymbol{r}) f_{Hxc}(\boldsymbol{r}, \boldsymbol{r}', \Omega)\Phi_q(\boldsymbol{r}') \tag{4.66}$$

provided the KS orbitals are chosen to be real. Discussion on the validity of these approximations and their use as tools to analyse full TDDFT spectra can be found in Appel et al. (2003). Generalized frequency-dependent, or "dressed" versions of these truncations have been used to derive approximations in certain cases, e.g. for double-excitations (see Chap. 8).

We return to finding a matrix formulation of Eq. 4.62 to yield exact excitations and oscillator strengths; we follow the exposition of Grabo et al. (2000). For a single KS transition from orbital i to a, we introduce the double-index $q = (i, a)$, with transition frequency

$$\omega_q = \varepsilon_a - \varepsilon_i \tag{4.67}$$

and transition density as in Eq. 4.64. Let $\alpha_q = f_i - f_a$ be the difference in their ground-state occupation numbers (e.g. $\alpha_q = 0$ if both orbitals are occupied or if both are unoccupied, $\alpha_q = 2$ if i is occupied while a unoccupied (a "forward" transition), $\alpha_q = -2$ if a is occupied while i unoccupied (a "backward" transition)). Reinstating the spatial dependence explicitly and defining

$$\zeta_q(\omega) = \int d^3r' \int d^3r'' \Phi_q^*(r') f_{Hxc}(r', r'', \omega)\xi(r'', \omega) \tag{4.68}$$

we can recast Eq. 4.62 as

$$\sum_q \frac{\alpha_q \Phi_q(r)}{\omega - \omega_q + i0^+}\zeta_q(\omega) = \lambda(\omega)\xi(r, \omega). \tag{4.69}$$

Solving this equation for $\xi(r, \omega)$, and reinserting the result on the r.h.s. of Eq. 4.68, we obtain

$$\sum_q \frac{M_{qq'}(\omega)}{\omega - \omega_{q'} + i0^+}\zeta_{q'}(\omega) = \lambda(\omega)\zeta_q(\omega) \tag{4.70}$$

where we have introduced the matrix elements

$$M_{qq'}(\omega) = \alpha_{q'} \int d^3r \int d^3r' \Phi_q^*(r) f_{Hxc}(r, r', \omega)\Phi_{q'}(r'). \tag{4.71}$$

Defining now $\beta_q = \zeta_q(\Omega)/(\Omega - \omega_q)$, and using the condition that $\lambda(\Omega) = 1$ at a true excitation energy, we can write, at the true excitations:

$$\sum_{q'} \left[M_{qq'}(\Omega) + \omega_q \delta_{qq'}\right] \beta_{q'} = \Omega \beta_q. \tag{4.72}$$

This eigenvalue problem rigorously determines the true excitation spectrum of the interacting system. The matrix is infinite-dimensional, going over all single-excitations of the KS system. In practice, it must be truncated. If only forward transitions are kept, this is known as the "Tamm–Dancoff" approximation.

The first matrix formulation of TDDFT linear response was derived in Casida (1995, 1996) by considering the response of the KS density matrix. Commonly known as "Casida's equations", these equations are similar in structure to TDHF, and are what is coded in most of the electronic structure codes today. By considering the poles and residues of the frequency-dependent polarizability, in Casida (1995) it was shown that the true frequencies Ω_I and oscillator strengths can be obtained from eigenvalues and eigenvectors of the following matrix equation:

$$RF_I = \Omega_I^2 F_I, \tag{4.73}$$

where

$$R_{qq'} = \omega_q^2 \delta_{qq'} + 4\sqrt{\omega_q \omega_{q'}} \int d^3r \int d^3r' \Phi_q(\mathbf{r}) f_{Hxc}(\mathbf{r}, \mathbf{r}', \Omega_I) \Phi_{q'}(\mathbf{r}') \tag{4.74}$$

The oscillator strength of transition I in the interacting system, defined as

$$f_I = \frac{2}{3}\Omega_I \left(|\langle \Psi_0 | \hat{x} | \Psi_I \rangle|^2 + |\langle \Psi_0 | \hat{y} | \Psi_I \rangle|^2 + |\langle \Psi_0 | \hat{z} | \Psi_I \rangle|^2 \right) \tag{4.75}$$

can be obtained from the eigenvectors F_I via

$$f_I = \frac{2}{3} \left(|x\mathbb{S}^{-1/2} F_I|^2 + |y\mathbb{S}^{-1/2} F_I|^2 + |z\mathbb{S}^{-1/2} F_I|^2 \right) / |F_I|^2 \tag{4.76}$$

where $\mathbb{S}_{ij,kl} = \delta_{i,k}\delta_{j,l}/\alpha_q \omega_q > 0$ with $q = (k, l)$ here.

The KS orbitals are chosen to be real in this formulation. Provided that real orbitals are also used in the secular equation (4.72), Casida's equations and Eq. 4.72 are easily seen to be equivalent. The SPA and SMA approximations, Eqs. 4.63 and 4.66 derived before, can be readily seen to result from keeping only the diagonal element in the matrix M, or in matrix R: neglecting off-diagonal terms, we immediately obtain Eq. 4.66. Assuming additionally that the correction to the bare KS transition is itself small, we take a square-root of Eq. 4.66 keeping only the leading correction, and find the single-pole-approximation Eq. 4.63.

The matrix equations are often re-written in the literature as

$$\begin{pmatrix} A & B \\ B^* & A^* \end{pmatrix} \begin{pmatrix} X \\ Y \end{pmatrix} = \omega \begin{pmatrix} -1 & 0 \\ 0 & 1 \end{pmatrix} \begin{pmatrix} X \\ Y \end{pmatrix} \tag{4.77}$$

where

$$A_{ia,jb} = \delta_{ij}\delta_{ab}(\varepsilon_a - \varepsilon_i) + 2 \int d^3r \int d^3r' \Phi_q^*(\mathbf{r}) f_{Hxc}(\mathbf{r}, \mathbf{r}') \Phi_{q'}(\mathbf{r}') \tag{4.78a}$$

$$B_{ia,jb} = 2 \int d^3r \int d^3r' \Phi_q^*(\mathbf{r}) f_{Hxc}(\mathbf{r}, \mathbf{r}') \Phi_{-q'}(\mathbf{r}'), \tag{4.78b}$$

which has the same structure as the eigenvalue problem resulting from time-dependent Hartree–Fock theory (Bauernschmitt and Ahlrich 1996a). However, we note that this form really only applies when an adiabatic approximation (see Sect. 4.7) is made for the kernel (but complex orbitals may be used).

We note that TDDFT linear response equations can be shown to respect the Thomas–Reiche–Kuhn sum-rule: the sum of the oscillator strengths equals the number of electrons in the system (see also Chap. 5). This of course is true for the KS oscillator strengths as well as the linear-response-corrected ones. As mentioned earlier, in the Tamm–Dancoff approximation, backward transitions are neglected, e.g. the B-matrix in Eq. 4.77 is set to zero, with the resulting equations resembling configuration-interaction singles (CIS) [see (Hirata 1999)]. In certain cases the Tamm–Dancoff approximation turns out to be nearly as good as (or sometimes "better" than (Casida et al. 2000, Hirata and Head-Gordon 1999)) full TDDFT, but it violates the oscillator strength sum-rule.

To summarize so far:

(i) The matrix formulations (Eqs. 4.72 and 4.73) are valid only for discrete spectra and hence are mostly used for finite systems, while the original Dyson-like integral equation, Eq. 4.58, is usually solved when dealing with the continuous spectra of extended systems. To obtain the continuous part of the spectra of finite systems (e.g. resonance widths and positions), the Sternheimer approach, described in Chap. 7 is often used.

(ii) To apply the TDDFT linear response formalism, there are evidently two ingredients. First, one has to find the elements of the non-interacting KS density-response function, i.e. use Eq. 4.51 to find all occupied and unoccupied KS orbitals living in the ground-state KS potential $v_{KS,0}$. An approximation is needed there for the ground-state xc potential. Second, one has to apply the xc kernel f_{xc}, for which in practice approximations are also needed. The next section discusses the kernel in a little more detail.

4.5.3 The xc Kernel

The central functional in linear response theory f_{xc} is simpler than that in the full theory, because instead of functionally depending on the density and its history as well as the initial-states as v_{xc} must, it depends only on the initial ground-state density. The kernel can be obtained from the functional derivative, Eq. 4.56, but often a more useful expression is to extract it from Eq. 4.58: one can isolate f_{xc} in Eq. 4.58 by applying the inverse response functions in the appropriate places, yielding

$$f_{xc}[n_{GS}](rt, r't') = \chi_{KS}^{-1}[n_{GS}](rt, r't') - \chi^{-1}[n_{GS}](rt, r't') - \frac{\delta(t - t')}{|r - r'|}, \quad (4.79)$$

where χ_{KS}^{-1} and χ^{-1} stand for the kernels of the corresponding inverse integral operators.

Note that the existence of the inverse density-response operators on the set of densities specified by Eqs. 4.45a–4.47 follows from Eq. 4.21 in the RG proof: the right-hand side of Eq. 4.21 is linear in the difference between the potentials. Consequently, the difference between $n(r, t)$ and $n'(r, t)$ is non-vanishing already in first order of $v(r, t) - v'(r, t)$. This result ensures the invertibility of linear response operators. The *frequency-dependent* response operators $\chi(r, r', \omega)$ and $\chi_{KS}(r, r', \omega)$, on the other hand, can be non-invertible at isolated frequencies (Mearns and Kohn 1987; Gross and Deb 1988). Recently, the numerical difficulties that the vanishing eigenvalues of $\chi_{KS}(r, r', \omega)$ cause for exact-exchange calculations of spectra have been highlighted in (Hellgren and von Barth 2009). The non-invertibility means that one can find a non-trivial (i.e. non-spatially constant) monochromatic perturbation that yields a vanishing density response. This might appear at first sight to be a counter-example to the density-potential mapping of the RG theorem, as it means that two different perturbations may be found which have the same density evolution (at least in linear response). However, this can only happen if the perturbation is truly monochromatic, having been switched on adiabatically from $t = -\infty$. If we instead think of the perturbation being turned on infinitely-slowly from $t = 0$, it must have an essential singularity in time: (e.g. $v \sim \lim_{\eta \to 0^+} e^{-\eta/t + i\omega t}$), i.e. the potential is not time-analytic about $t = 0$, and so is *a priori* excluded from consideration by the RG theorem.

Due to causality, $f_{xc}(rt, r't')$ vanishes for $t' > t$, i. e., f_{xc} is not symmetric with respect to an interchange of t with t'. Consequently, $f_{xc}(rt, r't')$ cannot be a second functional derivative $\delta^2 F_{xc}[n]/\delta n(r, t)\delta n(r', t')$ (Wloka 1971), and the exact $v_{xc}[n](r, t)$ cannot be a functional derivative, in contrast to the static case. (See also earlier Sect. 4.4.3).

Known exact properties of the kernel are given in Chap. 5. These include symmetry in exchange of r and r' and Kramers-Kronig relations for $f_{xc}(r, r', \omega)$. These relations make evident that frequency-dependence goes hand-in-hand with $f_{xc}(r, r', \omega)$ carrying a non-zero imaginary part.

The manipulations leading to the Dyson-like equation can be followed also in the ground-state Hohenberg-Kohn theory to yield static response equations. The frequency-dependent interacting and non-interacting response functions are replaced by the interacting and KS response functions to static perturbations, and the kernel reduces to the second functional derivative of the xc energy. It follows that

$$\lim_{\omega \to 0} f_{xc}[n_{GS}](r, r', \omega) = f_{xc}^{static}[n_{GS}](r, r') = \left. \frac{\delta^2 E_{xc}[n]}{\delta n(r)\delta n(r')} \right|_{n_{GS}}. \qquad (4.80)$$

The adiabatic approximation for the kernel used in almost all approximations today takes $f_{xc}[n_{GS}](r, r', \omega) = f_{xc}^{static}[n_{GS}](r, r')$. We return to this in Sect. 4.7.

Finally, we note that here we have only dealt with the linear response to time-dependent scalar fields at zero temperature. The corresponding formalism for systems at finite temperature in thermal equilibrium was developed in Ng and Singwi (1987) and Yang (1988). For the response to arbitrary electromagnetic fields, some early developments were made in Ng (1989), and, more recently, in the context of TDCDFT of the Vignale-Kohn functional, in van Faassen and de Baij (2004); Ullrich and Burke (2004).

4.5.4 Spin-Decomposed Equations

The linear-response formalisms presented above focussed on closed-shell singlet systems and their singlet excitations. To describe singlet-triplet splittings, linear response based on the spin-TDDFT of Liu and Vosko (1989) must be used. We distinguish two situations: first is when the initial ground-state is closed-shell and non-degenerate, and second when the initial state is an open-shell, degenerate, system. Here we focus on the first case, where the equations given above are straightforward to generalize, e.g. for the response of the spin-σ-density, we have

$$n_{1\sigma}(\mathbf{r}, \omega) = \sum_{\sigma'} \int d^3 r' \chi_{\sigma,\sigma'}(\mathbf{r}, \mathbf{r}', \omega) \delta v_{\text{ext},\sigma'}(\mathbf{r}', \omega) \qquad (4.81)$$

where $v_{\text{ext},\sigma}$ is the spin-dependent external potential and $\chi_{\sigma,\sigma'}(\mathbf{r}, \mathbf{r}', \omega)$ is the spin-decomposed density–density response function. The fundamental Dyson-like equation, Eq. 4.58, remains essentially the same, as do the matrix equations for the excitation energies, only spin-decomposed, with the spin-dependent xc kernel defined via

$$f_{\text{xc},\sigma\sigma'}[n_{0\uparrow}, n_{0\downarrow}](\mathbf{r}t, \mathbf{r}'t') = \left. \frac{\delta v_{\text{xc},\sigma}[n_\uparrow, n_\downarrow](\mathbf{r}, t)}{\delta n_{\sigma'}(\mathbf{r}', t')} \right|_{n_{0\uparrow} n_{0\downarrow}} \qquad (4.82)$$

The exact equations, Eq. 4.72, generalize to:

$$\sum_{\sigma'} \sum_{q'} \left[M_{q\sigma q'\sigma'}(\Omega) + \omega_{q\sigma} \delta_{qq'} \delta_{\sigma\sigma'} \right] \beta_{q',\sigma'} = \Omega \beta_{q\sigma} \qquad (4.83)$$

with the obvious spin-generalized forms of the terms, e.g.

$$M_{q\sigma q'\sigma'}(\omega) = \alpha_{q'\sigma'} \int d^3 r \int d^3 r' \Phi_{q\sigma}^*(\mathbf{r}) f_{\text{Hxc},\sigma\sigma'}(\mathbf{r}, \mathbf{r}', \omega) \Phi_{q'\sigma'}(\mathbf{r}') \qquad (4.84)$$

with $\alpha_{q\sigma} = f_{i\sigma} - f_{a\sigma}$ and $\Phi_{q,\sigma} = \varphi_{i\sigma}^*(\mathbf{r}) \varphi_{a\sigma}(\mathbf{r})$.

For spin-unpolarized ground-states, there are only two independent combinations of the spin-components of the xc kernel because the two parallel components are equal and the two anti-parallel are equal:

$$f_{\text{xc}} = \frac{1}{4} \sum_{\sigma\sigma'} f_{\text{xc},\sigma\sigma'} = \frac{1}{2}(f_{\text{xc},\uparrow\uparrow} + f_{\text{xc},\uparrow\downarrow}) \qquad (4.85a)$$

$$G_{\text{xc}} = \frac{1}{4} \sum_{\sigma\sigma'} f_{\text{xc},\sigma\sigma'} = \frac{1}{2}(f_{\text{xc},\uparrow\uparrow} - f_{\text{xc},\uparrow\downarrow}). \qquad (4.85b)$$

The spin-summed kernel, f_{xc} in Eq. 4.85a, is exactly the xc kernel that appeared in the previous section. For example, for the simplest approximation, adiabatic local spin-density approximation (ALDA) (see more shortly in Sect. 4.7), for spin-unpolarized ground-states

$$f_{xc}^{ALDA}[n](\boldsymbol{r}, \boldsymbol{r}') = \delta(\boldsymbol{r} - \boldsymbol{r}') \left. \frac{d^2[n e_{xc}^{hom}(n)]}{dn^2} \right|_{n=n(\boldsymbol{r})} \tag{4.86a}$$

$$G_{xc}^{ALDA}[n](\boldsymbol{r}, \boldsymbol{r}') = \delta(\boldsymbol{r} - \boldsymbol{r}') \frac{\alpha_{xc}^{hom}(n(\boldsymbol{r}))}{n(\boldsymbol{r})} \tag{4.86b}$$

where $e_{xc}^{hom}(n)$ is the xc energy per electron of a homogeneous electron gas of density n, and α_{xc}^{hom} is the xc contribution to its spin-stiffness. The latter measures the curvature of the xc energy per electron of an electron gas with uniform density n and relative spin-polarization $m = (n_\uparrow - n_\downarrow)/(n_\uparrow + n_\downarrow)$, with respect to m, at $m = 0$: $\alpha_{xc}^{hom}(n) = \delta^2 e_{xc}^{hom}(n, m)/\delta m^2|_{m=0}$.

For closed-shell systems, there is no singlet-triplet splitting in the bare KS eigenvalue spectrum: every KS orbital eigenvalue is degenerate with respect to spin. However, the levels spin-split when the xc kernel is applied. This happens even at the level of the SPA applied to Eq. 4.83 (Petersilka 1996b; Grabo et al. 2000): one finds the two frequencies

$$\Omega_{1,2} = \omega_q + \Re\left\{ M_{p\uparrow p\uparrow} \pm M_{p\uparrow p\downarrow} \right\}. \tag{4.87}$$

Using the explicit form of the matrix elements (Eq. 4.84) one finds, dropping the spin-index of the KS transition density, the singlet and triplet excitation energies within SPA,

$$\Omega^{singlet} = \omega_q + 2\Re \int d^3r \int d^3r' \Phi_q^*(\boldsymbol{r}) \left[\frac{1}{|\boldsymbol{r} - \boldsymbol{r}'|} + f_{xc}(\boldsymbol{r}, \boldsymbol{r}', \omega_q) \right] \Phi_q(\boldsymbol{r}') \tag{4.88a}$$

$$\Omega^{triplet} = \omega_q + 2\Re \int d^3r \int d^3r' \Phi_q^*(\boldsymbol{r}) G_{xc}(\boldsymbol{r}, \boldsymbol{r}', \omega_q) \Phi_q(\boldsymbol{r}'). \tag{4.88b}$$

This result shows that the kernel G_{xc} represents xc effects for the linear response of the frequency-dependent magnetization density $m(\boldsymbol{r}, \omega)$ (Liu and Vosko 1989). In this way, the SPA already gives rise to the singlet-triplet splitting in the excitation spectrum. For unpolarized systems, the weight of the pole in the spin-summed susceptibility (both for the Kohn–Sham and the physical systems) at $\Omega^{triplet}$ is exactly zero, indicating that these are the optically forbidden transitions to triplet states.

4.5.5 A Case Study: The He Atom

In this section, we take a break from the formal theory and show how TDDFT linear response works on the simplest system of interacting electrons found in nature, the helium atom.

Recall the two ingredients needed for the calculation: (i) the ground-state KS potential, out of which the bare KS response is calculated, and (ii) the xc kernel.

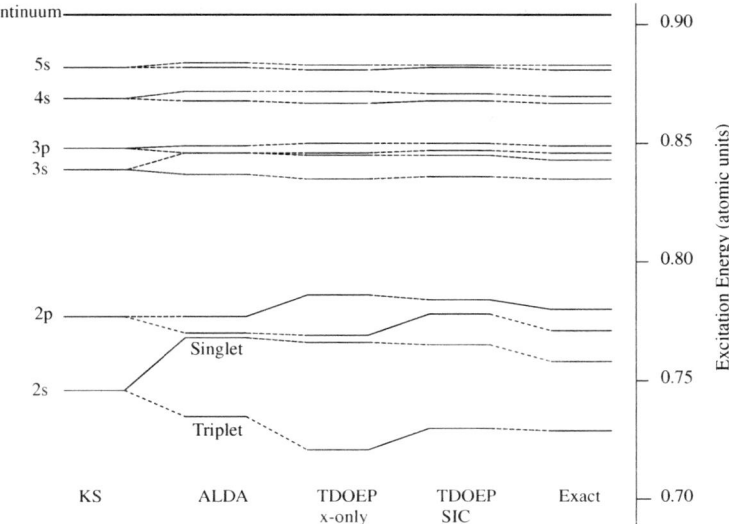

Fig. 4.2 Excitations in the Helium atom (from Petersilka (2000)): bare KS excitations out of the exact He ground-state potential (*left*), followed by TDDFT with the ALDA kernel, exact-exchange (TDOEP x-only) kernel, and self-interaction corrected LDA (TDOEP SIC) kernel, and finally the exact (*right*)

In usual practice, this means two approximate functionals are needed. However the helium atom is small enough that the essentially *exact* ground-state KS potential can be calculated in the following way. A highly-accurate wavefunction calculation can be performed for the ground-state, from which the density $n_{GS}(\mathbf{r}) = 2 \int d^3r' |\Psi_{GS}(\mathbf{r}, \mathbf{r}')|^2$ can be extracted. The corresponding KS system consists of a doubly-occupied orbital, $\varphi_0(\mathbf{r}) = \sqrt{n_{GS}(\mathbf{r})/2}$, so that the KS equation Eq. 4.51 can easily be inverted to find the corresponding ground-state KS potential $v_{KS}(\mathbf{r}) = \nabla^2 \varphi_0/(2\varphi_0) + \varepsilon_0$, where $\varepsilon_0 = -I$, the exact ionization potential.

First, we demonstrate the effect of the xc kernel, by utilizing the essentially exact ground-state KS potential, obtained by the above procedure beginning with a quantum monte carlo calculation for the interacting wavefunction, performed by Umrigar and Gonze (1994). The extreme left of Fig. 4.2 shows the bare KS excitations $\omega_q = \varepsilon_a - \varepsilon_i$. We notice that these are already very close to the exact spectrum, shown on the extreme right, and always lying in between the true singlet and triplet energies (Savin et al. 1998). The middle three columns show the correction due to the TDDFT xc kernel for which three approximations are shown. The first and simplest is the ALDA of Eq. 4.86a and the other two are orbital-dependent approximations which will be explained in Sect. 4.7. For now, we simply note that the bare KS excitations are good zeroth order approximations to the true excitations, providing an average over the singlet and triplet, while the approximate TDDFT corrections provide a good approximation to their spin-splitting.

Table 4.1 Singlet(s) and triplet (t) excitation energies of the helium atom (from Petersilka (2000)), in atomic units

Transition	Exact KS	EXX	LDASIC	TDEXX	TDLDASIC	Exact
1s → 2s	0.7460	0.7596	0.7838	0.7794 (s)	0.8039 (s)	0.7578 (s)
				0.7345 (t)	0.7665 (t)	0.7285 (t)
1s → 3s	0.8392	0.8533	0.8825	0.8591 (s)	0.8881 (s)	0.8425 (s)
				0.8484 (t)	0.8789 (t)	0.8350 (t)
1s → 4s	0.8688	0.8830	0.9130	0.8855 (s)	0.9154 (s)	0.8701 (s)
				0.8812 (t)	0.9117 (t)	0.8672 (t)
1s → 5s	0.8819	0.8961	0.9263	0.8974 (s)	0.9276 (s)	0.8825 (s)
				0.8953 (t)	0.9257 (t)	0.8811 (t)
1s → 6s	0.8888	0.9030	0.9333	0.9038 (s)	0.9341 (s)	0.8892 (s)
				0.9026 (t)	0.9330 (t)	0.8883 (t)
1s → 2p	0.7772	0.7905	0.8144	0.7981 (s)	0.8217 (s)	0.7799 (s)
				0.7819 (t)	0.8139 (t)	0.7706 (t)
1s → 3p	0.8476	0.8616	0.8906	0.8641 (s)	0.8930 (s)	0.8486 (s)
				0.8592 (t)	0.8899 (t)	0.8456 (t)
1s → 4p	0.8722	0.8864	0.9163	0.8875 (s)	0.9173 (s)	0.8727 (s)
				0.8854 (t)	0.9159 (t)	0.8714 (t)
1s → 5p	0.8836	0.8978	0.9280	0.8984 (s)	0.9285 (s)	0.8838 (s)
				0.8973 (t)	0.9278(t)	0.8832 (t)
1s → 6p	0.8898	0.9040	0.9343	0.9043 (s)	0.9346 (s)	0.8899 (s)
				0.9037 (t)	0.9342 (t)	0.8895 (t)

The exact results are from the variational calculation of Kono (1984). The second column shows the single-particle excitations obtained out of the exact KS potential, while the third and fourth columns show those of the approximate EXX and LDASIC potentials. The fifth and sixth columns then apply the respective xc kernels to get the TDDFT approximations

For most molecules of interest however, the exact ground-state KS potential is not available. Using LDA or semi-local GGA's can give results to within a few tenths of an eV for low-lying excitations. However, for higher excitations, (semi)local approximations run into problems because the LDA potential asympototically decays exponentially instead of as $-1/r$ as the exact potential does, so the higher lying bound-states become unbound. There is no Rydberg series in LDA/GGA atoms. For our simple helium atom, the situation is severe: *none* of the excitations are bound in LDA, and GGA does not improve this unfortunate situation. Use of a ground-state xc potential that goes as $-1/r$ at long-range pulls these excitations down from the continuum into the bound spectrum, and, as Table 4.1 shows, can be quite accurate. The table shows results using the exact-exchange approximation (EXX), and the self-interaction-corrected local density approximation (LDASIC); both bare KS excitations as well as the TDDFT values (i.e. corrected by the kernel) are shown. These approximations are discussed in detail in Sect. 4.7; to note for now, is that the ground-state KS potential in all cases has the correct long-range behavior. Notice also that the bare KS excitations are quite accurate; applying the kernel (second step) provides a small correction.

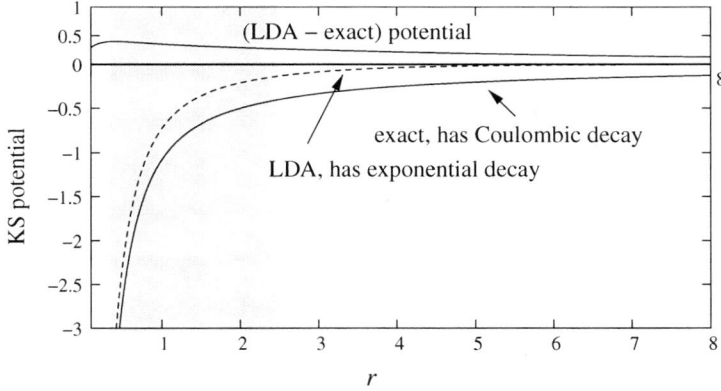

Fig. 4.3 The exact and LDA KS potentials for the He atom

Consider Fig. 4.3 which plots the true and LDA xc potentials for the case of the helium atom; similar pictures hold for any atom. In the shaded valence region, we notice that the LDA xc potential differs from the exact xc potential by nearly a constant. This effect is related to the derivative discontinuity, and it was argued in Perdew (1985) that this constant has a value $(I + A)/2$ where I is the ionization potential and A the electron affinity. The fact that the LDA xc potential runs almost parallel to the exact in this region, means that the valence orbitals are well-approximated in LDA, while their orbital energies are almost uniformly shifted up. This is why excitations, starting from the zeroth-order KS orbital energy *differences* in the valence region are generally approximated well in LDA. However, the rapid decay of v_{xc}^{LDA} to the zero-asymptote means that the higher-excitation energies are underestimated and eventually get squeezed into the continuum. (Unfortunately for the case of the He atom, this happens to even the lowest excitations.)

The top panel of Fig. 4.4 illustrates two effects of the too rapid decay of the LDA potential on the optical spectrum: (i) it pushes the valence levels up, so that the ionization potential is too low; the onset of the LDA continuum is red-shifted compared to the exact, and (ii) there is no Rydberg series in the LDA spectrum; instead their oscillator strengths appear in the LDA continuum, but in fact are not badly approximated (Wasserman et al. 2003). The reason for this accuracy is due to the LDA and true xc potentials running nearly parallel in the valence region: the LDA HOMO orbital, out of which the transitions are computed, very well-approximates the true HOMO, while the LDA continuum state at energy $E = \omega + I^{LDA}$ follows very closely the exact continuum state at energy $E = \omega + I^{exact}$ until a distance large enough away from the nucleus that the integrand does not contribute due to the decay of the HOMO. Noting that KS spectra are not true spectra, the lower panel shows the TDDFT-corrected spectrum using ALDA for the xc kernel; although not resolving the discrete part of the spectra, the overall oscillator strength envelope is not bad. For a detailed discussion, we refer the reader to Wasserman et al. (2003).

Fig. 4.4 Optical absorption spectrum in the He atom (from Wasserman et al. (2003)). The *top panel* shows the bare exact KS and LDA KS spectra. The *lower panel* shows the TDDFT ALDA spectra (*dashed line*, from Stener et al. (2001)), the exact calculations from Kono and Hattori (1984) and experimental results from Samson et al. (1994)

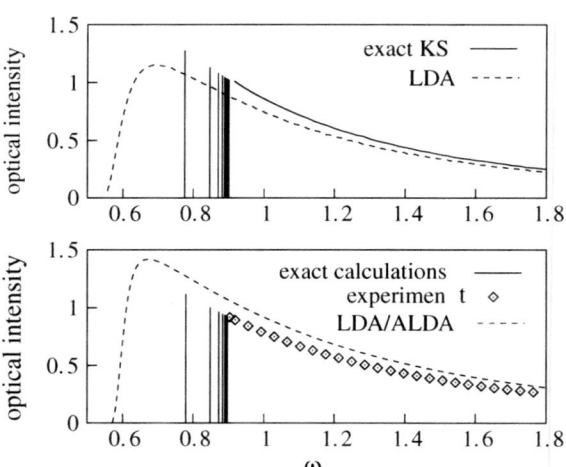

The purpose of this case study was to illustrate the workings of TDDFT response and investigate only the simplest functional on the simplest system. We note that most molecules have many more lower-lying excited states so that GGA's can do a much better job for more excitations. We return to the question of functional approximations in later sections throughout this book.

4.6 Higher-Order Response

Often the simplest way to calculate the non-linear response of a system to an external perturbation is via time-propagation. But, like in the linear-response case, it can be instructive to perform the non-linear response calculation directly in frequency-space. The higher-order terms in Eq. 4.46 can be expressed in terms of higher-order density–density response functions:

$$n_2(x) = \frac{1}{2!} \int dx' \int dx'' \chi^{(2)}(x, x', x'') \delta v(x') \delta v(x'') \tag{4.89a}$$

$$n_3(x) = \frac{1}{3!} \int dx' \int dx'' \int dx''' \chi^{(2)}(x, x', x'', x''') \delta v(x') \delta v(x'') \delta v(x''') \tag{4.89b}$$

$$\cdots$$

where we used the short-hand $x = (\mathbf{r}, t)$ and $\int dx = \int d^3r \int dt$. For quadratic response, the analogues of Eqs. 4.48 and 4.49 are

$$\chi^{(2)}(x, x', x'') = \left.\frac{\delta^2 n(x)}{\delta v_{ext}(x')\delta v_{ext}(x'')}\right|_{n_{GS}}$$

$$= -\sum_P \theta(t - t')\theta(t - t'')\langle\Psi_0|[[\hat{n}_H(x), \hat{n}_H(x')], \hat{n}_H(x'')]|\Psi_0\rangle$$

$$(4.90)$$

where the sum goes over all permutations of x, x', x''. Clearly, the interacting higher-order response functions are very difficult to calculate directly and instead we look to extract them from the KS response functions and xc kernels. Manipulations similar to those in the linear response case, but more complicated, lead to (Gross et al. 1996):

$$\chi^{(2)}(x, x', x'') = \int dy \int dy' \chi_{KS}^{(2)}(x, y, y') \left.\frac{\delta v_{KS}(y)}{\delta v(x')}\right|_{n_{GS}} \left.\frac{\delta v_{KS}(y')}{\delta v(x'')}\right|_{n_{GS}}$$

$$+ \int dy \chi_{KS}(x, y) \int dy' \int dy'' k_{xc}[n_{GS}](y, y', y'')\chi(y', x')\chi(y'', x'')$$

$$+ \int dy \chi_{KS}(x, y) \int dy' f_{Hxc}[n_{GS}](y, y')\chi^{(2)}(y', x', x'').$$

$$(4.91)$$

Here $\chi_{KS}^{(2)} = \delta^2 n(x)/\delta v_{KS}(x')\delta v_{KS}(x'')\big|_{n_{GS}}$ is the KS second-order density-response function, and

$$k_{xc}[n_{GS}](rt, r't', r''t'') = \left.\frac{\delta^2 v_{xc}[n](r, t)}{\delta n(r', t')\delta n(r'', t'')}\right|_{n=n_{GS}} \qquad (4.92)$$

is the dynamical second-order xc kernel. In the adiabatic approximation,

$$k_{xc}^{adia}[n](r, r', r'') = \frac{\delta^3 E_{xc}[n]}{\delta n(r)\delta n(r')\delta n(r'')} \qquad (4.93)$$

with $E_{xc}[n]$ a ground-state xc energy functional. Making Fourier-transforms with respect to $t - t'$ and $t - t''$, we arrive at the Dyson equation

$$n_2(r, \omega) = \frac{1}{2}\int d\omega' \int d^3r_1 d^3r_2 \chi^{(2)}(r, r_1, r_2, \omega, \omega - \omega')\delta v(r_1, \omega)\delta v(r_2, \omega - \omega')$$

$$= \frac{1}{2}\int d\omega' \int d^3r_1 d^3r_2 \left(\chi_{KS}^{(2)}(r, r_1, r_2, \omega, \omega - \omega')\delta v_{KS}(r_1, \omega)\delta v_{KS}(r_2, \omega - \omega')\right.$$

$$\left. + \int d^3r_3 \chi_{KS}(r, r_1, \omega)k_{xc}(r_1, r_2, r_3, \omega, \omega - \omega')n_1(r_2, \omega')n_1(r_3, \omega - \omega')\right)$$

$$+ \int d^3r_1 d^3r_2 \chi_{KS}(r, r_1, \omega)f_{Hxc}(r_1, r_2, \omega)n_2(r_2, \omega)$$

$$(4.94)$$

Likewise, one may work out Dyson-like response equations for the higher-order response functions, each time introducing a new higher-order xc kernel. These determine the frequency-dependent non-linear response. Sum-over-states expressions for the non-interacting KS density-response functions up to third-order may be found in Senatore and Subbaswamy (1987). We also point to Chap. 7, where higher-order response is discussed within a Sternheimer scheme.

Gross et al. (1996) pointed out a very interesting hierarchical structure that the TDDFT response equations have. At *any* order i,

$$n_i(\omega) = M_i(\omega) + \chi_{KS}(\omega) \star f_{Hxc}(\omega) \star n_i(\omega) \tag{4.95}$$

where M_i depends on *lower-order* density-response (and response-functions up to ith order). The last term on the right of Eq. 4.95 has the same structure for all orders. If we define the operator

$$L(\omega) = \hat{1} - \chi_{KS}(\omega) \star f_{Hxc}(\omega) \tag{4.96}$$

then

$$L(\omega) \star n_i(\omega) = M_i(\omega) \tag{4.97}$$

so $L(\omega)$ plays a significant role in determining what new poles are generated from electron-interaction effects in all orders of response (Elliott 2011).

4.7 Approximate Functionals

As noted earlier, the xc potential is a functional of the density, the initial true state, and the initial KS state. The exact functional has "memory", that is, it depends on the history of the density as well as these two initial states. This is discussed to some extent in Chap. 8. In fact these two sources of memory are intimately related, and often the elusive initial-state dependence can be replaced by a type of history-dependence. The xc kernel of linear response has simpler functional dependence, as it measures xc effects around the initial ground-state only. Functionally it depends only on the initial ground-state density, while memory-dependence appears as dependence on frequency in the arguments of f_{xc}.

It should be noted that when running response calculations, for formal consistency, the same approximation should be used for the xc kernel as is used for the ground-state potential, i.e. there must exist an approximate functional $v_{xc}^{app}[n](r, t)$ such that the initial potential $v_{xc}^{app}[n](r, t = 0)$ is used in the KS ground-state calculation, and such that $f_{xc}^{app}[n](r, r', t - t') = \delta v_{xc}^{app}[n](r, t)/\delta n(r', t')$ is used in the time-dependent response part. If different functionals were used for each step, one has left the framework of TDDFT response, since the calculation no longer is equivalent to computing the time-dependent response to an external perturbation. Nevertheless, this fact is often ignored in practical calculations.

We now will outline some of the different approximations people use today.

4.7.1 Adiabatic Approximations: ALDA, AGGA, AB3LYP, etc.

Almost all functionals in use today have no memory-dependence whatsoever. They are "adiabatic", meaning that the density at time t is plugged into a ground-state functional, i.e.

$$v_{xc}^{adia}[n](\boldsymbol{r}, t) = v_{xc}^{GS}[n(t)](\boldsymbol{r}).\tag{4.98}$$

For the xc kernel, we retrieve the static kernel of Eq. 4.80

$$f_{xc}^{adia}[n_{GS}](\boldsymbol{r}t, \boldsymbol{r}'t') = \left.\frac{\delta v_{xc}^{GS}[n(t)](\boldsymbol{r})}{\delta n(\boldsymbol{r}', t')}\right|_{n_{GS}} = \delta(t - t')\left.\frac{\delta^2 E_{xc}[n]}{\delta n(\boldsymbol{r})\delta n(\boldsymbol{r}')}\right|_{n_{GS}}.\tag{4.99}$$

If the external time-dependence is very slow (adiabatic) and the system begins in a ground-state, this approximation is justified. But this is not the usual case. Even if the density is reproducible by a system in its a ground-state, the wavefunction is usually not, so this appears to be quite a severe approximation. Nevertheless adiabatic approximations are the workhorse of the myriads of applications of TDDFT today, and work pretty well for most cases (but not all). Why this is so is still somewhat of an open question. Certainly adiabatic approximations trivially satisfy many exact conditions related to memory-dependence, so perhaps this is one reason. This is similar to the justification used for the success of LDA in the ground-state case, and the subsequent development of generalized gradient approximations (GGA) based on satisfaction of exact conditions. When considering excitation energies of systems, the bare KS orbital energy differences themselves are reasonably good approximations to the exact excitation energies. The kernel then just adds a small correction on top of this good zeroth order estimate, and hence even the simplest approximation such as an adiabatic one does a decent job. Many cases where the usual approximations fail, such as excited states of multiple-excitation character, or certain types of electronic quantum control problems, can be clearly understood to arise from lack of memory in the adiabatic approximation.

The adiabatic local density approximation, or ALDA is the simplest possible approximation in TDDFT. It is also often called TDLDA (for time-dependent LDA):

$$v_{xc}^{ALDA}[n](\boldsymbol{r}, t) = v_{xc}^{hom}(n(\boldsymbol{r}, t)) = \frac{d}{dn}[ne_{xc}^{hom}(n)]|_{n=n(\boldsymbol{r},t)}\tag{4.100}$$

where $e_{xc}^{hom}(n)$) is the xc energy per particle of the homogeneous electron gas. The corresponding xc kernel

$$f_{xc}^{ALDA}[n_{GS}](\boldsymbol{r}t, \boldsymbol{r}'t') = \delta(t - t')\delta(\boldsymbol{r} - \boldsymbol{r}')\frac{d^2}{dn^2}[ne_{xc}^{hom}(n)]|_{n=n_{GS}(\boldsymbol{r})}\tag{4.101}$$

which is completely local in both space and time, and its Fourier-transform, Eq. 4.86a is frequency-independent. Although it might appear justified only for slowly-varying systems in space and in time, it often gives reasonable results for systems far from this limit. Adiabatic GGA's and hybrid functionals are most commonly used for finite systems; hybrids in particular for the higher-lying excitations where it is important to catch the tail of the molecular potential.

4.7.2 Orbital Functionals

A natural way to break free of the difficulties in approximating functionals of the density alone, and still stay within TDDFT, is to develop functionals of the KS orbitals. The simplest functional of this kind is the exact-exchange functional, derived from the action:

$$
A_x[\{\phi_{i\alpha}\}] = -\frac{1}{2} \sum_\sigma \sum_{i,j}^{N_\sigma} \int_{-\infty}^t \mathrm{d}t' \int \mathrm{d}^3 r
$$
$$
\times \int \mathrm{d}^3 r' \frac{\varphi_{i\sigma}^*(r',t')\varphi_{j\sigma}(r',t')\varphi_{i\sigma}(r,t')\varphi_{j\sigma}^*(r,t')}{|r-r'|} \tag{4.102}
$$

Note that for more general functionals, the action needs to be defined on the Keldysh contour (van Leeuwen 1998) (see also Chap. 6). The exact exchange potential is then given by

$$
v_{x,\sigma}[\{\phi_{j\alpha}\}](r,t) = \frac{\delta A_x[\{\phi_{j\alpha}\}]}{\delta n_\sigma(r,t)} \tag{4.103}
$$

Orbital functionals are in fact implicit density functionals because orbitals are trivially functionals of the single-particle KS potential, which, by the RG theorem, is a functional of the density, $\varphi_j[v_{KS}][n](r,t)$. The xc potential is given by the functional derivative of the action with respect to the (spin-) density and the xc kernel is the second functional derivative. The equation satisfied by the xc potential is usually called the (time-dependent) Optimized Effective Potential (OEP) equation, and is discussed in Chap. 6. The exact-exchange functional is local in time when viewed as a functional of KS orbitals. However, viewed as an implicit functional of the density, it is non-local in time. The second-functional derivative with respect to the density then has a non-trivial dependence on $t - t'$.

Invoking a Slater-type approximation in each functional derivative of Eq. 4.102, (Petersilka 1996a, 1998) deduced,

$$
f_{x\sigma\sigma'}^{PGG}(\mathbf{r}, \mathbf{r}') = -\delta_{\sigma\sigma'} \frac{1}{|\mathbf{r} - \mathbf{r}'|} \frac{|\sum_k f_{k\sigma} \varphi_{k\sigma}(\mathbf{r})\varphi_{k\sigma}^*(\mathbf{r}')|^2}{n_\sigma(\mathbf{r})n_\sigma(\mathbf{r}')}. \tag{4.104}
$$

Evidently, with this approximation, the non-trivial $(t - t')$-dependence of the *exact* exchange-only kernel is not accounted for. However, it is clearly spatially non-local. For one and two electrons, Eq. 4.104 is exact for exchange.

The full numerical treatment of exact-exchange (TDEXX), including memory-dependence, has recently seen some progress. In Görling (1997b), the exact x kernel was derived from perturbation theory along the time-dependent adiabatic connection. One scales the electron-electron interaction by λ, defining a Hamiltonian

$$\hat{H}^\lambda = \hat{T} + \lambda \hat{V}_{ee} + \hat{V}^\lambda(t) \qquad (4.105)$$

such that the density $n^\lambda(\mathbf{r}, t) = n(\mathbf{r}, t)$ for any λ much like as is done in ground-state DFT. The initial state Ψ_0^λ is chosen to reproduce the same initial density and its first time-derivative, at any λ. For $\lambda = 0$, $\hat{V}^\lambda(t) = \hat{V}_{KS}(t)$ and for $\lambda = 1$, $\hat{V}^\lambda(t) = \hat{V}_{ext}(t)$. Performing perturbation theory to first order in λ yields the TDEXX potential and kernel, found in Görling (1997, 1998a, 1998b) while higher orders give correlation functionals. Until very recently there was only limited use of this kernel due to numerical instabilities (Shigeta et al. 2006). A series of papers reformulated the problem in terms of response of the KS potential itself, avoiding the calculation of numerically prohibitive inverse response functions (Hesselmann et al. 2009; Görling et al. 2010; Ipatov 2010), but needing a time-consuming frequency-iteration for each excitation energy. Most recently, a very efficient method has been derived that translates the problem onto a generalized eigenvalue problem (Hesselmann and Görling 2011). Although the results of full exact-exchange calculations for excitation energies are often numerically close to those of time-dependent Hartree–Fock (Hesselmann and Görling 2011) in the cases so far studied, there is a fundamental difference in the two methods: TDEXX operates with a local (multiplicative) potential, while that of time-dependent Hartree–Fock is non-local, i.e. an integral operator. Furthermore, the Hartree–Fock single-particle energy differences $\varepsilon_a^{HF} - \varepsilon_i^{HF}$ are usually too large and hence the Hartree–Fock kernel reduces the Hartree–Fock energy difference, while $\varepsilon_a^{KS} - \varepsilon_i^{KS}$ within EXX tend to be too small, so the x-kernel of TDDFT has to increase the KS excitation energy. Beyond the linear response regime, (Wijewardane and Ullrich 2008) computed nonlinear dynamics in semiconductor wells within TDEXX.

Another class of orbital-dependent functionals are self-interaction-corrected (SIC) functionals. An approximation at a similar level to Eq. 4.104 can be found in Petersilka (2000).

4.7.3 Hydrodynamically Based Kernels

The first proposal to incorporate memory-dependence was that of Gross and Kohn (1985), who suggested to use the frequency-dependent xc kernel of the homogeneous electron gas in the sense of an LDA:

$$f_{xc}^{LDA}[n_{GS}](\mathbf{r}, \mathbf{r}', \omega) = f_{xc}^{hom}(n_{GS}(\mathbf{r}), |\mathbf{r} - \mathbf{r}'|, \omega). \qquad (4.106)$$

and furthermore that the response $n_1(\mathbf{r}, \omega)$ is slowly varying enough on the length-scale of $f_{xc}^{hom}(n_{GS}(\mathbf{r}), |\mathbf{r} - \mathbf{r}'|; \omega)$ that only its uniform component contributes. That is, taking a spatial Fourier-transform with respect to $r - r'$, we include only the zeroth-Fourier component. This gives the Gross-Kohn kernel:

$$f_{xc}^{GK}[n_{GS}](r, r', \omega) = \delta(r - r') f_{xc}^{hom}(n_{GS}(r), q = 0, \omega) \qquad (4.107)$$

where q is the spatial Fourier transform variable. One requires the knowledge of the frequency-dependent response of a uniform electron gas, about which, indeed many exact properties are known, and parametrizations, believed to be accurate, exist (Conti 1999, 1997; Gross and Kohn 1985; Qian and Vignale- 2002, 2003). Chapter 24 discusses some of these.

Although the GK approximation has memory, it is completely local in space, a property which turns out to violate exact conditions, such as the zero-force rule and translational invariance (see Chaps. 5 and 24). Even ALDA does not violate these. To go beyond the adiabatic approximation consistently, both spatial and temporal non-locality must be included. This is perhaps not surprising in view of the fact that the density that at time t is at location r was at an earlier time $t' < t$ at a different location, i.e. memory is carried along with the fluid element. The development of memory-dependent functionals, often based on hydrodynamic schemes, is discussed further in Chaps. 8 and 24. These include the Dobson–Bünner–Gross (Dobson et al. 1997; Vignale and Kohn 1996; Tokatly 2005a, b), and (Kurzweil and Baer 2004) approaches. These are not commonly used; only the Vignale-Kohn functional has seen a few applications.

4.8 General Performance and Challenges

As shown in Sect. 4.2, TDDFT is an exact reformulation of non-relativistic time-dependent quantum mechanics. In principle, it yields exact electronic dynamics and spectra. In practice, its accuracy is limited by the functional approximations used. The simplest and computationally most efficient functional, ALDA, is local in both space and time, and it is perhaps surprising that it works as well as it does. We now discuss cases where it is essential to go beyond this simple approximation, and, further, beyond its adiabatic cousins. We organize this section into three parts: linear response in extended systems, linear response in finite systems, and real-time dynamics beyond the perturbative regime.

4.8.1 Extended Systems

We first ask, how well does ALDA perform for the response of solids? In simple metals, ALDA does well, and captures accurately the plasmon dispersion curves (Quong and Eguilez 1993). In fact, the ordinary plasmon is captured reasonably even by the Hartree potential alone, i.e. setting $f_{xc} = 0$, which is called the RPA. Applying f_{xc}^{ALDA} improves the description of its dispersion and linewidth.

The answer to the question above is rather more subtle for non-metallic systems. ALDA does a good job for electron energy loss (EEL) spectra, both when the impinging electron transfers finite momentum q and in the case of vanishing momentum-transfer. The EEL spectrum measures the imaginary part of the inverse dielectric function (see Chap. 3), which, in terms of the density-response function, is given in Eq. 3.56a (Onida et al. 2002; Botti et al. 2007). The optical response, which measures the imaginary part of the macroscopic dielectric function (see Chap. 3) is also quite well predicted by ALDA (Weissker et al. 2006) for *finite* wavevector q.

However, for optical response in the limit of vanishing wavevector, $q \rightarrow 0$, ALDA performs poorly for non-metallic systems. There are two main problems: (i) The onset of continuous absorption is typically underestimated, sometimes by as much as 30–50%. This problem is due to the fact that the KS gap in LDA is much smaller than the true gap. But even with the exact ground-state potential, there is very strong evidence (Knorr and Godlay 1992; Grüning et al. 2006, Niquet and Gonze 2004) that the exact KS gap is typically smaller than the true gap. To open the gap, the xc kernel must have an imaginary part (Giuliani and Vignale 2005). This follows from the fundamental Dyson equation (4.59)[2]: We know that the imaginary part of $\chi_{KS}(r, r', \omega) = 0$ for ω inside the KS gap. Then, for an approximate $f_{xc}(r, r', \omega)$ that is *real*, taking the imaginary part of Eq. 4.59 for ω inside the KS gap ($0 < \omega < E_g^{KS}$), yields

$$\Im m \chi_{KS}(\omega) = 0 \longrightarrow \left[\hat{1} - \Re e \chi_{KS}(\omega) \star f_{Hxc}(\omega) \right] \star \Im m \chi(\omega) = 0. \qquad (4.108)$$

Following the analogous procedure for ω inside the *true* gap ($0 < \omega < E_g$), where $\Im m \chi(r, r'\omega) = 0$, yields

$$\Im m \chi(\omega) = 0 \longrightarrow \Im m \chi_{KS}(\omega) \star \left[\hat{1} + f_{Hxc}(\omega) \star \Re e \chi(\omega) \right] = 0. \qquad (4.109)$$

In view of the fact that $f_{Hxc} = \chi_{KS}^{-1} - \chi^{-1}$, the expressions inside the square brackets in (4.108) and (4.109) cannot vanish identically in the full interval $0 < \omega < E_g^{KS}$ and $0 < \omega < E_g$, respectively. (The expressions may vanish at isolated frequencies corresponding to collective excitations). Hence we must conclude that wherever $\Im m \chi_{KS}(\omega) = 0$, then also $\Im m \chi(\omega) = 0$. That is, for frequencies inside the KS gap, the true response is also zero. Likewise, wherever $\Im m \chi(\omega) = 0$, then also $\Im m \chi_{KS}(\omega) = 0$. That is, for frequencies inside the true gap, the KS response is also zero. Putting the two together implies that the KS system and the true system must have the same gap when an approximation for f_{Hxc} is used that is purely real. This is clearly a contradiction, implying that f_{Hxc} must have a non-vanishing imaginary part. This, on the other hand, is equivalent to f_{xc} having a frequency-dependence, as mentioned in Sect. 4.5.3. Any adiabatic approximation however takes a kernel that is the second density-functional-derivative of a ground-state energy functional, and therefore is purely real. Further, as we shall shortly discuss, the kernel must have have

---

[2]   This argument is due largely to Giovanni Vignale.

a long-ranged part, that goes as $1/q^2$ as $q \to 0$, to get any non-vanishing correction on the gap. ALDA, on the other hand, is local in space, and so constant in $\boldsymbol{q}$.

(ii) The second main problem is that ALDA cannot yield any excitonic structure; again, one needs a long-ranged $1/q^2$ part in the kernel to get any significant improvement on RPA for optical response. To reproduce excitons, an imaginary part to f_{xc} is not required (Reining 2002).

Fundamentally, the long-ranged behaviour of f_{xc} can be deduced from exact conditions satisfied by the xc kernel, such as the zero-force theorem, that inextricably link time-nonlocality and space-non-locality. This is shown explicitly in Chap. 24. Lack of a long-ranged term in ALDA or AGGA for finite systems, or for EELs spectra, is not as critical as for the case of extended systems, since its contribution is much smaller there.

The need for this long-ranged behavior in the xc kernel is, interestingly, *not* a consequence of the long-rangedness of the Coulomb interaction. A simple way to see this is to consider the SPA Eq. 4.63 for a system of size L^3, where, for the extended system we consider $L \to \infty$. The transition densities Φ_q scale as $1/L^3$, so for a *finite*-ranged xc kernel, the xc-correction to the RPA value scales as $1/L^3$, and so vanishes in the extended-system limit (Giuliani and Vignale 2005).

An alternate way of seeing the need for the long-ranged kernel, is to note that the optical absorption measures the imaginary part of the macroscopic dielectric function, which can be written in terms of a modified density-response function (Botti et al. 2007) (and see Eqs. 3.51 and 3.55 in Chap. 3). Now $\chi_{\mathrm{KS}}(q \to 0) \sim q^2$ for infinite systems, so if f_{xc} is to have any significant non-vanishing effect on the optical response, it must have a component that diverges as $1/q^2$ as $q \to 0$.

Recent years have seen a tremendous effort to confront the problem of optical response in solids by including spatially-non-local dependence. Exact-exchange was shown in Kim and Görling (2002) to have apparent success in capturing the exciton. However, it was shown later in Bruneval (2006) that if done carefully, the excitonic structure predicted by exact-exchange is far too strong, essentially collapsing the entire spectrum onto the exciton. The earlier calculation of Kim and Görling (2002) fortuitously induced an effective screening of the interaction, since the long wavelength contributions were cut off in those calculations. TDCDFT has also been used (de Boeij 2001), with the motivation that local functionals of the current-density contain non-local information of the density, and this is discussed further in Chap. 24. Perhaps the most intense progress has been made in the development of kernels derived from many-body perturbation theory (MBPT), leading to what is now known as the "nanoquanta kernel". The latter is deduced from the Bethe-Salpeter approach of MBPT (Bruneval et al. 2005; Reining et al. 2002; Sottile et al. 2003; Adragna et al. 2003; Marini et al. 2003b; Stubner et al. 2004; von Barth et al. 2005). An important aspect is that the reference systems in the Bethe-Salpeter approach and the TDDFT approach are completely different: KS excitation energies and orbitals of the latter are *not* quasiparticle energies and wavefunctions that the former builds on. From the point of view of MBPT, the TDDFT xc kernel may be interpreted as having two roles: shifting the KS excitations to the quasiparticle ones (so-called $f_{xc}^{(1)}$,) and then accounting for the electron-hole interaction (so-called $f_{xc}^{(2)}$).

In practical uses of the nanoquanta kernel, explicit models for $f_{xc}^{(2)}$ are used on top of simply the quasiparticle energies (usually in GW approximation). We refer the reader to Botti et al. (2007) and Onida et al. (2002).

4.8.2 Finite Systems

In linear response calculations of optical spectra, use of local or semi-local functionals for low-lying excitations, or hybrid functionals for higher-lying ones, within the adiabatic approximation for the xc kernel yields results that are typically considerably better than those from TDHF or configuration interaction singles (CIS). These methods scale comparably to TDDFT, while the accuracy of TDDFT is, in most cases, far superior.

Most quantum chemistry applications use the B3LYP hybrid functional (Becke 1993a, b). While excitation energies are typically good to within a few tenths of an eV, structural properties fare much better (Furche 2002a; Elliott et al. 2009). For example, bond-lengths of excited states are within 1%, dipole moments and vibrational frequencies to within 5%. Chapter 16 discusses this more. Often the level of accuracy has been particularly useful in explaining, for the first time, mechanisms of processes in biologically and chemically relevant systems, e.g. the dual fluorescence in dimethyl-amino-benzo-nitrile (Rappoport and Furche 2004), and chiral identification of fullerenes (Furche 2002b).

In the following, we discuss several cases where the simplest approximations like ALDA and AGGA, perform poorly.

To be able to describe Rydberg excitations, it is essential that the ground-state potential out of which the bare KS excitations are computed has the correct $-1/r$ asymptotics. LDA and GGA do not have this feature, and as we have seen already in our case study of the He atom, the Rydberg excitations were absent in LDA/GGA. Solutions include exact-exchange methods, self-interaction corrected functionals, and hybrids. Step-like features in the ground-state potential as well as spatial non-locality in the xc kernel can also be essential: a well-known case is in the computation of polarizabilities of long-chain molecules (van Faassen 2002; van Gisbergen 1999b; Gritsenko 2000), and exact-exchange, as well as TDCDFT-methods have been explored for this problem. A more challenging problem is that of molecular dissociation: it is notoriously difficult to obtain accurate *ground-state* dissociation curves, since self-interaction errors in the usual functionals leads to fractional charges at large separation. As it dissociates, the exact ground-state potential for a molecule composed of open-shell fragments such as LiH, develops step and peak features in the bond-midpoint region (Perdew 1985; Gritsenko 1996; Helbig et al. 2009; Tempel et al. 2009), missed in GGAs and hybrids alike, but crucial for a correct description. To get even qualitatively correct excited state surfaces, frequency-dependence is crucial in the xc kernel (Maitra 2005b; Maitra and Tempel 2006a) (see also Chap. 8). The essential problem is that the true wavefunction has wandered far from a single-Slater-determinant, making the work of the ground-state exchange-correlation potential and the kernel very difficult. Such cases of strong correlation are one of the major motivators of time-dependent density-matrix functional theory, discussed in Chap. 26.

Another case where frequency-dependence is essential are states of double-excitation character. We will defer a discussion of this to Chap. 8.

A notorious failure of the usual approximations for finite systems is for charge-transfer excitations at large-separation (Dreuw 2003, 2004; Tozer 2003). To leading order in $1/R$, the exact answer for the lowest charge-transfer excitation frequency is:

$$\omega_{\text{CT}}^{\text{exact}} \to I^D - A^A - 1/R \qquad (4.110)$$

where I^D is the ionization energy of the donor, A^A is the electron affinity of the acceptor and $-1/R$ is the first electrostatic correction between the now charged species. Charge-transfer excitations calculated by TDDFT with the usual approximations however severely underestimate Eq. 4.110. Due to the exponentially small overlap between orbitals on the donor and acceptor, located at different ends of the molecule, f_{xc} must diverge exponentially with their separation in order to give any correction to the bare KS orbital energy difference (see e.g. Eq. 4.63). Semilocal functional approximations for f_{xc} give no correction, so their prediction for charge-transfer excitations reduces to the KS orbital energy difference, $\varepsilon_a - \varepsilon_i = \varepsilon_L(\text{acceptor}) - \varepsilon_H(\text{donor})$, where L, H subscripts indicate the KS LUMO and HOMO, respectively. This is a severe underestimate to the true energy, because the ionization potential is typically underestimated by the HOMO of the donor, due to the lack of the $-1/r$ asymptotics in approximate functionals (Sect. 4.5.5), while the LUMO of the acceptor lacks the discontinuity contribution to the affinity. The last few years have seen many methods to correct the underestimation of CT excitations, e.g. (Autschbach 2009; Tawada et al. 2004; Vydrov 2006; Zhao and Truhlar 2006; Stein et al. 2009a; Hesselmann et al. 2009; Rohrdanz 2009); most modify the ground-state functional to correct the approximate KS HOMO's underestimation of I using range-separated hybrids that effectively mix in some degree of Hartree–Fock, and most, but not all (Stein et al. 2009a; Hesselmann et al. 2009) determine this mixing via at least one empirical parameter. Fundamentally, staying within pure DFT, both the discontinuity contribution to A and the $-1/R$ tail in Eq. 4.110 come from f_{Hxc}, which must exponentially diverge with fragment separation (Gritsenko and Baerends 2004). Worse still, in the case of open-shell fragments, *not* covered by most of the recent fixes, additionally the exact f_{xc} is strongly frequency-dependent (Maitra 2005b).

We briefly mention two other challenges, which are further discussed later in this book. The difficulty that usual functionals have in capturing potential energy surfaces near a conical intersection is discussed in Chap. 14. This poses a challenge for coupled electron-dynamics using TDDFT, given that conical intersections are a critical feature on the potential energy landscape, funneling nuclear wavepackets between surfaces. The second challenge is the Coulomb blockade phenomenon in calculations of molecular transport. The critical need for functionals with a derivative discontinuity to describe this effect, and to obtain accurate conductances in nanostructures, is further discussed in Chap. 17.

4.8.3 Non-perturbative Electron Dynamics

In Chap. 18 of this book, we shall come back to the fascinating world of strong-field phenomena, several of which the usual approximations of TDDFT have had success in describing, and some of which the usual approximations do not capture well. Developments in attosecond laser science have opened up the possibility of electronic quantum control; recently the equations for quantum optimal control theory within the TDKS framework have been established and this is described in Chap. 13. Chaps. 14 and 15 discuss the difficult but extremely important question of coupling electrons described via TDDFT to nuclear motion described classically, in schemes such as Ehrenfest dynamics and surface-hopping. Here instead we discuss in general terms the challenges approximations in TDDFT face for real-time dynamics.

However first, we show how useful a density-functional picture of electron dynamics can be for a wide range of processes and questions, via the time-dependent electron localization function (TDELF). With the advent of attosecond lasers, comes the possibility of probing detailed mechanisms of electronic excitations and dynamics in a given process. For example, in chemical reactions, can we obtain a picture of bond-breaking and bond-forming? In Burnus (2005) it was shown how to generalize the definition of the electron localization function (ELF) used to analyze bonding in ground-state systems (Becke 1990), to time-dependent processes:

$$\text{TDELF}(\mathbf{r}, t) = \frac{1}{1 + \left[D_\sigma(\mathbf{r}, t) / D_\sigma^0(\mathbf{r}, t) \right]^2}, \qquad (4.111)$$

with

$$D_\sigma(\mathbf{r}, t) = \tau_\sigma(\mathbf{r}, t) - \frac{1}{4} \frac{|\nabla n_\sigma(\mathbf{r}, t)|^2}{n_\sigma(\mathbf{r}, t)} - \frac{j_\sigma^2(\mathbf{r}, t)}{n_\sigma(\mathbf{r}, t)} \qquad (4.112)$$

where j_σ is the magnitude of the KS current-density of spin σ, and $\tau_\sigma(\mathbf{r}, t) = \sum_{i=1}^{N_\sigma} |\nabla \varphi_{i\sigma}(\mathbf{r}, t)|^2$ is the KS kinetic energy-density of spin σ. In Eq. 4.111, $D_\sigma^0(\mathbf{r}, t) = \tau_\sigma^{\text{hom}}(n_\sigma(\mathbf{r}, t)) = \frac{3}{5}(6\pi^2)^{2/3} n_\sigma^{5/3}$ is the kinetic energy-density of the uniform electron gas. Using ALDA to evaluate the TDELF, this function has been useful for understanding time-resolved dynamics of chemical bonds in scattering and excitation processes (Burnus et al. 2005; Castro et al. 2007); features such as the temporal order of processes, and their time scales are revealed. As an example, in Fig. 4.5 we reproduce snapshots of the TDELF for laser-induced excitation of the $\pi \rightarrow \pi^*$ transition in the acetylene molecule, studied in Burnus et al. (2005). Many

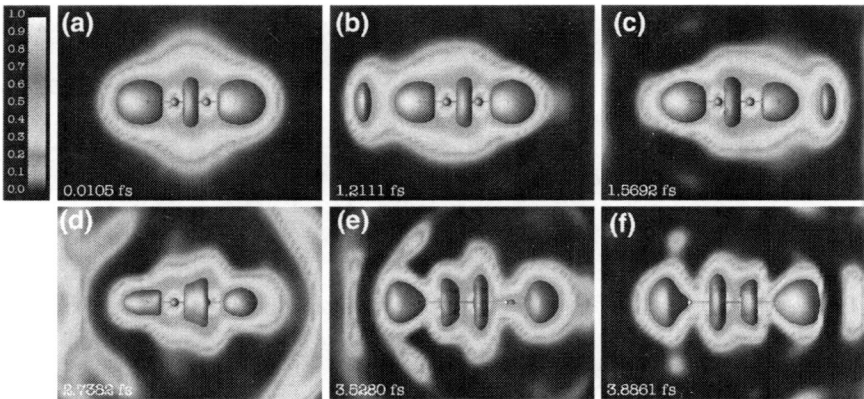

Fig. 4.5 Snapshots of the TDELF for the excitation of acetylene by a 17.5 eV laser pulse (from Burnus (2005)), polarized along the molecular axis. The pulse had a total length of 7 fs, an intensity of 1.2×10^{14} W/cm^2

interesting features can be observed from the TDELF. As the intensity of the laser field increases, the system begins to oscillate, and ionization is visible in the time slices at 1.2111 and 1.5692 fs. The figure clearly shows that after 3.5 fs, the transition from the ground-state to the antibonding state is complete: the original single torus signifying the triple bond in the ground-state has split into two separate tori, each around one carbon atom.

General success for dynamics in strong fields has been slower than for linear response applications. There are three main reasons. First, many of the observables of interest are not simply related to the time-dependent one-body density, so that, in addition to the approximation for the xc functional, a new ingredient is needed: approximate "observable functionals" to extract the properties of interest from the KS system. Sometimes these are simply the usual quantum mechanical operators acting directly on the KS system, e.g. high-harmonic generation spectra are measured by the dipole moment of the system, $\int d^3 r n(\mathbf{r}, \omega) z$. But if the observable is not simply related to the density, such as ionization probabilities (Ullrich and Gross 1997; Petersilka and Gross 1999), or cross-sections in atomic collisions (Henkel et al. 2009), in principle an observable functional is needed. Simply extracting double-ionization probabilities and momentum-densities using the usual operators acting on the KS wavefunction typically fails (Lappas and Van Leeuwen 1998; Wilken and Bauer 2006; Wilken and Bauer 2007; Rajam et al. 2009).

Second, lack of memory dependence in the usual xc approximations has been suggested to be often far more problematic than in the linear-response regime as is discussed in Chap. 8. We must deal with the full xc potential $v_{\mathrm{xc}}[n; \Psi_0, \Phi_0](\mathbf{r}, t)$, instead of the simpler xc kernel. The exact functional depends on the history of the density as well as on the initial state but almost all functionals used today are adiabatic. Third, a particularly severe difficulty is encountered when a system starting in a wavefunction dominated by a single Slater determinant evolves to a state that fundamentally needs at least two Slater determinants to describe it. This is the time-dependent (TD) analog of ground-state static correlation, and arises in electronic quantum control problems (Maitra et al. 2002b; Burke et al. 2005a), in ionization (Rajam 2009), and in coupled electron-ion dynamics (Levine et al. 2006). The TD KS system evolves the occupied orbitals under a one-body Hamiltonian, remaining in a single Slater determinant: the KS one-body density matrix is always idempotent (even with exact functionals), while, in contrast, that of the true system develops eigenvalues (natural occupation numbers) far from zero or one in these applications (Appel and Gross 2010). The exact xc potential and observable functionals consequently develop complicated structure that is difficult to capture in approximations. For example, in Rajam et al.(2009), a simple model of ionization in two-electron systems showed that the momentum distribution computed directly from the exact KS system contains spurious oscillations due to using a single, necessarily delocalized orbital, a non-classical description of the essentially classical two-electron dynamics.

Chapter 5
Exact Conditions and Their Relevance in TDDFT

Lucas O. Wagner, Zeng-hui Yang and Kieron Burke

5.1 Introduction

This chapter is devoted to exact conditions in time-dependent density functional theory. Many conditions have been derived for the exact ground-state density functional, and several have played crucial roles in the construction of popular approximations. We believe that the reliability of the most fundamental approximation of any density functional theory, the local density approximation (LDA), is due to the exact conditions that it satisfies. Improved approximations should satisfy at least those conditions that LDA satisfies, plus others. (Which others is part of the art of functional approximation).

In the time-dependent case, as we shall see, the adiabatic LDA (ALDA) plays the same role as LDA in the ground-state case, as it satisfies many exact conditions. But we do not have a generally applicable improvement beyond ALDA that includes nonlocality in time. For TDDFT, we have a surfeit of exact conditions, but that only makes finding those that are useful to impose an even more demanding task.

Throughout this chapter, we give formulas for pure DFT for the sake of simplicity (e.g. $E_{xc}[n]$), but in practice *spin* DFT is used (e.g. $E_{xc}[n_\uparrow, n_\downarrow]$). We use atomic units everywhere ($e^2 = \hbar = m_e = 1$), so energies are in units of Hartrees and distances are in Bohrs.

L. O. Wagner · Z. Yang
Department of Physics and Astronomy,
University of California,
Irvine, CA 92697, USA
e-mail: lwagner@uci.edu

K. Burke (✉)
Department of Chemistry, Department of Physics,
University of California,
Irvine, CA 92697, USA
e-mail: kieron@uci.edu

M. A. L. Marques et al. (eds.), *Fundamentals of Time-Dependent Density Functional Theory*, Lecture Notes in Physics 837, DOI: 10.1007/978-3-642-23518-4_5,
© Springer-Verlag Berlin Heidelberg 2012

5.2 Review of the Ground State

In ground-state DFT, the unknown exchange-correlation energy functional, $E_{xc}[n]$, plays a crucial role. In fact, it is this energy that we typically wish to approximate with some given level of accuracy and reliability, and *not* the density itself. Using such an approximation in a modern Kohn–Sham ground-state DFT calculation, we can calculate the total energy of any configuration of the nuclei of the system within the Born–Oppenheimer approximation. In this way we can extract the bond lengths and angles of molecules and deduce the lowest energy lattice structure of solids. We can also extract forces in simulations, and vibrational frequencies and phonons and bulk moduli. We can discover response properties to both external electric fields and magnetic fields (using spin DFT). The accuracy of the self-consistent density is irrelevant to most of these uses.

Given the central role of the energy, it makes sense to devote much effort to its study as a density functional. Knowledge of its behavior in various limits can be crucial to restraining and constructing accurate approximations, and to understanding their limitations. This task is greatly simplified by the fact that the total ground-state energy satisfies the variational principle. Many exact conditions use this in their derivation.

In this section we will review some of the more prominent exact conditions. They almost all concern the energy functional, which, as mentioned above, is crucial for good KS-DFT calculations. We also refer the interested reader to Perdew and Kurth (2003) for a thorough discussion. First, we will go over some of the formal definitions in DFT.

5.2.1 Basic Definitions

The xc energy as a functional of the density is written as (Levy 1979; Lieb 1983)

$$E_{xc}[n] = \min_{\Psi \to n} \langle \Psi | \hat{T} + \hat{V}_{ee} | \Psi \rangle - T_{KS}[n] - E_{H}[n], \qquad (5.1)$$

where Ψ is a correctly antisymmetrized electron wavefunction, the minimization of the kinetic and electron–electron repulsion energies is done over all such wavefunctions that yield the density $n(r)$, $T_{KS}[n]$ is the minimum (non-interacting) kinetic energy of a system with density $n(r)$, and

$$E_{H}[n] = \frac{1}{2} \int d^3 r \int d^3 r' \frac{n(r)n(r')}{|r - r'|} \qquad (5.2)$$

is the Hartree energy. The xc energy is usually split into an exchange piece, E_x, and a correlation piece, $E_c \equiv E_{xc} - E_x$. Exchange can be defined in a Hartree–Fock-like way in terms of the KS spin orbitals $\varphi_{i\sigma}(r)$:

$$E_x[n] = -\frac{1}{2} \sum_{i,j,\sigma}^{occ} \int d^3r \int d^3r' \frac{\varphi_{i\sigma}^*(r)\varphi_{j\sigma}^*(r')\varphi_{i\sigma}(r')\varphi_{j\sigma}(r)}{|r - r'|}. \tag{5.3}$$

To perform the self-consistent calculations in the non-interacting system, we need the functional derivative of the xc energy,

$$v_{xc}[n](r) = \frac{\delta E_{xc}[n]}{\delta n(r)}. \tag{5.4}$$

This is called the xc potential, and it is the essential part of the multiplicative KS potential $v_{KS}[n](r)$.

Orbital-dependent functionals. Some functionals are most naturally expressed in terms of the orbitals rather than the density. When varying the orbitals of these functionals, *nonlocal* potentials are obtained. For example, varying $\varphi_{i\sigma}$ in Eq. 5.3 leads to the nonlocal exchange term used in HF. There is a way to transform such orbital-dependent functionals into local potentials as in Eq. 5.4. This procedure is known as optimized effective potential (OEP) or optimized potential method (OPM) and is computationally expensive (Kümmel and Kronik 2008). Using OEP for E_x results in the exact exchange approximation (EXX) for E_{xc} in KS-DFT. The Krieger, Li, and Iafrate (KLI) approximation is a way to approximately solve EXX (Krieger et al. 1992a). An in-depth discussion of orbital-dependent functionals is found in Chap. 6.

Adiabatic connection. One can imagine smoothly connecting the interacting and non-interacting systems by multiplying the electron–electron repulsion term by λ, called the coupling-constant. Changing λ varies the strength of the interaction, and if we simultaneously change the external potential to keep the density fixed, we have a family of solutions for various interaction strengths. This makes all quantities (besides the density) functions of λ. When $\lambda = 0$, one has the non-interacting KS system, and when $\lambda = 1$, one has the fully interacting system. The following coupling-constant relations hold (Perdew and Kurth 2003):

- *xc energy λ dependence.* Altering the coupling-constant is simply related to scaling the density:

$$E_{xc}^\lambda[n] = \lambda^2 E_{xc}[n_{1/\lambda}], \tag{5.5}$$

where $n_{1/\lambda}(r)$ is the scaled density

$$n_\gamma(r) \equiv \gamma^3 n(\gamma r), \tag{5.6}$$

with $\gamma = 1/\lambda$.
- *Adiabatic connection formula.* By using the Hellmann–Feynman theorem, one can show:

$$E_{xc}[n] = \int_0^1 \frac{d\lambda}{\lambda} U_{xc}^\lambda[n], \tag{5.7}$$

where U_{xc}^{λ} is the potential contribution to exchange-correlation energy ($U_{xc} = V_{ee} - E_H$) at coupling-constant λ.

5.2.2 Standard Approximations

Despite a plethora of approximations (Perdew et al. 2005), no present-day approximation satisfies all the conditions mentioned in this chapter, as seen in tests on bulk solids and surfaces (Staroverov et al. 2004). With that the case, one must choose which conditions to impose on a given approximate form. Non-empirical (ab initio) approaches attempt to fix all parameters via exact conditions (Perdew et al. 1996a, b), while good empirical approaches might include one or two parameters that are fit to some data set (Becke 1988a; Lee et al. 1988; Becke 1993b).

There are two basic flavors of approximations: pure density functionals, which are often designed to meet conditions on the uniform gas, and orbital-dependent functionals (Grabo et al. 2000), which meet the finite-system conditions more naturally. The most sophisticated approximations being developed today use both (Tao et al. 2003). For a good discussion on what approximation is the right tool for the job, see Rappoport et al. (2009).

- *LDA.* The local density approximation is the bread and butter of DFT. It is the simplest, being derived from conditions on the uniform gas (Kohn and Sham 1965). Though it is too inaccurate for quantum chemistry (being off by about 1 eV or 30 kcal/mol), it is useful in solids and other bulk materials where the electrons almost look like a uniform gas. There can be only one LDA.
- *GGA.* The generalized gradient approximation came from trial and error when energies were allowed to depend on the gradient of the density. While more accurate than the LDA (getting errors down to 5 or 6 kcal/mol), and thus useful for quantum chemistry applications, there is no uniquely-defined GGA. BLYP is an empirical GGA that was designed to minimize the error in a particular data set. PBE is a non-empirical GGA designed to satisfy exact conditions.
- *Hybrid.* Hybrids have an exchange energy which is a mixture of GGA and HF, which attempts to get the best of both worlds:

$$E_{xc}^{hyb} = E_{xc}^{GGA} + a(E_x - E_x^{GGA}), \qquad (5.8)$$

where E_x is defined in (5.3). The parameter a was argued to be 0.25 for the non-empirical PBE0, but is fitted for the empirical B3LYP.

5.2.3 Finite Systems

The following conditions are derived for finite systems, just as the Hohenberg–Kohn theorem is. This list is by no means exhaustive; it is only meant to give an idea of some of the simpler and more useful conditions in ground-state DFT. As we will see, many of these conditions have analogs in TDDFT.

- *Signs of energy components.* From the variational principle and other elementary considerations, one can deduce

$$E_{xc}[n] \leq 0, \quad E_c[n] \leq 0, \quad E_x[n] \leq 0. \tag{5.9}$$

- *Zero xc force and torque theorem.* The xc potential cannot exert a net force or torque on the electrons (Levy and Perdew 1985):

$$\int d^3r n(\boldsymbol{r}) \, \nabla v_{xc}(\boldsymbol{r}) = 0 \tag{5.10a}$$

$$\int d^3r n(\boldsymbol{r}) \, \boldsymbol{r} \times \nabla v_{xc}(\boldsymbol{r}) = 0. \tag{5.10b}$$

- *xc virial theorem.*

$$E_{xc}[n] + T_c[n] = - \int d^3r n(\boldsymbol{r}) \, \boldsymbol{r} \cdot \nabla v_{xc}(\boldsymbol{r}), \tag{5.11}$$

where $T_c = T - T_{KS}$ is the kinetic contribution to the correlation energy. The xc virial theorem as well as the zero xc force and torque theorem are satisfied by all sensible approximate functionals.
- *Exchange scaling.* By using the scaled density (5.6), one can easily show

$$E_x[n_\gamma] = \gamma E_x[n]. \tag{5.12}$$

- *Correlation scaling.* The scaling of correlation is less simple than exchange, and will depend on whether one is in the high density limit (γ large) or low density limit (γ small) (Levy and Perdew 1985; Seidl et al. 2000):

$$E_c[n_\gamma] < \gamma E_c[n] \quad (\gamma < 1) \tag{5.13a}$$

$$E_c[n_\gamma] > \gamma E_c[n] \quad (\gamma > 1). \tag{5.13b}$$

It is also possible to derive the following results

$$E_c[n_\gamma] = E_c^{(2)}[n] + E_c^{(3)}[n]/\gamma + \cdots \quad (\gamma \to \infty) \tag{5.14a}$$

$$E_c[n_\gamma] = \gamma B[n] + \gamma^{3/2} C[n] + \cdots \quad (\gamma \to 0), \tag{5.14b}$$

where $E_c^{(2)}[n]$, $E_c^{(3)}[n]$, $B[n]$, and $C[n]$ are all scale-invariant functionals. These conditions are depicted in Fig. 5.1. Not all popular approximations satisfy these conditions.

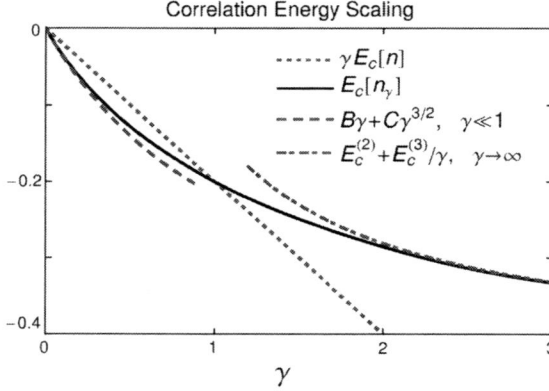

Fig. 5.1 Scaling of the correlation energy in ground state DFT, as well as the various conditions from Eq. 5.14a, b. The first two relations are illustrated with the *dotted line*. For $\gamma < 1$, the exact curve (*solid*) must lie below this *dotted line*, and for $\gamma > 1$ the exact curve must lie above—in both cases within the shaded region of the graph. The high density limit is shown with the *dot-dashed line*, and the low density limit with the *dashed line*. It is believed that not only is $E_c[n_\gamma]$ monotonic, but also its derivative with respect to γ

- *Self-interaction.* For any one-electron system (Perdew and Zunger 1981),

$$E_x[n] = -E_H[n], \quad E_c = 0 \quad (N = 1). \tag{5.15}$$

- *Lieb-Oxford bound.* For any density (Lieb and Oxford 1981),

$$E_{xc}[n] \geq 2.273\, E_x^{LDA}[n]. \tag{5.16}$$

In addition to conditions on E_{xc}, we also know some exact conditions on the xc potential and the KS eigenvalues.

- *Asymptotic behavior of potential.* Far from a Coulombic system

$$v_{xc}(\mathbf{r}) \to -1/r \quad (r \to \infty), \tag{5.17}$$

and

$$\varepsilon_{HOMO} = -I, \tag{5.18}$$

where ε_{HOMO} is the position of the highest occupied KS molecular orbital, and I the ionization potential. These results are intimately related to the self-interaction of one electron.

5.2.4 Extended Systems

The basic theorems of DFT (as discussed in the previous section) are proven for *finite* quantum mechanical systems, with densities that decay exponentially at large distances from the center. Their extension to extended systems, even those as simple as the uniform gas, requires careful thought. For ground-state properties, one can usually take results directly to the extended limit without change, but not always. For example, the high-density limit in Eq. 5.14a, b of the correlation energy for a finite system is violated by a uniform gas. With these things in mind, we will now discuss a set of conditions that involve the properties of the uniform or nearly uniform electron gas.

- *Uniform density.* When the density is uniform, $E_{xc} = n e_{xc}^{unif}(n)\mathcal{V}$, where $e_{xc}^{unif}(n)$ is the xc energy density per particle of a uniform electron gas of density n, and $\mathcal{V}$ is the volume. This forms the basis of LDA.
- *Slowly varying density.* For slowly varying densities, E_{xc} should recover the gradient expansion approximation (GEA):

$$
\begin{aligned}
E_{xc}[n] &= \int d^3 r\, n e_{xc}^{unif}(n) + \int d^3 r\, n \Delta e_{xc}^{GEA}(n, \nabla n) + \cdots \\
&= E_{xc}^{LDA}[n] + \Delta E_{xc}^{GEA}[n] + \cdots,
\end{aligned}
\tag{5.19}
$$

where $\Delta e_{xc}^{GEA}(n, \nabla n)$ is the leading correction to the LDA xc energy density for a slowly varying electron gas (Langreth and Perdew 1980). However, the GEA was found to give poor results and violate several important sum rules for the xc hole when applied to other systems (Burke et al. 1998). Fixing those sum-rules led to the development of ab initio GGAs. Though important in obtaining the energy for the ground-state, the xc hole rules have not been used in TDDFT and therefore will not be further discussed in this chapter.
- *Linear response of uniform gas.* Another generic limit is when a weak perturbation is applied to a uniform gas, and the resulting change in energy is given by the static response function, $\chi(q, \omega = 0)$. This function is known from accurate quantum Monte Carlo calculations (Moroni et al. 1995), and approximations can be tested against it.

5.3 Overview for TDDFT

The time-dependent problem is more complex than the ground-state problem, making the known exact conditions more difficult to classify. We make the basic distinction between general time-dependent perturbations, of arbitrary strength, and weak fields, where linear response applies. The former give conditions on $v_{xc}[n](r, t)$ for *all* time-dependent densities, the latter yield conditions directly on the xc kernel, which is a functional of the ground-state density alone. Of course, all of the former also yield conditions in the special case of weak fields.

In the time-dependent problem, the energy does not play a central role. Formally, the action plays an analogous role (see Chap. 9), but in practice, we never evaluate the action in TDDFT calculations (and it is identically zero on the real time evolution).

In TDDFT, our focus is truly the time-dependent density itself, and so, by extension, the potential determining that density. Thus many of our conditions are in terms of the potential.

Most pure *density* functionals for the ground-state problem produce poor approximations for the details of the potential. Such approximations work well only for quantities integrated over real space, such as the energy. Thus approximations that work well for ground-state energies are sometimes very poor as adiabatic approximations in TDDFT. Their failure to satisfy Eq. 5.17 leads to large errors in the KS energies of higher-lying orbitals (for example, consider the LDA potential for Helium in Fig. 4.3 of Chap. 4, which falls off exponentially rather than as $-1/r$), and (5.18) is often violated by several eV.

In place of the energy, there are a variety of physical properties that people wish to calculate. For example, quantum chemists are most often focused on the first few low-lying excitations, which might be crucial for determining the photochemistry of some biomolecule. Then the adiabatic generalization of standard ground-state approximations is often sufficient. At the other extreme, people who study matter in strong laser fields are often focused on ionization probabilities (see Chap. 18), and there the violation of Eq. 5.18 makes explicit density approximations too crude, and requires orbital-dependent approximations instead.

5.3.1 Definitions

In this section, we remind our readers of some of the basics from Chap. 4, the building blocks from which the exact conditions in TDDFT are proved. In contrast to the ground-state problem, the xc potential depends not only on the density but on the initial wavefunction $\Psi(0)$ and KS Slater determinant $\Phi(0)$, written symbolically as $v_{xc}[n; \Psi(0), \Phi(0)](r, t)$. This more complicated dependence comes about because two different wavefunctions, which are chosen to have the same density for all time, can come from completely different external potentials, which the xc potential accounts for. We can get rid of this initial wavefunction dependence if we start from a non-degenerate ground-state, where the wavefunction is a functional of the density alone, via the Hohenberg–Kohn theorem (Hohenberg and Kohn 1964). These things are further discussed in Chap. 8.

As the density evolves, the xc potential is determined not solely by the present density $n(r, t)$, but also by the history $n(r, t')$ for $0 \leq t' < t$. However, it is useful to break the xc potential up into two pieces, an *adiabatic* piece which only deals with the present density, and a *dynamic* piece which incorporates the memory dependence:

$$v_{xc}[n; \Psi(0), \Phi(0)](r, t) = v_{xc}^{adia}[n](r, t) + v_{xc}^{dyn}[n; \Psi(0), \Phi(0)](r, t). \quad (5.20)$$

The adiabatic piece of the potential,

$$v_{xc}^{adia}[n](r, t) = \left. \frac{\delta E_{xc}[n]}{\delta n(r)} \right|_{n(t)}, \quad (5.21)$$

is the xc potential for electrons as if their instantaneous density were a ground state. In the spirit of DFT, the dynamic piece is everything else.

In the linear response regime, small enough perturbations to the density will continuously change the xc potential:

$$v_{xc}[n + \delta n](r, t) - v_{xc}[n](r, t) = \int dt' \int d^3 r' \, f_{xc}[n](rt, r't') \delta n(r', t'), \quad (5.22)$$

where f_{xc} is the xc kernel, which can be written formally as the functional derivative:

$$f_{xc}[n_0](rt, r't') = \left. \frac{\delta v_{xc}[n](r, t)}{\delta n(r', t')} \right|_{n_0}. \quad (5.23)$$

The evaluation at n_0 reminds us that f_{xc} is used for the linear response of a density variation away from a ground-state density n_0.

Like the xc potential, the kernel can also be broken down into an adiabatic piece:

$$f_{xc}^{adia}(rt, r't') = \left. \frac{\delta^2 E_{xc}[n]}{\delta n(r) \delta n(r')} \right|_{n(t)} \delta(t - t'), \quad (5.24)$$

and a dynamic piece, which includes memory and everything else. The kernel is often Fourier-transformed from position space in the relative coordinate $(r - r')$ to momentum space (with wave-vector q), from the relative time $(t - t')$ to frequency (ω) domain, or both. Some conditions are more naturally expressed in momentum space and/or in the frequency domain. In the frequency domain, the adiabatic piece can be written as

$$f_{xc}^{adia}(r, r') = \lim_{\omega \to 0} f_{xc}(r, r'; \omega). \quad (5.25)$$

The kernel is discussed in more detail in Chap. 4.

5.3.2 Approximations

As we go through the various exact conditions, we will discuss whether the simplest approximations in present use satisfy them. We can divide all approximations into two classes based on whether or not the approximation neglects the dynamic term of Eq. 5.20; these classes are respectively adiabatic and non-adiabatic (i.e., memory) approximations. In the adiabatic approximation, familiar ground-state functionals (such as LDA, GGA, and hybrids) can produce xc potentials when one uses the approximate E_{xc} in Eq. 5.21. We mention two notable adiabatic approximations now.

- *ALDA*. The prototype of all TDDFT approximations is the adiabatic local density approximation, and it is the simplest pure density functional. The xc potential is as simple as can be:

$$v_{xc}^{ALDA}[n](r,t) = \frac{d\left[n e_{xc}^{unif}(n)\right]}{dn}\Bigg|_{n(r,t)}. \tag{5.26}$$

In linear response, the ALDA kernel is

$$f_{xc}^{ALDA}(rt,r't') = \frac{d^2\left[n e_{xc}^{unif}(n)\right]}{dn^2}\Bigg|_{n(r,t)} \delta(r-r')\delta(t-t'). \tag{5.27}$$

Like its ground-state inspiration, ALDA satisfies important sum rules by virtue of its simplicity, namely its locality in space and time. ALDA is commonly used in many calculations, and is described further in Chap. 4.

- *AA*. In the 'exact' adiabatic approximation, we use the exact E_{xc} in Eq. 5.21. This approximation is the best that an adiabatic approximation can do, unless there is some lucky cancellation of errors. Hessler et al. (2002) investigated AA applied to a time-dependent Hooke's atom system and found large errors in the instantaneous correlation energy. For the double ionization of a model Helium atom, Thiele et al. (2008) discovered that non-adiabatic effects were important only for high-frequency fields.

A key aim of today's methodological development is to build in correlation memory effects. Any attempt to build in memory goes beyond the adiabatic approximation, and thus belongs in the non-adiabatic class of approximations. The next three approximations belong to this dynamic class.

- *GK*. The Gross–Kohn approximation is simply to use the local frequency-dependent kernel of the uniform gas,

$$f_{xc}^{GK}(r,r';\omega) = \delta(r-r')f_{xc}^{unif}(n(r);\omega), \tag{5.28}$$

where

$$f_{xc}^{unif}(n;\omega) \equiv \lim_{q\to 0} f_{xc}^{unif}(n;q,\omega) \tag{5.29}$$

is the response of the uniform electron gas with density n. GK was the first approximation to go beyond the adiabatic approximation, but was found to violate translational invariance (see Sect. 5.4.6).

- *VK*. The Vignale–Kohn approximation sought to improve upon the shortcomings of GK. The VK approximation is simply the gradient expansion in the current density for a slowly-varying gas (see Chap. 24).

- *EXX*. Exact exchange, the orbital-dependent functional, is treated as an implicit density functional (see Chap. 6). When treated this way, EXX has some memory for more than two unpolarized electrons.

With the exception of EXX, non-adiabatic approximations are usually limited to the linear response regime and approximate the kernel, f_{xc}. There is now a major push to go beyond linear response for non-adiabatic approximations. The first such attempt was a bootstrap approach of Dobson et al. (1997). More recent attempts are described in Chap. 25 of this book and in Kurzweil and Baer (2004).

5.4 General Conditions

In this section, we discuss conditions that apply no matter how strong or how weak the time-dependent potential is. They apply to anything: weak fields, strong laser pulses, and everything in between. They apply also to the linear response regime, yielding the more specific conditions discussed in Sect. 5.5.

5.4.1 Adiabatic Limit

One of the simplest exact conditions in TDDFT is the adiabatic limit. For any finite system, or an extended system with a finite gap, the deviation from the instantaneous ground-state during a perturbation (of arbitrary strength) can be made arbitrarily small. This is the adiabatic theorem of quantum mechanics, which can be proven by slowing down the time-evolution, i.e., if the perturbation is $V(t)$, replacing it by $V(t/\tau)$ and making τ sufficiently large.

Similarly, as the time-dependence becomes very slow (or equivalently, as the frequency becomes small), for such systems the functionals reduce to their ground-state counterparts:

$$v_{xc}(\boldsymbol{r}, t) \rightarrow v_{xc}[n(t)](\boldsymbol{r}), \quad (\tau \rightarrow \infty) \tag{5.30}$$

where $v_{xc}[n](\boldsymbol{r})$ is the exact ground-state xc potential of density $n(\boldsymbol{r})$.

By definition, any adiabatic approximation satisfies this theorem, and so does EXX, by reducing to its ground-state analog for slow variations. On the other hand, if an approximation to $v_{xc}(\boldsymbol{r}, t)$ were devised that was not based on ground-state DFT, this theorem can be used in reverse to *define* the corresponding ground-state functional.

5.4.2 Equations of Motion

In this section, we discuss some elementary conditions that any reasonable TDDFT approximation should satisfy. Because these conditions are satisfied by almost all approximations, they are best applied to test the quality of propagation schemes.

For a scheme that does not automatically satisfy a given condition, then a numerical check of its error provides a test of the accuracy of the solution. A simple analog is the check of the virial theorem in ground-state DFT in a finite basis.

These conditions are all found via a very simple procedure. They begin with some operator that depends only on the time-dependent density, such as the total force on the electrons. The equation of motion for the operator in both the interacting and the KS systems are written down, and subtracted. Since the time-dependent density is the same in both systems, the difference vanishes. Usually, the Hartree term also separately satisfies the resulting equation, and so can be subtracted from both sides, yielding a condition on the xc potential alone. This procedure is well-described in the Chap. 24 for the zero xc force theorem.

Zero xc force and torque. These are very simple conditions saying that interaction among the particles cannot generate a net force (Vignale 1995a, b):

$$\int d^3r n(\boldsymbol{r}, t) \, \nabla v_{\mathrm{xc}}(\boldsymbol{r}, t) = 0 \tag{5.31a}$$

$$\int d^3r n(\boldsymbol{r}, t) \, \boldsymbol{r} \times \nabla v_{\mathrm{xc}}(\boldsymbol{r}, t) = \int d^3r \boldsymbol{r} \times \frac{\partial \boldsymbol{j}_{\mathrm{xc}}(\boldsymbol{r}, t)}{\partial t}, \tag{5.31b}$$

where $\boldsymbol{j}_{\mathrm{xc}}(\boldsymbol{r}, t)$ is the difference between the interacting current density and the KS current density (van Leeuwen 2001). The second condition says that there is no net xc torque, *provided* the KS and true current densities are identical. This is not guaranteed in TDDFT (but is in TDCDFT), as discussed in Sect. 4.4.4 of Chap. 4. The exchange-only KLI approximation, though incredibly accurate for ground state DFT, was found to violate the zero-force condition (Mundt et al. 2007). This is because KLI is not a solution to an approximate variational problem, but instead an approximate solution to the OEP equations. This means KLI also violates the virial theorem (Fritsche and Yuan 1998), which we describe next.

xc power and virial. By applying the same methodology to the equation of motion for the Hamiltonian, we find (Hessler et al. 1999):

$$\int d^3r \frac{dn(\boldsymbol{r}, t)}{dt} v_{\mathrm{xc}}(\boldsymbol{r}, t) = \frac{dE_{\mathrm{xc}}}{dt}. \tag{5.32}$$

while another equation of motion yields the virial theorem, which intriguingly has the exact same form as in the ground state, Eq. 5.11:

$$-\int d^3r n(\boldsymbol{r}, t)\boldsymbol{r} \cdot \nabla v_{\mathrm{xc}}[n](\boldsymbol{r}, t) = E_{\mathrm{xc}}[n](t) + T_{\mathrm{c}}[n](t). \tag{5.33}$$

These conditions are so basic that they are trivially satisfied by any reasonable approximation, including ALDA, AA, and EXX. Thus they are more useful as detailed checks on a propagation scheme, as mentioned earlier. The correlation contribution to Eq. 5.33 is very small, and makes for a very demanding test. But because the energy does not play the same central role as in the ground-state problem (and the action is *not* simply the time-integral of the energy—see Chap. 9), testing the propagation scheme is all they are used for so far.

5.4.3 Self-interaction

For any one-electron system,

$$v_x(\boldsymbol{r}, t) = -\int d^3 r' \frac{n(\boldsymbol{r}, t)}{|\boldsymbol{r} - \boldsymbol{r}'|}, \quad v_c(\boldsymbol{r}, t) = 0 \quad (N = 1) \tag{5.34}$$

These conditions are automatically satisfied by EXX. These conditions are instantaneous in time, so any adiabatic approximation that satisfies the ground-state conditions of Eq. 5.15 will also satisfy these time-dependent conditions, e.g. AA. On the other hand, LDA violates self-interaction conditions in the ground-state, so ALDA also violates these conditions in TDDFT.

5.4.4 Initial-State Dependence

There is a simple condition based on the principle that *any* instant along a given density history can be regarded as the initial moment (Maitra et al. 2002b; Maitra 2005a). This follows very naturally from the fact that the Schrödinger equation is first order in time. When applied to both interacting and non-interacting systems, we find:

$$v_{xc}[n; \Psi(t'), \Phi(t')](\boldsymbol{r}, t) = v_{xc}[n; \Psi(0), \Phi(0)](\boldsymbol{r}, t) \quad \text{for } t > t', \tag{5.35}$$

This is discussed in much detail in Chap. 8. Here we just mention that any adiabatic approximation, by virtue of its lack of memory and lack of initial-state dependence, automatically satisfies it. Interestingly, although EXX is instantaneous in the orbitals, it has memory (and so initial-state dependence) as a density functional (when applied to more than two unpolarized electrons).

This condition provides very difficult tests for any functional with memory. Consider any two evolutions of an interacting system, whose wavefunctions Ψ and Ψ' become equal after some time, t_c. This condition requires that the non-interacting systems have identical xc potentials at that time and forever after, even though they had different histories before then. This is illustrated in Fig. 5.2. An approximate functional with memory is unlikely, in general, to produce such identical potentials.

5.4.5 Coupling-Constant Dependence

Because of the lack of a variational principle for the energy, there are no definite results for various limits, as in Eq. 5.14a, b, nor is there a simple extension of the adiabatic connection formula (5.7), though Görling proposed an analog for time-dependent systems (Görling 1997). But there remains a simple connection between

Fig. 5.2 An illustration of the condition based on initial state dependence. The two wavefunctions Ψ and Ψ' become equal at time t_c, and therefore the KS potentials must become equal then and forever after

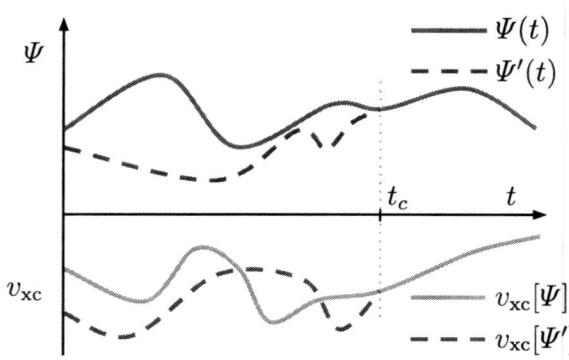

scaling and the coupling-constant for the xc potential (Hessler et al. 1999). For exchange, analogous to Eq. 5.12, the relation is linear:

$$v_x[n_\gamma; \Phi_\gamma(0)](rt) = \gamma v_x[n; \Phi(0)](\gamma r, \gamma^2 t), \qquad (5.36)$$

where

$$\Phi_\gamma(0) \equiv \gamma^{3N/2}\Phi(\gamma r_1, \ldots, \gamma r_N; t = 0) \qquad (5.37)$$

is the normalized initial state of the Kohn–Sham system with coordinates scaled by γ, and, for time-dependent densities,

$$n_\gamma(r, t) \equiv \gamma^3 n(\gamma r, \gamma^2 t). \qquad (5.38)$$

Though there is no simple expression for correlation scaling, we can relate it to the coupling-constant and find, analogous to Eq. 5.7:

$$v_c^\lambda[n; \Psi(0), \Phi(0)](r, t) = \lambda^2 v_c[n_{1/\lambda}; \Psi_{1/\lambda}(0), \Phi_{1/\lambda}(0)](\lambda r, \lambda^2 t), \qquad (5.39)$$

where $\Psi_{1/\lambda}(0)$ is the scaled initial state of the interacting system, defined as in Eq. 5.37, replacing γ with $1/\lambda$. For finite systems, it seems likely that taking the limit $\lambda \to 0$ makes the exchange term dominant (just as in the ground-state) (Hessler et al. 2002), but this has yet to be proven.

5.4.6 Translational Invariance

Consider a rigid boost $X(t)$ of a system starting in its ground state at $t = 0$, with $X(0) = dX/dt(0) = 0$. Then the xc potential of the boosted density will be that of the unboosted density, evaluated at the boosted point, i.e.,

$$v_{xc}[n'](r, t) = v_{xc}[n](r - X(t), t), \quad n'(r, t) = n(r - X(t), t). \qquad (5.40)$$

This condition is universally valid (Vignale 1995a). The GK approximation was found to violate this condition, which spurred on the development of the VK approximation. The harmonic potential theorem, a special case of translational invariance, is discussed in Chap. 24.

5.5 Linear Response

In the special case of linear response, all exchange-correlation information is contained in the kernel f_{xc}. Linear response is utilized in the great majority of TDDFT calculations, and the methods involved are thoroughly discussed in Chap. 7. As explained in Elliott et al. (2009), the chief use of linear response has been to extract electronic excitations. In this section, we shall discuss the exact conditions that pertain to f_{xc}, regardless of how it is employed.

5.5.1 Consequences of General Conditions

Each of the conditions listed below for f_{xc} can be derived from a general condition in Sect. 5.4.

Adiabatic limit. For any finite system, the exact kernel satisfies:

$$\lim_{\omega \to 0} f_{xc}(\boldsymbol{r}, \boldsymbol{r}'; \omega) = \frac{\delta^2 E_{xc}[n]}{\delta n(\boldsymbol{r}) \delta n(\boldsymbol{r}')}, \tag{5.41}$$

where E_{xc} is the exact xc energy. Obviously, any adiabatic functional satisfies this, with its corresponding ground-state approximation on the right.

Zero force and torque. The exact conditions on the potential of Sect. 5.4.2 also yield conditions on f_{xc}, when applied to an infinitesimal perturbation (see Chap. 24). Taking functional derivatives of Eq. 5.31a, b yields (Giuliani and Vignale 2005):

$$\int d^3 r \, n(\boldsymbol{r}) \nabla f_{xc}(\boldsymbol{r}, \boldsymbol{r}'; \omega) = -\nabla' v_{xc}(\boldsymbol{r}') \tag{5.42}$$

and

$$\int d^3 r \, n(\boldsymbol{r}) \boldsymbol{r} \times \nabla f_{xc}(\boldsymbol{r}, \boldsymbol{r}'; \omega) = -\boldsymbol{r}' \times \nabla' v_{xc}(\boldsymbol{r}'), \tag{5.43}$$

the latter assuming no xc transverse currents. Again, these are satisfied by ground-state DFT with the static xc kernel, so they are automatically satisfied by any adiabatic approximation. Similarly, in the absence of correlation, they hold for EXX. The general conditions employing energies, Eqs. 5.32 and 5.33, do not yield simple conditions for the kernel, because the functional derivative of the exact time-dependent xc energy is not the xc potential.

Self-interaction error. For one electron, functional differentiation of Eq. 5.34 yields:

$$f_x(r, r'; \omega) = -\frac{1}{|r - r'|}, \quad f_c(r, r'; \omega) = 0 \quad (N = 1). \tag{5.44}$$

These conditions are trivially satisfied by EXX, but violated by the density functionals ALDA, GK, and VK.

Initial-state dependence. The initial-state condition, Eq. 5.35, leads to very interesting restrictions on f_{xc} for arbitrary densities. But the information is given in terms of initial-state dependence, which is very difficult to find.

Coupling-constant dependence. The exchange kernel scales linearly with coordinates, as found by differentiating Eq. 5.36:

$$f_x[n_\gamma](r, r', \omega) = \gamma f_x[n](\gamma r, \gamma r', \omega/\gamma^2). \tag{5.45}$$

A functional derivative and Fourier-transform of Eq. 5.39 yields (Lein et al. 2000b)

$$f_c^\lambda[n](r, r', \omega) = \lambda^2 f_c[n_{1/\lambda}](\lambda r, \lambda r', \omega/\lambda^2). \tag{5.46}$$

These conditions are trivial for EXX. They can be used to test the derivations of correlation approximations in cases where the coupling-constant dependence can be easily deduced. More often, they can be used to *generate* the coupling-constant dependence when needed, such as in the adiabatic connection formula of Eq. 5.7.

A similar condition has also been derived for the coupling-constant dependence of the vector potential in TDCDFT (Dion and Burke 2005).

5.5.2 Properties of the Kernel

The kernel has many additional properties that come from its definition and other physical considerations.

Symmetry. Because the susceptibility is symmetric, so must also be the kernel:

$$f_{xc}(r, r'; \omega) = f_{xc}(r', r; \omega). \tag{5.47}$$

This innocuous looking condition is satisfied by any adiabatic approximation by virtue of the kernel being the second derivative of an energy, and is obviously satisfied by EXX.

Kramers–Kronig. The kernel $f_{xc}(r, r', \omega)$ is an analytic function of ω in the upper half of the complex ω-plane and approaches a real function $f_{xc}(r, r'; \infty)$ for $\omega \to \infty$. Therefore, defining the function

$$\Delta f_{xc}(\boldsymbol{r}, \boldsymbol{r}', \omega) = f_{xc}(\boldsymbol{r}, \boldsymbol{r}', \omega) - f_{xc}(\boldsymbol{r}, \boldsymbol{r}'; \infty), \tag{5.48}$$

we find

$$\Re e\, \Delta f_{xc}(\boldsymbol{r}, \boldsymbol{r}', \omega) = \mathcal{P} \int \frac{d\omega'}{\pi} \frac{\Im m\, f_{xc}(\boldsymbol{r}, \boldsymbol{r}', \omega')}{\omega' - \omega} \tag{5.49a}$$

$$\Im m\, f_{xc}(\boldsymbol{r}, \boldsymbol{r}', \omega) = -\mathcal{P} \int \frac{d\omega'}{\pi} \frac{\Re e\, \Delta f_{xc}(\boldsymbol{r}, \boldsymbol{r}', \omega')}{\omega' - \omega}, \tag{5.49b}$$

where $\mathcal{P}$ denotes the principle value of the integral. The kernel $f_{xc}(\boldsymbol{r}t, \boldsymbol{r}'t')$ is real-valued in the space and time domain, which leads to the condition in the frequency domain,

$$f_{xc}(\boldsymbol{r}, \boldsymbol{r}'; \omega) = f_{xc}^*(\boldsymbol{r}, \boldsymbol{r}'; -\omega). \tag{5.50}$$

The simple lesson here is that any adiabatic kernel (no frequency dependence) is purely real, and any kernel with memory has an imaginary part in the frequency domain (or else is not sensible). Many of the failures of current TDDFT approximations, e.g. the fundamental gap of solids, are linked to the lack of an imaginary part of the kernel (Giuliani and Vignale 2005). Because adiabatic approximations produce real kernels, we see that memory is required to produce complex kernels. Hellgren and von Barth (2009) showed that EXX has a complex kernel, since it has frequency-dependence (for more than two electrons). Both GK and VK have complex kernels satisfying the Kramers–Kronig conditions.

Adiabatic connection. A beautiful condition on the exact xc kernel is given simply by the adiabatic connection formula for the ground-state correlation energy (see Chap. 22)

$$-\frac{1}{2} \int d^3r \int d^3r' v_{ee}(\boldsymbol{r} - \boldsymbol{r}')$$

$$\times \int_0^\infty \frac{d\omega}{\pi} \int_0^1 d\lambda \Im m \left[\chi^\lambda(\boldsymbol{r}, \boldsymbol{r}'; \omega) - \chi_{KS}(\boldsymbol{r}, \boldsymbol{r}'; \omega) \right] = E_c. \tag{5.51}$$

Combined with the Dyson-like equation of Chap. 4 for χ^λ as a function of χ_{KS} and f_{xc}, this is being used to generate new and useful approximations to the ground-state correlation energy (Fuchs and Gonze 2002; Fuchs et al. 2005). Although computationally expensive, ways are being found to speed up the calculations (Eshuis et al. 2010).

Equation 5.51 provides an obvious exact condition on any approximate xc kernel for *any* system. Thus *every* system for which the correlation energy is known can be used to test approximations for f_{xc}. Note that, e.g. using ALDA for the kernel implicit in (5.51) does *not* yield the corresponding E_{xc}^{LDA}, but rather a much more sophisticated functional (Lein et al. 2000b). Even insertion of f_x yields correlation contributions to all orders in E_c. And lastly, even the exact adiabatic approximation, $f_{xc}[n_0](r, r'; \omega = 0)$, does not yield the exact $E_{xc}[n_0]$.

Functional derivatives. A TDDFT result ought to come from a TDDFT calculation, but this is not always the case. By a TDDFT calculation, we mean the result of an evolution of the TDKS equations of Chap. 4 with some approximation for the xc potential that is a functional of the density. This implies that the xc kernel should be the functional derivative of some xc potential, which also reduces to the ground-state potential in the adiabatic limit. All the approximations discussed here satisfy this rule. But calculations that intermix kernels with potentials in the solution of Casida's equations violate this condition, and will violate important underlying sum-rules, such as S_{-2} in Eq. 5.55.

5.5.3 Excited States

The following conditions have to do with the challenges of obtaining excited states in the linear response regime.

Infinite lifetimes of eigenstates. This may seem like an odd requirement. When TDDFT is applied to calculate a transition to an excited state, the frequency should be real. This is obviously true for ALDA and exact exchange, but not so clear when memory approximations are used. As mentioned in Sect. 5.5.2, the Kramers–Kronig relations mean that memory implies imaginary xc kernels, and these can yield imaginary contributions to the transition frequencies. Such effects were seen in calculations using the VK for atomic transitions (Ullrich and Burke 2004). Indeed, very long lifetimes were found when VK was working well, and much shorter ones occurred when VK was failing badly.

Single-pole approximation for exchange. This is another odd condition, in which two wrongs make something right. Using Görling–Levy perturbation theory (Görling and Levy 1993), one can calculate the exact exchange contributions to excited state energies (Filippi et al. 1997; Zhang and Burke 2004). To recover these results using TDDFT, one does *not* simply use f_x, and solve the Dyson-like equations. Like with Eq. 5.51, the infinite iteration yields contributions to all orders in the coupling-constant.

However, the single-pole approximation truncates this series after one iteration, and so drops all other orders. Thus the correct exact exchange results are recovered in TDDFT from the SPA solution to the linear response equations, and *not* by a full solution (Gonze and Scheffler 1999). This procedure can be extended to the next order (Appel et al. 2003).

Double excitations and branch cuts. Maitra et al. (Maitra et al. 2004; Cave et al. 2004) argued that a strong ω-dependence in f_{xc} allows double excitation solutions to Casida's equations, which effectively couples double excitations to single excitations. Similarly, the second ionization of the He atom implies a branch cut in its f_{xc} at the frequency needed (Burke et al. 2005a). Under limited circumstances, this frequency dependence can be estimated, but a generalization (Casida 2005) has been proposed. It would be interesting to check its compliance with the conditions listed in this chapter. A detailed discussion of double excitations can be found in Chap. 8.

Excitations in the adiabatic approximation. One misleading use of linear response has been to test the quality of different approximations to the ground-state E_{xc}. For instance, Jacquemin et al. (2010a) calculated the excitation energies for approximate E_{xc} functionals within adiabatic TDDFT and compared them to experimental values. However, even within AA—using the adiabatic approximation with the exact E_{xc}—the exact excitations would not be not obtained. Thus a good ground-state E_{xc} used in adiabatic linear response will not necessarily give good excitation energies.

Frequency sum-rules. In the limit that the wavelength of the incident light is much greater than the scale of the system, the linear response function determines the optical spectrum by Mahan and Subbaswamy (1990)

$$\sigma(\omega) = \frac{4\pi\omega}{3c} \Im m \left[\int d^3 r \int d^3 r' \boldsymbol{r} \cdot \boldsymbol{r}' \chi(\boldsymbol{r}, \boldsymbol{r}', \omega) \right], \tag{5.52}$$

where $\sigma(\omega)$ is the photoabsorption cross-section. In physics, the spectrum is usually described by the dimensionless oscillator strength $f(\omega)$. In atomic units, $f(\omega)$ is related to $\sigma(\omega)$ by (Friedrich 2006)

$$\sigma(\omega) = \frac{2\pi^2}{c} f(\omega). \tag{5.53}$$

The moments of the oscillator strength spectrum, S_n, are defined by

$$S_n = \sum_{\nu \in \text{excited states}} (E_\nu - E_0)^n f(E_\nu - E_0), \tag{5.54}$$

where the sum over continuum states will turn into an integral. Eqs. 5.52 and 5.53 show that S_n is related to the linear response function. Using basic quantum mechanics, S_n can be related to various general properties of the system, known as oscillator strength sum rules (Bethe and Salpeter 1957; Fano and Cooper 1968; Inokuti 1971); S_0 is the usual Thomas–Reich–Kuhn sum rule. In atomic units, the most used sum rules S_{-2} to S_2 are

$$S_{-2} = \alpha(\omega = 0), \quad S_{-1} = \frac{2}{3} \left\langle \left| \sum_j \boldsymbol{r}_j \right|^2 \right\rangle_0, \quad S_0 = N$$

$$S_1 = \frac{2}{3} \left\langle \left| \sum_j \mathbf{p}_j \right|^2 \right\rangle_0, \quad S_2 = \frac{4\pi}{3} \sum_{A \in \text{nuclei}} Z_A n(\boldsymbol{r} = \boldsymbol{R}_A), \tag{5.55}$$

where $\alpha(\omega = 0)$ is the static polarizability, Z_A is the charge of nucleus A and $\boldsymbol{R}_A$ its position. Sum rules for $n > 2$ do not converge due to the $\omega^{-7/2}$ asymptotic decay of

$f(\omega \to \infty)$ (Rau and Fano 1967), and $S_{n<-2}$ are related to various other properties of the system (Fano and Cooper 1968). For each sum rule, a new exact condition can be found for the kernel. Within the exact adiabatic (AA) approximation, the S_{-2} sum rule will be equal to the polarizability $\alpha(\omega = 0)$ of the real system, because the AA kernel is correct for $\omega \to 0$. By definition, the sums S_0 and S_2 for the exact Kohn–Sham ground state are equal to their counterparts in the interacting system, whereas the other sums are not. Since S_n is related to the linear response function, these differences yield exact conditions on f_{xc}, which connects the linear response functions of the Kohn–Sham system and the real system.

Scattering theory and real-time propagation. A vastly under-appreciated exact condition for TDDFT is the equivalence of time-dependent propagation and scattering theory. This can be particularly important in understanding the relation between bound and continuum states.

For example, much early work in TDDFT was performed by Yabana and Bertsch (1996), propagating ALDA for atoms and molecules in weak electric fields. By Fourier transformation of the time-dependent dipole moment, one can extract the photoabsorption spectrum. The fruitfly of such calculations is benzene, with a large $\pi \to \pi^*$ transition at about 6.5 eV, accurately given by ALDA. But closer inspection shows that the LDA ionization threshold is at about 5 eV, because the LDA xc potential is not deep enough. Thus this transition is in the LDA continuum, yet its position and area are given reasonably well by ALDA. This is no coincidence: ALDA describes the time-dependent density and its propagation for moderate times very well. All that has changed is the choice of a complete set of states onto which to project the results!

By following this logic, Wasserman et al. (2003) could capture the effect of Rydberg transitions using ALDA. However, ALDA puts many bound states in the continuum due to the exponential fall-off of the KS-LDA potential (as mentioned in Sect. 5.3). Thus the ionization potentials for the ALDA states are wrong, but the oscillator strength in the LDA continuum accurately approximates that of the true Rydberg transitions to the exact bound states. (However, it is *not* an exact condition that the KS oscillator strengths be correct, not even at the threshold where KS captures the right energy (Yang et al. 2009).) For an illustration, see Fig. 4.4 in Chap. 4 for the optical absorption for the helium atom. Using a trick due to Fano (1935), Wasserman showed (Wasserman and Burke 2005) that the Rydberg transition frequencies, could be extracted from ALDA. In van Faassen and Burke (2006a) it is shown the accuracy of this calculation for He, Be, and Ne, whereas van Faassen and Burke (2006b) shows the qualitative failure of ALDA for transitions to high angular momentum eigenstates (starting at the d orbitals).

One can go further, and even consider true continuum states. In scattering theory, the continuum states of the $N + 1$ particle system describe how a single electron scatters from an N particle system. Wasserman (2005) and van Faassen et al. (2007) developed methods to calculate scattering amplitudes and phase shifts based on time-propagation within TDDFT. With a given approximation, one can calculate the susceptibility of an atomic anion and deduce the scattering amplitude for an incident

Fig. 5.3 Electrons on a ring. A magnetic field $B(t)$ is turned on and steadily increases in (**b**); the resulting electric field $E(r)$ is uniform on a thin ring, accelerating electrons around the ring, producing the probability current $j(r, t)$. Note that in both (**a**) and (**b**) the densities are equal

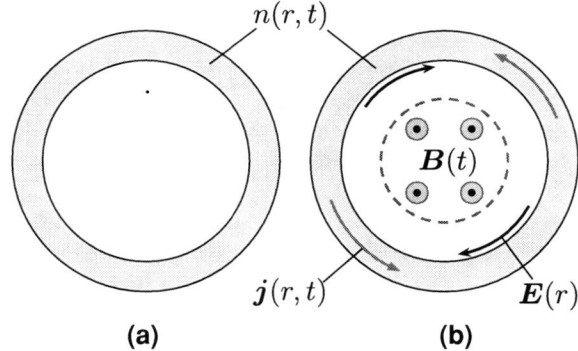

(a)   (b)

electron (Wasserman et al. 2005). Both these examples (the quantum defect and scattering) can be connected in the same framework (van Faassen and Burke 2009), and they illustrate that TDDFT fundamentally concerns time-propagation. Present-day approximations yield promising results; simple approximations like ALDA often yield accurate time-dependent densities, but their projection onto individual Kohn–Sham eigenstates may appear far more complicated.

5.6 Extended Systems and Currents

As mentioned in Sect. 5.2.4, care must be taken when extending exact ground-state DFT results to extended systems. This is even more so the case for TDDFT. The first half of the Runge–Gross theorem (see Chap. 4) provides a one-to-one correspondence between potentials and *current* densities, but a surface condition must be invoked to produce the necessary correspondence with densities. Without this condition, it can readily be seen that two periodic systems with completely different physics can have the same density (Maitra et al. 2003), as in Fig. 5.3. With hindsight, this is very suggestive that time-dependent functionals may contain a non-local dependence on the details at a surface. As such, they are more amenable to local approximations in the current rather than the density.

5.6.1 Gradient Expansion in the Current

As discussed elsewhere (Chap. 24) and first pointed out by Dobson (1994a), the frequency-dependent LDA (GK approximation) violates the translational invariance condition of Sect. 5.4.6. One can trace this failure back to the non-locality of the xc functional in TDDFT. But, by going to a current formulation, everything once again becomes reasonable. The gradient expansion in the current, for a slowly varying gas, was first derived by Vignale and Kohn (1996), and later simplified by Vignale et al. (1997), and is discussed in much detail in the Chap. 24.

For our purposes, the most important point is that, by construction, VK satisfies translational invariance. The frequency-dependence shuts off (it reduces to ALDA) when the motion is a rigid translation, but turns on when there is a true (non-translational) motion of the density (Vignale and Kohn 1996).

Any functional with memory should recover the VK gradient expansion in this limit, or justify why it does not. However, the VK approximation is *only* the gradient expansion, which for the ground-state was found to violate sum rules, as mentioned in Sect. 5.2.4. It is therefore likely that there exists something like a generalized gradient approximation, which is more accurate than VK.

5.6.2 Polarization of Solids

A decade ago, Gonze et al. (1995) pointed out that the periodic density in an insulating solid in an electric field is insufficient to determine the periodic one-body potential, in apparent violation of the Hohenberg–Kohn theorem (Hohenberg and Kohn 1964). In fact, this effect appears straightforwardly in the static limit of TDCDFT, and is even estimated by calculations using the VK approximation (van Faassen et al. 2003; Maitra et al. 2003). When translated back to TDDFT language, one finds a $1/q^2$ dependence in f_{xc}, where q is the wavevector corresponding to $r - r'$. This requires f_{xc} to have the same degree of nonlocality as the Hartree kernel, and this is missed by any local or semilocal approximation, such as ALDA, but *is* built in to EXX (Kim and Görling 2002) or AA. The need for a $1/q^2$ contribution in the optical response of solids led to much development (Onida et al. 2002) for a kernel that allows excitons (Reining et al. 2002; Sottile et al. 2007). Since the RG theorem can be proven for solids in electric fields of nonzero q, one can extract the $q \to 0$ (a constant E field) result at the end of the calculation (Maitra et al. 2003).

5.7 Summary

What lessons can we take away from this brief survey?

1. In the ground-state theory, the total xc energy is crucial for determining the energy of the system, and many conditions are proven for that functional. This is not so for TDDFT, for which only the time-dependent density matters. In the non-interacting system, the KS potential, and specifically its xc component, is what counts.
2. Approximate functionals depending explicitly on the density have poor-quality potentials, e.g. LDA and GGA. Thus successes in ground-state DFT do not translate directly into successes in TDDFT. One of the greatest challenges is that the potential is a far more sensitive functional of the density than vice versa. Though

we have enumerated many conditions on the xc potential, it is important to determine which conditions significantly affect the density, including those aspects of the density that are relevant to experimental measurements.

3. The adiabatic approximation satisfies many exact conditions by virtue of its lack of memory. Inclusion of memory may lead to violations of conditions that adiabatic approximations satisfy. This is reminiscent of the ground-state problem, where the gradient expansion approximation violates several key sum rules respected by the local approximation. Explicit imposition of those rules led to the development of generalized gradient approximations.

As shown in several chapters in this book, many people are presently testing the limits of our simple approximations, and very likely, these or other exact conditions will provide guidance on how to go beyond them.

Chapter 6
Orbital Functionals

Stephan Kümmel

Orbital functionals have developed into powerful tools of modern TDDFT as they allow to tackle two of the theory's most notorious problems: By explicitly using the orbitals, functionals that are free from electronic self-interaction and that incorporate particle number discontinuities in the ex-change–correlation potential can be constructed. This chapter presents an overview of why orbital functionals are needed and of the different ways in which they can be employed. The problem of electronic self-interaction and the advantages and drawbacks of the Kohn–Sham and generalized Kohn–Sham way of using orbital functionals are addressed. The problem of the time-dependent optimized effective potential is discussed in detail, and the chapter closes by looking at a few examples of orbital functionals which have been successfully used in practice.

6.1 Why Orbital Functionals are Needed

An orbital functional is a functional expression, e.g., for the energy or the action, that depends explicitly on a set of orbitals. Orbital functionals are one of the most successful concepts of density functional theory. Nevertheless, a discussion of the use of orbital functionals in TDDFT necessarily has to start on a sobering note: If there would be no need for the use of orbital functionals, TDDFT would be more fun to use in practice. Orbital functionals—at least the most commonly used ones which incorporate exact exchange or parts of it—are computationally considerably more expensive than semi-local functionals due to the many integral evaluations that Fock exchange requires. Additional complications—which will be discussed in this chapter—are associated with using such functionals in time-dependent Kohn–Sham calculations. Given these complications it may be surprising that nevertheless, orbital

S. Kümmel (✉)
Theoretical Physics IV, Department of Physics,
University of Bayreuth, 95440 Bayreuth, Germany
e-mail: stephan.kuemmel@uni-bayreuth.de

M. A. L. Marques et al. (eds.), *Fundamentals of Time-Dependent Density Functional Theory*, Lecture Notes in Physics 837, DOI: 10.1007/978-3-642-23518-4_6,
© Springer-Verlag Berlin Heidelberg 2012

functionals have become widely used in TDDFT. Their enormous success is based on decisive strengths that in many cases outweigh the increased computational cost and theoretical effort.

The advantages of orbital functionals have been demonstrated in many practical problems, but they are also easily seen on the formal level. To this end let us assume that we are staying within the realm of Kohn–Sham theory, i.e., our orbitals are Kohn–Sham orbitals. The discussion of other choices is defered to Sect. 6.2. It is well known (see Chaps. 4 and 5) that the unknown exact time-dependent exchange-correlation potential of Kohn–Sham theory is fully non-local with respect to space and time (Gross et al. 1996; Maitra et al. 2002b), i.e., $v_{xc}(r, t)$ depends not only on the density at spacepoint r at time t, but on the density at all points r' in space and at all times $t' \leq t$. These nonlocalities are explicitly visible in calculations that reconstruct the exact time-dependent $v_{xc}(r, t)$ (Thiele et al. 2008; Thiele and Kümmel 2009). Yet, incorporating such nonlocalities into explicit density functionals in a consistent manner is not easy at all. Kohn–Sham orbital functionals, on the other hand, include nonlocalities in a very natural way as the Kohn–Sham orbitals are nonlocal functionals of the density (Maitra et al. 2002b) (see Chap. 8).

This nonlocality is a straightforward consequence of the fact that the mapping between orbitals and density is provided by solving the Kohn–Sham equations, which is a nonlocal step. Thus, even a functional expression that is local or semilocal with respect to its dependence on the Kohn–Sham orbitals is nonlocal with respect to its dependence on the density. Whether the nonlocalities that are captured by a given orbital dependent approximation for v_{xc} are sufficient in the sense of capturing relevant physics that the ultimate exact functional would represent of course depends on the nature of the functional approximation.

Examples for functionals that are semilocal in the orbitals are the so called meta-GGAs (Tao et al. 2003) which use the kinetic energy density calculated from the orbitals, or self-interaction corrections based on iso-orbital indicators (Kümmel and Perdew 2003a). Such functionals can be used for TDDFT calculations in a spirit that one may call "adiabatic with respect to the orbitals", i.e., in a time-dependent Kohn–Sham (propagation) calculation, the instantaneous orbitals $\varphi_i(r, t)$ are used to calculate $v_{xc}(r, t)$. As stated above, such an approach is generally speaking non-adiabatic with respect to the density as the $\varphi_i(r, t)$ may depend on the entire history of the density. However, experience so far seems to indicate that the non-localities in meta-GGAs are not very pronounced. Such functionals will also not be discussed further here as most TDDFT calculations using orbital functionals fall into the different category of using expressions that are nonlocal in space with respect to the orbitals. The most prominent examples of this category of functionals are the ones that use exact Fock exchange or a fraction of it, e.g. hybrid functionals. Another one are self interaction correction (SIC) approaches. Both will be discussed in this chapter.

The main reason for why functionals incorporating exact exchange or a SIC have become popular is because, so far, this is the only known practical way of reducing or eliminating one-electron self interaction. The self-interaction problem, which is well known as one of the fundamental problems in ground-state DFT (Perdew and Zunger

1981; Kümmel and Kronik 2008), affects TDDFT as badly as it affects ground-state functionals. There are at least two reasons for this.

The first is that the asymptotic behavior of the Kohn–Sham potential is important in many TDDFT applications. Obvious examples are ionization processes or excitations to Rydberg states. The asymptotics of the Kohn–Sham potential is directly related to the self-interaction problem, as a v_{xc} that is free from one-electron self interaction will yield the correct $-1/r$ asymptotics of the Kohn–Sham potential. This can be seen from the following argument. At any given point in time, the Kohn–Sham potential of a finite, electrically neutral N-electron system can be written as (Hartree atomic units are used)

$$v_{KS}(\boldsymbol{r}, t) = v_{ext}(\boldsymbol{r}, t) + \int d^3 r' \frac{n(\boldsymbol{r}', t) - n_N(\boldsymbol{r}', t)}{|\boldsymbol{r} - \boldsymbol{r}'|}$$
$$+ \int d^3 r' \frac{n_N(\boldsymbol{r}', t)}{|\boldsymbol{r} - \boldsymbol{r}'|} + v_{xc}(\boldsymbol{r}, t), \qquad (6.1)$$

where the contribution of the N-th orbital density $n_N = |\varphi_N|^2$ to the Hartree potential has been written down separately. For the sake of notational transparency we here avoid a spin-dependent notation and assume a singly occupied highest orbital. One-electron self-interaction freeness means that for any one-electron density the total electron interaction must vanish, i.e., the correlation potential must be zero and exchange must be pure self exchange canceling the Hartree self interaction. The third term on the right-hand side of Eq. 6.1 is such a one-electron density contribution. In the far asymptotic region $|\boldsymbol{r}| \rightarrow \infty$ only the monopole contribution from each electrostatic term will remain, i.e., in this limit Eq. 6.1 becomes

$$v_{KS}(r, t) = -\frac{N}{r} + \frac{N-1}{r} + \frac{1}{r} + v_{xc}(r, t). \qquad (6.2)$$

For the sake of completeness one should note that the Kohn–Sham potential can go to a non-vanishing asymptotic constant on nodal surfaces of the highest occupied orbital (Della Sala and Görling 2002a, b; Kümmel and Perdew 2003b). If one considers the asymptotic potential on such a surface, the corresponding constant has to be taken into acccount in the asymptotics.

Let us assume that the total density is dominated by n_N in the far asymptotic region. This assumption will be discussed below. If it holds, then the third term on the right-hand side of Eq. 6.2 represents the Hartree interaction of the density n_N with itself. This unphysical self-interaction should be canceled by v_{xc}, and it is clear from Eq. 6.2 that, therefore, v_{xc} must fall of as $-1/r$. As a consequence of the first two terms the total Kohn–Sham potential will then fall of as $-1/r$ as well. This finding is in line with the straightforward argument that the Kohn–Sham potential is a one-particle potential, and one particle venturing out to infinity will feel the hole it left behind, thus experiencing an attractive $-1/r$ potential.

The above assumption is certainly fulfilled for ground-state densities of systems with an odd number of electrons, as the asymptotic exponential decay of a ground-state Kohn–Sham orbital is dominated by its eigenvalue (Kreibich et al. 1999).

This leads to a natural ordering of the orbitals with the highest occupied one decaying the slowest and thus asymptotically dominating the density. The far asymptotic region is therefore an iso-electron region. For ground-state densities with an even number of electrons, there may be two spin-orbitals having the same Kohn–Sham eigenvalue and therefore, the far-asymptotic region is an iso-orbital region (i.e., only one orbital shape can be detected), but not an iso-electron region. Thus, the correlation potential need not vanish. However, the Hartree self interaction must be canceled by the Fock self exchange, and exchange does not couple electrons of opposite spin. Consequently, an analogue of the above given argument holds for each spin channel separately. Turning to the general time-dependent case we note that time-dependent fields may destroy the initial orbital ordering. However, it still appears very likely— in particular if orbitals can be identified with electrons—that one (spin) orbital will asymptotically dominate the (spin) density. It need not be the one which was the highest occupied one in the ground state, i.e., in the above argument n_N may be replaced by some other orbital density—but otherwise, the argument remains. Thus, self-interaction and the asymptotics of the potential are directly related to each other.

The second reason why self-interaction is an important issue in TDDFT are the so-called step structures in v_{xc} that are related to particle number discontinuities in v_{xc} and derivative discontinuities in the exchange-correlation action functional (Mundt and Kümmel 2005). One-electron self interaction and missing particle number discontinuities and step structures are closely related problems (Perdew 1990). As electrons only come in integers, removing (or adding) an electron, i.e., distinguishing between a system of N electrons and two systems with $N - 1$ and 1 electrons, respectively, leads to a quantized change of certain properties. However, an electron that interacts with itself is never really a single electron. Therefore, self interaction "smears out" the discrete interaction effects that are associated with removing (or adding) a particle.

The existence of particle-number discontinuities in the ground-state v_{xc} has long been known (Perdew et al. 1982), and its enormous importance, e.g., in terms of the band gap question, is well understood (Kümmel and Kronik 2008). More recently it has been shown that similar discontinuities exist in TDDFT (Mundt and Kümmel 2005): As systems evolve continuously in time, the particle-number discontinuities are inherited by the time-dependent v_{xc}. These discontinuities are extremely important. As discussed in detail in Sect. 6.3 they lead to step structures in the potential which are decisive for the correct description of many electronic processes, and in particular such ones in which (long-range) charge-transfer is involved. Examples for such processes are charge-transfer excitations (Tozer 2003), correlated ionization (Lein and Kümmel 2005), dissociation of a system into well separated subsystems (Gritsenko and Baerends 1996; Tempel et al. 2009), or electronic transport (Toher et al. 2005). Figure 6.1 shows a simple example of how such step structures may look. It depicts snapshots taken at different times during the ionization process of a lithium model atom which was subjected to an electric field pulling density to the left, i.e., in -z direction [see (Mundt and Kümmel 2005) for details]. The figure shows the exchange potential for the two up-spin electrons. The system was initially (t = 0) in its ground state. Then a homogenous electrical field was linearly ramped up

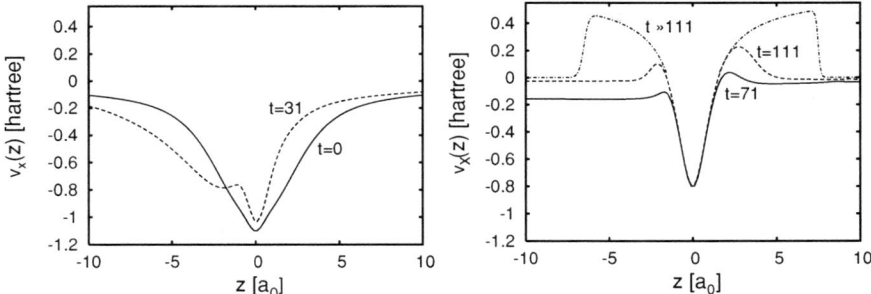

Fig. 6.1 A step structure building up in the exact exchange potential for one spin channel of a model lithium atom as the $2s$ electron is ionized to the left and the $1s$ electron remains bound. $t = 0$ shows the initial ground-state potential, plots labeled $t = 31$, 71 and 111 depict the potential at later times (in a.u.). $t \gg 111$ shows an extrapolated guess of how the potential will look at much later times

over a time of 40 a.u. to a strength of 0.06 a.u., and then held constant. At this field strength the $2s$ electron is ionized and the $1s$ electron remains bound. One can see that initially the exchange potential looks very smooth, but as the $2s$ electron escapes towards minus infinity a pronounced downward step in the potential builds up. This step is a reflection of the particle number discontinuity of the exact exchange potential. Similar steps are also seen in the exact time-dependent correlation potential that can be reconstructed from exact densities that are obtained by solving the correlated Schrödinger equation (Lein and Kümmel 2005).

In Sect. 6.3 we will investigate the "mechanism" by which the Kohn–Sham potential of orbital functionals achieves such step structures and particle number discontinuities. Before we do so, we however have to discuss some fundamental choices that the user of an orbital functional faces.

6.2 Using Orbital Functionals in TDDFT: Some Choices to Make

If one wants to use an orbital functional in an actual calculation one has to decide— just as for any other TDDFT calculation—whether one may either use linear response theory, or whether the excitation is nonlinear and beyond perturbation theory so that one may need to solve the time-dependent equations by propagating the orbitals in time (Marques et al. 2003; Mundt et al. 2007a). This decision depends on the nature of the excitation, e.g., how large the applied external fields are. The second choice one has to make is whether one would like to use—as we did in the previous section—time-dependent Kohn–Sham theory, i.e., use one multiplicative potential

$$v_{xc} = \frac{\delta A_{xc}\left[\{\varphi_j\}\right]}{\delta n} \tag{6.3}$$

for all orbitals [with A_{xc} denoting the exchange-correlation action functional (van Leeuwen 1998)], or whether one resorts to a time-dependent extension of the ground-state generalized Kohn–Sham theory (Seidl et al. 1996) and uses orbital specific potentials $u_{xc\,i}(r, t)$. The latter are defined in analogy to the ground-state definition (Grabo et al. 2000; Kümmel and Perdew 2003b) by replacing E_{xc} with a suitably chosen action functional A_{xc}, i.e.,

$$u_{xc\,i} = \frac{1}{\varphi_i^*} \frac{\delta A_{xc}\left[\{\varphi_j\}\right]}{\delta \varphi_i}. \tag{6.4}$$

The two different approaches—Kohn–Sham versus generalized Kohn–Sham— are typically used in different situations. On the one hand, explicitly linearized TDDFT calculations using exact Fock exchange, e.g., as part of hybrid functionals, mostly [though not exclusively (Ipatov et al. 2010)] employ orbital specific potentials. The first reason for this is that many quantum chemistry codes already had the linear-response Hartree–Fock equations coded. The orbital specific approach was therefore easy to implement. The second reason is that the exchange-correlation kernel f_{xc} that is needed in Kohn–Sham linear response theory is difficult to calculate for an orbital functional. Even without looking at the details of how the kernel is calculated this statement becomes understandable from Sect. 6.3. There it is shown that calculating v_{xc}, i.e., the first functional derivative defined in Eq. 6.3, is cumbersome and leads to an involved integro-differential equation. It is thus quite plausible that taking a further functional derivative to obtain $f_{xc} = \delta v_{xc}/\delta n$ is even more complicated.

On the other hand, calculations that explicitly propagate the Kohn–Sham orbitals in real time typically use the Kohn–Sham potential (Marques et al. 2003; Reinhard and Suraud 2003; Chu 2005; Mundt et al. 2007a). Real time propagation is mandatory for non-perturbative processes and, due to the just mentioned complications in calculating f_{xc}, also a viable option for linear-response Kohn–Sham calculations with orbital functionals (Marques et al. 2001; Mundt et al. 2007a).

There are pros and cons for both Kohn–Sham and generalized Kohn–Sham. One of the major advantages of the Kohn–Sham approach is that the theoretical concepts of its time-dependent version are well established. As the density [and initial conditions (Gross et al. 1996), which are suppressed in the notation of this chapter for notational clarity] determine the Kohn–Sham potential and as the same potential determines all orbitals, the orbitals are themselves functionals of the density (see Chap. 4). This is not the case for the generalized Kohn–Sham approach, where the mapping to one potential is not possible as each orbital is subjected to a different potential. Concepts such as the exchange-correlation kernel and the derivative discontinuity that are helpful for understanding and interpreting results are well defined only in the Kohn–Sham approach. This formal difference between the two approaches translates into very noticeable differences in practical calculations, e.g., in terms of how well eigenvalue differences approximate excitation energies (Körzdörfer and Kümmel 2010).

Yet, the generalized Kohn–Sham approach has a distinct advantage on the practical side. Taking the functional derivative of Eq. 6.4 is all that is required for obtaining the potentials. This is a straightforward calculation. Therefore, although formal arguments about the generalized Kohn–Sham approach have so far been mostly restricted to the ground-state context (Seidl et al. 1996), its linear-response version is in widespread use especially for hybrid functionals.

The practical advantanges of the generalized Kohn–Sham approach become obvious when one realizes that calculating the functional derivative in Eq. 6.3 is not straightforward at all, because the functional dependence of the orbitals on the density is well defined, but not explicitly known. Therefore, using the time-dependent Kohn–Sham v_{xc} for orbital functionals is a considerably more formidable task than using the orbital-specific potentials u_{xci}. The next section takes a closer look at this problem.

6.3 The Time-Dependent Optimized Effective Potential

The equation for the time-dependent Kohn–Sham exchange-correlation potential of an orbital functional is called the time-dependent optimized effective potential (TDOEP) equation for historical reasons. The first derivation of this equation straightforwardly employed the chain rule for functional derivatives (Ullrich et al. 1995a). A later second derivation employed the Keldysh time contour method to define an action that avoids causality problems (van Leeuwen 1998). For the example of exact exchange the action functional defined in this way is

$$
A_x[n] = -\frac{1}{2} \sum_{\substack{i,j=1 \\ \sigma=\uparrow,\downarrow}}^{N_\sigma} \int_C d\tau t'(\tau) \int d^3 r \int d^3 r'
$$
$$
\frac{\varphi_{i\sigma}^*(r',\tau)\varphi_{j\sigma}(r',\tau)\,\varphi_{i\sigma}(r,\tau)\varphi_{j\sigma}^*(r,\tau)}{|r - r'|}, \qquad (6.5)
$$

where τ is the Keldysh pseudotime, $\int_C$ denotes integration along the pseudotime contour, and $t'(\tau) = dt/d\tau$. For a given action functional the exchange-correlation potential $v_{xc\,\sigma}(r,t)$ for spin σ is defined as the functional derivative

$$
v_{xc\,\sigma}(r,t) = \left. \frac{\delta A_{xc}}{\delta n_\sigma(r,\tau)} \right|_{n_\sigma(r,t)}. \qquad (6.6)
$$

Evaluating this functional derivative after several steps (Ullrich et al. 1995a; van Leeuwen 1998) yields the TDOEP equation for $v_{xc\,\sigma}(r,t)$:

$$\sum_{j=1}^{N_\sigma} \frac{i}{\hbar} \int dt' \int d^3r' \left[v_{xc\,\sigma}(r',t') - u_{xc\,j\sigma}(r',t') \right] \varphi_{j\sigma}^*(r',t') \varphi_{j\sigma}(r,t)$$

$$\times \sum_{k=1}^{\infty} \varphi_{k\sigma}^*(r,t)\varphi_{k\sigma}(r',t')\theta(t-t') + \text{c.c.} = 0 \quad (6.7)$$

where $u_{xc\,j\sigma}(r,t)$ is defined by (cf. Eq. 6.4)

$$u_{xc\,j\sigma}(r,t) = \frac{1}{\varphi_{j\sigma}^*(r,t)} \left. \frac{\delta A_{xc}}{\delta \varphi_{j\sigma}(r,\tau)} \right|_{\varphi_{j\sigma}(r,t)}. \quad (6.8)$$

As the functional derivatives are taken at the orbitals and density in real time the pseudotime does not show up in the final equations and they can be interpreted in the usual way.

The TDOEP equation can be written more compactly by defining the so-called orbital shifts

$$\xi_{j\sigma}(r,t) = -\frac{i}{\hbar} \int dt' \int d^3r' \left[v_{xc\,\sigma}(r',t') - u_{xc\,j\sigma}^*(r',t') \right] \varphi_{j\sigma}(r',t')$$

$$\times \sum_{\substack{k=1 \\ k \neq j}}^{\infty} \varphi_{k\sigma}^*(r',t')\varphi_{k\sigma}(r,t)\theta(t-t') \quad (6.9)$$

Using this definition, Eq. 6.7 takes the short form

$$\sum_{j=1}^{N_\sigma} \xi_{j\sigma}^*(r,t)\varphi_{j\sigma}(r,t) + \text{c.c.} = g(r,t), \quad (6.10)$$

where

$$g(r,t) = \frac{i}{\hbar} \sum_{j=1}^{N_\sigma} |\varphi_{j\sigma}(r,t)|^2 \int_{-\infty}^{t} dt' \left[\bar{u}_{xc\,j\sigma}(t') - \bar{u}_{xc\,j\sigma}^*(t') \right] \quad (6.11)$$

and

$$\bar{u}_{xc\,j\sigma}(t) = \int d^3r \, \varphi_{j\sigma}^*(r,t) u_{xc\,j\sigma}(r,t)\varphi_{j\sigma}(r,t) \quad (6.12)$$

is an orbital average. Frequently, Eq. 6.10 takes a yet simpler form because the function $g(r,t)$ vanishes for functionals depending on $\{\varphi_{i\sigma}\}$ only through the combination $\varphi_{i\sigma}(r,t)\varphi_{i\sigma}^*(r',t')$ (Gross et al. 1996), a condition that is fulfilled by functionals such as exact exchange.

The orbital shift $\xi_{j\sigma}(r,t)$ is orthogonal to $\varphi_{j\sigma}(r,t)$ and obeys the equation

$$\left[i\hbar\frac{\partial}{\partial t} - \hat{H}_{\text{KS}\,\sigma}(\boldsymbol{r}, t)\right]\xi_{j\sigma}(\boldsymbol{r}, t) = \left\{v_{\text{xc}\,\sigma}(\boldsymbol{r}, t) - u^*_{\text{xc}\,j\sigma}(\boldsymbol{r}, t)\right.$$

$$\left.- \left[\bar{v}_{\text{xc}\,j\sigma}(t) - \bar{u}^*_{\text{xc}\,j\sigma}(t)\right]\right\}\varphi_{j\sigma}(\boldsymbol{r}, t), \quad (6.13)$$

where the Kohn–Sham Hamiltonian $\hat{H}_{\text{KS}\,\sigma}(\boldsymbol{r}, t)$ as usually is

$$\hat{H}_{\text{KS}\,\sigma}(\boldsymbol{r}, t) = -\frac{\hbar^2}{2m}\nabla^2 + v_{\text{ext}}(\boldsymbol{r}, t) + v_{\text{H}}(\boldsymbol{r}, t) + v_{\text{xc}\,\sigma}(\boldsymbol{r}, t) \quad (6.14)$$

and

$$\bar{v}_{\text{xc}\,j\sigma}(t) = \int d^3r\,\varphi^*_{j\sigma}(\boldsymbol{r}, t)v_{\text{xc}\,\sigma}(\boldsymbol{r}, t)\varphi_{j\sigma}(\boldsymbol{r}, t) \quad (6.15)$$

is another orbital average. Multiplying Eq. 6.10 by the Kohn–Sham potential and using the time-dependent Kohn–Sham equations together with Eq. 6.13 yields another form of the TDOEP equation:

$$v_{\text{xc}\,\sigma}(\boldsymbol{r}, t) = \frac{1}{2n_\sigma(\boldsymbol{r}, t)}\sum_{j=1}^{N_\sigma}\left\{\left|\varphi_{j\sigma}(\boldsymbol{r}, t)\right|^2\left\{u_{\text{xc}\,j\sigma}(\boldsymbol{r}, t) + \left[\bar{v}_{\text{xc}\,j\sigma}(t) - \bar{u}_{\text{xc}\,j\sigma}(t)\right]\right\}\right.$$

$$\left.- \frac{\hbar^2}{m}\nabla\cdot\left[\xi^*_{j\sigma}(\boldsymbol{r}, t)\nabla\varphi_{j\sigma}(\boldsymbol{r}, t)\right]\right\} + \text{c.c.} - f(\boldsymbol{r}, t), \quad (6.16)$$

where

$$f(\boldsymbol{r}, t) = \frac{1}{2n_\sigma(\boldsymbol{r}, t)}\left\{\frac{\hbar^2}{2m}\nabla^2 g(\boldsymbol{r}, t) + i\hbar\frac{\partial}{\partial t}\sum_{j=1}^{N_\sigma}\left[\xi^*_{j\sigma}(\boldsymbol{r}, t)\varphi_{j\sigma}(\boldsymbol{r}, t) - \text{c.c.}\right]\right\}. \quad (6.17)$$

Now all expressions needed to discuss the TDOEP equation are available.

We start by first looking not at the full TDOEP equation but at an approximation to it which has been used in most three dimensional propagation calculations so far (Marques et al. 2001; Chu 2005; Ullrich et al. 2000b). It is called the TDKLI approximation and is named after Krieger, Li, and Iafrate's corresponding approximation for the ground-state potential (Krieger et al. 1992b). In this approximation (which becomes exact in the one-orbital case) the terms in the second line on the right-hand side of Eq. 6.16 are omitted except for the complex conjugate. The remaining equation can readily be used in propagation schemes as all quantities on the right-hand side can be computed from the orbitals at a given time t. The TDKLI approximation is both ingeniously helpful and painfully misleading at the same time.

It is extremely helpful as it is computationally much easier than the full TDOEP, yet captures many of its crucial properties. The TDKLI approximation for v_{xc} does show, e.g., particle number discontinuities and the related step-like structures. Equation 6.16 is a good starting point for understanding the mechanism by which the TDKLI (and TDOEP) achieve this.

The particle number discontinuities are readily understandable by looking at the first line of Eq. 6.16 at a fixed time t. We make use of the arguments about the orbital ordering from Sect. 6.1, i.e., we assume a finite system and that one orbital density which we denote by $|\varphi_{N_\sigma\sigma}|^2$ dominates the density for $r \to \infty$. Then the decaying orbital densities $|\varphi_{j\sigma}|^2$ in the numerator will asymptotically suppress all terms except the ones from the N_σ orbital. Consequently, the orbital-average constants in the parenthesis on the right-hand side (first line) of Eq. 6.16 from the N_σ orbital will dominate the potential together with the asymptotic behavior that the $u_{\mathrm{xc}\,N_\sigma\sigma}$ term has (which is, e.g., $-1/r$ for exact exchange). To fulfill the requirement that the potential vanish at infinity one therefore must choose $\bar{v}_{\mathrm{xc}\,N_\sigma\sigma}$ such that the parenthesis vanishes, i.e.,

$$\bar{v}_{\mathrm{xc}\,N_\sigma\sigma} = \bar{u}_{\mathrm{xc}\,N_\sigma\sigma}. \tag{6.18}$$

This is always possible as it only amounts to adding a constant to the potential. However, if one now adds a particle to the system, one more orbital appears in the sum over orbitals in Eq. 6.16, and an additional constant term from these orbital's potential averages contributes. Thus, the potential jumps by a constant as soon as a fractional occupation of a new orbital is added—even if the occupation is extremely small. The same mechanism applies for the full TDOEP, but the magnitude of the potential jump may be different due to the additional terms in the second line of Eq. 6.16. This mechanism is in complete analogy to the ground-state case and detailed discussions of it can be found in Krieger et al. (1992b), Kümmel and Kronik (2008).

Let us now look at a time-dependent situation where a system is initially in the ground state and then some external time-dependent field is switched on that moves the highest orbital density far away from the system. This happens, e.g., if an electrical field that is strong enough to singly ionize the system is switched on in a real time calculation. Note that in such a calculation the particle number does not change: The highest orbital that is moved and the ionized core that remains behind still form one electronic system with the original number of electrons.[1] Therefore, one may think that the particle number discontinuity mechanism is not relevant in practice. Yet, think about how the spatial structure of $v_{\mathrm{xc}}(\mathbf{r}, t)$ changes as a function of time. To make the argument as transparent as possible we assume an idealized situation in which the initially highest occupied orbital is moved completely and all other $N - 1$ orbitals remain largely unaffected by the time-dependent field.

First consider the far asymptotic region. For very large r there is no qualitative change in the structure of v_{xc} because at any point in time during the ionization process the density is dominated by the highest orbital and the argument made above about the asymptotic form of the potential is always valid. However, now consider the intermediate region of space at a given time t. We define the intermediate region as all the points $\mathbf{r}$ that are so far away from the system's center that the lower $N - 1$

---

[1] Changes in particle number that may be due to numerical techniques such as absorbing boundaries are not relevant in this context because they do not represent physical changes of the particle number, but are a consequence of the numerical limitation that in practical calculations the grid size or basis set size is limited.

orbitals have decayed, but which are also so close to the system's center that at the given time t the highest orbital which is being ionized has already moved out of this region. Since the lower $N - 1$ orbitals more or less retained their ground-state shape and since the highest orbital is much further away, the potential in the intermediate region looks like the ground-state potential of an $N - 1$ electron system—except for the fact that it tends to a non-zero constant! The non-zero constant is an unavoidable consequence of the orbital average terms on the right-hand side of Eq. 6.16. We could only set one of them to zero via Eq. 6.18 as there is only one free constant in a potential—and the one free constant was already used to enforce that the potential decays to zero at infinity. Thus, v_{xc} in the intermediate region goes to a constant and only drops down again as one reaches the far asymptotic region in which the highest orbital density is appreciable. In other words, a "downward step" occurs in v_{xc} in the region of space where the highest orbital starts to dominate the density. As the system evolves, i.e., as we look at larger and larger times t, the intermediate region extends further and further outward. As a consequence, the potential step moves further and further outward as well. This is exactly the situation that is shown in Fig. 6.1.

One may now ask what the step structure has to do with the particle number discontinuity. The answer is that they are not the same, but that they are very closely related. Both are a consequence of the explicit sum over occupied orbitals and the orbital average terms in Eq. 6.16. One can further relate one to the other by taking the above described ionization process to the extreme limit that the highest orbital density "reaches infinity". Then, the step structure would also "reach infinity". The condition that v_{xc} must be zero at infinity would then force us to shift down the whole v_{xc} by exactly the height of the step. One may argue that "reaching infinity" in the above sense may be interpreted as being equivalent to taking the highest orbital away, i.e., removing one electron. Thus, one may argue that the shift in v_{xc} (i.e., the step height) has the same magnitude as and basically is the particle number discontinuity.

However, in the above description we brushed under the carpet a subtlety that comes in because we can hardly define in a precise way what we meant by "the orbital reaches infinity" in the above argument. In reality no orbital will reach infinity and, as said previously, the particle number in the propagation remains constant. The particle number discontinuity in v_{xc}, i.e., derivative discontinuity in A_{xc}, is however defined for particle number changes with all other variables kept fixed. This is not the same situation as the above described ionization process, because the step structure in v_{xc} may be influenced by dynamical and non-adiabatic effects. Therefore, we cannot exactly identify the step in v_{xc} with the derivative discontinuity, although the two are very closely related. The close relation has also been confirmed numerically in calculations that reconstructed the exact time-dependent $v_{xc}(\boldsymbol{r}, t)$ for an ionization process (Lein and Kümmel 2005). There, $v_{xc}(\boldsymbol{r}, t)$ developed a step of a height that corresponded to the magnitude of the ground-state derivative discontinuity.

With the particle number discontinuity the TDKLI potential incorporates one of the most complex properties of (TD)DFT. One may thus be tempted to believe that the KLI approximation is perfect. Unfortunately, this is not the case. In fact, the KLI approximation can be misleading because it accurately yields some properties while it fails badly for others. A prominent example of a KLI failure is found in molecular chains. The KLI approximation for the exact exchange functional yields quite accurate ground-state exchange energies for such systems, yet it badly over-estimates their static polarizability and hyperpolarizability (Gritsenko et al. 2000; Kümmel et al. 2004). With the static response already being seriously in error, one can hardly hope that the dynamical response is well described. For the self-interaction corrected LDA, i.e., another orbital functional (see Sect. 6.4), both the response prop-erties and the ground-state structure are poorly described by a straightforward KLI approximation (Körzdörfer et al. 2008). A further and quite fundamental shortcoming of the KLI approximation is that the KLI potential is not a functional derivative and violates the zero force theorem (Mundt and Kümmel 2007). The latter states that the density-averaged force exerted by v_{xc} must be zero (see Chap. 5). Violating this condition can lead to serious errors in the dynamics as the spurious forces exerted by v_{xc} may lead to unphysical effects such as self-excitation (Mundt and Kümmel 2007). Therefore, despite its many good properties, the TDKLI approximation is not a generally satisfactory option for the time-dependent use of orbital functionals.

A generally applicable scheme for solving the full TDOEP equation is thus highly desirable. Unfortunately, to date the TDOEP equation has only been solved for one-dimensional model systems (Wijewardane and Ullrich 2008). This solution is of great conceptual importance, as it proves that the equation can be solved in a stable numerical way. However, the scheme employed in Wijewardane and Ullrich (2008) is numerically so involved that hopes are low that it can be used to calculate the TDOEP for three dimensional molecules and solids.

Inspection of Eq. 6.16 suggests an idea for a generally applicable solution strategy. When one propagates the Kohn–Sham orbitals according to the time-dependent Kohn–Sham equations

$$i\hbar\frac{\partial}{\partial t}\varphi_{j\sigma}(\boldsymbol{r},t) = \hat{H}_{KS\,\sigma}(\boldsymbol{r},t)\varphi_{j\sigma}(\boldsymbol{r},t), \tag{6.19}$$

and the orbital shifts according to Eq. 6.13, one should be able to construct the full TDOEP from Eq. 6.16 at each time step from the sets of the $\{\varphi_j(\boldsymbol{r},t)\}$ and $\{\xi_j(\boldsymbol{r},t)\}$. Such a scheme would be the time-dependent analogue of the successful schemes which construct the ground-state OEP from the orbital shifts (Kümmel and Perdew 2003a, b; Horowitz et al. 2006, Cinal and Holas 2007) and would be readily applicable in three-dimensional problems. So far there has been one attempt to obtain the TDOEP along these lines (Mundt and Kümmel 2006). Unfortunately, it faced serious difficulties. The coupled numerical propagation of the φ_j and ξ_j as described above was not generally stable, and it could not be cleared up whether this was due to numerical issues or a fundamental problem.

One obvious difficulty with Eq. 6.16 is that the density appears in the denominator. For a finite system the density will decay exponentially in the asymptotic region, i.e., a numerical evaluation of Eq. 6.16 runs into the problem of dividing by numbers very close to zero. For the TDKLI approximation this problem is less severe because all terms in the numerator are multiplied by the orbital densities, i.e., numerator and denominator show similar decay behavior. The terms in the second line of Eq. 6.16 however do not benefit from such an effect, making the TDOEP more cumbersome

to evaluate. Similar problems have been observed in calculations of the ground-state
OEP, but can be sidestepped there by not using Eq. 6.16 for the construction of the
potential, but an error-correction scheme (Kümmel and Perdew 2003b). A similar
numerical trick for the time-dependent case is yet missing. Another uncertainty of
the TDOEP scheme rests in Eq. 6.17. It involves the imaginary part of the sum over
the orbitals and orbital shifts, whereas the TDOEP Eq. 6.10 defines its real part. In
the numerical scheme tested in Mundt and Kümmel (2006) the imaginary part was
set to zero. This is consistent with the other equations, but possibly not the only
choice. Finally, numerical tests revealed that numerical propagation of the φ_j with
analytically obtained ξ_j and the numerical propagation of the ξ_j with analytically
obtained φ_j were both stable. Thus, only the coupled equations of motion for orbitals
and orbital shifts were unstable, whereas each set of equations of motion separately
could readily be solved by numerical propagation.

We therefore have to conclude that a generally applicable solution scheme for the
TDOEP equation that would allow for obtaining the TDOEP for three-dimensional
systems is still missing. Developing such a scheme remains as one of the major
challenges in time-dependent Kohn–Sham theory.

6.4 A Few Examples

After having taken a close look at formal aspects of using orbital functionals in time-
dependent Kohn–Sham theory in the previous section, the final part of this chapter
is devoted to mentioning a few practical examples of situations in which orbital
functionals are successfully used.

Probably the most commonly used orbital functionals are the different hybrids
that mix a constant fraction of exact exchange with GGA exchange and correlation. For many organic molecules linear-response TDDFT using the orbital-specific
Hartree–Fock potentials for the exact exchange part yield quite reliable photoabsorption spectra. The great drawback of standard hybrid functionals is that with the
typically used constant fraction of about 20% exact exchange long-range charge-
transfer excitations are severely underestimated. This can lead to serious problems
of interpretation when both valence and charge-transfer excitations are present (Stein
et al. 2009a).

Range-separated hybrid functionals, which use Fock exchange but split the
Coulomb interaction into two parts with the decomposition

$$\frac{1}{|r-r'|} = \frac{\mathrm{erf}(\gamma|r-r'|)}{|r-r'|} + \frac{\mathrm{erfc}(\gamma|r-r'|)}{|r-r'|} \qquad (6.20)$$

and a range separation parameter γ, are promising candidates for overcoming this
problem. For example, a non-empirical way of fixing γ by exploiting the ionization
potential theorem has been put forward in Stein et al. (2009a). With thus determined values of γ one obtains quite accurate values for charge-transfer excitations

(Karolewski et al. 2011). The approach is not a panacea, though, as rather different values for γ are needed depending on whether one aims at a reliable description of ground-state binding energies or of charge-transfer excitations.

Exact exchange in Kohn–Sham TDDFT has been employed both in the explicitly linearized form (Ipatov et al. 2010) and in propagation calculations (Marques et al. 2001). An improved description of, e.g., excitonic effects (Kim and Görling 2002) has been reported, and naturally all problems related to one-electron self interaction are greatly reduced by using pure exact exchange. However, Fock exchange without correlation describes electronic binding only poorly, thus a correlation functional that is compatible with exact exchange is needed if one wants a universal functional that is useful for DFT and TDDFT.

The various time-dependent SIC schemes are examples for approaches that are self-interaction free and include correlation contributions (Chu 2005; Ullrich et al. 2000b). They go back to the ground-state DFT idea of correcting a given energy functional, e.g., LSDA, for self-interaction in an orbital-by-orbital fashion (Perdew and Zunger 1981), i.e.,

$$u^{\text{SIC}}_{\text{xc}\,j\sigma}[n_\uparrow, n_\downarrow] = v^{\text{LSDA}}_{\text{xc}\,\sigma}[n_\uparrow, n_\downarrow] - \left(v_{\text{H}}[|\varphi_{j\sigma}|^2] + v^{\text{LSDA}}_{\text{xc}\,\sigma}[|\varphi_{j\sigma}|^2, 0] \right). \qquad (6.21)$$

This approach can be extended to the time-domain by inserting the time-dependent orbitals in Eq. 6.21. Various flavors of TD-SIC in both Kohn–Sham and generalized Kohn–Sham variants have been developed, some with a particular focus on simplifications that reduce accuracy but ease computations (Legrand et al. 2002).

An interesting aspect of the SIC approach is that it is not unitarily invariant. Therefore, the usual OEP equation for the ground state has been replaced by a generalized OEP equation (Körzdörfer et al. 2008). Extending this concept to the time-domain and exploiting the unitary variance to stabilize the propagation scheme is a promising approach for obtaining a time-dependent Kohn–Sham SIC scheme (Hofmann and Kümmel 2009, unpublished). This may be a way to construct a functional that yields reasonably accurate results for both ground-state energetics and charge-transfer excitations.

In summary one can say that it will take further work to turn the relation between orbital functionals and TDDFT into a perfectly happy marriage—but already now it is certainly an exciting relationship.

Chapter 7
Response Functions in TDDFT: Concepts and Implementation

David A. Strubbe, Lauri Lehtovaara, Angel Rubio, Miguel A. L. Marques and Steven G. Louie

7.1 Introduction

Many physical properties of interest about solids and molecules can be considered as the reaction of the system to an external perturbation, and can be expressed in terms of response functions, in time or frequency and in real or reciprocal space. Response functions in TDDFT can be calculated by a variety of methods. Time-propagation

D. A. Strubbe (✉) · S. G. Louie
Department of Physics, University of California,
366 LeConte Hall MC 7300, Berkeley CA 94720-7300, USA
e-mail: dstrubbe@berkeley.edu

D. A. Strubbe · S. G. Louie
Materials Sciences Division, Lawrence Berkeley National Laboratory,
1 Cyclotron Road, Berkeley CA 94720, USA
e-mail: sglouie@berkeley.edu

M. A. L. Marques (✉)· L. Lehtovaara
Laboratoire de Physique de la Matière Condensée et Nanostructures,
Université Claude Bernard Lyon 1 et CNRS,
43 boulevard du 11 novembre 1918, 69622 Villeurbanne Cedex, France
e-mail: miguel.marques@tddft.org
e-mail: lauri.lehtovaaral@iki.fi

A. Rubio (✉)
Nano-Bio Spectroscopy Group and ETSF Scientific Development Centre,
Departamento Física de Materiales, Universidad del País Vasco,
Centro de Física de Materiales CSIC-UPV/EHU-MPC and DIPC,
Avenida Tolosa 72, E-20018 San Sebastián, Spain
e-mail: angel.rubio@ehu.es

A. Rubio
Fritz-Haber-Institut der Max-Planck-Gesellschaft, Faradayweg 4-6,
14195 Berlin, Germany

M. A. L. Marques et al. (eds.), *Fundamentals of Time-Dependent Density Functional Theory*, Lecture Notes in Physics 837, DOI: 10.1007/978-3-642-23518-4_7,
© Springer-Verlag Berlin Heidelberg 2012

is a non-perturbative approach in the time domain, whose static analogue is the method of finite differences. Other approaches are perturbative and are formulated in the frequency domain. The Sternheimer equation solves for the variation of the wavefunctions, the Dyson equation is used to solve directly for response functions, and the Casida equation solves for the excited states via an expansion in an electron-hole basis. These techniques can be used to study a range of different response functions, including electric, magnetic, structural, and $\mathbf{k} \cdot \mathbf{p}$ perturbations. In this chapter, we give an overview of the basic concepts behind response functions and the methods that can be employed to efficiently compute the response properties within TDDFT and the physical quantities that can be studied.

7.2 Response Functions

In this section, we will: (1) show how a response function maps an external field to a physical observable, (2) discuss how a specific response function is connected to a specific physical property, (3) link the fully interacting many-body density response function with the Kohn–Sham (KS) density response function, and (4) describe how the different orders of response functions form a hierarchy.

In spectroscopic experiments, an external field $F(\mathbf{r}, t)$ is applied to a sample. The sample, which is a fully interacting many-electron system from the theoretical point of view, responds to the external field. Then the response can be measured for some physical observable $\mathcal{P}$:

$$\Delta \mathcal{P} = \Delta \mathcal{P}_F[F]. \tag{7.1}$$

In general, the dependence of the functional $\Delta \mathcal{P}_F[F]$ on F is very complex, as it must reproduce the response for a field of any strength and shape. However, if the external field is weak, the response can be expanded as a power series with respect to the field strength (Bernard and Callen 1959; Peterson 1967). The first-order response, also called the linear response of the observable,

$$\delta \mathcal{P}^{(1)}(\mathbf{r}, t) = \int \mathrm{d}t' \int \mathrm{d}^3 r' \chi^{(1)}_{\mathcal{P} \leftarrow F}(\mathbf{r}, \mathbf{r}', t - t') \delta F^{(1)}(\mathbf{r}', t') \tag{7.2}$$

is a convolution of $\chi^{(1)}_{\mathcal{P} \leftarrow F}(\mathbf{r}, \mathbf{r}'; t - t')$, the linear response function, and $\delta F^{(1)}(\mathbf{r}'; t')$, the field expanded to first order in the field strength. The linear response function is nonlocal in space and in time, but the above time convolution simplifies to a product in frequency space:

$$\delta \mathcal{P}^{(1)}(\mathbf{r}; \omega) = \chi^{(1)}_{\mathcal{P} \leftarrow F}(\mathbf{r}, \mathbf{r}', \omega) \delta F^{(1)}(\mathbf{r}', \omega). \tag{7.3}$$

The linear response function $\chi^{(1)}_{\mathcal{P} \leftarrow F}(\mathbf{r}, \mathbf{r}', \omega)$ depends only on a single frequency ω, which is a consequence of the homogeneity of time.

At every order in the field strength, each observable/field pair has its own response function that is connected to a specific physical property. For example, the first-order response of the dipole moment to a dipole electric field is the polarizability $\alpha = \frac{\partial \mu}{\partial \mathcal{E}}$, the second-order response of the same pair provides the hyperpolarizability, and the first-order response of the magnetic moment to a homogeneous magnetic field is the magnetic susceptibility.

7.2.1 Linear Density Response

Perhaps the most important response function, from the TDDFT point of view, is the linear density response function $\chi(\boldsymbol{r}, \boldsymbol{r}', t - t') = \chi_{n \leftarrow v_{\text{ext}}}^{(1)}(\boldsymbol{r}, \boldsymbol{r}', t - t')$, as introduced in Eq. 4.47, which gives the linear response of the density $\delta n^{(1)}(\boldsymbol{r}, t)$ to an external scalar potential $\delta v_{\text{ext}}(\boldsymbol{r}', t')$. If the density response function $\chi(\boldsymbol{r}, \boldsymbol{r}', t - t')$ is obtained explicitly, it can then be used to calculate the first-order response of all properties derivable from the density with respect to any scalar field (e.g., polarizability, magnetic susceptibility).

The fully interacting many-body response function can be obtained from the corresponding Kohn–Sham system (Gross and Kohn 1985; Petersilka and Gross 1996) as described in Sect. 4.5.1. The Kohn–Sham system describes a non-interacting system of electrons subject to an external potential $v_{\text{KS}}(\boldsymbol{r}, t)$, which is the effective Kohn–Sham potential. Therefore, the so-called linear Kohn–Sham (density) response function measures how the density changes upon linear variation of the Kohn–Sham potential $v_{\text{KS}}(\boldsymbol{r}, t)$:

$$\delta n(\boldsymbol{r}, t) = \int dt' \int d^3 r' \chi_{\text{KS}}(\boldsymbol{r}, \boldsymbol{r}', t - t') \delta v_{\text{KS}}(\boldsymbol{r}', t'). \tag{7.4}$$

Note that, by virtue of the KS construction, the variation of the density $\delta n(\boldsymbol{r}, t)$ is the same as in the fully interacting system. In addition to the external potential $v_{\text{ext}}(\boldsymbol{r}, t)$, the effective Kohn–Sham potential has contributions from the Hartree and the exchange-correlation potentials:

$$\delta v_{\text{KS}}(\boldsymbol{r}', t') = \delta v_{\text{ext}}(\boldsymbol{r}', t') + \delta v_{\text{H}}[n](\boldsymbol{r}', t') + \delta v_{\text{xc}}[n](\boldsymbol{r}', t'), \tag{7.5}$$

where

$$\delta v_H[n](\boldsymbol{r}', t') = \int dt'' \int d^3 r'' \frac{\delta(t' - t'')}{|\boldsymbol{r}' - \boldsymbol{r}''|} \delta n(\boldsymbol{r}'', t''). \tag{7.6}$$

To calculate the variation of the exchange-correlation (xc) potential, one simply employs the chain-rule for functional derivatives:

$$\delta v_{\text{xc}}[n](\boldsymbol{r}', t') = \int dt'' \int d^3 r'' f_{\text{xc}}[n_{\text{GS}}](\boldsymbol{r}', \boldsymbol{r}'', t' - t'') \delta n(\boldsymbol{r}'', t''). \tag{7.7}$$

The exchange-correlation kernel

$$f_{xc}[n_{GS}](r', r'', t' - t'') = \left. \frac{\delta v_{xc}[n](r', t')}{\delta n(r'', t'')} \right|_{n=n_{GS}} \tag{7.8}$$

is the functional derivative of the exchange-correlation potential with respect to the density at the ground-state density n_{GS}. Note that the exchange-correlation kernel $f_{xc}[n_{GS}](r', r'', t' - t'')$ is a functional of the ground-state density and can be evaluated *before* any response calculation.

For the same external potential $v_{ext}(r, t)$, the fully interacting and the Kohn–Sham density responses must be the same. Therefore, we can set the right-hand-side of Eq. 4.47 equal to the right-hand-side of Eq. 7.4, and use Eq. 4.47 once more to replace δn in Eq. 7.7:

$$\int dt' \int d^3 r' \chi(r, r', t - t') \delta v_{ext}(r', t')$$

$$= \int dt' \int d^3 r' \chi_{KS}(r, r', t - t') \delta v_{ext}(r', t')$$

$$+ \int dt' \int d^3 r' \chi_{KS}(r, r', t - t')$$

$$\times \int dt'' \int d^3 r'' \left[\frac{\delta(t' - t'')}{|r' - r''|} + f_{xc}[n_{GS}](r', r'', t' - t'') \right]$$

$$\times \int dt''' \int d^3 r''' \chi(r'', r''', t'' - t''') \delta v_{ext}(r''', t'''). \tag{7.9}$$

As the density response function is an intrinsic property of the system, it cannot depend on the detailed form of the external potential. The terms multiplying the external potential $v_{ext}(r, t)$ must be point-wise equal, from which we obtain a Dyson-like equation for the density response function, which reads in frequency space [writing out the integrals in Eq. 4.59]:

$$\chi(r, r', \omega) = \chi_{KS}(r, r', \omega) + \int d^3 r'' \int d^3 r''' \chi_{KS}(r, r'', \omega)$$

$$\times \left[\frac{1}{|r'' - r'''|} + f_{xc}[n_{GS}](r'', r''', \omega) \right] \chi(r''', r', \omega). \tag{7.10}$$

The Kohn–Sham density response function $\chi_{KS}(r, r', \omega)$ is straightforward to obtain from first-order perturbation theory and has poles at the Kohn–Sham energy differences. Writing separate sums for occupied and unoccupied orbitals:

$$\chi_{KS}(r, r', \omega) = \lim_{\eta \to 0^+} \sum_{a,i} (n_i - n_a) \left[\frac{\varphi_i^*(r)\varphi_a(r)\varphi_i(r')\varphi_a^*(r')}{\omega - (\varepsilon_a - \varepsilon_i) + i\eta} - \frac{\varphi_i(r)\varphi_a^*(r)\varphi_a(r')\varphi_i^*(r')}{\omega - (\varepsilon_i - \varepsilon_a) + i\eta} \right],$$

$$\tag{7.11}$$

where $\varphi_i(\mathbf{r})$ and $\varphi_a(\mathbf{r})$ are occupied and unoccupied KS orbitals, respectively, and ε_i and ε_a are the corresponding KS eigenvalues.[1] The Eq. 7.10 can be formally written as

$$\chi = (1 - \chi_{KS} f_{Hxc})^{-1} \chi_{KS}, \tag{7.12}$$

where all terms on the right-hand-side are known from a ground-state Kohn–Sham calculation. This equation can be used directly to calculate the fully interacting density response function from a Kohn–Sham system (Hybertsen and Louie 1987).

Before introducing practical methods for calculating TDDFT response in Sect. 7.3, we will briefly discuss higher-order responses.

7.2.2 Higher-Order Density Response

In linear response, a system interacts only once with the external field and only with the field component which is first-order with respect to the field-strength parameter. For example, the magnetic field changes the kinetic-energy operator in the following way:

$$\frac{1}{2}\left(\hat{p} + \frac{\lambda}{c}\hat{A}\right)^2 = \frac{1}{2}\hat{p}^2 + \frac{\lambda}{2c}\hat{p}\cdot\hat{A} + \frac{\lambda}{2c}\hat{A}\cdot\hat{p} + \frac{\lambda^2}{2c^2}\hat{A}^2, \tag{7.13}$$

but only the terms $\hat{p}\cdot\hat{A}$ and $\hat{A}\cdot\hat{p}$ contribute to the linear response, because $\hat{A}^2$ term is second order with respect to the field strength parameter λ.

In second-order response, a system interacts twice with the linear component of the external field, but in addition, it also interacts once with the quadratic component of the external field $\delta v_{\text{ext}}^{(2)}$ (e.g., $\hat{A}^2$) if it exists. The second-order density response equation reads

$$\delta n^{(2)}(\mathbf{r}, t) = \frac{1}{2}\int dt' \int dt'' \int d^3r' \int d^3r'' \chi^{(2)}(\mathbf{r}, t, \mathbf{r}', t', \mathbf{r}'', t'')\delta v_{\text{ext}}^{(1)}(\mathbf{r}', t')\delta v_{\text{ext}}^{(1)}(\mathbf{r}'', t'')$$

$$+ \int dt' \int d^3r' \chi^{(1)}(\mathbf{r}, t, \mathbf{r}', t')\delta v_{\text{ext}}^{(2)}(\mathbf{r}', t'). \tag{7.14}$$

or in frequency space

$$\delta n^{(2)}(\mathbf{r}, \omega) = \frac{1}{2}\int d\omega' \int d\omega'' \int d^3r' \int d^3r'' \delta(\omega - (\omega' + \omega''))$$

$$\times \chi^{(2)}(\mathbf{r}, \mathbf{r}', \mathbf{r}'', \omega, \omega', \omega'')\delta v_{\text{ext}}^{(1)}(\mathbf{r}', \omega')\delta v_{\text{ext}}^{(1)}(\mathbf{r}'', \omega'')$$

$$+ \int d^3r' \chi^{(1)}(\mathbf{r}, \mathbf{r}', \omega)\delta v_{\text{ext}}^{(2)}(\mathbf{r}', \omega), \tag{7.15}$$

---

[1] This is the causal KS response function. The time-ordered KS response function would have $-i\eta$ in the second term.

where $\delta(\omega - (\omega' + \omega''))$ comes from the conservation of energy. The second-order response function $\chi^{(2)}(r, r', r'', \omega, \omega', \omega'')$ mixes two frequencies (which can be different, ω_1 and ω_2, or equal, $\omega_1 = \pm\omega_2$) yielding a new frequency. For example, if the field is monochromatic having only frequency ω_1, the second-order response generates second harmonics at frequency $2\omega_1$, and optical rectification at frequency $\omega = \omega_1 - \omega_1 = 0$.

The second-order Kohn–Sham response is fairly straightforward to obtain (Gross et al. 1996; Senatore and Subbaswamy 1987), and it reads[2]

$$
\begin{aligned}
\delta n^{(2)}(r, t) = {} & \frac{1}{2} \int dt' \int dt'' \int d^3r' \int d^3r'' \chi_{KS}^{(2)}(r, t, r', t', r'', t'') \delta v_{ext}^{(1)}(r', t') \delta v_{ext}^{(1)}(r'', t'') \\
& + \int dt' \int d^3r' \chi_{KS}^{(1)}(r, t, r', t') \delta v_{ext}^{(2)}(r', t') \\
& + \frac{1}{2} \int dt' \int dt'' \int dt''' \int d^3r' \int d^3r'' \int d^3r''' \chi_{KS}^{(1)}(r, t, r', t') \\
& \times k_{xc}(r', t', r'', t'', r''', t''') \delta n^{(1)}(r'', t'') \delta n^{(1)}(r''', t''') \\
& + \int dt' \int dt'' \int d^3r' \int d^3r'' \chi_{KS}^{(1)}(r, t, r', t') \\
& \times \left(\frac{\delta(t' - t'')}{|r' - r''|} + f_{xc}(r', t', r'', t'') \right) \delta n^{(2)}(r'', t''),
\end{aligned}
\tag{7.16}
$$

where

$$
k_{xc}(r', t', r'', t'', r''', t''') = \frac{\delta^2 v_{xc}(r', t')}{\delta n(r'', t'') \delta n(r''', t''')} \Big|_{n=n_{GS}}.
\tag{7.17}
$$

If $\chi^{(1)}$ is already solved for, the 2nd-order response equation can then be obtained by combining Eqs. 4.47, 7.16 and 7.14.

Higher-order responses are also straightforward to construct but become quickly cumbersome. The response equations form a hierarchic structure where the ith-order response requires the $(i - 1)$th- and lower-order responses. Note that all functional derivatives of the xc potentials are evaluated at the ground-state density.

7.3 Methods for Calculating Response Functions

In this section, we will briefly describe three different methods to calculate response from TDDFT: (i) time-propagation, (ii) Sternheimer, and (iii) Casida method. The time-propagation method (Yabana and Bertsch 1999) simply propagates a system

---

[2] Note this differs from (4.89a) in the last term: Equation 4.89a defined the nth-order response by the terms in the density expansion proportional to n factors of $\delta v_{ext}(r, t) = v_{ext}(r, t) - v_{ext}(r, 0)$. Here, on the other hand, we define the nth-order response as being proportional to the nth-power of the field strength.

under a given external field. The density response is obtained directly as the difference between the time-dependent density and the ground-state state density. As the method is nonperturbative, all orders of response are included in the calculation, and therefore, specific orders must be numerically extracted. The Sternheimer method (Sternheimer 1954; Baroni et al. 1987a; Gonze 1995a; Andrade et al. 2007) solves for a specific order of the response for a specific field in frequency space (i.e., it is a perturbative approach). The Sternheimer equations form a hierarchic structure, where higher-order responses can be calculated from lower-order responses. The Casida method (Casida 1995), instead of finding the response, finds the poles and residues of the first-order response function, which corresponds to finding the resonant transitions of a system. Note that physically all these techniques are equivalent as they are all based on Kohn–Sham DFT and are simply different ways to obtain the same quantities. Often the choice between them is done purely on numerical arguments, as each one is more adapted to certain numerical implementations (Marques and Rubio 2006b).

We will show that the three methods are connected to each other in a simple way. Since the purpose is to connect all these approaches, we will consider only weak perturbations. For pedagogical reasons, we make the following simplifying assumptions in the discussion below that can be easily generalized (see discussion at the end of this section): (i) the exchange-correlation functional does not have memory, i.e., we work within the adiabatic approximation, (ii) the system is spin-unpolarized, (iii) we have no fractional occupations, and (iv) we use no k-point sampling, i.e., only Γ-point or a non-periodic system, which allows us to use purely real (i.e., imaginary part is zero) ground-state Kohn–Sham wavefunctions. Assumption (i) is in practice not very restrictive, as a large majority of the functionals used in TDDFT are, indeed, adiabatic. Assumptions (ii–iv) are valid, for example, for closed-shell molecules. In any case it is fairly straightforward (but cumbersome) to remove the assumptions from the derivation. We will return to this topic at the end of the section.

7.3.1 Time-Propagation Method

In the time-propagation approach, the time-dependent Kohn–Sham equations are propagated in real-time, i.e., by solving the following nonlinear partial differential equation

$$i\frac{\partial}{\partial t}\varphi_k(\boldsymbol{r}, t) = \hat{H}_{\mathrm{KS}}[n](\boldsymbol{r}, t)\varphi_k(\boldsymbol{r}, t), \qquad (7.18)$$

starting from time $t = 0$ with the initial condition $\varphi_k(\boldsymbol{r}, t = 0) = \varphi_k^{(0)}(\boldsymbol{r})$, where $\varphi_k^{(0)}(\boldsymbol{r})$ are the ground-state Kohn–Sham wavefunctions. Here, we already have applied the adiabatic approximation by assuming that $\hat{H}_{\mathrm{KS}}[n](\boldsymbol{r}, t)$ has a functional dependence only on the instantaneous density $n(\boldsymbol{r}, t)$ instead of on its whole history.

If no perturbation is applied to the system, the system remains in the ground-state and the time-evolution of the KS wavefunctions is trivial: $\varphi_k(t) = \varphi^{(0)} e^{-i\varepsilon_k^{(0)} t}$. If we apply a weak time-dependent external perturbation with a given frequency ω, the time-evolution becomes nontrivial because of the nonlinearity of the Kohn–Sham Hamiltonian. A general form for a weak time-dependent external perturbation with a given frequency ω is

$$v_{\text{ext}}(\boldsymbol{r}, t) = \lambda v_{\text{ext}}^{\cos}(\boldsymbol{r}) \cos(\omega t) + \lambda v_{\text{ext}}^{\sin}(\boldsymbol{r}) \sin(\omega t) \tag{7.19}$$

or, rewriting in terms of the exponential

$$v_{\text{ext}}(\boldsymbol{r}, t) = \lambda v_{\text{ext}}^{+\omega}(\boldsymbol{r}) e^{+i\omega t} + \lambda v_{\text{ext}}^{-\omega}(\boldsymbol{r}) e^{-i\omega t}, \tag{7.20}$$

where λ is the strength of the perturbation. If we now insert this external potential to the TDKS equation and propagate in time, we can obtain physical observables from the time-dependent expectation values. For example, if we apply a weak delta pulse of a dipole electric field[3]

$$v_{\text{ext}}(\boldsymbol{r}, t) = -e\boldsymbol{r} \cdot \boldsymbol{K} \delta(t) = -e\boldsymbol{r} \cdot \boldsymbol{K} \frac{1}{2\pi} \int\limits_{-\infty}^{\infty} d\omega \exp(i\omega t), \tag{7.21}$$

we simply replace the ground-state wavefunctions (eigenfunctions of the Hamiltonian $\hat{H}_{\text{KS}}^{(0)}$) by

$$\varphi_k(\boldsymbol{r}, t = 0^+) = \exp\left\{ -\frac{i}{\hbar} \int\limits_{0^-}^{0+} dt \left[\hat{H}_{\text{KS}}^{(0)}(t) - e\boldsymbol{r} \cdot \boldsymbol{K} \delta(t) \right] \right\} \varphi_k(\boldsymbol{r}, t = 0^-)$$

$$= \exp\left(i e\boldsymbol{r} \cdot \boldsymbol{K}/\hbar\right) \varphi_k(\boldsymbol{r}, t = 0^-), \tag{7.22}$$

and propagate the free oscillations in time. Then the time-dependent dipole moment

$$\boldsymbol{\mu}(t) = -e \int d^3 r \boldsymbol{r} n(\boldsymbol{r}, t) \tag{7.23}$$

can be used to extract the dynamic polarizability tensor $\alpha(\omega)$. We Fourier-transform to obtain[4]

$$\alpha_{\gamma\delta}(\omega) = \frac{1}{K_\delta} \int\limits_{0}^{\infty} dt \left[\mu_\gamma(t) - \mu_\gamma(0^-) \right] e^{-i\omega t} + \mathcal{O}(K_\delta). \tag{7.24}$$

---

[3] Note that $\boldsymbol{K}$ has units of electric field times time.

[4] Note that the integration begins from $t = 0^-$ instead of $-\infty$, which basically corresponds to adding a Heaviside function $\theta(t - 0^-)$ inside the Fourier transform.

The imaginary part of the diagonal component of the dynamic polarizability $\mathcal{I}[\alpha_{\delta\delta}(\omega)]$ is proportional to the absorption spectrum. The above equation includes an integral over infinite time. Obviously, infinite time-propagation is not possible in practice, and we have to add an artificial lifetime to the equation by introducing a decay $e^{-\eta t}$:

$$\alpha_{\gamma\delta}(\omega) = \frac{1}{K_\delta} \int\limits_0^\infty dt \left[\mu_\gamma(t) - \mu_\gamma(0^-)\right] e^{-\mathrm{I}\omega t} e^{-\eta t} + \mathcal{O}(K_\delta). \tag{7.25}$$

which corresponds to forcing all excitations to decay back to the ground state with rate η.

Higher-order responses (e.g., hyperpolarizabilities) are automatically considered in the calculation. However, if the field is chosen weak enough, they have negligible contribution, as should be the case for a linear-response calculation. If the perturbation strength is increased, the nonlinear contributions begin to increase: 2nd order quadratically, 3rd order cubically, etc. In addition to the different scaling with respect to the perturbation strength, higher-order responses appear at combinations of existing frequencies, which will be discussed in the next section. To disentangle the different contributions from the time propagation scheme is not always a well-defined procedure.

7.3.2 Sternheimer Method

The time-propagation approach propagates the TDKS equations in real-time. If we apply time-dependent perturbation theory and transform the equations to frequency space, we obtain the Sternheimer method, which is also known as density-functional perturbation theory, particularly in its static form (Baroni et al. 2001), and as "coupled perturbed Kohn–Sham" in the quantum-chemistry literature.

As the time-dependent external perturbation was chosen weak, we can expand the TD-KS states and the TD-KS-Hamiltonian as a power series with respect to the perturbation strength λ. The perturbation expansion[5] of the TD-KS states reads

$$\varphi_k(\boldsymbol{r}, t) = \varphi_k^{(0)}(\boldsymbol{r}, t) + \lambda \varphi_k^{(1)}(\boldsymbol{r}, t) + \lambda^2 \varphi_k^{(2)}(\boldsymbol{r}, t) + \dots \tag{7.26}$$

where the zeroth-order response has trivial time-dependence $\varphi_k^{(0)}(\boldsymbol{r}, t) = \varphi_k^{(0)}(\boldsymbol{r}) e^{-i\varepsilon_k^{(0)} t}$. The perturbation expansion of the TD-KS Hamiltonian reads

$$\begin{aligned}\hat{H}_{\mathrm{KS}}(\boldsymbol{r}, t) = {} & \hat{H}_{\mathrm{KS}}^{(0)}[n^{(0)}](\boldsymbol{r}, t) + \lambda v_{\mathrm{ext}}^{(1)}(\boldsymbol{r}, t) + \lambda \hat{H}_{\mathrm{KS}}^{(1)}[n](\boldsymbol{r}, t) \\ & + \lambda^2 v_{\mathrm{ext}}^{(2)}(\boldsymbol{r}, t) + \lambda^2 \hat{H}_{\mathrm{KS}}^{(2)}[n](\boldsymbol{r}, t) + \dots\end{aligned} \tag{7.27}$$

---

[5]   Note that this expansion is not a Taylor expansion.

where $\hat{H}_{KS}^{(0)}[n^{(0)}](r)$ is the ground-state Hamiltonian. The $\hat{H}_{KS}^{(k)}[n](r, t)$ are the kth-order response Hamiltonians, i.e., kth derivatives of the Hamiltonian with respect to the magnitude of the bare external perturbation.[6] These response Hamiltonians arise from the nonlinearity of the TD-KS Hamiltonian: the Hartree and exchange-correlation potentials are affected too when the system is perturbed by the time-dependent external potential $v_{ext}(r, t)$.

The response Hamiltonians require the time-dependent density

$$n(r, t) = \sum_k n_k |\varphi_k(r, t)|^2 = n^{(0)}(r, t) + \lambda n^{(1)}(r, t) + \lambda^2 n^{(2)}(r, t) + \dots \quad (7.28)$$

Inserting the expansion for the KS wavefunctions

$$
\begin{aligned}
n(r, t) = \sum_k n_k \Big\{ & |\varphi_k^{(0)}(r, t)|^2 \\
& + \lambda \left\{ \left[\varphi_k^{(0)}(r, t) \right]^* \varphi^{(1)}(r, t) + \left[\varphi_k^{(1)}(r, t) \right]^* \varphi^{(0)}(r, t) \right\} \\
& + \lambda^2 \left\{ \left[\varphi_k^{(2)}(r, t) \right]^* \varphi^{(0)}(r, t) + \left[\varphi_k^{(0)}(r, t) \right]^* \varphi^{(2)}(r, t) + |\varphi_k^{(1)}(r, t)|^2 \right\} + \dots \Big\}
\end{aligned}
$$
$$(7.29)$$

where n_k is the occupation of the kth KS-state.

Each response Hamiltonian $\hat{H}_{KS}^{(k)}[n(r', t)](r, t)$ depends only on the response densities $n^{(j \leq k)}(r, t)$ which are of lower or equal order. For example, the zeroth-order response Hamiltonian is just the ground-state KS-Hamiltonian, which depends only on the ground-state density. The first-order response Hamiltonian

$$\hat{H}_{KS}^{(1)}[n](r, t) = \int d^3 r' f_{Hxc}\left[n^{(0)} \right](r, r') n^{(1)}(r', t) \quad (7.30)$$

has the first-order Hartree-exchange-correlation kernel $f_{Hxc}\left[n^{(0)} \right](r, r')$, which depends on the ground-state density $n^{(0)}(r)$, multiplied by the first-order density response $n^{(1)}(r, t)$.

Now, by equating different orders of λ in the TD-KS equation, we obtain in zeroth order

$$i\frac{\partial}{\partial t}\varphi_k^{(0)}(r, t) = \hat{H}_{KS}^{(0)}[n^{(0)}](r)\varphi_k^{(0)}(r, t), \quad (7.31)$$

in first order,

$$
\begin{aligned}
i\frac{\partial}{\partial t}\varphi_k^{(1)}(r, t) = & \hat{H}_{KS}^{(0)}[n^{(0)}](r)\varphi_k^{(1)}(r, t) \\
& + \left[\hat{H}_{KS}^{(1)}[n](r, t) + v_{ext}^{(1)}(r, t) \right]\varphi_k^{(0)}(r, t),
\end{aligned}
$$
$$(7.32)$$

---

[6] Remember that we are working within the adiabatic approximation here, and therefore, the TD-KS Hamiltonian has no memory.

in second order,

$$
\begin{aligned}
i\frac{\partial}{\partial t}\varphi_k^{(2)}(\boldsymbol{r},t) = {}& \hat{H}_{KS}^{(0)}[n^{(0)}](\boldsymbol{r})\varphi_k^{(2)}(\boldsymbol{r},t) \\
& + \left[\hat{H}_{KS}^{(1)}[n](\boldsymbol{r},t) + v_{ext}^{(1)}(\boldsymbol{r},t)\right]\varphi_k^{(1)}(\boldsymbol{r},t) \\
& + \left[\hat{H}_{KS}^{(2)}[n](\boldsymbol{r},t) + v_{ext}^{(2)}(\boldsymbol{r},t)\right]\varphi_k^{(0)}(\boldsymbol{r},t),
\end{aligned}
\tag{7.33}
$$

and so on. The equations form a hierarchy, where higher-order responses can be calculated from the lower-order ones (Gonze and Vigneron 1989; Gonze 1995).

The above equations still depend on time in a non-trivial way (except the zeroth order which is just the trivial time-propagation of the ground-state). Nevertheless, the only explicit time-dependence is in the time-dependent external potential. If the potential has only one frequency, the linear response will also have only one frequency. If the potential has two frequencies, the linear response has two. However, the second-order response will have frequencies which are sums and differences of the original frequencies. For example, in case of one frequency ω, the second-order response has frequency $\omega - \omega = 0$ and frequency $\omega + \omega = 2\omega$. Furthermore, in the case of two frequencies ω_1 and ω_2, the second-order response has frequencies 0, $2\omega_1$, $2\omega_2$, $\omega_1 + \omega_2$, and $|\omega_1 - \omega_2|$. The 3rd-order mixes three frequencies, and in addition to the frequencies of the field, it can also mix the frequencies generated by the 2nd-order response. Each new order brings new mixed frequencies.

From this point onward, we assume that we have only a single frequency ω in the external field:

$$
v_{ext}^{(1)}(\boldsymbol{r},t) = v_{ext}^{+\omega}(\boldsymbol{r})e^{+i\omega t} + v_{ext}^{-\omega}(\boldsymbol{r})e^{-i\omega t},
\tag{7.34}
$$

or, if we choose to use a cosine field,

$$
v_{ext}^{(1)}(\boldsymbol{r},t) = \frac{1}{2}v_{ext}^{\omega}(\boldsymbol{r})e^{+i\omega t} + \frac{1}{2}v_{ext}^{\omega}(\boldsymbol{r})e^{-i\omega t} = v_{ext}^{\omega}(\boldsymbol{r})\cos(\omega t).
\tag{7.35}
$$

A general first-order wavefunction in this case can be written as

$$
\begin{aligned}
\varphi(\boldsymbol{r},t) = {}& e^{-i\varepsilon^{(0)}t - i\lambda\Delta\varepsilon^{(1)}(t)} \\
& \times \left\{\varphi^{(0)}(\boldsymbol{r}) + \lambda\left[\varphi_{+\omega}^{(1)}(\boldsymbol{r})e^{i\omega t} + \varphi_{-\omega}^{(1)}(\boldsymbol{r})e^{-i\omega t}\right]\right\} + \mathcal{O}(\lambda^2),
\end{aligned}
\tag{7.36}
$$

where $\varphi_\omega(\boldsymbol{r})$ are now time-independent, and we have included a time-dependent level shift

$$
\Delta\varepsilon^{(1)}[n](t) = \int_{-\infty}^{t} dt\,\langle\varphi^{(0)}|\hat{H}_{KS}^{(1)}[n](t) + v_{ext}^{(1)}(t)|\varphi^{(0)}\rangle.
\tag{7.37}
$$

The first-order level shift $\Delta\varepsilon^{(1)}(t)$ is a first order correction to the phase of the zeroth-order wavefunction caused by the first-order Hamiltonian. By including it, we keep the first-order wavefunction $\varphi_{\pm\omega}^{(1)}$ orthogonal to the zeroth-order wavefunction $\varphi^{(0)}$. Otherwise, $\varphi_{\pm\omega}^{(1)}$ would be time-dependent and include a time-dependent $\varphi^{(0)}$ component in order to correct the phase (Langhoff et al. 1972).

From the wavefunction, we obtain the response densities. The zeroth-order density is just the ground-state density

$$n^{(0)}(\mathbf{r}, t) = \sum_k n_k |\varphi_k^{(0)}(\mathbf{r})|^2, \tag{7.38}$$

and the first-order response density

$$
\begin{aligned}
n^{(1)}(\mathbf{r}, t) &= \sum_k n_k \left\{ \left[\varphi_k^{(0)}(\mathbf{r})\right]^* \varphi_{k,+\omega}^{(1)}(\mathbf{r}) e^{i\omega t} + \left[\varphi_k^{(0)}(\mathbf{r})\right]^* \varphi_{k,-\omega}^{(1)}(\mathbf{r}) e^{-i\omega t} \right. \\
&\quad \left. + \left[\varphi_{k,+\omega}^{(1)}(\mathbf{r})\right]^* \varphi_k^{(0)}(\mathbf{r}) e^{-i\omega t} + \left[\varphi_{k,-\omega}^{(1)}(\mathbf{r})\right]^* \varphi_k^{(0)}(\mathbf{r}) e^{i\omega t} \right\} \\
&= \sum_k n_k \left\{ \left[\varphi_k^{(0)}(\mathbf{r})\right]^* \varphi_{k,+\omega}^{(1)}(\mathbf{r}) + \left[\varphi_{k,-\omega}^{(1)}(\mathbf{r})\right]^* \varphi_k^{(0)}(\mathbf{r}) \right\} e^{i\omega t} + \text{cc.} \tag{7.39}
\end{aligned}
$$

is oscillating at the frequency ω as expected.

Next, we insert the guess wavefunction to the TDKS equation and expand it up to first order. On the left-hand-side, we obtain

$$
\begin{aligned}
&i\frac{\partial}{\partial t} \left[\varphi_k^{(0)}(\mathbf{r}) + \lambda\varphi_{k,+\omega}^{(1)}(\mathbf{r}) e^{i\omega t} + \lambda\varphi_{k,-\omega}^{(1)}(\mathbf{r}) e^{-i\omega t}\right] e^{-i\varepsilon_k^{(0)} t - i\lambda\Delta\varepsilon_k^{(1)}(t)} \\
&= e^{-i\varepsilon_k^{(0)} t - i\lambda\Delta\varepsilon_k^{(1)}(t)} \left\{ \left[\varepsilon_k^{(0)} + \lambda\frac{\partial}{\partial t}\Delta\varepsilon_k^{(1)}(t)\right]\varphi_k^{(0)}(\mathbf{r}) + \lambda\left(-\omega + \varepsilon_k^{(0)}\right)\varphi_{k,+\omega}^{(1)}(\mathbf{r}) e^{i\omega t} \right. \\
&\quad \left. + \lambda\left(\omega + \varepsilon_k^{(0)}\right)\varphi_{k,-\omega}^{(1)}(\mathbf{r}) e^{-i\omega t} \right\} + \mathcal{O}(\lambda^2). \tag{7.40}
\end{aligned}
$$

On the right-hand-side, we obtain

$$
\begin{aligned}
&\left\{ \hat{H}_{\text{KS}}^{(0)}\left[n^{(0)}\right](\mathbf{r})\varphi_k^{(0)}(\mathbf{r}) + \lambda\hat{H}_{\text{KS}}^{(0)}\left[n^{(0)}\right](\mathbf{r})\left[\varphi_{k,+\omega}^{(1)}(\mathbf{r}) e^{i\omega t} + \varphi_{k,-\omega}^{(1)}(\mathbf{r}) e^{-i\omega t}\right] \right. \\
&\quad \left. + \lambda\left[\int d^3r' f_{\text{Hxc}}\left[n^{(0)}\right](\mathbf{r}, \mathbf{r}')n^{(1)}(\mathbf{r}', t) + v_{\text{ext}}^{(1)}(\mathbf{r}, t)\right]\varphi_k^{(0)}(\mathbf{r}) \right\} \\
&\quad \times e^{-i\varepsilon_k^{(0)} t - i\lambda\Delta\varepsilon_k^{(1)}(t)} + \mathcal{O}(\lambda^2). \tag{7.41}
\end{aligned}
$$

The first-order equation can now be written in matrix form by gathering terms proportional to the resonant part $e^{i\omega t - i\varepsilon_k^{(0)} t - i\lambda\Delta\varepsilon_k^{(1)}(t)}$ and to the anti-resonant part $e^{-i\omega t - i\varepsilon_k^{(0)} t - i\lambda\Delta\varepsilon_k^{(1)}(t)}$:

$$
\begin{aligned}
&\begin{pmatrix} \hat{H}_{\text{KS}}^{(0)} - \varepsilon_k^{(0)} + \omega & 0 \\ 0 & \hat{H}_{\text{KS}}^{(0)} - \varepsilon_k^{(0)} - \omega \end{pmatrix} \begin{pmatrix} \varphi_{k,+\omega}^{(1)} \\ \varphi_{k,-\omega}^{(1)} \end{pmatrix} \\
&= -\begin{pmatrix} \left(v_{\text{Hxc},+\omega}^{(1)} + v_{\text{ext},+\omega}^{(1)} - \varepsilon_{k,+\omega}^{(1)}\right)\varphi_k^{(0)} \\ \left(v_{\text{Hxc},-\omega}^{(1)} + v_{\text{ext},-\omega}^{(1)} - \varepsilon_{k,-\omega}^{(1)}\right)\varphi_k^{(0)} \end{pmatrix}, \tag{7.42}
\end{aligned}
$$

where

$$v_{\text{Hxc},\pm\omega}^{(1)} e^{\pm i\omega t} = \int d^3 r' f_{\text{Hxc}} \left[n^{(0)} \right] (r, r') n_{\pm\omega}^{(1)}(r', t), \qquad (7.43)$$

$$n_{\pm\omega}^{(1)}(r, t) = \sum_k n_k \left\{ \left[\varphi_k^{(0)}(r) \right]^* \varphi_{k,\pm\omega}^{(1)}(r) + \left[\varphi_{k,\mp\omega}^{(1)}(r) \right]^* \varphi_k^{(0)}(r) \right\} e^{\pm i\omega t}, \qquad (7.44)$$

and $\varepsilon_{k,\pm\omega}^{(1)}$ is the Fourier transform of $\frac{\partial}{\partial t} \Delta \varepsilon_k^{(1)}(t)$:

$$\varepsilon_{k,\pm\omega}^{(1)} = \left\langle \varphi_k^{(0)} | v_{\text{Hxc},\pm\omega}^{(1)} + v_{\text{ext},\pm\omega}^{(1)} | \varphi_k^{(0)} \right\rangle. \qquad (7.45)$$

In this form, the Sternheimer method looks like a set of linear equations, but in reality it is a nonlinear set of equations as the right-hand side depends on the solution through $v_{\text{Hxc},\pm\omega}^{(1)}$ which depends on $n_{\pm\omega}^{(1)}$ and therefore on $\varphi_{k,\pm\omega}^{(1)}$. The usual way is to introduce a self-consistent field (SCF) iteration for the response density $n_{\pm\omega}^{(1)}$, as for the ground-state DFT problem. This is the essence of the Baroni–Gianozzi–Testa method (Baroni et al. 1987a), originally used for static perturbations but equally applicable to TDDFT (Andrade et al. 2007).

By projecting the Sternheimer equation onto the unperturbed wavefunctions, one obtains the sum-over-states expression in second-order perturbation theory for the wavefunction:

$$\varphi_{k,\omega}^{(1)} = \sum_{m \neq k} |\varphi_m^{(0)}\rangle \frac{\langle \varphi_m^{(0)} | \hat{H}_\omega^{(1)} | \varphi_k^{(0)} \rangle}{\varepsilon_m^{(0)} - \varepsilon_k^{(0)} + \omega} \qquad (7.46)$$

where $\hat{H}_\omega^{(1)} = v_{\text{Hxc},\omega}^{(1)} + v_{\text{ext},\omega}^{(1)}$. Using the Sternheimer equation has the great advantage that it avoids the need for explicit calculation of the unoccupied states that would occur in this sum over states.

As with the time-propagation approach, we have to include an artificial lifetime. Otherwise, (i) the matrix is singular when ω corresponds to the KS-eigenvalue difference $\varepsilon_a - \varepsilon_i$ (an excitation in the independent particle picture), or (ii) the response becomes infinite when ω corresponds to a resonance (an excitation in the interacting picture). The former is simply a numerical issue, but the later one has physical meaning and will be used to derive Casida's equation in the next section. The artificial lifetime is introduced by multiplying the first-order wavefunction $\varphi_k^{(1)}(r, t)$ and the external potential $v_{\text{ext}}(r, t)$ by a decay term $e^{-\eta t}$. In the first order, the matrix equation then reads

$$\begin{pmatrix} \hat{H}_{\text{KS}}^{(0)} - \varepsilon_k^{(0)} + \omega + i\eta & 0 \\ 0 & \hat{H}_{\text{KS}}^{(0)} - \varepsilon_k^{(0)} - \omega + i\eta \end{pmatrix} \begin{pmatrix} \varphi_{k,+\omega}^{(1)} \\ \varphi_{k,-\omega}^{(1)} \end{pmatrix}$$
$$= -\hat{P}_c \begin{pmatrix} \left(v_{\text{Hxc},+\omega}^{(1)} + v_{\text{ext},+\omega}^{(1)} \right) \varphi_k^{(0)} \\ \left(v_{\text{Hxc},-\omega}^{(1)} + v_{\text{ext},-\omega}^{(1)} \right) \varphi_k^{(0)} \end{pmatrix} \qquad (7.47)$$

The matrix is no longer singular, and the resonances become Lorentzians with width η instead of delta functions. We also added a projector to the unoccupied space $\hat{P}_c = 1 - \hat{P}_{occ}$, which orthogonalizes the KS response wavefunctions with respect to the occupied KS ground-state wavefunctions. The components of the response wavefunctions in the occupied subspace are not needed because they cancel out in the density response. The projector avoids solving for these (possibly large) components, making the numerical solution more efficient and stable (Baroni et al. 2001; Andrade ety al. 2007). It also simplifies the equation by removing the level shift $\Delta\varepsilon_{\pm\omega}^{(1)}$. Finally, after the self-consistent solution is found, the linear density response is directly available from equation (7.44).

The Sternheimer equation is particularly suited to the calculation of higher-order responses, because solution of only the first-order problem can actually give access to up to third-order derivatives of the total energy (second-order response). In fact, the variational principle can be used to show that the $\{\phi_i^{(n)}\}$, the derivatives of order n of the KS wavefunctions with respect to a perturbation, can be used to construct all derivatives of the total energy up to order $2n + 1$ (this is the famous $2n + 1$ theorem for DFT (Gonze and Vigneron 1989)).

Consider a bare external perturbation $\hat{H}_{bare}^{(n)}$ and a total perturbation $\hat{H}^{(n)}$, including Hartree and exchange-correlation response (the "local fields" (Hybertsen and Louie 1987)). For $n = 0$, this theorem reduces to the familiar Hellman–Feynman Theorem, used in calculation of forces from only ground-state quantities:

$$E^{(1)} = \frac{\partial E}{\partial \lambda} = \sum_i^{occ} \langle \varphi_i | \hat{H}_{bare}^{(1)} | \varphi_i \rangle \tag{7.48}$$

At $n = 1$, the expression for the second derivative (first-order response) is

$$E^{(2)} = \sum_i^{occ} \left[\langle \varphi_i^{(0)} | \hat{H}^{(1)} | \varphi_i^{(1)} \rangle + \text{cc.} + \langle \varphi_i^{(0)} | \hat{H}_{bare}^{(2)} | \varphi_i^{(0)} \rangle \right] \tag{7.49}$$

and for the third derivative (second-order response) are

$$E^{(3)} = \sum_i^{occ} \left[\langle \varphi_i^{(1)} | \hat{H}^{(1)} | \varphi_i^{(1)} \rangle + \langle \varphi_i^{(1)} | \varphi_i^{(1)} \rangle \langle \varphi_i^{(0)} | \hat{H}^{(1)} | \varphi_i^{(0)} \rangle \right.$$
$$\left. + \langle \varphi_i^{(1)} | \hat{H}_{bare}^{(2)} | \varphi_i^{(0)} \rangle + \text{cc.} + \langle \varphi_i^{(0)} | \hat{H}_{bare}^{(3)} | \varphi_i^{(0)} \rangle \right]$$
$$+ \frac{1}{6} \int d^3 r \int d^3 r' \int d^3 r'' \frac{\delta^3 E[n^{(0)}]}{\delta n(r)\delta n(r')\delta n(r'')} n^{(1)}(r) n^{(1)}(r') n^{(1)}(r'') \tag{7.50}$$

where superscripts indicate the order of derivatives with respect to the perturbation (Gonze and Vigneron 1989). The bare perturbation $\hat{H}_{bare}^{(n)}$ is zero for $n > 1$ for electric fields, but not in general. The third functional derivative here is the second-order kernel K_{xc}.

Conveniently, it turns out even in these equations, only the projection of the wavefunction derivatives onto the unoccupied subspace $P_c\ \varphi_i^{(1)}$ are required for this formula (Debernardi and Baroni 1994), making the Sternheimer calculation more efficient. With this formula for $E^{(3)}$, the Sternheimer approach allows efficient access to phonon anharmonicities and nonlinear optical susceptibilities from solution of the first-order Sternheimer equation. This is true even for mixed derivatives with respect to perturbations in different directions or even entirely different perturbations. To get even higher orders, one can use the hierarchy of Sternheimer equations to solve for $\{\varphi_i^{(n)}\}$ from results at lower orders (Gonze and Vigneron 1989; Gonze 1995), with a somewhat more complicated calculation. The $2n + 1$ formulae for the energy derivatives at higher orders are straightforward but increasingly lengthy. For the time-dependent case, rather than total energies we use instead the action, or in the frequency domain, the Floquet quasi-energy (Langhoff et al. 1972).

The $2n + 1$ theorem actually also provides an alternate calculation approach for $\varphi^{(1)}$ (Gonze 1995; Dal Corso et al. 1996; Gonze 1997; Baroni et al. 2001). The formula for $E^{(3)}$ is variational with respect to $\varphi^{(1)}$, just as E is variational with respect to $\varphi^{(0)}$, as stated in the standard variational theorem of quantum mechanics. As a result, one can solve for $\varphi^{(1)}$ by direct minimization of the functional $E^{(3)}[\{\varphi_i^{(1)}\}]$. This approach is formally equivalent to solution by self-consistency, and the choice of technique is a question of numerical strategy.

7.3.3 Casida Method

From the Sternheimer method, we can continue to Casida's equation by writing the linear Sternheimer equation in the particle-hole basis, i.e., in the Kohn–Sham orbital basis including unoccupied states. First, we write the first-order response wavefunction as a linear combination of Kohn–Sham orbitals (i.e., sum-over-states expansion)

$$\varphi_k^{(1)}(\mathbf{r}) = \sum_a c_a^{(k)} \varphi_a^{(0)}(\mathbf{r}), \qquad (7.51)$$

where $c_a^{(k)}$ is the projection of the response of kth occupied state onto the ath unperturbed KS state $\varphi_a^{(0)}$. These coefficients represent excitations from state k to state a. As in Eq. 7.47, we are considering only the projection of $\varphi_k^{(1)}(\mathbf{r})$ into the unoccupied subspace, which will remove the level shift. We insert this linear combination into the first-order equation and multiply it from the left by $\langle \varphi_b^{(0)} |$, i.e., take an inner product with another basis function. The first-order TD-KS equation (7.42) now reads

$$\sum_a \left(-\omega + \varepsilon_k^{(0)}\right) \delta_{a,b} e^{i\omega t - i\varepsilon_k^{(0)} t} c_a^{(k,+\omega)} + \sum_a \left(\omega + \varepsilon_k^{(0)}\right) \delta_{a,b} e^{-i\omega t - i\varepsilon_k^{(0)} t} c_a^{(k,-\omega)}$$

$$= \sum_a \varepsilon_a^{(0)} \delta_{a,b} e^{i\omega t - i\varepsilon_k^{(0)} t} c_a^{(k,+\omega)} + \sum_a \varepsilon_a^{(0)} \delta_{a,b} e^{i\omega t - i\varepsilon_k^{(0)} t} c_a^{(k,-\omega)}$$

$$+ \left\langle \varphi_b^{(0)} | f_{\text{Hxc}}[n^{(0)}] n^{(1)}(t) | \varphi_k^{(0)} \right\rangle e^{-i\varepsilon_k^{(0)} t} + \left\langle \varphi_b^{(0)} | v_{\text{ext}}(t) | \varphi_k^{(0)} \right\rangle e^{-i\varepsilon_k^{(0)} t}. \qquad (7.52)$$

Writing the equations in a basis allows us to remove the nonlinearity caused by the Hartree-exchange-correlation kernel:

$$
\left\langle \varphi_b^{(0)} | \hat{f}_{\mathrm{Hxc}} \left[n^{(0)} \right] n^{(1)}(t) | \varphi_k^{(0)} \right\rangle e^{-i\varepsilon_k^{(0)} t}
$$
$$
= \int \mathrm{d}^3 r \int \mathrm{d}^3 r' f_{\mathrm{Hxc}} \left[n^{(0)} \right] (r, r') \times \sum_{k'} n_{k'} \left\{ \left\{ \left[\varphi_{k'}^{(0)}(r') \right]^* \varphi_{k',+\omega}^{(1)}(r') \right. \right.
$$
$$
+ \left[\varphi_{k',-\omega}^{(1)}(r') \right]^* \varphi_{k'}^{(0)}(r') \right\} e^{i\omega t} + \left\{ \left[\varphi_{k'}^{(0)}(r') \right]^* \varphi_{k',-\omega}^{(1)}(r') \right.
$$
$$
\left. + \left[\varphi_{k',+\omega}^{(1)}(r') \right]^* \varphi_{k'}^{(0)}(r') \right\} e^{-i\omega t} \right\} \varphi_b^{(0)}(r) \varphi_k^{(0)}(r) \tag{7.53}
$$

Inserting the expansion for $\varphi^{(1)}$ in terms of the unoccupied KS eigenfunctions gives

$$
\left\langle \varphi_b^{(0)} | f_{\mathrm{Hxc}} \left[n^{(0)} \right] n^{(1)}(t) | \varphi_k^{(0)} \right\rangle e^{-i\varepsilon_k^{(0)} t}
$$
$$
= \int \mathrm{d}^3 r \int \mathrm{d}^3 r' f_{\mathrm{Hxc}} \left[n^{(0)} \right] (r, r') \times \sum_{k',b'} n_{k'} \left\{ \left\{ \varphi_{k'}^{(0)}(r') c_{b'}^{(k',+\omega)} \varphi_{b'}^{(0)}(r') \right. \right.
$$
$$
+ \left[c_{b'}^{(k',-\omega)} \right]^* \varphi_{b'}^{(0)}(r') \varphi_{k'}^{(0)}(r') \right\} e^{i\omega t} + \left\{ \varphi_{k'}^{(0)}(r') c_{b'}^{(k',-\omega)} \varphi_{b'}^{(0)}(r') \right.
$$
$$
\left. + \left[c_{b'}^{(k',+\omega)} \right]^* \varphi_{b'}^{(0)}(r') \varphi_{k'}^{(0)}(r') \right\} e^{-i\omega t} \right\} \times \varphi_b^{(0)}(r) \varphi_k^{(0)}(r) \tag{7.54}
$$

This finally leads to

$$
\left\langle \varphi_b^{(0)} | f_{\mathrm{Hxc}} \left[n^{(0)} \right] n^{(1)}(t) | \varphi_k^{(0)} \right\rangle e^{-i\varepsilon_k^{(0)} t}
$$
$$
= \sum_{k',b'} n_{k'} K_{bk,b'k'} \times \left\{ \left\{ c_{b'}^{(k',+\omega)} + \left[c_{b'}^{(k',-\omega)} \right]^* \right\} e^{i\omega t} + \left\{ c_{b'}^{(k',-\omega)} + \left[c_{b'}^{(k',+\omega)} \right]^* \right\} e^{-i\omega t} \right\}, \tag{7.55}
$$

where

$$
K_{bk,b'k'} = \int \mathrm{d}^3 r \int \mathrm{d}^3 r' f_{\mathrm{Hxc}} [n^{(0)}](r, r') \varphi_{k'}^{(0)}(r') \varphi_{b'}^{(0)}(r') \varphi_b^{(0)}(r) \varphi_k^{(0)}(r) \tag{7.56}
$$

is the Hartree-exchange-correlation matrix element for interaction of excitations $b \leftarrow k$ and $b' \leftarrow k'$ (k and k' are occupied orbitals, b and b' unoccupied). This term couples independent-particle excitations (KS-eigenvalue differences) to interacting-particle excitations (TD-KS transition frequencies). Remember that we assumed the KS orbitals to be real functions.

The matrix form of the Sternheimer equation in the particle-hole basis reads as

$$
\begin{pmatrix} \Delta E + \omega I & 0 & -\eta & 0 \\ 0 & \Delta E - \omega I & 0 & -\eta \\ \eta & 0 & \Delta E + \omega I & 0 \\ 0 & \eta & 0 & \Delta E - \omega I \end{pmatrix} \begin{pmatrix} C^{(+\omega,re)} \\ C^{(-\omega,re)} \\ C^{(+\omega,im)} \\ C^{(-\omega,im)} \end{pmatrix}
$$
$$
= - \begin{pmatrix} K & K & 0 & 0 \\ K & K & 0 & 0 \\ 0 & 0 & K & -K \\ 0 & 0 & -K & K \end{pmatrix} N \begin{pmatrix} C^{(+\omega,re)} \\ C^{(-\omega,re)} \\ C^{(+\omega,im)} \\ C^{(-\omega,im)} \end{pmatrix} - N^{1/2} \begin{pmatrix} V_{+\omega,re} \\ V_{-\omega,re} \\ V_{+\omega,im} \\ V_{-\omega,im} \end{pmatrix}, \tag{7.57}
$$

where real and imaginary parts have been separated, $\Delta E_{bk,b'k'} = \delta_{k,k'}\delta_{b',b'}(\varepsilon_b - \varepsilon_k)$, $N_{bk,b'k'} = \delta_{k,k'}\delta_{b,b'}n_{k'}$, $V_{bk}^{\pm\omega} = \langle\varphi_b^{(0)}|v_{\text{ext},\pm\omega}|\varphi_k^{(0)}\rangle$ and K is the above Hartree-exchange-correlation kernel matrix. As one can easily see, the nonlinearity has been eliminated, i.e., the above equation is a linear equation if the first term on the right-hand side is moved to the left:

$$\left[\begin{pmatrix} -\Delta E - \mathcal{K} & -\mathcal{K} & \eta & 0 \\ \mathcal{K} & \Delta E + \mathcal{K} & 0 & -\eta \\ -\eta & 0 & -\Delta E - \mathcal{K} & \mathcal{K} \\ 0 & \eta & -\mathcal{K} & \Delta E + \mathcal{K} \end{pmatrix} - \omega I\right]$$
$$\times \begin{pmatrix} B^{(+\omega,re)} \\ B^{(-\omega,re)} \\ B^{(+\omega,im)} \\ B^{(-\omega,im)} \end{pmatrix} = -N \begin{pmatrix} -V_{+\omega,re} \\ V_{-\omega,re} \\ -V_{+\omega,im} \\ V_{-\omega,im} \end{pmatrix}, \tag{7.58}$$

where $\mathcal{K} = N^{1/2}K N^{1/2}$, $B = N^{1/2}C$, and we have already modified it slightly for convenience in the rest of the discussion.

In the limit when the lifetime parameter η goes to zero (i.e., infinite lifetime), the matrix has singularities at certain frequencies. As we included the Hartree-exchange-correlation kernel in the matrix, the response has poles only at the interacting resonance frequencies, and not at the noninteracting resonance frequencies as Eq. 7.47 did. Therefore, we can find the interacting resonance frequencies in the limit $\eta \to 0$ by finding the singularities of the matrix.

The matrix is a 2×2 block-diagonal system in the limit $\eta \to 0$, and the second diagonal block is the transpose of the first one. The blocks have the same eigenvalues, but the right and left eigenvectors of the blocks are swapped. We can focus on the first block and find the eigenvalues of the following equation:

$$\left[\begin{pmatrix} -\Delta E - \mathcal{K} & -\mathcal{K} \\ \mathcal{K} & \Delta E + \mathcal{K} \end{pmatrix} - \omega I\right]\begin{pmatrix} -B^{(+\omega,re)} \\ B^{(-\omega,re)} \end{pmatrix} = 0. \tag{7.59}$$

We apply an unitary transformation $Q = \frac{1}{\sqrt{2}}\begin{pmatrix} 1 & 1 \\ -1 & 1 \end{pmatrix}$, and multiply from left by $\Delta E^{\frac{1}{2}}$ to obtain

$$\left[\begin{pmatrix} 0 & \Delta E \\ \Delta E + 2\Delta E^{\frac{1}{2}}\mathcal{K}\Delta E^{-12} & 0 \end{pmatrix} - \omega I\right]\begin{pmatrix} \Delta E^{\frac{1}{2}}[\ B^{(+\omega,re)} + B^{(-\omega,re)}] \\ \Delta E^{\frac{1}{2}}[-B^{(+\omega,re)} + B^{(-\omega,re)}] \end{pmatrix} = 0. \tag{7.60}$$

The determinant of the matrix inside the square brackets can be easily calculated as ΔE and ωI are diagonal. Setting the determinant equal to zero gives us the eigenvalue equation

$$\Delta E^2 + 2\Delta E^{\frac{1}{2}}\mathcal{K}\Delta E^{\frac{1}{2}} = \omega^2 I \tag{7.61}$$

or, when we expand $\mathcal{K} = N^{\frac{1}{2}}K N^{\frac{1}{2}}$,

$$\Delta E^2 + 2\Delta E^{\frac{1}{2}} N^{\frac{1}{2}} K N^{\frac{1}{2}} \Delta E^{\frac{1}{2}} = \omega^2 I, \tag{7.62}$$

which is the well-known Casida's equation with one small difference: instead of differences of occupation numbers $(n_a - n_i)$, the actual occupation numbers appear. This is a consequence of our assumption of integral occupations. The extension to fractional occupations will be discussed in the next section.

The eigenvectors F of Casida's equation, Eq. 7.62, can be used to extract the strength of the response to the external field. After some algebra, for example, the polarizability can be written as

$$\alpha_{\gamma\delta}(\omega) = \mu_\gamma N^{\frac{1}{2}} \Delta E^{\frac{1}{2}} \sum_k F_k \left(\omega_k^2 - \omega^2\right)^{-1} F_k^\dagger \Delta E^{\frac{1}{2}} N^{\frac{1}{2}} \mu_\delta, \tag{7.63}$$

where μ_δ is the dipole-moment operator in direction δ, with matrix elements in the particle-hole basis $\mu_{\delta,ai} = \langle \varphi_a^{(0)} | r_\delta | \varphi_i^{(0)} \rangle$.

7.3.4 Generalizations and Discussion

In this section, we will discuss what changes if we do not make the assumptions of the beginning of the section. We start with the first assumption—the adiabatic approximation. Without the adiabatic approximation, the exchange-correlation functional has memory, i.e., the exchange-correlation functional depends on density at all previous times. In principle, it is trivial for the time-propagation method. We only have to store all previous densities and calculate the exchange-correlation potential from these. In practice, this is a very demanding task and often impossible beyond model systems.

In the Sternheimer method, memory will show up as a frequency dependence of the exchange-correlation kernels. At first order, the kernel depends only on one frequency, $f_{xc}[n_{GS}](r, r', \omega)$, but at higher orders it depends on multiple frequencies. Again, if explicit forms of the frequency-dependent kernels are known, it is straightforward to include memory (in principle). However, a practical implementation might not be easy and it will depend a lot on the actual form of the memory-dependence in the functionals, which remains an important unresolved theoretical issue (see Chap. 8).

In the case of the Casida method, the matrix becomes frequency-dependent (Casida 1995), which means that the linear eigenvalue problem becomes a nonlinear eigenvalue problem. A nonlinear eigenvalue problem is much harder to solve than a linear eigenvalue problem (e.g., SCF iterations may be required).

If a system is spin-polarized, each spin has its own exchange-correlation potential v_{xc}^α and v_{xc}^β. The exchange-correlation kernel is replaced by three exchange-correlation kernels $f_{xc}^{\alpha\alpha}$, $f_{xc}^{\beta\beta}$, and $f_{xc}^{\alpha\beta}$ (Casida 1995; Guan et al. 2000).

If a system has fractional occupation numbers, an excitation which happens from one partially occupied state i to another partially occupied state j will have an opposite excitation (or de-excitation) from j to i. The expressions from perturbation theory now contain the occupation difference $n_i - n_j$. The original presentation of the Casida method (Casida 1995) shows this general case. The situation for the Sternheimer method is significantly more complicated due to the question of what happens to the projection onto the unoccupied subspace. A computational scheme has been derived to solve the Sternheimer equation when the occupation function corresponds to the thermal Fermi–Dirac distribution or one of the related smearing methods, which is generally needed for reasonable convergence of the ground state of metals with respect to k-point sampling (de Gironcoli 1995). Introduction of some extra projectors into the Sternheimer equation allows the density response

$$n^{(1)}(\boldsymbol{r}) = \sum_{ij} \frac{\tilde{\theta}\left(\varepsilon_\mathrm{F} - \varepsilon_i\right) - \tilde{\theta}\left(\varepsilon_\mathrm{F} - \varepsilon_j\right)}{\varepsilon_i - \varepsilon_j} \varphi_i^*(\boldsymbol{r})\varphi_j(\boldsymbol{r})\langle\varphi_j|\hat{H}_\mathrm{bare}^{(1)}|\varphi_i\rangle \qquad (7.64)$$

to be written in the same form as Eq. 7.44 for the zero-temperature (semiconducting) case in which all states are full or empty, with the addition of terms corresponding to variation of the occupations and Fermi level in general (Baroni et al. 2001).

Finally, if we use k-points, the ground-state KS wavefunctions become complex and we cannot obtain the Casida's equation (7.62). However, we can still obtain a similar eigenvalue equation (Reining et al. 2002).

7.4 Applications of Linear Response

Having reviewed different methods of obtaining response functions, we will now consider the different perturbations that can be studied and how their response functions relate to physical quantities of interest. Electric, magnetic, structural, and $k \cdot p$ perturbations, as well as mixed perturbations, are commonly used to extract both static and dynamic response properties.

7.4.1 Response to Electric Perturbations

We will begin by considering electric perturbations, because they give access to optical properties and account for the vast majority of applications of TDDFT. In molecules, the basic quantity is the polarizability α, defined as the response of the dipole to an electric field $\mathcal{E}$, in the limit of zero applied field:

$$\alpha_{ij}(\omega) = \frac{\partial \mu_{i,\omega}}{\partial \mathcal{E}_{j,\omega}} = -\frac{\partial^2 E}{\partial \mathcal{E}_{i,-\omega}\partial \mathcal{E}_{j,\omega}} \qquad (7.65)$$

where ω denotes the frequency of the electric field. The cross-section for optical absorption (in the dipole approximation) can be calculated from the imaginary part:

$$\sigma_{ij}(\omega) = \frac{4\pi\omega}{c}\Im m\alpha_{ij}(\omega) \qquad (7.66)$$

The static polarizability (which is purely real) is commonly calculated by finite differences of applied field (Vila et al. 2010), and the dynamic polarizability can be computed by time-propagation (Yabana and Bertsch 1996), typically via application of an instantaneous pulsed electric field, which contains all frequencies. A Fourier transformation of the resulting free oscillations of the dipole moments yields the polarizability. The absorption spectrum is most often calculated by the Casida method (Casida 1995; Jamorski et al. 1996), which was designed for this problem; it calculates excited states, and a specific perturbation only enters in the calculation of oscillator strengths. It can be difficult to converge the real part of the dynamic polarizability below the optical gap in this method (Jamorski et al. 1996), in which case it is more efficient to do the calculation via the Sternheimer equation (Andrade et al. 2007). The electric field appears as a term $\mathcal{E}\cdot r$ in the Hamiltonian, so the perturbation used is $\partial H/\partial\mathcal{E} = r$. This is the response of the dipole moment to a homogeneous electric field, which couples to the dipole, so these are called dipole–dipole polarizabilities. Similar methodologies can be used for dipole–quadrupole polarizabilities (response to a field gradient) and other multipoles (Bishop 1990).

For solids, typically the susceptibility χ (polarizability per unit cell) and dielectric function ϵ are used instead of the polarizability, related via

$$\epsilon = 1 + 4\pi\chi = 1 + 4\pi\frac{\alpha}{\mathcal{V}} \qquad (7.67)$$

where $\mathcal{V}$ is the volume of the unit cell. The optical absorption is calculated just as for finite systems. There is a significant complication in applying a uniform electric field to a periodic system, because the operator r is not periodic. There are two ways to solve the problem: the original solution is to consider spatially modulated fields $\mathcal{E}(q) = \mathcal{E}_0 e^{iq\cdot r}$, which are periodic although not necessarily commensurate with the cell periodicity. In this case, one can consider the $q \to 0$ limit to obtain the response to a homogeneous electric field, which is used in TDDFT calculations in the sum-over-states (Hybertsen and Louie 1987; Levine and Allan 1989) and response-function approaches commonly used for crystals (Olevano et al. 1999; Sagmeister and Ambrosch-Draxl 2009). It is important to consider carefully the relation between microscopic/macroscopic and transverse/longitudinal responses in this method (see Chap. 3). Another solution is to use the quantum theory of polarization (Resta 1994; Vanderbilt and Resta 2006) to write the operator as $r = i\frac{\partial}{\partial k}$, which is periodic (Gonze 1997a). The k-point derivatives are obtained by finite differences, or again through perturbation theory. An equivalent approach is to calculate the polarization within a basis of Wannier functions (Dal Corso and Mauri 1994). In either method, we do not study the response of the dipole moment per unit cell, which is not a well-defined quantity, and instead use the polarization. To apply finite homogeneous electric fields in periodic systems, the electric–enthalpy approach can be used, in which a term $-\mu\cdot\mathcal{E}$ is added to the total energy functional to be minimized (Souza et al. 2002).

Armed with the dielectric function or polarizability, one can calculate many interesting properties. Inverting the dielectric matrix yields the loss function $\Im m \epsilon^{-1}(q, \omega)$, which describes the slowing of energetic electrons in a solid and is measured in electron energy-loss spectroscopy experiments (Onida et al. 2002; Marini et al. 2006a). Van der Waals interaction energies can be calculated too: the Hamaker coefficients in the expansion

$$\Delta E(R) = -\sum_{n=6}^{\infty} \frac{C_n}{R^n} \tag{7.68}$$

can be calculated from the Casimir–Polder relation as an integral over polarizabilities evaluated as a function of imaginary frequency. For example, the dominant C_6 term for interaction of molecules A and B is given by

$$C_6^{AB} = \frac{3}{\pi} \int_0^{\infty} du \, \alpha^{(A)}(iu) \, \alpha^{(B)}(iu) \tag{7.69}$$

Higher-order terms involve higher-order multipole polarizabilities. These coefficients have been calculated by TDDFT with molecular polarizabilities from time-propagation and Sternheimer methods, and surface susceptibilities from response functions, to study molecule-molecule (Marques et al. 2007) and molecule-surface interactions (Botti et al. 2008). Scaled interatomic C_6 coefficients from TDDFT can also be used to add Van der Waals interactions into DFT total energies as a post-processing step (Tkatchenko and Scheffler 2009).

Dielectric response can also be considered not for uniform fields but rather for point charges, giving $\epsilon(r, r', \omega)$ from a perturbation $1/|r - r'|$. Time-propagation has been used to study the spatially resolved plasmon response of liquid water (Tavernelli 2006). This form of the dielectric function can also be used as input for many-body perturbation theory via the GW approximation (Hedin and Lundqvist 1969) and Bethe–Salpeter equation. Typically these calculations use the RPA dielectric function, which is equivalent to using only the Hartree response and setting the kernel $f_{\rm xc} = 0$. However, as pointed out in the first practical implementation of this scheme (Hybertsen and Louie 1986), using instead the TDDFT ϵ is an approximate way of including the vertex Γ. This methodology has been used for quasiparticle and optical calculations on organic molecules (Tiago and Chelikowsky 2005). Recently progress has been made in replacing the expensive sums over states with solution of the time-dependent Sternheimer equation (Giustino et al. 2010), which can be done with RPA or including $f_{\rm xc}$.

Response to a related but more exotic perturbation can also be used to parametrize the DFT + U method, in which projectors on atomic-like orbitals are used to emulate Coulomb repulsion and correct the energies of localized d- and f-orbitals in strongly correlated materials (Anisimov et al. 1991). Ab initio values for U can be calculated from the screened response to a localized potential $\alpha_I P_I$, where P_I is an atomic-orbital projector, implemented via finite differences (Cococcioni and Gironcoli 2005).

Response to electric perturbations can be used to calculate nonlinear susceptibilities, describing nonlinear optical properties such as second-harmonic generation, optical rectification, and electrooptical effects (Shen 1984). The hyperpolarizability β of a molecule and second-order susceptibility $\chi^{(2)}$ of a solid are simply the derivatives with respect to field of α and χ, the next order in the Taylor expansion of the dipole moment:

$$\mu_i\,(\omega) = \mu_{i0} + \alpha_{ij}\,(\omega)\,\mathcal{E}_{j,\omega} + \frac{1}{2}\beta_{ijk}\,(\omega = \omega_1 + \omega_2)\,\mathcal{E}_{j,\omega_1}\mathcal{E}_{k,\omega_2} + \dots \quad (7.70)$$

though conventions can differ on what numerical factors may appear in this expansion (Willetts et al. 1992). With the $2n + 1$ theorem, solution of the Sternheimer equation can be used to calculate molecular hyperpolarizabilities (Andrade et al. 2007; Vila et al. 2010) as well as the nonlinear susceptibilites of semiconductors with the quantum theory of polarization (Dal Corso et al. 1996).

Finite differences are often also used to calculate static hyperpolarizabilities (Vila et al. 2010), and time-propagation can be used for dynamic hyperpolarizabilities; however, the advantage of being able to obtain the entire spectrum from a single calculation is lost, and separate calculations must be done for each set of input frequencies (Takimoto et al. 2007). The response-function technique has also recently been developed for $\chi^{(2)}$ in the $q \to 0$ limit, and applied to second-harmonic generation in zincblende semiconductors (Hübener et al. 2010).

7.4.2 Response to Magnetic Perturbations

Magnetic response offers a significant additional challenge compared to electric response because of the fact that the vector potential has to be formulated in a particular choice of gauge, which causes particular complications when localized-orbital bases or non-local pseudopotentials are used. The simplest quantity to consider is the magnetic susceptibility, the analogue of the electric susceptibility. The coupling in the Hamiltonian can be expressed with the vector potential A, field $B = \nabla \times A$, and spin magnetic moment $g\mu_B S$ (where μ_B is the Bohr magneton and S is the spin angular momentum), as

$$\hat{H} = \frac{1}{2}\left(p + \frac{1}{c}A\right)^2 + v + g\mu_B S \cdot B$$
$$= \hat{H}^{(0)} + \frac{1}{2c}(p \cdot A + A \cdot p) + \frac{A^2}{2c^2} + g\mu_B S \cdot B \quad (7.71)$$

The three perturbations are respectively the orbital paramagnetic, orbital diamagnetic, and spin paramagnetic contributions. Within the Coulomb gauge where $\nabla \cdot A = 0$ and p and A commute, the linear coupling to A can also be written in terms of the orbital angular momentum as $\frac{1}{2c}r \times p \cdot B = \frac{1}{2}L \cdot B$. In spin-unpolarized systems, the spin susceptibility is zero, so just the orbital perturbation is needed. The Sternheimer equation has been used with this perturbation to calculate static susceptibilities for boron fullerene molecules (Botti et al. 2009). There is actually the advantage, compared to electric perturbations, that the first-order response of the density is required to be zero if the ground state has time-reversal symmetry, which is the case in the absence of spin-polarization or magnetic fields, so that the magnetic Sternheimer equation does not require self-consistency (Mauri and Louie 1996).

To compute magnetic susceptibilities in solids, we have the same problem as for electric perturbations that the position operator is not periodic, which can similarly be handled either by taking the $q \rightarrow 0$ limit or by the quantum theory of polarization. The $q \rightarrow 0$ approach has been used for susceptibilities in crystals (Mauri and Louie 1996). It has also been used for chemical shifts in nuclear magnetic resonance (NMR) (Mauri et al. 1996; Pickard and Mauri 2001), which are ratios between the external field and the environment-dependent screened field at the position of the nuclei. The g-tensor of electron paramagnetic resonance (EPR), describing the direction-dependent spin susceptibility, has been calculated by a similar approach for radicals and defects, including spin-orbit and hyperfine effects (Pickard and Mauri 2002). The J coupling between nuclear spins in NMR can also be computed by the Sternheimer equation, via the magnetic field induced at one nucleus by the field of another (Joyce et al. 2007). Susceptibilities can also be studied by applying finite magnetic fields, but in the presence of non-local pseudopotentials the coupling in the Hamiltonian generates additional terms beyond those above to satisfy gauge-invariance, as can be handled with the ICL (Ismail-Beigi et al. 2001) or GIPAW methods (Pickard and Mauri 2003). When using pseudopotentials, it is important to note that core susceptibilities may be significant, unlike the electric case; they may be computed from separate atomic calculations (Mauri and Louie 1996), or handled via projector-augmented wave (PAW) methods (Pickard and Mauri 2001). Gauge-invariance for magnetic fields in localized-orbital calculations also requires special attention, and can be handled by the "gauge-including atomic orbitals" or "individual gauge for localized orbitals" methods. The susceptibilities of interest are usually static, and the NMR/EPR properties are treated as static since they are measured at radio frequencies.

To study spin waves in metals, dynamical susceptibilities have been calculated with the Sternheimer equation, where peaks in the spin susceptibility $\chi(q, \omega)$ show the magnon band structure (Savrasov 1998). Spinor wavefunctions are needed to allow spin rotations. Another kind of magnetic response is the spin-triplet optical excitation spectrum, inaccessible by electric perturbations which can only excite singlets. Time-propagation techniques have been used to calculate triplet states by applying an opposite kick to the up and down spins (Oliveira et al. 2008). A dynamic response that combines electric and magnetic response is circular dichroism, also known as optical rotation, in which a chiral molecule responds differently to left and right circularly polarized light. The rotatory strength as a function of frequency can be studied via the (orbital) magnetic moment induced by an electric field; the reverse is possible but more complicated to implement. These properties have been calculated for organic molecules with both time-propagation and Sternheimer approaches (Yabana and Bertsch 1999; Varsano et al. 2009).

7.4.3 Response to Structural Perturbations

There is a rich field of study regarding the response to perturbation of ionic positions and lattice parameters. Since it has been reviewed in great detail (Baroni et al. 2001), and is mostly concerned with static properties, we will consider only briefly most of these quantities and focus on those where TDDFT can be used. Forces on the ions and stresses on the unit cell (the diagonal part of which is the pressure) can be calculated via the Hellman–Feynman theorem, which is routinely done in static DFT for use in structural relaxation. These forces can additionally be used for calculation of dynamical matrices for vibrational properties by means of the frozen-phonon method, in which finite ionic displacements are used. However, only phonons with commensurate wavevectors q can be calculated, and large supercells may be required. Using the Sternheimer equation has the great advantage that dynamical matrices at arbitrary q may be calculated with effort comparable to that for zone-center phonons (Baroni et al. 2001). For displacement of ion α with potential v_α in direction i, the perturbation is $\partial v_\alpha / \partial R_{\alpha i}$.

The dynamical matrix is diagonalized to obtain phonon frequencies and eigenvectors in the harmonic approximation. This information, as function of cell volume, can also be used as input for the "quasi-harmonic approximation" which is used for free energies and other thermodynamic information about solids (Wallace 1972; Born and Huang 1954; Carrier et al. 2007). The phonon group velocities can be computed directly as analytic derivatives from the phonon perturbation calculation as well (Gonze and Vigneron 1989). Going beyond the harmonic approximation, the $2n + 1$ theorem gives access to anharmonic properties from Sternheimer calculations (Baroni et al. 2001). Finite-difference calculations have been used to calculate mechanical anharmonicity and electrical anharmonicity (second-order derivatives of force and polarization with respect to ionic displacement) for ionic contributions to the nonlinear susceptibility (Roman et al. 2006). Anharmonicities are needed for phonon linewidths in crystals, as well as to obtain vibrational frequencies in the presence of strong anharmonicity. Sternheimer phonon calculations also give the induced self-consistent potential, which is used to calculate the electron-phonon matrix elements between electronic states i and j and a phonon of wavevector q and branch ν:

$$ g_{ij\nu}(\mathbf{k}, \mathbf{q}) = \left\langle \varphi_{i\mathbf{k}+\mathbf{q}} \left| \frac{\mathrm{d}\hat{H}}{\mathrm{d}\lambda_{\mathbf{q}\nu}} \right| \varphi_{j\mathbf{k}} \right\rangle \tag{7.72} $$

With Wannier-function-based interpolation schemes, the electron-phonon coupling has been used to calculate the superconducting properties of boron-doped diamond (Giustino et al. 2007) and cuprates (Giustino et al. 2008).

Phonons are generally calculated from static response, an adiabatic approximation which is well justified when the phonon frequency is much less than the electronic band gap. For metals however this condition is not satisfied, and the system may not remain in the electronic ground state during phonon oscillation. Truly dynamical, or non-adiabatic, phonon calculations have been done for doped graphene (Lazzeri and Mauri 2006) and 2D intercalated compounds (Saitta et al. 2008), showing significant corrections in these systems. A TDDFT sum-over-states perturbation expression is used to find the dynamical matrix at a given frequency, and self-consistently iterated until the input and output frequencies coincide.

Besides these lattice-dynamics methods, another method for vibrational calculations is molecular dynamics. The system is evolved in time at finite temperature, and from the ionic trajectories, velocity autocorrelation functions are calculated, giving a power spectrum of vibrations as a function of frequency (Allen and Tildesley 1989). Such calculations can be done by empirical methods or *ab initio* MD, commonly in the DFT-based Car-Parrinello scheme (Car and Parrinello 1985). In systems such as liquids, the harmonic approximation fails completely and MD must be used to study the vibrational modes (Putrino and Parrinello 2002) and infrared spectra (Silvestrelli et al. 1997). Recently a new fast Ehrenfest dynamics method has been developed, in which TDDFT is used to propagate the wavefunctions between timesteps. This allows more efficient calculation of vibrational properties of large systems (Alonso et al. 2008; Andrade et al. 2009); the method will be discussed in detail in Chap. 21. TDDFT has also been used to study coherent excitation of phonons in Si by light, propagating the electronic system in the presence of an oscillating applied field, and analyzing the induced forces (Shinohara et al. 2010).

Bulk moduli and elastic constants can be calculated from the second derivative of the total energy with respect to pressure or shear, with finite dif-ferences (Lam and Cohen 1981) or the Sternheimer equation (Baroni et al. 1987b; Baroni et al. 2001). The strain perturbation involves a stretching of both the unit cell and the wavefunctions, and takes the tensorial form (Nielsen and Martin 1985)

$$T_{ij} = p_i p_j - r_i \frac{\partial v_{ext}}{\partial r_j} \qquad (7.73)$$

It is somewhat complicated to implement since it is not lattice-periodic in this form, but it can also be formulated in a lattice-periodic manner in terms of metric tensors (Hamann et al. 2005). Second-order elastic coefficients and Grüneisen parameters (variation of phonon frequencies with stress) can also be calculated from the $2n + 1$ theorem (Gonze and Vigneron 1989). The chain rule must be used to include ionic as well as electronic contributions.

7.4.4 Mixed Electric and Structural Response to Structural Perturbations

Raman spectroscopy measures vibrational frequencies by the energy gained or lost by a photon, and in the Placzek approximation the intensity of a vibrational peak is proportional to the Raman tensor (Lazzeri and Mauri 2003), the derivative of the polarizability with respect to ionic displacement:

$$\frac{\partial^3 E}{\partial \mathcal{E}_i \partial \mathcal{E}_j \partial \mathcal{R}_{ks}} = \frac{\partial \alpha_{ij}}{\partial \mathcal{R}_{ks}} \tag{7.74}$$

For solids, the susceptibility χ can be used instead. For off-resonant Raman, i.e., when the incident phonon is not resonant with an electronic excitation of the system, the static polarizability is generally used. There are many ways the response to the various perturbations could be calculated. Commonly finite differences are used for ionic response, with dielectric tensor calculated from sum over states (Baroni and Resta 1986b), finite differences (Roman et al. 2006), or the Sternheimer equation (Umari et al. 2001). Anharmonic Raman spectra of ice have been calculated with molecular dynamics by a Fourier transform of the autocorrelation function of the dielectric tensor ϵ_∞.

Purely perturbative methods have also been developed. In an earlier approach applied to silica (Lazzeri and Mauri 2003), the tensor is written as

$$\frac{\partial^3 E}{\partial \mathcal{E}_i \partial \mathcal{E}_j \partial \mathcal{R}_{ks}} = \int d^3 r \frac{\partial^2 \rho}{\partial \mathcal{E}_i \partial \mathcal{E}_j} \frac{d\hat{H}}{d\mathcal{R}_{ks}} \tag{7.75}$$

The first-order perturbations are calculated by the Sternheimer equation, and the second-order electric derivatives of the density matrix ρ are calculated via the second-order derivatives of the wavefunctions from a self-consistent sum-over-states expression. The $2n + 1$ theorem also makes it possible to do the computation from only first-order ionic and electric derivatives (Veithen et al. 2005). To study resonant Raman spectroscopy, dynamic polarizabilities must be used. This has been done in TDDFT with the complex polarization propagator approach to study the variation of the Raman spectrum with excitation energy for organic molecules (Mohammed et al. 2009); this method uses a relaxation toward the ground state in the equations of motion to broaden resonances and prevent divergences.

Another mixed response is Born effective charges, which can be used to calculate LO-TO splitting (Ghosez et al. 1998), infrared spectra (Pasquarello and Car 1997), and molecular dipole moments in liquids (Pasquarello and Resta 2003). They are defined by

$$Z^*_{\alpha i j} = \frac{\partial^2 E}{\partial \mathcal{E}_i \partial R_{\alpha j}} = \frac{\partial \mu_i}{\partial R_{\alpha j}} = \frac{\partial F_{\alpha j}}{\partial \mathcal{E}_i} \tag{7.76}$$

Born charges can be evaluated either as the response of the dipole moment (or polarization) in response to ionic displacement, which is natural in the context of a phonon calculation, or from the forces induced by an electric field, from the Sternheimer equation or finite differences (Gonze and Lee 1997).

A related quantity is the piezoelectric tensor γ, in which ionic displacement is replaced by strain e:

$$\gamma_{ijk}^* = \frac{\partial^2 E}{\partial \mathcal{E}_i \partial e_{jk}} = \frac{\partial \mu_i}{\partial e_{jk}} = \frac{\partial T_{jk}}{\partial \mathcal{E}_i} \tag{7.77}$$

The strain perturbation is not lattice-periodic, so piezoelectric tensors are most easily calculated by the stress T induced by an electric field (de Gironcoli et al. 1989). Both electronic and ionic contributions must be included.

A quite different quantity is the non-adiabatic coupling, which is used in molecular dynamics to govern the rate of hopping between the potential energy surfaces of the ground and excited states (Tully 1990). Going beyond the Born–Oppenheimer approximation, terms appear in the equation of motion containing $\langle \Phi_i | \partial/\partial R_{\alpha j} | \Phi_j \rangle$ (first-order coupling) and $\langle \Phi_i | \partial^2/\partial R_{\alpha j}^2 | \Phi_j \rangle$ (second-order coupling) (Hirai and Sugino 2009), with overlaps between many-body states i, j and their derivatives with respect to ionic displacement. The Casida method can be used for first-order non-adiabatic couplings, analogously to the calculation of oscillator strengths but where the dipole operator is replaced with the ionic perturbation (Hu et al. 2007). A time-propagation formula-tion has also been developed (Baer 2002) for the calculation. The second-order couplings cannot be calculated by these methods, but are negligible in simple cases (Hirai and Sugino 2009).

7.4.5 Response to $k \cdot p$ Perturbations

Response to an infinitesimal shift of k-point in a solid, often referred to as $k \cdot p$ perturbation theory, can be used to give various properties. These are by necessity static, not dynamic. Since the perturbation is applied to an individual state rather than to the whole system, it does not have an associated density response. With the Hellman–Feynman theorem, band velocities can be calculated as

$$v_{ik} = \frac{1}{\hbar} \frac{\partial \varepsilon_{ik}}{\partial k} = \frac{1}{\hbar} \left\langle u_{ik} | \frac{\partial H_k}{\partial k} | u_{ik} \right\rangle \tag{7.78}$$

where u_{ik} is the periodic part of the Bloch function and H_k is the effective Hamiltonian. The perturbation is

$$\frac{\partial H_k}{\partial k} = -i\nabla_k + k + \left[v_{pseudo}, r \right] \tag{7.79}$$

including a contribution from non-local pseudopotentials if they are used (Rohlfing and Louie 2000). Second-order perturbation theory with a sum over states can similarly give effective masses (Cardona and Pollak 1966; Yu and Cardona 1999), as frequently used in simple models of band structures and transport in semiconductors. $k \cdot p$ perturbation theory has also been used, in a finite-difference framework, for $q \to 0$ limits in GW (Hybertsen and Louie 1986) and Bethe–Salpeter (Rohlfing and Louie 2000) calculations. Additionally, $k \cdot p$ perturbations can be used to compute the $\partial/\partial k$ derivatives which are used in response calculations with the quantum theory of polarization (Olevano et al. 1999; Vanderbilt and Resta 2006). It is important, however, to note that perturbation theory cannot be used to compute the polarization itself, because it does not represent a consistent choice of gauge throughout the Brillouin zone (Resta 1994).

Chapter 8
Memory: History, Initial-State Dependence, and Double-Excitations

Neepa T. Maitra

8.1 Introduction

In ground-state DFT, the fact that the xc potential is a functional of the density is a direct consequence of the one-to-one mapping between ground-state densities and potentials. In TDDFT, the one-to-one mapping is between densities and potentials for a given initial state. This means that the potentials, most generally, are functionals of the initial state of the system, as well as of the density; and, not just of the instantaneous density, but of its entire history. These dependences are explicitly displayed in Eq. 4.28. Of particular interest is the xc potential, as that is the quantity that must be approximated. The Hartree potential has no memory, as the classical Coulomb interaction depends on the instantaneous density only, but since both the interacting and non-interacting mappings can depend on the initial state, the xc potential must be a functional of both the initial states and the density.

We use the term memory to refer to the dependence on quantities at earlier times: initial-state dependence and history-dependence of the density.

In a sense, memory arises because of the reduced nature of the density as a basic variable: if the wave function of the system was known, there would be no memory-dependence, since the wavefunction at time t contains the complete information about the system at time t, from which we can determine any observable. The density however traces out much of the information, desirably reducing the description involving $3N$ spatial variables plus time to a description using three variables plus time. Analogously to the theories of open systems, from this tracing out of degrees of freedom emerges memory dependence. In treating open systems bath degrees of freedom are traced out to get a reduced description in terms of system variables only (see Chaps. 10 and 11): the effect of the bath is embodied in an influence functional

Neepa T. Maitra (✉)
Hunter College and the Graduate Center,
City University of New York,
695 Park Avenue, New York, NY 10065, USA
e-mail: nmaitra@hunter.cuny.edu

M. A. L. Marques et al. (eds.), *Fundamentals of Time-Dependent Density Functional Theory*, Lecture Notes in Physics 837, DOI: 10.1007/978-3-642-23518-4_8, © Springer-Verlag Berlin Heidelberg 2012

that is non-local in time. Much like in open system theory with a low-dimensional bath, the TDDFT memory of early history persists at long times: time does not wash it away (as it would if we were tracing out a bath of a continuous spectrum).

In linear response, e.g. calculating spectra, (Chaps. 4 and 7) the system starts in its ground-state, which, by virtue of the Hohenberg–Kohn theorem, is itself a functional of its own density, assuming it is non-degenerate. Initial-state dependence is not explicitly needed in the functionals, provided that the functional space is reduced to that where the initial state is a non-degenerate ground state. The exact xc kernel has history-dependence, which translates into non-trivial frequency-dependence when a time-frequency Fourier transform is done.

In an adiabatic approximation, as discussed in Chap. 4, memory-dependence is completely neglected: the instantaneous density is input as a "ground-state" density into a ground-state xc potential approximation. In fact, even before Runge and Gross (RG) formally established their theory, adiabatic calculations of optical spectra were performed, that plugged the instantaneous density into the LDA (Ando 1977a, b; Zangwill and Soven 1980a, 1981); this is the ALDA, Eq. 4.100. The xc kernel in an adiabatic approximation is proportional to a $\delta(t-t')$ (Eq. 4.86a), which, upon Fourier transforming, yields a frequency-independent xc kernel. Since the inception of the RG theorem, there have been attempts to develop functionals with some memory dependence, with varying degrees of success and applicability. The earliest and simplest is the Gross–Kohn approximation (GK) for the xc kernel (Gross and Kohn 1985, 1990). Considering densities that are slowly varying in space, GK bootstraps the local density approximation to finite frequencies, i.e. the frequency-dependent kernel is approximated via the homogeneous electron gas response at finite frequency (Eq. 4.107), a spatially-local but time-nonlocal kernel. In the mid-nineties, it was realized, however, that a theory that depends on the density non-locally in time must also depend on it non-locally in space; otherwise, exact conditions, importantly the harmonic potential theorem, are violated (Vignale 1995a; Dobson 1994a) (see Chap. 24).

The idea that memory is locally carried by the electron "fluid", in a Lagrangian framework, was exploited by Dobson et al. (1997), who essentially applied the GK approximation in a frame moving along with the local velocity of the electron fluid. At about the same time, Vignale and Kohn showed that a theory local in space and non-local in time is possible instead in terms of the current density (Vignale and Kohn 1996; Vignale et al. 1997). Their functional has begun to be tested on a variety of systems with mixed successes (see Chap. 24). A fully spatially- and time-nonlocal hydrodynamic formulation using Landau Fermi-liquid theory was presented in 2003 (Tokatly and Pankratov 2003—see Chap. 25). Tokatly (2005a, b, 2007) further developed this, considering many-body dynamics in the co-moving Lagrangian frame, leading to time-dependent deformation functional theory. In this frame, xc is spatially local, and all complications, including memory, are contained in Green's deformation tensor characterizing the frame. A theory based on a Galilean-invariant "memory action functional" has also been formulated (Kurzweil and Baer 2004). Noting that functionals of the instantaneous KS orbitals incorporate infinite "KS memory" leads to another approach, e.g. time-dependent EXX displays memory effects near intersubband resonances in semiconductor quantum wells (Wijewardane and Ullrich 2008). We note none of the functionals proposed so far incorporate initial-state dependence, although orbital-dependent functionals do have memory of the KS initial state.

Today, however, almost all applications of TDDFT utilize an adiabatic approximation, absolutely memory-less. Certainly, an adiabatic approximation will work well if the system is slowly-varying enough that the system remains in a slowly-evolving ground-state, but this is hardly the typical case in dynamics.

Most of the rest of this chapter investigates memory properties of the *exact* functional, with general real-time dynamics and strong external fields in mind (Some specific phenomena are discussed in Chap. 18). Cases where exact results are available indicate that memory-dependence can play a vital role. Understanding how the exact functional behaves should prove a useful tool in constructing accurate approximations. We discuss history-dependence in the next section, followed by a section on initial-state dependence. We then show how these two sources of memory-dependence are entangled, and discuss an exact condition relating the two in Sect. 8.4. Implications of memory-dependence for quantum control type problems are then discussed. In the last section, we turn to the double-excitation problem in linear response, and discuss the frequency-dependent kernel that captures them. As in Chap. 4, atomic units are used throughout this chapter.

8.2 History Dependence: an Example

Consider a system in its ground state, assumed to be non-degenerate. As discussed in the introduction, we may then put aside initial-state dependence, and ask how far back does the system remember its past? How far back in time do observables at the present depend on the density in the past?

A useful tool to study this question is a time-dependent problem with at least two electrons, for which both the KS system and the interacting system are exactly, or exactly numerically, solvable. Two electrons in a Mathieu oscillator provides a good case (Hessler et al. 2002); the external potential has the form:

$$v_{ext}(\boldsymbol{r}, t) = \frac{1}{2}k(t)r^2, \quad \text{with } k(t) = \bar{k} - \epsilon \cos(\omega t), \tag{8.1}$$

with $\bar{k}$, ϵ, and ω appropriately chosen constants. The static version is often called the Hooke's atom; a paradigm for studies of exchange and correlation in the ground state (Taut 1993; Frydel et al. 2000), largely because, for some parameters, the interacting problem can be solved analytically. For the exact interacting solution of the time-dependent problem, transforming to center-of-mass and relative coordinates renders the Hamiltonian separable. Due to the spherical symmetry, one needs only to solve numerically two uncoupled one-dimensional time-dependent Schrödinger equations. From the evolving wavefunction, beginning in the ground state, the exact evolving density is obtained. Now, the KS wavefunction involves just one doubly-occupied spatial orbital, evolving in time. By requiring its density to yield half the density of the interacting wavefunction for all time, one can invert the KS equation to obtain the KS potential in terms of the evolving density (Hessler et al. 2002).

The exact interacting dynamics has the useful property that the evolving density breathes in and out while retaining the same (near Gaussian) profile: at each time t, it is essentially the density of a *ground state* of a certain Hooke's atom of spring constant k_{eff}. This spring constant is not equal to the actual spring constant in Eq. 8.1 at time t, except when the latter is modulated slowly enough such that the state remains an instantaneous ground state. In the general case, the state is *not* a ground state, but, at each instant in time, its *density* is that of a ground state of a Hooke's atom of spring constant $k_{\text{eff}}(t)$. This property allows us to compare the exact calculation with that of an exact adiabatic one in a relatively simple way.

Many interesting phenomena arise (Hessler et al. 2002); one typically finds significant differences between the adiabatic approximation and the exact KS case (except for very slow modulations). For example, the instantaneous correlation energy can become positive, which is impossible in any adiabatic approximation, since for ground states E_c is tied down below zero by the variational principle.

We now show that the correlation potential displays severe non-locality in time due to history dependence. It is convenient to define a type of density-weighted correlation potential via (Hessler et al. 1999)

$$\dot{E}_c(t) = \int d^3 r v_c(\mathbf{r}, t) \, \dot{n}(\mathbf{r}, t), \tag{8.2}$$

where the dot represents a time derivative. If $\dot{E}_c(t)$ depends not just on the density at and near time t, but also on its earlier history, then $v_c(t)$ must too. That is, non-locality in $\dot{E}_c$ directly implies non-locality in the correlation potential $v_c(t)$. The top panel of Fig. 8.1 plots the value of $k_{\text{eff}}(t)$, which, as discussed earlier, completely identifies the density profile. The density profiles within a time slice centered near $t = 4.8$ and one centered near $t = 28.9$ are almost the same, yet the values of $\dot{E}_c(t)$ near those times are significantly different. Other pairs of time-slices having this feature may also be found. The density at times near t is not enough to specify $v_c(t)$: in fact the exact correlation potential $v_c(t)$ is a highly non-local-in-time functional of the density, depending on its entire history. Any adiabatic approximation has no history dependence and fails to capture this effect.

This example, together with other studies (Ullrich 2006b; Wijewardane and Ullrich 2008) of dynamics in strong-fields (starting in a ground-state) suggest the exact functional typically has strong memory-dependence. However, not always: in strong-field double-ionization, for example, the xc potential appears not to be significantly non-local in time in a wide range of cases, although this depends on how the field is ramped on (Thiele et al. 2008). Likewise when a very high-frequency intense field is turned on very very slowly (Baer 2009).

Fig. 8.1 Non-locality in time: the *top panel* shows $k_{\text{eff}}(t)$, *middle panel* $\dot{E}_c$, *bottom panel* E_c. The parameters in Eq. 8.1 are $\bar{k} = 0.25$, $\omega = 0.75$ and $\epsilon = 0.1$

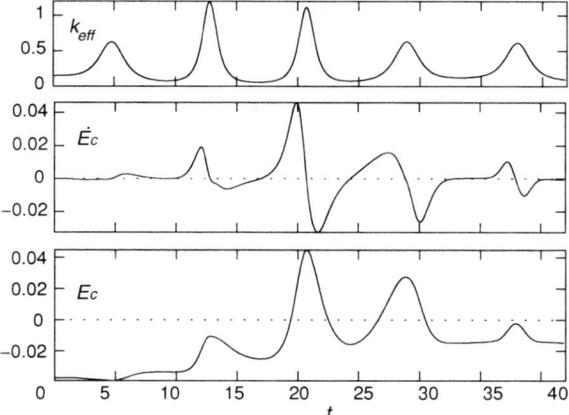

The instantaneous momentum-density in the Mathieu oscillator distinguishes time-slices where the instantaneous density is the same (Rajam et al. 2009). This suggests that memory-dependence is likely gentler in a theory that uses a joint position-momentum density, or density-matrix, as basic variable.

8.3 Initial-State Dependence

For a given time-dependent density how does the potential that yields this density depend on the choice of the initial wavefunction? Initial-state dependence has only begun to be explored (Maitra and Burke 2001, 2002b; Holas and Balawender 2002); unlike density-dependence, there is no precedent for initial-state dependence in ground-state DFT. For example, there is no analogue of the adiabatic approximation that could be used as a starting point for investigations.

One may wonder whether initial-state dependence actually exists. That is, if we constrain the density to evolve in a certain way, are the implicit constraints on the initial state enough to completely determine it? If this were the case, then there would be no initial-state dependence: knowing the history of the density would be enough to determine the functionals. We shall argue shortly that this is in fact the case for one electron, but not for more than one.

Let us first rephrase the question: consider a many-electron density $n(\boldsymbol{r}, t)$ evolving in time under an external time-dependent potential $v_{\text{ext}}(\boldsymbol{r}, t)$. Can we obtain the same density evolution by propagating a different initial state in a different potential?

One electron case. Consider one electron, evolving with density $n(\boldsymbol{r}, t)$. Let the electron's wavefunction be $\varphi(\boldsymbol{r}, t)$, where $n(\boldsymbol{r}, t) = |\varphi(\boldsymbol{r}, t)|^2$. An alternate candidate wavefunction $\tilde{\varphi}(\boldsymbol{r}, t)$ that evolves with identical density (in a different potential) must then be related to $\varphi(r, t)$ by a (real) phase $\alpha(r, t)$:

$$\tilde{\varphi}(r, t) = \varphi(r, t)e^{i\alpha(r,t)}. \tag{8.3}$$

The wavefunction at time t determines not just the density at time t but also its first time-derivative through the continuity equation:

$$\dot{n}(r, t) = -\nabla \cdot j(r, t), \tag{8.4}$$

where the current-density $j(r, t)$ is determined from:

$$j(r, t) = \frac{i}{2} \left[\varphi(r, t)\nabla\varphi^*(r, t) - \varphi^*(r, t)\nabla\varphi(r, t) \right]. \tag{8.5}$$

Because they evolve with the same density at all times, both $\varphi(r, t)$ and $\tilde{\varphi}(r, t)$ share the same $\dot{n}(r, t)$. From the continuity equation it follows that they have identical longitudinal currents, so:

$$0 = \dot{n}_\varphi(r, t) - \dot{n}_{\tilde{\varphi}}(r, t) = \nabla \cdot [n(r, t)\nabla\alpha(r, t)], \tag{8.6}$$

where on the right-hand side we have inserted the difference in the currents of $\tilde{\varphi}$ and φ, calculated using Eq. 8.5. Now if we multiply Eq. 8.6 by $\alpha(r, t)$ and integrate over all space, we obtain

$$0 = \int d^3r\alpha(r, t)\nabla \cdot [n(r, t)\nabla\alpha(r, t)] = -\int d^3rn(r, t)|\nabla\alpha(r, t)|^2 \tag{8.7}$$

In the last step, we integrated by parts, taking the surface term $\int_S d\sigma e_n \cdot (\alpha n \nabla\alpha)$, evaluated on a closed surface at infinity, to be zero. This will be true for any finite system, where the density decays at infinity, while the potential remains finite (or, if the potential grows, the density decays still faster).

Because the integrand in Eq. 8.7 cannot be negative anywhere, yet it integrates to zero, the integrand must be identically zero. Thus $\nabla\alpha(r, t) = 0$ everywhere. This is true even at nodes of the wavefunction, where $n(r_0, t) = 0$: if $\nabla\alpha$ was zero everywhere except at the nodes, then as a distribution it is equivalent to being zero everywhere, unless it was a delta-function at the node—but in that case the potential would be highly singular, and therefore unphysical. So, for physical potentials, $\alpha(r, t)$ must be constant in space, i.e. the wavefunctions $\varphi(r, t)$ and $\tilde{\varphi}(r, t)$ differ only by an irrelevant time-dependent phase. Thus, only one initial state (and one potential) can give rise to a particular density: the evolving density is enough to completely determine the potential and the initial states.

The vanishing of the surface term in Eq. 8.7 can be compared with the requirement on the potential in the Runge–Gross theorem, as discussed in Sect. 4.4.1. In Maitra and Burke (2001), an example of a pathological initial state is given, where the surface term does not vanish, even though the density decays exponentially at large distances: the potential in which it lives plummets to minus infinity at large distances, yielding wildly oscillatory behavior in the tails of the decaying wavefunction, embodying infinite kinetic energy and momentum. Such unphysical states are beyond consideration!

Many electrons. For many electrons, initial-state dependence is real and alive: one can find two or more different initial-states which evolve with identical density for all time in different external potentials.

A few simple examples of this are shown in Figs. 8.2 and 8.3. In Fig. 8.2, the density (thick solid line) of two non-interacting electrons in one dimension in an eigenstate (thin solid line) of the harmonic potential is considered (Maitra and Burke 2001; Holas and Balawender 2002). The two orbitals are the thin solid lines. If we keep this potential constant, the density will remain constant. We then ask, can we find another potential in which another non-interacting wavefunction evolves with this same, constant density for all time? There are in fact an infinite number of them, and one is shown here (dashed lines). The alternate potential was constructed using van Leeuwen's prescription (van Leeuwen 1999—see also Chap. 9), and is shown here at the initial time. It is not constant in time: both the alternate potential and the alternate orbitals evolve in time, in such a way as to keep the density constant at all times.

The significance of this for TDDFT comes to light when we imagine the density as the density of some *interacting* electronic system. For a KS calculation, we are free to choose any initial KS state which has this initial density: that is, both the potentials shown in the lower panel of Fig. 8.2, along with their respective orbitals, are fair game. The difference between these two KS potentials is exactly the difference in the xc potential, since the Hartree and external potentials are the same. So depending on this choice, the xc potentials are very different. Any functional without initial-state dependence would predict the same potential in both cases.

Figure 8.3 is another example of two different initial states that evolve with the same density for all time. This example, again of two non-interacting electrons, demonstrates that there is no one-to-one mapping between time-periodic densities of Floquet states and time-periodic potentials (see Maitra and Burke 2002a). Consider a periodically driven harmonic oscillator, containing two non-interacting electrons in a spin-singlet occupying two distinct quasi-energy orbitals. One can show that the density then periodically sloshes back and forth in the well. This is illustrated in the top panels of Fig. 8.3. The middle panel of Fig. 8.3 shows a doubly-occupied Floquet orbital (real and imaginary parts are the dashed lines) whose density (solid line) evolves identically to the density of the Floquet state in the top panel. This orbital sloshes back and forth in its potential, in a similar way to the orbitals of the driven oscillator. The lowest panel shows the potentials: the solid is the periodically driven harmonic potential corresponding to the Floquet state of the top panel, and the dashed is the periodically driven potential corresponding to that of the middle panel. Now, assuming there corresponds an interacting electron system whose density evolves exactly as shown, then both the Floquet state in the top panel and the middle panel

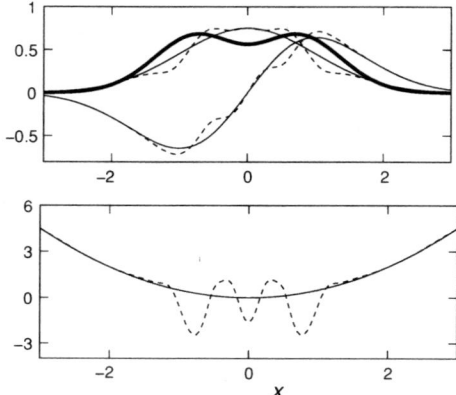

Fig. 8.2 An example of initial-state dependence for two non-interacting electrons: two different wavefunctions may evolve with the same density in different potentials. In the *top plot*, the *solid lines* are the two occupied orbitals of one wavefunction, which happens to be a stationary state of the harmonic oscillator potential, shown in the lower plot as a *solid line*. The density is shown as the *thick solid line* in the *top* figure. The *dashed lines* are the two orbitals of an alternative initial wavefunction, that evolves with the same density in the potential which, at the initial time, is shown as the *dashed line* in the *lower* figure

are possible KS wavefunctions, and both the solid and dashed potentials in the lower panels are possible KS potentials; again, their difference (the sloshing "bump" in the figure) is the difference in the xc potential.

Not only any adiabatic approximation, but any density-functional approximation that lacks initial-state dependence—even with history-dependence—would incorrectly predict the same potential for all choices of KS initial states that propagate with the same density. In the next section, we discuss how, in many cases, one can eliminate the need for initial-state dependence altogether, by transforming it into a history-dependence.

8.4 Memory: an Exact Condition

Part of what makes the memory dependence complex, is the intricate entanglement of initial state and history effects. This has consequences even for initial ground states. On the other hand, it allows the possibility for memory-dependence to be reduced to history-dependence alone.

Consider an interacting system, beginning with wavefunction $\Psi(0)$ at time 0, and evolving in time, with density $n(r, t)$. The xc potential at time t is determined by the density at all previous times, the initial interacting wavefunction, and the choice of the initial KS wavefunction $\Phi(0)$ for the KS calculation. Now say we can calculate the interacting wavefunction at a later time t', where $0 < t' < t$. Then, we may think of t', as the "initial" time for the inputs into the functional arguments of v_{xc}: that is,

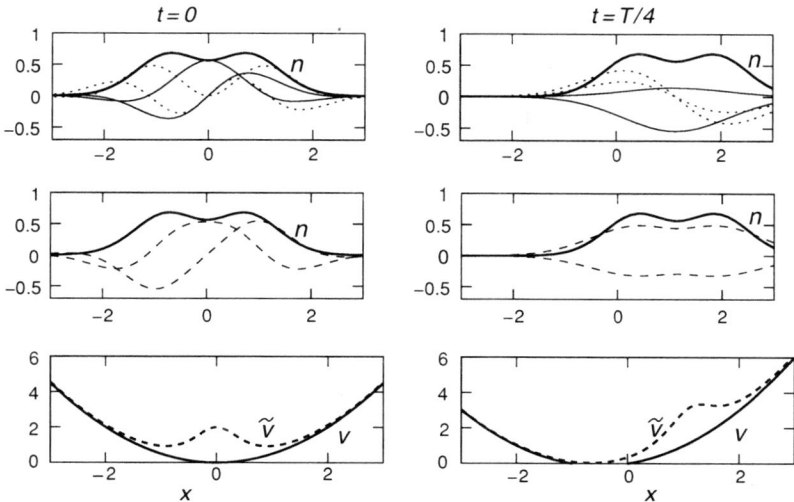

Fig. 8.3 *Top left panel*: the real and imaginary parts of the driven harmonic oscillator Floquet orbitals $\varphi^{(0)}(x, 0)$ (*solid*) and $\varphi^{(1)}(x, 0)$ (*dotted*) at time = 0, together with their density (*thick line*). *Middle left panel*: the real and imaginary parts of the alternative doubly-occupied Floquet orbital $\tilde{\varphi}(x, 0)$ (*dashed*), which has the same density shown (*thick line*). *Bottom left panel*: the two potentials, v is the *solid*, and $\tilde{v}$ is *dashed*. The right hand side shows the same quantities at $t = T/4$

$$v_{\mathrm{xc}} \left[n_{t'}, \Psi(t'), \Phi(t') \right] (r, t) = v_{\mathrm{xc}} \left[n, \Psi(0), \Phi(0) \right] (r, t) \quad \text{for} \quad t > t'. \quad (8.8)$$

Here, $\Psi(t') = \hat{U}(t')\Psi(0)$, where $\hat{U}(t)$ is the unitary evolution operator, and $\Phi(t') = \hat{U}_{\mathrm{KS}}(t')\Phi(0)$ where $\hat{U}_{\mathrm{KS}}(t')$ is the KS evolution operator. The subscript on the density means that the density is undefined for times earlier than the subscript, and it equals the evolving density $n(r, t)$ for times t greater than the subscript.

Equation 8.8 displays the relation between the memory effects: any dependence of the xc potential on the density at prior times may be transformed into an initial-state dependence and vice versa.

Like other exact conditions (discussed in Chap. 5), Eq. 8.8 may be used as a test for approximate functionals, but it is a very difficult condition to satisfy. For example, any of the recent attempts to include history-dependence, while ignoring initial-state dependence, must fail. If we restrict their application to systems beginning in the ground state, then Eq. 8.8 still produces a strict test of such functionals: imagine an exact time-dependent calculation beginning in the ground state of some system. Later, when $\Psi(t')$ is no longer a ground state, we evolve *backwards* in time in a different external potential, that leads us back to a different ground state at a different initial time. The history during the time before t' is different from the original history, but the xc potential for all times greater than t' should be the same for both the original evolution and the evolution along the alternative path. The extent to which these two differ is a measure of the error in a given history-dependent approximation, even applied only to initial ground states. Note that any adiabatic approximation ignoring initial-state dependence (such as those in the introduction) produces no difference. By ignoring both history dependence *and* initial-state dependence, the ALDA trivially satisfies Eq. 8.8.

A technical note: although the RG theorem was proven only for time-analytic potentials, i.e. those that equal their Taylor series expansions in t about the initial time for a finite time interval (Runge and Gross 1984), it holds also for piecewise analytic potentials, i.e. potentials analytic in each of a finite number of intervals (Maitra and Burke 2002b). This means that alternative allowed "pseudo-prehistories" can connect to the same wavefunction at some later time.

This raises the possibility of eliminating the initial-state dependence altogether: if we can evolve an initial interacting wavefunction that is not a ground state, backwards in time to a non-degenerate ground state, then the initial-state dependence may be completely absorbed into a history-dependence along this pseudo-prehistory.

As discussed in Chap. 4, one may choose any initial KS state that reproduces the density and divergence of the current of the interacting initial state (van Leeuwen 1999). In the procedure above, this choice is translated into the choice of which ground state the interacting wavefunction $\Psi(0)$ evolves back to, together with the pseudo-prehistory of the density thus generated. One can imagine that for a given wavefunction $\Psi(0)$ there may be many paths which evolve back to some ground state, each path generating a different pseudo-prehistory. Only for those which result in the same KS wavefunction $\Phi(0)$ [and of course interacting wavefunction $\Psi(0)$] will the xc potentials be identical after time 0.

In the linear response regime, the memory formula Eq. 8.8 yields an exact condition relating the xc kernel to initial-state variations (Maitra 2005a). We consider applying Eq. 8.8 in the perturbative regime, with the initial states at time 0 (on the right-hand-side) being ground-states. The initial states on the left-hand-side (i.e. the states at t') are not ground-states. We wish to express deviations from the ground-state values through functional derivatives with respect to the density and with respect to the initial states. Because the initial state determines the initial density and its first time-derivative, and puts constraints on higher-order time-derivatives of the density, the definition of a partial derivative with respect to the initial state is not trivial: what should be held fixed in the variation? The partial derivative with respect to the density, holding the initial state fixed, is simpler; for example, for the external potential this is a generalized inverse susceptibility, generalized to initial states which are not ground-states. Variations of the density at times greater than zero are included. In order to define an initial-state derivative, one considers an extension of the functionals to a higher space in which the initial-state variable and density variable are independent: one drops and Eqs. 4.6 and 4.20. Then one can show that

$$\sum_{\alpha} \int d^3 r_1 \left. \frac{\delta v_{KS}[n_{t'}, \Phi_{t'}](r, t)}{\delta \varphi_{t', \alpha}(r_1)} \right|_{(n_{GS}, \Phi_{GS}[n_{GS}])} \delta \varphi_{t', \alpha}(r_1)$$

$$- \int d^4 x_1 \cdots \int d^4 x_N \left. \frac{\delta v_{\text{ext}}[n_{t'}, \Psi_{t'}](r, t)}{\delta \Psi_{t'}(x_1, \ldots, x_N)} \right|_{(n_{GS}, \Psi_{GS}[n_{GS}])} \delta \Psi_{t'}(x_1, \ldots, x_N) + \text{c.c.}$$

$$= \int d^3 r_1 \int_0^{t'} dt_1 \ f_{xc}[n_{GS}](r, r_1, t - t_1) \delta n(r_1, t_1), \quad 0 < t' < t, \tag{8.9}$$

where the variables $x_i = (r_i, \sigma_i)$ represent spatial and spin coordinates, $\delta \varphi_{t', \alpha} = \delta \varphi_\alpha(t') = \varphi_\alpha(t') - \varphi_{\alpha, GS}[n_{GS}]$ represent the deviations at time t of the spin orbitals of the KS Slater determinant away from the ground-state values and $\delta \Psi$ is similarly the deviation of the interacting state away from its ground-state. This equation demonstrates the entanglement of initial-state dependence and history-dependence in the linear response regime: the expression for the xc kernel on the right is entirely expressed in terms of initial-state dependence on the left.

8.5 Memory in Quantum Control Phenomena

In recent years there have been huge advances in the control of chemical reactions, where nuclei are manipulated. The development of attosecond laser pulses opens the door to the possibility of manipulating electronic processes as well. Chapter 13 derives the equations for quantum optimal control theory within the KS framework. Here, instead, we present a couple of "gedanken" experiments to explore how the exact KS picture of the controlled dynamics looks compared to the exact true dynamics.

Let us say we are interested in driving a molecule from its ground state Ψ_{GS} in potential $v_{\text{ext}, GS}$ to its mth excited state Ψ_m. Let us say we are lucky enough to know the external time-dependent field that achieves this after a time $\tilde{t}$. The field is then turned off at time $\tilde{t}$ so that the molecule remains in the excited eigenstate. We now ask how this process is described in the corresponding KS system, i.e. what is the KS potential? Initially, this is the ground-state potential $v_{KS, 0}$ whose ground-state Φ_{KS} has density n_{GS}, the density of the interacting ground state of the molecule. The first observation is that the KS potential after time $\tilde{t}$ does *not* typically return to the initial KS potential, in contrast to the case of the interacting system. This is because, by definition, the density of the KS state equals the interacting density at all times; in particular, after time $\tilde{t}$ it is the density of the interacting excited state of potential $v_{\text{ext}, 0}$, but this is *not* guaranteed to be the density of the KS excited state of potential $v_{KS, 0}$. Only the ground-state density of an interacting system is shared by its KS counterpart, not the higher excited states; the final KS state of the molecule will not typically be an eigenstate of $v_{KS, 0}$.

There are two possibilities for the KS potential after time $\tilde{t}$. The first is that it becomes static, and the static final density, call it $\tilde{n}$, is that of an eigenstate of it.

From the above argument, the KS potential is however different from the initial, and does not equal the ground-state KS counterpart of the interacting case. For example, if we are exciting from the ground state of the helium atom to an excited state, the external potential of the interacting system is both initially and finally $-2/r$. The initial KS potential is the ground-state KS potential of helium, but the final is not; the final KS eigenstate has the same density as the excited state of interest in helium. Now we will argue that any adiabatic approximation, or indeed any potential that is not ultranonlocal in time, is unlikely to do well. Consider a time t beyond $\tilde{t}$. The density for times near t is constant, so any semi-local approximation for the potential will be any one of the potentials for which the density $\tilde{n}$ is the density of some eigenstate of it. In particular, for an adiabatic approximation, the potential is that for which the excited state density $\tilde{n}$ is the density of its *ground state*. There is no way for an approximate semi-local KS system to know that it should be the potential corresponding to the interacting system that has an mth excited state density of $\tilde{n}$. This information is encoded in the early history of the density, from times $0 < t < \tilde{t}$: the exact KS potential must be ultranonlocal in time, since, as time gets very large, it never forgets the early history. Alternatively, taking the "initial" time to be $\tilde{t}$ in the memory formula Eq. 8.8, this effect is an initial-state effect where the initial interacting (and non-interacting state) is not a ground-state.

The other possibility is that the KS (and xc) potential never becomes constant: it continues to change in time, with KS orbitals and orbital-densities changing in time in such a way that the total KS density remains static and equal to $\tilde{n}$. It is clear that any semi-local approximation will fail here, because for times beyond $\tilde{t}$, it will predict a constant potential since the density is constant. The exact xc (and KS) potential will be ultranonlocal in time; as time gets very large, one has to go way back in time, to times less than $\tilde{t}$, in order to capture any time dependence in the density.

This extreme non-locality is very difficult for a density functional approximation to capture: it may be that orbital functionals, which are implicit density functionals, provide a promising approach. Even so, there are cases where TDDFT faces a formidable challenge. Consider two electrons, beginning in a spin-singlet ground state (e.g. the ground state of helium). Imagine now finding an optimal control field that evolves the interacting state to a singlet singly-excited state (e.g. 1s2p of helium). Now, the KS ground state is a single Slater determinant with a doubly-occupied spatial orbital. This evolves under a one-body evolution operator, the KS Hamiltonian, so must remain a single Slater determinant. But a single excitation is a *double* Slater determinant, so can never be attained even with an orbital-dependent functional. This is a time-dependent analogue of static correlation, where a single Slater determinant is inadequate to describe a fundamentally multi-determinantal state (Maitra et al. 2002b). The KS description of the state is so far from the true description, that the exact xc potential and observable functionals consequently develop complicated structure difficult to capture in approximations. The KS state is not an eigenstate of the angular momentum operator, unlike that of the true state. If the overlap between the initial and final states is targetted, the maximum that can be achieved is 0.5 (Burke

Fig. 8.4 In the *top panel* is the density (*thick solid line*) of the two electron singlet excited state (*solid lines*) of the harmonic oscillator (*solid line* in the *lower panel*). The *dashed line* is the doubly-occupied orbital resulting from evolving the singlet ground-state to a state of the same density as the excited state. The potential in which this is an eigenstate—the ground-state—is shown as the *dashed line* in the *lower panel*

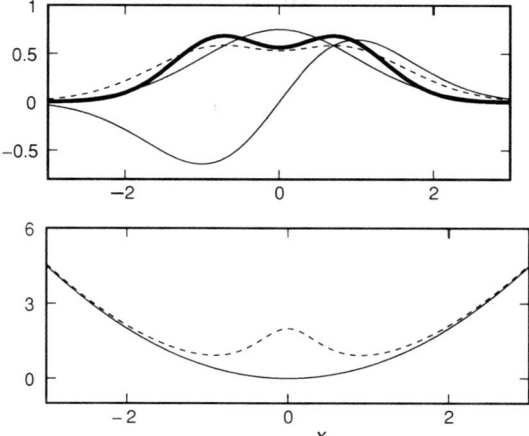

et al. 2005a), while close to 0.98 is achieved for the true interacting problem (but see also the last paragraph of this section).

A simplified model of this is shown in Fig. 8.4. Here the density of the first excited state of two electrons in a harmonic oscillator, considered to be the final KS potential, is shown. Attempting to evolve to this density from the ground state of the harmonic oscillator, which is a doubly-occupied orbital, the best KS can do is reach another doubly-occupied orbital (dashed), whose potential is shown in the lower panel (dashed).

We note that such problems do not arise in linear response regime, where we do not need to drive the system entirely into a single excited state: only perturbations of the ground-state are needed which have a small, non-zero projection on to the various excited states of the system.

It is very important to note that in the above examples the density is the target variable; the time-dependent density is the observable which the KS system is constructed to get exactly correct. However when quantum optimal control is performed directly with the TDKS system, then it is of course possible, and in most cases, more natural, instead to define the target variable directly related to the KS wavefunction. For example, one might instead consider targeting the density of a KS excited state, or an overlap with a KS excited state. The question is then whether the optimal field found for the KS system also achieves a good outcome for the true system. In most cases studied so far by A. Castro and E.K.U. Gross (personal communication), it fortunately does. Even in the case where the maximum overlap with the target KS state is as low as 0.5, it is quite possible that, with a clever choice of target functional, the optimal field found from the KS evolution applied to the interacting system yields a target overlap in the interacting system much closer to one.

8.6 Memory Effects in Excitation Spectra

The vast majority of applications of TDDFT today are in the linear response regime. Adiabatic approximations are used in the many successes here, but they are also the reason behind many of its notorious failures. This section discusses one of these: states of double-excitation character, for which frequency-dependence is essential to capture.

First, we clarify what is meant by a double- or multiple-excitation. The term is defined with respect to some single-particle picture, characterized by a set of single-particle orbitals that are solutions to a one-body Hamiltonian, N of which are occupied in the ground-state of this Hamiltonian. In TDDFT, this picture is naturally the KS system. A single-excitation swaps one of these occupied orbitals for an unoccupied orbital. A double-excitation instead swaps two occupied orbitals for two unoccupied orbitals, so represents an excited state where two electrons are excited with respect to the ground configuration. On the other hand, the true interacting eigenstates are linear combinations of this ground-state and all excitations, including double, triple, and higher (thinking of the determinants composed of the single-particle orbitals as a complete set of N-electron states). When we speak of a double-excitation in the true system, we mean it as a short-hand for a "state of double-excitation character", i.e. a state that has a significant fraction of a double-excitation with respect to a non-interacting single-particle picture. Clearly, in different theories that have different references (e.g. Hartree–Fock rather than KS), these fractions will differ.

Chapter 4 mentioned that often the KS excitations are themselves good zeroth-order approximations to the true excitations, with the xc kernel contributing a small enough correction that even the simplest adiabatic approximation does quite well. But this argument cannot apply to double-excitations, since in linear response of the KS system no double-excitations appear: to excite two electrons of a non-interacting system two photons would be required, beyond linear response. Only single-excitations of the KS system are available for an adiabatic kernel to mix. Indeed, if we consider the linear response function (Eq. 4.50) applied to the KS system, the numerator $\langle \Phi_{GS} | \hat{n}(r) | \Phi_I \rangle$ vanishes if Φ_I and Φ_{GS} differ by more than one orbital since the one-body operator $\hat{n}(r)$ cannot connect states that differ by more than one orbital. The true response function, on the other hand, retains poles at the true excitations which are mixtures of single, double, and higher-electron-number excitations, as the numerator $\langle \Psi_{GS} | \hat{n}(r) | \Psi_I \rangle$ remains finite due to the mixed nature of both Ψ_{GS} and Ψ_I. Within the adiabatic approximation, χ therefore contains more poles than χ_{KS}.

How does the exact kernel of TDDFT generate more poles, and capture states of multiple-excitations? One must go beyond the adiabatic approximation (Tozer and Handy 2000). In 2004, the exact frequency-dependence that is required when a double excitation mixes with a single excitation was demonstrated, and an approximate kernel based on this was derived (Maitra et al. 2004). We now discuss this.

A frequency-dependent kernel for double-excitations. Consider the simplest model: a two-by-two excitation subspace consisting of one KS single excitation

$\varphi_i \rightarrow \varphi_u$, of frequency $\omega_q = \epsilon_u - \epsilon_i$, and one double excitation (Φ_D) energetically close. We assume all other excitations lie far from these two levels, so that for frequencies close to ω_q :

$$\chi_{KS}(r, r', \omega) = \frac{A(r, r', \omega)}{\omega - \omega_q} \tag{8.10}$$

where the numerator only weakly depends on the frequency: $A(r, r', \omega) = \varphi_i^*(r)\varphi_u(r)\varphi_i(r')\varphi_u^*(r') + O(\omega - \omega_q)$. The KS double-excitation does not contribute to χ_{KS} from the argument above. Electron interaction mixes the KS single- and double-excitations, such that the true states have the form:

$$\Psi_a = m\Phi_D + \sqrt{1 - m^2}\Phi_q \text{ and } \Psi_b = \sqrt{1 - m^2}\Phi_D - m\Phi_q \tag{8.11}$$

where $0 < m < 1$ is a parameter to represent the fraction of double- and single-excitation character in the true interacting states. Inserting these into the expression for the true response function, Eq. 4.50, we obtain

$$\chi(r, r', \omega) = A(r, r', \omega)\left(\frac{1 - m^2}{\omega - \omega_a} + \frac{m^2}{\omega - \omega_b}\right), \tag{8.12}$$

where ω_a, ω_b are the true interacting excitation frequencies. Notice that the two interacting states share the oscillator strength of the KS single-excitation, in a ratio determined by the fraction of how much single-excitation each carries. Within this subspace, we then define a *dressed* (i.e. frequency-dependent) single-pole approximation (DSPA),

$$\omega = \omega_q + 2[q|f_{Hxc}(\omega)|q] \tag{8.13}$$

where the kernel on the right is derived from Eqs. 8.10 and 8.12, using

$$f_{Hxc}(\omega) = \chi_{KS}^{-1}(\omega) - \chi^{-1}(\omega). \tag{8.14}$$

(c.f. the SPA in Eq. 4.63; the square bracket notation in Eq. 8.13 indicates the doubleintegral of Eq. 4.63 but with the kernel evaluated at frequency ω, instead of at ω_q.)

Requiring that the DSPA recovers the exact frequencies ω_a, ω_b pins down the matrix element of $A^{-1}(r, r', \omega)$, and we find

$$2[q|f_{Hxc}(\omega)|q] = (\bar{\omega} - \omega_q) + \frac{\bar{\omega}'\bar{\omega} - \omega_a\omega_b}{(\omega - \bar{\omega}')}, \tag{8.15}$$

where $\bar{\omega}' = m^2\omega_a + (1 - m^2)\omega_b$ and $\bar{\omega} = (1 - m^2)\omega_a + m^2\omega_b$. Equation 8.15 gives the *exact* xc kernel matrix element for frequencies near the single and double of interest; we illustrate how it generates two poles in χ from the one in χ_{KS} in Fig. 8.5.

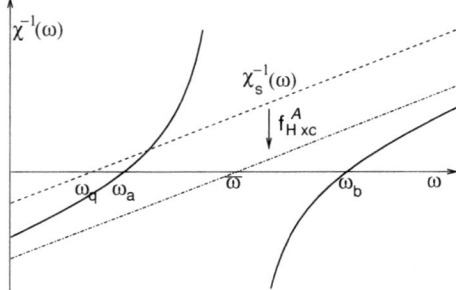

Fig. 8.5 Frequency-dependence near a double excitation (see text): near a single excitation, $\chi_{KS}^{-1}(\omega)$ (*upper dashed line*) has one zero at the KS transition ω_q, which an adiabatic kernel f_{Hxc}^A shifts to $\bar{\omega}$. Frequency-dependence of Eq. 8.15 renders two zeroes in the exact $\chi^{-1}(\omega)$ (*solid line*) at the transition frequencies ω_a, ω_b of the true mixed single and double states

The first term is adiabatic, while the second is strongly non-adiabatic (Maitra et al. 2004).

Equation 8.15 motivated the following practical approximation for the dressed kernel, when a single and double excitation lie near each other, in the limit of weak interaction (direct coupling to the rest of the KS excitations is neglected). Essentially one considers diagonalizing the many-body Hamiltonian in this two-by-two KS subspace, and requires that the kernel reduces to the adiabatic one (f_{xc}^A) in the limit that the single and double only weakly interact (see Maitra et al. 2004 for details). One obtains:

$$2\left[q|f_{xc}(\omega)|q\right] = 2\left[q|f_{xc}^A(\omega_q)|q\right] + \frac{|H_{qD}|^2}{\omega - (H_{DD} - H_{00})} \tag{8.16}$$

where the Hamiltonian matrix elements in the dynamical correction (second term) are those of the true interacting Hamiltonian, taken between the single (q) and double (D) KS Slater determinants of interest, as indicated, and H_{00} is the expectation value of the true Hamiltonian in the KS ground-state. The kernel is to be applied as an *a posteriori* correction to a usual adiabatic calculation: first, one scans over the KS orbital energies to see if the sum of two of their frequencies lies near a single excitation frequency, and then applies this kernel just to that pair. If a double-excitation mixes strongly with several single excitations, one performs the dressing Eq. 8.16 in a matrix spanned by those singles (see Cave et al. 2004; Mazur and Włodarczyk 2009, where this was done for polyenes).

Several alternate and more formal approaches have led essentially to Eq. 8.16. Casida (2005) derived Eq. 8.16 from a superoperator formalism, as a polarization propagator correction to adiabatic TDDFT. The Bethe–Salpeter equation (BSE) with a dynamically screened Coulomb interaction was used to derive a frequency-dependent kernel (Romaniello et al. 2009): here the frequency-dependence has two origins, one from folding the four-point BSE into the two-point TDDFT equation [as in the work on the optical response of solids, (e.g. Reining 2002; Gatti et al 2007a)], and the other (essential for double-excitations) from the frequency-dependence of the BSE kernel. The resulting kernel however yielded additional unphysical poles, attributed to the self-screening of GW. There are connections between this and the superoperator formalism (see Huix-Rotllant 2011), where explicit expressions for the kernel derived from algebraic diagrammatic construction with second-order polarization-propagator applied to KS are given. The spatial-dependence of the kernel was uncovered in an approach following more closely the original derivation but using the common energy denominator approximation to account for the effect of entire spectrum on the coupled single- and double-states (Gritsenko and Baerends 2009).

Where states of double-excitation character arise. In some systems, they lie even amongst the lowest-energy states, but we will also argue that they underlie several other failures of adiabatic TDDFT.

(i) In many conjugated molecules (e.g. polyenes), double-excitations infiltrate the low-lying excitations, which are as a result notoriously difficult to calculate (see Cave et al. 2004) for many references). For example, in butadiene, the HOMO $\rightarrow$ (LUMO + 1) and (HOMO − 1) $\rightarrow$ LUMO excitations are near-degenerate with a double-excitation of the HOMO to LUMO. If one runs an adiabatic calculation and simply assigns the energies according to an expected ordering, it may appear that one obtains a reasonable value for an expected state of double-excitation character (Hsu et al. 2001), however upon examining the make-up of the state, one will find it is instead a single-excitation (misplaced, for example due to basis-set issues). Cave et al (2004) applied dressed TDDFT to the dark $2^1 A_g$ state of butadiene and hexatriene, generalizing Eq. 8.16 to the case of two single excitations mixing with a double, obtaining results close to CASPT2. Linear polyenes were later studied in more detail (Mazur and Włodarczyk 2009; Mazur et al. 2011), analyzing more fully aspects such as self-consistent treatment of the kernel, and use of KS versus Hartree–Fock orbitals in the dressing, and successfully computing excited-state geometries with this dressed TDDFT. Dressed TDDFT was applied to low-lying excited states of 28 organic molecules (Huix-Rotllant et al. 2011b).

(ii) It is well known that charge-transfer excitations between fragments at large separations are severely underestimated with the usual approximations of TDDFT (see Sect. 4.8.2). If we are interested in charge-transfer between two distant open-shell species (e.g. LiH), the HOMO and LUMO are delocalized over the whole molecule. This is the case for the exact ground-state KS potential, for which a step appears in the bonding region that has exactly the size to re-align the two atomic HOMOs, as well as for local or semi-local approximations (Tempel et al. 2009). The HOMO–LUMO energy difference goes as the tunnel splitting, vanishing as the molecule is pulled apart; therefore *every* excitation out of the KS HOMO is near-degenerate with a double-excitation where a second electron goes from HOMO to LUMO (at almost zero KS cost). This yields a strong frequency-dependence of the kernel near *all* excitations, charge-transfer and local, for heteroatomic molecules composed of open-shell fragments at large separation (Maitra 2005b; Maitra and Tempel 2006a).

(iii) Double-excitations dog accurate calculations of coupled electron-nuclear dynamics (Levine et al. 2006): even when the vertical excitation does not contain much double-excitation character, the propensity for curve-crossing requires an accurate double-excitation description for accurate global potential energy surfaces. That same paper highlighted the difficulties approximate TDDFT has with obtaining conical intersections: in one example, the TDDFT dramatically exaggerated the shape of the intersection, while in another, its dimensionality was wrong, producing a seam rather than a point. Although the ground-state surface is not described well here with the approximate functionals, double-excitations are certainly relevant in the vicinity of conical intersections due to the near-degeneracy (see also Chap. 14).

(iv) In the He atom, the lowest double-excitation ($1s^2 \rightarrow 2s^2$) lies in the continuum, appearing as a resonance in the continuous spectrum. Autoionizing resonances that arise from bound core-valence single excitations are well-captured by the adiabatic kernels of TDDFT (see e.g. Stener et al. 2007; Hellgren and von Barth 2009) but those arising from double-excitations require a frequency-dependent kernel. An approximate kernel based on Fano's degenerate perturbation theory approach (Fano 1961) applied to the KS system can be derived (Krueger and Maitra 2009).

8.7 Outlook

Memory profoundly affects the structure of exact functionals in TDDFT. Here we have given some exact properties regarding initial-state dependence and history-dependence, and explored some memory effects on exactly solvable systems. Strong field dynamics is especially the regime where TDDFT may be the only feasible approach, as wavefunction methods for more than a few interacting electrons become prohibitively expensive. Yet it is in this regime that memory effects appear to be significant. Memory also influences the accuracy of linear response calculations, and we showed how a frequency-dependent kernel, derived from first principles, captures states of double-excitation character, missing in the usual memory-less adiabatic approximations. For TDDFT to be used for fully time-dependent phenomena as confidently as DFT is used for ground-state problems, and to build on its reliability for electronic spectra, further understanding and modeling of memory effects is required, along with further developments of memory-dependent approximations.

Part III
Advanced Concepts

Chapter 9
Beyond the Runge–Gross Theorem

Michael Ruggenthaler and Robert van Leeuwen

9.1 Introduction

The Runge–Gross theorem (Runge and Gross 1984) states that for a given initial
state the time-dependent density is a unique functional of the external potential. Let
us elaborate a bit further on this point. Suppose we could solve the time-dependent
Schrödinger equation (TDSE) for a given many-body system, i.e. we specify an
initial state $|\Psi_0\rangle$ at $t = t_0$ and evolve the wavefunction in time using the Hamiltonian
$\hat{H}(t)$. Then, from the wave function, we can calculate the time-dependent density
$n(\boldsymbol{r}, t)$. We can then ask the question whether exactly the same density $n(\boldsymbol{r}, t)$ can
be reproduced by an external potential $v'_{\text{ext}}(\boldsymbol{r}, t)$ in a system with a different given
initial state and a different two-particle interaction, and if so, whether this potential
is unique (modulo a purely time-dependent function). The answer to this question
is obviously of great importance for the construction of the time-dependent Kohn–
Sham equations. The Kohn–Sham system has no two-particle interaction and differs
in this respect from the fully interacting system. It has, in general, also a different
initial state. This state is usually a Slater determinant rather than a fully interacting
initial state. A time-dependent Kohn–Sham system therefore only exists if the question posed above is answered affirmatively. Note that this is a v-representability
question (see Sect. 4.4.2): is a density belonging to an interacting system also noninteracting v-representable? We will show in this chapter that, with some restrictions

M. Ruggenthaler · R. van Leeuwen (✉)
Department of Physics, Nanoscience Center,
University of Jyväskylä,
Survontie 9, P. O. Box 35,
40014 Jyväskylä,
Finland
e-mail: michael.ruggenthaler@jyu.fi
e-mail: robert.vanleeuwen@jyu.fi

M. A. L. Marques et al. (eds.), *Fundamentals of Time-Dependent Density Functional
Theory*, Lecture Notes in Physics 837, DOI: 10.1007/978-3-642-23518-4_9,
© Springer-Verlag Berlin Heidelberg 2012

on the initial states, potentials and densities, this question can indeed be answered affirmatively (van Leeuwen 1999, 2001; Giuliani and Vignale 2005; Ruggenthaler and van Leeuwen 2011b). We stress that we demonstrate here that the interacting-v-representable densities are also noninteracting-v-representable rather than aiming at characterizing the set of v-representable densities. The latter question has inspired much work in ground state density functional theory [for an extensive discussion see van Leeuwen (2003)] and has only been answered satisfactorily for quantum lattice systems (Chayes et al. 1985) and coarse grained approaches (Lammert 2010).

First we will introduce an extended version of the Runge–Gross theorem showing that time-analytic densities are v-representable in a different system provided we have an appropriate initial state. In the next paragraph using arguments from previous derivations we give a Runge–Gross type theorem where one treats the temporal restriction of the potentials for a spatial restriction. Then we show a one-to-one correspondence between densities and potentials provided we start from the initial state and have Laplace transformable switch-on potentials. Furthermore we will present the main ideas of a fixed-point proof for TDDFT which guarantees bijectivity of the density-potential mapping and v-representability of a broad class of potentials and densities. Finally we address certain problems of the v-representability of the quantum mechanical action and a recent elegant way to solve the resulting causality paradox.

9.2 The Extended Runge–Gross Theorem: Different Interactions and Initial States

We start by considering the Hamiltonian

$$\hat{H}(t) = \hat{T} + \hat{V}_{\text{ext}}(t) + \hat{V}_{\text{ee}}, \qquad (9.1)$$

where $\hat{T}$ is the kinetic energy, $\hat{V}_{\text{ext}}(t)$ the (in general time-dependent) external potential, and $\hat{V}_{\text{ee}}$ the two-particle interaction. In second quantization the constituent terms are, as usual, written as

$$\hat{T} = -\frac{1}{2} \sum_{\sigma} \int d^3r \, \hat{\psi}_{\sigma}^{\dagger}(r) \nabla^2 \hat{\psi}_{\sigma}(r), \qquad (9.2a)$$

$$\hat{V}_{\text{ext}}(t) = \sum_{\sigma} \int d^3r \, v_{\text{ext}}(r, t) \psi_{\sigma}^{\dagger}(r) \hat{\psi}_{\sigma}(r), \qquad (9.2b)$$

$$\hat{V}_{\text{ee}} = -\frac{1}{2} \sum_{\sigma\sigma'} \int d^3r \int d^3r' \, v_{\text{ee}} \left(|r - r'| \right) \hat{\psi}_{\sigma}^{\dagger}(r) \hat{\psi}_{\sigma'}^{\dagger}(r') \hat{\psi}_{\sigma'}(r') \hat{\psi}_{\sigma}(r). \qquad (9.2c)$$

where σ and σ' are spin variables. For the readers not used to second quantization we note that the first few basic steps in this chapter can also be derived in first quantization. For details we refer to (Giuliani and Vignale 2005; Vignale 2004). Good introductions to second quantization are found in Fetter and Walecka (1971) and Runge (1991).

The two-particle potential $v_{ee}(|r-r'|)$ in (9.2c) can be arbitrary, but will in practice almost always be equal to the repulsive Coulomb potential. We then consider some basic relations satisfied by the density and the current density. The time-dependent density is given as the expectation value of the density operator

$$\hat{n}(r) = \sum_\sigma \hat{\psi}_\sigma^\dagger(r)\hat{\psi}_\sigma(r), \qquad (9.3)$$

with the time-dependent many-body wavefunction, $n(r, t) = \langle \Psi(t)|\hat{n}(r)|\Psi(t)\rangle$. In the following we consider two continuity equations. If $|\Psi(t)\rangle$ is the state evolving from $|\Psi_0\rangle$ under the influence of Hamiltonian $\hat{H}(t)$ we have the usual continuity equation

$$\frac{\partial}{\partial t}n(r, t) = -\mathrm{i}\left\langle \Psi(t)\left|\left[\hat{n}(r), \hat{H}(t)\right]\right|\Psi(t)\right\rangle = -\nabla \cdot j(r, t), \qquad (9.4)$$

where the current operator is defined as

$$\hat{j}(r) = \frac{1}{2\mathrm{i}} \sum_\sigma \left\{ \hat{\psi}_\sigma^\dagger(r)\nabla\hat{\psi}_\sigma(r) - \left[\nabla\hat{\psi}_\sigma^\dagger(r)\right]\hat{\psi}_\sigma(r) \right\}, \qquad (9.5)$$

and has expectation value $j(r, t) = \langle \Psi(t)|\hat{j}(r)|\Psi(t)\rangle$. This continuity equation expresses, in a local form, the conservation of particle number. Using Gauss' law, the continuity equation says that the change of the number of particles within some volume can simply be measured by calculating the flux of the current through the surface of this volume.

As a next step, we can consider an analogous continuity equation for the current itself. We have

$$\frac{\partial}{\partial t}j(r, t) = -\mathrm{i}\left\langle \Psi(t)\left|\left[\hat{j}(r), \hat{H}(t)\right]\right|\Psi(t)\right\rangle. \qquad (9.6)$$

If we work out the commutator in more detail, we find the expression (Martin 1959)

$$\frac{\partial}{\partial t}j_\alpha(r, t) = -n(r, t)\frac{\partial}{\partial r_\alpha}v_{ext}(r, t) - \sum_\beta \frac{\partial}{\partial r_\beta}T_{\beta\alpha}(r, t) - V_{ee\,\alpha}(r, t). \qquad (9.7)$$

Here we have defined the momentum-stress tensor $\hat{T}_{\beta\alpha}$ (part of the energy-momentum tensor)

$$\hat{T}_{\beta\alpha}(r) = \frac{1}{2} \sum_\sigma \left\{ \frac{\partial}{\partial r_\beta}\hat{\psi}_\sigma^\dagger(r)\frac{\partial}{\partial r_\alpha}\hat{\psi}_\sigma(r) + \frac{\partial}{\partial r_\alpha}\hat{\psi}_\sigma^\dagger(r)\frac{\partial}{\partial r_\beta}\hat{\psi}_\sigma(r) \right.$$
$$\left. - \frac{1}{2}\frac{\partial^2}{\partial r_\beta \partial r_\alpha}\left[\hat{\psi}_\sigma^\dagger(r)\hat{\psi}_\sigma(r)\right] \right\}, \qquad (9.8)$$

and the quantity $\hat{V}_{\text{ee}\,\alpha}$ as

$$\hat{V}_{\text{ee}\,\alpha}(\boldsymbol{r}) = \sum_{\sigma,\sigma'} \int d^3r' \, \hat{\psi}_\sigma^\dagger(\boldsymbol{r}) \hat{\psi}_{\sigma'}^\dagger(\boldsymbol{r}') \frac{\partial}{\partial r_\alpha} v_{\text{ee}}(|\boldsymbol{r} - \boldsymbol{r}'|) \hat{\psi}_{\sigma'}(\boldsymbol{r}') \hat{\psi}_\sigma(\boldsymbol{r}). \tag{9.9}$$

The expectation values that appear in (9.7) are defined as

$$T_{\beta\alpha}(\boldsymbol{r}, t) = \left\langle \Psi(t) \left| \hat{T}_{\beta\alpha}(\boldsymbol{r}) \right| \Psi(t) \right\rangle \tag{9.10a}$$

$$V_{\text{ee}\,\alpha}(\boldsymbol{r}, t) = \left\langle \Psi(t) \left| \hat{V}_{\text{ee}\,\alpha}(\boldsymbol{r}) \right| \Psi(t) \right\rangle. \tag{9.10b}$$

The continuity equation (9.7) is a local quantum version of Newton's third law. Taking the divergence of (9.7) and using the continuity equation (9.4) we find

$$\frac{\partial^2}{\partial t^2} n(\boldsymbol{r}, t) = \nabla \cdot [n(\boldsymbol{r}, t) \nabla v_{\text{ext}}(\boldsymbol{r}, t)] + q(\boldsymbol{r}, t), \tag{9.11}$$

with $\hat{q}$ and $q(\boldsymbol{r}, t)$ being defined as

$$\hat{q}(\boldsymbol{r}) = \sum_{\alpha,\beta} \frac{\partial^2}{\partial r_\beta \partial r_\alpha} \hat{T}_{\beta\alpha}(\boldsymbol{r}) + \sum_\alpha \frac{\partial}{\partial r_\alpha} \hat{V}_{\text{ee}\,\alpha}(\boldsymbol{r}), \tag{9.12a}$$

$$q(\boldsymbol{r}, t) = \left\langle \Psi(t) | \hat{q}(\boldsymbol{r}) | \Psi(t) \right\rangle. \tag{9.12b}$$

Equation 9.11 will play a central role in our discussion of the relation between the density and the potential. This is because it represents an equation which directly relates the external potential and the electron density. From (9.11) we further see that $q(\boldsymbol{r}, t)$ decays exponentially at infinity when $n(\boldsymbol{r}, t)$ does, unless $v_{\text{ext}}(\boldsymbol{r}, t)$ grows exponentially at infinity. In the following we will, however, only consider finite systems with external potentials that are bounded at infinity [for a discussion of the set of allowed external potentials in TDDFT see Sect. 4.4.1 and in the case of ground state DFT we refer to Lieb (1983) and van Leeuwen (2003)].

Let us now assume that we have solved the time-dependent Schrödinger equation for the many-body system described by the Hamiltonian $\hat{H}(t)$ of (9.1) and initial state $|\Psi_0\rangle$ at $t = t_0$. We have thus obtained a many-body wavefunction $|\Psi(t)\rangle$ and density $n(\boldsymbol{r}, t)$. We further assume that $n(\boldsymbol{r}, t)$ is time-analytic at $t = t_0$. For our system, Eq. 9.11 is satisfied. We now consider a second system with Hamiltonian

$$\hat{H}'(t) = \hat{T} + \hat{V}'_{\text{ext}}(t) + \hat{V}'_{\text{ee}}. \tag{9.13}$$

The terms $\hat{V}'_{\text{ext}}(t)$ and $\hat{V}'_{\text{ee}}$ represent again the one- and two-body potentials. We denote the initial state by $|\Psi'_0\rangle$ at $t = t_0$ and the time-evolved state by $|\Psi'(t)\rangle$. The form of $\hat{V}'_{\text{ee}}$ is assumed to be such that its expectation value and its derivatives are finite. For the system described by the Hamiltonian $\hat{H}'$ we have an equation analogous to (9.11).

$$\frac{\partial^2}{\partial t^2} n'(\boldsymbol{r}, t) = \nabla \cdot \left[n'(\boldsymbol{r}, t) \nabla v'_{\text{ext}}(\boldsymbol{r}, t) \right] + q'(\boldsymbol{r}, t), \tag{9.14}$$

where $q'(\boldsymbol{r}, t)$ is the expectation value

$$q'(\boldsymbol{r}, t) = \langle \Psi'(t) | \hat{q}'(\boldsymbol{r}) | \Psi'(t) \rangle, \tag{9.15}$$

for which we defined

$$\hat{q}' = \sum_{\beta, \alpha} \frac{\partial^2}{\partial r_\beta \partial r_\alpha} \hat{T}_{\beta \alpha}(\boldsymbol{r}) + \sum_\alpha \frac{\partial}{\partial r_\alpha} \hat{V}'_{\text{ee}\,\alpha}(\boldsymbol{r}). \tag{9.16}$$

Our goal is now to choose v'_{ext} in (9.14) so that $n'(\boldsymbol{r}, t) = n(\boldsymbol{r}, t)$. We will do this by constructing v'_{ext} in such a way that for the k-th derivatives of the density at $t = t_0$ we have $\frac{\partial^k}{\partial t^k} n'(\boldsymbol{r}, t)|_{t=t_0} = \frac{\partial^k}{\partial t^k} n(\boldsymbol{r}, t)|_{t=t_0}$. First we need to discuss some initial conditions. As a necessary condition for the potential v'_{ext} to exist, we have to require that the initial states $|\Psi_0\rangle$ and $|\Psi'_0\rangle$ yield the same initial density, i.e.

$$n'(\boldsymbol{r}, t_0) = \langle \Psi'_0 | \hat{n}(\boldsymbol{r}) | \Psi'_0 \rangle = \langle \Psi_0 | \hat{n}(\boldsymbol{r}) | \Psi_0 \rangle = n(\boldsymbol{r}, t_0). \tag{9.17}$$

We now note that the basic Eq. 9.11 is a second order differential equation in time for $n(\boldsymbol{r}, t)$ and hence we need as additional requirement that $\frac{\partial}{\partial t} n'(\boldsymbol{r}, t) = \frac{\partial}{\partial t} n(\boldsymbol{r}, t)$ at $t = t_0$. The first time-derivative of the densities are determined by the initial states via the continuity equation (9.4):

$$\left. \frac{\partial}{\partial t} n'(\boldsymbol{r}, t) \right|_{t=t_0} = \langle \Psi'_0 | \nabla \cdot \hat{\boldsymbol{j}}(\boldsymbol{r}) | \Psi'_0 \rangle = \langle \Psi_0 | \nabla \cdot \hat{\boldsymbol{j}}(\boldsymbol{r}) | \Psi_0 \rangle = \left. \frac{\partial}{\partial t} n(\boldsymbol{r}, t) \right|_{t=t_0}. \tag{9.18}$$

This constraint also implies the weaker requirement that the initial state $|\Psi'_0\rangle$ must be chosen such that the initial momenta $\boldsymbol{P}(t_0)$ of both systems are the same:

$$\boldsymbol{P}(t) = \int \mathrm{d}^3 r \, \boldsymbol{j}(\boldsymbol{r}, t) = \int \mathrm{d}^3 r \, \boldsymbol{r} \frac{\partial}{\partial t} n(\boldsymbol{r}, t). \tag{9.19}$$

The equality of the last two terms in this equation follows directly from the continuity equation (9.4) and the fact that we are dealing with finite systems for which, barring pathological examples (van Leeuwen 2001, Maitra and Burke 2001(a,b)), currents and densities are zero at infinity. For notational convenience we first introduce the following notation for the k-th time-derivative at $t = t_0$ of a function f:

$$f^{(k)}(\boldsymbol{r}) = \left. \frac{\partial^k}{\partial t^k} f(\boldsymbol{r}, t) \right|_{t=t_0}. \tag{9.20}$$

Then our goal is to choose v'_{ext} in such a way that $n'^{(k)} = n^{(k)}$ for all k. Let us see how we can use (9.14) to do this. If we first evaluate (9.14) at $t = t_0$ we obtain, using the notation of (9.20), the expression

$$n'^{(2)}(r) = \nabla \cdot \left[n'^{(0)}(r) \nabla v_{\text{ext}}'^{(0)}(r) \right] + q'^{(0)}(r). \tag{9.21}$$

Since we want that $n'^{(2)} = n^{(2)}$ and have chosen the initial state $|\Psi_0'\rangle$ in such a way that $n'^{(0)} = n^{(0)}$ we obtain the following determining equation for $v_{\text{ext}}'^{(0)}$:

$$\nabla \cdot \left[n^{(0)}(r) \nabla v_{\text{ext}}'^{(0)}(r) \right] = n^{(2)}(r) - q'^{(0)}(r). \tag{9.22}$$

The right hand side is determined since $n^{(0)}$ and $n^{(2)}$ are given and $q'^{(0)}$ is calculated from the given initial state $|\Psi_0'\rangle$ as $q'^{(0)}(r) = \langle \Psi_0' | \hat{q}'(r) | \Psi_0' \rangle$. Equation 9.22 is of Sturm–Liouville type and has a unique solution for $v_{\text{ext}}'^{(0)}$ provided we specify a boundary condition. For a thorough discussion of this Sturm–Liouville equation see (Ruggenthaler et al. 2009; Penz and Ruggenthaler 2011). We will specify the boundary condition that $v_{\text{ext}}'^{(0)}(r) \to 0$ for $r \to \infty$. With this boundary condition we also fix the gauge of the potential. Having obtained $v_{\text{ext}}'^{(0)}$ let us now go on to determine $v_{\text{ext}}'^{(1)}$. To do this we differentiate (9.14) with respect to time and evaluate the resulting expression in $t = t_0$. Then we obtain the expression:

$$n'^{(3)}(r) = \nabla \cdot \left[n'^{(0)}(r) \nabla v_{\text{ext}}'^{(1)}(r) \right] + \nabla \cdot \left[n'^{(1)}(r) \nabla v_{\text{ext}}'^{(0)}(r) \right] + q'^{(1)}(r). \tag{9.23}$$

Since we want to determine $v_{\text{ext}}'^{(1)}$ such that $n'^{(3)} = n^{(3)}$ and the conditions on the initial states are such that $n'^{(0)} = n^{(0)}$ and $n'^{(1)} = n^{(1)}$, we obtain the following equation for $v_{\text{ext}}'^{(1)}$:

$$\nabla \cdot \left[n^{(0)}(r) \nabla v_{\text{ext}}'^{(1)}(r) \right] = n^{(3)}(r) - q^{(1)}(r) - \nabla \cdot \left[n^{(1)}(r) \nabla v_{\text{ext}}'^{(0)}(r) \right]. \tag{9.24}$$

Now all quantities on the right hand side of (9.24) are known. The initial potential $v_{\text{ext}}'^{(1)}$ was already determined from (9.22) whereas the quantity $q'^{(1)}$ can be calculated from

$$q'^{(1)}(r) = \left. \frac{\partial}{\partial t} q'(r, t) \right|_{t=t_0} = -i \left\langle \Psi_0' \left| \left[\hat{q}'(r), \hat{H}'(t_0) \right] \right| \Psi_0' \right\rangle. \tag{9.25}$$

From this expression we see that $q'^{(1)}$ can be calculated from the knowledge of the initial state and the initial potential $v_{\text{ext}}'^{(0)}$ which occurs in $\hat{H}'(t_0)$. Therefore, Eq. 9.24 uniquely determines $v_{\text{ext}}'^{(1)}$ (again with boundary conditions $v_{\text{ext}}'^{(1)} \to 0$ for $r \to \infty$). We note that in order to obtain (9.24) from (9.23) we indeed needed both conditions of (9.17) and (9.18). It is now clear how our procedure can be extended. If we take the k-th time-derivative of (9.14) we obtain the expression

$$n'^{(k+2)}(r) = q'^{(k)}(r) + \sum_{l=0}^{k} \binom{k}{l} \nabla \cdot \left[n'^{(k-l)}(r) \nabla v_{\text{ext}}'^{(l)}(r) \right]. \tag{9.26}$$

Demanding that $n'^{(k)} = n^{(k)}$ then yields

$$\nabla \cdot [n^{(0)}(\boldsymbol{r})\nabla v_{\text{ext}}'^{(k)}(\boldsymbol{r})] = n^{(k+2)}(\boldsymbol{r}) - q'^{(k)}(\boldsymbol{r}) - \sum_{l=0}^{k-1} \binom{k}{l} \nabla \cdot [n^{(k-l)}(\boldsymbol{r})\nabla v_{\text{ext}}'^{(l)}(\boldsymbol{r})].$$

(9.27)

The right hand side of this equation is completely determined since it only involves the potentials $v_{\text{ext}}'^{(l)}$ for $l = 1, \ldots, k - 1$ which were already determined. Similarly the quantities $q'^{(k)}$ can be calculated from multiple commutators of the operator $\hat{q}'$ and time-derivatives of the Hamiltonian $\hat{H}'(t_0)$ up to order $k - 1$ and therefore only involves knowledge of the initial state and $v_{\text{ext}}'^{(l)}$ for $l = 1, \ldots, k - 1$. We can therefore uniquely determine all functions $v_{\text{ext}}'^{(k)}$ from (9.27) (again taking into account the boundary conditions) and construct the potential $v_{\text{ext}}'(\boldsymbol{r}, t)$ from its Taylor series as

$$v_{\text{ext}}'(\boldsymbol{r}, t) = \sum_{k=0}^{\infty} \frac{1}{k!} v_{\text{ext}}'^{(k)}(\boldsymbol{r})(t - t_0)^k.$$

(9.28)

This determines $v_{\text{ext}}'(\boldsymbol{r}, t)$ completely within the convergence radius of the Taylor expansion. There is, of course, the possibility that the convergence radius is zero. However, this would mean that $v_{\text{ext}}'(\boldsymbol{r}, t)$ and hence $n(\boldsymbol{r}, t)$ and $v_{\text{ext}}(\boldsymbol{r}, t)$ are nonanalytic at $t = t_0$. Since the density of our reference system was supposed to be analytic we can disregard this possibility. If the convergence radius is non-zero but finite, we can propagate $|\Psi_0'\rangle$ to $|\Psi'(t_1)\rangle$ until a finite time $t_1 > t_0$ within the convergence radius and repeat the whole procedure above from $t = t_1$ by regarding $|\Psi'(t_1)\rangle$ as the initial state. This amounts to analytic continuation along the whole real time-axis and the complete determination of $v_{\text{ext}}'(\boldsymbol{r}, t)$ at all times. This completes the constructive proof of $v_{\text{ext}}'(\boldsymbol{r}, t)$.

Let us now summarize what we proved provided the densities and the potentials are both time-analytic. We specify a given density $n(\boldsymbol{r}, t)$ obtained from a many-particle system with Hamiltonian $\hat{H}$ and initial state $|\Psi_0\rangle$. If one chooses an initial state $|\Psi_0'\rangle$ of a second many-particle system with two-particle interaction $\hat{V}_{\text{ee}}'$ in such a way that it yields the correct initial density and initial time-derivative of the density, then, for this system, there is a unique external potential $v_{\text{ext}}'(\boldsymbol{r}, t)$ (determined up to a purely time-dependent function $c(t)$) that reproduces the given density $n(\boldsymbol{r}, t)$.

Let us now specify some special cases. If we take $\hat{V}_{\text{ee}}' = 0$ we can conclude that, for a given initial state $|\Psi_0'\rangle = |\Phi_0\rangle$ with the correct initial density and initial time derivative of the density, there is a unique potential $v_{\text{KS}}(\boldsymbol{r}, t)$ [modulo $c(t)$] for a noninteracting system that produces the given density $n(\boldsymbol{r}, t)$ at all times. This solves the noninteracting v-representability problem, provided we can find an initial state with the required properties. If the many-body system described by the Hamiltonian $\hat{H}$ is stationary for times $t < t_0$, the initial state $|\Psi_0\rangle$ at t_0 leads to a density with zero time-derivative at $t = t_0$. In that case, a noninteracting state with the required initial density and initial time-derivative of the density (namely zero) can be obtained via the so-called Harriman construction (Harriman 1981; Lieb 1983). Therefore a Kohn–Sham potential always exists for this kind of switch-on processes. The additional question whether this initial state can be chosen as a ground state of a noninteracting system is equivalent to the currently unresolved noninteracting v-representability question for stationary systems [Kohn 1983; Ullrich and Kohn 2002a; Dreizler and Gross 1990—for an extensive discussion see van Leeuwen (2003)].

We now take $\hat{V}'_{ee} = \hat{V}_{ee}$. We therefore consider two many-body systems with the same two-particle interaction. Our proof then implies that for a given v-representable density $n(r, t)$ that corresponds to an initial state $|\Psi_0\rangle$ and potential $v_{ext}(r, t)$, and for a given initial state $|\Psi'_0\rangle$ with the same initial density and initial time-derivative of the density, we find that there is a unique external potential $v'_{ext}(r, t)$ [modulo $c(t)$] that yields this given density $n(r, t)$. The case $|\Psi_0\rangle = |\Psi'_0\rangle$ (in which the constraints on the initial state $|\Psi'_0\rangle$ are trivially satisfied) corresponds to the well-known Runge–Gross theorem (see Sect. 4.4). Our results in this section therefore provide an extension of this important theorem with the additional assumption that the density is time-analytic. As a final note we mention that the proof discussed here has been extended in an elegant way by Vignale (2004) to time-dependent current-density functional theory. In that work it is shown that currents from an interacting system with some vector potential are also representable by a vector potential in a noninteracting system. This is, however, not true anymore if one considers scalar potentials. Interacting-v-representable currents are in general not noninteracting-v-representable (D'Agosta and Vignale 2005). Further extensions of the presented proof to open quantum systems (Yuen-Zhou et al. 2010) and to quantum electrodynamics (Ruggenthaler et al. 2011a) have only recently been given.

9.3 Runge–Gross Theorem for Dipole Fields

From the above considerations a straightforward formulation of the Runge–Gross theorem without any restrictions on the time-dependence can be found. The price one has to pay is that the spatial form of the time-dependent part of the external potential is restricted (Ruggenthaler et al. 2010). The most important example are the widely employed dipole fields. Hence we consider the set of external potentials of the form

$$\{v_{ext}(r, t) = v_0(r) + r \cdot E(t)\}, \qquad (9.29)$$

where $v_0(r)$ and $E(t)$ are arbitrary functions with $v_0(r) \to 0$ for $r \to \infty$ and we thus have implicitly chosen a gauge, i.e. the arbitrary time-dependent constant is assumed to be zero. Two systems differ only with respect to the external potentials $v_{ext}(r, t)$ and $v'_{ext}(r, t)$ and have the same initial state $|\Psi_0\rangle$ at $t = t_0$ and the same interaction. We will now deduce under which conditions two different systems lead to the same density. From (9.21) for the primed and unprimed system we can immediately deduce the condition

$$\nabla \cdot \left[n(r, t_0) \nabla \left(v_{ext}(r, t_0) - v'_{ext}(r, t_0) \right) \right] = 0. \qquad (9.30)$$

From similar considerations as in the original Runge–Gross proof for $v_0(r)$ and $v_0'(r)$ we can conclude $v_0'(r) = v_0(r)$, as for $r \to \infty$ the two dipole parts have to cancel each other. In a next step we notice from (9.19), that two systems with the same density have to lead to the same expectation value of the momentum, i.e. $P(t) - P'(t) = 0$. Therefore also the force expectation values of both systems have to be the same, i.e. $F(t) - F'(t) = \frac{\partial}{\partial t}[P(t) - P'(t)] = 0$. From (9.7), which is the local force equation, we can rewrite the force-expectation value in terms of the density as

$$F(t) = -\int d^3r\, n(r, t)\nabla v_{\text{ext}}(r, t) = -N E(t)\,. \tag{9.31}$$

Here the internal forces are zero as they can be written as a divergence of a second rank tensor (Tokatly 2005a) and $N > 0$ is the number of particles due to the integral of the density (for further discussion see Sect. 5.4.2). In order for both systems to have the same force expectation value we can conclude that they also need to have $E(t) = E'(t)$. Therefore the density uniquely determines the external potential and we have a Runge–Gross theorem for dipole fields. The existence of a noninteracting system producing the same density, however, needs due to the restriction on the spatial form of the potential further investigations.

9.4 Invertibility of the Linear Density Response Function

In this section we will address the question if we can recover the potential variation $\delta v_{\text{ext}}(r, t)$ from a given density variation $\delta n(r, t)$ that was produced by it. There is, of course, an obvious non-uniqueness since both $\delta v_{\text{ext}}(r, t)$ and $\delta v_{\text{ext}}(r, t)+c(t)$, where $c(t)$ is an arbitrary time-dependent function, produce the same density variation. However, this is simply a gauge of the potential and is easily taken care of. Thus, by an inverse we will always mean an inverse modulo a purely time-dependent function $c(t)$ and by different potentials we will always mean that they differ more than a gauge $c(t)$.

From the work of Mearns and Kohn (1987) we know that different potentials can yield the same density variations (see Sect. 4.5.3 for further details). However, in their examples these potentials are always potentials that exist at all times, i.e. there is no t_0 such that $\delta v_{\text{ext}} = 0$ for times $t < t_0$. On the other hand, we know from the Runge–Gross proof that a potential $\delta v_{\text{ext}}(r, t)$ (not purely time-dependent) that is switched on at $t = t_0$ and is analytic at t_0 always causes a nonzero density variation $\delta n(r, t)$. In this proof, the first nonvanishing time-derivative of δn at t_0 is found to be linear in the corresponding derivative of δv_{ext} and therefore the linear response function is invertible. Note that this conclusion holds even for an arbitrary initial state. The conclusion is therefore true for linear response to an already time-dependent system for which the linear response function depends on both t and t' separately, rather than on the time-difference $t - t'$. In the following we give an explicit proof for the invertibility of the linear response function for which the system is initially in its ground state. However, we will relax the condition that δv_{ext} is an analytic function in time, and we therefore allow for a larger class of external potentials than assumed in the Runge–Gross theorem. For clarification we further mention that it is sometimes assumed that the Dyson-type response equations of TDDFT are based on an adiabatic switch-on of the potential at all times. This is, however, not the case. The response functions can simply be derived by first order perturbation theory on the TDSE using a sudden switch-on of the external time-dependent potential (Fetter and Walecka 1971). The typical imaginary infinitesimals that occur in the denominator of the response functions result from the Fourier-representation of the causal Heaviside function (written as a complex contour integral) in the retarded density response function rather than from an adiabatically switched-on potential. The linear response equations of TDDFT are therefore in perfect agreement with a sudden switch-on of the potential.

We consider a many-body system in its ground state. At $t = 0$ (since the system is initially described by a time-independent Hamiltonian we can, without loss of generality, put the initial time $t_0 = 0$) we switch on an external field $\delta v_{\text{ext}}(\boldsymbol{r}, t)$ which causes a density response δn. We want to show that the linear response function is invertible for these switch-on processes. From simple first order perturbation theory on the TDSE we know that the linear density response is given by (Fetter and Walecka 1971)

$$\delta n(\boldsymbol{r}_1, t_1) = \int_0^{t_1} dt_2 \int d^3 r_2 \chi_R(\boldsymbol{r}_1 t_1, \boldsymbol{r}_2 t_2) \delta v_{\text{ext}}(\boldsymbol{r}_2, t_2), \qquad (9.32)$$

where

$$\chi_R(\boldsymbol{r}_1 t_1, \boldsymbol{r}_2 t_2) = -i\theta(t_1 - t_2) \langle \Psi_0 | [\Delta \hat{n}_H(\boldsymbol{r}_1, t_1), \Delta \hat{n}_H(\boldsymbol{r}_2, t_2)] | \Psi_0 \rangle, \qquad (9.33)$$

is the retarded density response function. Note that here, instead of the density operator $\hat{n}_H$ (in the Heisenberg picture with respect to the ground state Hamiltonian $\hat{H}$), we prefer to use the density fluctuation operator $\Delta \hat{n}_H = \hat{n}_H - \langle \hat{n}_H \rangle$ in the response function, where we use that the commutator of the density operators is equal to the commutator of the density fluctuation operators. Now we go over to a Lehmann representation of the response function and we insert a complete set of eigenstates of $\hat{H}$:

$$\delta n(\boldsymbol{r}_1, t_1) = i \sum_n \int_0^{t_1} dt_2 \int d^3 r_2 e^{i\Omega_n(t_1 - t_2)} f_n^*(\boldsymbol{r}_1) f_n(\boldsymbol{r}_2) \delta v_{\text{ext}}(\boldsymbol{r}_2, t_2) + \text{c.c.}, \quad (9.34)$$

where $\Omega_n = E_n - E_{\text{GS}} > 0$ are the excitation energies of the unperturbed system (we assume the ground state to be nondegenerate) and the functions f_n are defined as

$$f_n(\boldsymbol{r}) = \langle \Psi_{\text{GS}} | \Delta \hat{n}(\boldsymbol{r}) | \Psi_n \rangle. \qquad (9.35)$$

The density response can then be rewritten as

$$\delta n(r_1, t_1) = i \sum_n f_n^*(r_1) \int_0^{t_1} dt_2 a_n(t_2) e^{i\Omega_n(t_1 - t_2)} + \text{c.c.}, \qquad (9.36)$$

where we defined

$$a_n(t) = \int d^3r \, f_n(r) \delta v_{\text{ext}}(r, t). \qquad (9.37)$$

Now note that the time integral in (9.36) has the form of a convolution. This means that we can simplify this equation using Laplace transforms. The Laplace transform and its deconvolution property are given by

$$\hat{\mathcal{L}}f(s) = \int_0^\infty dt e^{-st} f(t), \qquad (9.38a)$$

$$\hat{\mathcal{L}}(f * g)(s) = \hat{\mathcal{L}}f(s)\hat{\mathcal{L}}g(s). \qquad (9.38b)$$

where the convolution product is defined as

$$(f * g)(t) = \int_0^t d\tau f(\tau) g(t - \tau). \qquad (9.39)$$

If we now take the Laplace transform of δn in (9.36) we obtain the equation:

$$\hat{\mathcal{L}}(\delta n)(r_1, s) = i \sum_n f_n^*(r_1) \frac{1}{s - i\Omega_n} \hat{\mathcal{L}}a_n(s) + \text{c.c.} \qquad (9.40)$$

If we multiply both sides with the Laplace transform $\hat{\mathcal{L}}(\delta v_{\text{ext}})$ of δv_{ext} and integrate over r_1 we obtain

$$\int d^3r_1 \hat{\mathcal{L}}(\delta v_{\text{ext}})(r_1, s)\hat{\mathcal{L}}(\delta n)(r_1, s) = i \sum_n \frac{1}{s - i\Omega_n} |\hat{\mathcal{L}}a_n(s)|^2 + \text{c.c.}$$

$$= -2 \sum_n \frac{\Omega_n}{s^2 + \Omega_n^2} |\hat{\mathcal{L}}a_n(s)|^2. \qquad (9.41)$$

This is the basic relation that we use to prove invertibility. If we assume that $\delta n = 0$ then also $\hat{\mathcal{L}}(\delta n) = 0$ and we obtain

$$0 = \sum_n \frac{\Omega_n}{s^2 + \Omega_n^2} |\hat{\mathcal{L}}a_n(s)|^2. \qquad (9.42)$$

However, since each prefactor of $|\hat{\mathcal{L}}a_n|^2$ in the summation is positive the sum can only be zero if $\hat{\mathcal{L}}a_n = 0$ for all n. This in its turn implies that $a_n(t)$ must be zero for all n. This means also that

$$
\begin{aligned}
\int d^3r \,\Delta\hat{n}(r)\delta v_{\text{ext}}(\boldsymbol{r}, t)|\Psi_0\rangle &= \sum_n |\Psi_n\rangle \int d^3r \,\langle\Psi_n|\Delta\hat{n}(r)|\Psi_0\rangle \,\delta v_{\text{ext}}(\boldsymbol{r}, t) \\
&= \sum_n a_n(t)|\Psi_n\rangle = 0.
\end{aligned}
\tag{9.43}
$$

Note that $a_0(t)$ is automatically zero since obviously $\langle\Psi_{\text{GS}}|\Delta\hat{n}(x)|\Psi_{\text{GS}}\rangle = 0$. If we write out the above equation in first quantization again we have

$$
\sum_{k=1}^{N} \Delta v_{\text{ext}}(\boldsymbol{r}_k, t)|\Psi_{\text{GS}}\rangle = 0,
\tag{9.44}
$$

where N is the number of electrons in the system and $\Delta v_{\text{ext}}(\boldsymbol{r}, t)$ is defined as

$$
\Delta v_{\text{ext}}(\boldsymbol{r}, t) = \delta v_{\text{ext}}(\boldsymbol{r}, t) - \frac{1}{N}\int d^3r\, n_{\text{GS}}(r)\delta v_{\text{ext}}(\boldsymbol{r}, t),
\tag{9.45}
$$

where n_{GS} is the density of the unperturbed system. Now (9.44) immediately implies that $\Delta v_{\text{ext}} = 0$ and, since the second term on the right hand side of (9.45) is a purely time-dependent function, we obtain

$$
\delta v_{\text{ext}}(\boldsymbol{r}, t) = c(t).
\tag{9.46}
$$

We have therefore proven that only purely time-dependent potentials yield zero density response. In other words, the response function is, modulo a trivial gauge, invertible for switch-on processes. Note that the only restriction we put on the potential $\delta v_{\text{ext}}(\boldsymbol{r}, t)$ is that it is Laplace-transformable. This is a much weaker restriction on the potential than the constraint that it be an analytic function at $t = t_0$, as required in the Runge–Gross proof. One should, however, be careful with what one means with an inverse response function. The response function defines a mapping $\chi : \delta\mathcal{V}_{\text{ext}} \to \delta\mathcal{N}$ from the set of potential variations from a nondegenerate ground state, which we call $\delta\mathcal{V}_{\text{ext}}$, to the set of first order density variations $\delta\mathcal{N}$ that are reproduced by it. We have shown that the inverse $\chi^{-1} : \delta\mathcal{N} \to \delta\mathcal{V}_{\text{ext}}$ is well-defined modulo a purely time-dependent function. However, there are density variations that can never be produced by a finite potential variation and are therefore not in the set $\delta\mathcal{N}$. An example of such a density variation is one which is identically zero on some finite volume.

Another consequence of the above analysis is the following. Suppose the linear response kernel has eigenfunctions, i.e. there is a λ such that

$$
\int dt_2 \int d^3r_2 \chi_R(\boldsymbol{r}_1 t_1, \boldsymbol{r}_2 t_2)\zeta(\boldsymbol{r}_2, t_2) = \lambda\zeta(\boldsymbol{r}_1, t_1).
\tag{9.47}
$$

Laplace transforming this equation yields

$$\int d^3 r_2 \, \Xi(\mathbf{r}_1, \mathbf{r}_2, s) \hat{\mathcal{L}} \zeta(\mathbf{r}_2, s) = \lambda \hat{\mathcal{L}} \zeta(\mathbf{r}_1, s), \tag{9.48}$$

where Ξ is the Laplace transform of χ explicitly given by

$$\Xi(\mathbf{r}_1, \mathbf{r}_2, s) = i \sum_n \frac{f_n^*(\mathbf{r}_1) f_n(\mathbf{r}_2)}{s - i\Omega_n} + \text{c.c.} \tag{9.49}$$

Since Ξ is a real Hermitian operator, its eigenvalues λ are real and its eigenfunctions $\hat{\mathcal{L}}\zeta$ can be chosen to be real. Then ζ is real as well and (9.41) implies (if we take $\delta v_{\text{ext}} = \zeta$ and $\delta n = \lambda \zeta$)

$$\lambda \int d^3 r \left[\hat{\mathcal{L}} \zeta(\mathbf{r}, s) \right]^2 < 0, \tag{9.50}$$

which implies $\lambda < 0$. We have therefore proven that if there are density variations that are proportional to the applied potential, then this constant of proportionality is negative. In other words, the eigenvalues of the density response function are negative. In this derivation we made again explicit use of Laplace transforms and therefore of the condition that $\zeta = 0$ for $t < 0$. The work of Mearns and Kohn shows that positive eigenvalues are possible when this restriction is not made. The same is true when one considers response functions for excited states (Gaudoin and Burke 2004). We finally note that similar results are readily obtained for the static density response function (van Leeuwen 2003) in which case the negative eigenvalues of the response function are an immediate consequence of the Hohenberg–Kohn theorem.

Let us now see what our result implies. We considered the density $n[v_{\text{ext}}]$ as a functional of v_{ext} and established that the response kernel $\chi[v_{\text{GS}}] = \delta n / \delta v_{\text{ext}}[v_{\text{GS}}]$ is invertible where v_{GS} is the potential in the ground state and that $\delta n / \delta v_{\text{ext}}[v_{\text{GS}}] < 0$ in the sense that its eigenvalues are all negative definite. We can now apply a fundamental theorem of calculus, the *inverse function theorem*. For functions of real numbers the theorem states that if a continuous function $y(x)$ is differentiable at x_0 and if $dy/dx(x_0) \neq 0$ then locally there exists an inverse $x(y)$ for y close enough to y_0, where $y(x_0) = y_0$. The theorem can be extended to functionals on function spaces [to be precise Banach spaces, for details see Choquet-Bruhat et al. (1991)]. For our case, this theorem implies that if the functional $n[v_{\text{ext}}]$ is differentiable at the ground state potential v_{GS} and the derivative $\chi[v_{\text{GS}}] = \delta n / \delta v_{\text{ext}}[v_{\text{GS}}]$ is an invertible kernel then for potentials v_{ext} close enough to v_{GS} (in Banach norm sense) the inverse map $v_{\text{ext}}[n]$ exists. Since we have shown that the linear response function $\chi[v_{\text{GS}}]$ is invertible this then proves the Runge–Gross theorem for Laplace transformable switch-on potentials for systems initially in the ground state.

9.5 Global Fixed-Point Proof of TDDFT

Here we reformulate and generalize the foundations of TDDFT. The central idea is to restate the fundamental one-to-one correspondence between densities and potentials as a global fixed-point question for potentials on a given time-interval. We prove that the unique fixed point, i.e. the unique potential generating a given density, is reached as the limiting point of an iterative procedure and show its convergence under some conditions. Assuming the existence of a certain density response function $\tilde{\chi}$ this approach avoids the usual restrictions of Taylor-expandability in time of the Runge–Gross proof and its extension. Here we will further assume certain smoothness properties of the potentials in space (finite second-oder spatial derivatives) in order to keep the presentation as simple as possible. This restriction as well as some others in the course of the presentation can be lifted at the expense of further mathematical details (Ruggenthaler and van Leeuwen 2011b; Penz and Ruggenthaler 2011). Before we outline the general proof we reinvestigate (9.11), which is of fundamental importance to TDDFT

$$-\nabla \cdot [n[v_{\text{ext}}](\boldsymbol{r}, t)\nabla v_{\text{ext}}(\boldsymbol{r}, t)] = q[v_{\text{ext}}](\boldsymbol{r}, t) - \frac{\partial^2}{\partial t^2} n[v_{\text{ext}}](\boldsymbol{r}, t). \quad (9.51)$$

Here we made the dependence on the external potential explicit. If we replace $n[v_{\text{ext}}](\boldsymbol{r}, t)$ in (9.51) by a *given* density n subject to similar conditions as (9.17) and (9.18) at time t_0 with the initial state $|\Psi_0\rangle$, i.e.

$$n(\boldsymbol{r}, t_0) = \langle \Psi_0 | \hat{n}(\boldsymbol{r}) | \Psi_0 \rangle \quad (9.52a)$$

$$\frac{\partial}{\partial t} n(\boldsymbol{r}, t_0) = -\langle \Psi_0 | \nabla \cdot \hat{\boldsymbol{j}}(\boldsymbol{r}) | \Psi_0 \rangle \quad (9.52b)$$

where $\hat{\boldsymbol{j}}(\boldsymbol{r})$ is the current-density operator (9.5), then (9.51) becomes

$$-\nabla \cdot [n(\boldsymbol{r}, t)\nabla v_{\text{ext}}(\boldsymbol{r}, t)] = q[v_{\text{ext}}](\boldsymbol{r}, t) - \frac{\partial^2}{\partial t^2} n(\boldsymbol{r}, t). \quad (9.53)$$

This is a nonlinear equation for v_{ext} which needs to be solved with specified boundary conditions (this amounts to fixing a gauge for v_{ext}). If we propagate the TDSE with initial state $|\Psi_0\rangle$ and with a potential v_{ext} that is a solution to (9.53) then for this potential clearly also the local force equation (9.51) will be satisfied with the same initial conditions (9.52a, b). Subtracting (9.53) from (9.51) then yields the equation

$$\frac{\partial^2}{\partial t^2} \tilde{n}(\boldsymbol{r}, t) - \nabla \cdot \left[\tilde{n}(\boldsymbol{r}, t)\nabla v_{\text{ext}}(\boldsymbol{r}, t) \right] = 0 \quad (9.54)$$

for the density difference $\tilde{n}(\boldsymbol{r}, t) = n[v_{\text{ext}}](\boldsymbol{r}, t) - n(\boldsymbol{r}, t)$ with initial conditions $\tilde{n}(\boldsymbol{r}, t_0) = \frac{\partial}{\partial t} \tilde{n}(\boldsymbol{r}, t_0) = 0$. The unique solution of (9.54) with these initial conditions is $\tilde{n}(\boldsymbol{r}, t) = 0$ and hence $n(\boldsymbol{r}, t) = n[v_{\text{ext}}](\boldsymbol{r}, t)$, i.e. the density in (9.53) is identical

to the one that is obtained from time-propagation of the TDSE with the solution of
(9.53). By making different choices for the density $n(r, t)$ in (9.53) we can deduce
some important consequences of this result. If we choose $n(r, t) = n[u_{ext}](r, t)$ to
be the density obtained by time-propagation of the TDSE with potential u_{ext} and the
same initial state $|\Psi_0\rangle$ then we must have $n[u_{ext}](r, t) = n[v_{ext}](r, t)$ where v_{ext}
is a solution of (9.53). The uniqueness of a potential v_{ext} for a given density n (the
Runge–Gross theorem) is thus equivalent to the uniqueness of the solution of (9.53),
i.e. $u_{ext} = v_{ext}$. If we choose $n(r, t)$ to be the density obtained by solving a TDSE for
a system with different two-particle interactions $\hat{V}'_{ee}$ and with different initial state
$|\Psi'_0\rangle$ then the existence of a solution to (9.53) implies that the same density can be
reproduced by a potential v_{ext} in our system with interaction $\hat{V}_{ee}$ and initial state $|\Psi_0\rangle$,
i.e. it is v-representable in our system. For the special case $\hat{V}_{ee} = 0$ this amounts
to reproducing the density of an interacting system within a noninteracting system,
which is known as the Kohn–Sham construction and forms the basis of virtually all
applications of TDDFT. The key question, which is crucial for the whole foundation
of TDDFT, is thus whether a solution to (9.53) is unique and exists. As we have
seen from the previous considerations existence and uniqueness have indeed been
established under the restrictions that the potential v_{ext} is Taylor-expandable around
the initial time as well as the density. Here our main goal is to present a proof that
lifts these restrictions.

Before going into details we first give the main idea. For a potential $v_{ext}^{(0)}(r, t)$
and initial state $|\Psi_0\rangle$ at time t_0 we can propagate the TDSE in a given time interval
$[t_0, t_1]$ and construct the function $q[v_{ext}^{(0)}](r, t)$ from (9.12b). This provides a mapping
$\mathcal{P} : v_{ext}^{(0)} \mapsto q[v_{ext}^{(0)}]$ (see Fig. 9.1). Let us fix a density $n(r, t)$ subject to the conditions
(9.52a, b) in terms of the initial state $|\Psi_0\rangle$. Then with the inhomogeneity $q[v_{ext}^{(0)}] - \frac{\partial^2}{\partial t^2} n$
we can solve (9.53) for a new potential $v_{ext}^{(1)}$, i.e.

$$-\nabla \cdot \left[n(r, t) \nabla v_{ext}^{(1)}(r, t) \right] = q\left[v_{ext}^{(0)} \right](r, t) - \frac{\partial^2}{\partial t^2} n(r, t) \qquad (9.55)$$

with given boundary conditions (this amounts to fixing a gauge for $v_{ext}^{(1)}$). This
provides us with a second map $\mathcal{W} : q\left[v_{ext}^{(0)} \right] \mapsto v_{ext}^{(1)}$ (see Fig. 9.1). The combined
map

$$\mathcal{F}\left[v_{ext}^{(0)} \right] = (\mathcal{W} \circ \mathcal{P})\left[v_{ext}^{(0)} \right] = v_{ext}^{(1)} \qquad (9.56)$$

maps our original potential $v_{ext}^{(0)}$ to a new one $v_{ext}^{(1)}$. If for some potential v_{ext} we
have $v_{ext} = \mathcal{F}[v_{ext}]$, i.e. v_{ext} is a fixed point of the mapping $\mathcal{F}$, then we satisfy
(9.53). Consequently, the question whether a solution to (9.53) exists and is unique
is equivalent to the question whether the mapping $\mathcal{F}$ has a unique fixed point. This
is exactly what we will prove in this section. Our proof is based on the following
inequality

$$\left\| \mathcal{F}\left[v_{ext}^{(1)} \right] - \mathcal{F}\left[v_{ext}^{(0)} \right] \right\|_\alpha \le a \left\| v_{ext}^{(1)} - v_{ext}^{(0)} \right\|_\alpha, \qquad (9.57)$$

Fig. 9.1 The
potential–potential mapping
$\mathcal{F}$ of (9.56) as composition
of the mappings $\mathcal{P}$ and $\mathcal{W}$

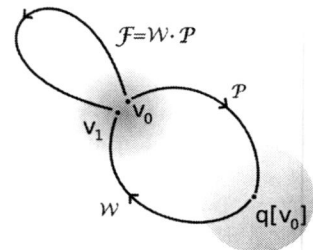

with $a < 1$ and where $\|\cdot\|_\alpha$ is a norm dependent on a positive parameter α on the
space of potentials. Comparable norms are commonly used in the solution of initial-
value problems (Evans 2010). This inequality can be derived in two steps. In the
first step we prove that if two potentials are close in norm then also the observables
$O[v_{\text{ext}}](t)$ calculated from them are close. In particular for $\hat{O} = \hat{q}(\boldsymbol{r})$ we prove

$$\left\| q\left[v_{\text{ext}}^{(1)}\right] - q\left[v_{\text{ext}}^{(0)}\right] \right\|_\alpha \leq \frac{C}{\sqrt{\alpha}} \left\| v_{\text{ext}}^{(1)} - v_{\text{ext}}^{(0)} \right\|_\alpha, \tag{9.58}$$

with C a positive constant. In the second step we prove that

$$\left\| \mathcal{F}\left[v_{\text{ext}}^{(1)}\right] - \mathcal{F}\left[v_{\text{ext}}^{(0)}\right] \right\|_\alpha \leq D \left\| q\left[v_{\text{ext}}^{(1)}\right] - q\left[v_{\text{ext}}^{(0)}\right] \right\|_\alpha \tag{9.59}$$

for a positive constant D. The combination of these two statements then yields (9.57)
where $a = CD/\sqrt{\alpha}$ and we can choose $\sqrt{\alpha} > CD$. It remains to prove (9.58)
and (9.59). Before we do so, we point out conditions on the set of potentials and
densities. In order for the boundary value problem of mapping $\mathcal{W}$ to be well defined
(Ruggenthaler et al. 2009; Penz and Ruggenthaler 2011) we have to restrict our
considerations on an arbitrarily large but finite region $\mathcal{V} \subset \mathbb{R}^3$ with boundary $\partial \mathcal{V}$.
The potentials are assumed to have finite second-order spatial derivatives, i.e. they
are generated by a Poisson equation $-\nabla^2 v_{\text{ext}}(\boldsymbol{r}, t) = 4\pi \rho_{\text{ext}}(\boldsymbol{r}, t)$ from a finite
external charge distribution ρ_{ext} that is only required to be piece-wise continuous
in the time-variable. This excludes the possibility of an external one-body potential
with a Coulombic singularity generated by a point charge but still includes potentials
generated by finite atomic nuclei (which are actually closer to the physical reality).
For such potentials the divergence of the force on the right hand side of (9.55) is finite.
The density $n(\boldsymbol{r}, t)$ in (9.55) is assumed to have a finite spatial derivative $\nabla n(\boldsymbol{r}, t)$
and finite second time-derivative $\frac{\partial^2}{\partial t^2} n(\boldsymbol{r}, t)$. These requirements make all terms in
(9.55) well-defined. Some of them (especially the exclusion of Coulombic potentials)
can be relaxed, however at the expense of an increase in mathematical technicalities
(Ruggenthaler and van Leeuwen 2011b; Penz and Ruggenthaler 2011). For the clarity
of our presentation we therefore restrict ourselves to these assumptions.

We start by deriving inequality (9.58) for the mapping $\mathcal{P}$ of Fig. 9.1. The main
ingredient is the fundamental theorem of calculus (Griffel 1985). Suppressing the
time-arguments we can write

$$O\left[v_{\text{ext}}^{(1)}\right] - O\left[v_{\text{ext}}^{(0)}\right] = \int_0^1 d\lambda \frac{dO}{d\xi}\left[v_{\text{ext}}^{(0)} + \xi\left(v_{\text{ext}}^{(1)} - v_{\text{ext}}^{(0)}\right)\right]\Big|_{\xi=\lambda}, \qquad (9.60)$$

for any operator expectation value $O[v_{\text{ext}}](t)$. In the case that we take $\hat{O} = \hat{q}(\boldsymbol{r})$ this equation yields

$$q\left[v_{\text{ext}}^{(1)}\right](\boldsymbol{r}, t) - q\left[v_{\text{ext}}^{(0)}\right](\boldsymbol{r}, t)$$

$$= \int_{t_0}^t dt' \int_{\mathcal{V}} d^3 r' \tilde{\chi}(\boldsymbol{r}t, \boldsymbol{r}'t')\left[v_{\text{ext}}^{(1)}(\boldsymbol{r}', t') - v_{\text{ext}}^{(0)}(\boldsymbol{r}', t')\right], \qquad (9.61)$$

where we defined

$$\tilde{\chi}(\boldsymbol{r}t, \boldsymbol{r}'t') = -i \int_0^1 d\lambda \left\langle \Psi_0 \left| \left[\hat{q}_{H_\lambda}(\boldsymbol{r}, t), \hat{n}_{H_\lambda}(\boldsymbol{r}', t')\right] \right| \Psi_0 \right\rangle \qquad (9.62)$$

and $\hat{O}_{H_\lambda}$ is the operator $\hat{O}$ in the Heisenberg representation with respect to Hamiltonian $\hat{H}_\lambda$ with potential $v_{\text{ext}}^\lambda = v_{\text{ext}}^{(0)} + \lambda\left(v_{\text{ext}}^{(1)} - v_{\text{ext}}^{(0)}\right)$. Equation 9.62 can be obtained directly from the TDSE by evaluating the expectation value $q\left[v_{\text{ext}}^\xi\right](\boldsymbol{r}, t)$ to first order in ξ around λ which amounts to linear response theory (Fetter and Walecka 1971). Equation 9.61 is of the form

$$f(\boldsymbol{r}, t) = \int_{t_0}^t dt' \int_{\mathcal{V}} d^3 r' \tilde{\chi}(\boldsymbol{r}t, \boldsymbol{r}'t')g(\boldsymbol{r}', t') = (\chi g)(\boldsymbol{r}, t). \qquad (9.63)$$

One can then derive that

$$\|f(t)\|^2 \leq [C(t)]^2 \int_{t_0}^t dt' \|g(t')\|^2, \qquad (9.64)$$

where we defined the norm $\|f(t)\|^2 = \int_{\mathcal{V}} d^3 r f(\boldsymbol{r}, t)^2$. In this expression the function $C(t)$ (the operator norm) is defined as

$$[C(t)]^2 = \sup_{g \neq 0} \frac{\|(\chi g)(t)\|^2}{\int_{t_0}^t dt' \|g(t')\|^2} \qquad (9.65)$$

Using a Cauchy–Schwarz inequality the function $C(t)$ can be shown to satisfy

$$[C(t)]^2 \leq \sup_{t' \in [t_0, t]} \int_{\mathcal{V}} d^3 r \int_{\mathcal{V}} d^3 r' \tilde{\chi}(\boldsymbol{r}t, \boldsymbol{r}'t')^2. \qquad (9.66)$$

The integral on the right hand side of inequality (9.64) can be manipulated as follows

$$
\int\limits_{t_0}^{t} dt' \|g(t')\|^2 = \int\limits_{t_0}^{t} dt' e^{-\alpha(t'-t_0)} e^{\alpha(t'-t_0)} \|g(t')\|^2
$$

$$
\leq \|g\|_{\alpha,t}^2 \int\limits_{t_0}^{t} dt' e^{\alpha(t'-t_0)} \leq \|g\|_{\alpha,t}^2 \frac{e^{\alpha(t-t_0)}}{\alpha}
$$

(9.67)

where α is an arbitrary positive number and where we defined

$$
\|g\|_{\alpha,t}^2 = \sup_{t' \in [t_0,t]} \left\{ \|g(t')\|^2 e^{-\alpha(t'-t_0)} \right\}.
$$

(9.68)

With this result inequality (9.64) can be written as

$$
\|f\|_{\alpha,t}^2 \leq \frac{[\tilde{C}(t)]^2}{\alpha} \|g\|_{\alpha,t}^2
$$

(9.69)

where $\tilde{C}(t) = \sup_{t' \in [t_0,t]} C(t')$. We will consider functions on an arbitrarily large but finite interval $[t_0, t_1]$ with $t_1 > t_0$. If we define $\|f\|_\alpha = \|f\|_{\alpha,t_1}$ and $C = \tilde{C}(t_1)$ and apply this to (9.61) we obtain (9.58). This concludes the first part of the proof. The discussion so far was related to the mapping $\mathcal{P}$. Let us now discuss the second mapping $\mathcal{W}$.

We consider the Sturm–Liouville operator $Q = -\nabla \cdot [n\nabla]$. By partial integration we find for two external potentials $v_{\text{ext}}^{(0)}$ and $v_{\text{ext}}^{(1)}$ within the standard inner product

$$
\left\langle v_{\text{ext}}^{(0)} | Q v_{\text{ext}}^{(1)} \right\rangle - \left\langle Q v_{\text{ext}}^{(0)} | v_{\text{ext}}^{(1)} \right\rangle =
$$
$$
\int\limits_{\partial \mathcal{V}} dS \cdot n(\mathbf{r}, t) \left[v_{\text{ext}}^{(1)}(\mathbf{r}, t) \nabla v_{\text{ext}}^{(0)}(\mathbf{r}, t) - v_{\text{ext}}^{(0)}(\mathbf{r}, t) \nabla v_{\text{ext}}^{(1)}(\mathbf{r}, t) \right].
$$

(9.70)

Hence, in order for Q to be self-adjoint the boundary term has to vanish. If $n(\mathbf{r}, t) \geq \epsilon > 0$ in $\mathcal{V}$ we know that we find solutions to the general inhomogeneous Sturm–Liouville problem (Ruggenthaler et al. 2009) with the boundary conditions $v_{\text{ext}} = 0$ on $\partial \mathcal{V}$, making Q a self-adjoint operator. We then have an orthonormal set of eigenfunctions (Griffel 1985). If the density is zero at the boundary then a boundedness condition at the edge singles out a unique set (a famous example is the Legendre equation which would correspond to an operator Q with density $n(x) = (1 - x^2)$ on $[-1, 1]$). For an extensive discussion of these issues in the one-dimensional case see (Bailey et al. 2001). In general we have a set of orthonormal eigenfunctions $\{\phi_i(\mathbf{r}, t)\}$ with $Q\phi_i = \lambda_i \phi_i$ and positive eigenvalues $0 \leq \lambda_0 < \lambda_1 \leq \cdots$, since $\lambda_i = \langle \phi_i | Q \phi_i \rangle = \langle \nabla \phi_i | n \nabla \phi_i \rangle \geq 0$. The eigenfunction to the eigenvalue zero is $\phi_0 = c(t)$. If we now subtract (9.55) for $v_{\text{ext}}^{(1)} = \mathcal{F}\left[v_{\text{ext}}^{(0)}\right]$ from the one for $v_{\text{ext}}^{(2)} = \mathcal{F}\left[v_{\text{ext}}^{(1)}\right]$ we obtain

$$-\nabla \cdot \left[n(\boldsymbol{r}, t)\nabla \left(v_{\text{ext}}^{(2)}(\boldsymbol{r}, t) - v_{\text{ext}}^{(1)}(\boldsymbol{r}, t)\right)\right] = q\left[v_{\text{ext}}^{(1)}\right](\boldsymbol{r}, t) - q\left[v_{\text{ext}}^{(0)}\right](\boldsymbol{r}, t). \quad (9.71)$$

We can then expand $v_{\text{ext}}^{(2)} - v_{\text{ext}}^{(1)}$ and $\zeta = q\left[v_{\text{ext}}^{(1)}\right] - q\left[v_{\text{ext}}^{(0)}\right]$ in terms of the eigenfunctions of Q, i.e. $\left(v_{\text{ext}}^{(2)} - v_{\text{ext}}^{(1)}\right)(\boldsymbol{r}, t) = \sum_{i=0}^{\infty} u_i(t)\phi_i(\boldsymbol{r}, t)$ and $\zeta(\boldsymbol{r}, t) = \sum_{i=0}^{\infty} \zeta_i(t)\phi_i(\boldsymbol{r}, t)$. Since ζ is a divergence $\int_V d^3r\,\zeta(\boldsymbol{r}, t) = 0$ and therefore the constant function ϕ_0 does not contribute to the expansion. Likewise the gauge fixing allows us to exclude ϕ_0 from the expansion of $v_{\text{ext}}^{(2)} - v_{\text{ext}}^{(1)}$. Inserting the expansions into (9.71) yields $u_i(t) = \zeta_i(t)/\lambda_i(t)$ $(\lambda_i(t) > 0)$ and

$$\begin{aligned}\left\|v_{\text{ext}}^{(2)}(t) - v_{\text{ext}}^{(1)}(t)\right\|^2 &= \sum_{i=1}^{\infty} \left|\frac{\zeta_i(t)}{\lambda_i(t)}\right|^2 \le \frac{1}{\lambda_1(t)^2} \sum_{i=1}^{\infty} |\zeta_i(t)|^2 \\ &= \frac{1}{\lambda_1(t)^2} \left\|q\left[v_{\text{ext}}^{(1)}\right](t) - q\left[v_{\text{ext}}^{(0)}\right](t)\right\|^2.\end{aligned} \quad (9.72)$$

If we now multiply (9.72) with $e^{-\alpha(t-t_0)}$ and take a maximum over the interval $[t_0, t_1]$ we arrive at (9.59) in which $D^2 = \max_{t\in[t_0,t_1]}\{\lambda_1(t)^{-2}\}$. This then together with (9.58) establishes our main inequality (9.57).

This equation can now be used to prove the uniqueness of a solution to (9.53). Suppose we would have two fixed-point solutions u_{ext} and v_{ext}, i.e. $u_{\text{ext}} = \mathcal{F}[u_{\text{ext}}]$ and $v_{\text{ext}} = \mathcal{F}[v_{\text{ext}}]$. Then by choosing $\sqrt{\alpha} = 2CD$ in (9.57) we find

$$\|v_{\text{ext}} - u_{\text{ext}}\|_\alpha = \|\mathcal{F}[v_{\text{ext}}] - \mathcal{F}[u_{\text{ext}}]\|_\alpha \le \frac{1}{2}\|v_{\text{ext}} - u_{\text{ext}}\|_\alpha \quad (9.73)$$

from which we conclude that $u_{\text{ext}} = v_{\text{ext}}$. Hence, if a solution to (9.53) exists then it is unique. This conclusion is equivalent to the Runge–Gross theorem. It states that a density $n[v_{\text{ext}}](\boldsymbol{r}, t)$ cannot be produced by another potential u_{ext} starting from the same initial state. This is now proven without assumptions on the Taylor-expandability in time of the potential. Suppose that the density of an interacting system is representable in a noninteracting system then this theorem guarantees that the effective potential producing the same density in this system is unique. This establishes the uniqueness of a Kohn–Sham scheme.

Let us now address the existence of a solution to (9.53). We see from (9.65) that the constant $C = C(t_1)$ in (9.58) is dependent on the response function $\tilde{\chi}$ and hence via Eq. 9.61 on potentials $v_{\text{ext}}^{(0)}$ and $v_{\text{ext}}^{(1)}$, i.e. $C = C\left[v_{\text{ext}}^{(0)}, v_{\text{ext}}^{(1)}\right]$. If a constant $C_{\text{sup}} = \sup_{v_{\text{ext}}^{(0)}, v_{\text{ext}}^{(1)}} C\left[v_{\text{ext}}^{(0)}, v_{\text{ext}}^{(1)}\right]$ exists when we range over all potentials $v_{\text{ext}}^{(0)}$ and $v_{\text{ext}}^{(1)}$ then (9.57) with $a = C_{\text{sup}}D/\sqrt{\alpha}$ and the choice $\sqrt{\alpha} > C_{\text{sup}}D$ is equivalent to the definition of a contractive mapping (Griffel 1985). Note that one can replace the assumption of the existence of a C_{sup} by a less severe assumption (Ruggenthaler and van Leeuwen 2011b). If we then let $v_{\text{ext}}^{(k)} = \mathcal{F}^k \left[v_{\text{ext}}^{(0)} \right]$ denote the k-fold application of the mapping $\mathcal{F}$ on a given initial potential $v_{\text{ext}}^{(0)}$ then (9.57) implies $\left\| v_{\text{ext}}^{(k+1)} - v_{\text{ext}}^{(k)} \right\|_\alpha \leq a^k \left\| v_{\text{ext}}^{(1)} - v_{\text{ext}}^{(0)} \right\|_\alpha$. Therefore the $v_{\text{ext}}^{(k)}$ are a Cauchy series which converges to a unique v_{ext}, i.e. $v_{\text{ext}}^{(k)} \rightarrow v_{\text{ext}}$ for $k \rightarrow \infty$. This is known as the contraction mapping theorem (Griffel 1985) and proves existence of a unique solution to Eq. 9.53, i.e. showing v-representability of the given density.

9.6 Consequences of v-Representability for the Quantum Mechanical Action

The role that is played by the energy functional in stationary DFT is played by the action functional in TDDFT. A correct form of the variational action principle for densities appears naturally within the framework of Keldysh theory (Marques et al. 2006a). However, historically the first action within TDDFT was defined by Peuckert (1978—who already made a connection to Keldysh theory) and later in the Runge–Gross paper (Runge and Gross 1984). However, as was discovered later (Gross et al. 1996; Burke and Gross 1998b) this form of the action and the usual variational principle leads to paradoxical results. Rajagopal (1996) attempted to introduce an action principle in TDDFT using the formalism of Jackiw and Kerman (1979) for deriving time-ordered n-point functions in quantum field theory. However, due to the time-ordering inherent in the work of Jackiw and Kerman the basic variable of Ragagopal's formalism is not the time-dependent density but a transition element of the density operator between a wavefunction evolving from the past and a wavefunction evolving from the future to a certain time t. Moreover, the action functional in this formalism suffers from the same difficulties as the action introduced by Runge and Gross. In this section we will sketch these difficulties and show that they arise due to a restriction of the variational freedom as a consequence of v-representability constraints. For two other recent discussions of these points we refer to van Leeuwen (2001) and Maitra et al. (2002). Finally we will introduce another resolution of these difficulties (Vignale 2008) using not the Keldysh formalism but modifying the variational principle.

We start with the following time-dependent action functional

$$A[\Psi] = \int\limits_{t_0}^{t_1} dt \, \left\langle \Psi | i\frac{\partial}{\partial t} - \hat{H}(t) | \Psi \right\rangle. \tag{9.74}$$

The usual approach is to require the action to be stationary under variations $\delta\Psi$ that satisfy $\delta\Psi(t_0) = \delta\Psi(t_1) = 0$. We then find after a partial integration

$$\delta A = \int\limits_{t_0}^{t_1} dt \; \left\langle \delta\Psi \,\big|\, i\frac{\partial}{\partial t} - \hat{H}(t) \,\big|\, \Psi \right\rangle + \text{c.c.} + i\langle \Psi | \delta\Psi \rangle \big|_{t_0}^{t_1}. \qquad (9.75)$$

With the boundary conditions and the fact that the real and imaginary part of $\delta\Psi$ can be varied independently we obtain the result that

$$\left[i\frac{\partial}{\partial t} - \hat{H}(t) \right] |\Psi\rangle = 0, \qquad (9.76)$$

which is just the time-dependent Schrödinger equation. We see that the variational requirement $\delta A = 0$, together with the boundary conditions is equivalent to the time-dependent Schrödinger equation.

A different derivation (Löwdin and Mukherjee 1972) which does not put any constraints on the variations at the endpoints of the time interval is the following. We consider again a first order change in the action due to changes in the wavefunction and require that the action is stationary. We have the general relation

$$0 = \delta A = \int\limits_{t_0}^{t_1} dt \; \left\langle \delta\Psi \,\left|\, i\frac{\partial}{\partial t} - \hat{H}(t) \,\right|\, \Psi \right\rangle + \int\limits_{t_0}^{t_1} dt \; \left\langle \Psi \,\left|\, i\frac{\partial}{\partial t} - \hat{H}(t) \,\right|\, \delta\Psi \right\rangle. \qquad (9.77)$$

We now choose the variations $\delta\Psi = \delta\tilde{\Psi}$ and $\delta\Psi = i\delta\tilde{\Psi}$ where $\delta\tilde{\Psi}$ is arbitrary. We thus obtain

$$0 = \delta A = \int\limits_{t_0}^{t_1} dt \; \left\langle \delta\tilde{\Psi} \,\left|\, i\frac{\partial}{\partial t} - \hat{H}(t) \,\right|\, \Psi \right\rangle + i \int\limits_{t_0}^{t_1} dt \; \left\langle \Psi \,\left|\, i\frac{\partial}{\partial t} - \hat{H}(t) \,\right|\, \delta\tilde{\Psi} \right\rangle \qquad (9.78a)$$

$$0 = \delta A = -i \int\limits_{t_0}^{t_1} dt \; \left\langle \delta\tilde{\Psi} \,\left|\, i\frac{\partial}{\partial t} - \hat{H}(t) \,\right|\, \Psi \right\rangle + i \int\limits_{t_0}^{t_1} dt \; \left\langle \Psi \,\left|\, i\frac{\partial}{\partial t} - \hat{H}(t) \,\right|\, \delta\tilde{\Psi} \right\rangle. \qquad (9.78b)$$

From (9.78a, b) we obtain

$$0 = \int\limits_{t_0}^{t_1} dt \; \left\langle \delta\tilde{\Psi} \,\left|\, i\frac{\partial}{\partial t} - \hat{H}(t) \,\right|\, \Psi \right\rangle. \qquad (9.79)$$

Since this must be true for arbitrary $\delta\tilde{\Psi}$ we again obtain the time-dependent Schrödinger equation

$$\left[i\frac{\partial}{\partial t} - \hat{H}(t) \right] |\Psi\rangle = 0. \qquad (9.80)$$

We did not need to put any boundary conditions on the variations at all. We only required that if $\delta\tilde{\Psi}$ is an allowed variation that then also $i\delta\tilde{\Psi}$ is an allowed variation.

Let us now derive and discuss the problems with the variational principle when one attempts to construct a time-dependent density-functional theory. The obvious definition of a density functional would be (Runge and Gross 1984)

$$A[n] = \int_{t_0}^{t_1} dt \left\langle \Psi[n] \left| i\frac{\partial}{\partial t} - \hat{H}(t) \right| \Psi[n] \right\rangle, \tag{9.81}$$

where $|\Psi[n]\rangle$ is a wavefunction which yields the density $n(\boldsymbol{r}, t)$ and evolves from a given initial state $|\Psi_0\rangle$ with initial density $n_0(\boldsymbol{r})$. By the Runge–Gross theorem the wavefunction is determined up to a phase factor. In order to define the action uniquely we have to make a choice for this phase factor. An obvious choice would be to choose the $|\Psi[n]\rangle$ that evolves in the external potential $v_{\text{ext}}(\boldsymbol{r}, t)$ that vanishes at infinity and yields the density $n(\boldsymbol{r}, t)$. This corresponds to choosing a particular kind of gauge. There are of course many more phase conventions possible. We can then rewrite the action functional (9.81) as

$$A[n] = A_0[n] - \int_{t_0}^{t_1} dt \int d^3r \, v_{\text{ext}}(\boldsymbol{r}, t) \, n(\boldsymbol{r}, t), \tag{9.82}$$

where the internal action is defined by

$$A_0[n] = \int_{t_0}^{t_1} dt \left\langle \Psi[n] \left| i\frac{\partial}{\partial t} - \hat{T} - \hat{V}_{\text{ee}} \right| \Psi[n] \right\rangle. \tag{9.83}$$

If we now make the action functional stationary, i.e. $\delta A = 0$, with respect to the density we immediately find

$$v_{\text{ext}}(\boldsymbol{r}, t) = \frac{\delta A_0[n]}{\delta n(\boldsymbol{r}, t)}. \tag{9.84}$$

This implies that

$$\frac{\delta v_{\text{ext}}(\boldsymbol{r}, t)}{\delta n(\boldsymbol{r}', t')} = \frac{\delta^2 A_0[n]}{\delta n(\boldsymbol{r}, t)\delta n(\boldsymbol{r}', t')}. \tag{9.85}$$

However, this equation is wrong as the left hand side is causal, i.e. it is nonzero only for $t > t'$, and the right hand side is symmetric with respect to t and t'. This finding was termed the causality paradox. Note that this problem arises for all possible forms of the action functional with the usual variational principle in real time, as the symmetry of the second functional derivative and the causality of the potential variations are independent of the form of the functional itself. Where does this causality paradox come from? Suppose we avoid the phase problem in the first place by defining a functional of the external potential rather than the density

$$A[v] = \int\limits_{t_0}^{t_1} dt \ \left\langle \Psi[v] \left| i\frac{\partial}{\partial t} - \hat{H}(t) \right| \Psi[v] \right\rangle. \tag{9.86}$$

Note that the potential v in the argument of the action is only used to parametrize the set of wavefunctions used in the action principle. This potential v is therefore not the same as the external potential in the Hamiltonian $\hat{H}(t)$ of (9.86) as this Hamiltonian is fixed. The state $|\Psi[v]\rangle$ is a state that evolves from a given initial state $|\Psi_0\rangle$ by solution of a time-dependent Schrödinger equation with potential v as its external potential. As the potential obviously defines $|\Psi[v]\rangle$ uniquely, including its phase, the action is well-defined. The question is now whether one can recover the time-dependent Schrödinger equation by making the action stationary with respect to potential variations δv. It is readily seen that this is not the case. The reason for this is that all variations $\delta\Psi$ of the wavefunction must now be caused by potential variations δv which lead to variations over a restricted set of wavefunctions. In other words, the variations $\delta\Psi$ must be v-representable. For instance, when deriving the Schrödinger equation from the variational principle one can not assume the boundary conditions $\delta\Psi(t_0) = \delta\Psi(t_1) = 0$. Since the time-dependent Schrödinger equation is first order in time, the variation $\delta\Psi(t)$ at times $t > t_0$ is completely determined by the boundary condition for $\delta\Psi(t_0)$. We are thus no longer free to specify a second boundary condition at a later time t_1. Moreover, we are not allowed to treat the real and imaginary part of $\delta\Psi$ as independent variations since both are determined simultaneously by the potential variation δv. This means that the first derivation of the TDSE that we presented in this section can not be carried out. It is readily seen that also the second derivation based on Eqs. 9.78a, b fails. If $\delta\Psi$ is a variation generated by some $\delta\hat{V}(t) = \int d^3 r \hat{n}(r)\delta v(r, t)$, then $\delta\Psi$ satisfies

$$\left[i\frac{\partial}{\partial t} - \hat{H}_v(t) \right] |\delta\Psi\rangle = \delta\hat{V}(t)|\Psi\rangle, \tag{9.87}$$

where $\hat{H}_v$ is a Hamiltonian with potential v and we neglected terms of higher order. Multiplication by the imaginary number "i" yields that the variation $i\delta\Psi$ must be generated by potential $i\delta v$. This potential variation is however imaginary and therefore not an allowed variation since all potential variations must be real. We therefore conclude that TDDFT can not be based on the usual variational principle.

A recent reformulation of the variational principle in TDDFT was given by Vignale (2008). From the above considerations we know that we are not allowed to pose the second boundary condition $\delta\Psi(t_1) = 0$ if we vary with respect to the density or the potential, respectively. Hence, in (9.75) we are no longer allowed to ignore the upper boundary term and have in terms of the density

$$\delta A[n] = i\langle\Psi[n](t_1)|\delta\Psi[n](t_1)\rangle. \tag{9.88}$$

In this approach the action functional is not stationary but its variation is equal to some density functional taking care of the upper boundary term. The functional derivative of the above equation together with (9.82) and (9.84) leads to

$$v_{\text{ext}}(\boldsymbol{r}, t) = \frac{\delta A_0[n]}{\delta n(\boldsymbol{r}, t)} - i \left\langle \Psi[n](t_1) \middle| \frac{\delta \Psi[n](t_1)}{\delta n(\boldsymbol{r}, t)} \right\rangle. \tag{9.89}$$

As shown in Vignale (2008) the additional boundary term is real and the dependence on the arbitrary upper limit t_1 is canceled. Actually, both terms on the right hand side of (9.89) have a noncausal dependence on $n(\boldsymbol{r}', t')$ for $t' > t$ which, however, is canceled exactly by the other term if combined. If we now calculate the redefined equation analogous to (9.85) we find

$$\frac{\delta v_{\text{ext}}(\boldsymbol{r}, t)}{\delta n(\boldsymbol{r}', t')} = \frac{\delta^2 A_0[n]}{\delta n(\boldsymbol{r}, t)\delta n(\boldsymbol{r}', t')}$$
$$- i \left\langle \Psi[n](t_1) \middle| \frac{\delta^2 \Psi[n](t_1)}{\delta n(\boldsymbol{r}, t)\delta n(\boldsymbol{r}', t')} \right\rangle - i \left\langle \frac{\delta \Psi[n](t_1)}{\delta n(\boldsymbol{r}', t')} \middle| \frac{\delta \Psi[n](t_1)}{\delta n(\boldsymbol{r}, t)} \right\rangle. \tag{9.90}$$

The first two terms on the right hand side are symmetric while the third term is antisymmetric. If we then subtract from (9.90) the same equation with $(\boldsymbol{r}, t)$ and $(\boldsymbol{r}', t')$ interchanged and note that due to causality and the convention $t > t'$ the term $\delta v_{\text{ext}}(\boldsymbol{r}', t')/\delta n(\boldsymbol{r}, t) = 0$, we arrive at

$$\frac{\delta v_{\text{ext}}(\boldsymbol{r}, t)}{\delta n(\boldsymbol{r}', t')} = 2\Im m \left\langle \frac{\delta \Psi[n](t_1)}{\delta n(\boldsymbol{r}', t')} \middle| \frac{\delta \Psi[n](t_1)}{\delta n(\boldsymbol{r}, t)} \right\rangle. \tag{9.91}$$

We note, that the right hand side has the structure of a Berry curvature. How the reformulated variational principle works in detail has been analyzed further in the same work (Vignale 2008) in terms of a simple example which can be solved exactly. This variational approach hence solves the causality paradox in real time. A solution of the paradox on the Keldysh time-contour (van Leeuwen 1998, 2001) is given in Chap. 3 of Marques et al. (2006a). There it is discussed how an extended type of action functional can be used as a basis from which the time-dependent Kohn–Sham equations can be derived. This has the immediate advantage that the action functional can then be directly related to the elegant formalism of nonequilibrium Green function theory which offers a systematic way of constructing time-dependent density functionals. Some examples of such functionals can be found in reference von Barth et al. (2005). With hindsight it is interesting to see that already the work of Peuckert (1978), which is one of the very first papers in TDDFT, makes a connection to Keldysh Green functions, and in fact several of his results (such as the adiabatic connection formula) are perfectly valid when interpreted in terms of the action formalism.

Chapter 10
Open Quantum Systems: Density Matrix Formalism and Applications

David G. Tempel, Joel Yuen-Zhou and Alán Aspuru-Guzik

10.1 Introduction

In its original formulation, TDDFT addresses the isolated dynamics of electronic systems evolving unitarily (Runge and Gross 1984). However, there exist many situations in which the electronic degrees of freedom are not isolated, but must be treated as a subsystem imbedded in a much larger thermal bath. The theory of open quantum systems (OQS) deals with precisely this situation, in which the bath exchanges energy and momentum with the system, but particle number is typically conserved. Several important examples include vibrational relaxation of molecules in liquids or impurities in a solid matrix, coupling to a photon bath in cavity quantum electrodynamics, photo-absorption of chromophores in a protein environment, electron–phonon coupling in single-molecule transport and exciton and energy transfer nanomaterials. Even with simple system–bath models, describing the reduced dynamics of many correlated electrons is computationally intractable. Therefore, applying TDDFT to OQS (OQS–TDDFT) offers a practical approach to the many-body open-systems problem.

David G. Tempel
Department of Physics, Harvard University,
17 Oxford Street, Cambridge, MA 02138, USA
e-mail: tempel@physics.harvard.edu

J. Yuen-Zhou
Department of Chemistry and Chemical Biology,
Harvard University, 12 Oxford Street, Room M138,
Cambridge, MA 02138, USA
e-mail: joelyuen@fas.harvard.edu

A. Aspuru-Guzik (✉)
Department of Chemistry and Chemical Biology,
Harvard University, 12 Oxford Street, Room M113,
Cambridge, MA 02138, USA
e-mail: aspuru@chemistry.harvard.edu

M. A. L. Marques et al. (eds.), *Fundamentals of Time-Dependent Density Functional Theory*, Lecture Notes in Physics 837, DOI: 10.1007/978-3-642-23518-4_10, © Springer-Verlag Berlin Heidelberg 2012

The formal development of the theory of OQS begins with the full unitary dynamics of the coupled system and bath, described by the Von Neumann equation for the density operator

$$\frac{d\hat{\rho}(t)}{dt} = -i[\hat{H}(t), \hat{\rho}(t)]. \qquad (10.1)$$

Here,

$$\hat{H}(t) = \hat{H}_S(t) + \alpha \hat{H}_{SB} + \hat{H}_B \qquad (10.2)$$

is the full Hamiltonian for the coupled system and bath and

$$\hat{H}_S(t) = -\frac{1}{2} \sum_{i=1}^{N} \nabla_i^2 + \sum_{i<j}^{N} v_{ee}(\mathbf{r}_i, \mathbf{r}_j) + \sum_i v_{ext}(\mathbf{r}_i, t) \qquad (10.3)$$

is the Hamiltonian of the electronic system of interest in an external potential $v_{ext}(\mathbf{r}, t)$. This potential generally consists of a static external potential due to the nuclei and an external driving field coupled to the system such as a laser field. For an interacting electronic system, $v_{ee}(\mathbf{r}, \mathbf{r}') = 1/|\mathbf{r} - \mathbf{r}'|$ is the two-body Coulomb repulsion. The system–bath coupling, $\hat{H}_{SB}$, acts in the combined Hilbert space of the system and bath and so it couples the two subsystems. Typically, for a single dissipation channel, the system–bath coupling is taken to have a bilinear form,

$$\hat{H}_{SB} = -\hat{S} \otimes \hat{B}, \qquad (10.4)$$

where $\hat{B}$ is an operator in the bath Hilbert space which generally couples to a local one-body operator $\hat{S} = \left[\sum_{i=1}^{N} \hat{s}(\hat{\mathbf{p}}_i, \hat{\mathbf{r}}_i) \right]$ in the system Hilbert space. Implicit in OQS is a weak interaction between the system and bath, so that one can treat the system–bath coupling perturbatively by introducing the small parameter α as in Eq. 10.2. $\hat{H}_B$ is the Hamiltonian of the bath, whose spectrum will typically consist of a dense set of bosonic modes such as photons or phonons. The density of states of $\hat{H}_B$ determines the structure of reservoir correlation functions, whose time-scale in turn determines the reduced system dynamics.

The goal of the theory of open quantum systems is to arrive at a reduced description of the dynamics of the electronic system alone, by integrating out the bosonic modes of the bath. In this way, one arrives at the quantum master equation, which describes the non-unitary evolution of the reduced system in the presence of its environment. In the next section, we derive the many-electron quantum master equation and discuss common approximations used to treat the system–bath interactions. We then formulate the master equation approach to TDDFT rigorously, by establishing a van Leeuwen construction for OQS. Next, we turn to a practical Kohn–Sham (KS) scheme for dissipative real-time dynamics and finally discuss the linear response version of OQS–TDDFT, giving access to environmentally broadened absorption spectra.

10.2 The Generalized Quantum Master Equation

We begin this section by deriving the formally exact many-electron quantum master equation using the Nakajima–Zwanzig projection operator formalism (Nakajima 1958; Zwanzig 1960, 2001). Using projection operators, the master equation can be systematically derived from first principles starting from the microscopic Hamiltonian in Eq. 10.2 (i.e. without phenomenological parameters). This is particularly amenable to TDDFT, in which the electronic degrees of freedom are treated using first principles as well. We will then discuss the Born–Markov approximation and the widely used Lindblad (1976) master equation.

10.2.1 Derivation of the Quantum Master Equation Using the Nakajima–Zwanzig Projection Operator Formalism

Our starting point is Eq. 10.1 for the evolution of the full density operator of the coupled system and bath,

$$\frac{d\hat{\rho}(t)}{dt} = -i[\hat{H}(t), \hat{\rho}(t)] \equiv -i\check{L}(t)\hat{\rho}(t), \tag{10.5}$$

where $\check{L}(t)$ is the Liouvillian superoperator for the full evolution defined by Eq. 10.5. It may be separated into a sum of Liouvillian superoperators as

$$\check{L}(t) = \check{L}_S(t) + \check{L}_{SB} + \check{L}_B, \tag{10.6}$$

where each term acts as a commutator on the density matrix with its respective part of the Hamiltonian. Our goal is to derive an equation of motion for the reduced density operator of the electronic system,

$$\hat{\rho}_S(t) = \text{Tr}_B\{\hat{\rho}(t)\}, \tag{10.7}$$

defined by tracing the full density operator over the bath degrees of freedom. To achieve this formally, we introduce the projection superoperators $\check{P}$ and $\check{Q}$. The operator $\check{P}$ is defined by projecting the full density operator onto a product of the system density operator with the equilibrium density operator of the bath,

$$\check{P}\hat{\rho}(t) = \hat{\rho}_B^{eq}\hat{\rho}_S(t). \tag{10.8}$$

$\check{Q} = 1 - \check{P}$ projects on the complement space. In this sense, $\check{P}$ projects onto the degrees of freedom of the electronic system we are interested in, while $\check{Q}$ projects onto irrelevant degrees of freedom describing the bath dynamics.

Using these projection operators, Eq. 10.5 can be written formally as two coupled equations:

$$\frac{d}{dt}\breve{P}\hat{\rho}(t) = -i\breve{P}\breve{L}\hat{\rho}(t) = -i\breve{P}\breve{L}\breve{P}\hat{\rho}(t) - i\breve{P}\breve{L}\breve{Q}\hat{\rho}(t) \tag{10.9a}$$

$$\frac{d}{dt}\breve{Q}\hat{\rho}(t) = -i\breve{Q}\breve{L}\hat{\rho}(t) = -i\breve{Q}\breve{L}\breve{P}\hat{\rho}(t) - i\breve{Q}\breve{L}\breve{Q}\hat{\rho}(t). \tag{10.9b}$$

If Eq. 10.9b is integrated and substituted into Eq. 10.9a, one obtains

$$\frac{d}{dt}\breve{P}\hat{\rho}(t) = -i\breve{P}\breve{L}\breve{P}\hat{\rho}(t) - \int_0^t d\tau\, \breve{P}\breve{L}e^{-i\int_\tau^t d\tau'\,\breve{Q}\breve{L}(\tau')}\breve{Q}\breve{L}\breve{P}\hat{\rho}(\tau)$$

$$- i\breve{P}\breve{L}e^{-i\int_0^t d\tau\,\breve{Q}\breve{L}(\tau)}\breve{Q}\hat{\rho}(0). \tag{10.10}$$

By performing a partial trace of both sides of Eq. 10.10 over the bath degrees of freedom, one arrives at the formally exact quantum master equation

$$\frac{d\hat{\rho}_S(t)}{dt} = -i[\hat{H}_S(t), \hat{\rho}_S(t)] + \int_0^t dt'\, \breve{K}(t, \tau)\hat{\rho}_S(\tau) + \Xi(t). \tag{10.11}$$

Here,

$$\breve{K}(t, \tau) = \mathrm{Tr}_B\left\{\breve{P}\breve{L}e^{-i\int_\tau^t d\tau'\breve{Q}\breve{L}(\tau')}\breve{Q}\breve{L}\hat{\rho}_B^{eq}\right\} \tag{10.12}$$

is the memory kernel.

$$\Xi(t) = \mathrm{Tr}_B\left\{-i\breve{P}\breve{L}e^{-i\int_0^t d\tau\breve{Q}\breve{L}(\tau)}\breve{Q}\hat{\rho}(0)\right\} \tag{10.13}$$

arises from initial correlations between the system and its environment (Meier and Tannor 1999). Equation 10.11 is still formally exact, as $\hat{\rho}_S(t)$ yields the exact expectation value of any operator depending on the electronic degrees of freedom. In practice, approximations to the memory kernel and initial correlation term are needed.

10.2.2 The Markov Approximation

One often invokes the Markov approximation, in which the memory kernel is local in time and the initial correlations vanish, i.e.

$$\int_0^t dt' \, \check{K}(t, t') \hat{\rho}_S(t') = \check{D} \hat{\rho}_S(t) \tag{10.14}$$

and

$$\Xi(t) = 0. \tag{10.15}$$

The Markov approximation is valid when $\tau_S \gg \tau_B$ is satisfied, where τ_S is the time-scale for the system to relax to thermal equilibrium and τ_B is the longest correlation time of the bath (Breuer and Petruccione 2002). Roughly speaking, the memory of the bath is neglected because the bath decorrelates from itself before the system has had a chance to evolve appreciably (Van Kampen 1992). The time-scale τ_S is inversely related to the magnitude of the system–bath coupling, and so a weak interaction between the electrons and the environment is implicit in this condition as well. The Lindblad form of the Markovian master equation,

$$\check{D} \hat{\rho}_S(t) = \hat{S} \hat{\rho}_S(t) \hat{S}^\dagger - \frac{1}{2} \hat{S}^\dagger \hat{S} \hat{\rho}_S(t) - \frac{1}{2} \hat{\rho}_S(t) \hat{S}^\dagger \hat{S}, \tag{10.16}$$

is constructed to guarantee positivity of the density matrix. This is desirable, since the populations of any physically sensible density matrix should remain positive during the evolution. The Lindblad equation is also trace preserving, which guarantees that the density matrix remains normalized as the system evolves. In the Lindblad equation, in addition to the Markov approximation, one performs second-order perturbation theory in the system–bath interaction (Eq. 10.4). These two approximations in tandem are referred to collectively as the Born–Markov approximation.

So far our discussion has focused on a system of interacting electrons coupled to a bath. We now turn to the formulation of OQS–TDDFT.

10.3 Rigorous Foundations of OQS–TDDFT

In order to formally establish an OQS–TDDFT starting from the many-body quantum master equation in Eq. 10.11, we must first establish the open-systems version of the van Leeuwen (1999) construction (see Chap. 9). This proves a one-to-one mapping between densities and potentials for non-unitary dynamics, as well as the existence of several different KS schemes (Yuen-Zhou et al. 2010, 2009).

10.3.1 The OQS–TDDFT van Leeuwen Construction

Our starting point is the master equation of Eq. 10.11, which evolves under the many-electron Hamiltonian in Eq. 10.3. We may now state a theorem concerning the construction of an auxiliary system.

Theorem *Let the original system be described by the density matrix $\hat{\rho}_S(t)$, which starting as $\hat{\rho}_S(0)$ evolves according to Eq. 10.11. Consider an auxiliary system associated with the density matrix $\hat{\rho}_S'(t)$ and initial state $\hat{\rho}_S'(0)$, which is governed by the master equation*

$$\frac{d\hat{\rho}_S'(t)}{dt} = -i[\hat{H}_S'(t), \hat{\rho}_S'(t')] + \int_0^t dt' \check{K}'(t, t')\hat{\rho}_S'(t') + \varXi'(t) \qquad (10.17)$$

and with $\check{K}'(t, t')$ and $\varXi'(t)$ fixed. Here,

$$\hat{H}_S'(t) = -\frac{1}{2}\sum_{i=1}^N \nabla_i^2 + \sum_{i<j}^N v_{ee}'(r_i, r_j) + \sum_i v_{ext}'(r_i, t), \qquad (10.18)$$

is the Hamiltonian of an auxiliary system with a different two-particle interaction $v_{ee}'(r, r')$. Under conditions discussed below, there exists a unique external potential $v_{ext}'(r, t)$ which drives the system in such a way that the particle densities in the original and the auxiliary systems are the same, i.e. $\langle \hat{n}(r)\rangle' = \langle \hat{n}(r)\rangle$ is satisfied for all times, where $\langle \hat{n}(r)\rangle \equiv (\text{Tr})\{\hat{\rho}_S(t)\hat{n}(r)\}$.

Proof The method we use closely parallels the van Leeuwen (1999) construction given for unitary evolution (see Chap. 9). By using Eq. 10.11 we can find an equation of motion for the second derivative of the particle density of the original system. This is done by first deriving the equation of motion for the particle density

$$\frac{\partial\langle \hat{n}(r)\rangle_t}{\partial t} = -\nabla \cdot \langle \hat{j}(r)\rangle_t + \text{Tr}\left\{\hat{n}(r)\left[\int_0^t dt' \check{K}(t, t')\hat{\rho}_S(t') + \varXi(t)\right]\right\}, \qquad (10.19)$$

as well as for the current density,

$$\frac{\partial\langle \hat{j}(r)\rangle_t}{\partial t} = -\frac{\langle \hat{n}(r)\rangle_t}{m}\nabla v_{ext}(r, t) + \mathcal{D}(r, t) + \frac{\mathcal{F}(r, t)}{m} + \mathcal{G}(r, t). \qquad (10.20)$$

We then differentiate both sides of Eq. 10.19 with respect to time and use Eq. 10.20 to eliminate the current. One thus arrives at,

$$\frac{\partial^2\langle \hat{n}(r)\rangle_t}{\partial t^2} = \nabla \cdot \{\langle \hat{n}(r)\rangle_t \nabla v_{ext}(r, t)/m$$
$$- \mathcal{D}(r, t) - \mathcal{F}(r, t)/m - \mathcal{G}(r, t)\} + \mathcal{J}(r, t), \qquad (10.21)$$

subject to the initial conditions

$$\langle \hat{n}(r)\rangle_{t=0}' = \langle \hat{n}(r)\rangle_{t=0}, \qquad (10.22a)$$

$$\frac{\partial \langle \hat{n}(r) \rangle_t'}{\partial t}\bigg|_{t=0} = \frac{\partial \langle \hat{n}(r) \rangle_t}{\partial t}\bigg|_{t=0}, \qquad (10.22b)$$

i.e. we demand that the densities and their first derivatives be the same at the initial time. In Eq. 10.21, each term has a clear physical interpretation. The quantity $\nabla v_{\text{ext}}(r, t)$ is proportional to the external electric field acting on the system,

$$\mathcal{D}(r, t) = -\frac{1}{4} \sum_{\alpha,\beta} \hat{\beta} \frac{\partial}{\partial \alpha} \left\langle \sum_i \{\hat{v}_{i\alpha}, \{\hat{v}_{i\beta}, \delta(r - \hat{r}_i)\}\} \right\rangle \qquad (10.23)$$

is the divergence of the stress tensor, where $\alpha, \beta = x, y, z$ label Cartesian indices, and

$$\mathcal{F}(r, t) = -\left\langle \sum_i \delta(r - \hat{r}_i) \sum_{j \neq i} \nabla_{r_i} v_{\text{ee}}(r_i - r_j) \right\rangle \qquad (10.24)$$

is the internal force density caused by the pairwise potential. In addition to these quantities which arise in usual TDDFT, we have defined two new quantities,

$$\mathcal{G}(r, t) = \text{Tr}\left\{ \hat{j}(r) \left[\int_0^t dt'\, \check{K}(t, t')\hat{\rho}_S(t') + \Xi(t) \right] \right\} \qquad (10.25a)$$

$$\mathcal{J}(r, t) = \text{Tr}\left\{ \hat{n}(r) \left[\int_0^t dt'\, \check{K}(t, t')\hat{\rho}_S(t') + \Xi(t) \right] \right\}, \qquad (10.25b)$$

which are unique to OQS–TDDFT and arise from forces induced by the bath.

We can now repeat the same procedure in the primed system, to arrive at the equation of motion for the second derivative of the density in terms of primed quantities,

$$\frac{\partial^2 \langle \hat{n}(r) \rangle_t'}{\partial t^2} = \nabla \cdot \{\langle \hat{n}(r) \rangle_t' \nabla v_{\text{ext}}'(r, t)/m$$
$$- \mathcal{D}'(r, t) - \mathcal{F}'(r, t)/m - \mathcal{G}'(r, t)\} + \mathcal{J}'(r, t). \qquad (10.26)$$

If we subtract Eq. 10.21 from Eq. 10.26 and demand that $\langle \hat{n}(r) \rangle_t' = \langle \hat{n}(r) \rangle_t$, we arrive at the equation

$$-\nabla \cdot \left[\frac{\langle \hat{n}(r) \rangle_t}{m} \nabla(\Delta v_{\text{ext}}'(r, t)) \right] = -\nabla \cdot \left[\mathcal{D}'(r, t) + \frac{\mathcal{F}'(r, t)}{m} + \mathcal{G}'(r, t) \right] + \mathcal{J}'(r, t)$$
$$+ \nabla \cdot \left[\mathcal{D}(r, t) + \frac{\mathcal{F}(r, t)}{m} + \mathcal{G}(r, t) \right] - \mathcal{J}(r, t), \qquad (10.27)$$

where we have defined $\Delta v'_{ext}(r, t) \equiv v'_{ext}(r, t) - v_{ext}(r, t)$. We now expand both sides of Eq. 10.27 in a Taylor series with respect to time to arrive at

$$
-\nabla \cdot \left[n_0(r) \nabla v'_{ext\, l}(r) \right] = - \nabla \cdot \left[n_0(r) \nabla v_{ext\, l}(r) \right] - \nabla \cdot \left[m\mathcal{D}'_l(r) + \mathcal{F}'_l(r) + m\mathcal{G}'_l(r) \right]
$$
$$
+ m\mathcal{J}'_l(r, t) + \nabla \cdot \left[m\mathcal{D}_l(r) + \mathcal{F}_l(r) + m\mathcal{G}_l(r) \right] - m\mathcal{J}_l(r, t)
$$
$$
+ \nabla \cdot \sum_{k=1}^{l} n_k(r) \nabla \left[\Delta v'_{ext\, l-k}(r) \right]. \tag{10.28}
$$

The left-hand side of Eq. 10.28 contains Taylor coefficients of $v'_{ext}(r, t)$ of order l, while the right-hand side depends only on Taylor coefficients of $v'_{ext}(r, t)$ of order $k < l$ and known quantities. Equation 10.28 can therefore be regarded as a *unique* recursion relation for constructing the Taylor coefficients of the auxiliary potential $v'_{ext}(r, t)$, once a suitable boundary condition is specified. We assume that $v'_{ext\, l}(r) \to 0$ sufficiently quickly as $|r| \to \infty$ for all l as in usual TDDFT. A more detailed discussion of this boundary condition is given in Chap. 4.

Several different KS schemes are now evident. If one sets $v'_{ee}(r, r') = 0$, but keeps the system open by setting $\check{K}'(t) = \check{K}^{KS}(t)$ and $\varXi'(t) = \varXi^{KS}(t)$, the auxiliary system is a non-interacting, but open KS system. This is similar to the construction used in (Burke et al. 2005c), but encompasses the non-Markovian case as well. However, one may also choose $v'_{ee}(r, r') = 0$ and $\check{K}'(t) = \varXi'(t) = 0$, whereby the density of the original open system is reproduced with a *closed* (unitarily evolving) and non-interacting KS system.

The OQS–TDDFT version of the Runge–Gross theorem, which is proven by setting $v'_{ee}(r, r') = v_{ee}(r, r')$, $\check{K}'(t) = \check{K}(t)$ and $\varXi'(t) = \varXi(t)$ in Eq. 10.28, requires only that the potential be time-analytic as in usual TDDFT. However, once one considers an auxiliary system with a different electron–electron interaction and/or system–bath coupling, all quantities appearing in Eq. 10.28 must be time-analytic, including the density, memory kernel and initial correlation terms in both the primed and unprimed systems. As discussed in Chap. 4 and (Maitra et al. 2010), it is possible for time-analytic potentials to generate densities that are not time-analytic. It seems plausible that a similar situation could arise in OQS–TDDFT, where certain potentials, initial states or memory kernels could produce densities that are not time-analytic and so the OQS–TDDFT van Leeuwen theorem might not hold. However, this still needs to be investigated more extensively. Also, as stated, the theorem assumes that the memory kernel and initial correlations do not depend on the external potential. In fact, this restriction is not essential as discussed in detail in (Yuen-Zhou et al. 2010).

10.3.2 The Double Adiabatic Connection

The content of our proof is conveniently summarized by parametrizing the auxiliary system's master equation with two coupling constants λ and β as,

Fig. 10.1 The relevant points on the double adiabatic connection square are: $(1,1)$: The original interacting OQS; $(1,0)$: The non-interacting yet open Kohn–Sham scheme; $(0,0)$: The non-interacting and closed Kohn–Sham scheme

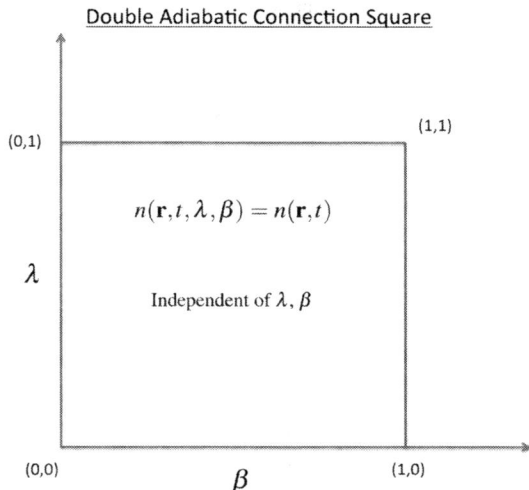

$$\frac{\mathrm{d}}{\mathrm{d}t}\hat{\rho}'_{\mathrm{S}}(\lambda, \beta, t) = -\,\mathrm{i}[\hat{H}'_{\mathrm{S}}(\lambda, \beta, t), \hat{\rho}'_{\mathrm{S}}(\lambda, \beta, t)]$$

$$+ \beta \left\{ \int_0^t \mathrm{d}\tau \; \check{K}'(t, \tau; \lambda)\hat{\rho}'_{\mathrm{S}}(\lambda, \beta, \tau) + \varXi'(t; \lambda) \right\}, \qquad (10.29)$$

where

$$\hat{H}'_{\mathrm{S}}(\lambda, \beta, t) = -\frac{1}{2}\sum_{i=1}^{N}\nabla_i^2 + \lambda \sum_{i<j}^{N} v_{\mathrm{ee}}(\mathbf{r}_i, \mathbf{r}_j) + \sum_i v'_{\mathrm{ext}}(\lambda, \beta, \mathbf{r}_i, t). \qquad (10.30)$$

Here, λ scales the electron–electron interaction and lies in the range $0 \leqslant \lambda \leqslant 1$. When $\lambda = 1$, we have a fully interacting system, while when $\lambda = 0$ we have a system of non-interacting electrons. The memory kernel and initial correlations are functions of λ as well. Similarly, β scales the non-unitary terms in the master equation and lies in the range $0 \leqslant \beta \leqslant 1$. When $\beta = 1$, we have a fully open system while when $\beta = 0$ the system evolves unitarily. The simple linear parameterization of Eq. 10.29 in terms of β is not unique, and one could consider a more complicated parameterization where $\check{K}'$ and $\varXi'$ depend on β as well.

The theorem of Sect. 10.3.1 guarantees the existence and uniqueness of a potential $v'_{\mathrm{ext}}(\lambda, \beta, \mathbf{r}, t)$ for all λ and β, which drives the auxiliary system in such a way that the true density is obtained independent of the values of λ and β. This can be viewed as a two-dimensional extension of the usual electron–electron adiabatic connection in closed-systems TDDFT (Görling 1997b). It is depicted graphically in Fig. 10.1. At the coordinate $(1,1)$, we have the original fully interacting and open system, while at the coordinate $(0,0)$ we have the non-interacting and closed KS scheme. Defining $\check{K}'(\lambda = 0) \equiv \check{K}^{\mathrm{KS}}$ and $\varXi'(\lambda = 0) \equiv \varXi^{\mathrm{KS}}$ as the memory kernel and initial correlations of an open system of non-interacting electrons, we see that the point (1,0) describes the open KS scheme. In the remainder of the chapter, we will focus on the points (0,0) and (1,0). However, our proof shows that any coordinate lying within the double adiabatic connection square represents a viable KS scheme. As in DFT and TDDFT, the double adiabatic connection provides a powerful tool for deriving exact conditions on OQS–TDDFT functionals and is currently being explored in more detail.

10.4 Simulating Real-Time Dissipative Dynamics with a Unitarily Evolving Kohn–Sham System

In the previous section, we saw that it is possible to take the KS system to be a non-interacting system evolving unitarily under a time-dependent driving field that will reproduce the density of the original interacting OQS. In this scheme, the KS potential v'_{ext} can be partitioned as

$$v'_{ext} = v_{ext} + v_H + v_{xc} + v_{bath}. \tag{10.31}$$

Here, v_{ext} is the original external potential acting on the real system. The electron–electron interaction is replaced by the sum of a Hartree term

$$v_H = \int d^3 r' \frac{\langle \hat{n}(r') \rangle_t}{|r - r'|}, \tag{10.32}$$

and a standard approximation to the exchange-correlation (xc) potential v_{xc}, such as an adiabatic functional (Furche and Ahlrichs 2002a). Finally, v_{bath} is a new term which represents a driving field that mimics the interactions of the system with the bath. This KS scheme places electron–electron and system–bath interactions on the same footing, so real-time TDDFT computer codes could in principle be easily modified to include the dissipative effects of an environment (Castro et al. 2006). Such a scheme is computationally desirable, since one would only need to propagate N orbital equations for an N-electron system as in usual TDDFT. This is in contrast to a density matrix approach, where one would need to propagate $M^2 - 1$ equations for the elements of the density matrix, with M being the (in principle infinite) dimensionality of the Hilbert space.

In usual TDDFT, the KS potential is a functional of the density and therefore the KS equations can be regarded as nonlinear Schrödinger equations (NLSE). In OQS, equations of motion for systems coupled to heat baths are often described by Langevin equations, where frictional forces are introduced through velocity-dependent potentials (Stokes law) (Zwanzig 2001). These frictional forces also give rise to a nonlinear Schrödinger equation, since velocity-dependent potentials can be regarded as functionals of the current or of time-derivatives of the density. Therefore, the search for approximations to v_{bath} could start by investigating work already done in the field of dissipative nonlinear Schrödinger equations (Kostin 1972, 1975; Bolivar 1998; Haas 2010) and time-dependent self consistent field (TD-SCF) methods (Makri and Miller 1987, López-López and Nest 2010, Martinazzo et al. 2006). In this section, we describe a simple Markovian bath functional (MBF) inspired by a NLSE suggested by Kostin (1972).

Consider a single particle in one dimension whose evolution is given by the NLSE,

$$i\frac{\partial \psi}{\partial t} = H\psi, \tag{10.33}$$

where

$$H = \frac{p^2}{2m} + v_{\text{ext}} + v_{\text{bath}}, \tag{10.34}$$

and p and v_{ext} are the momentum of the particle and the external potential respectively. The dissipative potential is chosen to have the form

$$v_{\text{bath}}(z, t) = \frac{\lambda}{2i} \ln \left[\frac{\psi(z, t)}{\psi^*(z, t)} \right]. \tag{10.35}$$

This NLSE has the very interesting property that it satisfies the zero-temperature Langevin equation for the expectation values of the particle's position and momentum. i.e.

$$\langle \dot{z} \rangle = \frac{\langle p \rangle}{M}, \tag{10.36a}$$

$$\langle \dot{p} \rangle = -\left\langle \frac{\partial v_{\text{ext}}(z, t)}{\partial z} \right\rangle - \lambda \langle p \rangle. \tag{10.36b}$$

Interestingly, v_{bath} in Eq. 10.35 can be written as a functional of the density and current as

$$v_{\text{bath}}[\langle \hat{n}(z') \rangle_t, \langle \hat{j}(z') \rangle_t] = \lambda \int_{-\infty}^{z} dz' \frac{\langle \hat{j}(z') \rangle_t}{\langle \hat{n}(z') \rangle_t}. \tag{10.37}$$

This identification is very appealing, since the frictional force is proportional to the space integral of the velocity field of the particle, $\langle \hat{j}(z') \rangle_t / \langle \hat{n}(z') \rangle_t$. Furthermore, the friction coefficient λ can be derived from a microscopic model of harmonic bath modes (Zwanzig 2001; Nitzan 2006; Tuckerman 2010). Note that v_{bath} at a given time only depends on the momentum of the particle at the same instant, implying that this NLSE is Markovian. This situation can be obtained in the limit where the dynamics of the bath can be described as white noise (Peskin and Steinberg 1998).

Although the discussion above has been given for a single particle, we can heuristically propose Eq. 10.37 as a MBF for TDDFT. In practice, we can re-express v_{bath} in terms of the orbitals of a time-dependent single Slater determinant KS wavefunction (orbital-dependent functional),

$$\Phi_{KS}(t) = \frac{1}{\sqrt{N!}} \det[\varphi_i(z_j, t)] \qquad (10.38)$$

as

$$v_{\text{bath}}(z, t) = \lambda \int\limits_{-\infty}^{z} dz' \frac{\sum_i |\varphi_i(z', t)|^2 \nabla \alpha_i(z', t)}{\sum_i |\varphi_i(z', t)|^2}, \qquad (10.39)$$

where $\alpha_i(z', t)$ is the phase of the ith orbital. The extension of the functional to more dimensions follows analogously, although the limits of integration must be studied with care. In the higher dimensional case, the Kohn–Sham current may differ from the physical current by a purely transverse term and one must resort to a formulation in terms of vector potentials using TDCDFT (see Chap. 24). Equation 10.39 is easy to implement in a real-time propagation, and has been implemented for a model Helium system interacting with a heat bath (Yuen-Zhou et al. 2010). Non-Markovian extensions, as well as functionals where several timescales of relaxation and dephasing exist, are currently under development.

Neuhauser and Lopata (2008) have recently reported an important result, which could also be considered a MBF in our formalism. Their functional is inspired by an optimal control approach, where they demand that the energy in the KS system decays monotonically. They show that

$$v_{\text{bath}}[\langle \hat{j}(z') \rangle_t] = \int dz' a(z') \frac{\partial \langle \hat{j}(z') \rangle_t}{\partial t} \hat{j}(z) \qquad (10.40)$$

achieves such goal. This functional couples the time-derivative of the current-density to the current operator with a spatially dependent proportionality constant $a(z')$. Their studies of a jellium cluster also show numerical robustness and provide a practical scheme to include dissipation in a real-time KS calculation.

10.5 OQS–TDDFT in the Linear Response Regime Using the Open Kohn–Sham Scheme

In addition to real-time dynamics, one can also consider OQS–TDDFT in the linear response regime, which gives access to environmentally broadened spectra (Tempel et al. 2011a). The starting point is the density–density response function of an interacting OQS evolving according to Eq. 10.11:

$$\chi(\mathbf{r}, \mathbf{r}'; \omega) = \text{Tr}_S \left\{ \hat{n}(\mathbf{r}) \frac{1}{\omega + \check{L}_S - i\check{K}(\omega)} \left[\hat{\rho}_S^n(\mathbf{r}', 0) + \Xi(\omega) \right] \right\}, \qquad (10.41)$$

where

$$\hat{\rho}_S^n(\mathbf{r}, 0) = \mathrm{Tr}_R\{[\hat{n}(\mathbf{r}), \hat{\rho}^{\mathrm{eq}}]\} \tag{10.42}$$

is the commutator of the particle density operator with the full equilibrium density matrix of the combined system and reservoir, traced over the reservoir degrees of freedom. From Eq. 10.41, one sees that the primary effect of the bath is to introduce a frequency-dependent self-energy $\check{K}(\omega)$, which shifts the poles of the response function into the complex plane. In the absence of coupling to the bath, the poles of Eq. 10.41 would lie at the excitation frequencies of the isolated system, which are the eigenvalues of $\check{L}_S$. The real part of $\check{K}(\omega)$ can be interpreted as an excited-state lifetime while the imaginary part is a Lamb shift of the energy.

In order to access the poles of the many-body response function in Eq. 10.41, one introduces an auxiliary *open*, but non-interacting KS system with a density–density response function given by

$$\chi^{\mathrm{KS}}(\mathbf{r}, \mathbf{r}', \omega) = \mathrm{Tr}_S\left\{\hat{n}(\mathbf{r})\frac{1}{\omega + \check{L}^{\mathrm{KS}} - i\check{K}^{\mathrm{KS}}(\omega)}([\hat{n}(\mathbf{r}'), \hat{\rho}_S^{\mathrm{KS}}(0)] + \Xi^{\mathrm{KS}}(\omega))\right\}. \tag{10.43}$$

Here, $\check{L}^{\mathrm{KS}}$ is the Liouvillian for the ground or equilibrium-state Kohn–Sham–Mermin Hamiltonian (Kohn and Sham 1965) and $\hat{\rho}_S^{\mathrm{KS}}(0)$ is the corresponding KS density matrix. $\check{K}^{\mathrm{KS}}(\omega)$ is a KS self-energy which describes coupling of non-interacting electrons to the environment. It is chosen to be a one-body superoperator and easily constructed in terms of KS orbitals and eigenvalues of $\check{L}_{ks}$. As mentioned in Sect. 10.3, the existence and uniqueness of such a KS system is guaranteed by setting $\check{K}' = \check{K}^{\mathrm{KS}}$, $\Xi' = \Xi^{\mathrm{KS}}$ and $v'_{ee}(\mathbf{r}, \mathbf{r}') = 0$ in the OQS–TDDFT van Leeuwen construction. For this scheme, we partition the potential as

$$v'_{\mathrm{ext}} = v_{\mathrm{ext}} + v_H + v_{\mathrm{xc}}^{\mathrm{open}}, \tag{10.44}$$

where $v_{\mathrm{xc}}^{\mathrm{open}}$ not only accounts for electron–electron interaction within the system, but must also correct for the difference between $\check{K}^{\mathrm{KS}}$ and $\check{K}$ in the system–bath interaction. This scheme is better suited to response theory than the closed Kohn–Sham scheme discussed in Sect. 10.4, since relaxation and dephasing is already accounted for in the KS system through $\check{K}^{\mathrm{KS}}$. The unknown (OQS–TDDFT) exchange-correlation functional only needs to correct the relaxation and dephasing in the KS system to that of the interacting system, rather than needing to explicitly account for the entire effect of the environment.

Reminiscent of usual TDDFT, Eqs. 10.41 and 10.43 are related by the Dyson-like equation

$$\chi(\mathbf{r}, \mathbf{r}', \omega) = \chi^{\mathrm{KS}}(\mathbf{r}, \mathbf{r}', \omega) + \int d^3y \int d^3y' \chi^{\mathrm{KS}}(\mathbf{r}, \mathbf{y}, \omega)$$
$$\left\{\frac{1}{|\mathbf{y} - \mathbf{y}'|} + f_{\mathrm{xc}}^{\mathrm{open}}[n^{\mathrm{eq}}, \check{K}, \check{K}^{\mathrm{KS}}](\mathbf{y}, \mathbf{y}', \omega)\right\} \times \chi(\mathbf{y}', \mathbf{r}', \omega), \tag{10.45}$$

where

$$f_{\text{xc}}^{\text{open}}[n^{\text{eq}}, \check{K}, \check{K}^{\text{KS}}](r, r', \omega) = \left.\frac{\delta v_{\text{xc}}^{\text{open}}(r, \omega)}{\delta n(r', \omega)}\right|_{n=n_{\text{eq}}} \tag{10.46}$$

is the OQS–TDDFT exchange correlation kernel. It is a functional of the equilibrium density as well as the memory kernel in both the interacting and Kohn–Sham systems. For Markovian environments, it is straightforward to reformulate Eq. 10.45 as a Casida-type equation (Casida 1996),

$$\left\{\omega^2 - \bar{\Omega}(\omega)\right\} F = 0, \tag{10.47}$$

where the frequency-dependent operator $\bar{\Omega}(\omega)$ can be expressed in a basis of Kohn–Sham-Mermin orbitals as

$$\bar{\Omega}_{ijkl}(\omega) = \delta_{ik}\delta_{jl}\left\{(\omega_{lk}^{\text{KS}} + \Delta_{kl}^{\text{KS}})^2 + (\Gamma_{kl}^{\text{KS}})^2 - 2i\omega\Gamma_{kl}^{\text{KS}}\right\}$$
$$+ 4\sqrt{(f_i - f_j)(\omega_{ji}^{\text{KS}} + \Delta_{ij}^{\text{KS}})}K_{ijkl}(\omega)\sqrt{(f_k - f_l)(\omega_{lk}^{\text{KS}} + \Delta_{kl}^{\text{KS}})}. \tag{10.48}$$

Here, Γ_{kl}^{KS} and Δ_{kl}^{KS} arise from matrix elements of the real and imaginary parts of $\check{K}^{\text{KS}}$, respectively.

$$K_{ijkl}(\omega) = \int d^3r \int d^3r' \varphi_i^*(r)\varphi_j^*(r)$$
$$\times \left\{\frac{1}{|r - r'|} + f_{\text{xc}}^{\text{open}}[n^{\text{eq}}, \check{K}, \check{K}^{\text{KS}}](r, r', \omega)\right\} \varphi_k(r')\varphi_l(r') \tag{10.49}$$

are matrix elements of the OQS Hartree-exchange-correlation kernel. Equation 10.48 is a non-hermitian and explicitly frequency-dependent operator yielding complex eigenvalues. The real part of the eigenvalues are interpreted as excitation energies while the imaginary parts give the linewidths. Since the KS system is open, the bare KS spectrum is already broadened at zeroth-order. $f_{\text{xc}}^{\text{open}}$ has the task of not only shifting the location of the bare KS absorption peaks to that of the interacting system, but it must also correct the linewidths.

As a simple example, we solved the OQS Casida equations in Eq. 10.47 to obtain the absorption spectrum of a C^{2+} cation interacting with the modes of the electromagnetic field in vacuum, giving rise to radiative natural linewidths. The electromagnetic field acts as a photon bath, while the C^{2+} cation can be treated as an OQS in our formalism (Cohen-Tannoudji 2004). As a crude first approximation, we used an adiabatic functional (ATDDFT) for $f_{\text{xc}}^{\text{open}}$ in Eq. 10.48, and solved Eq. 10.47 for the three lowest dipole allowed transitions ($2s \rightarrow 2p, 3p, 4p$). The local density approximation (LDA) with the modified Perdew–Zunger (PZ) parametrization was used for the groundstate functional as well as the adiabatic exchange-correlation kernel. From Fig. 10.2, we see that the adiabatic functional places the location of the absorption peaks in essentially the correct place as in usual TDDFT, but leaves

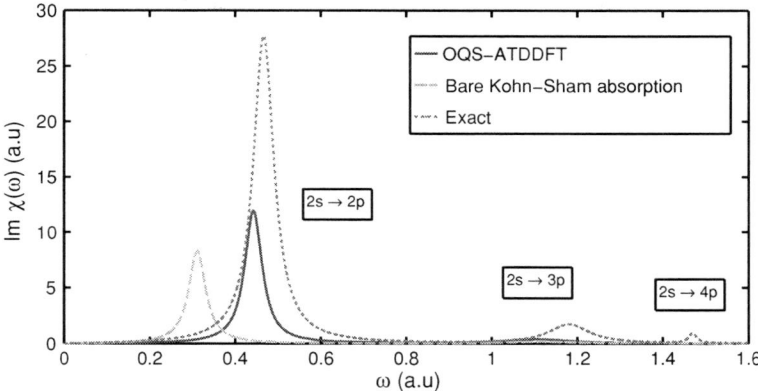

Fig. 10.2 Absorption Spectrum of C^{2+} including the three lowest dipole allowed transitions. The curves shown are: *a* The bare Kohn–Sham spectrum (*green-dashed*). *b* The spectrum obtained by solving Eq. 10.47 with an adiabatic exchange-correlation kernel (*blue*). *c* The numerically exact spectrum obtained using experimental data (*red-dashed*). For visualization, all linewidths have been scaled by a factor of c^3 since the radiative lifetime in vacuum is extremely small

the linewidths unchanged relative to their bare KS value. To correct the linewidths, one needs a frequency-dependent bath functional yielding additional broadening. Such a functional with the correct frequency-dependence to first-order in Görling–Levy perturbation theory (Görling and Levy 1993; Görling 1998a) was discussed in (Tempel et al. 2011a) for the $2s \rightarrow 2p$ transition. The frequency-dependent kernel matrix element in Eq. 10.49 was found to be

$$K^{bath}_{2s2p,2s2p}(\omega) = -\frac{i}{2(\varepsilon_{2s} - \varepsilon_{2p})} \left(\omega + i\Gamma^{KS}_{2p,2s} \right) \left(\Gamma^1_{2p,2s} \right), \qquad (10.50)$$

where $\Gamma^{KS}_{2p,2s}$ is the bare KS linewidth and $\Gamma^1_{2p,2s}$ is a correction derived from first-order Görling–Levy perturbation theory. In Fig. 10.3, we see that including Eq. 10.50 when solving the OQS Casida equations yields a large correction to the linewidth, although the oscillator strength is unchanged. To correct the oscillator strength as well, one needs higher-order corrections. A similar formalism can be used to capture line broadening due to vibrational relaxation in molecules and electron–phonon scattering and is currently being explored.

10.6 Positivity of the Lindblad Master Equation for Time-Dependent Hamiltonians

The Lindblad form of the master equation ensures preservation of trace and positivity throughout the time-evolution. However, in the usual formulation, one assumes that both the system Hamiltonian and jump operators are time-independent.

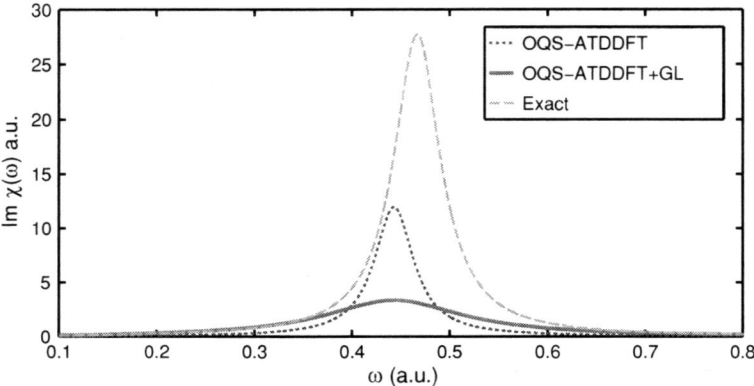

Fig. 10.3 The curves shown are: *a* The correction to the bare Kohn–Sham linewidth using the frequency-dependent bath functional to first-order in GL perturbation theory (Eq. 10.50) (*green*). *b* The spectrum obtained by solving Eq. 10.47 with an adiabatic exchange-correlation kernel (*blue-dashed*). *c* The numerically exact spectrum obtained using experimental data (*red-dashed*). All linewidths are scaled by a factor of c^3

In this section we prove that the Lindblad equation maintains positivity and is trace preserving, even if the Hamiltonian and jump operators are time-dependent.

Let us begin by generalizing Eq. 10.16 to include N dissipation operators, which may be time-dependent as well as an explicitly time-dependent Hamiltonian:

$$\frac{d\hat{\rho}_S}{dt} = -i[\hat{H}_S(t), \hat{\rho}_S(t)] + \sum_{i=1}^{N} \hat{S}_i(t)\hat{\rho}_S(t)\hat{S}_i^\dagger(t) - \frac{1}{2}\hat{S}_i^\dagger(t)\hat{S}_i(t)\hat{\rho}_S(t)$$

$$- \frac{1}{2}\hat{\rho}_S(t)\hat{S}_i^\dagger(t)\hat{S}_i(t). \tag{10.51}$$

We now show that this master equation preserves the positivity and trace of the density matrix under its evolution, *irrespective* of the time-dependence of $\hat{H}$ and $\hat{S}_i$. To verify this, we expand $\hat{\rho}_S(t + \Delta t)$ for small Δt :

$$\hat{\rho}_S(t + \Delta t) = \hat{\rho}_S(t) + \frac{\partial \hat{\rho}_S}{\partial t}\Delta t + O(\Delta t^2)$$

$$\approx \sum_{i=0}^{N} M_i(\Delta t)\hat{\rho}_S(t)M_i(\Delta t)^\dagger. \tag{10.52}$$

Using Eq. 10.51, we can define the so-called Kraus operators $M_i(\Delta t)$ by:

$$M_0(\Delta t) = I + \left(-i\hat{H} - \frac{1}{2}\sum_{i=1}^{N}\hat{S}_i^\dagger\hat{S}_i\right)\Delta t, \tag{10.53}$$

and

$$M_i(\Delta t) \equiv \hat{S}_i \sqrt{\Delta t}, \tag{10.54}$$

for $i > 0$. An important property of the Kraus operators is that they satisfy

$$\sum_{i=0}^{N} M_i(\Delta t)^\dagger M_i(\Delta t) = I, \tag{10.55}$$

as can be easily checked from Eqs. 10.53 and 10.54 . Notice that, in general, $M_i(\Delta t)$ are time dependent if $\hat{S}_i$ are as well.

Let us write $\hat{\rho}_S(t) = \sum_j p_j |\xi_j(t)\rangle\langle\xi_j(t)|$, where $\{|\xi_i(t)\rangle\}$ is the basis that diagonalizes $\hat{\rho}_S(t)$ at each instant in time and $p_j \geq 0$ for all j. Notice that up to $O(\Delta t)$,

$$\hat{\rho}_S(t + \Delta t) = \sum_j \sum_{i=0}^{N} p_j \left[M_i(\Delta t)|\xi_j(t)\rangle \right]\left[\langle\xi_j(t)|M_i(\Delta t)^\dagger \right], \tag{10.56}$$

which shows that $\hat{\rho}_S(t + \Delta t)$ is positive semidefinite if $\hat{\rho}_S(t)$ is. The preservation of the trace can also be readily shown, i.e.

$$\text{Tr}\{\hat{\rho}_S(t + \Delta t)\} = \text{Tr}\left\{ \sum_{i=0}^{N} M_i(\Delta t)\hat{\rho}_S(t)M_i(\Delta t)^+ \right\} = \text{Tr}\{\hat{\rho}_S(t)\}. \tag{10.57}$$

The proofs above are an adaptation of the discussion of the Lindblad equation in the textbook by Schumacher and Westmoreland[1](2010).

For completeness, we now introduce the concept of a *semigroup*. Consider the integrated form of the equation of motion for $\hat{\rho}_S(t)$ in the form of a dynamical map, $\hat{\rho}_S(t) = \Phi_{t,0}\hat{\rho}_S(0)$, where $\Phi_{t,0}$ is a dynamical map that propagates the density matrix from the initial time 0 to the final time t. The semigroup property is expressed as the following identity for the composition map (Breuer and Petruccione 2002):

$$\Phi_{s,0}\Phi_{t,0} = \Phi_{s+t,0}. \tag{10.58}$$

Notice that, on the one hand, the semigroup property will not be satisfied in general for time-dependent $\hat{H}$ or $\hat{S}_i$. On the other hand, for the case where both types of operators are time-independent, it can be shown that the Liouvillian $\check{L}$ is the most general form of the generator of a quantum dynamical semigroup, with the dynamical map being the exponential map: $\Phi_{s,0} = e^{\check{L}s}$ (Gorini et al. 1976). However, it is clear from the above discussion that although the semigroup property may not hold, neither positivity nor trace-preservation are contingent upon the time-independence of $\hat{H}$ or $\hat{S}_i$.

---

[1] The authors discuss the derivation of the Lindblad equation as the generator of a completely positive map.

10.7 Comparison of OQS–TDDFT in the Stochastic Schrödinger Equation and Master Equation Approaches

The next chapter presents a different approach to OQS–TDDFT based on the method of stochastic Schrödinger equations (SSE) rather than master equations. In this section, we briefly give an overview of the connection between the two approaches.

It is well known that if $\hat{H}_S$ does not depend on the state of the system, Eq. 10.51 can be "unraveled" as the evolution of an ensemble of Stochastic Schrödinger Equations which reconstructs the density matrix upon appropriate manipulation of the calculated trajectories [see Chap. 7 and (Dalibard et al. 1992; Diosi 1986)]. Hence, under these circumstances the SSE yields the same density matrix dynamics as the Lindblad equation. There are several interesting features of the SSE, such as its lower numerical cost (Breuer et al. 1997) as well as the novel conceptual insights it provides based on the theory of weak measurements, and more importantly, the monitoring of single quantum systems (Plenio 1998).

When considering OQS–TDDFT, the Kohn–Sham (KS) potential is a functional of the ensemble-averaged density. In this case, a KS-SSE will still yield an unraveling of the KS master equation, provided the numerical propagation of the SSE is performed carefully as explained in Chap. 11.

The numerical KS-SSE scheme which unravels the KS master equation proceeds by constructing the KS potential as a functional of the ensemble-averaged density, $\langle \hat{n}(r, t) \rangle$ and then using the *same* KS potential for each stochastic trajectory. In this way, one arrives at a closed equation of motion for the statistical operator, since the KS Hamiltonian depends only on ensemble averaged quantities and is therefore not a stochastic field. Such a numerical scheme would proceed by propagating all the SSE trajectories in the ensemble simultaneously, at each time-step computing the ensemble-averaged $\langle \hat{n}(r, t) \rangle$ and from this density constructing the KS potential to be used in the next time step. This is the KS-SSE scheme presented in the next chapter and is perfectly consistent with the master equation approach presented in this chapter.

The reader should, however, be aware of a KS-SSE propagation scheme presented by Di Ventra and coworkers, which differs from that discussed here and in Chap. 11 (D'Agosta and Ventra 2008). In the scheme of DiVentra and coworkers, one uses a different KS potential for each trajectory and then performs the ensemble averaging afterwards. Such a scheme does not yield a closed equation of motion for the statistical operator, since in this case the KS potential is a stochastic field. To highlight the differences between DiVentra and coworkers' KS-SSE scheme and the one presented in this book, it is useful to recognize an analogy between OQS–TDDFT which describes non-equilibrium behavior and static DFT of systems in equilibrium at a finite temperature as formulated by Kohn, Sham and Mermin (KSM) (Kohn and Sham 1965; Mermin 1965). In the static equilibrium case, one also considers a density matrix representing an ensemble of different microstates. However, in the usual KSM formulation of DFT one considers a *single* Kohn–Sham potential which is a functional of the ensemble averaged density. This is analogous to the master equation formalism presented in this chapter and its KS-SSE unraveling presented in the next chapter. One does not construct different Kohn–Sham potentials corresponding to different microstates of the system, which would be analogous the KS-SSE method of ensemble averaging proposed in the literature by DiVentra and coworkers.

10.8 Conclusions and Outlook

We have discussed the formal foundations of OQS–TDDFT in the density matrix representation starting from a many-electron quantum master equation and established a van Leeuwen construction which allows for a variety of different Kohn–Sham schemes.

The first scheme we discussed uses a non-interacting and closed (unitarily evolving) Kohn–Sham system to reproduce the dynamics of an interacting OQS. With suitable functionals, this scheme is remarkably useful for dissipative real-time dynamics, since it can be easily implemented in existing real-time codes. We presented the simple yet practical Markovian bath functional, which has shown promising results for a model Helium system (Yuen-Zhou et al. 2010). Future research will focus on understanding exact conditions with the goal of developing more sophisticated functionals. In (Tempel et al. 2011b), a systematic study of the exact OQS–TDDFT functional was carried out for a one-electron model OQS. The exact functional was shown to have memory dependence (Maitra and Burke 2001) and share some features with existing dissipation functionals in time-dependent current DFT (TDCDFT) (Vignale and Kohn 1996; Vignale et al. 1997; Ullrich and Vignale 2002). However, in OQS–TDDFT dissipation arises from coupling to a dense bosonic bath, which differs from TDCDFT where dissipation arises as an intrinsic feature of the interacting electron liquid.

The second scheme we discussed uses an open KS system to calculate broadened absorption spectra in linear response TDDFT. By using an open KS system, the bare KS spectrum is already broadened, while the OQS–TDDFT exchange-correlation kernel generates additional line broadening and shifts. The development of more sophisticated frequency-dependent functionals to capture additional broadening and asymmetric lineshapes due to non-Markovian effects is currently being explored as well.

In the next chapter, an alternative formulation of OQS–TDDFT based on stochastic wavefunctions rather than density matrices will be presented.

Chapter 11
Open Quantum Systems: A Stochastic Perspective

Heiko Appel

11.1 Introduction

Time-dependent density-functional theory (TDDFT) provides an efficient approach for the study of excited state properties as well as the real-time dynamics of many-particle systems (Marques et al. 2006a). The original formulation of TDDFT was designed to treat the unitary time-evolution of a closed quantum system with a fixed number of particles (Runge and Gross 1984). On the other hand, most experimental situations have in common that the respective system of interest is coupled to some surrounding which influences it in a non-negligible way. Already a single atom in vacuum can not be regarded as completely isolated, since the atom is embedded in the surrounding photon field. This results in, for example, spontaneous emission which can not be described by the original TDDFT approach. Other examples where the coupling to the surrounding plays a prominent role include hot electron relaxation in bulk systems and surfaces after laser irradiation, thermalization due to electron–phonon coupling, decoherence in pump–probe experiments, exciton propagation and relaxation in biological chromophores, vibrational relaxation in nanomaterials and molecular systems.

Even if we were able to prepare a perfectly isolated quantum system, we would need to regard a measurement of the system as bringing the system into contact with an environment, i.e. the measurement apparatus itself constitutes an environment. Such a measurement can be regarded as a non-unitary projection of the system state onto states of the observables. This results in a relative loss of information and increase of entropy in the system. As a consequence, the time-evolution of our system of interest on time-scales of a laboratory experiment can no longer be regarded as unitary.

H. Appel (✉)
Fritz-Haber-Institut der Max-Planck-Gesellschaft,
Faradayweg 4–6, 14195 Berlin-Dahlem, Germany
e-mail: appel@fhi-berlin.mpg.de

M. A. L. Marques et al. (eds.), *Fundamentals of Time-Dependent Density Functional Theory*, Lecture Notes in Physics 837, DOI: 10.1007/978-3-642-23518-4_11, © Springer-Verlag Berlin Heidelberg 2012

Given all these aspects which are ubiquitous in real-world experiments, it would therefore be desirable to extend the range of applicability of TDDFT also to situations where environment induced dissipation and decoherence play an important role for the dynamical evolution of the system. In recent years, several attempts have been made to also consider open quantum systems in a density-functional formalism. A TDDFT approach based on a Kohn–Sham master equation was pioneered by Burke et al. (2005c) and in recent work this has been pursued by the group of Aspuru-Guzik (Yuen-Zhou et al. 2009; Tempel et al. 2011a; Tempel and Aspuru-Guzik 2011b). This group also proposed a description of open quantum systems in terms of a unitarily evolving closed Kohn–Sham system (Yuen-Zhou et al. 2009, 2010). A computational TDDFT study of a dissipative molecular device has been performed by Chen and collaborators (Zheng et al. 2007). The authors justified their approach by proposing a time-dependent extension of the holographic electron density theorem for ground states (Riess and Münch 1981). Friction in TDDFT real-time propagations was considered by Neuhauser and Lopata (2008) and an analogy to system-bath studies was drawn. An alternative approach to treat open quantum systems in TDDFT has been developed by Di Ventra and D'Agosta (2007; D'Agosta and Di Ventra 2008) and Appel and Di Ventra (2009, 2011) and is based on stochastic Kohn–Sham equations.

In the present chapter we will give a brief overview of this stochastic density functional approach to open quantum systems. The material of the present text is organized as follows. We start in Sect. 11.2 with some general remarks on choices for the partitioning of a physical scenario into the system of interest and the environment and discuss typical physical assumptions which are frequently employed in open quantum system theories. In Sect. 11.3, we illustrate how to integrate out the bath degrees of freedom directly in terms of wavefunctions. In general, this leads to a non-Markovian stochastic Schrödinger equation for the system dynamics. In order to illustrate the relation between different stochastic Schrödinger equations and quantum master equations, we show in Sect. 11.4 how to derive the Lindblad master equation from the stochastic Schrödinger equation in the Born–Markov limit. This example also illuminates the possible starting points that can be used for a TDDFT approach to open quantum systems. In Sect. 11.5, we introduce the formal aspects of stochastic current density functional theory. We discuss numerical aspects for a solution of the stochastic time-dependent Kohn–Sham (TDKS) equations and show some examples for an application of stochastic current DFT to molecular systems. Finally, we conclude the chapter with a summary and outlook for future prospects of TDDFT for open quantum systems.

11.2 General Remarks on Open Quantum Systems

11.2.1 Partitioning into System and Environment

As motivated in the introduction, we have to face a situation where our system of interest is coupled to a typically much larger quantum system which we regard as the environment (or bath). To describe such a situation quantum mechanically, let us consider the total Hamiltonian for the combined system and bath and their mutual interaction in the form

$$\hat{H}(t) = \hat{H}_S(t) \otimes \hat{I}_B + \hat{I}_S \otimes \hat{H}_B + \alpha \hat{H}_{SB}, \tag{11.1}$$

where $\hat{H}_S(t)$ denotes the system Hamiltonian, $\hat{H}_B$ describes the bath degrees of freedom and $\hat{I}_S$, $\hat{I}_B$ denote unit operators in the respective system and bath manifolds. In general, we allow for time-dependent system Hamiltonians, $\hat{H}_S(t)$, which e.g. can include the interaction with external electromagnetic fields. To lowest order, the interaction between the two subsystems is typically assumed to have the following bilinear form

$$\hat{H}_{SB} = \sum_{j=1}^{m} \hat{S}_j \otimes \hat{B}_j, \tag{11.2}$$

which connects the system and the bath with a coupling strength α. This bilinear expression covers the most common types of baths. As an example, photon and phonon baths have such interaction terms when they are coupled to fermionic degrees of freedom (e.g. Fröhlich interaction in the case of a phonon bath). In general, one can consider m different types of baths coupled to the system, each describing different relaxation channels.

Clearly, the partitioning of the total Hamiltonian in Eq. 11.1 is not unique. Depending on the physical situation, different partitionings of $\hat{H}(t)$ might be of interest. For instance, if we aim at describing dissipation and dephasing during the dynamics of a molecule on a surface, we could include the electronic and ionic degrees of the molecule as well as the electronic structure of the surface in the system Hamiltonian $\hat{H}_S(t)$. The ionic degrees of freedom of the surface and the corresponding bulk would then constitute one possible choice for the bath, i.e. would represent a phonon bath which provides relaxation channels for the electronic excitations in the system. On the other hand, for reasons of simplicity and computational cost, we could also include the ionic degrees of freedom of the molecule in the bath. This is of course less flexible when, for example, conformational changes of the molecule are of interest. In that case, the first form of the partitioning would be more appropriate but also more difficult for an explicit treatment. This simple example already shows that a predictive theory for open quantum systems not only relies on the description of the electronic many-body effects in the subsystem S alone (e.g. within a TDDFT approach), but importantly also requires as input a microscopic picture of the relevant decoherence and dissipation channels. Even if we could solve the effective system dynamics exactly, we would still need to rely on the knowledge of thermodynamic properties of the bath, the form of bath-correlation functions and the microscopic form of the system-bath interaction. It is precisely these macroscopic thermodynamic properties of the bath, the bath-correlation functions, and the microscopic form of the system-bath interaction term, $\hat{H}_{SB}$, that determine the relaxation rates for the dynamics of the system. Therefore, we have to keep in mind that both an effective description of the system dynamics as well as a given partitioning of the total Hamiltonian enter in a theory for open quantum systems.

11.2.2 Physical Assumptions

In principle, in order to describe a bath with true thermodynamic properties, e.g. with Poincaré recurrence times pushed to infinity, an infinite amount of degrees of freedom in the bath is required. This implies that we can not simulate the dynamics of the full Hamiltonian $\hat{H}(t)$ on a finite computer.[1] The goal of open quantum system theories is therefore to find an effective description for the dynamics of the system degrees of freedom only. In order to reach this goal, there are several physical assumptions that typically enter a description of systems coupled to external environments. In general, the interaction between the system and the bath is assumed to be weak. It is therefore possible to treat the system-bath coupling perturbatively in the coupling constant α. The bath is generally modelled as a large set of bosonic modes with a dense energy spectrum. Typical examples of such environments include, as mentioned above, a photon bath or a phonon bath coupled to the system. Since we do not know the microscopic state of all the infinitely many degrees of freedom of the environment, in general only macroscopic thermodynamic properties of the bath, like e.g. temperature or pressure are known. In the most common case, one assumes that the bath is in thermal equilibrium at a given temperature T. A further assumption that enters most open quantum system descriptions is that the system and bath are initially uncorrelated. This allows one to work with factorized initial states but it may be a severe assumption for some cases.

11.3 Stochastic Schrödinger Equations

In this section, we aim at finding an effective description for the dynamics of the system S in the presence of the bath. The common textbook way of integrating out the bath degrees of freedom is performed by "tracing" the statistical operator

---

[1] This also implies that it is difficult to establish exact reference systems for interacting Fermions coupled to an environment. Analytical solutions are available only for very simple system-bath models.

$\hat{\rho}$ (N-body density matrix for mixed states) of the combined system and bath over all bath degrees of freedom, i.e. one considers the following partial trace $\hat{\rho}_S = \mathrm{Tr}_B\{\hat{\rho}\}$, where $\hat{\rho}_S$ is called the reduced statistical operator of the system S (Breuer and Petruccione 2002; Weiss 2007). One can then find approximate equations of motion, so-called quantum master equations, for this reduced statistical operator $\hat{\rho}_S$. A frequently used method to derive approximate quantum master equations is the Nakajima–Zwanzig projection operator technique (Nakajima 1958; Zwanzig 1960), cf. also Chap. 10 in the present volume. Although many different forms for approximate master equations can be derived with such approaches, they often share the problem that the reduced density matrix $\hat{\rho}_S$ does not maintain positivity when evolved with approximate master equations. A loss of positivity implies that the eigenvalues of $\hat{\rho}_S$, which describe occupation probabilities, can become negative. This precludes a statistical interpretation of physical observables and renders the approximate time-evolution of the density matrix unphysical. The loss of positivity is a known problem for, e.g. the Redfield equations or also for the Caldeira–Leggett equation (Suarez et al. 1992). In some cases, slippage factors for the initial conditions can be employed to curtail the problem (Gaspard and Nagaoka 1999a).

In the present section, we summarize an alternative approach that has been developed over the last few decades by Diosi (1988), Dalibard et al. (1992), Zoller and Gardiner et al. (Dum et al. 1992a, b; Gardiner et al. 1992) as well as Carmichael et al. (1989). In this approach, instead of using the density matrix, the bath degrees of freedom are integrated out directly at the level of the wavefunction. As will be illustrated, this provides an alternative route to describe open quantum systems entirely in terms of wavefunctions. The advantage of this approach is that one directly works with a statistical ensemble of state vectors which ensures by construction that positivity is maintained throughout the time-evolution of the system. We will briefly recall here the basic steps of this wavefunction based approach before we turn our attention in Sect. 11.5 to a density-functional formulation.

As discussed in Sect. 11.2, we start with the total Hamiltonian for system and bath and their mutual interaction as given by (11.1). Since the combined set of system S and bath B can be regarded as a closed quantum system, it follows a unitary time evolution given by the time-dependent Schrödinger equation

$$i\frac{d}{dt}|\Psi(t)\rangle = \hat{H}(t)|\Psi(t)\rangle. \tag{11.3}$$

As a first step towards an effective equation of motion for the system S, we consider the many-body eigenstates of the bath Hamiltonian resulting from the static Schrödinger equation of the bath

$$\hat{H}_B \Psi_{B,n}(x_B) = \varepsilon_n \Psi_{B,n}(x_B). \tag{11.4}$$

Since the eigenstates, $\Psi_{B,n}(x_B)$, of the Hermitian Hamiltonian, $\hat{H}_B$, form a complete orthonormal set, we can expand the total wavefunction in this time-independent basis of bath eigenmodes[2]

$$\Psi(x_S, x_B; t) = \sum_q \Psi_{S,q}(x_S; t)\Psi_{B,q}(x_B). \tag{11.5}$$

This gives rise to expansion coefficients $\Psi_{S,q}(x_S; t)$, which depend on the system coordinates and on time. Let us consider in the following a representative coefficient, $\Psi_{S,q}(x_S; t)$. The square $|\Psi_{S,q}(x_S; t)|^2$ of such a coefficient describes the amplitude for finding the system in bath mode q. In order to project on a given bath mode, we define the following Feshbach projection operators (Feshbach 1958; Nordholm and Rice 1975)

$$\begin{aligned}
\hat{P}_q &= \hat{I}_S \otimes |\Psi_{B,q}\rangle\langle\Psi_{B,q}|, \\
\hat{Q}_q &= \hat{I}_S \otimes \sum_{p \neq q} |\Psi_{B,p}\rangle\langle\Psi_{B,p}|.
\end{aligned} \tag{11.6}$$

By applying these projection operators to the time-dependent Schrödinger equation of the combined system, Eq. 11.3, we arrive at

$$\begin{aligned}
\mathrm{i}\partial_t \hat{P}_q \Psi(t) &= \hat{P}_q \hat{H} \hat{P}_q \Psi(t) + \hat{P}_q \hat{H} \hat{Q}_q \Psi(t), \\
\mathrm{i}\partial_t \hat{Q}_q \Psi(t) &= \hat{Q}_q \hat{H} \hat{Q}_q \Psi(t) + \hat{Q}_q \hat{H} \hat{P}_q \Psi(t).
\end{aligned} \tag{11.7}$$

Solving formally for the second equation and inserting the result back into the first equation yields an effective Schrödinger equation projected on the system manifold

$$\mathrm{i}\partial_t \hat{P}_q \Psi(t) = \hat{P}_q \hat{H} \hat{P}_q \hat{P}_q \Psi(t) + \overbrace{\hat{P}_q \hat{H} \hat{Q}_q \mathrm{e}^{-\mathrm{i}\hat{Q}_q \hat{H} \hat{Q}_q t} \hat{Q}_q \Psi(0)}^{\text{Source Term}}$$

$$\underbrace{-\mathrm{i}\int_0^t \hat{P}_q \hat{H} \hat{Q}_q \mathrm{e}^{\mathrm{i}\hat{Q}_q \hat{H} \hat{Q}_q (t-\tau)} \hat{Q}_q \hat{H} \hat{P}_q \hat{P}_q \Psi(\tau)\, \mathrm{d}\tau}_{\text{Memory Term}} . \tag{11.8}$$

So far, no approximations have been made. This equation is still fully coherent and describes the time-evolution of a representative system wave function, $\Psi_S(t) = \hat{P}_q \Psi(t)$, in the presence of the bath. The source term takes the initial conditions into account, whereas the memory term records the past history of the interaction between system and bath.

We emphasize that Eq. 11.8 has a formal similarity to the quantum transport formulation of Kurth et al. (2005), which is also discussed in the present volume (see Chap. 17). However, there are also notable physical differences. In the work of Kurth and Stefanucci, the projection operators project onto real-space regions (lead and

---

[2] One could also consider a mixed state of the overall system as the initial state. In the present discussion, this changes only the weights of the states in the statistical ensemble at the initial time.

device regions) and in addition, the bath is fermionic since it refers to the electronic
degrees of freedom in the leads. Hence, in the quantum transport formulation, not only
momentum and energy transfer have to be taken into account, but also an exchange
of particles. This exchange of particles implies that the projected wavefunction (e.g.
in the central device region) is in general not normalized. In contrast, in the present
chapter, we consider only bosonic baths, so that energy and momentum can be
transfered, but particle number is conserved.

The effective Schrödinger equation, Eq. 11.8, for the system degrees of freedom
is still a many-body equation and, due to the many-body operators in the source and
memory terms, is quite involved to solve in practice. It is therefore common to introduce at this point several approximations which all rest on physical assumptions for
the interaction between the system and the bath and the macroscopic thermodynamic
properties of the bath. As discussed in Sect. 11.2.2, we assume that the bath has a
dense energy spectrum and always remains in thermal equilibrium. Since the phases
of all the bath degrees of freedom (in principle infinitely many) are not known, we
perform a random-phase approximation.[3] Furthermore, it is assumed that the interaction between the system and the bath is weak so that a perturbative expansion up
to second order in the coupling constant α is sufficiently accurate (Born approximation). With these assumptions, the fully coherent Eq. 11.8 can be turned into a
non-Markovian stochastic Schrödinger equation of the form (Gaspard and Nagaoka
1999b)

$$
i\partial_t \Psi_S(t) = \hat{H}_S \Psi_S(t) + \alpha \sum_q l_q(t) \hat{S}_q \Psi_S(t)
$$

$$
- i\alpha^2 \sum_{pq} \int_0^t C_{pq}(t - \tau) \hat{S}_p^\dagger e^{-i\hat{H}_S(t-\tau)} \hat{S}_q \Psi_S(\tau) \, d\tau + \mathcal{O}(\alpha^3),
$$

(11.9)

where $C_{pq}(t - \tau)$ are bath correlation functions. In this equation, the source term of
Eq. 11.8 appears as stochastic forcing term, where $l_q(t)$ denotes a stochastic process
with zero ensemble average $\overline{l_q(t)} = 0$, and correlation functions

$$
\overline{l_p(t)l_q(t')} = 0,
$$

(11.10a)

$$
\overline{l_p^*(t)l_q(t')} = C_{pq}(t - t'),
$$

(11.10b)

where $\overline{l}$ denotes the statistical average of l. Due to the memory integral, Eq. 11.9
is still rather involved to solve. Typically, the Markov approximation is invoked at
this point. This amounts to assuming a δ-correlated bath, which in terms of the bath
correlation functions takes the form

$$
\overline{l_p^*(t)l_q(t')} \propto \delta_{pq}\delta(t - t').
$$

(11.11)

---

[3] The physical content of this approximation becomes clear in the next chapter, where we consider
the derivation of the Lindblad equation from the stochastic Schrödinger equation.

Physically, this means that the relaxation timescales inside the bath are much faster than the relaxation timescales in the system. As a result, the bath retains no memory and behaves always the same way as seen from the system perspective. Finally, we arrive with these assumptions at the Born–Markov limit of the stochastic Schrödinger equation

$$i\partial_t \Psi_S(t) = \hat{H}_S \Psi_S(t) + \alpha \sum_q l_q(t) \hat{S}_q \Psi_S(t) - \frac{i\alpha^2}{2} \sum_q \hat{S}_q^\dagger \hat{S}_q \Psi_S(t) + \mathcal{O}(\alpha^3). \quad (11.12)$$

We note, in passing, that the Markov approximation has been performed here mainly since it is a reasonable approximation for many types of environments. In addition, the equations become much more tractable computationally in the Markov limit. If this approximation is physically too severe for the considered application, e.g. if the timescales of the system are comparable to the typical timescales of the bath, we have to resort to the more involved solution of the non-Markovian stochastic Schrödinger equation in Eq. 11.9.

The second term on the rhs of Eq. 11.12 describes the fluctuations induced by the presence of the bath and the third term is responsible for dissipation. Both terms are not independent of each other, but are connected by a fluctuation–dissipation relation which ensures that the norm of the wavefunction is preserved on average (Van Kampen 1992).

For a given and fixed initial state, the solution of the stochastic Schrödinger equation for different stochastic processes, $l_q(t)$, results in different stochastic trajectories in the Hilbert space of the system. These solutions form a statistical ensemble, $\{|\Psi_{Sj}\rangle\}$, and are usually termed stochastic realizations. The ensemble, $\{|\Psi_{Sj}\rangle\}$, describes the properties of the subsystem, S, in the presence of the bath. To see this, consider an expectation value of some system observable $\hat{O}_S$. Using Eq. 11.5 and the orthonormality of the bath eigenstates, we have

$$\langle \Psi(x_S, x_B, t) | \hat{O}_S | \Psi(x_S, x_B, t) \rangle = \sum_n \langle \Psi_{S,n}(x_S, t) | \hat{O}_S | \Psi_{S,n}(x_S, t) \rangle \quad (11.13)$$

and by normalizing the functions, $\Psi_{S,n}(x_S, t)$, according to

$$\widetilde{\Psi}_{S,n}(x_S, t) = \Psi_{S,n}(x_S, t) / \sqrt{p_n(t)}, \quad (11.14)$$

where

$$p_n(t) = \langle \Psi_{S,n}(x_S, t) | \Psi_{S,n}(x_S, t) \rangle, \quad (11.15)$$

we can define the following statistical operator

$$\hat{\rho}_S = \sum_n p_n(t) | \widetilde{\Psi}_{S,n}(x_S, t) \rangle \langle \widetilde{\Psi}_{S,n}(x_S, t) |$$
$$= \overline{| \widetilde{\Psi}_S(t) \rangle \langle \widetilde{\Psi}_S(t) |}, \quad (11.16)$$

and immediately recognize that the average (11.13) can be rewritten as

$$\langle \Psi(x_S, x_B, t) | \hat{O}_S | \Psi(x_S, x_B, t) \rangle = \text{Tr}\{\hat{\rho}_S \hat{O}_S\}. \tag{11.17}$$

Because of the interaction with the bath, the system, S, is necessarily in a *mixture* of states. This mixture is defined by the macrostate

$$\{p_n(t), |\widetilde{\Psi}_{Sn}(x_S, t)\rangle\}. \tag{11.18}$$

We emphasize, at this point, that not only first-order moments of observables can be computed from the macrostate $\{p_n(t), |\widetilde{\Psi}_{Sn}(x_S, t)\rangle\}$. Since we have the full statistical ensemble at hand, we can also directly compute higher-order moments or cumulants of the distribution for any observable. Situations where the full statistical ensemble turns out to be useful include, for example, the calculation of shot noise which arises from the autocorrelation function of the current operator. In addition, working directly with wavefunctions in the statistical ensemble has the distinct advantage that we always deal with physical states. The average in Eq. 11.16 thus always maintains positivity by construction.

11.4 Derivation of Master Equations from Stochastic Schrödinger Equations

In classical statistical systems, a trajectory approach is based on Langevin equations and describes, for example, the Brownian motion of a particle in a thermal surrounding. The corresponding time-evolution of the probability distribution is given by the Fokker–Planck equation and can be obtained from the Langevin trajectories by averaging over a statistical ensemble of trajectories. The quantum case can be viewed similarly. The stochastic Schrödinger equation describes the motion of particles embedded in a surrounding in terms of stochastic trajectories for the state vectors in the Hilbert space of the system. Similar to the Fokker–Planck equation in the classical case, the quantum master equation then describes the probability distribution for these stochastic state vector "trajectories".

To illustrate this relation in the quantum case, let us consider the derivation of the Lindblad master equation from the stochastic Schrödinger equation in the Born–Markov limit. This exemplifies for which cases we can actually arrive at a *closed* equation of motion for the statistical operator and for which cases this is not possible using statistical averages.

For notational convenience, let us denote in the following by $|\psi\rangle$ a single member of the statistical ensemble, $\{|\psi\rangle_j\}$, and with an overline the statistical average over this ensemble. If we denote the stochastic integral for the stochastic process $l(t)$ by

$$W(t) = \int_0^t dt'\, l(t') \tag{11.19}$$

and observe that $W(t)$ is a Wiener process (Van Kampen 1992), we can write the differential increment of the stochastic Schrödinger equation (11.12) for a single bath operator $\hat{S}$ in the form [4]

$$d|\Psi_S\rangle = \left[-i\hat{H}_S|\Psi_S\rangle - \frac{1}{2}\hat{S}^\dagger\hat{S}|\Psi_S\rangle \right] dt - i\hat{S}|\Psi_S\rangle dW. \qquad (11.20)$$

Using Ito stochastic calculus, let us now consider the following differential

$$d|\Psi_S\rangle\langle\Psi_S| = (d|\Psi_S\rangle)\langle\Psi_S| + |\Psi_S\rangle(d\langle\Psi_S|) + (d|\Psi_S\rangle)(d\langle\Psi_S|). \qquad (11.21)$$

Note, that unlike in normal calculus, we also have to keep the third term in the product rule above, since a statistical average over the Wiener increment $dW^\dagger dW$ is proportional to dt. This will cause terms quadratic in dW to contribute to *first* order in dt. By inserting Eq. 11.20 and its Hermitian conjugate into Eq. 11.21 we find

$$\begin{aligned}
d|\Psi_S\rangle\langle\Psi_S| = &- \left(i\hat{S}|\Psi_S\rangle\langle\Psi_S|dW + \text{h.c.} \right) - i\left[\hat{H}_S, |\Psi_S\rangle\langle\Psi_S| \right]dt \\
&- \frac{1}{2}\left\{ \hat{S}^\dagger\hat{S}, |\Psi_S\rangle\langle\Psi_S| \right\}dt + \hat{S}|\Psi_S\rangle\langle\Psi_S|\hat{S}^\dagger dW^\dagger dW \\
&+ \left(\hat{S}|\Psi_S\rangle\langle\Psi_S|\hat{H}_S dWdt + \text{h.c.} \right) + \left(\frac{i}{2}\hat{S}|\Psi_S\rangle\langle\Psi_S|\hat{S}^\dagger\hat{S}dWdt + \text{h.c.} \right) \\
&+ \hat{H}_S|\Psi_S\rangle\langle\Psi_S|\hat{H}_S dt^2 + \frac{1}{4}\hat{S}^\dagger\hat{S}|\Psi_S\rangle\langle\Psi_S|\hat{S}^\dagger\hat{S}dt^2 + \frac{i}{2}\left\{ \hat{H}_S, |\Psi_S\rangle\langle\Psi_S|\hat{S}^\dagger\hat{S} \right\}dt^2.
\end{aligned}$$
$$(11.22)$$

In order to construct the statistical operator from the state vectors, we perform in the next step the statistical average over all members of the stochastic ensemble, i.e.

$$d\hat{\rho}_S = d\overline{|\Psi_S\rangle\langle\Psi_S|}. \qquad (11.23)$$

Taking the properties $\overline{dW} = 0$ and $\overline{dW^\dagger dW} = dt$ of the stochastic process $l(t)$ into account, we notice that only the second, third and fourth term on the rhs of Eq. 11.22 contribute to first order in dt and we arrive at

$$d\hat{\rho}_S = -i\overline{\left[\hat{H}_S, |\Psi_S\rangle\langle\Psi_S| \right]}dt - \frac{1}{2}\overline{\left\{ \hat{S}^\dagger\hat{S}, |\Psi_S\rangle\langle\Psi_S| \right\}}dt + \overline{\hat{S}|\Psi_S\rangle\langle\Psi_S|\hat{S}^\dagger}dt + O(dt^2). \qquad (11.24)$$

Only if the Hamiltonian $\hat{H}_S$ of the system and the bath operator $\hat{S}$ do not explicitly depend on the state vectors or similarly do not depend on some external stochastic field [like they would for the case when a stochastic thermostat for the nuclei is employed (Bussi et al. 2007)], we have

$$\overline{\left[\hat{H}_S, |\Psi_S\rangle\langle\Psi_S| \right]} = \left[\hat{H}_S, \overline{|\Psi_S\rangle\langle\Psi_S|} \right] \qquad (11.25a)$$

---

[4] To simplify the notation, the coupling parameter α has been absorbed in $\hat{S}$ in the following discussion.

$$\overline{\left\{ \hat{S}^\dagger \hat{S}, |\Psi_S\rangle\langle\Psi_S| \right\}} = \left\{ \hat{S}^\dagger \hat{S}, \overline{|\Psi_S\rangle\langle\Psi_S|} \right\} \tag{11.25b}$$

$$\overline{\hat{S}|\Psi_S\rangle\langle\Psi_S|\hat{S}^\dagger} = \hat{S}\overline{|\Psi_S\rangle\langle\Psi_S|}\hat{S}^\dagger \tag{11.25c}$$

and only in that case Eq. 11.24 can be written in the form

$$\frac{\mathrm{d}}{\mathrm{d}t}\hat{\rho}_S = -\mathrm{i}\left[\hat{H}_S, \hat{\rho}_S \right] - \frac{1}{2}\left\{ \hat{S}^\dagger \hat{S}, \hat{\rho}_S \right\} + \hat{S}\hat{\rho}_S\hat{S}^\dagger, \tag{11.26}$$

which is the well-known master equation in the Born–Markov limit. Furthermore, if we consider the limit where the bath operators and the Hamiltonian do not depend on time, this is the prominent Lindblad master equation (Lindblad 1976; Gardiner 1985; Breuer and Petruccione 2002; Weiss 2007).[5]

If, on the other hand, the Hamiltonian does depend on the state vectors we have

$$\overline{\hat{H}_S[|\Psi_S\rangle]} \neq \hat{H}_S[\overline{|\Psi_S\rangle}], \tag{11.27}$$

or if the Hamiltonian or the bath operators depend explicitly on some stochastic field, we have to stay with Eq. 11.24 and perform explicit averages over *stochastic Hamiltonians*. In summary, this implies that we do not necessarily have a closed equation of motion for the statistical operator $\hat{\rho}_S$ at hand. Only if we can prove that the Hamiltonian depends exclusively on averages over the state vectors (like e.g. the ensemble averaged N-body density $\rho_S = \overline{|\Psi_S\rangle\langle\Psi_S|}$,) we can use Eq. 11.25a and only in that case we arrive at a closed equation of motion for the statistical operator $\hat{\rho}_S$.

A similar situation already arises for classical statistical systems. If, for example, the Langevin equation depends on internal degrees of freedom of the system, such as the velocity of a Brownian particle, then there is no closed Fokker–Planck equation which describes the probability distribution for that case. We have to stay with the Langevin trajectories and perform averages over an ensemble of such trajectories.

11.5 Stochastic Current Density Functional Theory

11.5.1 Formal Aspects of Stochastic Current Density Functional Theory

The stochastic Schrödinger equations, Eqs. 11.9 and 11.12, provide an approximate way (Born and Born–Markov limit respectively) of integrating out all the bath degrees

---

[5] Let us emphasize at this point that the Lindblad theorem (Lindblad 1976) guarantees a semigroup property for $\hat{\rho}_S$ which ensures completely positive evolution (i.e. all probabilities computed from $\hat{\rho}_S$ stay positive). However, this semigroup property can *only* be established in the Lindblad framework if the Hamiltonian does not depend on time. The Lindblad theorem is therefore not useful in TDDFT since the Hamiltonian is generally time-dependent due to the dependence on the time-evolving density.From this derivation, we can also see that in the Markov limit the physical content of the random-phase approximation, that we introduced in the previous section, leads to a form of fluctuation and dissipation which is exactly equivalent to the Lindblad equation.

of freedom. However, we are still left with the many-body problem of the system Hamiltonian, $\hat{H}_S$, itself. The operator, $\hat{H}_S$, so far still contains the full electron–electron interaction term, and it is desirable to find an effective way to describe exchange and correlation effects within the system S. In the present section we will comment on how to construct a time-dependent current density functional approach based on the interacting many-electron stochastic Schrödinger equation in Eq. 11.12. Using a stochastic approach for the open system problem in terms of wavefunctions has several advantages. The derivation of the stochastic Schrödinger equation remains valid for time-dependent system Hamiltonians, and it does not rely on a semigroup property. As illustrated in the previous section, the positivity of the density matrix (11.16) is ensured since we average only over physical states, i.e. the probabilities obtained from (11.16) remain always positive. Hence, this approach is a sound starting point to formulate a TDDFT approach for open quantum systems.

Stochastic time-dependent current DFT rests on the following basic theorem, which establishes a one-to-one mapping between the statistically averaged current density $\overline{j(r,t)}$ and the vector potential $A(r,t)$ acting on the system.

Theorem *For fixed dissipation (bath operators $\hat{S}_j$), two-body interaction $v_{ee}(r, r')$ and fixed initial states, there is, under reasonable physical assumptions, a one-to-one mapping between the open-system current density $\overline{j(r,t)}$ and the external vector potential $A(r,t)$.*

For details on the proof of the theorem, we refer to Di Ventra and D'Agosta (2007). The one-to-one mapping, which is established by the proof, allows us to introduce a non-interacting auxiliary stochastic Kohn–Sham system

$$\mathrm{d}\left|\Phi_S^{KS}\right\rangle = \left(-\mathrm{i}\hat{H}_{KS} - \frac{1}{2}\sum_j \hat{S}_j^\dagger \hat{S}_j\right)\left|\Psi_S^{KS}\right\rangle \mathrm{d}t - \mathrm{i}\sum_j \hat{S}_j \left|\Psi_S^{KS}\right\rangle \mathrm{d}W, \quad (11.28)$$

which is coupled to the environment and evolves with an effective vector potential $A^{\mathrm{eff}}(r,t)$ and yields the same statistically averaged current density $\overline{j(r,t)}$ as the interacting many-electron system. As in normal time-dependent current DFT we have to assume non-interacting A-representability. It is important to realize that this Kohn–Sham equation is in general still a many-body equation, since the bath operators $\hat{S}_j$ are not necessarily single-particle operators. Only if we assume that the system-bath interaction can be expressed in terms of single-particle bath operators we can formulate Eq. 11.28 as a set of single-particle stochastic Kohn–Sham equations. We will assume for the following that this is a reasonably good approximation. We also emphasize that the basic variable of this stochastic Kohn–Sham scheme is the *ensemble averaged* current density $\overline{j(r,t)}$. Due to the ensemble average this current density is not stochastic anymore. Since the one-to-one mapping in the theorem is established for the variable $\overline{j(r,t)}$, the *exact* TDKS potential is a functional of this ensemble averaged current density (and initial states, etc.). Provided no external stochastic fields are considered and no loss of information occurs due to approximations in the functional, the Hamiltonian of the stochastic Kohn–Sham system is therefore also *not* stochastic and the *very same* Hamiltonian is used for all stochastic realizations in the statistical ensemble. The stochastic nature of the Kohn–Sham orbitals arises in this case only because of the coupling to the bath. In other words, the first two terms in (11.28) are deterministic and only the third term is stochastic due to the stochastic process dW. If on the other hand there is a loss of information present due to an approximate reduced, or traced description of the electronic degrees of freedom or the system is exposed to external stochastic potentials, we have a situation which requires to consider an ensemble of different stochastic Hamiltonians.

11.5.2 Practical Aspects of a Stochastic Simulation: Quantum Jump Algorithm

A common way of solving a stochastic Schrödinger equation of the form of Eq. 11.28 relies on adapted integration schemes, like, for example, modified Runge–Kutta methods (Kloeden and Platen 1992; Breuer and Petruccione 2002). Although elementary in their form, they turn out not to be the most stable choice for systems with a large number of states or for very long integration periods in time, as e.g. the case when timescales up to several picoseconds have to be reached with electronic time steps of a few attoseconds. An alternative to sample the stochastic process given in Eq. 11.28 consists in a piecewise deterministic evolution of the system which is alternating with interactions with the bath (Breuer and Petruccione 2002). Since only a deterministic time-evolution is required as an ingredient for such an approach, the known propagation algorithms employed in closed quantum system simulations (Castro et al. 2004a) can be used readily. In the following, we briefly illustrate the basic concept of this so-called quantum jump algorithm.

Consider the deterministic time-evolution given by the following norm-preserving non-linear Schrödinger equation

$$\frac{d}{dt}\varphi_j(t) = -i\left(\hat{\mathcal{H}}_S + \frac{i}{2}||\hat{S}\psi||^2\right)\varphi_j(t), \tag{11.29}$$

where the non-Hermitian Hamiltonian, $\hat{\mathcal{H}}_S$, is given by

$$\hat{\mathcal{H}}_S = \hat{H}_S - \frac{i}{2}\hat{S}^\dagger\hat{S}. \tag{11.30}$$

The main idea of the algorithm is to propagate the physical states of interest with Eq. 11.29 and to introduce the fluctuations, due to the interaction with the bath operator $\hat{S}$, in terms of instantaneous jumps of the wavefunction according to

$$\varphi_j(t_k^+) = \frac{\hat{S}\varphi_j(t_k^-)}{||\hat{S}\varphi_j(t_k^-)||}. \tag{11.31}$$

The waiting times between two successive jumps at times t_k and t_{k+1} are drawn from a waiting-time distribution. It can be shown (Breuer and Petruccione 2002), that this distribution can be sampled with a Monte-Carlo procedure by propagating auxiliary states ξ_j^{aux} with the non-Hermitian Hamiltonian in Eq. 11.30. The action of the bath operator, Eq. 11.31, is taking place when the norm of the auxiliary state ξ_j^{aux} drops below a uniform random number, $\eta_k \in [0, 1]$.

In summary, the algorithm works as follows:

1. Draw a uniform random number, $\eta_k \in [0, 1]$, for the Kohn–Sham Slater determinant

2. Propagate N auxiliary orbitals, ξ_j^{aux}, under the non-Hermitian dynamics

$$i\partial_t \xi_j^{\text{aux}} = \left[\hat{H}_{\text{KS}} - \frac{i}{2} \hat{S}^\dagger \hat{S} \right] \xi_j^{\text{aux}}, \quad j = 1 \ldots N$$

3. Propagate the orbitals, φ_j^{KS} with $j = 1 \ldots N$, of the Kohn–Sham system with a norm-conserving dynamics according to

$$i\partial_t \varphi_j^{\text{KS}} = \left[\hat{H}_{\text{KS}} - \frac{i}{2} \hat{S}^\dagger \hat{S} + i \| \hat{S} \psi_j^{\text{KS}} \|^2 \right] \varphi_j^{\text{KS}}$$

4. If the norm of the Slater determinant formed with auxiliary orbitals drops below the drawn random number η_k, act with the bath operator on the Kohn–Sham orbitals and update the auxiliary orbitals

$$\| \text{Det}\{\xi_j^{\text{aux}}(t_k^-)\} \| \le \eta_k \rightarrow \begin{cases} \varphi_j^{\text{KS}}(t_k^+) = \hat{S} \varphi_j^{\text{KS}}(t_k^-) / \| \hat{S} \varphi_j^{\text{KS}}(t_k^-) \| \\ \xi_j^{\text{aux}}(t_k^+) = \varphi_j^{\text{KS}}(t_k^+) \end{cases}$$

5. Go to step 1.

In Fig. 11.1, we illustrate the piecewise deterministic time-evolution that is generated by this algorithm. By averaging over an ensemble of trajectories that arise from this Monte-Carlo sampling, we can then compute at any point in time, according to Eq. 11.17, expectation values for physical observables of interest. We recall here, that a single trajectory should not be used for a probabilistic interpretation. In order to perform averages, we always have to consider a statistical ensemble of states.

11.5.3 Stochastic Quantum Molecular Dynamics

The original formulation of stochastic time-dependent current DFT (Di Ventra and D'Agosta 2007) was restricted to electronic degrees of freedom only, without allowing for ionic motion. This covers situations where the ionic motion is considered as part of the bath, e.g. a purely electronic system coupled to a phonon bath. However, there are many situations where one would like to treat some of the ionic degrees of freedom explicitly. As mentioned in the introduction, an example of such a situation is the study of molecules on surfaces. In this case one would like to

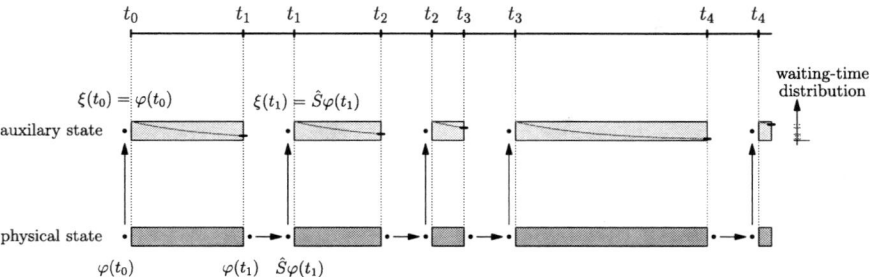

Fig. 11.1 Schematic illustration of the quantum jump algorithm that is employed for the simulation of the stochastic process associated with the stochastic Kohn–Sham equations. The time evolution of the Kohn–Sham state vector (lower track) is punctuated by "quantum jumps" at which the wavefunction changes discontinuously in time. The points in time where the jumps appear are determined by a waiting-time distribution which is sampled with the help of an auxiliary propagation (upper track)

include the ionic motion of the molecule in the Hamiltonian of the system, whereas the ionic motion of the surface atoms provide a dissipation and dephasing channel which can be taken into account in terms of a phonon bath. Other examples, where an explicit description of ionic motion is desirable, include chromophores embedded in a protein environment or molecules solvated in a liquid environment. To account for such situations, stochastic quantum molecular dynamics (MD) has been introduced recently (Appel and Di Ventra 2009, 2011). This molecular dynamics scheme is an extension of stochastic time-dependent current DFT, which allows one, in principle, to include quantum nuclei in the system Hamiltonian. Unlike standard MD approaches, like Born–Oppenheimer MD, Car–Parrinello MD, or Ehrenfest MD, which always consider a closed quantum system for the electrons, this stochastic quantum MD approach allows to couple both the electronic and the ionic degrees of freedom to a thermal bath.

Let us consider this approach in the limit of classical nuclei. In Fig. 11.2, we illustrate the dynamics of a neon dimer with soft-Coulomb interaction (Su and Eberly 1991) that is coupled to a thermal bath and compare to a standard TDDFT/Ehrenfest propagation. In both cases the time-dependent adiabatic LDA was used for the exchange-correlation functional. As initial state of the dimer we use in both cases the same stretched configuration for the ionic coordinates. The time evolution shows that the vibrational oscillations of the dimer continue indefinitely in the closed quantum system case (left panel), since no energy dissipation takes place. On the other hand, in the open quantum system case (right panel), energy is dissipated into the thermal environment and we observe a damping of the ionic vibrations (Appel and Di Ventra 2011). Note that we have not used a thermostat for the classical ions here. Instead, the damping of the ionic motion arises due to the coupling of the electrons to the thermal bath. In other words, the Ehrenfest forces, that electrons exert on the classical ions, differ *qualitatively* between standard closed-system Ehrenfest TDDFT and stochastic quantum molecular dynamics. This simple example shows that the environment has

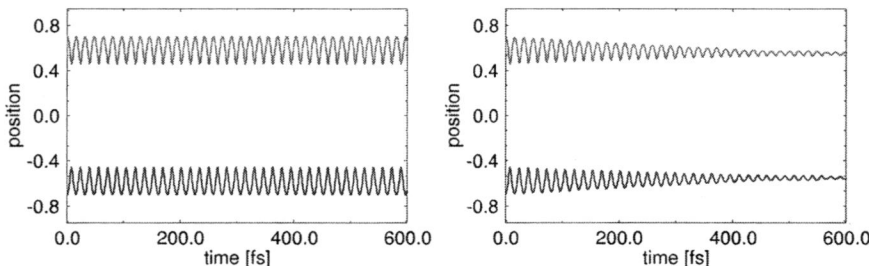

Fig. 11.2 Both panels show the atomic positions of a Neon dimer with soft-Coulomb interaction as function of time. In both cases, the same stretched configuration of the dimer has been used as initial condition. *Left panel*: normal TDDFT/Ehrenfest propagation for a closed quantum system. *Right panel*: Open quantum system simulation using stochastic quantum molecular dynamics for a bath temperature of 290 K. A relaxation time of $\tau = 300$ fs has been used for the stochastic quantum molecular dynamics simulation and 100 stochastic trajectories have been used for the statistical average

a non-negligible effect on the system dynamics and that electronic dissipation and dephasing can play an important role for the dynamics of molecules embedded in some surrounding.

11.6 Open Questions in TDDFT for Open Quantum Systems and Outlook

In summary, we presented a basic introduction to the concepts of stochastic current density functional theory. We have illustrated how to trace out the bath degrees of freedom directly in terms of wavefunctions. In general, this leads to a non-Markovian stochastic Schrödinger equation which still contains the electron–electron interaction for the system degrees of freedom. Based on such a stochastic many-body equation which takes the bath interaction into account we have then shown how to establish a stochastic current density functional theory. We have discussed numerical aspects for the solution of the stochastic Kohn–Sham equations and an application of the recently introduced stochastic quantum molecular dynamics has been provided.

The description of open quantum systems in terms of time-dependent density functional approaches still has to be regarded as a very young field of research. Basic theorems have been proposed for the quantum master and stochastic approaches. A few applications indicate the potential of the theories and the questions that can be answered. Examples include relaxation in quantum transport (Burke et al. 2005c), broadening of molecular spectra (Tempel et al. 2011a) or quantum MD of electrons and ions coupled to an environment (Appel and Di Ventra 2009, 2011).

In TDDFT for open quantum systems many questions remain open and provide a large and exciting field for future studies. Most prominent is the development of functionals. So far, in both the quantum master and stochastic TDKS approaches, primarily standard functionals from closed quantum system TDDFT have been employed, most notably the adiabatic local density approximation. So far, orbital functionals and extensions of the TDOEP integral equations (see Chap. 6) have not been considered at all in the context of open quantum systems. At present, it is unclear for most cases how the functionals in the respective approaches depend on the bath operators/dissipation. The memory and initial state dependence in the density functionals of open quantum systems most likely will also play a different role as in closed system TDDFT, since the dynamics is irreversible and has dissipative decay and decoherence contributions dictated by the bath. Another interesting route for future research is optimal control of dissipative systems. How much of the dynamics is still controllable when the system is connected to an environment? What is the role of exchange and correlation within the system in this case?

We also have emphasized the importance of modelling the microscopic form of the system-bath interaction and the properties of the bath (e.g. bath correlation functions). Even if the exact functional would be available to describe the electron exchange and correlation effects in the system, the starting point of an open quantum system formulation still rests on a given partitioning of the total Hamiltonian of the system and the bath and the corresponding microscopic form of their mutual interaction. So far, primarily model assumptions for this interaction have been employed. For a predictive theory, it would be desirable to have a microscopic form of the bath operators at hand. This would allow one to deduce relaxation rates from the microscopic interaction between system and bath. One example along these lines is a system coupled to a phonon bath. The relaxation rates for a system coupled to such a bath would then be related to the electron–phonon coupling matrix elements. Further work is required to reach such a predictive level with open quantum system approaches in TDDFT.

Chapter 12
Multicomponent Density-Functional Theory

Robert van Leeuwen and Eberhard K. U. Gross

12.1 Introduction

The coupling between electronic and nuclear motion plays an essential role in a wide range of physical phenomena. A few important research fields in which this is the case are superconductivity in solids, quantum transport where one needs to take into account couplings between electrons and phonons, the polaronic motion in polymer chains, and the ionization-dissociation dynamics of molecules in strong laser fields. Our goal is to set up a time-dependent multicomponent density-functional theory (TDMCDFT) to provide a general framework to describe these diverse phenomena. In TDMCDFT the electrons and nuclei are treated completely quantum mechanically from the outset. The basic variables of the theory are the electron density n, which will be defined in a body-fixed frame attached to the nuclear framework, and the diagonal of the nuclear N-body density matrix Γ, which will depend on all nuclear coordinates. The chapter is organized as follows: we start out by defining the coordinate transformations to obtain a suitable Hamiltonian for defining our densities to be used as basic variables in the theory. We then discuss the basic one-to-one correspondence between TD potentials and TD densities, and subsequently, the resulting TD Kohn–Sham equations, the action functional, and linear response theory. As an example we discuss a diatomic molecule in a strong laser field.

R. van Leeuwen
Department of Physics, Nanoscience Center,
University of Jyväskylä, PO Box, 35 Survontie 9,
40014 Jyväskylä, Finland
e-mail: robert.vanleeuwe@jyu.fi

E. K. U. Gross (✉)
Max-Planck Institut für Mikrostrukturphysik,
Weinberg 2, 06120 Halle, Germany
e-mail: hardy@mpi-halle.mpg.de

M. A. L. Marques et al. (eds.), *Fundamentals of Time-Dependent Density Functional Theory*, Lecture Notes in Physics 837, DOI: 10.1007/978-3-642-23518-4_12, © Springer-Verlag Berlin Heidelberg 2012

12.2 Fundamentals

We consider a system composed of N_e electrons with coordinates $\{r\}$ and N_n nuclei with masses $M_1 \ldots M_{N_n}$, charges $Z_1 \ldots Z_{N_n}$, and coordinates denoted by $\{R\}$. By convention, the subscripts "e" and "n" refer to electrons and nuclei, respectively, and atomic units are employed throughout this chapter. In non-relativistic quantum mechanics, the system described above is characterized by the Hamiltonian

$$\hat{H}(t) = \hat{T}_n(\{R\}) + \hat{V}_{nn}(\{R\}) + \hat{U}_{ext,n}(\{R\}, t) + \hat{T}_e(\{r\}) + \hat{V}_{ee}(\{r\})$$
$$+ \hat{U}_{ext,e}(\{r\}, t) + \hat{V}_{en}(\{r\}, \{R\}), \tag{12.1}$$

where

$$\hat{T}_n = -\frac{1}{2} \sum_{\alpha=1}^{N_n} \frac{\nabla_\alpha^2}{M_\alpha} \quad \text{and} \quad \hat{T}_e = -\frac{1}{2} \sum_{j=1}^{N_e} \nabla_j^2 \tag{12.2}$$

denote the kinetic-energy operators of the nuclei and electrons, respectively,

$$\hat{V}_{nn} = \frac{1}{2} \sum_{\substack{\alpha,\beta=1 \\ \alpha \neq \beta}}^{N_n} \frac{Z_\alpha Z_\beta}{|R_\alpha - R_\beta|}, \quad \hat{V}_{ee} = \frac{1}{2} \sum_{\substack{i,j=1 \\ i \neq j}}^{N_e} \frac{1}{|r_i - r_j|}, \tag{12.3}$$

and

$$\hat{V}_{en} = -\sum_{j=1}^{N_e} \sum_{\alpha=1}^{N_n} \frac{Z_\alpha}{|r_j - R_\alpha|} \tag{12.4}$$

represent the interparticle Coulomb interactions. Truly external potentials representing, e.g., a laser pulse applied to the system, are contained in

$$\hat{U}_{ext,n}(t) = \sum_{\alpha=1}^{N_n} u_{ext,n}(R_\alpha, t) \tag{12.5a}$$

$$\hat{U}_{ext,e}(t) = \sum_{j=1}^{N_e} u_{ext,e}(r_j, t). \tag{12.5b}$$

Defining electronic and nuclear single-particle densities conjugated to the true external potentials (12.5a) and (12.5b), a multicomponent density-functional theory (MCDFT) formalism can readily be formulated on the basis of the above Hamiltonian (Capitani 2000; Gidopoulos 1998). However, as discussed in Kreibich (2000), Kreibich and Gross (2001a), Kreibich et al. (2008), such a MCDFT is not useful in practice because the single-particle densities necessarily reflect the symmetry of the true external potentials and are therefore not characteristic of the internal properties of the system. In particular, for all isolated systems where the external potentials (12.5a) and (12.5b) vanish, these densities are constant, as a consequence of the translational invariance of the respective Hamiltonian. A suitable MCDFT is obtained by defining the densities with respect to internal coordinates of the system (Kreibich and Gross 2001a; van Leeuwen 2004b). To this end, new electronic coordinates are introduced according to

$$r'_j = \mathcal{R}(\alpha, \beta, \gamma)\, (r_j - R_{\mathrm{CMN}}) \qquad j = 1 \ldots N_{\mathrm{e}}, \tag{12.6}$$

where the center-of-mass of the nuclei (CMN) is defined as

$$R_{\mathrm{CMN}} = \frac{1}{M_{\mathrm{nuc}}} \sum_{\alpha=1}^{N_{\mathrm{n}}} M_\alpha R_\alpha, \quad \text{where} \quad M_{\mathrm{nuc}} = \sum_{\alpha=1}^{N_{\mathrm{n}}} M_\alpha. \tag{12.7}$$

The quantity $\mathcal{R}$ is a three-dimensional orthogonal matrix representing the Euler rotations. The Euler angles (α, β, γ) are functions of the nuclear coordinates $\{R\}$ and specify the orientation of the body-fixed coordinate frame. They can be determined in various ways. One way is by requiring the inertial tensor of the nuclei to be diagonal in the body-fixed frame. The conditions that the off-diagonal elements of the inertia tensor are zero in terms of the rotated coordinates $\mathcal{R}(R_\alpha - R_{\mathrm{CMN}})$ then give three determining equations for the three Euler angles in terms of the nuclear coordinates $\{R\}$ (Villars and Cooper 1970). A common alternative to determine the orientation of the body-fixed system is provided by the so-called Eckart conditions (Eckart 1935; Louck 1976; Bunker and Jensen 1998) [for recent reviews see (Sutcliffe (2000) and Meyer 2002)] which are suitable to describe small vibrations in molecules and phonons in solids (van Leeuwen 2004b). A general and very elegant discussion on the various ways the body-fixed frame can be chosen is given in Littlejohn and Reinsch (1997). In this work we will not make a specific choice, as our derivations are independent of such choice. The most important point is that, by virtue of Eq. 12.6, the electronic coordinates are defined with respect to a coordinate frame that is attached to the nuclear framework and rotates as the nuclear framework rotates. The nuclear coordinates themselves are not transformed any further at this point, i.e.,

$$R'_\alpha = R_\alpha \qquad \alpha = 1 \ldots N_{\mathrm{n}}. \tag{12.8}$$

Of course, introducing internal nuclear coordinates is also desirable. However, the choice of such coordinates depends strongly on the specific situation to be described: If near-equilibrium situations in systems with well-defined geometries are considered, normal or—for a solid—phonon coordinates are most appropriate, whereas fragmentation processes of molecules are better described in terms of Jacobi coordinates (Meyer 2002; Schinke 1993). Therefore, to keep a high degree of flexibility, the nuclear coordinates are left unchanged for the time being and are transformed to internal coordinates only prior to actual applications of the final equations that we will derive.

As a result of the coordinate changes of Eq. 12.6, the Hamiltonian (12.1) transforms into

$$
\hat{H}(t) = \hat{T}_n(\{\boldsymbol{R}\}) + \hat{V}_{nn}(\{\boldsymbol{R}\}) + \hat{U}_{ext,n}(\{\boldsymbol{R}\}, t) + \hat{T}_e(\{\boldsymbol{r}'\}) + \hat{V}_{ee}(\{\boldsymbol{r}'\})
$$
$$
+ \hat{T}_{MPC}(\{\boldsymbol{r}'\}, \{\boldsymbol{R}\}) + \hat{V}_{en}(\{\boldsymbol{r}'\}, \{\boldsymbol{R}\}) + \hat{U}_{ext,e}(\{\boldsymbol{r}'\}, \{\boldsymbol{R}\}, t). \qquad (12.9)
$$

Since we transformed to a noninertial coordinate frame a mass-polarization and Coriolis (MPC) term

$$
\hat{T}_{MPC} = \sum_{\alpha=1}^{N_n} -\frac{1}{2M_\alpha} \left[\nabla_{\boldsymbol{R}_\alpha} + \sum_{j=1}^{N_e} \frac{\partial \boldsymbol{r}'_j}{\partial \boldsymbol{R}_\alpha} \nabla_{\boldsymbol{r}'_j} \right]^2 - \hat{T}_n(\{\boldsymbol{R}\}) \qquad (12.10)
$$

appears. Obviously, this MPC term is not symmetric in the electronic and nuclear coordinates. However, this was not expected since only the electrons refer to a noninertial coordinate frame, whereas the nuclei are still defined with respect to the inertial frame. Therefore, all MPC terms arise solely from the electronic coordinates, representing fictitious forces due to the electronic motion in noninertial systems (for a detailed form of these terms within the current coordinate transformation see van Leeuwen (2004b)). The kinetic-energy operators $\hat{T}_n$ and $\hat{T}_e$, the electron–electron and nuclear–nuclear interactions, as well as the true external potential $\hat{U}_{ext,n}$ acting on the nuclei are formally unchanged in Eq. 12.9 and therefore given in Eqs. 12.2 and 12.3 with the new coordinates replacing the old ones, whereas the electron–nuclear interaction now reads

$$
\hat{V}_{en}(\{\boldsymbol{r}'\}, \{\boldsymbol{R}\}) = -\sum_{j=1}^{N_e} \sum_{\alpha=1}^{N_n} \frac{Z_\alpha}{\left| \boldsymbol{r}'_j - \mathcal{R}(\alpha, \beta, \gamma)(\boldsymbol{R}_\alpha - \boldsymbol{R}_{CMN}) \right|}. \qquad (12.11)
$$

The quantity

$$
\boldsymbol{R}''_\alpha = \mathcal{R}(\alpha, \beta, \gamma)(\boldsymbol{R}_\alpha - \boldsymbol{R}_{CMN}) \qquad (12.12)
$$

that appears in Eq. 12.11 is a so-called shape coordinate (Littlejohn 1997; van Leeuwen 2004b), i.e., it is invariant under rotations and translations of the nuclear framework. This is, of course, precisely the purpose of introducing a body-fixed frame: The attractive nuclear Coulomb potential (12.11) that the electrons in the body-fixed frame experience is invariant under rotations or translations of the nuclear framework. As a further consequence of the coordinate transformation (12.6), the true external potential acting on the electrons now not only depends on the electronic coordinates, but also on all the nuclear coordinates:

$$
\hat{U}_{ext,e}(\{\boldsymbol{r}'\}, \{\boldsymbol{R}\}, t) = \sum_{j=1}^{N_e} u_{ext,e}(\mathcal{R}^{-1}\boldsymbol{r}'_j + \boldsymbol{R}_{CMN}, t). \qquad (12.13)
$$

Therefore, in the chosen coordinate system, the electronic external potential is not a one-body operator anymore, but acts as an effective interaction. Consequences of this fact are discussed later.

12.2.1 Definition of the Densities

As already mentioned above, it is not useful to define electronic and nuclear single-particle densities in terms of the inertial coordinates r and R, since such densities necessarily reflect the symmetry of the corresponding true external potentials, e.g., they are constant for vanishing external potentials. Instead, we use the diagonal of the nuclear N_n-body density matrix

$$\Gamma(\{R\}, t) = \sum_{\{s\},\{\sigma\}} \int d^3 r'_1 \cdots \int d^3 r'_{N_e} \left| \Psi_{\{s\}\{\sigma\}}(\{R\}, \{r'\}, t) \right|^2, \qquad (12.14)$$

and the electronic single-particle density referring to the body-fixed frame:

$$n(r', t) = N_e \sum_{\{s\},\{\sigma\}} \int d^3 R'_1 \cdots \int d^3 R'_{N_n}$$
$$\times \int d^3 r'_1 \cdots \int d^3 r'_{N_e-1} \left| \Psi_{\{s\}\{\sigma\}}(\{R\}, \{r'\}, t) \right|^2. \qquad (12.15)$$

Here $\Psi_{\{s\}\{\sigma\}}(\{R\}, \{r'\}, t)$ represents the full solution of the TD Schrödinger equation with the Hamiltonian (12.9). The quantities $\{s\}$ and $\{\sigma\}$ denote the nuclear and electronic spin coordinates. The electronic density (12.15) represents a conditional density. It is proportional to the probability density of finding an electron at position r' as measured from the nuclear center-of-mass, given a certain orientation of the nuclear framework. Therefore the electronic density calculated through (12.15) reflects the internal symmetries of the system, e.g., the cylindrical symmetry of a diatomic molecule, instead of the Galilean symmetry of the underlying space.

12.3 The Runge–Gross Theorem for Multicomponent Systems

In order to set up a density-functional framework, our next task is to prove the analogue of the Runge–Gross theorem (Runge 1984) for multicomponent systems. To this end, we slightly modify the Hamiltonian (12.9) to take the form

$$\hat{H}(t) = \hat{T}_n(\{R\}) + \hat{V}_{nn}(\{R\}) + \hat{T}_e(\{r'\}) + \hat{V}_{ee}(\{r'\})$$
$$+ \hat{T}_{MPC}(\{r'\}, \{R\}) + \hat{V}_{en}(\{r'\}, \{R\}) + \hat{U}_{ext,e}(\{r'\}, \{R\}, t)$$
$$+ \hat{V}_{ext,n}(\{R\}, t) + \hat{V}_{ext,e}(\{r'\}, t). \qquad (12.16)$$

The potentials $\hat{V}_{ext,n}(\{R\}, t)$ and $\hat{V}_{ext,e}(\{r'\}, t)$, given by

$$\hat{V}_{ext,n}(\{R\}, t) = v_{ext,n}(\{R\}, t) \quad \text{and} \quad \hat{V}_{ext,e}(\{r'\}, t) = \sum_{j=1}^{N_e} v_{ext,e}(r'_j, t), \quad (12.17)$$

are potentials conjugate to the densities $\Gamma(\{R\}, t)$ and $n(\{r'\}, t)$ and are introduced to provide the necessary mappings between potentials and densities. In the special case $\hat{V}_{\text{ext, n}}(\{R\}, t) = \hat{U}_{\text{ext, n}}(\{R\}, t)$ and $\hat{V}_{\text{ext, e}}(\{r'\}, t) = 0$, the external potentials reduce to those of the Hamiltonian (12.9). It is important to note that the potential $\hat{U}_{\text{ext, e}}(\{r'\}, \{R\}, t)$ depends on both the electronic and nuclear coordinates and is therefore treated as a fixed many-body term in Hamiltonian (12.16). The mass-polarization and Coriolis terms in $\hat{T}_{\text{MPC}}$ are complicated many-body operators. They are treated here as additional electron–nuclear interactions which ultimately enter the exchange-correlation functional. For Hamiltonians of the form (12.16) we can apply the proof of the basic one-to-one correspondence along the same lines as Li and Tong (1985). Two sets of densities $\{\Gamma(\{R\}, t), n(r', t)\}$ and $\{\Gamma'(\{R\}, t), n'(r', t)\}$, which evolve from a common initial state Ψ_0 at $t = t_0$ under the influence of two sets of potentials $\{v_{\text{ext, n}}(\{R\}, t), v_{\text{ext, e}}(r', t)\}$ and $\{v'_{\text{ext, n}}(\{R\}, t), v'_{\text{ext, e}}(r', t)\}$ always become different infinitesimally after t_0 provided that at least one component of the potentials differs by more than a purely time-dependent function:

$$v_{\text{ext, n}}(\{R\}, t) \neq v'_{\text{ext, n}}(\{R\}, t) + C(t) \quad \text{or} \quad v_{\text{ext, e}}(r', t) \neq v'_{\text{ext, e}}(r', t) + C(t).$$
(12.18)

Consequently a one-to-one mapping between time-dependent densities and external potentials,

$$\{v_{\text{ext, n}}(\{R\}, t), v_{\text{ext, e}}(r', t)\} \leftrightarrow \{\Gamma(\{R\}, t), n(r', t)\}$$
(12.19)

is established for a given initial state Ψ_0. We again stress that since the external potential acting on the electrons $\hat{U}_{\text{ext, e}}(\{r'\}, \{R\}, t)$ in the body-fixed frame attains the form of an electron–nuclear interaction, the one-to-one mapping is still functionally dependent on $u_{\text{ext, e}}(r', \{R\}, t)$.

12.4 The Kohn–Sham Scheme for Multicomponent Systems

On the basis of the multicomponent Runge–Gross theorem we can set up the Kohn–Sham equations. For this we consider an auxiliary system with Hamiltonian

$$\hat{H}_{\text{KS}}(t) = \hat{T}_{\text{n}}(\{R\}) + \hat{T}_{\text{e}}(\{r'\}) + \hat{V}_{\text{KS, n}}(\{R\}, t) + \hat{V}_{\text{KS, e}}(\{r'\}, t),$$
(12.20)

where we introduced the potentials

$$\hat{V}_{\text{KS, n}}(\{R\}, t) = v_{\text{KS, n}}(\{R\}, t) \quad \text{and} \quad \hat{V}_{\text{KS, e}}(\{r'\}, t) = \sum_{j=1}^{N_e} v_{\text{KS, e}}(r'_j, t).$$
(12.21)

This represents a system in which the interelectronic interaction as well as the inter-action between the nuclei and the electrons have been switched off. According to the multicomponent Runge–Gross theorem there is at most one set of potentials

$\left\{ \hat{V}_{KS,n}(\{R\},t),\ \hat{V}_{KS,e}(\{r'\},t) \right\}$ (up to a purely time-dependent function) that reproduces a given set of densities $\left\{ \Gamma(\{R\},t), n(r',t) \right\}$. The potentials determined in this way are therefore functionals of the densities n and Γ and will henceforth be denoted as the Kohn–Sham potentials for the multicomponent system. The corresponding Hamiltonian of Eq. 12.20 will be denoted as the multicomponent Kohn–Sham Hamiltonian. In the Kohn–Sham Hamiltonian the electronic and nuclear motion have become separated. If we therefore choose the initial Kohn–Sham wavefunction $\Phi_{KS,0}$ to be a product of a nuclear and an electronic wavefunction then the time-dependent Kohn–Sham wavefunction will also be such a product, i.e.,

$$\Phi_{KS,\{s\}\{\sigma\}}(\{R\},\{r'\},t) = \Phi_{e,\{\sigma\}}(\{r'\},t)\,\Phi_{n,\{s\}}(\{R\},t) \tag{12.22}$$

and the corresponding densities are given by

$$\Gamma(\{R\},t) = \sum_{\{s\}} \left| \Phi_{n,\{s\}}(\{R\},t) \right|^2 \tag{12.23a}$$

$$n(r',t) = N_e \sum_{\{\sigma\}} \int d^3r'_1 \cdots \int d^3r'_{N_e-1} \left| \Phi_{e,\{\sigma\}}(\{r'\},t) \right|^2. \tag{12.23b}$$

The electronic and nuclear Kohn–Sham wavefunctions satisfy the equations

$$\left[i\frac{\partial}{\partial t} - \hat{T}_n(\{R\}) - \hat{V}_{KS,n}[n,\Gamma](\{R\},t) \right] \Phi_{n,\{s\}}(\{R\},t) = 0 \tag{12.24a}$$

$$\left[i\frac{\partial}{\partial t} - \hat{T}_e(\{r'\}) - \hat{V}_{KS,e}[n,\Gamma](\{r'\},t) \right] \Phi_{e,\{\sigma\}}(\{r'\},t) = 0. \tag{12.24b}$$

Note that the potential $\hat{V}_{KS,n}$ in the nuclear Kohn–Sham equation (12.24a) is an N_n-body interaction, whereas the electronic Kohn–Sham potential $\hat{V}_{KS,e}$ is a one-body operator. Hence, by choosing the initial electronic Kohn–Sham wavefunction as a Slater determinant consisting of orbitals φ_j, the electronic Kohn–Sham equation (12.24b) attains the usual form

$$\left\{ i\frac{\partial}{\partial t} - \left[-\frac{1}{2}\nabla'^2 + v_{KS,e}[n,\Gamma](r',t) \right] \right\} \varphi_j(r',t) = 0 \tag{12.25a}$$

$$n(r',t) = \sum_{j}^{N_e} |\varphi_j(r',t)|^2. \tag{12.25b}$$

The nuclear equations (12.23a) and (12.24a), together with the electronic equations (12.25a) and (12.25b), provide a formally exact scheme to calculate the electronic density n and the N_n-body nuclear density Γ. For practical applications it remains to obtain good approximations for the potentials $v_{KS,n}[n,\Gamma]$ and $v_{KS,e}[n,\Gamma]$. More insight into this question is obtained from the multicomponent action functional to be discussed in the next paragraph.

12.5 The Multicomponent Action

We start by defining a multicomponent action functional

$$\tilde{A}[v_{\text{ext, e}}, v_{\text{ext, n}}] = i \ln \langle \Psi_0 | \hat{T}_C \exp\{-i \int_C dt\, \hat{H}(t)\} | \Psi_0 \rangle. \qquad (12.26)$$

The Hamiltonian in this expression is the one of Eq. 12.16. Furthermore Ψ_0 is the initial state of the system and $\hat{T}_C$ denotes time-ordering along the Keldysh time contour C running along the real time-axis from t_0 to t and back to t_0. The time-dependent potentials $v_{\text{ext,e}}$ and $v_{\text{ext,n}}$ are correspondingly defined on this contour. The case discussed here is for an initial pure state. In case the initial system is in thermodynamic equilibrium the expectation value with respect to Ψ_0 can be replaced by a thermodynamic trace and the contour can be extended to include a final vertical stretch running from t_0 to $t_0 - i\beta$, where β is the inverse temperature of the initial ensemble. In that case the functional is closely related to the grand potential as is extensively discussed in Chap. 5. The main property of the action (12.26) which is important for multicomponent density-functional theory is that

$$\frac{\delta \tilde{A}}{\delta v_{\text{ext, e}}(\boldsymbol{r}, t)} = n(\boldsymbol{r}, t) \quad \text{and} \quad \frac{\delta \tilde{A}}{\delta v_{\text{ext, n}}(\{\boldsymbol{R}\}, t)} = \Gamma(\{\boldsymbol{R}\}, t). \qquad (12.27)$$

(From now on, for ease of notation, we will remove the prime from the electronic coordinate.) We now do a Legendre transform to obtain a functional of n and Γ and we define

$$A[n, \Gamma] = -\tilde{A}[v_{\text{ext, e}}, v_{\text{ext, n}}] + \int_C dt \int d^3 r\, n(\boldsymbol{r}, t) v_{\text{ext, e}}(\boldsymbol{r}, t)$$

$$+ \int_C dt \int d^3 R_1 \cdots \int d^3 R_{N_{\text{n}}} \Gamma(\{\boldsymbol{R}\}, t) v_{\text{ext, n}}(\{\boldsymbol{R}\}, t), \qquad (12.28)$$

where in this equation $v_{\text{ext,e}}$ and $v_{\text{ext,n}}$ (by virtue of the multicomponent Runge–Gross theorem) are now regarded as functionals of n and Γ. From the chain rule of differentiation we then easily obtain

$$\frac{\delta A}{\delta n(\boldsymbol{r}, t)} = v_{\text{ext, e}}(\boldsymbol{r}, t) \quad \text{and} \quad \frac{\delta A}{\delta \Gamma(\{\boldsymbol{R}\}, t)} = v_{\text{ext, n}}(\{\boldsymbol{R}\}, t). \qquad (12.29)$$

For the Hamiltonian $\hat{H}_{KS}(t)$ of Eq. 12.20 we can now further define an action functional analogous to Eq. 12.26

$$\tilde{A}_{KS}[v_{KS, e}, v_{KS, n}] = i \ln \langle \Phi_0 | \hat{T}_C \exp\{-i \int_C dt\, \hat{H}_{KS}(t)\} | \Phi_0 \rangle \qquad (12.30)$$

where Φ_0 is the initial state of the auxiliary system. By a Legendre transform we then obtain the functional $A_{KS}[n, \Gamma]$. With $A[n, \Gamma]$ and $A_{KS}[n, \Gamma]$ well-defined we now can define the exchange-correlation part $A_{xc}[n, \Gamma]$ of the action through the equation

$$
\begin{aligned}
A[n, \Gamma] = A_{KS}[n, \Gamma] &- \frac{1}{2} \int_C dt \int d^3 r_1 \int d^3 r_2 v_{ee}(r_1, r_2) n(r_1, t) n(r_2, t) \\
&- \int_C dt \int d^3 r \int d^3 R_1 \cdots \int d^3 R_{N_n} \left[v_{en}(r, \{R\}) \right. \\
&\left. + u_{ext, e}(r, \{R\}, t) \right] \times n(r, t) \Gamma(\{R\}, t) \\
&- \int_C dt \int d^3 R_1 \cdots \int d^3 R_{N_n} v_{nn}(\{R\}) \Gamma(\{R\}, t) - A_{xc}[n, \Gamma],
\end{aligned}
$$

(12.31)

where we subtracted the Hartree-like parts of the electron–electron and electron–nuclear interaction and the internuclear repulsion, using the definitions

$$
v_{en}(r, \{R\}) = -\sum_{\alpha=1}^{N_n} \frac{Z_\alpha}{|r - \mathcal{R}(R - R_{CMN})|}
\tag{12.32a}
$$

$$
u_{ext, e}(r, \{R\}, t) = u_{ext, e}(\mathcal{R}^{-1} r + R_{CMN}, t).
\tag{12.32b}
$$

These Hartree terms are treated separately because they are expected to be the dominant potential-energy contributions whereas the remainder, $A_{xc}[n, \Gamma]$, is expected to be smaller. No such dominant contributions arise from the mass-polarization and Coriolis terms which are usually rather small. The contributions coming from $\hat{T}_{MPC}$ are therefore completely retained in $A_{xc}[n, \Gamma]$. Differentiation of (12.31) with respect to n and Γ then yields

$$
\begin{aligned}
v_{KS, e}(r, t) = v_{ext, e}(r, t) &+ \int d^3 r' v_{ee}(r, r') n(r', t) + \int d^3 R_1 \cdots \\
&\times \int d^3 R_{N_n} \left[v_{en}(r, \{R\}) + u_{ext, e}(r, \{R\}, t) \right] \Gamma(\{R\}, t) + v_{xc, e}(r, t),
\end{aligned}
$$

(12.33)

and

$$
\begin{aligned}
v_{KS, n}(\{R\}, t) = v_{ext, n}(\{R\}, t) &+ v_{nn}(\{R\}) \\
&+ \int d^3 r \left[v_{en}(r, \{R\}) + u_{ext, e}(r, \{R\}, t) \right] \\
&\times n(r, t) + v_{xc, n}(\{R\}, t),
\end{aligned}
$$

(12.34)

where we have defined the electronic and nuclear exchange-correlation potentials as

$$v_{xc,e}(\boldsymbol{r}, t) = \frac{\delta A_{xc}[n, \Gamma]}{\delta n(\boldsymbol{r}, t)} \quad \text{and} \quad v_{xc,n}(\{\boldsymbol{R}\}, t) = \frac{\delta A_{xc}[n, \Gamma]}{\delta \Gamma(\{\boldsymbol{R}\}, t)}. \tag{12.35}$$

The main question is now how to obtain explicit functionals for the exchange-correlation potentials. One of the most promising ways of obtaining these may be the development of orbital functionals as in the OEP approach. Such functionals can be deduced by a diagrammatic expansion of the action functionals.

12.6 Linear Response and Multicomponent Systems

We will now consider the important case of linear response in the multicomponent formalism. Such approach will, for instance, be very useful in weak field problems such as electron–phonon coupling in solids. For convenience we first introduce the notation $i = (\boldsymbol{r}_i, t_i)$ and $\underline{i} = (\{\boldsymbol{R}\}, t_i)$. Let us then define the set of response functions:

$$\boldsymbol{\chi}_{12} = \begin{pmatrix} \chi_{ee}(1, 2) & \chi_{en}(1, \underline{2}) \\ \chi_{ne}(\underline{1}, 2) & \chi_{nn}(\underline{1}, \underline{2}) \end{pmatrix} = \begin{pmatrix} \dfrac{\delta n(1)}{\delta v_e(2)} & \dfrac{\delta n(1)}{\delta v_n(\underline{2})} \\ \dfrac{\delta \Gamma(\underline{1})}{\delta v_e(2)} & \dfrac{\delta \Gamma(\underline{1})}{\delta v_n(\underline{2})} \end{pmatrix}. \tag{12.36}$$

Similarly for the Kohn–Sham system we have

$$\boldsymbol{\chi}_{KS,12} = \begin{pmatrix} \chi_{KS,ee}(1, 2) & 0 \\ 0 & \chi_{KS,nn}(\underline{1}, \underline{2}) \end{pmatrix} = \begin{pmatrix} \dfrac{\delta n(1)}{\delta v_{KS,e}(2)} & \dfrac{\delta n(1)}{\delta v_{KS,n}(\underline{2})} \\ \dfrac{\delta \Gamma(\underline{1})}{\delta v_{KS,e}(2)} & \dfrac{\delta \Gamma(\underline{1})}{\delta v_{KS,n}(\underline{2})} \end{pmatrix}, \tag{12.37}$$

in which the mixed response functions $\chi_{KS,en} = \chi_{KS,ne} = 0$ since in the Kohn–Sham system the nuclear and electronic systems are decoupled. The two sets of response functions are related by an equation that is very similar to that of ordinary TDDFT

$$\boldsymbol{\chi}_{12} = \boldsymbol{\chi}_{KS,12} + \boldsymbol{\chi}_{KS,13} \cdot (v_{34} + \boldsymbol{f}_{xc,34}) \cdot \boldsymbol{\chi}_{42} \tag{12.38}$$

where "·" denotes a matrix product and integration over the variables 3 and 4, respectively. The matrices $\boldsymbol{f}_{xc}$ and v are defined as

$$\boldsymbol{f}_{xc,12} = \begin{pmatrix} f_{xc,ee}(1, 2) & f_{xc,en}(1, \underline{2}) \\ f_{xc,ne}(\underline{1}, 2) & f_{xc,nn}(\underline{1}, \underline{2}) \end{pmatrix} = \begin{pmatrix} \dfrac{\delta v_{xc,e}(1)}{\delta n(2)} & \dfrac{\delta v_{xc,e}(1)}{\delta \Gamma(\underline{2})} \\ \dfrac{\delta v_{xc,n}(\underline{1})}{\delta n(2)} & \dfrac{\delta v_{xc,n}(\underline{1})}{\delta \Gamma(\underline{2})} \end{pmatrix} \tag{12.39a}$$

$$v_{12} = \begin{pmatrix} v_{ee}(1,2) & v_{en}(1,\underline{2}) + u_{ext,e}(1,\underline{2}) \\ v_{en}(2,\underline{1}) + u_{ext,e}(2,\underline{1}) & 0 \end{pmatrix}. \tag{12.39b}$$

The Eq. 12.38 is the central equation of the multicomponent response theory and is readily derived by application of the chain rule of differentiation. As an example we calculate

$$\begin{aligned} \chi_{ee}(1,2) &= \frac{\delta n(1)}{\delta v_e(2)} \\ &= \int d3 \frac{\delta n(1)}{\delta v_{KS,e}(3)} \frac{\delta v_{KS,e}(3)}{\delta v_{ext,e}(2)} + \int d\underline{3} \frac{\delta n(1)}{\delta v_{KS,n}(\underline{3})} \frac{\delta v_{KS,n}(\underline{3})}{\delta v_{ext,e}(2)} \\ &= \frac{\delta n(1)}{\delta v_{KS,e}(2)} + \int d3 \frac{\delta n(1)}{\delta v_{KS,e}(3)} \frac{\delta v_{Hxc,e}(3)}{\delta v_{ext,e}(2)}, \end{aligned} \tag{12.40}$$

From which readily follows

$$\chi_{ee}(1,2) = \chi_{KS,ee}(1,2) + \int d3 \chi_{KS,ee}(1,3)$$
$$\times \left\{ \int d4 \frac{\delta v_{Hxc,e}(3)}{\delta n(4)} \chi_{ee}(4,2) + \int d\underline{4} \frac{\delta v_{Hxc,e}(3)}{\delta \Gamma(\underline{4})} \chi_{ne}(\underline{4},2) \right\}, \tag{12.41}$$

where $v_{Hxc,e} = v_{KS,e} - v_e$. We further have

$$\frac{\delta v_{Hxc,e}(3)}{\delta n(4)} = v_{ee}(3,4) + f_{xc,ee}(1,2) \tag{12.42a}$$

$$\frac{\delta v_{Hxc,e}(3)}{\delta \Gamma(\underline{4})} = v_{en}(3,\underline{4}) + u_{ext,e}(3,\underline{4}) + f_{xc,en}(3,\underline{4}). \tag{12.42b}$$

Inserting these expressions into (12.41) we have established one entry in the matrix equation (12.38). The other entries can be verified analogously. We finally note that Eq. 12.39b still contains the term $u_{ext,e}$, which is inconvenient in practice. However, to calculate the linear response to the true external field we anyway need to expand further in powers of $u_{ext,e}$. If we do this we obtain Eq. 12.38 with $u_{ext,e} = 0$ in Eq. 12.39b and two additional equations for the response functions $\delta n/\delta u_{ext,e}$ and $\delta \Gamma/\delta u_{ext,e}$ which will not be discussed here (Butriy et al. 2007). From the structure of the linear response equation (12.38) it is readily seen that electronic Kohn–Sham excitations (poles of $\chi_{ee,KS}$) and nuclear vibrational Kohn–Sham excitations (poles of $\chi_{nn,KS}$) will in general mix. The exchange-correlation kernels in f_{xc} will then have to provide the additional shift such that the true response functions in χ will contain the true excitations of the coupled electron–nuclear system.

12.7 Example

As an application of the formalism we discuss the case of a diatomic molecule in a strong laser field. The Hamiltonian of this system in laboratory frame coordinates is given by

$$\hat{H}(t) = -\frac{1}{2M_1}\nabla_{\boldsymbol{R}_1}^2 - \frac{1}{2M_2}\nabla_{\boldsymbol{R}_2}^2 - \frac{1}{2}\sum_{i=1}^{N_e}\nabla_i^2 + v_{en} + v_{ee} + v_{nn} + v_{laser}(t) \quad (12.43)$$

where

$$v_{nn}(\{R\}) = \frac{Z_1 Z_2}{|R_1 - R_2|} \quad (12.44a)$$

$$v_{en}(\{r\}, \{R\}) = -\sum_{i=1}^{N_e}\left\{\frac{Z_1}{|r_i - R_1|} + \frac{Z_2}{|r_i - R_2|}\right\} \quad (12.44b)$$

$$v_{laser}(\{r\}, \{R\}, t) = \left\{\sum_{i=1}^{N_e} r_i - Z_1 R_1 - Z_2 R_2\right\} \cdot E(t) \quad (12.44c)$$

and where $E(t)$ represents the electric field of the laser.

We now have to perform a suitable body-fixed frame transformation to refer the electron coordinates to a nuclear frame. For the diatomic molecule a natural choice presents itself: we determine the Euler angles by the requirement that the internuclear axis be parallel to the z-axis in the body-fixed frame, i.e., $\mathcal{R}(\boldsymbol{R}) = R e_z$, where $\boldsymbol{R} = \boldsymbol{R}_1 - \boldsymbol{R}_2$ and $R = |\boldsymbol{R}|$. For the special case of the diatomic molecule only two Euler angles are needed to specify the rotation matrix $\mathcal{R}$. From (12.6) and (12.7) we see that the electron–nuclear interaction and the external laser field transform to

$$v_{en}(\{r'\}, \{R\}) = -\sum_{i=1}^{N_e}\left\{\frac{Z_1}{|r'_i - \frac{M_2}{M_1+M_2}R e_z|} + \frac{Z_2}{|r'_i + \frac{M_1}{M_1+M_2}R e_z|}\right\} \quad (12.45a)$$

$$v_{laser}(t) = \left\{N_e \boldsymbol{R}_{CMN} - Z_1 \boldsymbol{R}_1 - Z_2 \boldsymbol{R}_2 + \sum_{i=1}^{N_e}\mathcal{R}^{-1}r'_i\right\} \cdot E(t). \quad (12.45b)$$

With these expressions the Kohn–Sham potentials of Eqs. 12.33 and 12.34 attain the form

$$v_{KS, e}(\boldsymbol{r}, t) = \int d^3 r' v_{ee}(\boldsymbol{r}, \boldsymbol{r}')n(\boldsymbol{r}', t) + \int d^3 R_1 \cdots$$
$$\times \int d^3 R_{N_n}\left[v_{en}(\boldsymbol{r}, \{R\}) + \mathcal{R}^{-1}\boldsymbol{r}\cdot E(t)\right]\Gamma(\{R\}, t) + v_{xc, e}(\boldsymbol{r}, t)$$
$$(12.46)$$

and

$$v_{\text{KS, n}}(\{\boldsymbol{R}\}, t) = [N_e \boldsymbol{R}_{\text{CMN}} - Z_1 \boldsymbol{R}_1 - Z_2 \boldsymbol{R}_2] \cdot \boldsymbol{E}(t) + v_{\text{nn}}(\{\boldsymbol{R}\})$$
$$+ \int d^3 r \left[v_{\text{en}}(\boldsymbol{r}, \{\boldsymbol{R}\}) + \mathcal{R}^{-1} \cdot \boldsymbol{E}(t) \right] n(\boldsymbol{r}, t) + v_{\text{xc, n}}(\{\boldsymbol{R}\}, t).$$

(12.47)

Since the rotation matrix $\mathcal{R}$ only depends on $\boldsymbol{R}$, the nuclear Kohn–Sham potential is readily seen to be separable in terms of the coordinates $\boldsymbol{R}$ and $\boldsymbol{R}_{\text{CMN}}$. The nuclear Kohn–Sham wavefunction can then be written as

$$\Phi_{n, s_1, s_2}(\boldsymbol{R}_1, \boldsymbol{R}_2, t) = \Upsilon(\boldsymbol{R}_{\text{CMN}}, t)\xi(\boldsymbol{R}, t)\theta(s_1, s_2) \qquad (12.48)$$

where θ is a nuclear spin function of the nuclear spin coordinates s_1 and s_2 and Υ and ξ satisfy the equations

$$\left\{ i\partial_t - \left[-\frac{1}{M_{\text{nuc}}} \nabla^2_{\boldsymbol{R}_{\text{CMN}}} + Q_{\text{tot}} \boldsymbol{R}_{\text{CMN}} \cdot \boldsymbol{E}(t) \right] \right\} \Upsilon(\boldsymbol{R}_{\text{CMN}}, t) = 0 \qquad (12.49a)$$

$$\left\{ i\partial_t - \left[-\frac{1}{2\mu} \nabla^2_{\boldsymbol{R}} + \bar{v}_{\text{KS, n}}[n, \Gamma](\boldsymbol{R}, t) \right] \right\} \xi(\boldsymbol{R}, t) = 0, \qquad (12.49b)$$

where we defined the total nuclear mass $M_{\text{nuc}} = M_1 + M_2$, the total charge $Q_{\text{tot}} = N_e - Z_1 - Z_2$ and the reduced mass $\mu = M_1 M_2 / (M_1 + M_2)$. The potential $\bar{v}_{\text{KS, n}}$ has the form

$$\bar{v}_{\text{KS, n}}(\boldsymbol{R}, t) = [-q_n \boldsymbol{R} + \boldsymbol{d}(\boldsymbol{R}, t)] \cdot \boldsymbol{E}(t) + \frac{Z_1 Z_2}{R} - \int d^3 r n(\boldsymbol{r}, t)$$
$$\times \left(\frac{Z_1}{|\boldsymbol{r} - \frac{M_2}{M_1 + M_2} R \boldsymbol{e}_z|} + \frac{Z_2}{|\boldsymbol{r} + \frac{M_1}{M_1 + M_2} R \boldsymbol{e}_z|} \right) + v_{\text{xc, n}}(\boldsymbol{R}, t),$$

(12.50)

where we have defined

$$q_n = \frac{M_2 Z_1 - M_1 Z_2}{M_1 + M_2} \qquad (12.51a)$$

$$\boldsymbol{d}(\boldsymbol{R}, t) = \mathcal{R}^{-1} \int d^3 r n(\boldsymbol{r}, t) \boldsymbol{r}. \qquad (12.51b)$$

We see that the nuclear center-of-mass motion has been decoupled from the nuclear relative motion. The nuclear center-of-mass wavefunction corresponds to a so-called Volkov plane wave. If it is normalized to a volume $\mathcal{V}$ then the nuclear density matrix can be written as

$$\Gamma(\boldsymbol{R}_1, \boldsymbol{R}_2, t) = \frac{1}{\mathcal{V}} N(\boldsymbol{R}, t), \qquad (12.52)$$

Fig. 12.1 Time-evolution (in units of optical cycles τ) of the nuclear density $N(R, t)$ obtained for a one-dimensional model H_2^+-molecule in a $\lambda = 228$ nm, $I = 5 \times 10^{13}$ W/cm^2 laser field from the exact solution, the time-dependent Hartree approximation and a time-dependent correlated variational approach

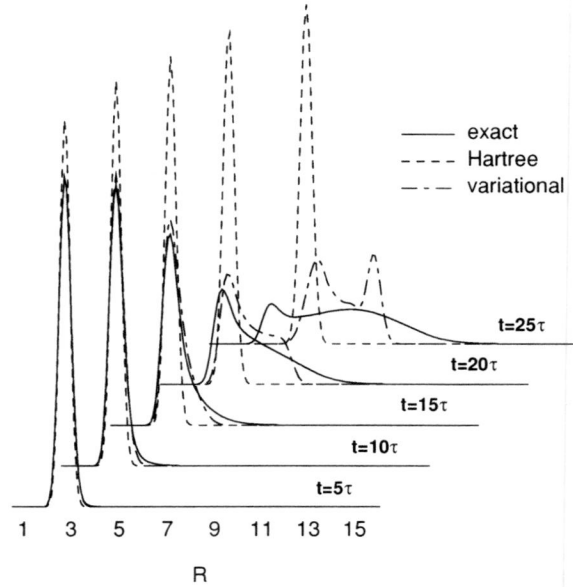

where we defined the density of the relative nuclear "particle" as

$$N(R, t) = |\xi(R, t)|^2. \tag{12.53}$$

In terms of this quantity, the electronic Kohn–Sham potential (12.46) attains the form

$$v_{KS, e}(r, t) = D^{-1} r \cdot E(t) + \int d^3 r' \frac{n(r', t)}{|r - r'|} - \int d^3 R N(R, t)$$

$$\times \left\{ \frac{Z_1}{|r - \frac{M_2}{M_1 + M_2} Re_z|} + \frac{Z_2}{|r + \frac{M_1}{M_1 + M_2} Re_z|} \right\} + v_{xc,e}(R, t), \tag{12.54}$$

where

$$D^{-1} = \int d^3 R N(R, t) \mathcal{R}^{-1} \tag{12.55}$$

We have now completely defined the multicomponent Kohn–Sham equations for a diatomic molecule in a laser field in the dipole approximation. The next task is to develop appropriate functionals for the exchange-correlaton potentials, particularly for the electron–nuclear correlation. The simplest approach is to treat the electron–nuclear correlation in the Hartree approach where we put $v_{xc, n} = 0$ in Eq. 12.50. This approach has been tested (Kreibich et al. 2004) in a one-dimensional model system for H_2^+ which is a suitable testcase since (1) it can be compared to the exact solution of the Schrödinger equation and (2) there are no electron–electron correlations. When the model molecule is exposed to a strong laser field, the nuclear density $N(R, t)$ shows a time-dependence as shown in Fig. 12.1. In this plot we can clearly see that the exact nuclear wavepacket splits, and part of the nuclear wavepacket moves away and decribes a dissociating molecule. However, within the Hartree approach to the electron–nuclear correlation, the nuclear wavepacket remains sharply peaked around the molecular equilibrium bond distance. This means that electron–nuclear correlation beyond the Hartree approximation is very important [for a more extensive discussion see Kreibich and Gross (2001a) and Kreibich et al. (2004)]. This is corroborated by the fact that a variational Ansatz for the time-dependent wavefunction in terms of correlated orbitals (denoted as "variational" in Fig. 12.1) does yield the qualitatively correct splitting of the nuclear wavepacket.

12.8 Conclusions

We showed how to set up a multicomponent density-functional scheme for general systems of electrons and nuclei in time-dependent external fields. The basic quantities in this theory are the electron density referred to a body-fixed frame and the nuclear density matrix. Important for future applications will be the development of functionals for electron–nuclear correlations. The first steps in this direction have already been taken in the MCDFT for stationary systems (Kreibich and Gross 2001a; Kreibich et al. 2004, 2008). The first application of the linear response approach of the TDMCDFT formalism was published recently using the simple Hartree approximation for the electron–nuclear interaction (Butriy et al. 2007). The development of improved functionals beyond this Hartree aproximation is an important goal for the future. For completeness, we finally note that also another approach to constructing a density functional theory for time-dependent systems of electrons and nuclei has been developed on the basis of the equations of motion of mixed-electron nuclear density matrices (Krishna 2009). This formalism allows also for the straightforward construction of semi-classical approximations. We are, however, not aware of any applications of this formalism. We further like to mention some further recent work on electron–proton functionals derived from the electron–proton pair density (Chakraborty et al. 2009).

Chapter 13
Quantum Optimal Control

Alberto Castro and Eberhard K. U. Gross

13.1 Introduction

All applications of time-dependent density-functional theory (TDDFT) until now have attempted to describe the response of many-electron systems to external fields. Given its success in this task, it seems timely, therefore, to address the inverse problem: given a prescribed goal (e.g., the transfer of electronic charge to a given region in space, or the population of a given excited state), what is the external perturbation that achieves this goal in an optimal way? This is the problem studied by quantum optimal control theory (QOCT). The essentials of this theory make no assumption on the nature of the quantum system whose behaviour is being engineered, or on the particular methodology used to model the system. It must, therefore, be complemented with a suitable model for describing the dynamics of the quantum system. In this chapter, we describe how this model can be TDDFT.

QOCT (Shapiro and Brumer 2003; Rice and Zhao 2000; Werschnik and Gross 2007; Werschnik 2006; Brif et al. 2010; Rabitz 2009) provides the necessary tools to theoretically design driving fields capable of controlling a quantum system towards a given state or along a prescribed path in Hilbert space. This discipline has grown steadily in the last two decades. The reason is that it is the most powerful theoretical counterpart to the exponentially growing field of experimental quantum coherent control. The key aspect for the success of this area is that the laser pulse creation and

A. Castro (✉)
Instituto de Biocomputación y Física de Sistemas Complejos,
Universidad of Zaragoza, Mariano Esquillor s/n, Edificio I+D,
50018 Zaragoza, Spain
e-mail: acastro@bifi.es

E. K. U. Gross
Max-Planck Institut für Mikrostrukturphysik, Weinberg 2,
06120 Halle, Germany
e-mail: hardy@mpi-halle.mpg.de

M. A. L. Marques et al. (eds.), *Fundamentals of Time-Dependent Density Functional Theory*, Lecture Notes in Physics 837, DOI: 10.1007/978-3-642-23518-4_13, © Springer-Verlag Berlin Heidelberg 2012

shaping techniques have improved impressively over the last decades (Weiner 2000), and thus the area of experimental optimal control has become a well established field. An outstanding number of possible technological applications are to be expected: the quest for systems able to perform quantum computing (Tesch et al. 2001; Palao and Kosoff 2002), the synthesis of design-molecules by laser-induced chemical reactions (Laarmann et al. 2007), or the control of electron currents in molecular switches using light (Geppert et al. 2004), to name a few examples, may benefit from these recent advances.

The essential concepts of QOCT can be formulated for arbitary quantum systems, that are to be described either via the wave function, the density matrix, or in some cases the propagator of the process. Since the full quantum dynamics of interacting particles cannot be treated exactly for more than a few degrees of freedom, the application of QOCT, therefore, relies on the previous existence of a methodology capable of constructing a sufficiently good model for the relevant process. In fact, the solution of the necessary equations typically requires multiple propagations, both forwards and backwards, for the system under study. We know that this is especially challenging if this is a many-electron system. In fact, in most applications, few-level simplifications and models are typically postulated when handling the QOCT equations. Unfortunately, these simplifications are not always accurate enough: strong pulses naturally involve many electronic levels, and normally perturbative treatments are not useful. Non-linear laser-matter interaction must sometimes be described *ab initio*.

As it is shown elsewhere in this book, TDDFT is a viable alternative to computationally more expensive approaches based on the wave function for the description of this type of processes. We are referring to many-electron systems irradiated with femtosecond pulses, with intensities typically ranging from 10^{11} to 10^{15} Wcm^{-2}, a non-linear regime that nevertheless allows for a non-relativistic treatment. This may lead to a number of interesting phenomena, e.g. above-threshold or tunnel ionization, bond hardening or softening, high harmonic generation, photo-isomerization, photo-fragmentation, Coulomb explosion, etc (Protopapas et al. 1997; Brabec and Krausz 2000; Scrinzi et al. 2006).

It is therefore necessary to inscribe TDDFT into the general QOCT framework. In this chapter, we describe the manner in which this inscription can be done. We present here generalized equations, valid for non-collinear spin configurations as well as providing detailed derivations of the key equations.

13.2 The Essential QOCT Equations

For the sake of completeness, we start by recalling the essential QOCT equations (by which we mean those that provide the gradient or numerical derivative of the target functional with respect to the control function). We do so for a generic quantum system, before establishing the results for TDDFT in the next section. For alternative presentations of the following results, we refer the reader, for example, to (Peirce et al. 1988; Tersigni et al. 1990; Serban et al. 2005).

We depart from a generic Schrödinger equation in the form:

$$\dot{\varphi}(t) = -i\hat{H}[u, t]\varphi(t), \tag{13.1a}$$

$$\varphi(t_0) = \varphi_0, \tag{13.1b}$$

where $\varphi(t)$ is the state of the system at time t (that lives in the Hilbert space $\mathcal{H}$), and $\hat{H}[u, t]$ is the Hamiltonian, that dependes on some parameter u (or perhaps a set of parameters) that we will call hereafter *control*. We wish to find the control u that makes the evolution optimal in some manner. The goodness of the process is encoded into a functional that depends on how the system evolves:

$$F : \overline{\mathcal{H}}^0 \times \mathcal{P} \to \mathbb{R} \tag{13.2a}$$

$$(\varphi, u) \to \mathbb{R}, \tag{13.2b}$$

The set $\overline{\mathcal{H}}^0$ is defined in the following way: if we consider that the process occurs in the time interval $[0, T]$, the possible evolutions of the sytem can be contained in the space $\overline{\mathcal{H}} = \mathcal{H} \times [0, T]$. However, we restrict the domain of F to those evolutions whose initial condition is consistent with Eq. 13.1b:

$$\overline{\mathcal{H}}^0 = \{\varphi \in \overline{\mathcal{H}} \text{ such that } \varphi(t_0) = \varphi_0\}. \tag{13.3}$$

The set $\mathcal{P}$ contains the allowed control parameters u. The explicit dependence of F on the control u is typically due to the presence of a *penalty* function, whose role is to avoid undesirable regions of the control search space (e.g., too high frequency components of a laser pulse, unrealistical intensities, etc.). In many cases, the functional F is split as:

$$F[\varphi, u] = J_1[\varphi] + J_2[u], \tag{13.4}$$

where J_1 is the real objective, and J_2 is the penalty function. The objective may depend on the full evolution of the system during the time interval $[0, T]$, and/or on the state of the system at the final time T:

$$J_1[\varphi] = J_1^i[\varphi] + J_1^f[\varphi(T)], \tag{13.5}$$

where the functional J_1^i depends on the full evolution of the system, whereas J_1^f is a functional of the final state. In most cases, these functionals are expectation values of some observable:

$$J_1^i[\varphi] = \int dt \langle \varphi(t) | \hat{O}^i(t) | \varphi(t) \rangle, \tag{13.6a}$$

$$J_1^f[\varphi] = \int dt \langle \varphi(T) | \hat{O}^f | \varphi(T) \rangle. \tag{13.6b}$$

The specification of a particular control u determines the evolution of the system; in mathematical terms, we have a mapping:

$$\varphi : \mathcal{P} \to \overline{\mathcal{H}}^0 \tag{13.7a}$$
$$u \to \varphi[u]. \tag{13.7b}$$

Therefore, the optimization problem can be formulated as the problem of finding the extrema of the function G:

$$G[u] = F[\varphi[u], u]. \tag{13.8}$$

There are many optimization algorithms capable of maximizing functions utilizing solely the knowledge of the function values ("gradient-free algorithms"). We have recently employed one of them in this context (Castro et al. 2009). However, QOCT provides the solution to the problem of computing the gradient of G—or, properly speaking, the functional derivative if u is a function.

The equations that we are seeking, therefore, are those that provide the gradient of G, that will permit us to efficiently perform its optimization. We start by noting that not all the elements of the domain $\overline{\mathcal{H}}^0$ are possible evolution of the system: the search for maxima of F must be constrained to the subset of solutions of Schrödinger's equation. This is achieved by introducing a Lagrangian functional in the form:

$$L[\varphi, \lambda, u] = -2\Re \int_0^T dt \ \langle \lambda(t) | \dot{\varphi}(t) + i\hat{H}[u, t]\varphi(t) \rangle, \tag{13.9}$$

and defining a new "total" functional as the sum of F and this new Lagrangian term:

$$J[\varphi, \lambda, u] = F[\varphi, u] + L[\varphi, \lambda, u]. \tag{13.10}$$

Note that, for a possible evolution of the system $\varphi[u]$, the Lagrangian function is zero, and therefore $J[\varphi[u], \lambda, u] = F[\varphi[u], u] = G[u]$. And note that setting to zero the functional derivative of J with respect to λ gives precisely Schrödinger's equation:

$$\frac{\delta J[\varphi, \lambda, u]}{\delta \lambda^*(x, t)} = -\dot{\varphi}(x, t) - i\hat{H}[u, t]\varphi(x, t) = 0. \tag{13.11}$$

In other words, nullifying this functional derivative is equivalent to establishing the map $u \to \varphi[u]$. Analogously, a $u \to \lambda[u]$ mapping can be established by nullifying the functional derivative of J with respect to φ:

$$\frac{\delta J[\varphi, \lambda, u]}{\delta \varphi^*(x, t)} = \dot{\lambda}(x, t) + i\hat{H}[u, t]\lambda(x, t) - \delta(t - T)\lambda(x, t) + \frac{\delta J_1^i}{\delta \varphi^*(x, t)}$$

$$+ \delta(t - T)\frac{\delta J_1^f}{\delta \varphi^*(x, T)}. \tag{13.12}$$

The $u \rightarrow \lambda[u]$ mapping is thus established by prescribing the following equations of motion:

$$\dot{\lambda}(x, t) = -i\hat{H}[u, t]\lambda(x, t) + \frac{\delta J_1^i}{\delta \varphi^*(x, t)}, \tag{13.13a}$$

$$\lambda(x, T) = \frac{\delta J_1^f}{\delta \varphi^*(x, T)}. \tag{13.13b}$$

Note that (i) if J_1^i is not zero (i.e. the target functional depends on the evolution of the system during the time interval $[0, T]$, or as it is sometimes put, there is a *time dependent target*), the equation of motion is an inhomogeneous Schrödinger equation. If J_1^i is zero, λ follows the same Schrödinger equation as y; (ii) the boundary condition given by Eq. 13.13b is given at time T. Therefore, Eq. 13.13a must be propagated *backwards*.

Once we have prepared these ingredients, we may compute the gradient of $G[u]$ by utilizing the chain rule in the identity:

$$G[u] = J[\varphi[u], \lambda[u], u]. \tag{13.14}$$

Since the functional derivatives of J with respect to φ and λ, applied at $\varphi[u]$ and $\lambda[u]$ are zero, we are left with the expression:

$$\nabla_u G[u] = \nabla_u J_2[u] + 2\Im m \int_0^T dt \, \langle \lambda[u](t)|\nabla_u H[u, t]|\varphi[u](t)\rangle. \tag{13.15}$$

Often, the controlling field is not specified by a discrete set of parameters, but rather by a real function $\varepsilon(t)$. For example, the Hamiltonian may be:

$$\hat{H}[\epsilon, t] = \hat{H}_0 + \varepsilon(t)\hat{V}. \tag{13.16}$$

The search space, in this case, is the space of real continuous functions in $[0, T]$. The function G is then a functional $G = G[\varepsilon]$, and consequently the gradient of Eq. 13.15 must then be replaced by a functional derivative with respect to ε, and the result is:

$$\frac{\delta G}{\delta \varepsilon(t)} = \frac{\delta J_2}{\delta \varepsilon(t)} + 2\Im m \langle \lambda[\varepsilon](t)|\hat{V}|\varphi[\varepsilon](t)\rangle. \tag{13.17}$$

Finally, let us give an example of a very common choice of penalty function: let the Hamiltonian given by Eq. 13.16 be the one that describes an atom or molecule irradiated by a laser pulse, whose temporal shape is determined by $\varepsilon(t)$. The total irradiated energy or *fluence* will be proportional to $\int_0^T dt \, \varepsilon^2(t)$, and it is usually considered an undesired feature to require very high fluences. A natural choice for the penalty functional is therefore:

$$J_2[\varepsilon] = -\alpha \int_0^T dt\; \varepsilon^2(t), \tag{13.18}$$

for α some positive weighting constant. The critical points of G in this case can then easily found by substituting this expression in Eq. 13.17, and equating the functional derivative to zero:

$$\varepsilon(t) = \frac{1}{\alpha}\Im\langle\lambda[\varepsilon](t)|\hat{V}|\varphi[\varepsilon](t)\rangle. \tag{13.19}$$

In this case, the simplest algorithm to find the optimum consists of the following steps: (i) Propagate the Schrödinger equation 13.11 forward in time with some given initial condition; (ii) Propagate the Schrödinger equation (13.13a) with the initial condition (13.13b); (iii) From the solutions $\varphi(t)$ and $\lambda(t)$ of steps (i) and (ii), calculate a new control field from Eq. 13.19 and then perform again steps (i) and (ii) using the new control field, etc. This algorithm has been shown to be monotonically convergent (Zhu et al. 1998; Ohtsuki et al. 2004; Serban et al. 2005).

13.3 Optimization for the TDKS System

In the previous section, we have summarized some of the fundamental aspects of QOCT for a generic quantum system. Equation 13.15, or alternatively, Eq. 13.17, permits to utilize some optimization algorithm that may lead to a satisfactory maximum. The use of those equations, however, requires the solution of Eqs. 13.1a and 13.1b, and Eqs. 13.13a and 13.13b which are forwards and backwards propagation of Schrödinger's equation for the process being studied. Depending on the nature of the process and on the complexity of the model used to describe it, this task may be anywhere from trivial to unfeasible. In particular, for many-electron systems, an exact numerical solution is not possible, and thus TDDFT may be used to tackle the problem.

At this point, let us recall that TDDFT is not a scheme that permits to solve the Schrödinger equation in some approximate manner, but rather a methodology aimed at *bypassing* the use of the many-electron wave function, and avoiding the Schrödinger equation fulfilled by this wave function. The theory provides us with a method to obtain the time-dependent electron density (by means of solving the equations of the associated Kohn–Sham system), but the many electron wave function is not available. However, the previous description of the QOCT methodology works directly with the wave function, and, in particular, it is assumed that the target functional is defined in terms of this object, i.e. the objective is to find the control u that maximizes a functional of the many-electron wave function,

$$F_{\text{int}} = F_{\text{int}}[\Psi, u]. \tag{13.20}$$

As usual, we allow for the possibility of an explicit dependence of the functional on the control u in order to include, if necessary, a penalty function. The label "int" means that this functional is defined in terms of the wave function of the interacting system of electrons. Every observable, however, is known to be a functional of the time-dependent density by virtue of the fundamental one-to-one correspondence of TDDFT, and therefore, in principle it should be possible to find a functional of the density $\tilde{F}$, such that:

$$F_{\text{int}}[\Psi, u] = \tilde{F}[n[\Psi], u], \tag{13.21}$$

where $n[\Psi]$ is the electronic density corresponding to the many-electron wave function Ψ (we consider hereafter a system of N electrons):

$$n[\Psi](r, t) = \langle \Psi(t)| \sum_{i=1}^{N} \delta(r - \hat{r}_i)|\Psi(t)\rangle, \tag{13.22}$$

This is the object provided by TDDFT, and so we may substitute the problem of formulating QOCT in terms of the real interacting system, by the formulation of the optimization problem for the non-interacting system of electrons. The equations of motion for the single-particle orbitals of this system, also known as time-dependent Kohn–Sham equations, are:

$$i\frac{\partial}{\partial t}\varphi_i(r\sigma, t) = \left[-\frac{\nabla^2}{2} + \hat{V}_0 + v_{\text{H}}[n(t)](r) + \hat{V}_{\text{xc}}[n_{\sigma\sigma'}(t)] + \hat{V}_{\text{ext}}[u]\right]\varphi_i(r\sigma, t) \tag{13.23}$$

with the (spin-) densities

$$n_{\sigma\sigma'}(r, t) = \sum_{i=1}^{N} \varphi_i^*(r\sigma, t)\varphi_i(r\sigma', t), \tag{13.24}$$

and

$$n(r, t) = \sum_{\sigma} n_{\sigma\sigma}(r, t), \tag{13.25}$$

for $i = 1, \ldots, N$ orbitals. The greek indexes σ, σ' run over the two spin configurations, up and down. The densities are, by construction, equal to that of the *real*, interacting system of electrons. $\hat{V}_0$ represents the internal, time independent fields— usually a nuclear Coulomb potential $V_n(r)$, and may include as well a spin-orbit coupling term of the form $\sigma \cdot \nabla V_n \times \hat{\alpha}$ (where σ is the vector of Pauli matrices). The term

$$v_{\text{H}}[n(t)](r) = \int d^3r' \frac{n(r', t)}{|r - r'|} \tag{13.26}$$

one must consider its functional derivatives. From the condition $\delta J / \delta \underline{\varphi}^* = 0$, we obtain

$$i\underline{\dot{\chi}}(t) = \left[\underline{\underline{\hat{H}}}_{\mathrm{KS}}[n_{\sigma\sigma'}[u](t), u, t] + \underline{\underline{\hat{K}}}[\varphi[u](t)] \right] \underline{\chi}(t) - i\frac{\delta F}{\delta \underline{\varphi}^*}, \tag{13.35a}$$

$$\underline{\chi}(T) = \underline{0}. \tag{13.35b}$$

The presence of the non-diagonal operator matrix $\underline{\underline{\hat{K}}}[\varphi[u](t)]$ is the main difference with respect to the normal QOCT equations for a linear quantum system. Its origin is the non-linear dependence of the Kohn–Sham Hamiltonian with respect to the propagating orbitals. It is defined by:

$$\hat{K}_{ij}[\varphi[u](t)]\psi(r\sigma) = -2i \sum_{\sigma'} \varphi_i[u](r\sigma', t) \times \Im\mathrm{m}$$

$$\times \left[\sum_{\alpha\beta} \int \mathrm{d}^3 r' \psi^*(r'\alpha) f_{\mathrm{Hxc}}^{\alpha\beta,\sigma\sigma'}[n_{\sigma\sigma'}[u](t)](r, r')\varphi_j[u](r'\beta, t) \right], \tag{13.36}$$

where $f_{\mathrm{Hxc}}^{\alpha\beta,\sigma\sigma'}$ is the *kernel* of the Kohn–Sham Hamiltonian, defined as:

$$f_{\mathrm{Hxc}}^{\alpha\beta,\sigma\sigma'}[n_{\sigma\sigma'}](r, r') = \frac{\delta_{\alpha\beta}\delta_{\sigma\sigma'}}{|r - r'|} + \frac{\delta v_{\mathrm{xc}}^{\alpha\beta}[n_{\sigma\sigma'}](r)}{\delta n_{\sigma\sigma'}(r')}. \tag{13.37}$$

If we now note that $G[u] = J[\varphi[u], \varphi[u], u]$, we arrive to:

$$\nabla_u G[u] = \nabla_u F[\varphi, u]\Big|_{\underline{\varphi}=\underline{\varphi}[u]} + 2\Im\mathrm{m} \left[\sum_{j=1}^{N} \int_0^T \mathrm{d}t \, \langle \chi_j[u](t)|\nabla_u \hat{V}_{\mathrm{ext}}[u](t)|\varphi_j[u](t)\rangle \right]. \tag{13.38}$$

Several aspects deserve further discussion:

(i) Equation 13.35a represents a set of first-order differential equations, whose solution, as it happened in the generic case, must be obtained by *backwards* propagation, since the boundary condition, Eq. 13.35b is given at the end of the propagation interval, T. Note that this propagation depends on the Kohn–Sham orbitals $\varphi[u]$. Therefore, the numerical procedure consists of a forward propagation to obtain $\varphi[u]$, followed by a backwards propagation to obtain $\chi[u]$.

(ii) These backwards equations are inhomogeneous, due to the presence of the last term in Eq. 13.35a, the functional derivative of F with respect to φ. The reason is that the previous Eqs 13.35a and 13.35b refer to a general "time-dependet target" case, i.e. the target functional depends on the full evolution of the system. In many cases of interest, however, the target functional F takes a "static" form, which can be expressed as:

$$J_1[\underline{\varphi}, u] = O[\varphi(T), u],\tag{13.39}$$

for some functional O whose argument is not the full evolution of the Kohn–Sham system, but only its value at the end of the propagation. In this case, the inhomogeneity in Eq. 13.35a vanishes, and instead we obtain a different final-value condition:

$$\chi_i[u](r\sigma, T) = \frac{\delta O[\underline{\varphi}[u], u]}{\delta \varphi_i^*(r\sigma)}.\tag{13.40}$$

(iii) The previous Eq. 13.38 assumes that u is a set of N parameters, $u \in \mathbb{R}^N$ that determines the control function. As we discussed in the previous section, if the control is in fact a time-dependent function $\varepsilon(t)$, the gradient has to be substituted by a functional derivative, and the result is:

$$\frac{\delta G}{\delta \varepsilon(t)} = \left.\frac{\delta F[\underline{\varphi}, \varepsilon]}{\delta \varepsilon(t)}\right|_{\underline{\varphi}=\underline{\varphi}[\varepsilon]} + 2\Im m \left[\sum_{j=1}^N \langle \chi_j[\varepsilon](t)|\hat{D}|\varphi_j[\varepsilon](t)\rangle\right].\tag{13.41}$$

We have assumed here that the external potential $\hat{V}_{\text{ext}}$ is determined by the function u by a linear relationship:

$$\hat{V}_{\text{ext}}[\varepsilon](t) = \varepsilon(t)\hat{D}.\tag{13.42}$$

This is the most usual case ($\hat{D}$ would be the dipole operator, and $\varepsilon(t)$ the amplitude of an electric field), but of course it would be trivial to generalize this to other possibilities.

The previous scheme permits therefore to control the Kohn–Sham system. It is important to reflect on the original assumption given by Eq. 13.21, which permits to identify this control with that of the real system of interest. Although the existence of a density functional equivalent to the wave function functional is known to exist *in principle*, we will not always have explicit knowledge of this density functional. We do have this knowledge immediately, for example, if the target functional F_{int} is given by the expectation value of some one-body local operator $\hat{A}$:

$$F_{\text{int}}[\Psi] = \langle \Psi(T)|\hat{A}|\Psi(T)\rangle = \int d^3r \, n(r, T)a(r),\tag{13.43}$$

where $\hat{A} = \sum_{i=1}^N a(\hat{r}_i)$. In this case, Eq. 13.40 is simply:

$$\chi_i[u](r\sigma, T) = a(r)\varphi_i[u](r\sigma, T).\tag{13.44}$$

Unfortunately, in some cases the situation is not that favourable. For example, a very common control goal is the transition from an initial state to a target state. In other words, the control operator $\hat{A}$ is the projection operator onto the target state $\hat{A} = |\Psi_{\text{target}}\rangle\langle\Psi_{\text{target}}|$. We have no exact manner to substitute, in this case, the F_{int}

functional by a functional $\tilde{F}$ defined in terms of the density, and afterwards in terms of the KS determinant. It can be approximated, however, by an expression in the form:

$$F[\underline{\varphi}] = |\langle \underline{\varphi}(T)| \sum_I c_I |\varphi^I \rangle|^2, \tag{13.45}$$

where $\varphi(T)$ is the TDKS determinant at time T, and we compute its overlap with a linear combination of Slater determinants φ^I, weighted with some coefficients c_I. These Slater determinants would be composed of occupied and unoccupied ground state KS orbitals, $\varphi^I = \det[\varphi_1^I, \ldots, \varphi_N^I]$. In this case, Eq. 13.40 takes the form:

$$\chi_i[u](r\sigma, T) = \sum_{IJ} \lambda_{IJ}(r\sigma)\langle \underline{\varphi}(T)|\varphi^I \rangle \langle \varphi^J |\underline{\varphi}(T)\rangle, \tag{13.46a}$$

$$\lambda_{IJ}(r\sigma) = c_I c_J^* \mathrm{Tr}\{(M^I)^{-1} A_I^i(r\sigma)\}, \tag{13.46b}$$

where $M_{mn}^I = \langle \varphi_m | \varphi_n^I \rangle$ and $A_I^i(r\sigma)_{mn} = \delta_{mi} \varphi_n^I(r\sigma)$.

In conclusion, we have shown how TDDFT can be combined with QOCT, and we have demonstrated how the resulting equations are numerically tractable. This provides a scheme to perform QOCT calculations from first principles, in order to obtain tailored function-specific laser pulses capable of controlling the electronic state of atoms, molecules, or quantum dots. Most of the previous applications of QOCT were targeted to control, with femto-second pulses, the motion of the nuclear wave packet on one or few potential energy surfaces, which typically happens on a time scale of hundreds of femtoseconds or picoseconds. The approach shown here, on the other hand, is particularly suited to control the motion of the electronic degrees of freedom which is governed by the sub-femto-second time scale. The possibilities that are open thanks to this technique are numerous: shaping of the high harmonic generation spectrum (i.e. quenching or increasing given harmonic orders), selective excitation of electronic excited states that are otherwise difficult to reach with conventional pulses, control of the electronic current in molecular junctions, and many more.

Part IV
Real-Time Dynamics

Chapter 14
Non-Born–Oppenheimer Dynamics and Conical Intersections

Mark E. Casida, Bhaarathi Natarajan and Thierry Deutsch

14.1 Introduction

The area of excited state dynamics is receiving increasing attention for a number of reasons, including the the importance of photochemical processes in basic energy sciences, improved theoretical methods and the associated theoretical understanding of photochemical processes, and the advent of femtosecond (and now attosecond) spectroscopy allowing access to more detailed experimental information about photochemical processes. Since photophysical and chemical processes are more complex than thermal (i.e., ground state) processes, simulations quickly become expensive and even unmanageable as the model system becomes increasingly realistic. With its combination of simplicity and yet relatively good accuracy, TDDFT has been finding an increasingly important role to play in this rapidly developing field. After reviewing some basic ideas from photophysics and photochemistry, this chapter will cover some of the strengths and weaknesses of TDDFT for modeling photoprocesses. The emphasis will be on going beyond the Born–Oppenheimer approximation.

There are distinct differences between how solid-state physicists and chemical physicists view photoprocesses. We believe that some of this is due to fundamental

M. E. Casida (✉) · B. Natarajan
Laboratoire de Chimie Théorique,
Département de Chimie Moléculaire (DCM, UMR CNRS/UJF 5250),
Institut de Chimie Moléculaire de Grenoble (ICMG, FR2607),
Université Joseph Fourier (Grenoble I),
38041 Grenoble, France
e-mail: mark.casida@ujf-grenoble.fr
e-mail: bhaarathi.natarajan@ujf-grenoble.fr

T. Deutsch · B. Natarajan
CEA, INAC, SP2M, L_Sim,
38054 Grenoble Cedex 9, France
e-mail: theirry.deutsch@cea.fr

M. A. L. Marques et al. (eds.), *Fundamentals of Time-Dependent Density Functional Theory*, Lecture Notes in Physics 837, DOI: 10.1007/978-3-642-23518-4_14,
© Springer-Verlag Berlin Heidelberg 2012

differences in the underlying phenomena being studied but that much is due to the use of different approximations and the associated language. Ultimately anyone who wants to work at the nanointerface between molecules and solids must come to terms with these differences, but that is not our objective here. Instead we will adopt the point of view of a chemical physicist (or physical chemist)—see, e.g. (Michl and Bonačić-Koutecký 1990).

The usual way to think about molecular dynamics is in terms of the potential energy surfaces that come out of the Born–Oppenheimer separation. In thermal processes, vibrations are associated with small motions around potential energy surface minima. Chemical reactions are usually described as going over passes (transition states) on these hypersurfaces as the system moves from one valley (reactants) to another (products). Photoprocesses are much more complicated (see Fig. 14.1). Traditionally they include not only process that begin by absorption of a photon, but also any process involving electronically-excited states, such as chemiluminescence (e.g., in fireflies and glow worms) where the initial excitation-energy is provided by a chemical reaction. The Franck–Condon approximation tells us that the initial absorption of a photon will take us from one vibronic state to another without much change of molecular geometry, thus defining a Franck–Condon region on the excited-state potential energy surfaces. The molecule can return to the ground state by emitting a photon of the initial wavelength or, depending upon vibronic coupling and perturbations from surrounding molecules, the molecule may undergo radiationless relaxation to a lower energy excited state before emitting or it may even decay all the way to the ground state without emitting. If emission takes place from a long-lived excited state of the same spin as the ground state, then we speak of fluorescence. If emission takes place from an excited state with a different spin due to intersystem crossing, then we speak of phosphorescence. If it is unsure whether the emission is fluorescence or phosphorescence, then we just say the molecule luminesces. Because of the large variety of de-excitation processes, excited molecules usually return too quickly to their ground state for the molecular geometry to change much. We then speak of a photophysical process because no chemical reaction has taken place. Thus fluorescence is usually described as an excited molecule relaxing slightly to a nearby minimum on the excited state potential energy surface where it is momentarily trapped before it emits to the ground state. It follows that the photon emitted during fluorescence is Stokes shifted to a lower energy than that of the photon initially absorbed.

Photochemical reactions occur when the excited molecule decays to a new minimum on the ground state surface, leading to a new chemical species (product.) This may have positive value as a way for synthesizing new molecules or negative value because of photodegradation of materials or because of photochemically-induced cancers. Either way the photochemical reaction must occur quickly enough that it can compete with other decay processes. Photochemical reactions almost always occur via photochemical funnels where excited state and ground-state surfaces come together, either almost touching (avoided crossing) or crossing (conical intersection.) These funnels play a role in photochemical reactions similar to transition states for thermal reactions. However it must be kept in mind that these funnels

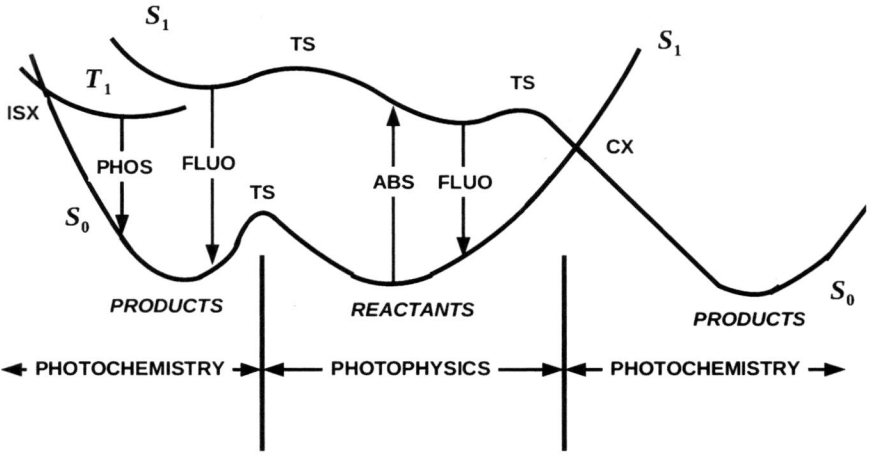

Fig. 14.1 Schematic representation of potential energy surfaces for photophysical and photochemical processes: S_0 ground singlet state; S_1 lowest excited singlet state; T_1 lowest triplet state; *ABS* absorption; *FLUO* fluorescence; *PHOS* phosphorescence; *ISX* intersystem crossing; *CX* conical intersection; *TS* transition state

may be far from the Franck–Condon region on the excited state potential energy surface, either because there is an easy energetically-"downhill" process or because, unless the absorption wavelength can be carefully tuned to a known vertical excitation energy, the system will typically arrive in an electronically-excited state with excess dynamical energy which can be used to move from one excited state potential energy surface valley over a transition state to funnels in another basin of the excited state potential energy surface. While conical intersections are forbidden in diatomic molecules, they are now believed to be omnipresent in the photochemistry of polyatomic molecules. Traditional simple models involve symmetry constraints which correspond to a potential energy surface cut, typically revealing an avoided crossing rather than the nearby conical intersection corresponding to a less symmetric geometry. A particularly striking example is provided by experimental and theoretical evidence that the fundamental photochemical reaction involved in vision passes through a conical intersection (Polli et al. 2010). For these reasons, modern photochemical modeling often involves some type of dynamics and, when this is not possible, at least focuses on finding conical intersections that can explain the reaction.

While a single-reference electronic structure method may be adequate for describing photophysical processes, the usual standard for describing photochemical processes is a multireference electronic structure method such as the complete active space self-consistent field (CASSCF) method. [See (Helgaker et al. 2000) for a review of modern quantum chemical methods.] This is because the first approximation to the wave function along a reaction pathway is as a linear combination of the wave functions of the initial reactants and the final products. Since CASSCF is both computationally heavy and requires a high-level of user intervention, a simpler method such as TDDFT would be very much welcome. Early work in TDDFT in quantum chemistry foresaw increasing applications of TDDFT in photochemical modeling. For example, avoided crossings between cross-sections of excited state potential energy surfaces may be described with TDDFT because of the multireference-like nature of TDDFT excited states (Casida 1998). However great attention must also be paid to problems arising from the use of approximate functionals (Casida 2002). In particular, the TDDFT Tamm-Dancoff approximation (TDA) (Hirata 1999) was found to give improved shapes of excited-state potential energy surfaces (Casida 2000; Cordova et al. 2007), albeit at the price of losing oscillator strength sum rules. A major advance towards serious investigations of TDDFT for describing photoprocesses came with the implementation of analytical derivatives for photochemical excited states in many electronic structure programs [see Chap. 16 and (Van Caillie and Amos 1999, 2000; Furche and Ahlrichs 2002a; Hutter 2003; Rappoport and Furche 2005; Doltsinis and Kosov 2005; Scalmani et al. 2006).] This made it possible to relax excited state geometries and to calculate Stokes shifts within the framework of TDDFT. In fact, TDDFT has become a standard part of the photochemical modeler's toolbox. It is typically used for calculating absorption spectra and exploring excited state potential energy surfaces around the Franck–Condon region. TDDFT also serves as a rapid way to gain the chemical information needed to carry out subsequent CASSCF calculations. [See, e.g., (Diau et al. 2001a, b, 2002; Sølling et al. 2002; Diau and Zewail 2003) for some combined femtosecond spectroscopy/theoretical studies of photochemical reactions which make good use of TDDFT.] It would be nice to be able to use a single method to model entire photochemical processes. The advent of mixed TDDFT/classical surface-hopping Tully-type dynamics (Tapavicza et al. 2007; Werner et al. 2008; Tapavicza et al. 2008; Tavernelli et al. 2009a, c; Barbatti et al. 2010) is giving us a way to extend the power of TDDFT to the exploration of increasingly complicated photochemical processes.

The rest of this chapter is organized as follows: The next section reviews non-Born–Oppenheimer phenomena from a wave-function point of view, with an emphasis on mixed quantum/classical dynamics. This sets the stage for our discussion of TDDFT for non-Born–Oppenheimer dynamics and conical intersections in Sect. 14.3. We sum up in Sect. 14.4.

14.2 Wave-Function Theory

Most likely anyone who has made it this far into this chapter has seen the Born–Oppenheimer approximation at least once, if not many times. However, it is relatively rare to find good discussions that go beyond the Born–Oppenheimer approximation (Doltsinis and Marx 2002; Cederbaum 2004). This section tries to do just this from a wave-function point of view, in preparation for a discussion of TDDFT approaches to the same problems in the following section. We first begin by reviewing (again!) the Born–Oppenheimer approximation, but this time with the point of view of identifying the missing terms. We then discuss mixed quantum/classical approximations, and end with a discussion of the pathway method and ways to find and characterize conical intersections. (Mixed quantum/quantum and quantum/semiclassical methods are also interesting, but have been judged beyond the scope of this chapter.) We shall use Hartree atomic units ($\hbar = m_e = e = 1$) throughout and adapt the convention in this section that electronic states are labeled by small Latin letters, while nuclear degrees of freedom are labeled by capital Latin letters.

14.2.1 Born–Oppenheimer Approximation and Beyond

As is well-known, the Born–Oppenheimer approximation relies on a separation of time scales: Since electrons are so much lighter and so move so much faster than nuclei, the electrons may be thought of as moving in the field of nuclei which are "clamped" in place and the nuclei move in a field which is determined by the mean field of the electrons. The Born–Oppenheimer approximation provides a precise mathematical formulation of this physical picture. Our interest here is in where the Born–Oppenheimer approximation breaks down and what terms are needed to describe this breakdown.

Consider a molecule composed of M nuclei and N electrons. Denote the nuclear coordinates by $\bar{R} = (R_1, R_2, \ldots, R_M)$ and electronic coordinates by $\bar{r} = (r_1, r_2, \ldots, r_N)$. The full Hamiltonian, $\hat{H}(\bar{R}, \bar{r}) = \hat{T}_n(\bar{R}) + \hat{H}_e(\bar{r}; \bar{R}) + V_{nn}(\bar{R})$, is the sum of an electronic Hamiltonian, $\hat{H}_e(\bar{r}; \bar{R}) = \hat{T}_e(\bar{r}) + V_{en}(\bar{r}; \bar{R}) + V_{ee}(\bar{r})$, with its electronic kinetic energy, $\hat{T}_e$, electron-nuclear attraction, V_{en}, and electron–electron repulsion, V_{ee}, with the missing nuclear terms—namely the nuclear kinetic energy, $\hat{T}_n$, and the nuclear–nuclear repulsion, V_{nn}. Solving the time-dependent Schrödinger equation,

$$\hat{H}(\bar{R}, \bar{r})\Phi(\bar{R}, \bar{x}, t) = i\frac{d}{dt}\Phi(\bar{R}, \bar{x}, t), \tag{14.1}$$

is a formidable $(N + M)$-body problem. ($\bar{x}$ denotes inclusion of electron spin. We have decided to omit nuclear spin for simplicity. Note, however, that explicit inclusion of nuclear spin can sometimes be important—for example, the properties of *ortho-* and *para*-hydrogen.) That is why the Born–Oppenheimer expansion (which is not yet the Born–Oppenheimer approximation!),

$$\Phi(\bar{R}, \bar{x}, t) = \sum_j \Psi_j(\bar{x}; \bar{R})\chi_j(\bar{R}, t), \tag{14.2}$$

is used, where the electronic wave functions are solutions of the time-independent electronic problem in the field of clamped nuclei,

$$\hat{H}_e(\bar{r}; \bar{R})\Psi_j(\bar{x}; \bar{R}) = E_j^e(\bar{R})\Psi_j(\bar{x}; \bar{R}). \tag{14.3}$$

Inserting the Born–Oppenheimer expansion (Eq. 14.2) into the full Schrödinger equation (Eq. 14.1), left multiplying by $\Psi_i^*(\bar{x}; \bar{R})$, and integrating over $\bar{x}$ gives the time-dependent Schrödinger equation for the nuclear degrees of freedom,

$$\left[\hat{T}_{\mathrm{n}}(\bar{R}) + V_i(\bar{R})\right] \chi_i(\bar{R}, t) + \sum_j \hat{V}_{i,j}(\bar{R}) \chi_j(\bar{R}, t) = \mathrm{i}\frac{\partial}{\partial t}\chi_i(\bar{R}, t). \qquad (14.4)$$

Here, $V_i(\bar{R}) = E_i^{\mathrm{e}}(\bar{R}) + V_{\mathrm{nn}}(\bar{R})$, is the *adiabatic* potential energy surface for the ith electronic state. [Notice that this is a different use of the term "adiabatic" than in the TDDFT "adiabatic approximation" for the exchange-correlation (xc) functional.] The remaining part, $\hat{V}_{i,j}(\bar{R})$, is the hopping term which couples the ith and jth potential energy surfaces together. It should be kept in mind that the Born–Oppenheimer expansion (Eq. 14.2) is exact and hence so is Eq. 14.4. As is well known, the Born–Oppenheimer approximation neglects the hopping terms,

$$\left[\hat{T}_{\mathrm{n}}(\bar{R}) + V_i(\bar{R})\right] \chi_i(\bar{R}, t) = \mathrm{i}\frac{\partial}{\partial t}\chi_i(\bar{R}, t). \qquad (14.5)$$

We, on the other hand, are interested in precisely the terms neglected by the Born–Oppenheimer approximation. The hopping term is given by

$$\hat{V}_{i,j}(\bar{R})\chi_j(\bar{R}, t) = -\sum_I \frac{1}{2m_I}\left[G_{i,j}^{(I)}(\bar{R}) + 2F_{i,j}^{(I)}(\bar{R}) \cdot \nabla_I\right]\chi_j(\bar{R}, t), \qquad (14.6)$$

where,

$$G_{i,j}^{(I)}(\bar{R}) = \int \mathrm{d}^3\bar{x}_1 \cdots \int \mathrm{d}^3\bar{x}_N \Psi_i^*(\bar{x}; \bar{R})\left[\nabla_I^2\Psi_j(\bar{x}; \bar{R})\right]$$
$$= \langle i|\nabla_I^2|j\rangle, \qquad (14.7)$$

is the scalar coupling matrix and,

$$F_{i,j}^{(I)}(\bar{R}) = \int \mathrm{d}^3\bar{x}_1 \cdots \int \mathrm{d}^3\bar{x}_N \Psi_i^*(\bar{x}; \bar{R})\left[\nabla_I\Psi_j(\bar{x}; \bar{R})\right]$$
$$= \langle i|\nabla_I|j\rangle, \qquad (14.8)$$

is the derivative coupling matrix (Cederbaum 2004). Note that the derivative coupling matrix is also often denoted $d_{i,j}^I$ and called the nonadiabatic coupling vector (Doltsinis and Marx 2002). Here we have introduced a compact notation for some complicated objects: Both the scalar and derivative coupling matrices are simultaneously a function of the nuclear coordinates, a matrix in the electronic degrees of freedom, and a vector in the nuclear degrees of freedom, and a matrix in the electronic degrees of freedom. However the derivative coupling matrix is also a vector in the three spatial coordinates of the Ith nucleus.

Interestingly the scalar coupling matrix and derivative coupling matrix are not independent objects. Rather, making use of the resolution of the identity for the electronic states, it is straightforward to show that,

$$\sum_k \left(\delta_{i,k}\nabla_I + F_{i,k}^{(I)}(\bar{R})\right) \cdot \left(\delta_{k,j}\nabla_I + F_{k,j}^{(I)}(\bar{R})\right) = \nabla_I^2 + G_{i,j}^{(I)}(\bar{R}) + 2F_{i,j}^{(I)}(\bar{R}) \cdot \nabla_I.$$
$$(14.9)$$

We may then rewrite the time-dependent nuclear equation (14.4) as,

$$
-\left\{ \sum_I \frac{1}{2m_I} \left[\sum_k \left(\delta_{i,k} \nabla_I + F_{i,k}^{(I)}(\bar{\boldsymbol{R}}) \right) \cdot \left(\delta_{k,j} \nabla_I + F_{k,j}^{(I)}(\bar{\boldsymbol{R}}) \right) \right] \right\} \chi_j(\bar{\boldsymbol{R}}, t)
$$
$$
+ V_i(\bar{\boldsymbol{R}}) \chi_i(\bar{\boldsymbol{R}}, t) = i \frac{\partial}{\partial t} \chi_j(\bar{\boldsymbol{R}}, t), \tag{14.10}
$$

which is known as the group Born–Oppenheimer equation (Cederbaum 2004). Evidently this is an equation which can be solved within a truncated manifold of a few electronic states in order to find fully quantum mechanical solutions beyond the Born–Oppenheimer approximation.

More importantly for present purposes is that Eq. 14.10 brings out the importance of the derivative coupling matrix. The derivative coupling matrix can be rewritten as,

$$
F_{i,j}^{(I)}(\bar{\boldsymbol{R}}) = \frac{\langle i | \left[\nabla_I \hat{H}_e(\bar{\boldsymbol{R}}) \right] | j \rangle - \delta_{i,j} \nabla_I E_i^e(\bar{\boldsymbol{R}})}{E_j^e(\bar{\boldsymbol{R}}) - E_i^e(\bar{\boldsymbol{R}})}. \tag{14.11}
$$

Since this equation is basically a force-like term, divided by an energy difference, we see that we can neglect coupling between adiabatic potential energy surfaces when (i) the force on the nuclei is sufficiently small (i.e., the nuclei are not moving too quickly) and (ii) when the energy difference between potential energy surfaces is sufficiently large.

These conditions often break down in funnel regions of photochemical reactions. There is then a tendency to follow diabatic surfaces, which may be defined rigorously by a unitary transformation of electronic states (when it exists) to a new representation satisfying the condition, $F_{i,j}^{(I)}(\bar{\boldsymbol{R}}) \approx 0$. The advantage of the diabatic representation (when it exists, which is not always the case) is that it eliminates the off-diagonal elements of the derivative coupling matrix in the group Born–Oppenheimer equation (Eq. 14.10), hence eliminating the need to describe surface hopping. At a more intuitive level, the character of electronic states tends to be preserved along diabatic surfaces because

$$
\left\langle i \Big| \frac{\mathrm{d}j}{\mathrm{d}t} \right\rangle = \dot{\boldsymbol{R}} \cdot \langle i | \nabla j \rangle = \dot{\boldsymbol{R}} \cdot F_{i,j} \approx 0 \tag{14.12}
$$

in this representation. For this reason, it is usual to trace diabatic surfaces informally in funnel regions by analyzing electronic state character, rather than seeking to minimize the nonadiabatic coupling vector. Avoided crossings of adiabatic surfaces are then described as due to configuration mixing of electronic configurations belonging to different diabatic surfaces.

14.2.2 Mixed Quantum/Classical Dynamics

Solving the fully quantum-mechanical dynamics problem of coupled electrons and nuclei is a challenge for small molecules and intractable for larger molecules. Instead it is usual to use mixed quantum/classical methods in which the nuclei are described by Newtonian classical mechanics while the electrons are described by quantum mechanics. Dividing any quantum system into two parts and then approximating one using classical mechanics is the subject of on-going research (Kapral 2006). In general, no rigorous derivation is possible and wave-function phase information (e.g., the Berry phase) is lost which may be important in some instances. Nevertheless mixed quantum/classical approximations are intuitive: Most nuclei (except perhaps hydrogen) are heavy enough that tunneling and other quantum mechanical effects are minor, so that classical dynamics is often an *a priori* reasonable first approximation. Of course, rather than thinking of a single classical trajectory for the nuclear degree of freedom, we must expect to think in terms of ensembles (or "swarms") of trajectories which are built to incorporate either finite temperature effects or to try to represent quantum mechanical probability distributions or both. The purpose of this subsection is to introduce some common mixed quantum/classical methods.

The most elementary mixed quantum/classical approximation is Ehrenfest dynamics. According to Ehrenfest's (1927) theorem Newton's equations are satisfied for mean values in quantum systems, $d\langle \hat{r} \rangle / dt = \langle \hat{p} \rangle / m$ and $d\langle \hat{p} \rangle / dt = -\langle \nabla V \rangle$. Identifying the position of the nuclei with their mean value, we can then write an equation, $m_I \ddot{\bar{R}}_I(t) = -\nabla_I V(\bar{R}(t))$, whose physical interpretation is that the nuclei are moving in the mean field of the electrons. Here

$$V(\bar{R}(t)) = \langle \Psi(\bar{R}, t) | \hat{H}_e(\bar{R}(t)) | \Psi(\bar{R}, t) \rangle + V_{nn}(\bar{R}(t)), \qquad (14.13)$$

where the electronic wave function is found by solving the time-dependent equation,

$$\hat{H}_e(\bar{x}, \bar{R}(t)) \Psi(\bar{x}; \bar{R}, t) = i \frac{\partial}{\partial t} \Psi(\bar{x}; \bar{R}, t). \qquad (14.14)$$

While Ehrenfest dynamics has been widely and often successfully applied, it suffers from some important drawbacks. The first drawback is that the nuclei always move on average potential energy surfaces, rather than adiabatic or diabatic surfaces, even when far from funnel regions where the nuclei would be expected to move on the surface of a single electronic state. While this is serious enough, since it suggests errors in calculating branching ratios (i.e., relative yields of different products in a photoreaction), a more serious drawback is a loss of microscopic reversibility. That is, the temporal variation of the mean potential energy surface depends upon past history and can easily be different for forward and reverse processes.

A very much improved scheme is the fewest switches method of Tully (1990) (Hammes-Schiffer and Tully 1994). Here the nuclei move on well-defined adiabatic potential energy surfaces,

$$m_I \ddot{\boldsymbol{R}}_I(t) = -\nabla_I V_i(\bar{\boldsymbol{R}}(t)), \tag{14.15}$$

and the electrons move in the field of the moving nuclei,

$$\hat{H}_e(\bar{\boldsymbol{r}}; \bar{\boldsymbol{R}}(t))\Psi(\bar{\boldsymbol{x}}, t) = i\frac{d}{dt}\Psi(\bar{\boldsymbol{x}}, t). \tag{14.16}$$

To determine the probability that a classical trajectory describing nuclear motion hops from one electronic potential energy surface to another, we expand

$$\Psi(\bar{\boldsymbol{x}}, t) = \sum_m \Psi_m(\bar{\boldsymbol{x}}; \bar{\boldsymbol{R}}(t))C_m(t), \tag{14.17}$$

in solutions of the time-independent Schrödinger equation,

$$\hat{H}(\bar{\boldsymbol{r}}; \bar{\boldsymbol{R}}(t))\Psi_m(\bar{\boldsymbol{x}}; \bar{\boldsymbol{R}}(t)) = E_m(\bar{\boldsymbol{R}}(t))\Psi_m(\bar{\boldsymbol{x}}; \bar{\boldsymbol{R}}(t)). \tag{14.18}$$

The probability of finding the system on surface m is then given by, $P_m(t) = |C_m(t)|^2$. The coefficients may be obtained in a dynamics calculation by integrating the first-order equation,

$$\dot{C}_m(t) = -iE_m(t)C_m(t) - \sum_n \left\langle m \left| \frac{dn}{dt} \right. \right\rangle C_n(t). \tag{14.19}$$

A not unimportant detail is that the nonadiabatic coupling elements need not be calculated explicitly, but instead can be calculated using the finite difference formula,

$$\langle m(t + \Delta t/2) | \dot{n}(t + \Delta t/2) \rangle = \frac{\langle m(t)|n(t + \Delta t) \rangle - \langle m(t + \Delta t)|n(t) \rangle}{2\Delta t}. \tag{14.20}$$

In practice, it is also important to minimize the number of surface hops or switches in order to keep the cost of the dynamics calculation manageable. [Tully also suggests (p. 1066 of Ref. (Tully 1990)) that too rapid switching would lead to trajectories behaving incorrectly *as if* they were on an average potential energy surface.] Tully accomplished this by introducing his fewest-switches algorithm which is a type of Monte Carlo procedure designed to correctly populate the different potential energy surfaces with a minimum of surface hopping. Briefly, the probability of jumping from surface m to surface n in the interval $(t, t + \Delta t)$ is given by $g_{m \to n}(t, \Delta t) = \dot{P}_{m,n}(t)\Delta t / P_{m,m}(t)$ where $P_{m,n}(t) = C_m(t)C_n^*(t)$. A random number ξ is generated with uniform probability on the interval $(0,1)$ and compared with $g_{m \to n}(t, \Delta t)$. The transition $m \to n$ occurs only if $P_n^{(m-1)} < \xi < P_n^{(m)}$ where $P_n^{(m)} = \sum_{l=1,m} P_{n,l}$ is the sum of the transition probabilities for the first m states. Additional details of the algorithm, beyond the scope of this chapter, involve readjustment of nuclear kinetic energies and the fineness of the numerical integration grid for the electronic part of the calculation with respect to that of the grid for the nuclear degrees of freedom.

It is occasionally useful to have a simpler theory for calculating the probability of potential energy surface hops which depends only on the potential energy surfaces and not on the wave functions. Such a theory was suggested by Landau (1932) and Zener (1932) (see also Wittig 2005). Their work predates the modern appreciation of the importance of conical intersections and so focused on surface hopping at avoided crossings. The Landau–Zener model assumes that surface hopping occurs only on the surface where the two diabatic surfaces cross that give rise to the avoided crossing where the surface hopping occurs. After some linearizations and an asymptotic limit, it is possible to arrive at a very simple final formula,

$$P = \exp\left[-\frac{\pi^2 \Delta E_{\text{adia}}^2}{h(\text{d}|\Delta E_{\text{dia}}|/\text{d}t)} \right], \tag{14.21}$$

for the probability of hopping between two potential energy surfaces. This formula is to be applied at the point of closest approach of the two potential energy surfaces where the energy difference is ΔE_{adia}. However $\text{d}|\Delta E_{\text{dia}}|/\text{d}t$ is evaluated as the maximum of the rate of change of the *adiabatic* energy difference as the avoided crossing is approached. While not intended to be applied to conical intersections, it is still applicable in photodynamics calculations in the sense that trajectories rarely go exactly through a conical intersection.

14.2.3 Pathway Method

Dynamics calculations provide a swarm of reaction trajectories. The "pathway method" provides an alternative when dynamics calculations are too expensive or a simplified picture is otherwise desired, say, for interpretational reasons. The pathway method consists of mapping out minimum energy pathways between the initial Franck–Condon points obtained by vertical excitations and excited-state minima or conical intersections. Although analogous to the usual way of finding thermal reaction paths, it is less likely to be a realistic representation of true photoprocesses except in the limit of threshold excitation energies since excess energy is often enough to open up alternative pathways over excited-state transition states. While the necessary ingredients for the photochemical pathway method are similar to those for thermal reactions, conical intersections are a new feature which is quite different from a thermal transition state. This section provides a brief review for finding and characterizing conical intersections.

The notion of a conical intersection arises from a relatively simple argument (Yarkony 2001). The potential energy surface of a molecule with f internal degrees of freedom is an f-dimensional hypersurface in an $(f + 1)$-dimensional space (the extra dimension is the energy axis.) If two potential energy surfaces simply cross "without seeing each other", then the crossing space is characterized by the constraint

$$E_i(\bar{\boldsymbol{R}}) = E_j(\bar{\boldsymbol{R}}), \tag{14.22}$$

making the crossing space $(f-1)$-dimensional. However in quantum mechanics, we also have the additional constraint,

$$H_{i,j}(\bar{R}) = 0. \tag{14.23}$$

This makes the crossing space $(f-2)$-dimensional. This means that there will be two independent directions in hyperspace in which the two potential energy surfaces will separate. These two directions define a branching plane. Within the 3-dimensional space defined by the energy and the branching plane, the conical intersection appears to be a double cone (see Fig. 14.6), the point of which represents an entire $(f-2)$-dimensional space. Of course, $f=1$ for a diatomic and no conical intersection is possible. This is the origin of the well-known avoided crossing rule for diatomics. Here we are interested in larger molecules where the low dimensionality of the branching space in comparison with the dimensionality of the parent hyperspace can make the conical intersection hard to locate and characterize.

In the pathway method, the system simply goes energetically downhill until two potential energy surfaces have the same energy (Eq. 14.22). The resultant intersection space must be analyzed and the branching plane extracted so that the surface crossing region can be properly visualized and interpreted. In order to do so, let us recall a result from elementary calculus. Imagine a trajectory, $\bar{R}(\tau)$, depending upon some parameter τ within the conical intersection surface. Then $\nabla C(\bar{R})$ must be perpendicular to the conical intersection for any constraint function $C(\bar{R}) = 0$ because,

$$0 = \frac{dC(\bar{R}(\tau))}{d\tau} = \nabla C(\bar{R}) \cdot \frac{d\bar{R}}{d\tau}. \tag{14.24}$$

and we can always choose $d\bar{R}/d\tau \neq 0$. Taking the gradient of Eq. 14.23 defines the derivative coupling vector, $f_{i,j} = \nabla H_{i,j}(\bar{R})$, while taking the gradient of Eq. 14.22 defines the gradient difference vector, $g_{i,j} = \nabla E_i(\bar{R}) - \nabla E_j(\bar{R})$. Together the derivative coupling vector and gradient difference vector are referred to as the branching vectors which characterize the branching plane. [Note that the derivative coupling vector is essentially the numerator of the derivative coupling matrix expression given in Eq. 14.11. This confusion of nomenclature is unfortunate but present in the literature.]

The condition that $d\bar{R}/d\tau$ be perpendicular to the branching plane provides a constraint for use in the exploration of the conical intersection hyperspace when seeking the minimum energy conical intersection or the first-order saddle point in conical intersection. In particular, there has been considerable effort devoted to the problem of developing efficient algorithms for finding minimum energy points within the conical intersection space (Koga and Morokuma 1985; Atchity et al. 1991; Ragazos et al. 1992; Yarkony 1996; Domcke and Stock 1997; Izzo and Klessinger 2000). Furthermore, an automated systematic exploration method for finding minimum energy conical intersections has very recently developed (Maeda et al. 2009). First-order saddle points and the corresponding minimum energy pathways both within the conical intersection hypersurface may be useful reference points when mapping out a surface, and an optimization method was developed for such high-energy points within the conical intersection hypersurface (Sicilia et al. 2008). Some of the minimum energy conical intersection optimizers use the branching plane conditions explicitly to keep the degeneracy of the two adiabatic states during optimizations (Manaa and Yarkony 1993; Bearpark et al. 1994; Anglada and Bofill 1997), making explicit use of both the derivative coupling vector and gradient difference vector at every step. Most well-established optimization algorithms assume smoothness of the function to be optimized. Since the potential energy surface necessarily has a discontinuous first derivative in the vicinity of a conical intersection, the above-mentioned algorithms for finding minimum energy conical intersections have required access to the gradient difference vector and derivative coupling vectors. The gradient difference vector can easily be obtained from analytical gradients, if available, or by numerical energy differentiation if analytical gradients are not yet available. However ways for finding the derivative coupling vector are not yet available for all methods since implementation of an analytical derivative method is often regarded as a prerequisite (Ciminelli et al. 2004; Maeda et al. 2010). Some approaches make use of a penalty function to get around the need to calculate the derivative coupling vector and these have proven very useful for finding minimum energy conical intersection regions without the need for the derivative coupling vector (Levine et al. 2008). This is especially important for methods such as renormalized coupled-cluster theories and TDDFT or free-energy methods for which the electronic wave function is not completely defined, considerably complicating the problem of how to calculate derivative coupling vector matrix elements. However, convergence of penalty function methods is in general slower than methods which make explicit use of the branching plane constraints, especially if tight optimization of the energy difference, $(E_i - E_j)$, is desired (Keal et al. 2007).

14.3 TDDFT

The last section discussed the basic theory of non-Born–Oppenheimer dynamics and conical intersections from a wave-function point of view. We now wish to see to what extent we can replace wave-function theory with what we hope will be a simpler DFT approach. As usual in DFT, we seek both the guiding light of formal rigor and pragmatic approximations that work. We will take a more or less historical approach to presenting this material. In this section, upper case Latin indices designate electronic states, while lower case Latin indices designate orbitals.

One of the early objectives of TDDFT was to allow simulations of the behavior of atoms and clusters in intense laser fields, well beyond the linear-response regime and too complex to be handled by comparable wave-function methods. The closely related topic of ion-cluster collisions was studied early on using TDDFT in a very simplified form (Yabana et al. 1998). The Ehrenfest method was the method of choice for TDDFT simulations coupling electronic and nuclear degrees of freedom in this area. The gradient of the potential (14.13) is calculated with the help of the

Hellmann-Feynman theorem as,

$$\nabla_I V(\bar{\boldsymbol{R}}(t)) = \langle \Phi_s(\bar{\boldsymbol{R}}, t)|\nabla_I \hat{H}_s(\bar{\boldsymbol{R}}(t))|\Phi_s(\bar{\boldsymbol{R}}, t)\rangle + \nabla_I V_{nn}(\bar{\boldsymbol{R}}(t)). \qquad (14.25)$$

Note that the first integral on the right hand side only involves the (time-dependent) charge density—at least in the usual TDDFT adiabatic approximation. Among the notable work done with this approximation is early studies of the dynamics of sodium clusters in intense laser fields (Calvayrac et al. 1998), the development of the time-dependent electron localization function (Burnus et al. 2005), and (more recently) the study of electron-ion dynamics in molecules under intense laser pulses [see Chap. 18 and (Kawashita et al. 2009).] Besides limitations associated with the TDDFT adiabatic approximation, the TDDFT Ehrenfest method suffers from the same intrinsic problems as its wave-function sibling—namely that it is implicitly based on an average potential energy surface and so does not provide state-specific information, and also suffers from problems with microscopic irreversibility.

To our knowledge, the first DFT dynamics on a well-defined excited state potential energy surface was not based upon TDDFT but rather on the older multiplet sum method of Ziegler et al. (1977) (Daul 1994). This was the work of restricted open-shell Kohn–Sham (ROKS) formalism of Irmgard Frank et al. (1998) who carried out Car-Parinello dynamics for the open-shell singlet excited state $^1(i, a)$ using the multiplet sum method energy expression,

$$E_s = 2E\left[\Phi_{i\uparrow}^{a\uparrow}\right] - E\left[\Phi_{i\downarrow}^{a\uparrow}\right], \qquad (14.26)$$

where $\Phi_{i\sigma}^{a\tau}$ is the Kohn–Sham determinant with the $i\sigma$ spin-orbital replaced with the $a\tau$ spin-orbital. Such a formalism suffers from all the formal difficulties of the multiplet sum method, namely that it is just a first-order estimate of the energy using a symmetry-motivated zero-order guess for the excited state wave function and assumes that DFT works best for states which are well-described by single determinants. Nevertheless appropriate use of the multiplet sum method can yield results similar to TDDFT. A recent application of this method is to the study of the mechanism of the electrocyclic ring opening of diphenyloxirane (Friedrichs and Frank 2009).

The implementation of TDDFT excited state derivatives in a wide variety of programs not only means that excited state geometry optimizations may be implemented, allowing the calculation of the Stokes shift between absorption and fluorescence spectra, but that the pathway method can be implemented to search for conical intersections in TDDFT. Unless nonadiabatic coupling matrix elements can be calculated within TDDFT (*vide infra*), then a penalty method should be employed as described in the previous section under the pathway method. This has been done by Levine et al. (2006) using conventional TDDFT and by Minezawa and Gordon (2009) using spin-flip TDDFT. We will come back to these calculations later in this section.

The most recent approach to DFT dynamics on a well-defined excited state potential energy surface is Tully-type dynamics (Tully 1990; Hammes-Schiffer and Tully

1994; Tully 1998a) applied within a mixed TDDFT/classical trajectory surface-hopping approach. Surface-hopping probabilities can be calculated from potential energy surfaces alone within the Landau–Zener method (Eq. 14.21), however a strict application of Tully's method requires nonadiabatic coupling matrix elements as input. Thus a key problem to be addressed is how to calculate nonadiabatic coupling matrix elements within TDDFT. Initial work by Craig, Duncan, and Prezhdo used a simple approximation which neglected the xc-kernel (Craig et al. 2005). A further approximation, commented on by Maitra (2006), has been made by Craig and co-workers (Craig et al. 2005; Habenicht et al. 2006) who treated the electronic states as determinants of Kohn–Sham orbitals which are propagated according to the time-dependent Kohn–Sham equation. This means that neither the excitation energies nor the associated forces could be considered to be accurate.

The first complete mixed TDDFT/classical trajectory surface-hopping photo-dynamics method was proposed and implemented by Tapavicza, Tavernelli, and Röthlisberger (Tapavicza et al. 2007) in a development version of the CPMD code. It was proposed that the nonadiabatic coupling matrix elements be evaluated within Casida's *ansatz* (Casida 1995) which was originally intended to aid with the problem of assigning excited states by considering a specific functional form for an approximate excited state wave function. Note that numerical integration of Eq. 14.19 to estimate the coefficients, $C_m(t)$, for the true system of interacting electrons also involves making assumptions about the initial interacting excited state. Casida's *ansatz* is a more logical choice for this than is a simple single determinant of Kohn–Sham orbitals. For the TDA, the Casida *ansatz* takes the familiar form, $\Psi_I = \sum_{ia\sigma} \Phi_{i\sigma}^{a\sigma} X_{ia\sigma}$.

In fact, matrix elements between ground and excited states may be calculated exactly in a Casida-like formalism because of the response theory nature of Eq. 14.11 (Chernyak Mukamel 2000a, Hu 2007b, Send 2010). Test results show reasonable accuracy for nonadiabatic coupling matrix elements as long as conical intersections are not approached too closely (Baer 2002; Hu et al. 2007b; Tavernelli 2009c; Tavernelli et al. 2009a; Send and Furche 2010). One likely reason for this is the divergence of Eq. 14.11 when $E_I = E_J$. Hu and Sugino attempted to further improve the accuracy of nonadiabatic coupling matrix elements by using average excitation energies (Hu Sugino 2007a). The problem of calculating nonadiabatic coupling matrix elements between two excited states is an open problem in TDDFT, though the ability to calculate excited state densities (Furche and Ahlrichs 2002a) suggests that such matrix elements could be calculated from double response theory using Eq. 14.11. The idea is that adding a second time-dependent electric field *in addition* to the first perturbation which allows the extraction of excited state densities, should allow the extraction of the excited state absorption spectrum using linear response theory in much the same way that this is presently done for the ground state. To our knowledge, this has never yet been done. But were it to be done, the extension to the derivative coupling matrix through Eq. 14.11 should be trivial.

Soon after the implementation of mixed TDDFT/classical trajectory surface-hopping photodynamics in CPMD, a very similar method was implemented in TURBOMOL and applied (Werner et al. 2008, Mirić et al. 2008, Barbatti et al. 2010).

A version of TURBOMOL capable of doing mixed TDDFT/classical trajectory surface-hopping photodynamics using analytic nonadiabatic coupling matrix elements (see Chap. 16) has recently appeared (Send and Furche 2010) and has been used to study the photochemistry of vitamin-D (Tapavicza 2010). Time-dependent density-functional tight-binding may be regarded as the next step in a multiscale approach to the photodynamics of larger systems. From this point of view, it is interesting to note that mixed TDDFT-tight binding/classical trajectory surface-hopping photo-dynamics is also a reality (Mitrić et al. 2009). Given the increasingly wide-spread nature of implementations of mixed TDDFT/classical trajectory surface-hopping photodynamics, we can only expect the method to be increasingly available to and used by the global community of computational chemists.

Before going further, let us illustrate the state-of-the-art for TDDFT when applied to non-Born–Oppenheimer dynamics and conical intersections. We will take the example of the photochemical ring opening of oxirane (structure I in Fig. 14.2.) While this is not the "sexy application" modeling of some biochemical photoprocess, the photochemistry of oxiranes is not unimportant in synthetic photochemistry and, above all, this is a molecule where it was felt that TDDFT "ought to work" (Cordova et al. 2007). A first study showed that a main obstacle to photodynamics is the presence of triplet and near singlet instabilities which lead to highly underesti-mated and even imaginary excitation energies as funnel regions are approached. This is illustrated in Fig. 14.3 for C_{2v} ring opening. While the real photochemical process involves asymmetric CO ring-opening rather than the symmetric C_{2v} CC ring-opening, results for the symmetric pathway have the advantage of being easier to analyze. The figure shows that applying the TDA strongly attenuates the insta-bility problem, putting most curves in the right energy range. Perhaps the best way to understand this is to realize that, whereas time-dependent Hartree–Fock (TDHF), is a nonvariational method and hence allows variational collapse of excited states, TDA TDHF is the same as configuration interaction singles (CIS) which is varia-tional. There is however still a cusp in the ground state curve as the ground state configuration changes from σ^2 to $(\sigma^*)^2$. According to a traditional wave-function picture, these two states, which are each double excitations relative to each other should be included in configuration mixing in order to obtain a proper description of the ground state potential energy surface in the funnel region [see Chap. 8 and (Cordova et al. 2007; Huix-Rotllant et al. 2010).]

Figure 14.4 shows an example of mixed TDA TDPBE/classical trajectory surface-hopping calculations for the photochemical ring-opening of oxirane with the initial photoexcitation prepared in the $^1(n, 3p_z)$ state. Part (b) of the figure clearly shows that more than one potential energy surface is populated after about 10 fs. The Landau–Zener process is typical of the dominant physical process which involves an excitation from the HOMO nonbonding lone pair on the oxygen initially to a $3p_z$ Rydberg orbital. As the reaction proceeds, the ring opens and the target Rydberg orbital rapidly changes character to become a CO σ^* antibonding orbital (Fig. 14.5.) Actual calculations were run on a swarm of 30 trajectories, confirming the mecha-nism previously proposed Gomer–Noyes mechanism (Gomer and Noyes 1950) (Fig. 14.2), but also confirming other experimental by-products and giving unprecedented

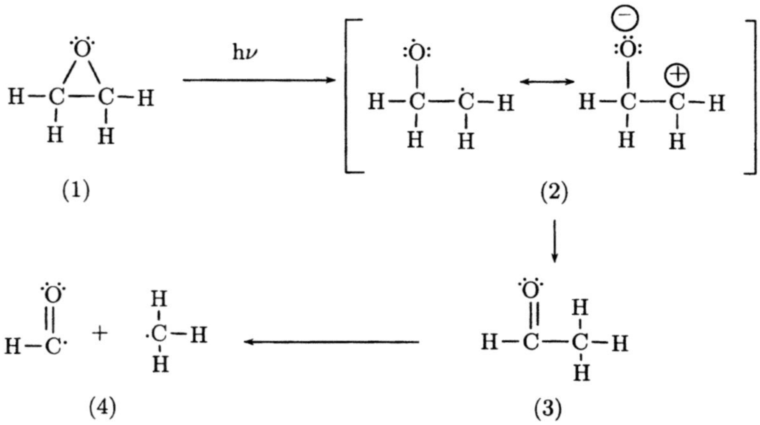

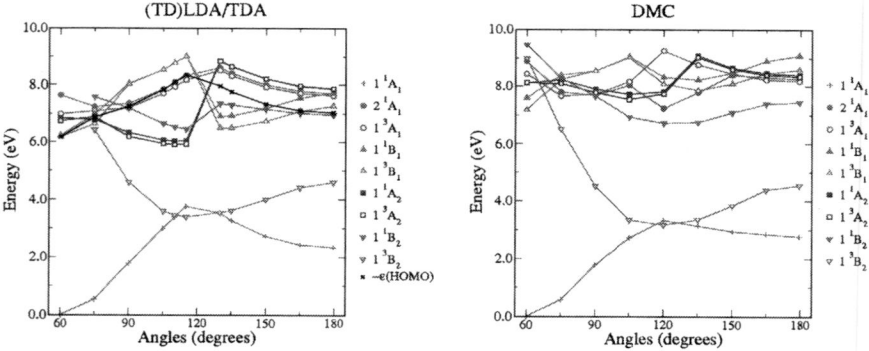

Fig. 14.2 Mechanism proposed by Gomer and Noyes in 1950 for the photochemical ring opening of oxirane. Reprinted with permission from (Tapavicza et al. 2008). Copyright 2008, American Institute of Physics

Fig. 14.3 Comparison of TDA TDLDA and diffusion Monte Carlo curves for C_{2v} ring opening of oxirane. Reprinted with permission from Cordova et al. (2007). Copyright 2007, American Institute of Physics

state-specific reaction details such as the orbital description briefly described above.

The oxirane photochemical ring-opening passes through a conical intersection, providing a concrete example of a conical intersection to study with TDDFT. We now return to the study by Levine, Ko, Quenneville, and Martinez of conical intersections using conventional TDDFT Levine et al. (2006) who noted that strict conical intersections are forbidden by the TDDFT adiabatic approximation for the simple reason that there is no coupling matrix element (Eq. 14.23) to zero out between the ground and excited states. Figure 14.6 shows a CASSCF conical intersection close to the oxirane photochemical funnel. Also shown are the TDA TDDFT surfaces

Fig. 14.4 **a** Cut of potential energy surfaces along reaction path of a Landau–Zener (*dashed line*) and a fewest-switches (*solid line*) trajectory (*black*, S_0; *blue*, S_1; *green*, S_2; *magenta*, S_3.) Both trajectories were started by excitation into the $^1(n, 3p_z)$ state, with the same geometry and same initial nuclear velocities. The running states of the Landau–Zener and the fewest-switches trajectory are indicated by the *red crosses* and *circles*, respectively. The geometries of the Landau–Zener trajectory are shown at time *a* 0, *b* 10, and *c* 30 fs. **b** State populations (*black*, S_0; *blue*, S_1; *green*, S_2; *magenta*, S_3) as a function of the fewest-switches trajectory in (**a**). Reprinted with permission from Tapavicza et al. (2008). Copyright 2008, American Institute of Physics

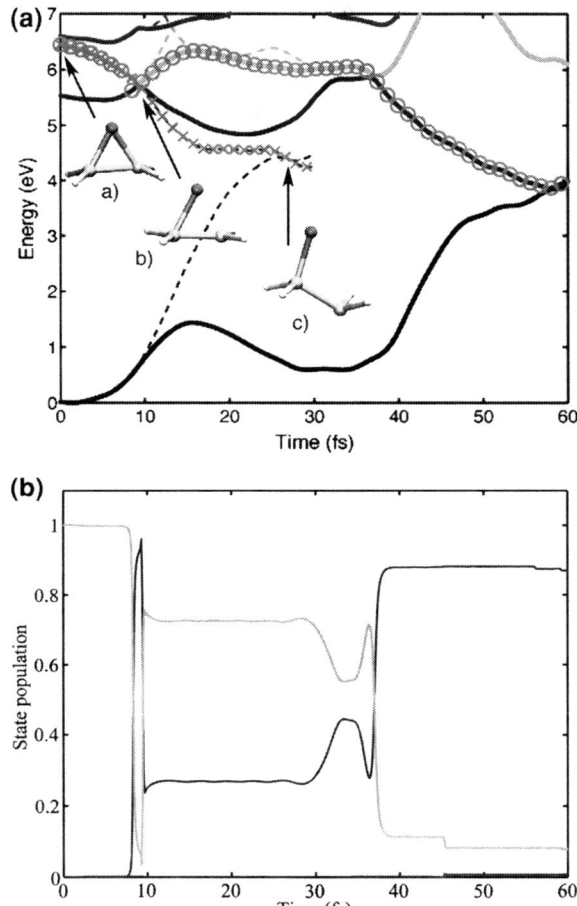

calculated with the same CASSCF branching coordinates. Interestingly the CASSCF and TDDFT conical intersections look remarkably similar. However closer examination shows that the TDDFT "conical intersection" is actually two *intersecting* cones rather than a true conical intersection, confirming the observation of Levine *et al.* This was analyzed in detail in Tapavicza et al. (2008) where it was concluded that the problem is that we are encountering effective noninteracting v-representability. True noninteracting v-representability means that there is no noninteracting system whose ground state gives the ground state density of the interacting system. This only means that there is some excited state of the noninteracting system with integer occupation number which gives the ground state density of the interacting system. What we call effective noninteracting v-representability is when the LUMO falls below the HOMO (or, in the language of solid-state physics, there is a "hole below the Fermi level".) This is exactly what frequently happens in the funnel region.

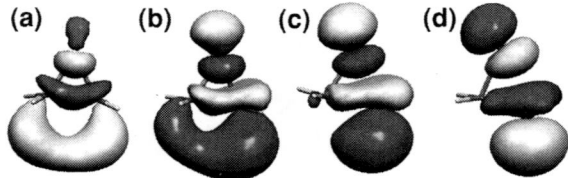

Fig. 14.5 Change of character of the active state along the reactive Landau–Zener trajectory, shown in Fig. 14.1. Snapshots were taken at times **a** 2.6, **b** 7.4, **c** 12.2, and **d** 19.4 fs. For **a** and **b**, the running state is characterized by a transition from the highest occupied molecular orbital (HOMO) to the lowest unoccupied molecular orbital (LUMO) plus one (LUMO + 1), while for **c** and **d** it is characterized by a HOMO–LUMO transition due to orbital crossing. Note that the HOMO remains the same oxygen nonbonding orbital throughout the simulation. Reprinted with permission from Tapavicza et al. (2008). Copyright 2008, American Institute of Physics

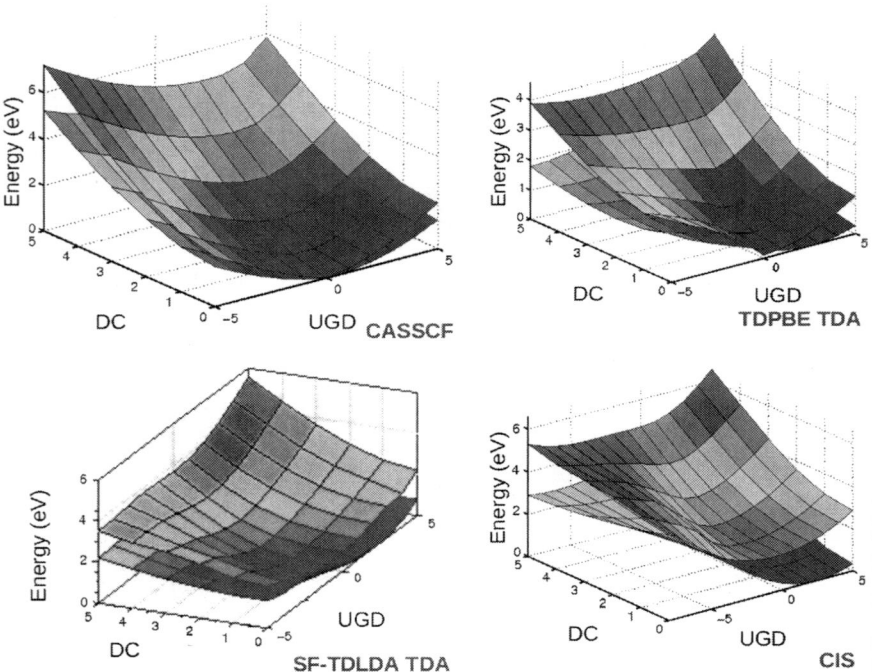

Fig. 14.6 Comparison of the S_0 and S_1 potential energy surfaces calculated using different methods for the CASSCF branching coordinate space. Reproduced from Huix-Rotllant et al. (2010) by permission of the PCCP Owner Societies

Spin-flip (SF) TDDFT (Slipchenko and Krylov 2003; Shao et al. 2003; Wang and Ziegler 2004) offers one way to circumvent some of the problems of effective noninteracting v-representability in funnel regions. This is because we can start from the lowest triplet state which has fewer effective noninteracting v-representability

Fig. 14.7 C_{2v} potential energy curves: full calculation (*solid lines*), two-orbital model (*dashed lines*.) Reproduced from Huix-Rotllant et al. (2010) by permission of the PCCP Owner Societies

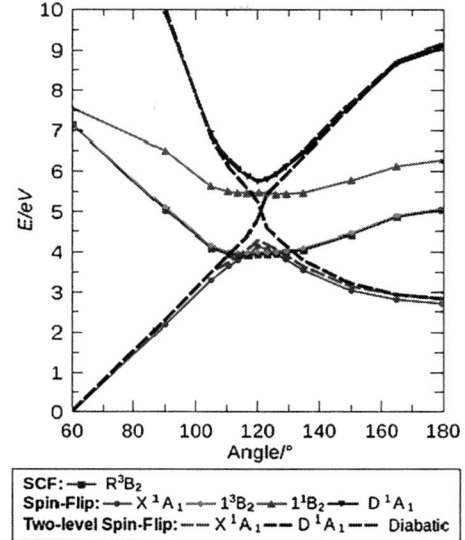

problems and then use SFs to obtain both the ground state and a doubly-excited state (see Chap. 8). Analytic derivatives are now available for some types of SF-TDDFT (Seth et al. 2010). Figure 14.7 shows that SF-TDDFT works fairly well for treating the avoided crossing in the C_{2v} ring-opening pathway of oxirane. Minezawa and Gordon also used SF-TDDFT to locate a conical intersection in ethylene (Minezawa and Gordon 2009). However Huix-Rotllant, Natarajan, Ipatov, Wawire, Deutsch, and Casida found that, although SF-TDDFT does give a true conical intersection in the photochemical ring opening of oxirane, the funnel is significantly shifted from the position of the CASSCF conical intersection (Huix-Rotllant 2010). The reason is that the key funnel region involves an active space of over two orbitals which is too large to be described accurately by SF-TDDFT.

There are other ways to try to build two- and higher-excitation character into a DFT treatment of excited states. Let us mention here only multireference configuration interaction (MRCI)/DFT (Grimme and Waletzke 1999), constrained density functional theory-configuration interaction (CDFT-CI) (Wu et al. 2007), and mixed TDDFT/many-body theory methods based upon the Bethe-Salpeter equation (Romaniello 2009) or the related polarization propagator approach (Casida 2005; see also: Huix-Rotllant, M. and M. E. Casida, "Formal Foundations of Dressed Time-Dependent Density-Functional Theory for Many-Electron Excitations", Condensed Matter ArXiv, arxiv.org/abs/1008.1478) or the simpler dressed TDDFT approach (Maitra et al. 2004; Cave et al. 2004; Gritsenko and Baerends 2009, Mazur and Włodarczyk 2009, Mazur 2010, Huix-Rotllant et al. 2011) .

All of these may have the potential to improve the DFT-based description of funnel regions in photochemical reactions. Here however we must be aware that we may be in the process of building a theory which is less automatic and requires the high amount of user intervention typical of present day CASSCF calculations. This is certainly the case with CDFT-CI which has already achieved some success in describing conical intersections (Kaduk and Van Voorhis 2010).

14.4 Perspectives

Perhaps the essence of dynamics can be captured in a simple sentence: "You should know from whence you are coming and to where you are going." Of course this rather deterministic statement must be interpreted differently in classical and quantum mechanics. Here however we would like to think about its meaning in terms of the development of DFT for applications in photoprocesses. Theoretical developments in this area have been remarkable in recent years, opening up the possibility for a more detailed understanding of femtosecond (and now also attosecond) spectroscopy. In this chapter we have tried to discuss the past, the present, and a bit of the future.

The past treated here has been the vast area of static investigation and dynamic simulations of photophysical and photochemical processes. We have first described more traditional wave-function techniques. We have also mentioned and made appropriate references to important work on early DFT work involving Ehrenfest TDDFT and restricted open-shell Kohn–Sham DFT dynamics. Our emphasis has been on photochemical processes involving several potential energy surfaces, partly because of our own personal experiences, but also because photo*chemical* processes start out as photo*physical* processes in the Franck–Condon region and then rapidly become more complicated to handle.

The present-day status of DFT photodynamics is perhaps best represented by the recent availability of mixed TDDFT and TDDFTB/classical surface-hopping dynamics codes as well as serious efforts to investigate and improve the quality of the TDDFT description of photochemical funnel regions. First applications have already shown the utility of this theory and we feel sure that other applications will follow as programs are made broadly available to computational scientists. Finally we have ended the last section with some speculations about the future concerning the need for explicit double- and higher-excitations to correctly describe funnel regions.

As expected, we could not treat everything of relevance to the chapter title. Roi Baer's recent work indicating that Berry phase information is somehow included in the ground-state charge density is most intriguing (Baer 2010a). Also on-going work on multicomponent DFT capable of treating electrons and nuclei on more or less the same footing (Kreibich and Gross 2001a, Kreibich et al. 2008) would seem to open up new possibililties for developing useful non-Born–Oppenheimer approximations within a DFT framework. We are sure that still other potentially relevant work has been unfortunately omitted either because of space limitations or for other reasons.

Do we know where this field is going? Certainly non-Born–Oppenheimer photo-dynamics using some form of DFT is currently a hot and rapidly evolving area. Exactly what lies in store may not yet be clear, but what we do know is that we are going to have fun getting there!

Chapter 15
On the Combination of TDDFT with Molecular Dynamics: New Developments

José L. Alonso, Alberto Castro, Pablo Echenique and Angel Rubio

15.1 Introduction

In principle, we should not need the time-dependent extension of density-functional theory (TDDFT) for excitations, and in particular for most equilibrium and non-equilibrium molecular dynamics (MD) studies of closed systems: the theorem by

J. L. Alonso · P. Echenique
Departamento de Física Teórica, Universidad de Zaragoza,
Pedro Cerbuna 12, 50009 Zaragoza, Spain
e-mail: alonso.buj@gmail.com

J. L. Alonso · P. Echenique · A. Castro
Instituto de Biocomputación y Física de Sistemas Complejos, Universidad of Zaragoza,
Mariano Esquillor s/n, Edificio I+D, 50018 Zaragoza, Spain
e-mail: acastro@bifi.es

P. Echenique (✉)
Instituto de Química Física "Rocasolano", CSIC,
Serrano 119, 28006 Madrid, Spain
e-mail: echenique.p@gmail.com

A. Rubio
Nano-Bio Spectroscopy Group and ETSF Scientific Development Centre,
Departamento de Física de Materiales, Universidad del País Vasco
Centro de Física de Materiales CSIC-UPV/EHU-MPC and DIPC,
Avenida Tolosa 72, 20018 San Sebastián, Spain
e-mail: angel.rubio@ehu.es

A. Rubio
Fritz-Haber-Institut der Max-Planck-Gesellschaft,
Faradayweg 4–6, 14195 Berlin-Dahlem, Germany

P. Echenique
Unidad Asociada IQFR-BIFI,
Zaragoza, Spain

M. A. L. Marques et al. (eds.), *Fundamentals of Time-Dependent Density Functional Theory*, Lecture Notes in Physics 837, DOI: 10.1007/978-3-642-23518-4_15,
© Springer-Verlag Berlin Heidelberg 2012

Hohenberg and Kohn (1964) teaches us that for any observable that we wish to look at (including observables dependent on excited states) there is a corresponding functional of the ground-state density. Yet the unavailability of such magic functionals in many cases (the theorem is a non-constructive existence result) demands the development and use of the alternative exact reformulation of quantum mechanics provided by TDDFT. This theory defines a convenient route to electronic excitations and to the dynamics of a many-electron system subject to an arbitrary time-dependent perturbation (discussed in previous chapters of this book, e.g., in Chap. 4). This is, in fact, the main purpose of inscribing TDDFT in a MD framework—the inclusion of the effect of electronic excited states in the dynamics. However, as we will show in this chapter, it may not be the only use of TDDFT in this context.

The term "ab initio molecular dynamics" (AIMD) has been exclusively identified in the past with the Car–Parrinello (CP) technique (Car and Parrinello 1985). This method combines ground-state DFT with MD, providing an efficient reformulation of ground-state Born-Oppenheimer MD (gsBOMD) (Marx and Hutter 2000). However, the "AIMD" words have broader meaning, and should include all the possible MD techniques that make use of a first principles approach to tackle the many-electron problem. For example, Ehrenfest MD can also be one AIMD scheme if TDDFT is used to propagate the electronic subsystem. This is the most common manner in which TDDFT and MD have been combined in the past: as a means to study fast out-of-equilibrium processes, typically intense laser irradiations or ionic collisions (Saalmann and Schmidt 1996, 1998; Reinhard and Suraud 1999; Kunert and Schmidt 2001; Castro et al. 2004b).

Nevertheless, there are other possibilities. In this chapter we review two recent proposals: In Sect. 15.2, we show how TDDFT can be used to design efficient gsBOMD algorithms (Alonso et al. 2008; Andrade et al. 2009), even if the electronic excited states are not relevant in this case. The work described in Sect. 15.3 addresses the problem of mixed quantum-classical systems at thermal equilibrium (Alonso et al. 2010).

Atomic units are used throughout the document in order to get rid of constant factors such as $\hbar$ or $1/4\pi\varepsilon_0$.

15.2 Fast Ehrenfest Molecular Dynamics

In order to derive the quantum-classical MD known as Ehrenfest molecular dynamics (EMD) from the time-dependent Schrödinger equation for a molecular system, one starts with a separation ansatz for the wave function of the molecular system between the electrons and the nuclei (Gerber et al. 1982), leading to the so-called time-dependent self-consistent-field (TDSCF) equations (Tully 1998b; Bornemann et al. 1996). The next step is to approximate the nuclei as classical point particles via short wave asymptotics or Wentzel–Kramers–Brillouin (WKB) approximation (Tully 1998b; Bornemann et al. 1996; Wentzel 1926). The resultant Ehrenfest MD scheme is contained in the following system of coupled differential equations (Bornemann et al. 1996):

$$M_\alpha \frac{d^2}{dt^2} \mathbf{R}_\alpha(t) = -\langle \Phi(t) | \nabla_{\mathbf{R}_\alpha} \hat{H}_e(\bar{\mathbf{R}}(t), t) | \Phi(t) \rangle, \tag{15.1}$$

$$i \frac{d}{dt} | \Phi(t) \rangle = \hat{H}_e(\bar{\mathbf{R}}(t), t) | \Phi(t) \rangle, \tag{15.2}$$

where $\Phi(t)$ is the state of the quantum subsystem (we will assume that this is a set of N electrons), and $\{ \mathbf{R}_\alpha \}_{\alpha=1}^M$ are the position coordinates of M classical particles (a set of M nuclei of masses M_α and charges Z_α). The quantum (or *electronic*) Hamiltonian operator $\hat{H}_e(\bar{\mathbf{R}}, t)$ depends on these classical coordinates, and is usually given by:

$$\hat{H}_e(\bar{\mathbf{R}}, t) = \sum_{i=1}^N \frac{1}{2} \hat{p}_i^2 + \sum_{i,j<i} \frac{1}{|\hat{r}_i - \hat{r}_j|} + \sum_{\beta<\alpha} \frac{Z_\alpha Z_\beta}{|\mathbf{R}_\alpha - \mathbf{R}_\beta|}$$
$$- \sum_{\alpha,i} \frac{Z_\alpha}{|\mathbf{R}_\alpha - \mathbf{r}_i|} + \sum_i v_{ext}^e(\mathbf{r}_i, t) + \sum_\alpha v_{ext}^n(\mathbf{R}_\alpha, t), \tag{15.3}$$

where v_{ext}^e and v_{ext}^n are external potentials acting on the electrons and nuclei, respectively (Echenique and Alonso 2007) (see also Chap. 14 for more about EMD).

Given this definition, one can show that Eq. 15.1 can be rewritten as:

$$M_\alpha \frac{d^2}{dt^2} \mathbf{R}_\alpha(t) = - \int d^3 r n(\mathbf{r}, t) \nabla_{\mathbf{R}_\alpha} v_0(\mathbf{r}, \bar{\mathbf{R}}(t)), \tag{15.4}$$

where

$$v_0(\mathbf{r}, \bar{\mathbf{R}}) = - \sum_\alpha \frac{Z_\alpha}{|\mathbf{R}_\alpha - \mathbf{r}|} + \frac{1}{N} \sum_\alpha v_{ext}^n(\mathbf{R}_\alpha, t) + \frac{1}{N} \sum_{\beta<\alpha} \frac{Z_\alpha Z_\beta}{|\mathbf{R}_\alpha - \mathbf{R}_\beta|}; \tag{15.5}$$

a result which is known as the "electrostatic force theorem" in the quantum chemistry literature (Levine 2000), and which is based in the fact that the gradient $\nabla_{\mathbf{R}_\alpha} \hat{H}_e(\bar{\mathbf{R}}(t), t)$ is a one-body local multiplicative operator (as far as the electrons are concerned), i.e., it is a sum of one-electron operators whose action amounts to a multiplication in real space (Eschrig 2003; Von Barth 2004).

Equation 15.4 shows that the knowledge of the time-dependent electronic density $n(\mathbf{r}, t)$ suffices to obtain the nuclear movement. This fact is the basis for TDDFT-based Ehrenfest MD (Ehrenfest-TDDFT): instead of solving Eq. 15.2, we solve the corresponding time-dependent Kohn–Sham system, which provides $n(\mathbf{r}, t)$ (see also Chap 4):

$$i \frac{\partial}{\partial t} \varphi_j(\mathbf{r}, t) = -\frac{1}{2} \nabla^2 \varphi_j(\mathbf{r}, t) + v_{KS}[n](\mathbf{r}, t) \varphi_j(\mathbf{r}, t), \quad j = 1, \ldots, N, \tag{15.6}$$

being

$$v_{KS}[n](\mathbf{r}, t) = \sum_\alpha \frac{-Z_\alpha}{|\mathbf{R}_\alpha(t) - \mathbf{r}|} + v_H[n](\mathbf{r}, t) + v_{xc}[n](\mathbf{r}, t) + v_{ext}^e(\mathbf{r}, t). \tag{15.7}$$

and

$$n(\boldsymbol{r}, t) = 2 \sum_{j=1}^{N} |\varphi_j(\boldsymbol{r}, t)|^2, \qquad (15.8)$$

where $v_{KS}[n](\boldsymbol{r}, t)$ is the time-dependent Kohn–Sham potential, and $v_H[n](\boldsymbol{r}, t)$ and $v_{xc}[n](\boldsymbol{r}, t)$ are the Hartree and exchange-correlation potential, respectively. For simplicity, we assume an even number of electrons in a spin-compensated configuration, and omit initial-state dependence in the functionals.

The equations of motion for Ehrenfest-TDDFT (15.4, 15.6 and 15.8) can be derived from the following Lagrangian (assuming an adiabatic approximation for the exchange and correlation potential, as it is commonly done in practical implementations of TDDFT):

$$L_\mu[\varphi, \dot{\varphi}, \bar{\boldsymbol{R}}, \dot{\bar{\boldsymbol{R}}}] = \mu \frac{i}{2} \sum_j \left(\langle \varphi_j | \dot{\varphi}_j \rangle - \langle \dot{\varphi}_j | \varphi_j \rangle \right) + \sum_\alpha \frac{1}{2} M_\alpha \dot{\bar{\boldsymbol{R}}}_\alpha^2 - E_{KS}[\varphi, \bar{\boldsymbol{R}}],$$

$$(15.9)$$

for $\mu = 1$ (the reason for including this parameter μ will become clear in what follows). We use a dot to denote time-derivatives.

The term E_{KS} is the Kohn–Sham ground-state energy functional:

$$E_{KS}[\varphi, \bar{\boldsymbol{R}}] = 2 \sum_j \left\langle \varphi_j \left| \frac{\hat{p}^2}{2} \right| \varphi_j \right\rangle - \int d^3r \sum_\alpha \frac{Z_\alpha}{|\boldsymbol{R}_\alpha - \boldsymbol{r}|} n(\boldsymbol{r})$$

$$+ \frac{1}{2} \int d^3r v_H[n](\boldsymbol{r}) n(\boldsymbol{r}) + E_{xc}[n] + \sum_{\beta < \alpha} \frac{Z_\alpha Z_\beta}{|\boldsymbol{R}_\alpha - \boldsymbol{R}_\beta|}. \qquad (15.10)$$

Note that, when the time-dependent orbitals are introduced into this expression (as it is done in Ehrenfest-TDDFT), it becomes a functional of the Kohn–Sham orbitals at each time, and only an implicit functional of the time-dependent density. Also, from here on, we assume that there are no external potentials v_{ext}^e and v_{ext}^n, since they do not add anything to the following discussion.

It is worth remarking now that Ehrenfest MD differs from gsBOMD, and it is instructive to see in which way. We do so in the initial formulation of Eqs. 15.1 and 15.2 using the N-electron wavefunction for simplicity, i.e., we forget for a moment the TDDFT formalism.

To illustrate the main concepts we start by projecting the Ehrenfest MD equations into the adiabatic basis, formed at each nuclear configuration by the set of eigenfunctions of the electronic Hamiltonian:

$$\hat{H}_e(\bar{\boldsymbol{R}})|\Psi_m(\bar{\boldsymbol{R}})\rangle = E_m(\bar{\boldsymbol{R}})|\Psi_m(\bar{\boldsymbol{R}})\rangle, \qquad (15.11a)$$

$$|\Phi(t)\rangle = \sum_m c_m(t)|\Psi_m(\bar{\boldsymbol{R}}(t))\rangle. \qquad (15.11b)$$

The result is:

$$M_\alpha \frac{d^2}{dt^2} R_\alpha(t) = -\sum_m |c_m(t)|^2 \nabla_{R_\alpha} E_m(\bar{R}(t))$$

$$-\sum_{mn} c_m^*(t) c_n(t) \left[E_m(\bar{R}(t)) - E_n(\bar{R}(t)) \right] d_\alpha^{mn}(\bar{R}(t)) \quad (15.12)$$

and

$$i \frac{d}{dt} c_m(t) = E_m(\bar{R}(t)) c_m(t) - i \sum_n c_n(t) \left[\sum_\alpha \dot{R}_\alpha \cdot d_\alpha^{mn}(\bar{R}(t)) \right] \quad (15.13)$$

where the "non-adiabatic couplings" are defined as:

$$d_\alpha^{mn}(\bar{R}) = \langle \Psi_m(\bar{R}) | \nabla_{R_\alpha} \Psi_n(\bar{R}) \rangle. \quad (15.14)$$

If the nuclear velocities are small (compared to the electronic one), the last term in Eq. 15.13 can be neglected, and if we assume that the electronic system starts from the ground state ($c_m(0) = \delta_{m0}$), EMD reduces to gsBOMD:

$$M_\alpha \frac{d^2}{dt^2} R_\alpha(t) = \nabla_{R_\alpha} E_0(\bar{R}(t)), \quad (15.15a)$$

$$c_m(t) = \delta_{m0}. \quad (15.15b)$$

Now, in order to integrate the gsBOMD equations, one can make use of ground-state DFT, since the only necessary ingredient is the ground-state energy $E_0(\bar{R}(t))$. One could thus precompute this hyper-surface, in order to propagate the nuclear dynamics *a posteriori*, or else only compute the energies at the $\bar{R}$ points visited by the dynamics (a procedure normally known as "on-the-fly"). However, Car and Parrinello (1985) proposed an alternative, based on the following Lagrangian:

$$L_\lambda^{CP}[\varphi, \dot{\varphi}, \bar{R}, \dot{\bar{R}}] = \lambda \frac{i}{2} \sum_j \langle \dot{\varphi}_j | \dot{\varphi}_j \rangle + \sum_\alpha \frac{1}{2} M_\alpha \dot{R}_\alpha^2$$

$$- E_{KS}[\varphi, \bar{R}] + \sum_{ij} \Lambda_{ij} \left(\langle \varphi_i | \varphi_j \rangle - \delta_{ij} \right). \quad (15.16)$$

Note the presence of a fictional mass λ, and of a set of Lagrange multipliers Λ_{ij} associated to the constraints that keep the KS orbitals orthonormal along the evolution. The Car–Parrinello (CP) equations that stem from this Lagrangian are, for the nuclei,

$$M_\alpha \frac{d^2}{dt^2} R_\alpha(t) = -\nabla_{R_\alpha} E_{KS}[\varphi(t), \bar{R}(t))], \quad (15.17)$$

while the electronic wavefunctions follow

$$\lambda \ddot{\varphi}_j(\boldsymbol{r}, t) = -\frac{1}{2} \nabla^2 \varphi_j(\boldsymbol{r}, t) + v_{\text{KS}}[n](\boldsymbol{r}, t) \varphi_j(\boldsymbol{r}, t) + \sum_k \Lambda_{jk} \varphi_k(\boldsymbol{r}, t), \quad (15.18)$$

and finally, the orthogonality conditions

$$\langle \varphi_i(t) | \varphi_j(t) \rangle = \delta_{ij}. \tag{15.19}$$

The first of these three sets of equations ensures that CP molecular dynamics (CPMD) is (approximately) equivalent to gsBOMD if the KS orbitals stay close to the ground-state ones; the second equation is an auxiliary, *fictional* electronic propagation that enforces this proximity to the ground state for a certain range of values of the "mass" λ; whereas the last equation demands the constant orthonormality of the electronic orbitals. Another role of the fictional mass λ; is to accelerate the fake electronic dynamics, and as a consequence to improve the numerical efficiency. This efficiency (in addition to the success of DFT in the calculation of total energies with chemical accuracy) has made CPMD the method of choice for performing ab initio gsBOMD during the last decades.

When attempting simulations of very large systems, the calculations must be done using the massive parallel architectures presently available, therefore one must ensure a good scalability of the computational algorithms with respect to the number of processors and the size of the systems (i.e., number of atoms). The CPMD technique at a given point has to face the problem posed by the need of orthonormalization, as required by Eq. 15.19. This is a very non-local process (regardless of the algorithm used), and therefore very difficult to parallelize efficiently. Linear-scaling methods and other approaches have been proposed recently (Kühne et al. 2007) to improve the speed of the CP technique.

One possibility to circumvent the orthonormalization issue is to do Ehrenfest-TDDFT (which automatically conserves the orthonormality) instead of CPMD, for those cases in which the coupling to higher electronic excited states is weak, and therefore Ehrenfest-TDDFT is almost equivalent to gsBOMD. This fact was first realized by Theilhaber (1992). Unfortunately, the required time step for Ehrenfest-TDDFT is very small (two to three orders of magnitude smaller than the CPMD time-step), which makes it very inefficient computationally. The reason is that the simulation must follow the real electronic motion, which is very fast (in contrast to the fictional electronic motion used in CPMD). In (Alonso et al. 2008) and (Andrade et al. 2009), however, it was shown how the time-step can be increased by modifying the μ parameter in the definition of the Lagrangian function given in Eq. 15.9, which for normal Ehrenfest dynamics should be $\mu = 1$.

For any μ, the equations of motion derived from this Lagrangian function are:

$$i\mu \frac{\partial}{\partial t} \varphi_j(\boldsymbol{r}, t) = -\frac{1}{2} \nabla^2 \varphi_j(\boldsymbol{r}, t) + v_{\text{KS}}[n](\boldsymbol{r}, t) \varphi_j(\boldsymbol{r}, t), \tag{15.20a}$$

$$M_\alpha \frac{d^2}{dt^2} \boldsymbol{R}_\alpha(t) = -\nabla_{\boldsymbol{R}_\alpha} E_{\text{KS}}[\varphi(t), \bar{\boldsymbol{R}}(t)]. \tag{15.20b}$$

The only difference with respect to Ehrenfest-TDDFT is the appearance of the μ parameter multiplying the time-derivative of the time-dependent KS equations. The most relevant features of this dynamics are:

1. The orthogonality of the time-dependent KS orbitals is automatically preserved along the evolution, so that there is no need to perform any orthonormalization procedure.
2. The "exact" total energy of the system, defined as

$$E = \frac{1}{2} \sum_\alpha M_\alpha \dot{\boldsymbol{R}}_\alpha^2 + E_{KS}[\varphi, \bar{\boldsymbol{R}}], \tag{15.21}$$

is also preserved along the evolution. Note that it is independent of μ and it coincides with the same exact energy that is preserved along the gsBOMD evolution. In contrast, the preserved energy in CPMD is given by:

$$E_{CP} = E + \frac{1}{2}\lambda \sum_j \langle \dot{\varphi}_j | \dot{\varphi}_j \rangle, \tag{15.22}$$

where E, given by Eq. 15.21, is now time-dependent. It can be seen how the new constant of motion E_{CP} actually depends on λ, which is the fictional electronic mass introduced in the CP formulation.
3. To illustrate the effect of μ in the coupled Eqs. 15.20a, b, let us use in (15.20a) [but not in (15.20b), this is not a change of variables] $t = \mu t_e$,

$$i\mu \frac{\partial \varphi_j}{\partial t} = i \frac{\partial \varphi_j}{\partial t_e}. \tag{15.23}$$

The 'new' Eq. 15.20a can be seen as a standard Ehrenfest-TDDFT ($\mu = 1$) propagation in time t_e. Consequently, in the 'old' Eq. 15.20a, the maximum Δt is $\Delta t = \mu \Delta t_e$, where Δt_e is the time-step needed in Ehrenfest-TDDFT ($\mu = 1$). Therefore, our method is μ times faster than standard Ehrenfest-TDDFT.
4. This dynamics has also the effect of scaling the TDDFT excitation energies by a factor $1/\mu$. Hence, we may open or close the electronic gap by using a smaller or larger than one value of μ. Obviously, if $\mu \to 0$, then the gap becomes infinite and we retrieve the adiabatic (gsBOMD) regime.
5. Taking into account the two previous points and recalling that the purpose of this modified Ehrenfest dynamics is to reproduce, albeit approximately, the gsBOMD results, it becomes clear that there is a tradeoff affecting the optimal choice for the value of μ: low values (but still larger than one) will give physical accuracy, while large values will produce a faster propagation. The optimal value is the maximum value that still keeps the system near the adiabatic regime. It is reasonable to expect that this value will be given by the ratio between the electronic gap and the highest vibrational frequency of the nuclei. For many systems, like some molecules or insulators, this ratio is large and we can expect large improvements with respect to standard Ehrenfest MD. For other systems, like metals, this ratio is small or zero and the new method will not work.

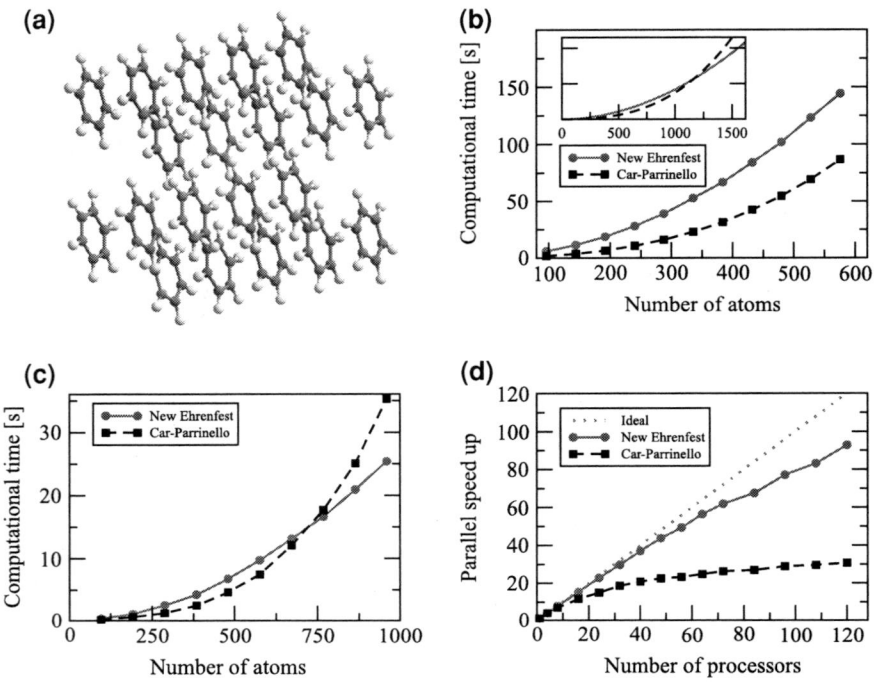

Fig. 15.1 a Scheme of the benzene molecule array. **b** Single processor computational cost for different system sizes. (*Inset*) Polynomial extrapolation for larger systems. **c** Parallel computational cost for different system sizes. **d** Parallel scaling with respect to the number of processors for a system of 480 atoms. In both cases, a mixed states-domain parallelization is used to maximize the performance. This figure has been taken from Andrade et al. (2009), with permission by ACS

6. Regarding the scaling with the system size, the modified Ehrenfest dynamics evidently inherits the main advantage of the original one: since propagation preserves the orthonormality of the KS orbitals, it needs not be imposed and the numerical cost is proportional to $N_W N_C$ (with N_W the number of orbitals and N_C the number of grid points or basis set coefficients). For CPMD, a reorthogonalization has to be done each time step, so the cost is proportional to $N_W^2 N_C$. From these scaling properties, we can predict that for large enough systems the Ehrenfest method will be less costly than CP. For smaller systems, however, this gain will not compensate for the fact that the time-step, despite being increased by the μ factor, will still need to be one or two orders of magnitude smaller than the time-step utilized in CPMD.

Some numerical examples, performed with the octopus code (Marques et al. 2003; Castro et al. 2006), that give an idea of the performance of this modified Ehrenfest dynamics were shown in (Alonso et al. 2008) and (Andrade et al. 2009). We reproduce here one case: the vibrational spectrum of an artificial benzene crystal (see

Fig. 15.1). Essentially, the calculations consist of the time-propagation of the system, either with the standard CP technique or with the modified Ehrenfest dynamics, for an interval of time departing from a Boltzmann distribution of velocities at a given temperature. Then, the vibrational frequencies are obtained from the Fourier transform of the velocity autocorrelation function.

Panels (b) and (c) display the serial and parallel computational cost, respectively, of the two methods, defined as the computer time needed to propagate one atomic unit of time. In the serial case, it can be seen how, for the system sizes studied, CP is more efficient; a different scaling can already be guessed from the curve; indeed, if these curves are extrapolated (inset), one can predict a crossing point where the new Ehrenfest technique starts to be advantageous. This is more patent in the parallel case, as can be seen in panel (c). Panel (d) displays the different scalability of the two methods: for a fixed system size, the system is equally divided in a variable number of processors, and the figure displays the different speed-ups obtained.

The key conclusion is that the lack of the orthonormalization step permits a new efficient parallelization layer, on top of the usual ones that are commonly employed in CPMD (domain decomposition, and Brillouin zone k-points): since the propagation step is independent for each orbital, it is natural to parallelize the problem by distributing the KS states among processors. Communication is only required once per time-step to calculate quantities that depend on a sum over states: the time dependent densities and the forces over the ions.

15.3 MD at Finite Electronic Temperature

The previous section has addressed algorithmic alternatives to the solution of the gsBOMD Eqs. 15.15a, b. These represent the evolution of the classical nuclei, interacting all-to-all through the potential $E_0(\bar{R})$. The resulting dynamics can be used to calculate equilibrium averages at a given finite temperature, by assuming ergodicity and computing time averages over a number of trajectories, once the system has been appropriately coupled to a thermostat. The resulting marginal equilibrium density in the nuclear positions space is, in the canonical ensemble:

$$p_{\text{gsBO}}(\bar{R}) = \frac{e^{-\beta E_0(\bar{R})}}{\int d^3 R'_1 \cdots \int d^3 R'_M e^{-\beta E_0(\bar{R}')}}, \tag{15.24}$$

with $\beta = 1/k_B T$, or $\beta = 1/RT$ if per-mole units are used.

However, this scheme ignores completely the dynamics of the electrons, by assuming that, even at a finite temperature, they are continuously tied to their ground state. This assumption is legitimate if the electronic gap is large compared to $k_B T$ at the temperature of interest. Indeed, in many physical, chemical or biological processes the dynamical effects arising from the presence of low lying electronic excited states have to be taken into account. For instance, in situations where the hydrogen bond is weak, different states come close to each other and non-adiabatic proton transfer transitions become rather likely at normal temperature (May and Kühne 2004). In these circumstances, the computation of ensemble averages cannot be based on a model that assumes the nuclei moving on the ground-state BO surface.

In the DFT realm, the inclusion of electronic excited states in the dynamics is very often done by working with partial occupation numbers to account for the electronic excitations (Grumbach et al. 1994; Alavi et al. 1994, 1995; Marzari et al. 1997), ideally making use of temperature-dependent exchange and correlation functionals (Mermin 1965; Prodan 2010; Eschrig 2010). This scheme is however tied to DFT, and is hindered by the difficulty of realistically approximating this functional. Other alternative options are Ehrenfest dynamics and surface hopping (Tully 1990) (for more on recent progress in non-adiabatic electronic dynamics in mixed quantum-classical dynamics, see, for example, (Zhu et al. 2005)).

Recently, Alonso et al. (2010) have proposed a new alternative, which can make use of the ability of TDDFT to compute electronic excited states. In the following, we make a summary of the new technique.

In order to arrive to a general quantum-classical formalism, and to a suitable expression for the quantum-classical equilibrium distribution that is considered to be the correct one in the literature, it is preferable in this case to follow the partial Wigner transformation route (Wigner 1932), as done by Kapral (1999) and Nielsen (2001). Let us assume a quantum system of two particles of masses m and M ($M > m$) living both in one dimension, whose canonical position and momentum operators are $(\hat{z}, \hat{p})$ and $(\hat{Z}, \hat{P})$, respectively. The generalization to more particles and higher dimension is straightforward. Given an operator $\hat{A}$, its partial Wigner transform $\hat{A}_W$ with respect to the large-mass coordinate is defined as:

$$\hat{A}_W(Z, P) = (2\pi)^{-1} \int dz' e^{iPz'} \langle Z - z'/2 | \hat{A} | Z + z'/2 \rangle. \tag{15.25}$$

The operator $\hat{A}_W(Z, P)$ acts on the Hilbert space of the *light* particle, and depends on the two real numbers (Z, P). It is possible to reformulate all quantum theory in terms of these partial Wigner transforms; in particular, if the Hamiltonian for the two particles is given by:

$$\hat{H} = \frac{\hat{P}^2}{2M} + \frac{\hat{p}^2}{2m} + v(\hat{z}, \hat{Z}), \tag{15.26}$$

its transformation is:

$$\hat{H}_W(Z, P) = \frac{P^2}{2M} + \frac{\hat{p}^2}{2m} + v(\hat{z}, Z), \tag{15.27}$$

i.e., one just has to substitute the quantum operators of the heavy particle by the real numbers (Z, P).

If the state of the system is described by the density matrix $\hat{\rho}(t)$, its evolution will be governed by von Neumann's equation,

$$\frac{d}{dt}\hat{\rho}(t) = -i\left[\hat{H}, \hat{\rho}(t)\right],$$ (15.28)

which can be cast into its partial Wigner-transformed form:

$$\frac{\partial}{\partial t}\hat{\rho}_W = -i\left(\hat{H}_W e^{\Lambda/2i}\hat{\rho}_W - \hat{\rho}_W e^{\Lambda/2i}\hat{H}_W\right),$$ (15.29)

where Λ is the "Poisson bracket operator",

$$\Lambda = \frac{\overleftarrow{\partial}}{\partial P}\frac{\overrightarrow{\partial}}{\partial Z} - \frac{\overleftarrow{\partial}}{\partial Z}\frac{\overrightarrow{\partial}}{\partial P},$$ (15.30)

and the arrows indicate the direction in which each derivative acts.

Note that up to now, this is an exact reformulation of quantum mechanics (no classical or semiclassical limit has been taken). However, this is also a convenient departure point to take the classical limit for the heavy particle. After an appropriate change of coordinates (Kapral 1999), if we retain only the first order terms in $\eta = (m/M)^{1/2}$, Eq. 15.29 is transformed into:

$$\frac{\partial}{\partial t}\hat{\rho}_W = -i\left[\hat{H}_W, \hat{\rho}_W\right] + \frac{1}{2}\left(\{\hat{H}_W, \hat{\rho}_W\} - \{\hat{\rho}_W, \hat{H}_W\}\right),$$ (15.31)

where $\{\cdot, \cdot\}$ is the Poisson bracket with respect to the canonical conjugate coordinates (Z, P),

$$\{\hat{H}_W, \hat{\rho}_W\} = \frac{\partial\hat{H}_W}{\partial Z}\frac{\partial\hat{\rho}_W}{\partial P} - \frac{\partial\hat{H}_W}{\partial P}\frac{\partial\hat{\rho}_W}{\partial Z},$$ (15.32)

and both $\hat{\rho}_W$ and $\hat{H}_W$ are functions of (Z, P).

The equilibrium density matrix in the partial Wigner representation at the classical limit for the heavy particle, denoted by $\hat{\rho}_W^{eq}$ should be stationary with respect to the evolution at first order in $\eta = (m/M)^{1/2}$ in Eq. 15.31. If we use this property and expand the equilibrium density matrix in powers of η:

$$\hat{\rho}_W^{eq}(Z, P) = \sum_{n=0}^{\infty} \eta^n \hat{\rho}_W^{eq\,(n)}(Z, P),$$ (15.33)

it can then be proved (Nielsen 2001) that the zero-th order term is given by:

$$\hat{\rho}_W^{eq\,(0)}(Z, P) = \frac{1}{\mathcal{Z}}e^{-\beta\hat{H}_W(Z,P)},$$ (15.34)

with

$$\mathcal{Z} = \text{Tr}_Q\left[\int dZ \int dP e^{-\beta\hat{H}_W(Z,P)}\right],$$ (15.35)

the symbol $\mathrm{Tr_Q}$ meaning trace over the quantum degrees of freedom.

Note that (15.34) corresponds, at fixed classical variables (Z, P), to the equilibrium density matrix *for the electronic states*. However, it is only an *approximation* to the true quantum-classical equilibrium density matrix, since it is not a stationary solution to the quantum-classical Liouvillian given in Eq. 15.31. This distribution is often regarded, however, as the correct equilibrium distribution of the canonical ensemble for a mixed quantum-classical system (Mauri et al. 1993; Parandekar and Tolly 2005; Parandekar and Tully 2006; Schmidt et al. 2008, Bastida et al. 2007), and the average of observables is computed as:

$$\langle \hat{O}(\hat{z}, \hat{p}, Z, P) \rangle = \mathrm{Tr_Q} \int \mathrm{d}Z \int \mathrm{d}P \, \hat{O}(\hat{z}, \hat{p}, Z, P) \hat{\rho}_{\mathrm{W}}^{\mathrm{eq}(0)}(Z, P). \tag{15.36}$$

As mentioned, the careful analysis described in (Kapral and Ciccotti 1999; Nielsen and Martin 1985), shows that this is a first order approximation in the square root of the quantum-classical mass ratio $\eta = (m/M)^{1/2}$, and therefore an acceptable approximation if this ratio is small.

In the remaining part of this chapter, and following (Alonso et al. 2010), we will write a system of dynamic equations for the classical particles such that the equilibrium distribution in the space of classical variables is in fact given by Eq. 15.34. This is also a goal of surface hopping methods (Tully 1990), although it is not fully achieved since these methods do not exactly yield this distribution (Schmidt et al. 2008). We will do this by deriving a temperature-dependent effective potential for the classical variables, which differs from the ground-state potential energy surface used in gsBOMD. It is straightforward, however, to write an equation that gives the expression for the effective potential in terms of this potential energy surface together with the BO surfaces corresponding to the excited states of the electronic Hamiltonian. Despite this property, it is worth remarking that the approach described here is based on the assumption that the full system of electrons and nuclei is in thermal equilibrium at a given temperature, and not on the assumption that electrons immediately follow the nuclear motion (i.e., the adiabatic approximation), which is at the core of the BO scheme.

Let us assume that we are only interested in the average of observables that depend explicitly only on the degrees of freedom of the heavy, classical particle, $A = A(Z, P)$. It is a matter of algebra [using Eqs. 15.34 and 15.36] to prove that this average can be written as:

$$\langle A(Z, P) \rangle = \frac{1}{\mathcal{Z}} \int \mathrm{d}Z \int \mathrm{d}P \, A(Z, P) \mathrm{e}^{-\beta H_{\mathrm{eff}}(Z, P; \beta)}, \tag{15.37}$$

where we have introduced an *effective* Hamiltonian H_{eff}, defined as:

$$H_{\mathrm{eff}}(Z, P; \beta) = -\frac{1}{\beta} \ln \mathrm{Tr_Q} \mathrm{e}^{-\beta H_{\mathrm{W}}(Z, P)}. \tag{15.38}$$

The partition function $\mathcal{Z}$ can also be written in terms of the effective Hamiltonian:

$$\mathcal{Z} = \int dZ \int dP e^{-\beta H_{\mathrm{eff}}(Z,P;\beta)}, \tag{15.39}$$

Hence, the quantum subsystem has been "integrated out", and does not appear explicitly in the equations any more (of course, it has not disappeared, being hidden in the definition of the effective Hamiltonian). In this way, the more complicated quantum-classical calculations have been reduced to a simpler classical dynamics with an appropriate effective Hamiltonian, which produces the same equilibrium averages of classical observables (Eq. 15.37) as the one we would obtain using Eq. 15.34 in 15.36, and hence incorporates the quantum back-reaction on the evolution of the classical variables, at least at the level of equilibrium properties.

In the case of a molecular system, the total (partially Wigner transformed) Hamiltonian reads:

$$\hat{H}(\bar{R}, \bar{P}) = T_{\mathrm{n}}(\bar{P}) + \hat{H}_{\mathrm{e}}(\bar{R}), \tag{15.40}$$

where $\bar{R}$ denotes collectively all nuclear coordinates, $\bar{P}$ all nuclear momenta, $T_n(\bar{P})$ is the total nuclear kinetic energy, and $\hat{H}_{\mathrm{e}}(\bar{R})$ is the electronic Hamiltonian in Eq. 15.3, that includes the electronic kinetic term and all the interactions. The effective Hamiltonian, defined in Eq. 15.38 in general, is in this case of a molecular system given by:

$$
\begin{aligned}
H_{\mathrm{eff}}(\bar{R}, \bar{P}; \beta) &= T_{\mathrm{n}}(\bar{P}) - \frac{1}{\beta} \ln \mathrm{Tr}_{\mathrm{Q}} e^{-\beta \hat{H}_{\mathrm{e}}(\bar{R})} \\
&= T_{\mathrm{n}}(\bar{P}) + v_{\mathrm{eff}}(\bar{R}; \beta),
\end{aligned} \tag{15.41}
$$

where the last equality is a definition for the *effective* potential $v_{\mathrm{eff}}(\bar{R}; \beta)$.

Now, making use of the adiabatic basis, defined in Eq. 15.11a as the set of all eigenvectors of electronic Hamiltonian $\hat{H}_{\mathrm{e}}(\bar{R})$, we can rewrite $v_{\mathrm{eff}}(\bar{R}; \beta)$ as:

$$v_{\mathrm{eff}}(\bar{R}; \beta) = E_0(\bar{R}) - \frac{1}{\beta} \ln \left[1 + \sum_{n>0} e^{-\beta E_{n0}(\bar{R})} \right], \tag{15.42}$$

where $E_{n0}(\bar{R}) = E_n(\bar{R}) - E_0(\bar{R})$. It is for the computation of these excitation energies that TDDFT can be employed. The proposed dynamics would be, therefore, *based* on TDDFT. Of course, any other many-electron technique can also be used.

This equation permits to see explicitly how the ground state energy E_0 differs from v_{eff}, and in consequence how a MD based on v_{eff} is going to differ from a gsBOMD. In particular, notice that $v_{\mathrm{eff}}(\bar{R}; \beta) \leq E_0(\bar{R})$, and compare the marginal probability density in the gsBOMD case in Eq. 15.24 to the one produced using the new dynamics:

$$p_{\mathrm{eff}}(\bar{R}) = \frac{\left[1 + \sum_{n>0} e^{-\beta E_{n0}(\bar{R})} \right] e^{-\beta E_0(\bar{R})}}{\int d^3 R_1' \cdots \int d^3 R_M' \left[1 + \sum_{n>0} e^{-\beta E_{n0}(\bar{R}')} \right] e^{-\beta E_0(\bar{R}')}}. \tag{15.43}$$

Finally, note that to the extent that nuclei do not have quantum behavior near conical intersections or spin crossings (see Yarkony (1996) and also Chap. 14), nothing prevents us to use this equation also in these cases.

The definition of the classical, effective Hamiltonian for the nuclear coordinates in Eq. 15.41 allows us now to use any of the well-established techniques available for computing canonical equilibrium averages in a classical system. Of course, since H_{eff} in Eq. 15.41 depends on T, any Monte Carlo or dynamical method must be performed at the same T that H_{eff} was computed in order to produce consistent results, given in this case by the convenient expression (15.37). For example, we could use (classical) Monte Carlo methods, or, if we want to perform MD simulations, we could propagate the stochastic Langevin dynamics associated to the Hamiltonian (15.41):

$$M_J \ddot{\vec{R}}_J(t) = -\nabla_J v_{\text{eff}}(\bar{\vec{R}}(t); \beta) - M_J \gamma \dot{\vec{R}}_J(t) + M_J \vec{\Xi}(t), \tag{15.44}$$

where $\vec{\Xi}$ is a vector of stochastic fluctuations, obeying $\langle \Xi_i(t) \rangle = 0$ and $\langle \Xi_i(t_1) \Xi_j(t_2) \rangle = 2\gamma k_B T \delta_{ij} \delta(t_1 - t_2)$ which relates the dissipation strength γ and the temperature T to the fluctuations (fluctuation–dissipation theorem).

Indeed, it is well-known that this Langevin dynamics is equivalent to the Fokker–Planck equation for the probability density $W(\bar{\vec{R}}, \bar{\vec{P}})$ in the classical phase space (Van Kampen 2007):

$$\frac{\partial W(\bar{\vec{R}}, \bar{\vec{P}}; t)}{\partial t} = \{H_{\text{eff}}(\bar{\vec{R}}, \bar{\vec{P}}; \beta), W(\bar{\vec{R}}, \bar{\vec{P}}; t)\}$$
$$+ \gamma \sum_J \partial_{P_J}(P_J + M k_B T \partial_{P_J}) W(\bar{\vec{R}}, \bar{\vec{P}}; t). \tag{15.45}$$

Any solution to Eq. 15.45 approaches at infinite time a distribution $W_{\text{eq}}(\bar{\vec{R}}, \bar{\vec{P}})$ such that $\partial_t W_{\text{eq}}(\bar{\vec{R}}, \bar{\vec{P}}) = 0$. This stationary solution is unique and equal to the Gibbs distribution, $W_{\text{eq}}(\bar{\vec{R}}, \bar{\vec{P}}) = \mathcal{Z}^{-1} e^{-\beta H_{\text{eff}}(\bar{\vec{R}}, \bar{\vec{P}}; \beta)}$ (Van Kampen 2007). Thus, the long-time solutions of Eq. 15.45, and hence those of Eq. 15.44 reproduce the canonical averages in Eq. 15.37. This property, which is also satisfied by other dynamics like the one proposed by Nosé (1984, 1991) if the H_{eff} in Eq. 15.41 is used, comes out in a very natural way from the present formalism while it is yet unclear of other ab initio MD candidates for going beyond gsBOMD (Mauri et al. 1993; Parandekar and Tully 2005; Schmidt et al. 2008; Bastida et al. 2007).

When would this new MD scheme be useful? The approach introduced in this section is particularly suited to the case of conical intersections or spin-crossing (see Yarkony (1996) and also Chap. 14), since it does not assume that the electrons or quantum variables immediately follow the nuclear motion, in contrast to any adiabatic approach. Another interesting application pertains the debated issue of quantum effects in proton transfer (Iyengar et al. 2008). It is a matter of current debate to what extent protons behave "quantum-like" in biomolecular systems (e.g. is there any trace of superposition, tunneling or entanglement in their behavior?). Recently, McKenzie et al. (Bothma et al. 2010) have carefully examined the issue, and concluded that "tunneling well below the barrier only occurs for temperatures less than a temperature T_0 which is determined by the curvature of the potential energy surface at the top of the barrier." In consequence, the correct determination of this curvature is of paramount importance.

The curvature predicted by the temperature-dependent effective potential introduced here is smaller than the one corresponding to the ground state potential energy surface, in the cases in which the quantum excited surfaces approach, at the barrier top, the ground state one. Therefore, T_0 would be smaller than that corresponding to the ground state potential energy surface (see Eq. 8 in Bothma et al. (2010)), and hence the conclusion in this reference "that quantum tunneling does not play a significant role in hydrogen transfer in enzymes" is reinforced by the results of the new dynamics.

Chapter 16
Excited-State Properties and Dynamics

Dmitrij Rappoport and Jürg Hutter

Structures and dynamics of electronically excited states of molecules play a central role in our understanding and modeling of molecular photophysics and photochemistry. Given the enormous success of density functional based methods for molecular ground-state properties, it is desirable to have methods at our disposal for computing excited-state forces and other first-order properties in the framework of DFT.

However, DFT is, with notable exceptions (Theophilou 1979; Levy and Nagy 1999; Görling 1999c; Frank et al. 1998; Filatov and Shaik 1999), essentially a ground-state theory due to the density variational principle (Perdew and Kurth 2003). The ground state (of a given spin and spatial symmetry) plays a unique role, distinct from higher-lying electronic states in density functional based approaches, in contrast to state-specific methods such as multi-configuration self-consistent field (MCSCF) method and related theories. As a consequence, methods suitable for computing ground-state properties cannot simply be carried over to properties of excited states. Instead, excited-state first-order properties, e.g., gradients or dipole moments, have to be expressed as sums of derivatives of excitation energies and the respective ground-state first-order properties. Non-adiabatic coupling matrix elements between the ground state and an electronic excited state formally involve derivatives of ground- and excited-state wavefunctions and their derivation in the density functional framework is not immediately obvious. However, as shown by Chernyak and Mukamel (2000a) by using the off-diagonal analog of the Hellmann–Feynman theorem, they can be expressed in terms of transition densities and are thus computable from TDDFT linear response.

D. Rappoport (✉)
Department of Chemistry and Chemical Biology, Harvard University,
12 Oxford Street, Cambridge, MA 02138, USA
e-mail: rappoport@chemistry.harvard.edu

J. Hutter
Institute of Physical Chemistry, University of Zurich,
Winterthurerstrasse 190, 8057 Zurich, Switzerland
e-mail: hutter@pci.uzh.ch

M. A. L. Marques et al. (eds.), *Fundamentals of Time-Dependent Density Functional Theory*, Lecture Notes in Physics 837, DOI: 10.1007/978-3-642-23518-4_16,
© Springer-Verlag Berlin Heidelberg 2012

The availability of excited-state gradients makes structure optimizations and systematic investigations of excited-state potential energy surfaces possible in the framework of TDDFT. Together with non-adiabatic coupling matrix elements, they provide the necessary input quantities for performing excited-state dynamics simulations.

16.1 Derivatives of Excited-State Energies in TDDFT

Electronic excitation energies are obtained in the framework of the TDDFT by solving the matrix form of the TDDFT linear response equation (Casida's equation), (4.77). An equivalent formulation is given by the variational problem for the functional

$$
G[X, Y, \omega] = \frac{1}{2} \left[(X + Y)^{\dagger} (A + B)(X + Y)(X - Y)^{\dagger}(A - B)(X - Y) \right]
$$
$$
+ \frac{\omega}{2} \left[(X + Y)^{\dagger}(X - Y) + (X - Y)^{\dagger}(X + Y) - 2 \right]. \tag{16.1}
$$

Here, X, Y are the expansion coefficients of the transition or first-order response density in terms of the ground-state KS orbitals,

$$
\rho^{(1)}(\boldsymbol{r}, \boldsymbol{r}') = \frac{1}{2} \sum_{i\sigma a\tau} \left(X_{ia\sigma} \varphi_{a\tau}(\boldsymbol{r})\varphi_{i\sigma}(\boldsymbol{r}') + Y_{ia\sigma} \varphi_{i\sigma}(\boldsymbol{r})\varphi_{a\tau}(\boldsymbol{r}') \right). \tag{16.2}
$$

The stationarity conditions for the functional G,

$$
\frac{\partial G}{\partial (X + Y)_{ia\sigma}} = \sum_{jb\sigma'} (A + B)_{ia\sigma jb\sigma'} (X + Y)_{jb\sigma'} - \omega(X - Y)_{ia\sigma} = 0, \tag{16.3a}
$$

$$
\frac{\partial G}{\partial (X - Y)_{ia\sigma}} = \sum_{jb\sigma'} (A - B)_{ia\sigma jb\sigma'} (X - Y)_{jb\sigma'} - \omega(X + Y)_{ia\sigma} = 0, \tag{16.3b}
$$

$$
\frac{\partial G}{\partial \omega} = \sum_{ia\sigma} (X + Y)_{ia\sigma} (X - Y)_{ia\sigma} - 1 = 0, \tag{16.3c}
$$

yield the original TDDFT linear response equation and the normalization condition for the coefficients X, Y (Furche 2001; Furche and Ahlrichs 2002a; Hutter 2003).

In the following, we use the spin-orbital formulation and choose all spin orbitals to be real. In keeping with the usual notation, $i, j, \ldots$ denote occupied KS orbitals, $a, b, \ldots$ are virtual (unoccupied) orbitals, while $p, q, \ldots$ stand for general KS orbitals. We denote the total number of KS orbitals as N. The spin indices σ, τ can take the values (α, β). Generally, two types of excitations are distinguished: spin-conserving excitations with $\sigma = \tau$ and spin-flip excitations, for which $\sigma \neq \tau$.

The matrices $(A + B)$ and $(A - B)$ are known as orbital rotation Hessians and were first introduced in the framework of time-dependent Hartree–Fock (TDHF) theory (Thouless 1972). The matrix elements of A and B were given in (4.78) for pure TDDFT. Here, we use a generalized formulation, which allows us to interpolate between pure TDDFT and TDHF,

$$(A + B)_{ia\sigma jb\sigma'} = (F_{ab\sigma}\delta_{ij} - F_{ij\sigma}\delta_{ab})\delta_{\sigma\sigma'}$$
$$+ K^{\mathrm{H}}_{ia\sigma jb\sigma'} + K^{\mathrm{xc}+}_{ia\sigma jb\sigma'} - K^{\mathrm{HFX}+}_{ia\sigma jb\sigma'}, \tag{16.4a}$$

$$(A - B)_{ia\sigma jb\sigma'} = (F_{ab\sigma}\delta_{ij} - F_{ij\sigma}\delta_{ab})\delta_{\sigma\sigma'} + K^{\mathrm{xc}-}_{ia\sigma jb\sigma'} - K^{\mathrm{HFX}-}_{ia\sigma jb\sigma'}. \tag{16.4b}$$

$F_{pq\sigma}$ denotes the KS matrix in the basis of ground-state KS orbitals and is diagonal for canonical KS orbitals, $F_{pq\sigma} = \varepsilon_{p\sigma}\delta_{pq}$. The matrix elements of the Hartree kernel K^{H}, the exchange-correlation kernels $K^{\mathrm{xc}\pm}$, and the nonlocal HF exchange kernels $K^{\mathrm{HFX}\pm}$ depend on the excitation type and the form of the approximate exchange-correlation functional and are defined in Table 16.1. The original formulation of TDDFT was given for spin-conserving excitations and pure density functionals (non-hybrids), in which case only symmetric Hartree and xc kernels contribute (Casida 1995; Bauernschmitt and Ahlrichs 1996a),

$$K^{\mathrm{HFX}\pm} = K^{\mathrm{xc}-} = 0, \tag{16.5}$$

and we recover (4.78).

Inclusion of nonlocal HF exchange in DFT has been pursued in various ways for over 15 years. While the addition of nonlocal, orbital-dependent HF exchange to xc potential is, strictly speaking, an extension of Hohenberg–Kohn DFT (Seidl et al. 1996; Görling and Levy 1997a), the successes of the functionals constructed in this way are indisputable. Global hybrids employ a linear combination of the local xc potential and nonlocal HF exchange potential, controlled by the hybrid mixing parameter c_{x} (Becke 1993a, b). The global hybrid construction includes the limiting cases of pure or non-hybrid TDDFT ($c_{\mathrm{x}} = 0$) and TDHF ($c_{\mathrm{x}} = 1$, $K^{\mathrm{xc}\pm}_{ia\sigma jb\sigma'} = 0$). A more recent class of long-range corrected (LRC) functionals is based on a range separation of electron repulsion (Stoll and Savin 1985; Savin 1996; Toulouse et al. 2005; Iikura et al. 2001; Tawada et al. 2004; Yanai et al. 2004; Vydrov and Scuseria 2006; Baer et al. 2010b),

$$\frac{1}{|\mathbf{r} - \mathbf{r}'|} = \frac{1 - [c_{\mathrm{x}} + c_{\mathrm{LR}}\,\mathrm{erf}(\mu|\mathbf{r} - \mathbf{r}'|)]}{|\mathbf{r} - \mathbf{r}'|} + \frac{c_{\mathrm{x}} + c_{\mathrm{LR}}\,\mathrm{erf}(\mu|\mathbf{r} - \mathbf{r}'|)}{|\mathbf{r} - \mathbf{r}'|}, \tag{16.6}$$

where μ is the range separation parameter. The first term in (16.6) represents the short-range electron repulsion, which is captured by a modified semilocal xc functional, while the second term contributes to the nonlocal HF exchange kernels of (16.4a, b) . Long-range corrected functionals remedy some of the problems in TDDFT and have been employed for computing excitation energies (Tawada et al. 2004; Yanai et al. 2004; Chai and Head-Gordon 2008) and excited-state gradients (Chiba et al. 2006; Nguyen et al. 2010b).

Table 16.1 Interaction kernels for electron repulsion in (16.4a, b)

Interaction kernel	Excitation type	
	Spin-conserving	Spin-flip
Hartree kernel		
$K^{\mathrm{H}}_{ia\sigma jb\sigma'}$	$2(ia\sigma\|jb\sigma')$	0
xc kernels		
$K^{\mathrm{xc}+}_{ia\sigma jb\sigma'}$	$2f^{\mathrm{xc}}_{ia\sigma jb\sigma'}$	$2f^{\mathrm{xc,SF}}_{ia\sigma jb\sigma'}$
$K^{\mathrm{xc}-}_{ia\sigma jb\sigma'}$	0	$(1-2\delta_{\sigma\sigma'})2f^{\mathrm{xc,SF}}_{ia\sigma jb\sigma'}$
Nonlocal HF exchange kernels		
Global hybrid functionals		
$K^{\mathrm{HFX}\pm}_{ia\sigma jb\sigma'}$	$c_{\mathrm{x}}\delta_{\sigma\sigma'}[(ab\sigma\|ij\sigma)\pm(ja\sigma\|ib\sigma)]$	$c_{\mathrm{x}}[\delta_{\sigma\sigma'}(ab\tau\|ij\sigma)$ $\pm(1-\delta_{\sigma\sigma'})(ja\tau\|ib\sigma)]$
Long-range corrected (LRC) functionals		
$K^{\mathrm{HFX}\pm}_{ia\sigma jb\sigma'}$	$\delta_{\sigma\sigma'}[(c_{\mathrm{x}}(ab\sigma\|ij\sigma)+c_{\mathrm{LR}}(ab\sigma\|ij\sigma)_{\mathrm{LR}})$ $\pm(c_{\mathrm{x}}(ja\sigma\|ib\sigma)+c_{\mathrm{LR}}(ja\sigma\|ib\sigma)_{\mathrm{LR}})]$	$\dots$

See (16.4a, b) and subsequent text for details

Definitions:

$$(pq\sigma|rs\sigma') = \int \mathrm{d}^3 r \int \mathrm{d}^3 r' \varphi_{p\sigma}(\boldsymbol{r})\varphi_{q\sigma}(\boldsymbol{r})\frac{1}{|\boldsymbol{r}-\boldsymbol{r}'|}\varphi_{r\sigma'}(\boldsymbol{r}')\varphi_{s\sigma'}(\boldsymbol{r}')$$

$$(pq\sigma|rs\sigma')_{\mathrm{LR}} = \int \mathrm{d}^3 r \int \mathrm{d}^3 r' \varphi_{p\sigma}(\boldsymbol{r})\varphi_{q\sigma}(\boldsymbol{r})\frac{\mathrm{erf}(\mu|\boldsymbol{r}-\boldsymbol{r}'|)}{|\boldsymbol{r}-\boldsymbol{r}'|}\varphi_{r\sigma'}(\boldsymbol{r}')\varphi_{s\sigma'}(\boldsymbol{r}')$$

$$f^{\mathrm{xc}}_{pq\sigma rs\sigma'} = \int \mathrm{d}^3 r \int \mathrm{d}^3 r' \varphi_{p\sigma}(\boldsymbol{r})\varphi_{q\sigma}(\boldsymbol{r})\frac{\delta^2 E_{\mathrm{xc}}}{\delta n_\sigma(\boldsymbol{r})\delta n_{\sigma'}(\boldsymbol{r}')}\varphi_{r\sigma'}(\boldsymbol{r}')\varphi_{s\sigma'}(\boldsymbol{r}')$$

$$f^{\mathrm{xc,SF}}_{pq\sigma rs\sigma'} = \int \mathrm{d}^3 r \varphi_{p\sigma}(\boldsymbol{r})\varphi_{q\tau}(\boldsymbol{r})\frac{1}{m(\boldsymbol{r})}\frac{\delta E_{\mathrm{xc}}}{\delta m(\boldsymbol{r})}\varphi_{r\sigma'}(\boldsymbol{r})\varphi_{s\tau'}(\boldsymbol{r}) \quad (\sigma\neq\tau, \sigma'\neq\tau')$$

$$m(\boldsymbol{r}) = n_\alpha(\boldsymbol{r}) - n_\beta(\boldsymbol{r})$$

Spin-flip excitations involve the change of the S_z spin vector component by ±1. Besides applications in computing electronic excitations in open-shell systems, they have been proposed as a tool for describing open-shell singlet molecules (biradicals) using the corresponding triplet state as a reference (Shao et al. 2003). Exchange-correlation interaction kernels for spin-flip excitations have been formulated in the framework of noncollinear xc potentials (Wang and Ziegler 2004; Wang et al. 2005b; Seth et al. 2010; Vahtras and Rinkevicius 2007; Rinkevicius et al. 2010).

For the purpose of computing derivatives of excitation energies we assume that the ground-state KS problem and the TDDFT linear response equation have been solved for a specific excitation with energy ω and transition densities $(X + Y)$ and $(X - Y)$. In other words, the functional G and its derivatives are to be evaluated at the stationary point (X, Y, ω). Excited-state properties are defined as derivatives of excited-state energy with respect to an external perturbation. We will denote the coupling strength parameter of the perturbation by ξ, and the corresponding derivatives will be indicated by a superscript. In the following we will deal exclusively with first-order properties, specifically with excited-state gradients (forces), i.e., first-order derivatives of excited-state energies with respect to nuclear displacements. However, the formalism outlined here is also applicable to other one-electron perturbations, e.g., external electric fields giving rise to excited-state dipole moments.

The dependence of the KS orbitals on the perturbation, e.g., the nuclear configuration, is taken into account by considering the Lagrangian (Furche and Ahlrichs 2002a; Hutter 2003)

$$L[X, Y, \omega, \varphi, Z, W] = G[X, Y, \omega] + \sum_{ia\sigma} Z_{ia\sigma} F_{ia\sigma} - \sum_{\substack{rs\sigma \\ r \le s}} W_{rs\sigma}(S_{rs\sigma} - \delta_{rs}), \quad (16.7)$$

where $S_{rs\sigma}$ are overlap integrals of KS orbitals. The Lagrange multipliers $Z_{ia\sigma}$ and $W_{rs\sigma}$ provide N^2 constraints in the variation of G. They ensure that the KS orbitals $\varphi_{p\sigma}$ are solutions of the ground-state KS equations and remain orthonormal, respectively, for $\xi \ne 0$. The explicit inclusion of Lagrangian multipliers Z and W simplifies the derivation since derivatives of the matrix elements $(A + B)_{ia\sigma jb\sigma'}$ and $(A - B)_{ia\sigma jb\sigma'}$ with respect to KS orbitals can be avoided from the beginning. This approach is known in quantum chemistry as the Z-vector method (Handy and Schaefer 1984; Helgaker and Jørgensen 1989) and is connected to the Dalgarno–Sternheimer interchange theorem (Hirschfelder et al. 1964).

For almost all practical calculations, the KS orbitals are expressed as linear combinations of basis functions

$$\varphi_{p\sigma}(\boldsymbol{r}) = \sum_{\mu p\sigma} C_{\mu p\sigma} \phi_\mu(\boldsymbol{r}). \quad (16.8)$$

From the stationarity of the Lagrangian with respect to the expansion coefficients $C_{\mu p\sigma}$,

$$\frac{\delta L}{\delta C_{\mu p\sigma}} = 0, \quad (16.9)$$

we can derive a linear equation for the Lagrange multipliers $Z_{ia\sigma}$ (the Z-vector equation)

$$\sum_{jb\sigma'} (A + B)_{ia\sigma jb\sigma'} Z_{jb\sigma'} = -R_{ia\sigma}. \quad (16.10)$$

The right-hand side R of this equation takes the form

$$R_{ia\sigma} = \frac{1}{2} \sum_b \left\{ (X + Y)_{ib\sigma} H^+_{ab\sigma}[X + Y] + (X - Y)_{ib\sigma} H^-_{ab\sigma}[X - Y] \right\}$$

$$- \frac{1}{2} \sum_j \left\{ (X + Y)_{ja\sigma} H^+_{ji\sigma}[X + Y] + (X - Y)_{ja\sigma} H^-_{ji\sigma}[X - Y] \right\}$$

$$+ H^+_{ia\sigma}[T] + 2 \sum_{jb\sigma'kc\sigma''} g^{xc}_{ia\sigma jb\sigma'kc\sigma''}(X + Y)_{jb\sigma'}(X + Y)_{kc\sigma''}.$$

$$(16.11)$$

Here, the results of linear operators H^+ and H^- acting on a general vector V are given by

$$H^+_{pq\sigma}[V] = \sum_{rs\sigma'} \left(K^{\mathrm{H}}_{pq\sigma rs\sigma'} + K^{\mathrm{xc}+}_{pq\sigma rs\sigma'} - K^{\mathrm{HFX}+}_{pq\sigma rs\sigma'} \right) V_{rs\sigma'}, \qquad (16.12a)$$

$$H^-_{pq\sigma}[V] = -\sum_{rs\sigma'} K^{\mathrm{HFX}-}_{pq\sigma rs\sigma'} V_{rs\sigma'}, \qquad (16.12b)$$

and the matrix elements of $K^{\mathrm{HFX}\pm}$, $K^{\mathrm{xc}+}$ are of the spin-conserving type irrespective of the considered excitation. The unrelaxed difference density matrix T reads

$$T_{ij\sigma} = -\frac{1}{2} \sum_a \left[(X+Y)_{ia\sigma}(X+Y)_{ja\sigma} + (X-Y)_{ia\sigma}(X-Y)_{ja\sigma} \right], \quad (16.13a)$$

$$T_{ab\sigma} = \frac{1}{2} \sum_i \left[(X+Y)_{ia\sigma}(X+Y)_{ib\sigma} + (X-Y)_{ia\sigma}(X-Y)_{ib\sigma} \right], \qquad (16.13b)$$

$$T_{ia\sigma} = T_{ai\sigma} = 0. \qquad (16.13c)$$

The right-hand side of (16.10) also includes matrix elements of the third functional derivative of the xc functional, which for spin-conserving excitations is of the form

$$g^{\mathrm{xc}}_{\sigma\sigma'\sigma''}(r, r', r'') = \frac{\delta^3 E_{\mathrm{xc}}}{\delta n_\sigma(r)\delta n_{\sigma'}(r')\delta n_{\sigma''}(r'')}, \qquad (16.14)$$

while for spin-flip excitations it is given by

$$g^{\mathrm{xc,SF}}_{\sigma\sigma'\sigma''}(r, r', r'') = \delta_{\sigma\sigma'} \left[-\frac{1}{n_s^2(r)} \frac{\delta m(r)}{\delta n_{\sigma''}(r'')} \frac{\delta E_{\mathrm{xc}}}{\delta m(r)} + \frac{1}{m(r)} \frac{\delta^2 E_{\mathrm{xc}}}{\delta m(r)\delta n_{\sigma''}(r'')} \right], \qquad (16.15)$$

where $m(r) = n_\alpha(r) - n_\beta(r)$ is the spin polarization density.

Once the Z-vector equation has been solved, the relaxed difference density matrix P of the excitation is obtained by

$$P = T + Z, \qquad (16.16)$$

and the Lagrange multipliers W are calculated from

$$W_{ij\sigma} = \frac{1}{1+\delta_{ij}} \Bigg\{ \sum_a \omega[(X+Y)_{ia\sigma}(X-Y)_{ja\sigma} + (X-Y)_{ia\sigma}(X+Y)_{ja\sigma}]$$

$$- \sum_a \varepsilon_{a\sigma}[(X+Y)_{ia\sigma}(X+Y)_{ja\sigma} + (X-Y)_{ia\sigma}(X-Y)_{ja\sigma}]$$

$$+ H^+_{ij\sigma}[P] + 2 \sum_{kc\sigma'ld\sigma''} g^{\mathrm{xc}}_{ij\sigma kc\sigma'ld\sigma''}(X+Y)_{kc\sigma'}(X+Y)_{ld\sigma''} \Bigg\}, $$

$$(16.17a)$$

$$W_{ab\sigma} = \frac{1}{1+\delta_{ab}} \Big\{ \sum_i \omega[(X+Y)_{ia\sigma}(X-Y)_{ib\sigma} + (X-Y)_{ia\sigma}(X+Y)_{ib\sigma}]$$

$$+ \sum_i \varepsilon_{i\sigma}[(X+Y)_{ia\sigma}(X+Y)_{ib\sigma} + (X-Y)_{ia\sigma}(X-Y)_{ib\sigma}]\Big\},$$

$$(16.17b)$$

and

$$W_{ia\sigma} = \frac{1}{2}\sum_j \{(X+Y)_{ja\sigma} H_{ji\sigma}^+[X+Y] + (X-Y)_{ja\sigma} H_{ji\sigma}^-[X-Y]\}$$

$$+ \varepsilon_{i\sigma} Z_{ia\sigma}.$$

$$(16.17c)$$

With the knowledge of Lagrange multipliers Z and W, first-order derivatives of excitation energies may be computed by straightforward differentiation of the Lagrangian L with respect to the coupling strength parameter ξ. By virtue of stationarity of L, no derivatives of its variational parameters occur in the final expression. Therefore, only Hellmann–Feynman terms and derivatives of basis functions with respect to the perturbation contribute to the gradient, as indicated by superscript $^{(\xi)}$,

$$\omega^\xi = L^\xi = \sum_{pq\sigma} F_{pq\sigma}^{(\xi)} P_{pq\sigma} - \sum_{pq\sigma} S_{pq\sigma}^{(\xi)} W_{pq\sigma}$$

$$+ \frac{1}{2}\sum_{ia\sigma jb\sigma'} \left(K_{ia\sigma jb\sigma'}^{H(\xi)} + K_{ia\sigma jb\sigma'}^{xc+(\xi)} - K_{ia\sigma jb\sigma'}^{HFX+(\xi)} \right)(X+Y)_{ia\sigma}(X+Y)_{jb\sigma'}$$

$$+ \frac{1}{2}\sum_{ia\sigma jb\sigma'} \left(K_{ia\sigma jb\sigma'}^{xc-(\xi)} - K_{ia\sigma jb\sigma'}^{HFX-(\xi)} \right)(X-Y)_{ia\sigma}(X-Y)_{jb\sigma'}.$$

$$(16.18)$$

The resulting gradient expression includes derivative KS matrix elements $F_{pq\sigma}^{(\xi)}$ and derivatives of overlap integrals $S_{pq\sigma}^{(\xi)}$. Their evaluation is completely analogous to the ground-state energy gradient (Pople et al. 1992), with density matrix P and energy-weighted density matrix W given by (16.16) and (16.17a–c), respectively. In addition, derivatives of interaction kernels give rise to further terms, which involve derivative Coulomb integrals $(ia\sigma|jb\sigma')^{(\xi)}$ and derivatives of matrix elements of the xc kernel $f_{ia\sigma jb\sigma'}^{xc(\xi)}$ (Furche and Ahlrichs 2002a).

16.2 Implementation of Excited-State Energy Derivatives

16.2.1 Atom-Centered Basis Sets

The most popular atom-centered basis sets include Gaussian-type (Davidson and Feller 1987) and Slater-type basis sets (van Lenthe and Baerends 2003). The use of Gaussian-type basis sets in molecular calculations makes efficient evaluation of matrix elements over basis functions possible. Several analytical integration schemes exist for Coulomb integrals over Gaussian basis functions (Helgaker et al. 2000), while matrix elements of exchange-correlation potential and its functional derivatives can be efficiently computed by numerical quadrature (Becke 1988b; Treutler and Ahlrichs 1995). The matrix element-driven implementation of excited-state energy derivatives includes repeated transformations between the KS orbital representation, which is required in the matrix multiplication steps in (16.11), (16.13a–c), and (16.17a–c), and the basis set representation, which enters the matrix element evaluation (Furche and Rappoport 2005b). An analogous approach is employed to compute the off-diagonal contributions to the matrix-vector products arising in the iterative solution of the TDDFT linear response equation (Bauernschmitt and Ahlrichs 1996a; Chernyak et al. 2000b; Furche and Rappoport 2005b). In both cases, the similarity to the Fock matrix construction of ground-state KS calculations (Weiss et al. 1993) makes efficient algorithms of ground-state DFT transferable to excited-state calculations.

Atom-centered basis functions depend on the positions of the nuclei. Therefore, the gradient expression (16.18) includes not only matrix elements of the perturbation (Hellmann–Feynman terms) but also terms depending on derivatives of basis functions, which are known as Pulay forces in quantum chemistry (Pulay 1987). In the basis set representation, the gradient (16.18) can be expressed in terms of derivative matrix elements and auxiliary one- and two-particle density matrices. For spin-conserving excitations and global hybrid functionals, the gradient of the excitation energy takes the following form (Furche and Ahlrichs 2002a):

$$\omega^\xi = L^\xi = \sum_{\mu\nu\sigma} h_{\mu\nu}^\xi P_{\mu\nu\sigma} - \sum_{\mu\nu\sigma} S_{\mu\nu}^\xi W_{\mu\nu\sigma} + \sum_{\mu\nu\sigma} v_{\mu\nu\sigma}^{xc(\xi)} P_{\mu\nu\sigma}$$
$$+ \sum_{\mu\nu\kappa\lambda\sigma'} (\mu\nu|\kappa\lambda)^\xi \Gamma_{\mu\nu\sigma\kappa\lambda\sigma'} + \sum_{\mu\nu\kappa\lambda\sigma'} f_{\mu\nu\sigma\kappa\lambda\sigma'}^{xc(\xi)} (X+Y)_{\mu\nu\sigma} (X+Y)_{\kappa\lambda\sigma'}.$$

(16.19)

Here, indices μ, ν, κ, λ denote basis functions. The contributing terms include derivative core Hamiltonian matrix elements $h_{\mu\nu}^\xi$, derivatives of overlap integrals $S_{\mu\nu}^\xi$, derivatives of matrix elements of xc potential $v_{\mu\nu\sigma}^{xc(\xi)}$, and derivative Coulomb integrals $(\mu\nu|\kappa\lambda)^\xi$. Their evaluation is analogous to the ground-state DFT energy gradient (Pople et al. 1992) and can be combined with the latter. The effective two-particle density matrix,

$$\Gamma_{\mu\nu\sigma\kappa\lambda\sigma'} = P_{\mu\nu\sigma} D_{\kappa\lambda\sigma'} + (X+Y)_{\mu\nu\sigma} (X+Y)_{\kappa\lambda\sigma'} - \frac{1}{2} c_x \delta_{\sigma\sigma'} \big(P_{\mu\lambda\sigma} D_{\kappa\nu\sigma'}$$
$$+ P_{\mu\kappa\sigma} D_{\lambda\nu\sigma'} + (X+Y)_{\mu\lambda\sigma} (X+Y)_{\kappa\nu\sigma'} + (X+Y)_{\mu\kappa\sigma} (X+Y)_{\lambda\nu\sigma'}$$
$$- (X-Y)_{\mu\lambda\sigma} (X-Y)_{\kappa\nu\sigma'} + (X-Y)_{\mu\kappa\sigma} (X-Y)_{\lambda\nu\sigma'} \big),$$

(16.20)

separates into products of one-particle quantities, where $D_{\mu\nu\sigma} = \sum_i C_{\mu i\sigma} C_{\nu i\sigma}$ is the ground-state KS density matrix.

Flexible Gaussian basis sets developed for ground states usually perform well in excited-state calculations. The smallest recommendable basis sets are of split-valence quality and have polarization functions on all atoms except H, e.g., SV(P) (Schäfer et al. 1992; Weigend and Ahlrichs 2005) or 6-31G* (Hariharan and Pople 1973; Francl et al. 1982; Rassolov et al. 1998). Especially in larger systems, these basis sets can give useful accuracy, however, excitation energies may be overestimated by 0.2–0.5 eV for valence excitations. Triple-zeta valence basis sets with two sets of polarization functions, e.g., cc-pVTZ (Dunning 1989; Woon and Dunning 1993; Wilson et al. 1999) or TZVPP (Schäfer et al. 1994; Weigend and Ahlrichs 2005), usually lead to basis set errors well below the functional error; larger basis sets (Weigend et al. 2003; Dunning 1989; Woon and Dunning 1993; Wilson et al. 1999) may be used to benchmark. Diffuse augmentation may be crucial for higher excitations and Rydberg states. Diffuse augmentation schemes based on ground-state optimization of anions (Kendall et al. 1992; Woon and Dunning 1993) and extrapolation (Woon and Dunning 1994) have been developed. Property-optimized diffuse augmented basis sets are derived from variational principle for static electric polarizabilities (Rappoport and Furche 2010).

16.2.2 Plane-Wave Basis Sets

Plane-wave basis sets are independent of the positions of the nuclei, and their convergence towards the basis set limit is controlled by a single cutoff parameter. Plane-wave codes are particularly efficient in the absence of nonlocal HF exchange, when transition densities and excited-state difference densities can be expressed in the real space as

$$n_{X+Y}(\boldsymbol{r}) = \sum_{\mu\nu\sigma}(X+Y)_{\mu\nu\sigma}\phi_\mu(\boldsymbol{r})\phi_\nu(\boldsymbol{r}), \qquad (16.21)$$

and analogously for T, Z, and P. Only Hellmann–Feynman terms contribute to the gradient and (16.18) reduces to (Hutter 2003)

$$L^\xi = \sum_{\mu\nu\sigma}h^\xi_{\mu\nu}P_{\mu\nu\sigma} = \int d^3r\, n_P(\boldsymbol{r})\frac{\partial v_{\text{ext}}(\boldsymbol{r})}{\partial\xi} \qquad (16.22)$$

where n_P is the relaxed difference density of the excitation.

16.2.3 Tamm–Dancoff Approximation

An often used approximation to TDDFT excitation energies is the Tamm–Dancoff approximation (TDA) (Grimme 1996; Hirata and Head-Gordon 1999; Hutter 2003), which amounts to restricting $Y_{ia\sigma} = 0$ in the variation of the Lagrange functional L. As a result, the TDDFT linear response equation reduces to the symmetric eigenvalue problem

$$AX^{\text{TDA}} = \omega^{\text{TDA}} X^{\text{TDA}}. \tag{16.23}$$

While the TDA excitation energies are generally quite close to the full TDDFT excitation energies, transition moments do not satisfy the common sum rules (Furche 2001) and are generally less accurate than the full TDDFT transition moments. However, even though the TDA is hardly less expensive than the full TDDFT treatment, it can be more robust with respect to triplet instabilities (Čížek and Paldus 1967, Bauernschmitt and Ahlrichs 1996b).

16.2.4 Resolution-of-the-Identity Approximation

The resolution-of-the-identity (RI-J) approximation is an efficient approximation to the Hartree contributions in (16.4a, b) (16.12a, b) and (16.18) (Eichkorn et al. 1995, Bauernschmitt et al. 1997). It is based on an expansion of transition densities and difference densities in a basis of auxiliary atom-centered basis functions. The error in the Hartree contribution is minimized by choosing the expansion coefficients with respect to the Coulomb metric (Dunlap et al.1979; Eichkorn et al. 1995). In the RI-J approximation, four-center Coulomb integrals can be completely avoided. The evaluation of Hartree contributions in (16.4a, b), (16.12a, b) and (16.18) is instead performed in three low-scaling steps and involves only three-center and two-center Coulomb integrals.

The quality of the auxiliary basis sets is critical for the accuracy of the RI-J approximation. Standard auxiliary basis sets developed for ground-state RI-J calculations (Eichkorn et al. 1995, 1997; Weigend 2006) are also sufficiently accurate in excited-state calculations, although diffuse augmentation is sometimes necessary. The errors introduced by the RI-J approximation are usually less than 10% of the basis set errors for excitation energies (Bauernschmitt et al. 1997; Rappoport and Furche 2005) and even less for excited-state structures and first-order properties (Rappoport and Furche 2005).

16.3 Performance of TDDFT for Excited-State Energies and their Derivatives

16.3.1 Singlet Excitations

Prediction of electronic excitation energies in molecular systems with good accuracy is among the greatest successes of TDDFT and continues to drive methodological and algorithmic development. TDDFT has evolved into an established method for simulations of optical spectra, and, increasingly, excited-state properties and dynamics. It provides a useful level of accuracy for many medium-size and large molecular systems at a computational cost that is similar to that of ground-state DFT calculations. Performance of TDDFT for electronic excitation energies and excited-state properties has been the subject of several recent reviews (Grimme 2004; Rosa et al. 2004; Dreuw and Head-Gordon 2005; Furche and Rappoport 2005b; Furche and Burke 2005c; Elliott et al. 2009; Rappoport et al. 2009). In this section we give an overview of the typical performance of TDDFT for electronic excitation energies and excited-state properties, discuss some problem cases, and outline several new developments.

Several benchmark studies on small and medium-size molecules have demonstrated that average accuracy of 0.2–0.5 eV can be achieved with TDDFT for vertical excitation energies of singlet valence excitations (Bauernschmitt and Ahlrichs 1996a; Parac and Grimme 2002; Dierksen and Grimme 2004; Silva-Junior et al. 2008; Jacquemin et al. 2009; Goerigk et al. 2009; Caricato et al. 2010). However, the accuracy of vertical excitation energies remains markedly dependent on the type of the approximate xc functional and kernel, and given the large number of available approximations, the choice of the "best" functional might seem a daunting task. Fortunately, the type of the approximate xc functional (LDA, GGA, global hybrid, LRC functional) often gives a good indication of its performance for a particular class of excitations. In Table 16.2 we show a statistical evaluation of vertical excitation energies for singlet valence excitations in small and medium-size organic molecules from Jacquemin et al. (2009) and Goerigk et al. (2009). We consider some of the established approximations for xc functionals that represent LDA, GGA, meta-GGA, and global hybrid functionals. In addition, we include more recent developments such as LRC functionals, and double hybrids (Grimme 2006; Goerigk et al. 2009). For comparison, excitation energies from TDHF are given. Theoretical estimates from Schreiber et al. (2008) obtained from high-level wave function methods serve as a reference. Mean errors, mean unsigned errors, and root-mean-square (RMS) deviations as well as largest positive and negative deviations from the reference are shown.

Vertical excitation energies are severely underestimated by the LDA, for which deviations of up to 1 eV are not unusual. Gradient-corrected functionals such as Becke–Perdew 1986 (BP86) (Becke 1988a; Perdew et al. 1986a) and PBE (Perdew et al. 1996b) offer a modest improvement over LDA with respect to the mean unsigned errors and to the width of the error distribution. In general, results obtained with different GGA functionals are very similar to each other. The meta-GGA functional of Tao, Perdew, Staroverov, and Scuseria (TPSS) (Tao et al. 2003) performs somewhat better than GGAs. Table 16.2 also illustrates the well-known tendency of TDHF to overestimate vertical excitation energies of singlet valence excitations by about 1 eV on average. Global hybrid functionals interpolate between the overestimation of TDHF and the underestimation of GGAs. It was found that global hybrid functionals containing 20–25% nonlocal HF exchange such as B3LYP and PBE0 are quite accurate for singlet valence excitations. On the other hand, global hybrids containing a higher proportion of nonlocal HF exchange, such as Becke's half-and-half hybrid (BHLYP), tend to give too high vertical excitation energies.

Table 16.2 Mean errors, mean unsigned errors, root-mean-square (RMS) deviations, and maximum negative and positive deviations in electron volt from theoretical best estimates of vertical excitation energies in 28 molecules (103 singlet valence excited states)

	TDHF	LDA	BP86	PBE	TPSS	B3LYP
Mean	0.99	−0.48	−0.44	−0.45	−0.29	−0.08
Mean uns.	1.05	0.57	0.52	0.53	0.42	0.26
RMS dev.	1.23	0.68	0.62	0.64	0.51	0.32
Max neg.	−0.61	−1.42	−1.36	−1.38	−1.17	−0.75
Max pos.	2.61	0.70	0.66	0.64	0.74	0.75
	PBE0	BHLYP	LC-ω PBE	CAM-B3LYP	B2PLYP	
Mean	0.05	0.42	0.41	0.22	0.01	
Mean uns.	0.24	0.49	0.46	0.31	0.19	
RMS dev.	0.32	0.60	0.61	0.42	0.25	
Max neg.	−0.66	−0.56	−0.68	−0.64	−0.70	
Max pos.	0.92	1.57	2.16	1.46	0.88	

TDHF and TDDFT results for the following functionals are shown: LDA in Vosko-Wilk-Nusair V parametrization (Vosko et al. 1980), GGA functionals BP86 (Becke 1988a; Perdew 1986a) and PBE (Perdew et al. 1996b), TPSS meta-GGA functional (Tao et al. 2003), global hybrids B3LYP (Becke 1993a, b), PBE0 (Perdew et al. 1996c), and BHLYP (Becke 1993b), LRC functionals LC-ω PBE (Vydrov and Scuseria 2006; Vydrov et al. 2006b) and CAM-B3LYP (Yanai et al. 2004) and double hybrid functional B2PLYP (Grimme 2006; Grimme and Neese 2007). The data are from Jacquemin et al. (2009), Goerigk et al. (2009).

LRC functionals are represented by the LC-ω PBE functional of Scuseria and co-workers ($\mu = 0.40$ a.u.$^{-1}$, $\alpha = 0$, $\beta = 1$) (Vydrov and Scuseria 2006; Vydrov et al. 2006b) and the Coulomb-attenuated B3LYP functional CAM-B3LYP ($\mu = 0.33$ a.u.$^{-1}$, $\alpha = 0.19$, $\beta = 0.46$) (Yanai et al. 2004), cf. (16.6). The available results indicate that LRC functionals show performance similar to global hybrids for singlet $\pi \to \pi^*$ and $n \to \pi^*$ excitations (Silva-Junior et al. 2008; Jacquemin et al. 2009, 2007b). Considerable dependence of the results on the range separation parameter μ was observed (Peach et al. 2006; Rohrdanz and Herbert 2008; Rohrdanz et al. 2009; Wong et al. 2009).

Double hybrid functionals combine a GGA part, nonlocal HF exchange as well as a nonlocal orbital-dependent correlation functional (Grimme 2006). The computational cost of double hybrid functionals is therefore higher than that of global hybrids and corresponds to that of Møller–Plesset second-order perturbation theory (MP2). An approximate procedure for computing excitation energies within the double hybrid scheme was developed, in which the linear response is computed using the GGA and nonlocal HF exchange kernels, while the nonlocal correlation is added as a perturbative correction (Grimme and Neese 2007). Initial benchmark studies using the double hybrid functional B2PLYP showed encouraging results, which are somewhat better than those for global hybrids. For a detailed discussion, we refer to Jacquemin et al. (2009), Goerigk et al. (2009).

Table 16.3 summarizes the performance of configuration interaction singles (CIS) method, LDA, BP86 and PBE functionals of GGA type and the B3LYP global hybrid

Table 16.3 Mean unsigned errors compared to experiment for adiabatic excitation energies, excited-state bond distances, dipole moments, and vibrational frequencies with CIS, LDA in the Perdew–Wang parameterization (Perdew and Wang 1992a), GGA functionals BP86 (Becke 1988a; Perdew 1986a) and PBE (Perdew et al. 1996b), and B3LYP global hybrid (Becke 1993a,b)

Property	Exp. mean	CIS	LDA	BP86	PBE	B3LYP
34 ad. exc. energies (eV)	4.5	0.6[a]	0.3[b]	0.3	0.3	0.3
40 bond distances (pm)	142.2	3.5[a]	1.5[b]	1.3	1.3	1.3
10 dipole moments (D)	1.3	0.4[a]	0.1[a]	0.1	0.1	0.2
80 vib. frequencies (cm^{-1})	1258	169[c]	62[b]	49	49	61

The data are from (Furche and Ahlrichs 2002a)
[a] Excludes the $1^2\Sigma^+$ state of NO (instability)
[b] Excludes the 1^1B_1 state of CCl_2 and the $1^2\Sigma^+$ state of NO (instabilities)
[c] Excludes the $1^2\Sigma^+$ state of NO (instability) and the ν_{13} ($1a_2$) frequency of the 1^1B_2 state of pyridine (saddle point)

for adiabatic excitation energies and excited-state bond distances, dipole moments, and vibrational frequencies. The most striking observation is that structural parameters and other properties of excited states are generally predicted with better accuracy than excitation energies (Furche and Ahlricks 2002a; Rappoport and Furche 2005). This is understandable since excited-state properties depend only on relative energies and thus benefit from significant error cancellation. As a consequence, excited-state bond distances may be reproduced with an accuracy of 1–2 pm with LDA, GGA, and global hybrids, which is comparable to the accuracy of ground-state DFT. Another consequence of the error cancellation is that adiabatic excitation energies are generally predicted with an accuracy similar to that of vertical excitation energies. Similarly, excited-state energy differences are predicted with better accuracy than transition energies of excitations from the ground state (Silva-Junior et al. 2008). It should be noted, however, that the shapes of excited-state potential energy surfaces may be considerably affected by the shortcomings of present approximate TDDFT such as lack of double excitations or spurious charge-transfer states, see below. Larger and less systematic errors or even qualitatively wrong results may be obtained for adiabatic excitation energies and excited-state properties in these cases.

The performance of TDDFT for the calculation of transition moments and oscillator strengths of dipole-allowed transitions was recently assessed in several benchmark studies (Miura et al. 2007; Silva-Junior et al. 2008; Caricato et al. 2011). It was found that LDA and GGA functionals tend to underestimate oscillator strengths compared to high-level wave function methods, while inclusion of nonlocal HF exchange leads to some improvement. LRC functionals predict oscillator strengths with fairly good accuracy.

16.3.2 Charge-Transfer Excitations in TDDFT

The accuracy of present TDDFT methodology is often sufficient for predicting electronic absorption spectra and properties of optically allowed excited states. However, detailed studies of electronic excited-state manifolds and the corresponding potential energy surfaces reveal a number of drawbacks, which can be traced back to limitations of the approximate xc potentials and kernels currently in use. Well-known problem cases are charge-transfer excited states, Rydberg transitions, spin-forbidden excitations, and excitations with considerable doubles character. Less obvious weaknesses include the well-documented underestimation of excitation energies of L_a transitions in polycyclic aromatic hydrocarbons (Grimme and Parac 2003; Parac and Grimme 2003; Wong and Hsieh 2010) or the erroneous chain length dependence of excitation energies in cyanine dyes (Jacquemin et al. 2007a; Send et al. 2011). These shortcomings have spurred broad research activity aimed to understand their causes and develop practicable solutions. In the following, we will briefly discuss some typical problem cases and outline some proposed solutions. We also refer to Chap. 4 for theoretical background.

The problem of long-range charge transfer (CT) is perhaps the most prominent drawback of the present TDDFT methodology and has recently received a great deal of attention. Excited states exhibiting complete or partial charge transfer are frequently encountered in photochemical studies and play a central role in photosynthesis and organic photovoltaics. The amount of charge transfer often depends on the molecular geometry and may increase upon structural relaxation in the excited state. In fact, excited states such as twisted intramolecular charge-transfer (TICT) states are essentially of valence type at the Franck–Condon region but gain strong charge-transfer character at the excited-state energy minimum (Rettig 1986). In TDDFT, excitation energies of CT states are commonly underestimated by a significant amount. Indeed, errors in excitation energies of CT states are often 1 eV or more (Tozer et al. 1999; Peach et al. 2008). In some cases, the relative energies and thus structures and properties of charge-transfer excited states may be predicted with satisfactory accuracy (Rappoport and Furche 2004). However, CT errors of approximate TDDFT often generate a number of spurious low-lying CT states, which may have a profound effect on excited-state dynamics (Magyar and Tretiak 2007). The underestimation might even be strong enough to cause spurious crossings with ground-state potential energy surfaces to appear (Kozak et al. 2010). Particularly striking is the occurrence of spurious low-lying CT states in molecular aggregates. Instead of following the $-1/R$ dependence on the intermolecular distance R, the excitation energies of CT states are grossly underestimated (Dreuw et al. 2003; Bernasconi et al. 2003) and essentially distance-independent (Dreuw et al. 2003; Dreuw and Head-Gordon 2004). This behavior has consequences for calculations of energy transfer rates and Davydov splittings in dimers (Sagvolden et al. 2009). A useful diagnostic for the CT character of a given excitation was proposed by Tozer and co-workers (Peach et al. 2008). It is based on spatial overlap between the occupied and the virtual orbitals involved in the excitation; values of the diagnostic close to zero indicate that the excitation has significant CT or Rydberg character, while valence excitations correspond to larger values of the diagnostic.

The failure of present approximate TDDFT to accurately describe long-range CT excitations is attributed to the wrong asymptotic shape of ground-state xc potentials, which leads to errors in orbital energies. Moreover, the essentially local character of the non-hybrid xc kernels of (16.4a, b) has the consequence that their matrix elements vanish at large separations between the donor and acceptor orbitals (Dreuw et al. 2003). We refer to Autschbach (2009) and Chap. 4 for a more in-depth discussion. A host of different solutions for the CT problem in TDDFT has been proposed. They range from asymptotic corrections to xc potentials (Tozer and Handy 1998; Tozer et al. 1999) a posteriori energy corrections using methods exhibiting the correct asymptotic behavior, such as CIS (Dreuw et al. 2003; Dreuw and Head-Gordon 2004), to introducing nonlocality in the xc kernel using the global hybrid or LRC constructions (Tawada et al. 2004; Yanai et al. 2004; Vydrov and Scuseria 2006; Baer et al. 2010b). Global hybrids were found to alleviate the CT problem, however, a high percentage of about 50% of nonlocal HF exchange is required to obtain satisfactory accuracy for CT states. This high proportion of HF exchange leads to a strong overestimation of singlet valence excitation energies, cf. Table 16.2. LRC functionals showed promise to systematically improve the description of CT excited-states in TDDFT, and strong impetus for the development of LRC functionals came from these applications. Several benchmark studies showed that LRC functionals systematically increase excitation energies of CT states (Peach et al. 2006, 2008; Plötner et al. 2010; Wong and Hsieh 2010). However, the choice of the optimal range-separation parameter μ remains an open question; it was shown that the parameter μ is strongly system-dependent and the requirements of ground-state calculations and excitation energies with respect to its choice are inherently different (Peach et al. 2006; Rohrdanz and Herbert 2008; Rohrdanz et al. 2009; Wong et al. 2009). A variational procedure for finding the optimal value of the range-separation parameter for a given excitation was developed and successfully applied to CT excitations (Stein et al. 2009a, b). A combination of global exchange and LRC exchange was also proposed for treating CT excitations (Rohrdanz et al. 2009).

16.3.3 Rydberg, Triplet Excitations and Excitations with Doubles Character

The wrong asymptotic shape of semilocal xc potentials and the lack of derivative discontinuity are responsible for systematic underestimation of Rydberg excitation energies (Casida et al. 1998b; Tozer and Handy 1998; Della Sala and Görling 2002a, b; Grüning et al. 2002; Caricato et al. 2010). Introduction of a nonlocal HF exchange via the global hybrid scheme reduces the errors in excitation energies; however, a much higher percentage of nonlocal HF exchange is typically required for Rydberg excitations than for valence excitations (Caricato et al. 2010). LRC functionals show slightly better performance than global hybrids (Peach et al. 2008). It should, however, be noted that using special techniques, accurate Rydberg excitation energies may be computed even within the LDA (Wasserman and Burke 2005c). The diffuse character of Rydberg excited states makes the use of augmented basis sets absolutely necessary (Wiberg et al. 2002; Ciofini and Adamo 2007). In the absence of diffuse basis functions, Rydberg excitations appear extremely blue-shifted or are entirely absent from excitation spectra.

The accuracy of spin-forbidden excitations to triplet states has been recently addressed in several benchmark studies (Silva-Junior et al. 2008; Jacquemin et al. 2010b). The excitation energies of triplet excitations are significantly underestimated by LDA and GGA functionals; however, in contrast to singlet excitations, inclusion of nonlocal HF exchange barely affects the quality of the results.

Excited-states with double-excitation character are often optically forbidden (dark) but are assumed to play an important role in photochemical reactions. Of particular interest are the dark $2^1 A_g$ states of polyenes (Hsu et al. 2001; Cave et al. 2004; Levine et al. 2006; Huix-Rotllant et al. 2011). Within adiabatic TDDFT, only single excitations are obtained from linear response TDDFT calculations. Double and higher excitations can only be described by introducing frequency dependence into the xc kernel, see Chap. 8. However, is was observed that double excitations are often strongly coupled to single excitations of similar energies. The absence of this coupling in the adiabatic TDDFT has a profound effect on molecular electronic excitation spectra. As was shown for the $2^1 A_g$ excited state of trans-butadiene, the character of $2^1 A_g$ state predicted by TDDFT methods is qualitatively different from that obtained by multi-reference wave function methods such as CASSCF (Levine et al. 2006). A dressed TDDFT approach was developed for including the frequency dependence for the model case of a double excitation coupled to an individual single excitation, which showed promising results for dark states of polyenes (Maitra et al. 2004; Cave et al. 2004). Several generalizations of this approach were recently proposed (Gritsenko and Baerends 2009; Huix-Rotllant et al. 2011).

16.4 Non-Adiabatic Coupling Matrix Elements

In cases where the adiabatic approximation to the separation of nuclear and electronic quantum effects fails, the non-adiabatic coupling vector or first-order nonadiabatic coupling matrix elements (NACMEs),

$$d_{fi} = \langle \Psi_i \mid \frac{\partial}{\partial \xi} \Psi_f \rangle, \tag{16.24}$$

also called derivative coupling between the adiabatic states Ψ_i and Ψ_f, becomes the central quantity linking the change of the electronic state with nuclear motion. The variable ξ in Eq. 16.24 stands for a nuclear coordinate. Examples of nonadiabatic effects include radiationless decay of electronically excited states by internal conversion or through conical intersections, predissociation, and excitation energy transfer (Domcke et al. 2004; Keal et al. 2007; Sagvolden et al. 2009).

In TDDFT the wavefunctions are not directly accessible and the calculation of d_{fi} has to be pursued via a time-dependent response approach. Similar to other properties the coupling matrix elements are extracted from the time evolution of observables. The theory of NACMEs in TDDFT was pioneered by Chernyak and

Mukamel (2000a). In recent years several approaches ranging from finite differences to approximate finite basis set response equations have been published (Baer 2002; Billeter and Curioni 2005; Hu et al. 2007b, 2008, 2009; Tapavicza et al. 2007; Tavernelli et al. 2009a, b, c). A complete derivation that includes also Pulay terms following the gradient theory presented in Sect. 16.1 was given by Send and Furche (2010). They derive an expression for first-order NACMEs from time-dependent response theory by considering the time evolution of

$$C^\xi(t) = \left\langle \Psi(t) \mid \frac{\partial}{\partial \xi} \Psi(t) \right\rangle, \tag{16.25}$$

under the influence of an external perturbation. First order couplings between the ground state and an excited state n are extracted from residues of the first order response of C^ξ,

$$d_{n0}^\xi = -\sum_{ia}(X - Y)_{nia}\langle \varphi_a(r) \mid \varphi_i^\xi(r')\rangle, \tag{16.26}$$

where $\varphi_i^\xi(r)$ is the derivative of KS orbital i with respect to an atomic coordinate. Couplings between excited states may be calculated from nonlinear responses of C^ξ (Tavernelli et al. 2010a). The presence of the derivative of KS orbitals in Eq. 16.26 seems to imply that the coupled perturbed KS equations for all nuclear coordinates have to be solved in order to calculate a complete coupling vector. This would result in computational costs similar to second derivatives of the ground state energy. However, Send and Furche (2010) were able to recast Eq. 16.26 into a form reminiscent of the ground state and excited state nuclear gradients

$$d_{n0}^\xi = \sum_{\mu\nu\sigma}h_{\mu\nu}^\xi P_{\mu\nu\sigma}^n - \sum_{\mu\nu\sigma}S_{\mu\nu}^\xi W_{\mu\nu\sigma}^n + \sum_{\mu\nu\sigma}v_{\mu\nu\sigma}^{xc(\xi)} P_{\mu\nu\sigma}^n$$

$$+ \sum_{\mu\nu\sigma\kappa\lambda\sigma'}(\mu\nu|\kappa\lambda)^\xi \Gamma_{\mu\nu\sigma\kappa\lambda\sigma'}^n - \frac{1}{2}\sum_{\mu\nu\sigma}T_{\mu\nu}^\xi (X - Y)_{n\mu\nu\sigma}. \tag{16.27}$$

In Eq. 16.27 we have made use of the following alternative definitions of the one- and two-particle density matrices

$$P_{ia\sigma}^n = \omega_n^{-1}(X + Y)_{nia\sigma} \tag{16.28a}$$

$$W_{ia\sigma}^n = \varepsilon_{i\sigma}P_{ia\sigma}^n + \frac{1}{2}(X - Y)_{nia\sigma} \tag{16.28b}$$

$$W_{ij\sigma}^n = \frac{1}{1 + \delta_{ij}}H_{ij\sigma}^+[P^n] \tag{16.28c}$$

$$\Gamma_{\mu\nu\sigma\kappa\lambda\sigma'}^n = P_{\mu\nu\sigma}^n P_{\kappa\lambda\sigma'}^n - \frac{1}{2}c_x\delta_{\sigma\sigma'}\left[P_{\mu\lambda\sigma}^n D_{\kappa\nu\sigma} + P_{\nu\lambda\sigma}^n D_{\kappa\mu\sigma}\right] \tag{16.28d}$$

$$T_{\mu\nu}^\xi = \langle \chi_\mu^\xi \mid \chi_\nu\rangle - \langle \chi_\mu \mid \chi_\nu^\xi\rangle \tag{16.28e}$$

where $D_{\mu\nu\sigma}$ is an element of the ground state density matrix and ω_n is the excitation energy of state n.

16.5 Excited-State Dynamics

The Born–Oppenheimer separation between the motions of nuclei and electrons is at the heart of most electronic structure calculations. This separation allows the description of the dynamics of the nuclei with quantum or classical mechanics while the electronic structure remains in an adiabatic eigenstate of the system. For systems in the electronic ground state, the coupling of Newtonian dynamics with DFT has become a very successful simulation method. Using a standard integration scheme, Newton's equations of motion are solved starting from an initial configuration. When combined with DFT, this requires the solution of the KS equations and the calculation of the first derivative with respect to atomic positions at each time step.

The method of adiabatic dynamics can easily be applied to other states than the ground state. Having defined an excited state energy surface as the sum of the ground state energy and an excitation energy from linear response within TDDFT, we only need the nuclear forces to follow the system in time. Applying Newton's equations of motion restricts the dynamics to the initially chosen adiabatic state. However, this restriction together with the neglect of the quantum mechanical behavior of the nuclei often limits the scope of the simulations. Quantum effects such as tunneling, interference and level quantization may be important for many systems. The single state approximation will be much more severe for excited states than it is for the ground state, and transitions between electronic states have to be taken into account for most problems of interest. When such transitions occur, the forces experienced by the nuclei may change drastically. This effect has to be properly incorporated into the dynamics and is crucial for describing many dynamical effects like photochemistry, electron transfer in molecules, or radiationless transitions (see Chap. 14).

In Chap. 14 advanced methods for excited-state dynamics are derived from the non-relativistic time-dependent Schrödinger equation. If the total wavefunction at each set of nuclear positions is expanded into the time-independent adiabatic electronic wavefunctions one find that the nuclear wavefunctions are solutions to the following equation

$$
i\hbar\frac{\partial}{\partial t}\chi_n(\bar{\boldsymbol{R}}, t) = \left[-\sum_\alpha \frac{1}{2M_\alpha}\nabla^2_{\boldsymbol{R}_\alpha} + E_n(\bar{\boldsymbol{R}})\right]\chi_n(\boldsymbol{R}, t)
$$
$$
- \sum_\alpha \frac{1}{M_\alpha}\sum_m \left[\boldsymbol{d}_{nm}^{\alpha(1)}\cdot\nabla_{\boldsymbol{R}_\alpha} + \frac{1}{2}\boldsymbol{d}_{nm}^{\alpha(2)}\right]\chi_m(\bar{\boldsymbol{R}}, t), \qquad (16.29)
$$

where $E_n(\boldsymbol{R})$ is the adiabatic potential energy surface and $\boldsymbol{d}_{nm}^{\alpha(1)}$ and $\boldsymbol{d}_{nm}^{\alpha(2)}$ are the first- and second-order nonadiabatic coupling vectors. Equation 16.29 is used in wave packet dynamics and in connection with TDDFT response calculations uses the nuclear forces and NACMEs derived in the previous sections. The second-order NACMEs are often either neglected or approximated using a resolution of the identity method that allows to reduce second-order NACMEs to first-order NACMEs (Hirai and Sugino 2009).

An alternative to wave package dynamics that replaces nuclear quantum dynamics by a classical equation but retains the coupling between different adiabatic potential energy surfaces is Tully's surface hopping (TSH) method (Tully and Preston 1971; Tully 1990). In this method a set of trajectories is propagated independently according to Newton's equation of motion. Nuclear quantum correlations are neglected and transfer of amplitude between different potential energy surfaces is achieved by a stochastic procedure (surface hopping). The time-dependent quantities of interest are the complex-valued amplitudes $C_n(t)$ that are propagated according to (Tapavicza et al. 2007; Barbatti et al. 2007; Tavernelli et al. 2010b)

$$i\hbar \dot{C}_m(t) = \sum_n C_n(t) \left[E_m \delta_{nm} - i\hbar \sum_\alpha \dot{R}_\alpha \cdot d_{nm}^{\alpha(1)} \right]. \qquad (16.30)$$

The classical trajectories of the nuclear positions are propagated adiabatically until a switch to another adiabatic potential energy surface is initiated. In the fewest switches algorithm (Tully 1990), the probability for a transition from state m to state n within dt from time t is

$$g_{nm}(t, dt) = -2 \int_t^{t+dt} d\tau \frac{1}{|C_n(\tau)|^2} \Re e \left[C_m(\tau) C_n^*(\tau) \sum_\alpha \dot{R}_\alpha \cdot d_{nm}^{\alpha(1)}(\tau) \right]. \qquad (16.31)$$

A switch is initiated if and only if

$$\sum_{i \leq m-1} g_{ni} < \eta < \sum_{i \leq m} g_{ni}, \qquad (16.32)$$

where η is a uniform random number in the interval $[0, 1]$. The combination of surface hopping and TDDFT has been successfully applied to a series of problems from photochemistry (Tapavicza et al. 2007, 2008; Aquino et al. 2009; Hirai and Sugino 2009).

16.6 Solvation Effects and Coupling to Classical Force Fields

Many chemical reactions take place in condensed phase and therefore an accurate theoretical description of solvent effects is of utmost importance in computational quantum chemistry. Excited electronic states are generally more susceptible to environment effects, specifically solvation. Effective solvation models in the framework of the polarizable continuum model (PCM) have been applied to simulate bulk solvation effects on structures and dynamics of electronically excited molecules in solution

(Tomasi et al. 2005; Caricato et al. 2006; Scalmani et al. 2006; Mennucci et al. 2009). Other applications to solvation effects in the excited state are using explicit solvent models. One route to the simulation of large condensed systems is the combination of electronic structure theory methods with empirical force fields. The idea behind this combined quantum mechanics/molecular mechanics (QM/MM) (Sherwood 2000) methods is to describe a part of the molecule/system quantum mechanically and the rest within the empirical force field approach. In this QM/MM approach we can write the total energy of the systems as

$$E_{\text{tot}} = E_{\text{QM}} + E_{\text{MM}} + E_{\text{QM-MM}}, \tag{16.33}$$

where E_{QM} is the energy of the QM part of the system, E_{MM} the energy of the MM system, and $E_{\text{QM-MM}}$ describes the coupling of the two parts. We will be concerned here with the case where E_{QM} is given by the TDDFT energy and we don't have to discuss E_{MM} further. However, the coupling term $E_{\text{QM-MM}}$ has to be investigated. The force field parameters, as well as van der Waals parameters were optimized to describe the interaction of a ground state system. It is therefore important to ensure that in the case of the description of excited states with TDDFT, the actual excitation region is well separated from the QM–MM boundary. The important interaction term in our case is the electrostatic interaction of the classical point charges with the electronic charge density of the quantum system

$$E_{\text{QM-MM}}^{\text{el}} = \sum_I \int d^3 r \frac{n_{\text{QM}}(r) q_I(R_I)}{|r - R_I|}. \tag{16.34}$$

The sum in Eq. 16.34 runs over all point charges q_I at position R_I. If we now define the charge density of the QM system as the derivative of the total energy with respect to the external potential

$$n^\omega(r) = \frac{\partial E_\omega}{\partial v_{\text{ext}}(r)} = n(r) + n_P(r), \tag{16.35}$$

we see that it is the relaxed density that acts as the true charge density of the system in the excited state. Replacing $n_{\text{QM}}(r)$ with $n^\omega(r)$ in the definition of $E_{\text{QM-MM}}^{\text{el}}$ also ensures that the calculation of the excitation energy and nuclear forces is still consistent. If we add the interaction term to the Lagrangian (16.7) the additional external potential from the classical point charges appears in the definition of A (16.4a, b) and the KS matrix F. This coupling scheme has been applied in a series of applications to spectroscopy in solution (Sulpizi et al. 2003, 2005).

The scheme outlined above, based on an interaction energy given by Eq. 16.34 applies also to other situations where a subsystem treated by TDDFT is embedded into a charge density. A prominent case of such a method is the KS method with constrained electron density embedding by Wesolowski and Warshel (1993). An extension of this method for TDDFT has been formulated (Casida and Wesolowski 2004) and has been used to calculate the solvatochromic shift of excitation energies for molecules in solution (Neugebauer et al. 2005).

Chapter 17
Electronic Transport

Stefan Kurth

17.1 Introduction

In this chapter we will tackle the problem of describing electron transport at the nanoscale. Transport spectroscopy, i.e., the measurement of physical quantities related to electron transport such as, e.g., the conductivity, provides an important experimental tool to investigate mesoscale or nanoscale system. In recent years, transport measurements through single molecules attached to two metallic leads have become feasible. In fact, electronic devices based on single molecules form the basis for the vision of a "molecular electronics" (Cuniberti et al. 2005).

In order to properly describe electron transport at the nanoscale, a quantum theory of transport is of paramount importance. This theory should ideally be able to take into account the atomistic details of the junction since transport properties of molecules certainly depend strongly on the details of, e.g., the geometry of the junction or the chemical bonding between molecule and metallic leads. Ground state DFT (Hohenberg and Kohn 1964; Kohn and Sham 1965) provides a useful tool to describe the atomistic details of a molecular junction with reasonable accuracy. Transport properties, however, do not fall in the domain of ground-state DFT because transport is a non-equilibrium phenomenon.

S. Kurth (✉)
Nano-Bio Spectroscopy Group and ETSF Scientific Development Centre,
Dpto. de Física de Materiales, Universidad del País Vasco UPV/EHU,
Centro Física de Materiales CSIC-UPV/EHU, Av. Tolosa 72,
20018 San Sebastian, Spain
e-mail: stefan_kurth@ehu.es

S. Kurth
IKERBASQUE,
Basque Foundation for Science, 48011 Bilbao, Spain

M. A. L. Marques et al. (eds.), *Fundamentals of Time-Dependent Density Functional Theory*, Lecture Notes in Physics 837, DOI: 10.1007/978-3-642-23518-4_17, © Springer-Verlag Berlin Heidelberg 2012

At present, the standard methodology for atomistic modelling of electron transport is based on a combination of non-equilibrium Green function theory (NEGF) (Keldysh 1965) with DFT (Lang 1995; Di Ventra et al. 2000; Brandbyge et al. 2002; Evers et al. 2004; Koentopp et al. 2008). In this approach, transport is viewed as a scattering problem: the device region (i.e., the molecule or nanostructure) is coupled to two metallic leads at different chemical potentials and the electrons are scattered from the self-consistent Kohn–Sham potential in the device region. By construction, this approach is concerned with the steady state only, i.e., a constant, time-independent current flowing as a consequence of the two leads being kept at different chemical potentials.

Instead of directly aiming for the steady state, one may also adopt a time-dependent point of view. Starting from the contacted lead-molecule-lead system initially in its ground state (Cini 1980), one then drives the system out of equilibrium by applying a bias in the leads and follows the time evolution of the system towards a steady state (if this steady state is actually achieved). This approach faces the problem that one essentially has to treat a macroscopic system (the leads coupled to the device), although in the end one is interested in the transport properties of the nanoscale device only.

Of course, also the transport problem is inherently a many-body problem of interacting electrons. Therefore it is crucial to treat the electron-electron interaction properly. Not surprisingly, here we choose a TDDFT approach to deal with the interaction. The locality of the KS potential is actually quite advantageous when partitioning the system into left and right leads and central nanodevice region.

In the steady-state limit, TDDFT has been shown to lead to an exchange-correlation correction of the zero-bias conductance (Koentopp et al. 2006, Stefanucci et al. 2007a) which, however, vanishes for the typical local and adiabatic approximation to the xc kernel. This correction is more easily accessible in a current-DFT framework (Koentopp et al. 2006; Vignale and Di Ventra 2009).

In the present chapter we will review three TDDFT approaches to time-dependent transport: (i) modelling the leads as large but finite system, (ii) using an embedding technique to take into account the coupling of the nanodevice to semi-infinite leads and (iii) a quantum kinetic approach which describes the coupling to an environment via a master equation.

Of course, besides TDDFT there are also other approaches for time-dependent transport through interacting nanodevices which, however, will not be covered here. These include non-equilibrium many-body techniques propagating the Kadanoff-Baym equations (Myöhänen et al. 2008; Myöhänen et al. 2009; von Friesen et al. 2010), the time-dependent renormalization group (Branschädel et al. 2010; Heidrich-Meisner et al. 2009), real-time path integral (Mühlbacher and Rabani 2008) and Monte Carlo approaches (Werner et al. 2010).

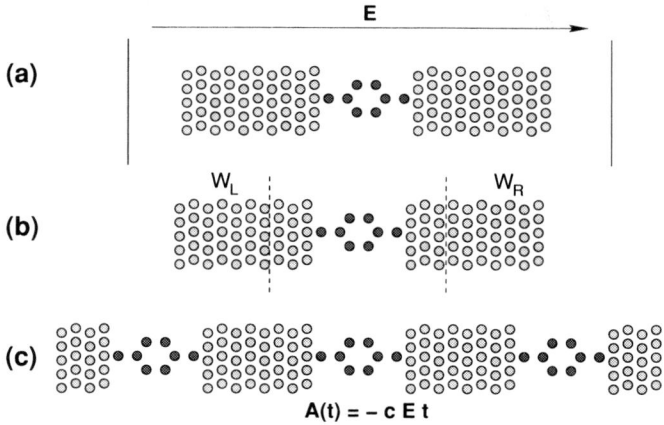

Fig. 17.1 Schematic sketch of the three different approaches to transport discussed in the text: **a** the nanosystem or molecule attached to two large but finite leads is placed in an external electric field; **b** the molecule is attached to two semi-infinite leads with applied bias $\Delta w = w_L - w_R$; **c** the molecule plus a finite portion of the metallic leads are repeated periodically in a time-dependent vector potential describing the applied external electric field

17.2 TDDFT Approaches to Transport

17.2.1 Finite Systems

A technical difficulty in the theoretical treatment of (time-dependent) electron transport stems from the fact that one wants to describe a nanoscopic system coupled to macroscopic leads. In fact, for non-interacting electrons it has been shown (Stefanucci and Almbladh 2004b) that a system perturbed from its ground state by application of a DC bias will evolve towards a steady state provided that the density of states is a smooth function of energy which implies that the system lead-device-lead has to be (infinitely) extended. Of course, this is a theoretical idealization since in the real world the leads, though macroscopic, are still finite. It is therefore natural to model the system by the device connected to large but finite leads (see Fig. 17.1a). Of course, then one can at best achieve a quasi-steady current with a finite lifetime which crucially depends on the system size. This finite-system approach has been used in conjunction with TDDFT to model transport through nanoscale systems (Bushong et al. 2005; Sai et al. 2007; Cheng et al. 2006; Evans et al. 2008; Evans and Van Voorhis 2009; Evans et al. 2009).

In the approach of Di Ventra and coworkers (Bushong et al. 2005; Sai et al. 2007) the (finite) system consists of two large reservoirs connected by a constriction with cross section much smaller than the cross section of the reservoirs. Initially, the system is exposed to an external potential of the form

$$v(\boldsymbol{r}) = v_0 \left[\Theta(z - z_0) - \Theta(-z - z_0) \right] \tag{17.1}$$

which mimics the bias. Here, $\pm z_0$ are in the right and left leads, respectively, which bear a potential offset from the central constriction (located at $z = 0$). The initial state is calculated from a standard self-consistent Kohn–Sham scheme in the presence of the potential $v(\boldsymbol{r})$ which induces a charge imbalance between left and right leads. At time $t = 0$, the external potential $v(\boldsymbol{r})$ is switched off and the system starts to discharge across the constriction. The time evolution is followed with TDDFT using the adiabatic LDA for exchange and correlation. The time-dependent current in the middle of the junction is monitored and one can see how it evolves towards a current plateau (interpreted as the steady state current) before the finite size of the system leads to a decaying current due to reflections at the far ends of the leads. The methodology has been validated for a tight-binding model of noninteracting electrons (Bushong et al. 2005). It has then been applied with TDDFT for monoatomic gold chains of various chain length (Bushong et al. 2005) as well as for the study of current flow patterns in jellium model junctions (Sai et al. 2007) and two dimensional gold leads connected through a single gold atom (Sai et al. 2007).

In Refs. (Cheng et al. 2006; Evans et al. 2008; Evans and Van Voorhis 2009; Evans et al. 2009) a similar approach is taken. One difference is the use of Gaussian basis functions in these works. Two different setups are used for the time-dependent simulations. In one case, Löwdin atomic populations (Löwdin 1955) are used to build the initial state (Cheng et al. 2006) with different chemical potentials in left and right leads. This "bias" is switched off at $t = 0$ and the time evolution is calculated within TDDFT. In the second case, the simulation starts from the ground state with a bias being suddenly switched on at $t = 0$. One problem of this approach is the presence of significant fluctuations in the resulting time-dependent currents. A steady-state current is extracted by time-averaging the current over a time interval significantly larger than the time step of the propagation. In (Evans and Van Voorhis 2009) where a benzenedithiol attached to gold leads is studied, charge oscillations on the molecule are found and the system appears not to evolve towards a steady state.

17.2.2 Infinite Systems via Embedding Technique

In this section we will describe an embedding approach to take into account semi-infinite leads coupled to the central device, see Fig. 17.1b).

We write the time-dependent KS Hamiltonian of the coupled system left lead (L)—central device (C)—right lead (R) in a localized basis as

$$\hat{H}_{\text{KS}}(t) = \begin{pmatrix} \hat{H}_{\text{LL}} & \hat{H}_{\text{LC}} & \hat{H}_{\text{LR}} \\ \hat{H}_{\text{CL}} & \hat{H}_{\text{CC}} & \hat{H}_{\text{CR}} \\ \hat{H}_{\text{RL}} & \hat{H}_{\text{RC}} & \hat{H}_{\text{RR}} \end{pmatrix} \tag{17.2}$$

where $\hat{H}_{\alpha\alpha}$ is the part of the Hamiltonian which describes the isolated region $\alpha = \text{L, C, R}$ and $\hat{H}_{\alpha\alpha'}$ ($\alpha \neq \alpha'$) describes the coupling between the different regions.

(In a grid representation, the $\hat{H}_{\alpha\alpha'}$ ($\alpha \neq \alpha'$) contain the off-diagonal parts of the kinetic energy operator stemming from the discretization of the Laplacian, in a tight-binding picture they contain the hopping matrix elements.) In general, the Hamiltonian blocks $\hat{H}_{\alpha\alpha'}$ are time-dependent but for notational convenience we here do not write this time dependence explicitly.

The time-dependent KS equation for KS orbital φ_k reads

$$\left[i\partial_t - \hat{H}_{ks}(t) \right] \varphi_k(t) = 0 \tag{17.3}$$

where the KS orbital is $\varphi_k(t) = (\varphi_{k,L}(t), \varphi_{k,C}(t), \varphi_{k,R}(t))^T$ with $\varphi_{k,\alpha}(t)$ being the projection of $\varphi_k(t)$ onto region α. We now assume that the Hamiltonian blocks coupling the left and right leads vanish identically,

$$\hat{H}_{LR} = 0 \quad \text{and} \quad \hat{H}_{RL} = 0. \tag{17.4}$$

Then one can derive from Eq. 17.3 the following equation of motion for the KS orbital projected onto the central region $\varphi_{k,C}(t)$

$$\left[i\partial_t - \hat{H}_{CC}(t) \right] \varphi_{k,C}(t) = \int_0^t dt' \hat{\Sigma}_{emb}^R(t, t') \varphi_{k,C}(t') + \sum_{\alpha=L,R} \hat{H}_{C\alpha} \hat{G}_{\alpha\alpha}^R(t, 0) \varphi_{k,\alpha}(0). \tag{17.5}$$

Here $\hat{G}_{\alpha\alpha}^R(t, t')$ is the retarded Green function for lead α which satisfies the equation of motion

$$\left(i\partial_t - \hat{H}_{\alpha\alpha} \right) \hat{G}_{\alpha\alpha}^R(t, t') = \delta(t - t') \tag{17.6}$$

with initial conditions $\hat{G}_{\alpha\alpha}^R(t, t^+) = 0$ and $\hat{G}_{\alpha\alpha}^R(t, t^-) = -i$.

$$\hat{\Sigma}_{emb}^R(t, t') = \sum_{\alpha=L,R} \hat{H}_{C\alpha} \hat{G}_{\alpha\alpha}^R(t, t') \hat{H}_{\alpha C} \tag{17.7}$$

is the embedding self energy. Equation 17.5 describes the time evolution of the KS orbital $\varphi_{k,C}(t)$ projected onto the central device in the presence of the leads. The first term on the r.h.s. of this equation we call the memory term because it depends on the history of $\varphi_{k,C}(t)$. Physically, it describes how the KS electron leaving the device region is scattered back from the leads. The second term on the r.h.s. of Eq. 17.5 depends on the the KS orbital projected on the leads at initial time $t = 0$. This term we call the source term because it describes how the leads act as particle sources for the central device.

We have suggested a practical algorithm to solve Eq. 17.5 (Kurth et al. 2005). We work with a real-space or tight-binding representation of the KS Hamiltonian or, in other words, our basis functions are completely localized on a given site (grid point). Since the KS potential is a local, multiplicative potential in real space, in this basis the time dependence of the KS Hamiltonian (17.2) is restricted to the diagonal elements of the matrix while all off-diagonal matrix elements are independent of time. For convenience we assume that the time-dependent potential in the leads is independent of position or site, i.e.,

$$\hat{H}_{\alpha\alpha}(t) = \hat{H}_{\alpha\alpha}(0) + w_\alpha(t)\hat{1}_\alpha \qquad (17.8)$$

where the time-dependent bias $w_\alpha(t)$ in lead α ($\alpha = $ L, R) is an arbitrary, externally controlled function of time with $w_\alpha(0) = 0$. Physically, this assumption describes perfect metallic screening, i.e., any external electric field is completely screened inside the lead at each instant of time. This allows for a significant simplification in the calculation of the memory and source terms, mainly because the lead Green function $\hat{G}^R_{\alpha\alpha}(t, t')$ for the time-dependent case can trivially be related to that for time-independent leads.

In order to solve Eq. 17.5, we also need to calculate the initial KS orbital $\varphi_k(0)$. In the general case, the (self-consistent) calculation of a scattering state for a system without spatial periodicity is nontrivial. However, for the simple model systems studied below, they can be obtained by standard means. The actual time propagation in our algorithm is performed with a modified version of the Crank-Nicholson algorithm. For details on the implementation we refer the reader to the original work (Kurth et al. 2005; Stefanucci et al. 2008a). The original algorithm has been generalized to describe spin dynamics in quantum dots (Stefanucci et al. 2008b) and to the case of superconducting leads (Stefanucci et al. 2010). An alternative but similar algorithm to implement the transparent boundary conditions in transport has been suggested in (Zheng et al. 2007).

In the following section we will discuss two applications of the formalism described above to rather simple model systems. In both cases, the assumption that the biased system evolves towards a steady state will turn out to be unjustified. First we will consider non-interacting electrons and show that the presence of bound states inhibits the evolution towards a steady current for a system under DC bias. In the second case, we will show that a discontinuous TDDFT xc potential can have a profound effect on the time evolution of a biased system and may lead not to a steady state but to a dynamical state of correlation-induced density oscillations. This dynamical state can be shown to be closely related to the Coulomb blockade phenomenon.

Bound state oscillations. We consider an infinitely extended system of non-interacting electrons. The system is partitioned into a nanodevice region connected to two semi-infinite, metallic leads which is initially in its ground state and then driven out of equilibrium by applying a DC bias. It is reasonable to expect that the time-evolution of the system will eventually lead to the formation of a steady state current. It has been show theoretically (Stefanucci and Almbladh 2004a, b) that such a steady state is achieved if the local density of states in the device region is a smooth function of energy. However, if the biased system supports two or more bound states in the device region (i.e., the local density of states has delta peaks), a system of non-interacting electrons under a DC bias does not evolve towards a steady state (Dhar and Sen 2006, Stefanucci 2007). Instead, it can be shown that in the long-time limit the density consists of two contributions, a steady-state contribution and a dynamical contribution which reads

$$n^{\text{dyn}}(\boldsymbol{r}, t) = \sum_{b,b'} f_{b,b'} \cos((\varepsilon_b^\infty - \varepsilon_{b'}^\infty)t)\varphi_b^*(\boldsymbol{r})\varphi_{b'}(\boldsymbol{r}). \qquad (17.9)$$

Here, $\varphi_b(\boldsymbol{r})$ and ε_b^∞ are the bound states and eigenenergies of the biased Hamiltonian. The coefficients $f_{b,b'}$ can be shown to depend on the history of the system (Stefanucci 2007) and their diagonals $f_{b,b}$ may be interpreted as occupation numbers of bound state $\varphi_b(\boldsymbol{r})$ in the long-time limit (Khosravi et al. 2008, 2009).

We take a one-dimensional model system as an example. We start with electrons in the ground state of a constant potential $v(z) = 0$ and the plane-wave eigenstates are occupied up to the Fermi energy $\varepsilon_F = 0.2$ a.u. At time $t = 0$, we suddenly switch on both a bias $w_L(t) = 0.05\Theta(t)$ a.u. in the left lead (i.e. for $z < -1$ a.u.) and a gate potential $v_{\text{gate}}(z, t) = v_g(z)\Theta(t)$ with $v_g(z) = -1.8$ a.u. for $|z| \leq 1$ a.u. and zero otherwise. The biased Hamiltonian then has two bound states with energies $\varepsilon_1^\infty = -1.291$ a.u. and $\varepsilon_2^\infty = -0.114$ a.u.

In Fig. 17.2 we show the modulus of the Fourier transform of the current $I(t)$ for $0 \leq t \leq 500$ a.u. at position $z = 0$. We recognize a number of well defined peaks which can clearly be identified with characteristic transitions of the system. The main peak occurs, as expected, at the transition frequency between the energy levels of the Hamiltonian at $t \to \infty$. In addition there are smaller peaks which correspond to transitions from the two bound levels to the Fermi energy shifted by the bias in the leads. A time-frequency analysis reveals (Khosravi et al. 2008) that only the peak with the transition frequency $\omega = \varepsilon_2^\infty - \varepsilon_1^\infty$ corresponds to truly persistent oscillations in the current. The transitions towards the continuum of the leads appear only during a transient period after switch-on of the bias.

In the inset of Fig. 17.2 we also show the time-dependent current for the situation described above with initial potential $v(z)$. In addition, the time-dependent current is shown for a different inital potential, $v(z) = v_g(z)$, chosen such that after switching on the bias in the leads the potential landscape is exactly the same in both cases, i.e., in both situations the biased Hamiltonian supports the same two bound states. One can see that in the first case [$v(z) = 0$] the current oscillates with a rather large amplitude. In the second case [$v(z) = v_g(z)$], although not visible on the scale of this plot, the current oscillates with the same frequency but with a much smaller amplitude. This example shows explicitly the dependence of the coefficients $f_{b,b'}$ in Eq. 17.9 on the initial state of the system. A dependence on the history of the system can also be observed when gate and/or bias are not switched on suddenly but smoothly with a switching time T_g. Then the amplitude of the current oscillations depends on T_g and the oscillations become small in the limit of slow switching (Khosravi et al. 2008).

Dynamical Coulomb blockade. In the present subsection we will show, again for a simple model system, that also when the interaction is taken into account via TDDFT,

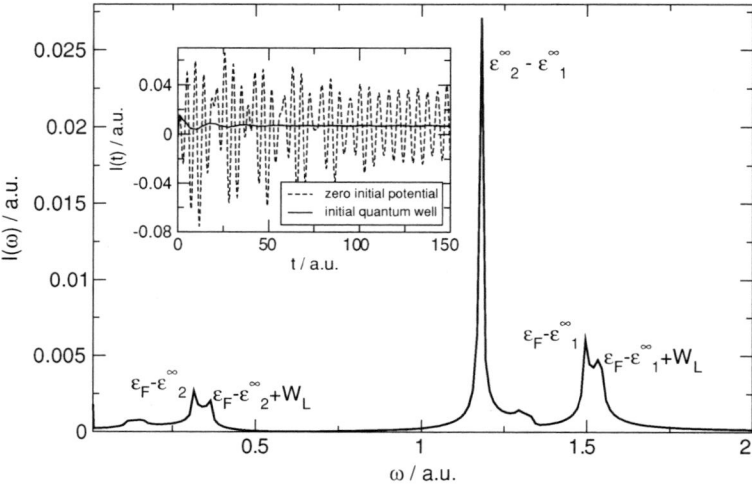

Fig. 17.2 Modulus of the discrete Fourier transform of the current for a system of non-interacting electrons in one dimension in the initial potential $v(z) = 0$. At $t = 0$, a bias $w_L = 0.05$ a.u. and a gate potential $v_g(z)$ are suddenly switched on. Here the gate potential is $v_g(z) = -1.8$ a.u. for $|z| \leq 1$ a.u. and zero otherwise. The *inset* shows the time-dependent current for this situation (*dashed line*) as well as for the case where the initial potential is $v(z) = v_g(z)$. At time $t = 0$ the bias $w_L = 0.05$ a.u. is suddenly switched such that the potential for times $t > 0$ is the same in both cases

a biased system not always evolves towards a steady state. Here the crucial ingredient is the use of an exchange-correlation functional which has a derivative discontinuity at integer particle number. In static DFT, Perdew et al. (Perdew et al. 1982) have established the existence of such a derivative discontinuity for the exact exchange-correlation energy functional. In a TDDFT context, the derivative discontinuity has been investigated in Refs. (Lein and Kümmel 2005; Mundt and Kümmel 2005) (see also Chap. 6).

In the present section we will use an exchange-correlation potential which is a local, but discontinuous function of the density. In certain parameter regimes, which can be identified with the regime of Coulomb blockade, this discontinuity leads to the biased system approaching a dynamical state of persistent density and current oscillations in the long-time limit (Kurth et al. 2010).

We consider the model of a single interacting impurity connected to two tight-binding leads of non-interacting electrons described by the Hamiltonian

$$\hat{H}(t) = \hat{H}_C + \hat{H}_L(t) + \hat{H}_R(t) + \hat{H}_T .\tag{17.10}$$

Here the Hamiltonian of the impurity with a Hubbard-like interaction reads

$$\hat{H}_C = \sum_{\sigma=\uparrow,\downarrow} \varepsilon_C \hat{d}_\sigma^\dagger \hat{d}_\sigma + U \hat{d}_\uparrow^\dagger \hat{d}_\uparrow \hat{d}_\downarrow^\dagger \hat{d}_\downarrow \tag{17.11}$$

and $\hat{d}_\sigma^\dagger$ ($\hat{d}_\sigma$) is the creation (annihilation) operator for an electron with spin σ at the impurity. The Hamiltonian for lead $\alpha = L, R$ is

$$\hat{H}_\alpha(t) = \sum_{i=1}^{\infty}\sum_\sigma [\varepsilon_\alpha + w_\alpha(t)]\hat{c}_{i\sigma\alpha}^\dagger\hat{c}_{i\sigma\alpha} - \sum_{i=1}^{\infty}\sum_\sigma \left(v_\alpha\hat{c}_{i\sigma\alpha}^\dagger\hat{c}_{i+1\sigma\alpha} + \text{H.c.}\right) \quad (17.12)$$

with the time-dependent bias $w_\alpha(t)$ in lead α and $\hat{c}_{i\sigma\alpha}^\dagger$ ($\hat{c}_{i\sigma\alpha}$) creates (annihilates) an electron with spin σ at site i in lead α. Finally, the tunneling Hamiltonian connecting the impurity site with the leads is given by

$$\hat{H}_T = -\sum_\sigma \sum_{\alpha=L,R} \left(v_{\text{link}}\hat{d}_\sigma^\dagger\hat{c}_{1\sigma\alpha} + \text{H.c.}\right). \quad (17.13)$$

In a TDDFT framework, the interacting Hamiltonian (17.10) is mapped onto an effective single-particle Hamiltonian. In a local and adiabatic approximation, the KS Hamiltonian for the combined system reads $\hat{H}^{ks}(t) = \hat{H}_C^{KS}(t) + \sum_{\alpha=L,R}\hat{H}_\alpha(t) + \hat{H}_T$ with the KS Hamiltonian for the impurity

$$\hat{H}_C^{KS}(t) = \sum_\sigma v_{KS}(n_0(t))\hat{d}_\sigma^\dagger\hat{d}_\sigma \quad (17.14)$$

where the KS potential

$$v_{KS}(n_0(t)) = \varepsilon_0 + \frac{1}{2}Un_0(t) + v_{xc}(n_0(t)) \quad (17.15)$$

depends only on the local density $n_0(t)$ at the impurity. For the xc potential we use a modified version (Kurth et al. 2010; Kurth and Stefanucci 2011) of the Bethe-ansatz LDA (BALDA) (Lima et al. 2003) and its adiabatic extension to TDDFT (Verdozzi 2008). Just as the usual LDA is based on the xc energy per particle of the uniform electron gas, the BALDA is based on the xc energy per site of the uniform one-dimensional Hubbard model. This model contains two parameters: the Hubbard interaction U present at each site and the hopping parameter v connecting each pair of neighboring sites and the xc energy per site is given in terms of these parameters (Lima et al. 2003). In our present model, however, the interaction is only present at the impurity which is connected to the leads by the hopping parameter v_{link}. Therefore we use a modification of the original BALDA by replacing v in the original parametrization by v_{link}. The explicit form of the modified BALDA xc potential then is

$$v_{xc}(n) = \Theta(1-n)v_{xc}^{(<)}(n) - \Theta(n-1)v_{xc}^{(<)}(2-n) \quad (17.16)$$

where

$$v_{xc}^{(<)}(n) = -\frac{1}{2}Un - 2v_{\text{link}}\left[\cos\left(\frac{\pi n}{2}\right) - \cos\left(\frac{\pi n}{\xi}\right)\right]. \quad (17.17)$$

The parameter ξ is given as solution of the equation

$$\frac{2\xi}{\pi} \sin(\pi/\xi) = 4 \int\limits_0^\infty dz\, \frac{J_0(z)J_1(z)}{z[1 + \exp(Uz/(2v_{\text{link}}))]} \tag{17.18}$$

where $J_{i=0,1}(z)$ are Bessel functions. A very important property of the BALDA xc potential is its discontinuity at half filling (Lima et al. 2002):

$$v_{\text{xc}}(n = 1^+) - v_{\text{xc}}(n = 1^-) = U - 4v_{\text{link}} \cos\left(\frac{\pi}{\xi}\right). \tag{17.19}$$

We *assume* now that under application of a DC bias a steady state exists. Then, using non-equilibrium Green's functions techniques, one can derive a self-consistency condition for the steady-state density $n_0^\infty := n_0(t \to \infty)$ at the impurity. Under the additional assumption of the absence of localized (bound) states in the system, this condition reads

$$n_0^\infty = 2 \sum_{\alpha=\text{L, R}} \int\limits_{-\infty}^{\varepsilon_F+w_\alpha} \frac{d\omega}{2\pi} \Gamma(\omega - w_\alpha)|G(\omega)|^2 \tag{17.20}$$

where ε_F is the Fermi energy of the contacted system in the ground state, w_α ($\alpha = \text{L, R}$) is the constant bias applied in lead α and

$$G(\omega) = \left[\omega - v_{\text{KS}}(n_0^\infty) - \sum_{\alpha=\text{L,R}} \Sigma_\alpha(\omega - w_\alpha)\right]^{-1}. \tag{17.21}$$

$\Sigma_\alpha(\omega) = \Lambda_\alpha(\omega) - \frac{i}{2}\Gamma_\alpha(\omega)$ is the analytically known embedding self energy for non-interacting tight-binding leads with real and imaginary parts $\Lambda_\alpha(\omega)$ and $\Gamma_\alpha(\omega)$, respectively, and the total width function is $\Gamma(\omega) = \sum_\alpha \Gamma_\alpha(\omega)$. Note that the KS potential in Eq. 17.21 is the local and adiabatic KS potential of Eq. 17.15 evaluated at the non-equilibrium steady-state density n_0^∞.

Taking the discontinuity (17.19) of the BALDA xc potential seriously, one finds that there are parameter values for which Eq. 17.20 *does not have any solution*. Nevertheless, the steady-state analysis can still provide useful insight into the physical nature of these situations if, instead of using the xc potential of Eq. 17.16, we use a modified xc potential with a smoothened discontinuity (Kurth et al. 2010)

$$v_{\text{xc}}(n) = f(n)v_{\text{xc}}^{(<)}(n) - [1 - f(n)]v_{\text{xc}}^{(<)}(2 - n) \tag{17.22}$$

where $f(n) = [\exp((n - 1)/a) + 1]^{-1}$ with a smoothening parameter a. Such a smoothening is also physically reasonable: while the exact xc potential for the Hubbard model is certainly discontinuous, in our case the coupling of the impurity to the leads introduces some broadening of the isolated impurity levels which are expected to lead to a smoothened xc potential.

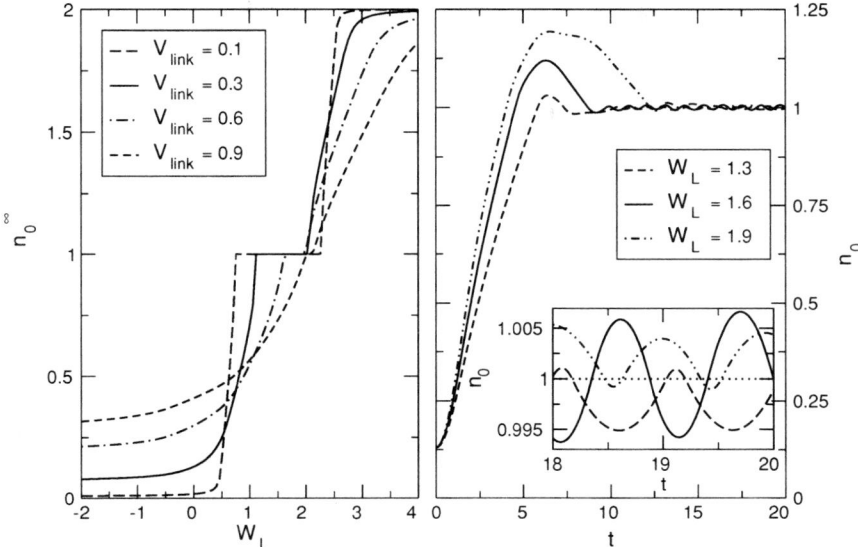

Fig. 17.3 *Left panel*: steady state density at impurity as function of the appliied bias w_L for different values of v_{link}. *Right panel*: time evolution of the density at the impurity for $v_{link} = 0.3$ for different biases in the Coulomb blockade regime. *Inset*: magnification for large times showing the density oscillations around unity. The dotted straight line at unity serves as guide for the eye

The BALDA ground state densities at the impurity as function of the on-site energy ε_0 have been found (Kurth and Stefanucci 2011) to be in rather good agreement with Quantum Monte Carlo results (Wang et al. 2008). In the left panel of Fig. 17.3 we show the steady-state densities as function of the applied bias w_L in the left lead for different values v_{link} of the hopping between leads and impurity. The other parameters are $\varepsilon_F = 1.5$, $\varepsilon_C = 2$, $\varepsilon_\alpha = 0$, and $U = 2$ (all energies given in units of $v_L = v_R = v$). For the smoothening parameter we chose $a = 10^{-4}$. As the coupling v_{link} becomes smaller, a clear step of the steady-state density develops around unity. The width of this step increases and approaches U in the limit $v_{link} \to 0$. It reflects the energy cost a second electron entering the impurity has to pay if the impurity is already occupied by one electron. This is nothing but the Coulomb blockade effect. We therefore conclude that the discontinuity in the xc potential is a necessary ingredient for a DFT description of Coulomb blockade.

For a truly discontinuous xc potential, the steady state self-consistency condition for certain parameter values does not have a solution. We have therefore looked at the time evolution of systems in this parameter regime, the regime of Coulomb blockade (i.e., the regime of the plateau in the steady-state density for the smoothened potential, Fig. 17.3). We consider a system initially in its ground state. At $t = 0$, we suddenly switch on a bias in the left lead and follow the time evolution of the system. The right panel of Fig. 17.3 shows the time-dependent density at the impurity for $v_{link} = 0.3$

and three different values for w_L in the Coulomb blockade regime. After application of the bias, the density rises from its initial value and seems to settle around the value of unity, i.e., single occupancy of the impurity. However, closer inspection of the time-dependent density at large times (see inset in right panel of Fig. 17.3) reveals that the density shows oscillations around unity with rather small amplitude of the order of 5×10^{-3} and, in accordance with our analysis of the steady-state condition, does not settle towards a steady state. This behaviour can be understood (Kurth et al. 2010; Kurth and Stefanucci 2011) in the following way: when the density is away from the critical value of unity (where the discontinuity occurs), the KS potential changes at the same rate as the density. On the other hand, when the density crosses unity, the KS potential changes discontinuously (or very rapidly for the smoothened potential). For instance, if the density at some instant in time crosses unity from below, the potential suddenly jumps up by the discontinuity and tends to push the density towards lower values. However, the rate of change of the density is limited by the inertia of the electrons and for some time the density will continue to grow before it decreases and eventually crosses the value unity from above. Now the potential suddenly decreases and attracts more electrons but again the resulting change in the density will not be immediate. In this way, the system eventually reaches a dynamical state of oscillating density but never a steady state, establishing a dynamical picture of Coulomb blockade as a sequence of charging and discharging of the impurity.

17.2.3 Quantum Kinetic Approach

In this final section, we would like to discuss another approach to time-dependent transport which combines quantum kinetic theory with TDDFT (Gebauer and Car 2004a, b, 2005; Gebauer et al. 2005; Burke and Car 2005).

This approach is somewhat different from the ones discussed in the previous sections in the sense that is concerned with the electronic system coupled to a bath, typically the phononic degrees of freedom of the system. Similarly to Chap. 10, we here discuss the theory of open systems (i.e., systems coupled to a bath) in terms of master equations. Alternatively, open system can also be described with an approach based on the stochastic Schrödinger equation (see Chap. 11).

We write the total (electronic + bath) Hamiltonian of the system as

$$\hat{H}_{tot} = \hat{H}_{el} + \hat{H}_{bath} + \hat{H}_{coup} \qquad (17.23)$$

where $\hat{H}_{el}$ and $\hat{H}_{bath}$ are the Hamiltonians for the electrons and the bath, respectively, while $\hat{H}_{coup}$ describes the coupling between them. The density operator of the combined system

$$\hat{S}_{tot}(t) = |\Psi(t)\rangle\langle\Psi(t)| \qquad (17.24)$$

satisfies the equation of motion

$$\frac{\mathrm{d}}{\mathrm{d}t}\hat{S}_{\mathrm{tot}}(t) = -\mathrm{i}\left[\hat{H}_{\mathrm{tot}}, \hat{S}_{\mathrm{tot}}\right].\tag{17.25}$$

The reduced density operator for the electronic subsystem is defined by tracing out the bath degrees of freedom

$$\hat{S}_{\mathrm{red}}(t) = \mathrm{Tr}_{\mathrm{bath}}\left\{\hat{S}_{\mathrm{tot}}\right\}.\tag{17.26}$$

An equation of motion for the reduced density operator can then be derived under two assumptions (Breuer and Petruccione 2002): (i) the coupling between electrons and bath is assumed to be weak and therefore it is sufficient to treat $\hat{H}_{\mathrm{coup}}$ only up to second order and (ii) the Markov approximation is justified, i.e., the time scale on which the electronic system varies is large compared to the time scale over which bath (phonon) correlation functions decay. Under these conditions, the equation of motion for $\hat{S}_{\mathrm{red}}(t)$ reads

$$\frac{\mathrm{d}}{\mathrm{d}t}\hat{S}_{\mathrm{red}}(t) = -\mathrm{i}\left[\hat{H}_{\mathrm{el}}, \hat{S}_{\mathrm{red}}(t)\right] + \check{C}[\hat{S}_{\mathrm{red}}(t)]\tag{17.27}$$

where $\check{C}$ is a superoperator whose explicit form depends on the bath (for a phonon bath this form has been given in Refs. (Gebauer and Car 2004b; Burke and Car 2005)).

Similar to standard TDDFT, Burke and Car (Burke and Car 2005b) showed that, given the electron-electron interaction, the superoperator $\check{C}$, and the initial reduced density matrix $\hat{S}_{\mathrm{red}}(0)$, two different time-dependent one-body potentials always lead to different time-dependent densities. In a second step, they then mapped the problem of interacting electrons in contact with a reservoir onto a problem of non-interacting Kohn–Sham electrons also in contact with a reservoir with the same time-dependent density as the interacting system. In practice, this leads to a KS version of the master equation (17.27).

In the finite system approach to transport (Sect. 17.2.1), the boundary condition of vanishing wavefunctions at the boundaries of the simulation box are used. In the infinite system approach via the embedding techniques (Sect. 17.2.2) open (or transparent) boundary conditions are used. In contrast, in practical applications of the KS master equation approach periodic boundary conditions are used (see Fig. 17.1c): the central device together with a finite portion of the metallic leads is repeated periodically which may also be viewed as a ring geometry. A spatially constant electric field E is applied throughout the system by using a time-dependent, spatially constant vector potential $A = -cEt$ where c is the velocity of light. The setup of the transport problem with periodic boundary conditions has the advantage that it allows for the use of plane waves as computationally efficient basis set.

The presence of the external electric field throughout the system (ring) constantly accelerates the electrons and prevents them from reaching a steady state. It is the task of the collison term, i.e., the last term on the r.h.s. of Eq. 17.27 (or its KS analogue), to dissipate the energy injected into the system by the presence of the electric field.

The quantum kinetic approach to transport described here has been applied to simple model systems (Gebauer and Car 2004a, 2004b, 2005) and to an atomistic model of a self-assembled monolayer of benzene dithiolate molecules contacted between two gold electrodes (Gebauer et al. 2005).

17.3 Conclusions

We have sketched three different TDDFT approaches to time-dependent transport: (i) a finite system approach where the leads are mimicked by large but finite electrodes; (ii) an approach describing an infinite system using an embedding technique to implement the transparent boundary conditions and (iii) a quantum kinetic approach solving a KS master equation for the electronic system in contact with a bath. The explicitly time-dependent view on transport taken here naturally is not restricted to the steady state which, in fact, in some situations is not achieved at all during the time evolution. However, the natural target of study for these approaches are truly dynamical phenomena in transport beyond the steady state. As we have seen, TDDFT can help to gain physical insights into time-dependent transport even for model systems but it is probably the only method which can become (or already is, to some extent) practical for simulations of realistic systems.

Chapter 18
Atoms and Molecules in Strong Laser Fields

Carsten A. Ullrich and André D. Bandrauk

18.1 Introduction: New Light Sources for the Twenty-first Century

The interactions of superstrong and ultrashort laser pulses with atoms and molecules have been a subject of great interest over the past two decades, as reflected in many books (Gavrila 1992; Piraux et al. 1993; Delone and Krainov 2000; Batani et al. 2000; Brabec 2008; Mulser and Bauer 2010) and review articles (Mainfray and Manus 1991; Freeman and Bucksbaum 1991; Burnett and Reed 1993; Protopapas et al. 1997; Salières et al. 1999; Joachain et al. 2000; Brabec and Krausz 2000; Dörner et al. 2002; Becker et al. 2002; Krausz and Ivanov 2009). The beginning of the twenty-first century is witnessing the development of several large- and medium-scale experimental facilities dedicated to the generation of laser light with unprecedented capabilities. The frequency spectrum covered by these new light sources ranges from the infrared up to the extreme ultraviolet and soft x-ray [produced in the FLASH free-electron laser facility at DESY (Richter et al. 2009)]. This calls for the development of new theoretical and computational tools to simulate laser-matter interactions at extreme conditions.

C. A. Ullrich (✉)
Department of Physics,
University of Missouri-Columbia,
Columbia, MO 65211, USA
e-mail: ullrichc@missouri.edu

A. D. Bandrauk
Département de Chimie, Faculté de Sciences,
Université de Sherbrooke,
Sherbrooke QC J1K 2R1, Canada
e-mail: Andre.Bandrauk@USherbrooke.ca

M. A. L. Marques et al. (eds.), *Fundamentals of Time-Dependent Density Functional Theory*, Lecture Notes in Physics 837, DOI: 10.1007/978-3-642-23518-4_18   351
© Springer-Verlag Berlin Heidelberg 2012

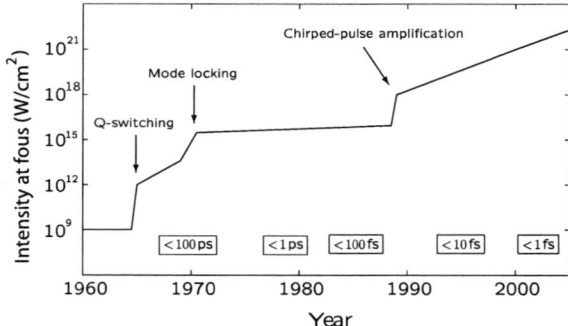

Fig. 18.1 The attainable pulsed-laser peak intensity has increased by 12 orders of magnitude over the last 40 years. Ultrashort pulses can be generated with durations of a few femtoseconds and, of late, even down to the attosecond regime (Drescher et al. 2001; Paul et al. 2001; Hentschel et al. 2001)

In this chapter we review some of the phenomena taking place in the ultrafast and ultrastrong regime; see also Chap. 1. This presents us with the opportunity to discuss the successes and challenges of TDDFT, which were outlined earlier in Chap. 4, in the context of specific applications.[1]

As shown in Fig. 18.1, attainable laser intensities have increased by 12 orders of magnitude since the invention of the laser in 1960, as a result of a series of technological advances. Today, petawatt pulses with focused intensities in excess of 10^{21} W/cm^2 can be produced from both large-scale and lab-scale lasers. By comparison, the atomic unit of intensity is $I_0 = 3.52 \times 10^{16}$ W/cm^2, which, by the relation $I = c\mathcal{E}^2/8\pi$, corresponds to an electric field strength $\mathcal{E}_0 = e/a_0^2 = 5.14 \times 10^{11}$ V/m, i.e. the electric field which an electron experiences in the $1s$ orbital of a hydrogen atom. Thus, the forces produced by the laser field match and even exceed the Coulomb forces that attract the electrons to the nucleus, or that bind the atoms together in a molecule. For intensities above 10^{18} W/cm^2 the electronic motion in the laser focus becomes relativistic, and nonrelativistic TDDFT ceases to be applicable.

On the other hand, the minimum pulse lengths have dramatically decreased. Whereas at the end of the twentieth century the focus was on femtosecond photochemistry and photophysics, culminating with the 1999 Nobel Prize to A. H. Zewail for "femtochemistry", major efforts are now underway to develop and apply attosecond optical pulses (Drescher et al. 2001; Paul et al. 2001; Hentschel et al. 2001; Krausz and Ivanov 2009). Passing the attosecond frontier makes it possible to image and to manipulate the dynamics of electrons on their natural inneratomic time scales (Kienberger et al. 2002; Drescher et al. 2002; Kienberger et al. 2004; Wickenhauser et al. 2005; Haessler et al. 2010), defined by the period of the $1s$ hydrogen orbit, 24 attoseconds.

---

[1] This chapter is an updated version of Ullrich and Bandrauk (2006c) with many additional references and new examples in the area of direct double ionization of helium and molecular strong-field processes.

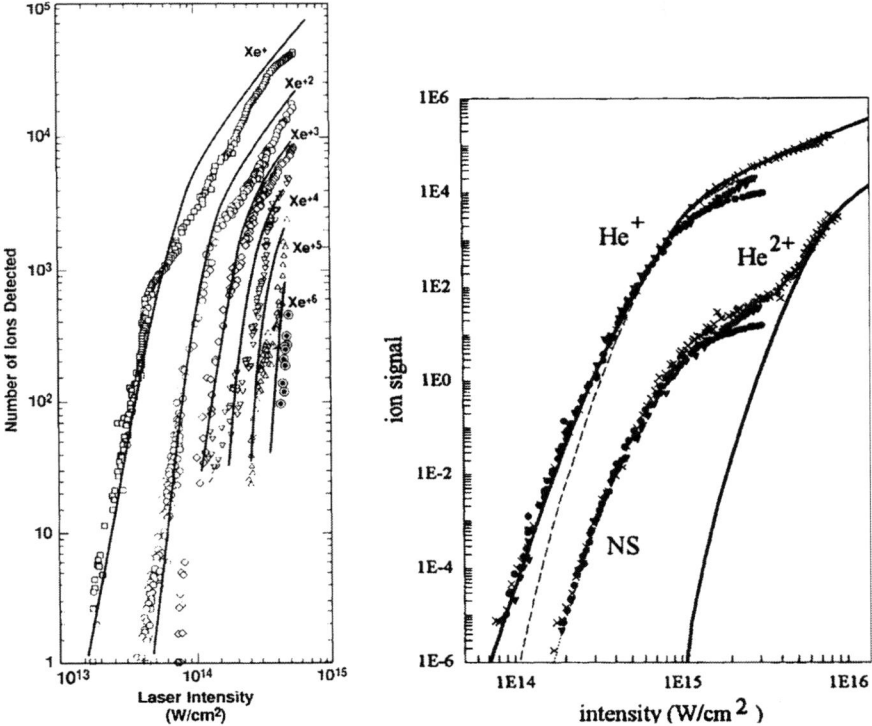

Fig. 18.2 *Left*: Sequential multiphoton ionization of Xe by 1 ps laser pulses of wavelength 585 nm [reproduced from Perry et al. (1988)]. *Right*: Non-sequential double ionization of He by 160 fs laser pulses of wavelength 780 nm [reproduced from Walker et al. (1994)]. In both panels, the full lines follow from rate equation models including only sequential processes

18.2 Atoms in Strong Laser Fields: an Overview

18.2.1 Multiphoton Ionization

An intense laser pulse can ionize an atom even when the photon energy is much smaller than the ionization potential. Figure 18.2 shows experimental data for the number of xenon ions as a function of laser intensity, at a wavelength of 585 nm and a pulse duration of 1 ps (Perry et al. 1988). At this wavelength, six photons are necessary to ionize the Xe atom, whereas the ionization process leading from, e.g. Xe^{+5} to Xe^{+6} requires already 34 photons. Notice that if a certain charge state is reached, increasing the laser intensity by about 50% is enough to remove a further electron, while the number of photons which should be absorbed at each step rapidly grows with the degree of ionization. Recently, multiphoton ionization up to Xe^{+24} was observed (Yamakawa et al. 2004; DiChiara et al. 2010).

Fig. 18.3 Schematic
representation of three
nonlinear phenomena of
atoms in intense laser fields
(L'Huillier 2002): sequential
and non-sequential
multiphoton ionization,
above-threshold ionization
(ATI), and high-harmonic
generation (HHG). A simple
interpretation is based on a
semiclassical recollision
model

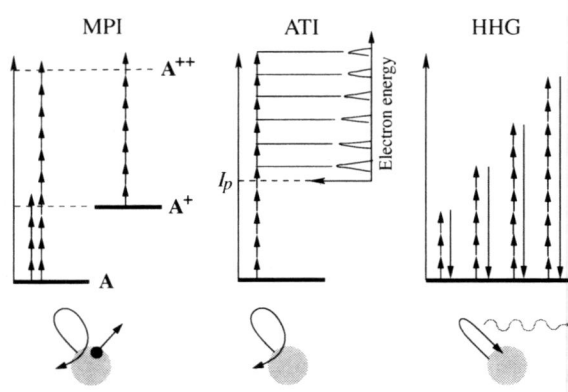

For relatively long pulse lengths and not too high intensities (L'Huillier et al. 1983; Perry et al. 1988), ionization proceeds via stepwise removal of the electrons. Ionization yields such as shown in the left panel of Fig. 18.2 can be theoretically explained by kinetic rate-equation models employing ionization rates obtained from lowest-order perturbation theory. For each ion species, the ion yield rises very steeply with intensity ($\sim I^N$, where N is the minimum number of photons required). At the saturation intensity, a marked change appears in the slope of the curves, associated with a depletion of an ion species when the ionization probability reaches unity.

For higher intensities (around 10^{15} W/cm^2) and shorter pulses (around 100 fs), this simple picture of multiphoton ionization becomes more complicated, and highly correlated ionization mechanisms come into play (Dörner and Weber 2002). Since the first observations in the early and mid-1990s (Fittinghoff et al. 1992; Walker et al. 1994; Larochelle et al. 1998), non-sequential double ionization of helium has been a hot field of research. The right panel of Fig. 18.2 shows the famous "helium-knee", indicating an enhancement of the He^{2+} yield by several orders of magnitude over what sequential ionization models would predict.

After a lot of initial controversy, the question of the non-sequential double ionization mechanism has now been settled. Experimental observations such as the measurement of recoil ion momentum (Weber et al. 2000; Moshammer et al. 2000), along with the suppression of the enhancement for elliptically polarized light (Fittinghoff et al. 1992), find their explanation in a simple three-step recollision model (Corkum 1993; Kulander et al. 1993; Lewenstein et al. 1994; Yudin and Ivanov 2001; Liu et al. 2004). In the first (bound-free) step, an electron is set free from its parent atom by tunnelling or (at higher intensities) over-the-barrier ionization. In the second (free-free) step, the driving laser field dominates the electron dynamics, and the ionic Coulomb force can be ignored. As the phase of the laser field reverses, the electron is driven back to the atomic core. In the third step, the electron can then scatter off the core and in the process knock out another electron, or produce harmonic generation through radiative recombination (see below). These processes are illustrated schematically in Fig. 18.3.

Correlated multiple ionization is by no means limited to helium, but has been observed in other rare-gas atoms as well. For instance, neon atoms in 25 fs laser pulses show simultaneous emission of up to four photoelectrons due to recollision (Rudenko et al. 2004).

At even higher intensities and at very high frequencies, theory predicts the surprising phenomenon of stabilization against ionization (Pont et al. 1998, 1990; Kulander et al. 1991a; Eberly and Kulander 1993; Bauer and Ceccherini 1999; Gavrila 2002): as the laser intensity is increased, atomic ionization rates pass through a maximum and then *decrease*. Experimental evidence of stabilization has been reported for atomic Rydberg states (de Boer et al. 1993, 1994; van Druten et al. 1997).

18.2.2 Above-Threshold Ionization

Another highly nonlinear phenomenon known as above-threshold ionization (ATI) (Becker et al. 2002) occurs when electrons absorb a large number of extra photons in addition to those needed to overcome the ionization barrier.[2] One detects a whole sequence of equally spaced peaks in the kinetic-energy distribution of the photoelectrons (Agostini et al. 1979; Kruit et al. 1983), with energies $n\hbar\omega - I_p - U_p$. Here, I_p is the atomic ionization potential, and $U_p = e^2\mathcal{E}^2/4m\omega^2$ is the ponderomotive potential associated with the wiggle motion of a free electron in a laser field. Most of the early work in the 1980s concentrated on the low-energy part of the ATI spectrum, investigating the role of the ponderomotive potential, the AC-Stark shifted resonant excited states, and the transition from multiphoton to tunnelling regime.

In the mid-1990s the experimental precision in recording photoelectron spectra increased significantly, and it was found that ATI spectra extend over many tens of eV, with a decrease for the first orders up to $\sim 2U_p$, followed by a large plateau extending up to $\sim 10U_p$ (Paulus et al. 1994, 2001; Grasbon et al. 2003), see Fig. 18.4. This can again be explained in the semiclassical recollision model, where electrons are lifted into the continuum at some phase of the laser's electric field and start from the atom with zero velocity. The energy $2U_p$ is the resulting classical maximum kinetic energy for electrons leaving the atom without rescattering. Free electrons that reencounter the ion and elastically rescatter may acquire a maximal classical energy of $10U_p$. With linear polarization, electrons are generated along the polarization direction. However, angular distributions may exhibit a much more complex off-axis structure at the edge of the plateau ("scattering rings"), which can be explained as a consequence of the rescattering of the electron wavepacket on the parent ion.

---

[2] Strictly speaking, ATI belongs to the general class of multiphoton ionization phenomena. However, in practice it is understood that "multiphoton ionization" refers to the counting of various ionized states of atoms or molecules produced by the laser field, whereas ATI specifically refers to the photoelectron spectra.

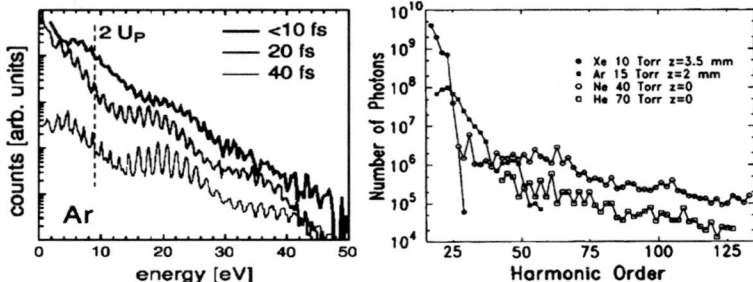

Fig. 18.4 *Left*: ATI spectrum for argon, for various pulse lengths, at 800 nm and 0.8×10^{14} W/cm^2 [reproduced from Grasbon et al. (2003)]. *Right*: HHG in rare-gas atoms driven by 1 ps laser pulses of wavelength 1053 nm and intensity 1×10^{15} W/cm^2 [reproduced from L'Huillier and Balcou (1993)]

18.2.3 Harmonic Generation

When an atom interacts with a laser field, a dipole moment is induced which in turn acts as a source of radiation. At high laser intensities, the atomic response becomes extremely nonlinear. As a result, pronounced signals at multiples of the driving frequency appear in the photoemission spectrum [high-harmonic generation or HHG (Salières et al. 1999; Brabec and Krausz 2000)]. Because atoms have inversion symmetry, only odd multiples of the driving frequency are emitted. In contrast to the perturbative picture prevailing at weak laser fields, strong laser pulses can yield a very large number of harmonics (McPherson et al. 1987; Ferray et al. 1988; Li et al. 1989; L'Huillier et al. 1992; Sarukura et al. 1991; Miyazaki and Sakai 1992; Wahlström et al. 1993; Kondo et al. 1994; L'Huillier and Balcou 1993; Macklin et al. 1993).

Figure 18.4 shows experimental data on harmonic generation of rare-gas atoms (L'Huillier and Balcou 1993), exhibiting the typical rapid decrease over the first few harmonics followed by an extended plateau. The highest harmonic observed here is the 135th harmonic of 1053 nm (which corresponds to 7.8 nm) with He as target atom. We find that the width of the plateau decreases going from He to Xe, while at the same time the absolute intensity of the observed harmonics becomes larger. This behavior is linked to differences in the static polarizabilities of the target atoms (Liang et al. 1994; Chin and Golovinski 1995). Harmonic orders of around 300 have been observed using ultrashort, high-intensity laser pulses, where the atoms experience only a few laser cycles (Zhou et al. 1996; Spielmann et al. 1997; Schnürer et al. 1998). Under these conditions, harmonic frequencies extend beyond 500 eV, reaching into the water window with wavelengths around 2.7 Å.

Theoretically, the cutoff of the HHG plateaus is predicted to occur at $\hbar\omega_c = I_p + 3.2 U_p$, following the three-step recollision model (Corkum 1993). Here, the ponderomotive potential U_p refers to the saturation intensity of the respective atomic species, i.e. that intensity at which the atom gets ionized. This intensity is usually significantly lower than the peak intensity of the laser pulse. The data shown in Fig. 18.4 is in excellent agreement with this simple rule (L'Huillier and Balcou 1993).

18.2.4 Theoretical Methods

There exists a large body of theoretical work devoted to the various aspects of the nonlinear physics of atoms in strong fields, for a review see (Joachain et al. 2000). From the experimental phenomenology it is clear that perturbative and semi-perturbative methods fail to capture the extremely high degree of nonlinearity, and one must resort to non-perturbative theories. Many approaches, most prominently Floquet (Chu and Telnov 2004) and R-matrix Floquet (Burke et al. 1990, 1991) theories, are based on the assumption that the Hamiltonian of the atom-laser field system is periodic in time. Although this is not true for a realistic laser pulse, one can nevertheless incorporate some pulse shape effect into (R-matrix) Floquet calculations, provided that the atom remains in a Floquet eigenstate that is adiabatically connected to the initial state. Clearly, such an assumption can be expected to break down for ultrashort, femtosecond pulses. The same is true for the many-body S-matrix theory (Becker and Faisal 2005), which can be viewed as a low-frequency approach.

In general, therefore, one needs to resort to a direct integration of the time-dependent Schrödinger equation (TDSE). Since the pioneering work by Kulander (1987), and Kulander et al. (1992, 1993), most activity focused on the hydrogen atom, where the TDSE can be numerically solved without restrictions on large grids (Cormier and Lambropoulos 1997; Tong and Chu 1997). This strategy becomes of course tremendously involved for atoms having more than one electron; a propagation of the full two-electron wave function of helium in all three spatial dimensions was carried out in Parker et al. (1996, 1998), using a massively parallel supercomputer. In general, however, the TDSE for two-electron systems can be solved only on a restricted basis (Lambropoulos et al. 1998). As an alternative one can treat one- and two-electron atoms and molecules as one-dimensional (1D) model systems, which has been particularly useful to elucidate the mechanism of non-sequential multiphoton ionization (Eberly et al. 1992; Bauer 1997; Lappas and van Leeuwen 1998; Lein et al. 2000a; Dahlen and van Leeuwen 2001; Dahlen 2002).

All studies dealing with many-electron atoms in strong laser fields have made use of more or less severe approximations to reduce the problem to a tractable size. Most conspicuously, in the single-active electron (SAE) model (Kulander 1988; Kulander et al. 1991b; Tang et al. 1991; Awasthi et al. 2008) the TDSE is solved for only one "active" electron while the remaining electrons are frozen in their initial configuration, their influence on the active electron being simulated by a *static* model potential. This strategy successfully models the screening of the nuclear charge by the inner electrons, but cannot describe collective effects arising from electronic correlation.

18.3 TDDFT for Atoms in Strong Laser Fields

The first TDDFT studies of atoms in strong laser fields were carried out in the mid-1990s (Ullrich et al. 1995b, 1996; Ullrich and Gross 1997; Erhard and Gross 1997; Tong and Chu 1998), solving the TDKS equations for He, Be, and Ne atoms in the presence of time-dependent potentials of the form

$$v_{\text{laser}}(\boldsymbol{r}, t) = \mathcal{E} z f(t) \sin(\omega t), \tag{18.1}$$

describing a laser field in the length gauge (Joachain et al. 2000). Here, $\mathcal{E}$ is the electric-field amplitude, ω is the laser frequency, and $f(t)$ is a function between 0 and 1 which describes the switching-on or pulse envelope of the laser.

We now discuss the Ne atom in intense laser fields of wavelength 248 nm in some detail (Ullrich and Gross 1997). Due to the linear polarization of the laser field, the rotational symmetry of the atom around the z-axis is preserved at all times. It is thus appropriate to solve the TDKS equations using cylindrical coordinates. For this example, the complete outer shell is propagated, consisting of $2s$, $2p_0$ and $2p_1$ orbitals, each doubly occupied. Here, the $2p_0$ orbital is oriented along the laser polarization axis, whereas the two $2p_1$ orbitals are perpendicular, and have thus an identical time evolution. The inner $1s$ orbital is kept frozen in its initial state. The TDKS equations are solved on a grid with a finite-difference representation, and using the Crank-Nicholson algorithm for the time propagation. The calculations are done in exchange-only, treated in ALDA and TDKLI (Ullrich et al. 1995a).

Under the influence of the intense driving field, the entire valence shell gets strongly excited, and the KS orbitals acquire substantial continuum contributions. To prevent electronic flux from being reflected back from the edges of the numerical grid, absorbing boundary conditions are introduced in the form of a so-called mask function (a complex potential would be an alternative). Over the course of time, the norm of the KS orbitals $N_j(t)$ thus *decreases*, even though the time propagation algorithm is unitary. This allows us to describe ionization in a straightforward manner, by calculating the number of electrons remaining in a finite volume:

$$N_j(t) = \int_{\mathcal{V}} \mathrm{d}^3 r |\phi_j(\boldsymbol{r}, t)|^2, \quad N(t) = \sum_j N_j(t) = \int_{\mathcal{V}} \mathrm{d}^3 r\, n(\boldsymbol{r}, t). \tag{18.2}$$

Here, $\mathcal{V}$ refers to a volume centered about the nucleus which contains essentially the entire wavefunction at $t = t_0$. The total number of escaped electrons, $N_{\text{esc}}(t) = N_0 - N(t)$, is thus a simple functional of the time-dependent density. The left panel of Fig. 18.5 shows $N_{2s}(t)$, $N_{2p_0}(t)$, and $N_{2p_1}(t)$ for the Ne atom. As expected, the $2s$ orbital is the least ionized of the three orbitals. The $2p_0$ and $2p_1$ orbital differ by about an order of magnitude in their degree of ionization, which is a typical observation and due to the fact that the $2p_0$ orbital is oriented along the polarization axis which makes it easier for the electrons to escape. The difference between ALDA and TDKLI can be understood from the differences of the KS orbital eigenvalues (the electrons are more weakly bound in LDA than in KLI).

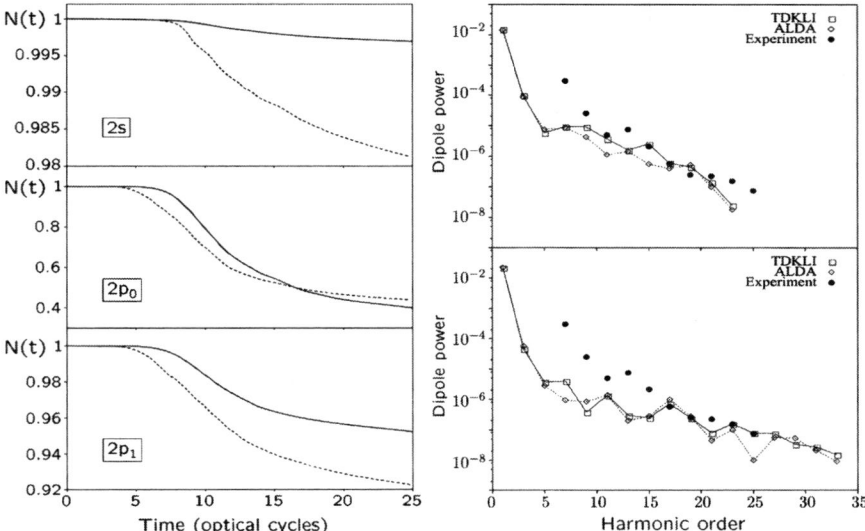

Fig. 18.5 *Left*: time-dependent norm $N(t)$ of the Ne $2s$, $2p_0$ and $2p_1$ orbitals, calculated with TDKLI (*full lines*) and ALDA (*dashed lines*). The laser parameters are $\lambda = 248$ nm, $I = 3 \times 10^{15}$ W/cm^2, and a 10-cycle linear ramp switching-on. *Right*: Harmonic distributions for Ne at $I = 3 \times 10^{15}$ W/cm^2 (*top*) and $I = 5 \times 10^{15}$ W/cm^2 (*bottom*). Experimental data from Sarukura et al. (1991)

Another observable which is a straightforward density functional is the dipole moment $d(t) = \int d^3 r z n(\mathbf{r}, t)$. The associated power spectrum, $|d(\omega)|^2$, yields the contribution to the HHG spectrum of a single atom, which is proportional to the experimentally observed HHG spectra to within a reasonable approximation [in general, however, one needs to include propagation effects within the interaction volume (Salières et al. 1999)]. The right panel of Fig. 18.5 shows the calculated HHG spectra for Ne, at the two intensities I=3 and 5×10^{15} W/cm^2. Comparison with experimental data (Sarukura et al. 1991) taken with a laser pulse of peak intensity 4×10^{17} W/cm^2 shows that the observed HHG spectra must have been produced at the rising edge of the pulse.

To connect with multiphoton ionization experiments, one needs to calculate the probability of finding the atom in one of the possible charge states to which it can ionize, $P^{+n}(t)$. In the context of TDDFT, this is a difficult problem and has been extensively discussed in the literature (Lappas and van Leeuwen 1998; Dahlen and van Leeuwen 2001; Dahlen 2002; Petersilka and Gross 1999; Ullrich 2000). A rigorous definition of the $P^{+n}(t)$ involves the time-dependent many-body wave function (Ullrich 2000a):

$$P^0(t) = \int_{\mathcal{V}} d^4x_1 \cdots \int_{\mathcal{V}} d^4x_N |\Psi(\mathbf{x}_1, \ldots, \mathbf{x}_N, t)|^2 \qquad (18.3a)$$

$$P^{+1}(t) = \binom{N}{1} \int_{\overline{\mathcal{V}}} d^4x_1 \int_{\mathcal{V}} d^4x_2 \cdots \int_{\mathcal{V}} d^4x_N |\Psi(\mathbf{x}_1, \ldots, \mathbf{x}_N, t)|^2 \qquad (18.3b)$$

and similarly for all other P^{+n}, where $\overline{\mathcal{V}}$ refers to the region outside the integration volume $\mathcal{V}$, and $\mathbf{x}$ denotes spatial and spin coordinates. In the case of a two-electron system, one can rewrite these formulas by introducing the pair correlation function $g(\mathbf{r}_1, \mathbf{r}_2, t) = 2|\Psi(\mathbf{r}_1, \mathbf{r}_2, t)|^2 / n(\mathbf{r}_1, t) n(\mathbf{r}_2, t)$:

$$P^0(t) = \frac{1}{2} \int_{\mathcal{V}} d^3r_1 \int_{\mathcal{V}} d^3r_2 \, n(\mathbf{r}_1, t) n(\mathbf{r}_2, t) g(\mathbf{r}_1, \mathbf{r}_2, t) \qquad (18.4a)$$

$$P^{+1}(t) = \int_{\mathcal{V}} d^3r \, n(\mathbf{r}, t) - \int_{\mathcal{V}} d^3r_1 \int_{\mathcal{V}} d^3r_2 \, n(\mathbf{r}_1, t) n(\mathbf{r}_2, t) g(\mathbf{r}_1, \mathbf{r}_2, t) \qquad (18.4b)$$

$$P^{+2}(t) = 1 - P^{+1}(t) - P^0(t). \qquad (18.4c)$$

The task is thus twofold: one needs to find an accurate xc functional for the TDKS equation, as well as an accurate expression for the pair correlation function. A straightforward approximation is to set the latter equal to its x-only limit of 1/2, which amounts to an independent-particle approximation. For a two-electron system with a doubly occupied TDKS orbital, this gives

$$P^0(t) = N(t)^2, \quad P^{+1}(t) = 2N(t)[1 - N(t)], \quad P^{+2}(t) = [1 - N(t)]^2, \quad (18.5)$$

where $N(t) = \frac{1}{2} \int_{\mathcal{V}} d^3r \, n(\mathbf{r}, t)$. Figure 18.6 shows results for the Ne atom, where up to Ne^{+4} ionic states are found. However, it turns out that the independent-particle approximation for $P^{+n}(t)$ leads to significant errors for the non-sequential double ionization probabilities (Lappas and van Leeuwen 1998; Dahlen and van Leeuwen 2001; Dahlen 2002), even if one uses the *exact* TDDFT orbital densities, as we will demonstrate below for the case of a 1D H_2 molecule. Attempts to improve this situation by employing a local-density approximation for the pair correlation function $g_c[n](\mathbf{r}_1, \mathbf{r}_2)$ met with little success (Petersilka and Gross 1999). This is indicative of a sizable degree of "correlation" of the two-electron dynamics, in particular for the case of longer wavelengths. At 248 nm, on the other hand, the independent-particle approximation works much better (Dahlen 2002). Similar findings have been reported by Bauer (1997), Bauer and Ceccherini (2001).

An important breakthrough in the TDDFT description of multiphoton ionization was made by Lein and Kümmel (2005), who realized that the key property of the xc potential is the discontinuity upon change of particle number (Mundt and Kümmel 2005; De Wijn et al. 2008). In the case of the helium atom, x-only theories fail to

Fig. 18.6 Population of the ionized states of Ne (TDKLI), using the independent-particle approximation for $P^{+n}(t)$. Laser: $\lambda = 248$ nm, $I = 5 \times 10^{15}$ W/cm^2

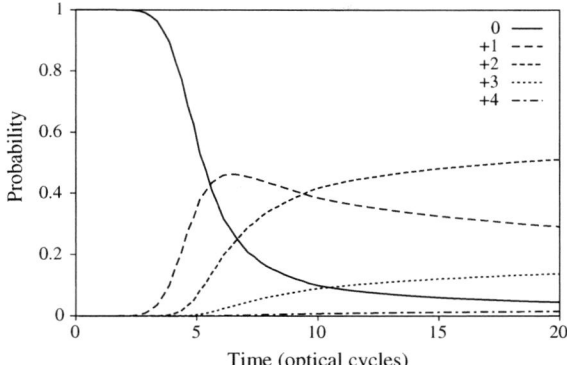

capture the enhancement of double ionization: the helium knee is purely due to correlation effects. Lein and Kümmel proposed the following simple model correlation potential with a discontinuity:

$$v_c(z, t) = [s(t) - 1]\left[v_H(z, t) + v_X(z, t)\right], \tag{18.6}$$

where $s(t)$ is a step-like function of the number of bound electrons, depending on a positive, sufficiently large constant C (e.g. $C=50$):

$$s(t) = \frac{2/N(t)}{1 + \exp[C(2/N(t) - 2)]}. \tag{18.7}$$

Initially, $N=2$ so that $s=1$ and the correlation potential vanishes. As the number of bound electrons decreases and approaches 1, $s(t)$ grows smoothly to $s=2$ and then suddenly jumps to zero as $N(t)$ passes through 1. The correlation potential then cancels out the Hartree plus exchange potentials, as it should for a single-particle system. Using the correlation potential (18.6) and the simple, uncorrelated ion probabilities (18.5) one indeed finds a plateau structure, as shown in the middle panel of Fig. 18.7 . However, the knee is still too high; the remaining error lies with the formulas to calculate the ion probabilities. Wilken and Bauer (2006) developed a model expression for the pair correlation function in Eqs. 18.4a, b, c, using an adiabatic approximation:

$$g = \begin{cases} -\dfrac{1}{2} & 0 \leq N \leq 1 \\[2mm] \dfrac{\rho^A(z, z', t)}{n^A(z, t)n^A(z', t)} - \dfrac{1}{2} & 1 \leq N \leq 2, \end{cases} \tag{18.8}$$

where n^A is an approximation for the density of the helium atom, using linear combinations of ground-state densities, and ρ^A is a similar approximation for the pair density. In this way, the helium knee is very well reproduced, as shown in the right-hand panel of Fig. 18.7 .

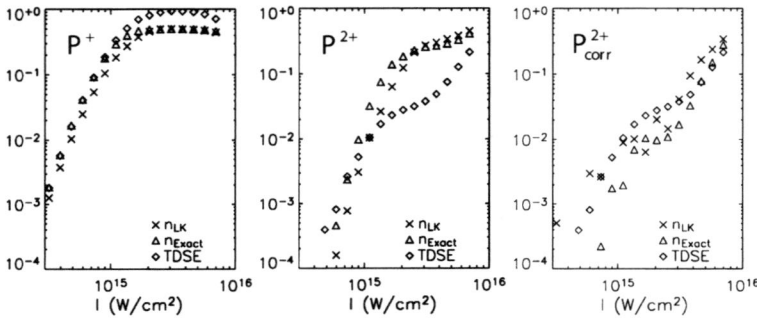

Fig. 18.7 Single and double ionization probabilities for a one-dimensional helium atom in a 780 nm laser pulse. *Left* and *middle*: ion probabilities calculated with the uncorrelated expressions (18.5), using the exact density and the density obtained from the Lein-Kümmel correlation potential (18.6), compared to the exact ion probabilities following from the two-electron TDSE. *Right*: P^{2+} using the pair correlation function (18.8). (Adapted from Wilken and Bauer, 2006)

Similar issues arise for the calculation of ATI spectra, which are rigorously defined in terms of the many-body wavefunction and can therefore in principle be expressed as functionals of $n(\mathbf{r}, t)$. In practice, however, all TDDFT approaches calculate the kinetic-energy spectra associated with the KS single-particle orbitals (Pohl et al. 2000; Nguyen et al. 2004), despite the fact that the KS wave function has no strict physical meaning. One obtains the kinetic energy distribution $P_{KS}(E)$ by recording the KS orbitals $\varphi_j(\mathbf{r}, t)$ over time at a point $\mathbf{r}_b$ near the grid boundary and subsequent Fourier transformation, so that

$$P_{KS}(E) = \sum_j |\varphi_j(\mathbf{r}_b, E)|^2. \tag{18.9}$$

Alternatively (Vèniard et al. 2003), one can obtain $P_{KS}(E)$ by propagating the TDKS equation for an additional Δt after the end of the laser pulse, and calculating the probability of finding an electron of energy E in the spatial region between $r_\pm = \Delta t \sqrt{2(E \pm \Delta E)}$, where ΔE is the energy resolution.

18.4 Molecules in Strong Fields

18.4.1 Overview

The interaction of molecules with intense laser pulses introduces new challenges due to the presence of the extra degrees of freedom of the nuclear motion, and the associated additional time scales. The shortest nuclear motion period is that of the proton, 10–15 fs, which is comparable to the ionization times at intensities of the order of 10^{14} W/cm^2 at wavelengths of 800–1064 nm. Thus, the nuclear motion should be treated on an equal footing with the radiative processes induced by intense laser fields, in order to study the photochemical dynamics in the nonperturbative multiphoton regime. This section summarizes a few highlights of the physics of molecules in strong fields; more details can be found in recent review articles (Bandrauk and Kono 2003; Marangos 2004). The topic of molecular alignment in strong fields has been reviewed in Corkum et al. (1999) and Stapelfeldt and Seideman (2003).

The dressed-molecule representation (Bandrauk 1994a, b) has been a successful approach at intensities below 10^{13} W/cm^2 where ionization is negligible. Here, resonant processes are pictured as crossings of "dressed" molecular potentials. At higher intensities, the dynamics of photophysical processes can be conveniently described using laser-induced molecular potentials (LIMPs). In the intensity range 10^{14}–10^{15} W/cm^2, rapid ionization starts to set in, accompanied by considerable distortions of intermolecular potentials, creating LIMPs that can lead to bond softening via laser-induced avoided crossings. The molecular ions can also undergo above-threshold dissociation, the equivalent of ATI in atoms.

Compared to atoms, molecules offer new perspectives of laser-induced electron recollision (LIERC) with parent molecular ions, such as "diffraction" from more than one nuclear center. This leads to a new molecular phenomenon, laser-induced electron diffraction (LIED) (Zuo et al. 1996), a new tool for probing molecular geometry changes on ultrashort time scales (Bandrauk and Chelkowski 2001; Itatani et al. 2004). Much of the theoretical understanding of LIERC and LIED in molecules is based on exact solutions of the TDSE for the one-electron H_2^+ and H_3^{++} molecules for static nuclei (Bandrauk and Chelkowski 2001, Itatani et al. 2004) as well as moving nuclei (non-Born-Oppenheimer) (Bandrauk et al. 2003b).

For multielectron atoms, double ionization in intense laser fields is a highly correlated process. As we discussed above, LIERC is a dominant mechanism for the nonsequential steps, transforming sequential to double ionization at higher intensities. Molecules differ from atoms by the multicenter Coulomb nature of the electron recollision process, leading to diffraction (Zuo and Bandrauk 1996b; Itatani et al. 2004) and even collision with neighboring ions (Bandrauk and Yu 1998, 1999). In particular, at large internuclear distances, tunnelling ionization, which is the first step in atomic LIERC, becomes more complicated as a consequence of charge transfer and charge resonance effects, first predicted in Mulliken (1939) due to the existence of excited ion-pair states (Martin and Hepburn 1997) which cross the ground state at high intensities in both 1D (Kawata et al. 2000) and recent 3D (Harumiya et al. 2002) simulations of H_2. Similar effects were found in 1D models of H_3^+ at high intensities (Kawata et al. 2001).

The H_2 molecule is the prototype model of the two-electron chemical bond with bonding and anti-bonding molecular orbitals. Earlier 3D calculations of the nonlinear response of H_2 in an intense laser field were performed using a frozen core approximation (Krause et al. 1991), and TDHF using finite-element basis sets (Yu and Bandrauk 1995). Exact 1D (Kawata et al. 2001) and 3D (Harumiya et al. 2002) TDSE numerical solutions of H_2 were obtained on large finite grids at equilibrium and large intermolecular distances in order to confirm the universal molecular phenomena of charge resonance enhanced ionization (CREI), first discovered in 3D

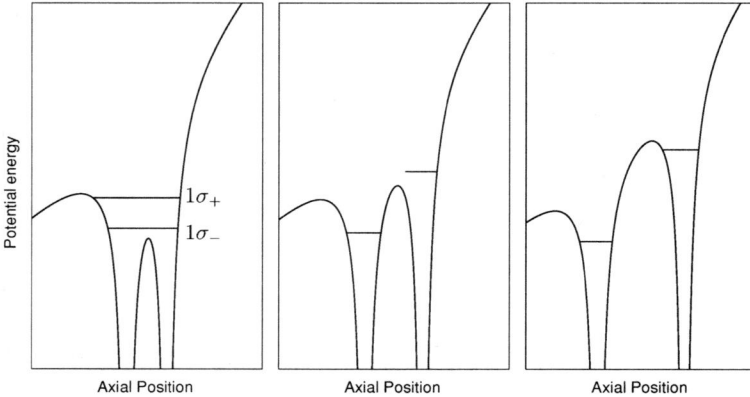

Fig. 18.8 Schematic illustration of CREI in a diatomic molecule. For certain internuclear separations above a critical R_c, the LUMO is raised above the inner barrier. Ionization proceeds via radiative coupling between the HOMO $1\sigma_-$ and LUMO $1\sigma_+$ ("essential" or "doorway" states) (Bandrauk and Kono 2003a)

(Zuo and Bandrauk 1995) and 1D (Zuo and Bandrauk 1995, 1996b; Chelkowski and Bandrauk 1995; Seideman et al. 1995) one-electron molecular model simulations. As illustrated in Fig. 18.8 , an electron in a diatomic molecule experiences essentially a double potential well, which becomes distorted in the presence of a laser field by the gradient of the optical potential across the molecule. There exists a critical separation between the two nuclei, R_c, at which the bound state in the upper well is raised to the point that it barely sees any barrier to tunnel through into the second well and from then on to ionization. As a result, for certain nuclear distances the ionization rate of the molecule can be an order of magnitude higher than for the individual atoms.

A number of recent experiments have addressed the issue of sequential and nonsequential ionization in diatomic molecules such as H_2 (Alnaser et al. 2004b), D_2 (Sakai et al. 2003; Lègarè et al. 2003), O_2, N_2, NO, and I_2 (Guo and Gibson 2001; Eremina et al. 2004; Alnaser et al. 2004a; Suzuki et al. 2004). These studies have confirmed the validity of the CREI and LIERC mechanisms for multielectron systems, and have explored the phenomenon of Coulomb explosion following rapid ionization with few-cycle pulses, combined with a possible imaging of the fragments. Other recent studies have focused on multiphoton ionization of larger, polyatomic molecules (Markevitch et al. 2003, 2004).

18.4.2 A 1D Example: H_2 with Fixed Nuclei

1D models have proven to be useful to recover the essential physics of laser-molecular interactions at high intensities, i.e. tunneling ionization and laser control of the ionized electron trajectory. For molecules such as H_2 and H_3^+ in linear configuration, the arrangement of nuclei enhances refocusing of the ionized electron along the internuclear axis (Bandrauk and Yu 1999; Kawata et al. 2001; Villeneuve et al. 1996).

For H_2 with static nuclei, the 1D Born-Oppenheimer Hamiltonian for two electrons with coordinates z_1, z_2 with respect to the center of mass of two protons situated at positions $\pm R/2$ is written as

$$\hat{H}_0 = -\frac{1}{2} \sum_{i=1}^{2} \left[\frac{\partial^2}{\partial z_i^2} - \frac{1}{\sqrt{(z_i \pm R/2)^2 + c}} \right] + \frac{1}{\sqrt{(z_1 - z_2)^2 + d}}. \qquad (18.10)$$

The softening parameters c and d remove Coulomb singularities and allow the use of high-order split-operator methods for solving the TDSE (Bandrauk and Shen 1993; 1994c). Propagating the TDSE with $\hat{H}_0$ (18.10) in imaginary time yields the ground state electronic energies and molecular potentials; choosing $c = 0.7$ and $d = 1.2375$ reproduces accurately the first three electronic states of H_2. The TDSE for H_2 in an intense laser field then becomes

$$i\frac{\partial \Psi(z_1, z_2, t)}{\partial t} = \left[\hat{H}_0 + v(z_1, z_2, t) \right] \Psi(z_1, z_2, t), \qquad (18.11)$$

where $v(z_1, z_2, t) = -(z_1 + z_2)\mathcal{E}(t)\cos(\omega t)$ is the dipolar (long wavelength) form of the electron-laser interaction for a laser pulse of frequency ω and electric-field envelope $\mathcal{E}(t)$.

We now discuss the ionization of the ground $X^1\Sigma_g^+$ state of H_2 at laser wavelength $\lambda = 800$ nm and intensities $10^{13} < I < 10^{15}$ W/cm^2. Single and double ionization probabilities, P^{+1} and P^{+2}, are obtained by numerical integration of the total two-electron probabilities $|\Psi(z_1, z_2, t)|^2$, see Eqs. 18.3a, b, c, where the integration volume $\mathcal{V}$ is the region $|z_i| \leq 6$ a.u. The numerical integration procedure is verified by projecting the total final wavefunction $\Psi(z_1, z_2, T)$ on complete sets of field free bound and continuum states of H_2^+ (Pindzola et al. 1997). The grid is large enough ($|z_i| \leq 1000$) to capture all of the first ionized electrons and identify double ionization due to recollision processes.

Figure 18.9 shows a comparison of P^{+1} and P^{+2} obtained from the exact one-electron density, but evaluated using the independent-particle expressions (18.5), versus the exact results. The one-electron ionization probabilities, $H_2 \rightarrow H_2^+ + e^-$, and the sequential double ionization, $H_2^+ \rightarrow H_2^{++} + e^-$, agree very well in both methods. Both P^{+2} by density and exact integration show a knee (plateau) at the saturation of the first ionization, $I \sim 10^{15}$ W/cm^2. The knee coincides with 100% ($P^{+1} = 1$) of the first ionization. The exact double ionization P^{+2} is parallel to P^{+1} at low intensities, confirming its source from recollision of the first electron with the ion core. The efficiency of the double ionization through recollision is $P^{+2}/P^{+1} \sim 10^{-2}$. Notice that the ratio is underestimated at low intensities and overestimated in the knee region from the exact one-electron density, thus confirming the necessity of introducing the correlation factor in (18.4a, b).

Figure 18.10 compares TDEHF and ALDA versus the exact probabilities. In the TDEHF method, one uses two different, non-orthogonal initial orbitals, $1\sigma_g$ and $1\sigma_g'$,

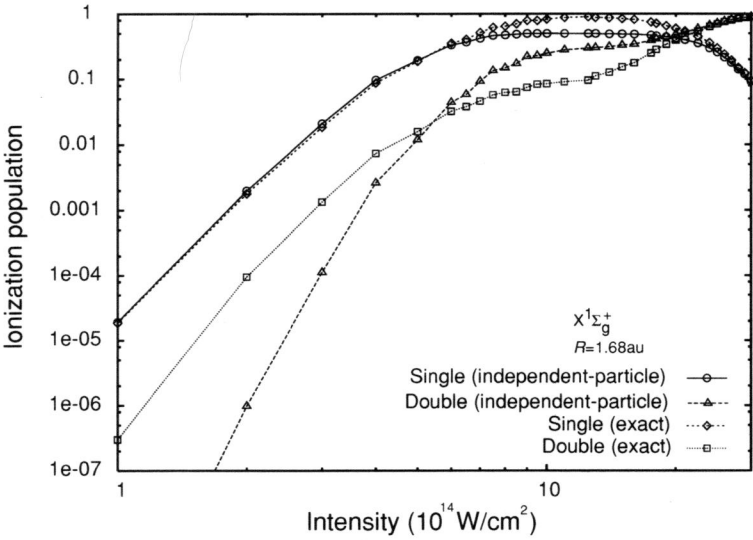

Fig. 18.9 Population of singly and doubly ionized states of 1D H_2, $\lambda = 800$ nm, comparing exact results and independent-particle results based on Eq. 18.5, but evaluated with the exact density (Adapted from Bandrauk and Lu, 2006)

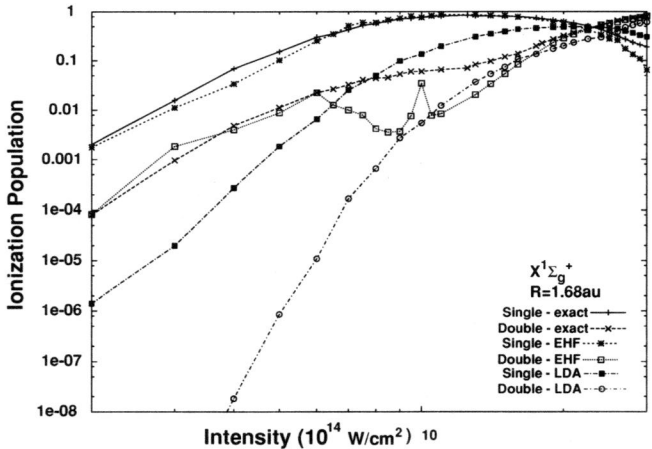

Fig. 18.10 Population of singly and doubly ionized states of 1D H_2, $\lambda = 800$ nm, comparing exact results and TDEHF (Adapted from Nguyen and Bandrauk, 2006)

which represent the propensity for electrons to be inequivalent in the ground state, i.e. inner and outer electrons. This distinction becomes amplified upon ionization and reflects better the physics than TDHF-like theories where both electrons are initially equivalent. Figure 18.10 shows that TDEHF agrees fairly well with the exact P^{+1}

and P^{+2} below the knee. Some anomalies appear in the knee region where the first electron has already ionized and cannot recollide. Similar effects take place in 1D models of He (Dahlen and van Leeuwen 2001; Dahlen 2002). The ALDA single and double ionization are grossly underestimated, reflecting the improper asymptotic behavior of electronic wavefunction and ionization potential, as well as insufficient screening.

18.4.3 TDDFT for Molecules in Strong Fields

Even in 1D, exact numerical calculations of strong-field molecular processes are presently limited to at most two electrons plus moving nuclei. To describe the dynamics of multielectron molecular systems without making the SAE approximation, TDDFT is unrivalled due to its computational simplicity. We will now review some recent applications.

In Chu and Chu (2001a, b, 2004b), HHG and multiphoton ionization were studied for various dimers with fixed nuclei in strong laser fields, using xc potentials with the correct asymptotic behavior. The role of the binding energy and orientation of individual molecular orbitals was explored, and a variety of correlation and interference effects between these orbitals in HHG spectra is discussed, as well as the role of inner valence electrons in determining the total ionization. Along similar lines, Dundas and Rost (Dundas and Rost 2005) used an all-electron, x-only ALDA approach to investigate the suppression of single ionization in N_2, O_2, and F_2 due to destructive interference of outgoing electron waves from the ionized electron orbitals. This again points to the insufficiency of the SAE model in describing multielectron systems.

As an example, consider the ionization of CO_2, a linear molecule whose highest occupied molecular ground-state orbitals ($1\pi_u$, $3\sigma_g$, and $1\pi_g$) are shown in Fig. 18.11 (Fowe and Bandrauk 2009, 2010a, b). As long as the molecule is oriented parallel to the field ($\theta = 0$), the π-orbitals along the x and y direction are equivalent; for other orientations, differences develop. Figure 18.11 compares the ionization of the molecular TDKS orbitals for two different xc functionals, LDA and LB94 (van Leeuwen and Baerends 1994). The latter has the correct asymptotics and yields excellent ionization energies [see also Awasthi et al. (2008)]. In LDA, the electrons are too weakly bound and ionize too easily. The main finding is that the ionization of individual orbitals depends strongly on their orientation, and the highest orbital is not necessarily the one that ionizes most rapidly. This is further illustrated in Fig. 18.12 using the TDELF. It can be seen that the dominant response comes in both cases from the lone pair region of the molecule ($1\pi_u$), but at $\theta = 90°$ the C-O bond region (involving inner orbitals) is significantly affected as well.

Baer et al. (2003) considered ionization and HHG in benzene by short circularly-polarized pulses, propagating 15 TDKS orbitals within ALDA on a 3D grid with pseudopotentials. The interplay between bound-bound and bound-continuum

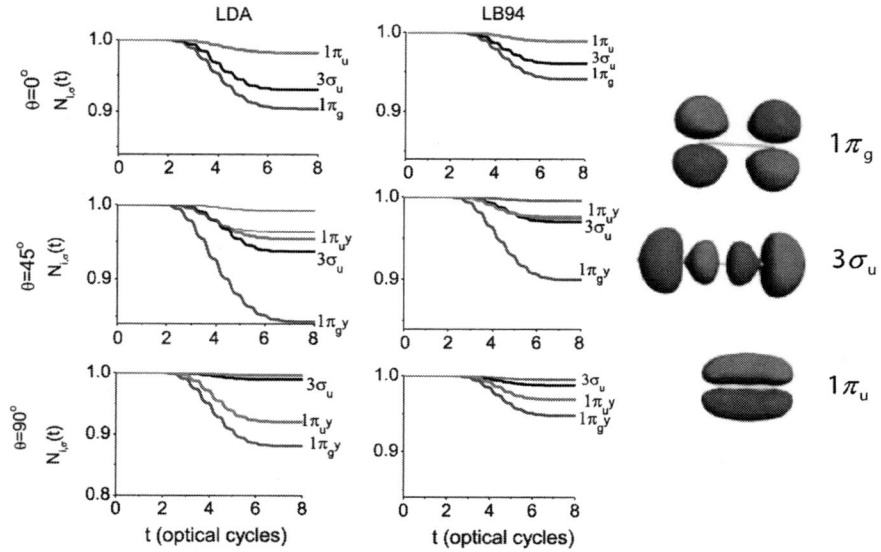

Fig. 18.11 *Right*: the three highest occupied ground-state molecular orbitals of CO_2. *Left*: orbital population of CO_2 subject to 800 nm laser pulses with $3.5 \times 10^{14} W/cm^2$, oriented at different angles with respect to the field and calculated with adiabatic LDA and LB94 xc functionals. (Adapted from Fowe and Bandrauk, 2010a)

transitions as well as multielectron dynamics was found to cause some unique features in the HHG spectra, specific to the ring geometry of the molecule and the circularly polarized light. Similar studies for circular carbon chains and nanotubes were carried out in Liu et al. (2010).

Several TDDFT studies have explored the route leading from molecules to clusters, e.g. using models based on chains of rare-gas atoms (Vèniard et al. 2002), or small silver molecules with fixed nuclear positions (Nobusada and Yabana 2004). A massively parallel TDKS calculation combining the dynamics of the highly excited electron cloud with the (classical) motion of the nuclei was carried out by Calvayrac et al. (Calvayrac et al. 1998), simulating the Coulomb explosion of a Na_{12} cluster. A more recent study of two-color photoionization in Na_4 and Na_4^+ clusters (Nguyen et al. 2004) found counterintuitive asymmetries and unexpectedly large plateaus in the ATI spectra.

The fully quantum mechanical, non-Born-Oppenheimer description of correlated electron and nuclear dynamics is one of the biggest challenges for TDDFT (Kreibich and Gross 2001a; Kreibich et al. 2001b, 2003, 2004; Butriy et al. 2007). A comparison of different forms of the electron-nuclear wavefunctions of H_2^+ shows that a simple Hartree mean-field approximation for the nuclear wavefunction is unable to describe dissociation processes; much better results are achieved with a correlated

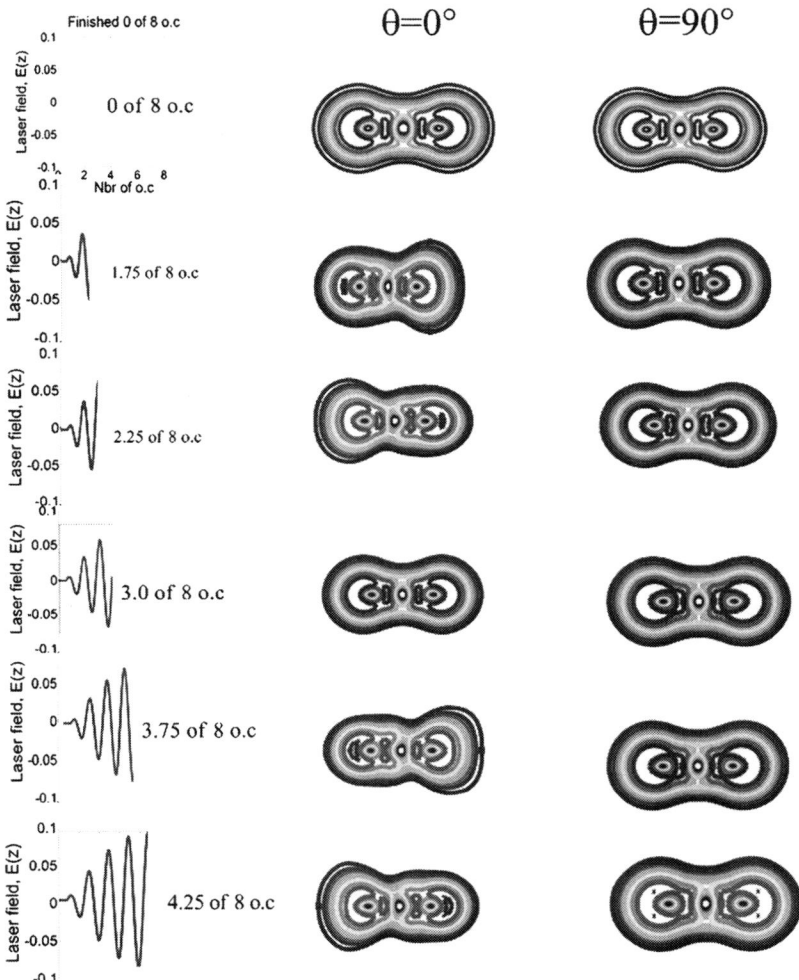

Fig. 18.12 Snapshots of isosurfaces of the TDELF as the CO_2 molecule evolves in the laser field of Fig. 18.11 (Adapted from Fowe and Bandrauk, 2010b)

variational ansatz for the electron-nuclear wavefunction. This is in line with other recent calculations for the ionization dynamics of multielectron systems based on TDEHF approaches (Dahlen and van Leeuwen 2001; Dahlen 2002; Kitzler et al. 2004; Zanghellini et al. 2004, 2005).

18.5 Conclusion and Perspectives

In this chapter, we have given a (by no means exhaustive) review of the rich phenomenology and the often counterintuitive effects of intense laser-matter interaction. Experimental and theoretical research continues to move at a rapid pace towards ever increasing intensities and decreasing pulse lengths, exploring new regimes of electronic and nuclear dynamics. For example, Coulomb explosions of molecular clusters, induced by table-top lasers, have been observed to trigger nuclear fusion reactions (Ditmire et al. 1999). Another example: when an electron is suddenly removed, the remaining xc hole is filled on a 50 attosecond time scale, which appears to be a universal phenomenon (Breidbach and Cederbaum 2005). Thus, the nonlinear dynamics of atoms and molecules in short, intense laser pulses provides many fascinating and challenging opportunities for TDDFT.

The true power of TDDFT emerges when dealing with multielectron systems, and there are many applications for ionization and harmonic generations of atoms and molecules in strong laser pulses. However, it is clear that we need xc functionals that do a better job in capturing highly correlated dynamical processes. Simple semi-classical models give a good intuitive, and often quantitative, understanding of many important strong-field phenomena, such as double photoionization via laser-induced recollision. Unfortunately, these simple scenarios appear to be extremely hard to capture with TDDFT using traditional xc functionals such as the ALDA. Progress in this direction is likely to come through xc functionals that are orbital-based (TDOEP), since these functionals exhibit discontinuities upon change of particle number. This discontinuity is a crucial feature for a correct description of correlated ionization processes (Lein and Kümmel 2005; Mundt and Kümmel 2005).

A particularly tough problem for TDDFT is the proper description of molecular dissociation and of CREI processes that occur when molecules are stretched. The fundamental reason for these difficulties is that in such situations the true molecular state is described by two or more Slater determinants (because of near-degeneracy, also known as static correlation), whereas TDDFT always works with single Slater determinants. Thus, the xc functional has to carry a heavy extra burden, as discussed in Chap. 8.

The problem of stretched H_2 has recently been addressed with an xc orbital functional based on the RPA (Fuchs et al. 2005). Another promising direction is time-dependent density-matrix functional theory (Giesbertz et al. 2008, 2009) (see Chap. 26). However, a full TDDFT formulation of coupled electron-nuclear dynamics (with non-classical nuclei), although possible in principle (Butriy et al. 2007), has not yet become practical.

Finally, even if one solves the TDKS equations with extremely good xc functionals, the problem remains that there are observables that are very hard to extract from the density, such as ionization probabilities or ATI photoelectron spectra. Often, the best one can do is to evaluate these observables using the TDKS orbitals, thus committing one of the four "deadly sins" of TDDFT (Burke et al. 2005a). However, recent efforts (Rohringer et al. 2006; Wilken and Bauer 2007) to construct "read-out functionals" beyond the independent-particle approximation show that there is continuing progress towards new applications of TDDFT for strong fields.

Part V
Numerical Aspects

Chapter 19
The Liouville-Lanczos Approach to Time-Dependent Density-Functional (Perturbation) Theory

Stefano Baroni and Ralph Gebauer

Most current implementations of time-dependent density-functional theory are designed to deal with the lowest-lying portion of the spectrum (often just a few of the very first discrete lines) of systems consisting of up to a few tens of atoms. We introduce a method that allows for the simulation of extended portions of the spectrum of systems virtually of the same size as possibly treatable with state-of-art ground-state DFT techniques.

19.1 Introduction

Current implementations of time-dependent (TD) density-functional theory (DFT) (Runge and Gross 1984) fall into three broad categories. In the first, the TDDFT charge susceptibility is obtained from the independent-electron susceptibility using a Dyson-like linear equation (Gross and Kohn 1985; Petersilka et al. 1996a); in the second, the poles of the susceptibility, which correspond to excitations energies, are addressed as the eigenvalues of a suitable linear (super-) operator equation (Casida 1995; Petersilka et al. 1996a); finally, the full spectrum of a system can be obtained by Fourier analyzing the time series generated by the expectation value of some observable (such as, e.g., the dipole) calculated along the perturbed time evolution of the TDDFT molecular orbitals (Yabana and Bertsch 1996; Marques et al. 2003; Castro et al. 2006).

S. Baroni (✉)
SISSA – Scuola Internazionale Superiore di Studi Avanzati,
via Bonomea 265, 34136 Trieste, Italy
e-mail: baroni@sissa.it

R. Gebauer
The Abdus Salam International Centre for Theoretical Physics,
Strada Costiera 11, 34151 Trieste, Italy
e-mail: rgebauer@ictp.it

M. A. L. Marques et al. (eds.), *Fundamentals of Time-Dependent Density Functional Theory*, Lecture Notes in Physics 837, DOI: 10.1007/978-3-642-23518-4_19,
© Springer-Verlag Berlin Heidelberg 2012

The first approach allows for a straightforward conceptual juxtaposition of TDDFT and many-body perturbation theory (MBPT) (Onida et al. 2002). Computationally much lighter than MBPT, this approach still requires the manipulation (inversion, multiplication) of large matrices, which is hard to accomplish for large systems or basis sets. Also, many unoccupied (virtual) eigenstates of the unperturbed Kohn–Sham (KS) Hamiltonian need to be calculated, a task which again may become critical for large systems or basis sets, as it is easily the case with plane waves or real-space grids.

The second approach results in a non-Hermitian eigenvalue equation that has the same structure as in the TD Hartree–Fock theory (Thouless 1960; McLachlan and Ball 1964), and the dimension of the resulting matrix (the Liouvillian) is twice the product of the number of occupied states with the number of virtual states. The calculation of a few eigenstates of such a large matrix can be accomplished using iterative techniques (Stratmann et al. 1998), possibly in conjunction with the Tamm–Dancoff approximation (TDA), which amounts to enforcing Hermiticity by neglecting the anti-Hermitian component of the Liouvillian (Tamm 1945; Dancoff 1950; Hirata and Head-Gordon 1999). Many existing molecular applications of TDDFT have been performed within such a framework, which is probably near to optimal when a small number of excited states is addressed. In a large system, however, the number of quantum states in any given energy range grows with the system size. The number of pseudo-discrete states in the continuum grows with the basis-set size even in a small system. For these reasons, a method to model the absorption spectrum directly, without calculating individual excited states, would be highly desirable.

The aforementioned third approach does provide such a method, where the TD KS equations are solved in the time domain, and various susceptibilities can be obtained by Fourier analyzing the linear response of the system to appropriate perturbations. This scheme has the same numerical complexity as ground-state DFT iterative methods, and it also gives easy access to nonlinear optical properties. Because of this, real-time methods have recently gained popularity in conjunction with the use of real-space grids (Marques et al. 2003). The main limitation here is that stable integration of the TD KS equations requires a time step as small as 10^{-3}, fs in typical pseudopotential applications—which decreases as the number of grid points or plane waves increases—resulting in rather long simulation runs.

Walker et al. (2006) and Rocca et al. (2008) introduced an alternative approach to the calculation of optical spectra within TDDFT—named the Liouville-Lanczos method—which keeps and enhances the advantages of the previous methods, yet avoiding most of their drawbacks. By using a super-operator formulation of linearized TDDFT, the dynamical polarizability of an interacting-electron system is first represented as an off-diagonal matrix element of the resolvent of the Liouvillian super-operator. One-electron operators and density matrices are treated using a representation borrowed from time-independent density-functional perturbation theory (DFPT) (Baroni et al. 1987a, Baroni et al. 2001), which permits to avoid the calculation of virtual KS orbitals. The resolvent of the Liouvillian is then evaluated using a generalization of the *recursion method* by Haydock, Heine, and Kelly (Bullet et al. 1980), based on the Lanczos bi-orthogonalization algorithm (see Chap. 7 of

Saad (2003)). Each step of the Lanczos recursion essentially requires twice as many operations as a single step of the iterative diagonalization of the unperturbed KS Hamiltonian. Suitable extrapolation of the Lanczos coefficients (Rocca et al. 2008) allows for a dramatic reduction of the number of Lanczos steps necessary to obtain well converged spectra, bringing such number down to hundreds or a few thousands, at worst, in typical plane-wave pseudopotential applications. The resulting numerical workload is only a few times larger than that needed by a ground-state KS calculation for a same system.

19.2 Statement of the Problem, Minimal Theoretical Background, and Notation

In the dipole approximation, the response of molecular systems to electromagnetic radiation is described by the dynamical polarizability tensor, $\alpha_{ij}(\omega)$, whose elements are defined as the dipole moment linearly induced along the ith Cartesian direction by a perturbing electric field of unit strength, polarized along the jth axis, and oscillating at the frequency ω. The absorption coefficient is essentially the product of the frequency times the imaginary part of the diagonal elements (or trace) of the polarizability (Bassani and Altarelli 1983). We address here the dynamical polarizability of a molecular system at *clamped nuclei*, and we use atomic (Hartree) units throughout: $\hbar = 1$; $e = 1$; $m_e = 1$.

The polarizability of a system of interacting electrons can be expressed as:

$$\alpha_{ij}(\omega) = \text{Tr}\left[\hat{r}_i \hat{\rho}'_j(\omega)\right], \tag{19.1}$$

where carets indicate quantum mechanical operators, $\hat{r}_i$ is the ith component of the dipole (or position) operator, $\hat{\rho}'_j(\omega) = \hat{\rho}_j(\omega) - \hat{\rho}^{(0)}$, is the response density matrix, $\hat{\rho}_j(\omega)$ being the one-electron density matrix of the system perturbed by an external homogeneous electric field of unit strength polarized along the jth cartesian axis and oscillating at frequency ω, and $\hat{\rho}^{(0)}$ is its unperturbed counterpart. In TDDFT the response density matrix can be expressed as the solution of the linearized quantum Liouville equation (Walker et al. 2006; Rocca et al. 2008):

$$(\omega - \check{L}) \cdot \hat{\rho}'_j(\omega) = [\hat{r}_j, \hat{\rho}^{(0)}], \tag{19.2}$$

where $\check{L}$ is the TDDFT Liouvillian of the system, defined as:

$$\check{L} \cdot \hat{\rho}' = \left[\hat{H}^{(0)}, \hat{\rho}'\right] + \left[\hat{V}'_{\text{Hxc}}[\hat{\rho}'], \hat{\rho}^{(0)}\right], \tag{19.3}$$

$\hat{H}^{(0)}$ is the unperturbed KS Hamiltonian, $\hat{V}'_{\text{Hxc}}[\hat{\rho}']$ is the linear correction to the Hartree-plus-exchange-correlation (xc) potential, whose coordinate representation reads:

$$v'_{\text{Hxc}}(\boldsymbol{r}, \omega) = \int d^3r' \left[\frac{1}{|\boldsymbol{r} - \boldsymbol{r}'|} + f_{\text{xc}}(\boldsymbol{r}, \boldsymbol{r}', \omega)\right] \rho'(\boldsymbol{r}, \boldsymbol{r}', \omega), \tag{19.4}$$

and f_{xc} is the so-called xc kernel (Gross and Kohn 1985) that, in the adiabatic DFT approximation (Zangwill and Soven 1980a, Bauernschmitt and Ahlrichs 1996a), is independent of ω. Traces of products of operators, such as in Eq. 19.1, have the same algebra as scalar products in linear spaces, $\mathrm{Tr}\left[\hat{A}^\dagger B\right] = (\hat{A}, \hat{B})$, and this property is instrumental in expressing the polarizability as an *off-diagonal matrix element of the resolvent of the Liouvillian*. By solving Eq. 19.2, we can express the polarizability in Eq. 19.1 as

$$\alpha_{ij}(\omega) = -\left(\hat{r}_i, (\omega - \check{L})^{-1} \cdot \left[\hat{r}_j, \hat{\rho}^{(0)}\right]\right), \tag{19.5}$$

where $(\hat{A}, \hat{B})$ indicates the scalar product between operators $\hat{A}$ and $\hat{B}$, in the sense as defined above. Of course, in order to give a well-defined meaning to Eq. 19.5, a well-defined representation must be given to operators, *super-operators* (i.e., operators acting on the linear space of quantum mechanical operators), and for scalar products defined in this linear space. This will be done in Sect. 19.2.1.

19.2.1 Representation of the Response Density Matrix and of Other Operators

The coordinate representation of the response density matrix is:

$$\rho'(\mathbf{r}, \mathbf{r}', \omega) = 2\sum_{v=1}^{N_v}\left[\widetilde{\varphi}'_v(\mathbf{r}, \omega)\varphi_v^{(0)*}(\mathbf{r}') + \varphi_v^{(0)}(\mathbf{r})\widetilde{\varphi}_v'^*(\mathbf{r}', -\omega)\right], \tag{19.6}$$

where $\varphi_v^{(0)}$ are unperturbed KS orbitals, $\widetilde{\varphi}'_v(\mathbf{r}, \omega) = \varphi_v(\mathbf{r}, \omega) - \varphi_v^{(0)}(\mathbf{r})$ denotes the Fourier transform of the first-order correction to the vth KS orbital, $N_v = N_e/2$ is the number of occupied KS states (N_e being the number of electrons), and the factor 2 accounts for the spin degeneracy of KS states in a system that is assumed to be non-magnetic.

As shown in Eq. 19.6, the response density matrix at any given frequency ω is uniquely determined by the two sets of response orbitals $\{\widetilde{\varphi}'_v(\mathbf{r}, \omega)\}$ and $\{\widetilde{\varphi}_v'^*(\mathbf{r}, -\omega)\}$. Standard time-dependent perturbation theory indicates that each response orbital $\widetilde{\varphi}'_v$ can be chosen to be orthogonal to the KS occupied-state manifold. For this reason the response density matrix of Eq. 19.6 has vanishing matrix elements between pairs of occupied and virtual states, namely $\left\langle\varphi_c^{(0)}|\hat{\rho}'|\varphi_{c'}^{(0)}\right\rangle = \left\langle\varphi_v^{(0)}|\hat{\rho}'|\varphi_{v'}^{(0)}\right\rangle = 0\ \forall(v, v')$ and (c, c'), where v and v' denotes generic occupied (*valence*) states and c and c' generic virtual (*conduction*) states. This is to say that in the representation of the unperturbed KS states (the KS representation) the response density matrix has the block structure:

$$\hat{\rho}' \rightarrow \begin{pmatrix} 0 & \rho'_{vc} \\ \rho'_{cv} & 0 \end{pmatrix}. \tag{19.7}$$

Note that in the frequency domain the density matrix is *not* Hermitian. Being the Fourier transform of a Hermitian operator, it satisfies the relation: $\hat{\rho}(\omega)^{\dagger} = \hat{\rho}(-\omega)$. The right-hand side of Eq. 19.2, $[\hat{r}_j, \hat{\rho}^{(0)}]$, has a similar block structure in the KS representation. Therefore, the effective dimension of the linear system, Eq. 19.2, is $2N_v(N - N_v)$, N being the size of the one-electron basis set. Unfortunately, this representation does require the explicit knowledge of all the virtual KS orbitals, whose calculation is generally avoided, or even impossible, in modern electronic-structure methods, particularly when (very) large basis sets (such as plane waves or real-space grids) are used.

Inspired by these considerations, we define the standard batch representation (SBR) of the response density matrix as:

$$\hat{\rho}' \xrightarrow{SBR} \begin{pmatrix} \{q_v\} \\ \{p_v\} \end{pmatrix}, \tag{19.8}$$

where $\{q_v\}$ and $\{p_v\}$ indicate the sets (*batches*) of orbitals:

$$q_v(r) = \frac{1}{2}\left(\widetilde{\varphi}'_v(r, \omega) + \widetilde{\varphi}'^{*}_v(r, -\omega)\right) \tag{19.9a}$$

$$p_v(r) = \frac{1}{2}\left(\widetilde{\varphi}'_v(r, \omega) - \widetilde{\varphi}'^{*}_v(r, -\omega)\right). \tag{19.9b}$$

The 45° rotation of the $\widetilde{\varphi}'$ orbitals in Eq. 19.9 is introduced for further notational convenience. According to Eq. 19.9 the most general response density matrix can be represented in terms of two batches of N_v orbitals orthogonal to the manifold of occupied KS states. This orthogonality condition can be enforced either implicitly, by expressing each orbital as a linear combination of unoccupied KS states, or explicitly by acting upon each of them with the projector over the virtual KS manifold, $\hat{Q}$. The first option would eventually lead to the KS representation of Eq. 19.7, which requires the explicit calculation of virtual KS states. By expressing the virtual-state projector as $\hat{Q} = \hat{I} - \hat{P}$ (where $\hat{I}$ is the identity operator and $\hat{P} = \sum_{v=1}^{N_v}\left|\varphi_v^{(0)}\right\rangle\left\langle\varphi_v^{(0)}\right|$ is the projector over the occupied-state KS manifold), instead, the second option allows one to enforce the orthogonality condition without any explicit reference to any virtual states, following a practice that was introduced in the framework of (time-independent) DFPT (Baroni et al. 1987a). Note that for a time-reversal invariant system, whose KS orbitals can be chosen to be real, the SBR of the response charge density can be expressed in terms of the $\{q_v\}$ orbitals alone:

$$n'(r, \omega) = 4\sum_{v=1}^{N_v}\varphi_v^{(0)}(r)q_v(r). \tag{19.10}$$

General one-particle quantum mechanical operators can be given a similar representation. The SBR of a general operator is defined as:

$$\hat{A} \xrightarrow{SBR} \begin{pmatrix} \{a_v^q\} \\ \{a_v^p\} \end{pmatrix} = a, \tag{19.11}$$

where the orbitals $a_v^q(\boldsymbol{r})$ and $a_v^p(\boldsymbol{r})$ are defined as

$$a_v^q(\boldsymbol{r}) = \frac{1}{2}\left[\hat{Q}\hat{A}\varphi_v^{(0)}(\boldsymbol{r}) + \left(\hat{Q}\hat{A}^\dagger\varphi_v^{(0)}(\boldsymbol{r})\right)^*\right] \tag{19.12a}$$

$$a_v^p(\boldsymbol{r}) = \frac{1}{2}\left[\hat{Q}\hat{A}\varphi_v^{(0)}(\boldsymbol{r}) - \left(\hat{Q}\hat{A}^\dagger\varphi_v^{(0)}(\boldsymbol{r})\right)^*\right]. \tag{19.12b}$$

The $\{a_v^q\}$ and $\{a_v^p\}$ functions of Eqs. 19.12a, b will be referred to as the *upper* (or *q*-like) and *lower* (or *p*-like) components of the SBR of the $\hat{A}$ operator. If $\hat{A}$ is a Hermitian operator, then its SBR is given by

$$\hat{A} = \hat{A}^\dagger \xrightarrow{SBR} \begin{pmatrix} \{\hat{Q}\hat{A}\varphi_v^{(0)}\} \\ \{0\} \end{pmatrix}, \tag{19.13}$$

where we have assumed again that the ground-state orbitals $\varphi_v^{(0)}$ are real because of time-reversal invariance. Other operators appearing in Eq. 19.2 are represented as:

$$\left[\hat{H}^{(0)}, \hat{\rho}'\right] \xrightarrow{SBR} \begin{pmatrix} \{(\hat{H}^{(0)} - \varepsilon_v^{(0)})p_v\} \\ \{(\hat{H}^{(0)} - \varepsilon_v^{(0)})q_v\} \end{pmatrix}, \tag{19.14a}$$

$$\left[\hat{A}, \hat{\rho}^{(0)}\right] \xrightarrow{SBR} \begin{pmatrix} \{0\} \\ \{\hat{Q}\hat{A}\varphi_v^{(0)}\} \end{pmatrix}, \tag{19.14b}$$

where $\varepsilon_v^{(0)}$ are unperturbed KS orbital energies. Clearly the SBR of a general operator is incomplete because it misses the information contained in the matrix blocks that vanish in the KS-state representation of the response density matrix, Eq. 19.7. It is however sufficient to calculate traces of products of any operator with any response density matrix having the block structure of Eq. 19.7. By using the SBR, the polarizability in Eq. 19.5 can be expressed as:

$$\alpha_{ij}(\omega) = -\left(r_i, (\omega - L)^{-1} \cdot y_j\right), \tag{19.15}$$

where r_i, y_j, and L are the SBR representations of $\hat{r}_i$, $[\hat{r}_j, \hat{\rho}^{(0)}]$, and of the Liouvillian, respectively:

$$\hat{r}_i \xrightarrow{SBR} \begin{pmatrix} \{r_{i,v}\} \\ \{0\} \end{pmatrix} = r_i \tag{19.16a}$$

$$\left[\hat{r}_j, \hat{\rho}^{(0)}\right] \xrightarrow{SBR} \begin{pmatrix} \{0\} \\ \{r_{j,v}\} \end{pmatrix} = y_j \tag{19.16b}$$

$$\breve{L} \xrightarrow{SBR} \begin{pmatrix} 0 & \breve{D} \\ \breve{D} + 2\breve{K} & 0 \end{pmatrix} = L, \tag{19.16c}$$

the $r_{i,v}$ orbitals are defined as:

$$r_{i,v}(\boldsymbol{r}) = \hat{Q}\hat{r}_i \varphi_v^{(0)}(\boldsymbol{r}),$$

(19.17)

and the $\check{D}$ and $\check{K}$ super-operators are defined as:

$$\check{D} \cdot \{u_v(\boldsymbol{r})\} = \left\{\left(\hat{H}^{(0)} - \varepsilon_v^{(0)}\right) u_v(\boldsymbol{r})\right\}$$

(19.18a)

$$\check{K} \cdot \{u_v(\boldsymbol{r})\} = \left\{4\varphi_v^{(0)}(\boldsymbol{r}) \sum_{v'=1}^{N_v} \int \mathrm{d}^3 r' f_{\mathrm{xc}}(\boldsymbol{r}, \boldsymbol{r}') \varphi_{v'}^{(0)}(\boldsymbol{r}') u_{v'}(\boldsymbol{r}')\right\}.$$

(19.18b)

Finally, the SBR of scalar products (traces of products of operators) reads:

$$\mathrm{Tr}\left(\hat{A}^\dagger \hat{B}\right) \xrightarrow{SBR} \sum_{v=1}^{N_e} \left(\langle a_v^q | b_v^q \rangle + \langle a_v^p | b_v^p \rangle\right) = (a, b),$$

(19.19)

where $\left(\{a_v^q\}, \{a_v^p\}\right)$, $\left(\{b_v^q\}, \{b_v^p\}\right)$ are the SBRs of $\hat{A}$ and $\hat{B}$, respectively, and brackets $\langle \phi | \psi \rangle$ indicate standard quantum-mechanical scalar products between one-electron orbitals $|\phi\rangle$ and $|\psi\rangle$. Note that, according to these definitions, the two vectors that bracket the resolvent of the Liouvillian in Eqs. 19.5 and 19.15 are orthogonal because the commutator of two Hermitian operators is anti-Hermitian and the trace of the product of a Hermitian and an anti-Hermitian operator vanishes.

19.2.2 Dipole Operator in Periodic Boundary Conditions

In order to obtain the SBR of the dipole operator and of its commutator with the unperturbed density matrix, Eqs. 19.16a, b, one needs to evaluate the orbitals in Eq. 19.17. In periodic boundary conditions the position operator $\hat{r}_i$ is ill defined, since it is both non-periodic and not bound from below. As a consequence it is not possible to compute the expectation value of $\hat{r}_i$ on Bloch states. However in the calculation of $\hat{Q}\hat{r}_i \varphi_v^{(0)}$ only off-diagonal matrix elements of $\hat{r}_i$ are required:

$$\hat{Q}\hat{r}_i \varphi_v^{(0)} = \sum_c \left|\varphi_c^{(0)}\right\rangle \left\langle \varphi_c^{(0)} \left| \hat{r}_i \right| \varphi_v^{(0)} \right\rangle,$$

(19.20)

which are well defined in periodic boundary conditions (Baldereschi and Tosatti 1978). Indeed, one has

$$\langle \varphi_c^{(0)} | \hat{r}_i | \varphi_v^{(0)} \rangle = \frac{1}{\varepsilon_c^{(0)} - \varepsilon_v^{(0)}} \langle \varphi_c^{(0)} | [\hat{H}^{(0)}, \hat{r}_i] | \varphi_v^{(0)} \rangle$$

(19.21)

and, if the potential operator in the unperturbed Hamiltonian is purely local, then the commutator in Eq. 19.21 is simply proportional to the momentum operator,

$$[\hat{H}^{(0)}, \hat{r}_i] = -\hat{p}_i. \tag{19.22}$$

When the potential acting on electrons has non-local contributions (which is the case for the vast majority of pseudopotentials), an explicit correction due to those non-local terms must be added to the momentum operator in Eq. 19.22 (Baroni and Resta 1986a; Tobik and Dal Corso 2004).

In practice, the Liouville-Lanczos approach to TDDFT, as well as to DFPT, is designed so as to avoid any explicit reference to virtual eigenpairs of the KS Hamiltonian, so that Eq. 19.21 cannot be used directly. However the relevant orbitals $\hat{Q}\hat{r}_i\varphi_v^{(0)}$ can be obtained by solving a set of linear systems, as proposed in Baroni et al. (1987a) and thoroughly explained in Baroni et al. (2001).

19.3 Algorithm

According to the discussion in the previous section, any component of the polarizability tensor can be expressed as an off-diagonal element of the resolvent of the Liouvillian (super-) operator. At first sight, it may seem that the calculation of such a matrix element, Eq. 19.15, would require the solution of a $n \times n$ linear system to invert $(\omega - L)$ for each different value of the frequency ω, a daunting task as the system size and/or the number of frequencies grow large. As an expedient alternative, the Liouville-Lanczos approach to TDDFT uses a generalization of the recursion method by Haydock, Heine, and Kelly (Bullet et al. 1980), based on the Lanczos bi-orthogonalization algorithm (LBOA) (Saad 2003, Chap. 7). The LBOA allows for the bulk of the numerical work to be done once for all the frequencies, while the linear system is inexpensively solved in an approximate representation where the matrix to be inverted is both tridiagonal and of much smaller size (Walker et al. 2006; Rocca et al. 2008; Baroni et al. 2010).

19.3.1 Lanczos Bi-orthogonalization Algorithm

Given a pair of vectors, u^1 and v^1 normalized by the condition $(u^1, v^1) = 1$ (although not strictly necessary, it is convenient to assume that both vectors coincide with y_j in the present case), the LBOA amounts to the following recursion:

$$\gamma^1 v^0 = \beta^1 u^0 = 0 \tag{19.23a}$$

$$v^1 = u^1 = y_j \tag{19.23b}$$

$$\beta^{l+1} v^{l+1} = L v^l - \alpha^l v^l - \gamma^l v^{l-1} \tag{19.23c}$$

$$\gamma^{l+1} u^{l+1} = L^\top u^l - \alpha^l u^l - \beta^l u^{l-1}, \tag{19.23d}$$

where β^{l+1} and γ^{l+1} are scaling factors for the newly generated u^{l+1} and v^{l+1} vectors, chosen so as to enforce bi-normalization:

$$(u^{l+1}, v^{l+1}) = 1, \qquad (19.24)$$

and

$$\alpha^l = (u^l, Lv^l). \qquad (19.25)$$

Because of the special block structure of the Liouvillian, Eq. 19.16c, and of the starting vector y_j, Eq. 19.16b, all the α's vanish: $\alpha^l = 0$; one also has $|\beta^l| = |\gamma^l|$. In exact arithmetics, it is known that the two sequences of vectors generated by the recursion in Eq. 19.23 are bi-orthogonal, i.e., $(u^k, v^l) = \delta_{kl}$, where δ_{kl} is the Kronecker symbol. The resulting algorithm is described in detail, e.g., in Chap. 7 of Saad (2003). Let us now define $^mV = [v^1, v^2, \ldots, v^m]$ and $^mU = [u^1, u^2, \ldots, u^m]$ as the two $n \times m$ rectangular matrices whose columns are the elements of the bi-orthogonal set of vectors generated by m steps of the LBOA, Eq. 19.23 (the left and right iterates of the Lanczos recursion). The following Lanczos factorization holds in terms of the quantities calculated from the LBOA:

$$L\,^mV = {}^mV\,{}^mT + \beta^{m+1} v^{m+1}\,{}^me_m^\top, \qquad (19.26a)$$

$$L^\top\,{}^mU = {}^mU\,{}^mT^\top + \gamma^{m+1} u^{m+1}\,{}^me_m^\top, \qquad (19.26b)$$

$${}^mU^\top\,{}^mV = {}^mI, \qquad (19.26c)$$

where ^mT is the $m \times m$ tridiagonal matrix made out of the LBOA coefficients,

$$
{}^mT =
\begin{pmatrix}
0 & \gamma^2 & & & 0 \\
\beta^2 & 0 & \gamma^3 & & \\
& \beta^3 & 0 & \ddots & \\
& & \ddots & \ddots & \gamma^m \\
0 & & & \beta^m & 0
\end{pmatrix}, \qquad (19.27)
$$

^me_l is the l-th unit vector in an m-dimensional space, "$\top$" indicates matrix transposition, and ^mI is the $m \times m$ unit matrix.

19.3.2 Calculation of the Polarizability

Let us now rewrite Eq. 19.26 as:

$$(\omega - L)^mV = {}^mV(\omega - {}^mT) - \beta^{m+1} v^{m+1}\,{}^me_m^\top. \qquad (19.28)$$

By multiplying Eq. 19.28 by $r_i^\top (\omega - L)^{-1}$ on the left and by $(\omega - T_M)^{-1}\, {}^m e_1$ on the right, we obtain:

$$
\begin{aligned}
r_i^\top (\omega - L)^{-1}\, {}^m V\, {}^m e_1 &= r_i^\top\, {}^m V (\omega - {}^m T)^{-1}\, {}^m e_1 \\
&\quad + \beta^{m+1} r_i^\top (\omega - L)^{-1} v^{m+1}\, {}^m e_m^\top (\omega - {}^m T)^{-1}\, {}^m e_1.
\end{aligned}
\tag{19.29}
$$

Using (19.29) and taking the relation ${}^m V\, {}^m e_1 = v^1 = y_j$ into account, Eq. 19.15 can be cast into:

$$
\alpha_{ij}(\omega) = - \left({}^m z_{ij}, {}^m w_j(\omega) \right) + \varepsilon_m(\omega),
\tag{19.30}
$$

where ${}^m z_{ij}$ and ${}^m w_j(\omega)$ are m-dimensional arrays defined as

$$
{}^m z_{ij} = {}^m V_j r_i
\tag{19.31}
$$

and as the solution of the tridiagonal linear system:

$$
\left(\omega - {}^m T_j \right) {}^m w_j(\omega) = {}^m e_1,
\tag{19.32}
$$

respectively,

$$
\varepsilon_m(\omega) = -\beta^{m+1} \left(r_i, (\omega - L)^{-1} v^{m+1} \right) \left({}^m e_m, (\omega - {}^m T)^{-1}\, {}^m e_1 \right)
\tag{19.33}
$$

is the error made when truncating the Lanczos chain to its first m terms, and a j suffix has been appended to the ${}^m T$ and ${}^m V$ arrays, so as to indicate that different Lanczos chains are generated for different polarizations of the perturbing electric field, r_j. Note that the components of the ${}^m z_{ij}$ array can be calculated on the fly at each Lanczos iteration (i.e., without storing the Lanczos iterates) as:

$$
z_{ij}^l = \left(r_i, v_j^l \right) = \sum_{v=1}^{N_v} \left\langle r_{i,v} | v^{q,l}_{j,v} \right\rangle,
\tag{19.34}
$$

where the orbitals $\{r_{i,v}\}$ are defined as in Eq. 19.17, $v^{q,l}_{j,v}$ is the vth *upper* (q-like) component of the l-th Lanczos right iterate generated from $u^1 = v^1 = y_j$ (see Eq. 19.16b), and a j suffix has also been appended to v^l for notational uniformity. The size of the error term ε_m Eq. 19.33 generally decreases by increasing the number m of steps in the Lanczos chain. A manageable number of Lanczos steps is found to be sufficient to achieve the accuracy needed for spectroscopic applications, as demonstrated by the numerical applications discussed in the following sections.

In practice, the algorithm outlined above is performed in two steps, for any given external perturbation (such as, e.g., different polarizations j of the perturbing electric field). The first, time consuming but frequency-independent, step consists in the LBOA factorization, yielding the tridiagonal matrix ${}^m T$ and the ${}^m z$ array, whose components are calculated on the fly at each Lanczos iteration using Eq. 19.34. The calculation of several response functions (such as, e.g., different components of the molecular dipole, corresponding to different rows of the polarizability tensor) implies the simultaneous calculation of different $^m z$ arrays. In the second, inexpensive, step, the response functions are calculated from Eq. 19.30 upon solution of Eq. 19.32, for any different frequency one is interested in.

19.3.3 Extrapolating the Lanczos Recursion

The components of the $^m z$ array in Eq. 19.30 decrease rather rapidly when the number of iterations grows large, so that only a relatively small number of components have to be explicitly calculated. A much larger number of β and γ coefficients is necessary, however, to obtain well converged solutions of Eq. 19.32. Rocca et al. (2008) showed that, for large iteration counts, β and γ oscillate around two distinct values for odd and even counts, whose average roughly equals one fourth the width of the spectrum of the Liouvillian (which extends from minus to plus the maximum excitation energy), and whose difference is of the order of one half the Liouvillian gap (which extends from minus to plus the minimum excitation energy). As an example, in Fig. 19.1 we display the typical behavior of the elements of the $^m z$ array and of the Lanczos β coefficients as functions of the Lanczos iteration count, in the case of a fullerene C_{60} molecule. The average value of the β's for large iteration counts is roughly one half the kinetic-energy cutoff for the plane-wave basis sets (30 Ry in this case), which in turn is of the order of the maximum excitation energy. The difference between the averages for odd and even counts is here 3.2 eV, to be compared with a calculated optical gap of 3.5 eV.

These results can be understood in terms of an analogy with the continued-fraction expansion of the local density of states (LDOS) for tight-binding Hamiltonians, a problem that has been the main motivation for the development of recursion methods and their application to electronic-structure theory in the seventies (Bullet et al. 1980). In particular, Turchi et al. (1982) have shown that the coefficients of the continued-fraction expansion of a *connected* LDOS asymptotically tend to a constant—which equals one fourth of the band width—whereas they oscillate between two values in the presence of a gap: in the latter case the average of the two limits equals one fourth of the total band width, whereas their difference equals one half the energy gap. Further details on the analogy between the spectral properties of TB Hamiltonians and of the TDDFT Liouvillian in a plane-wave representation can be found in Rocca et al. (2008).

The rapid decrease of the components of the $^m z$ array, together with the observed asymptotic behavior of the coefficients of the tridiagonal matrix suggest an effective strategy to enhance the accuracy of the Liouville-Lanczos TDDFT algorithm, by extrapolating the results obtained from a relatively small number of iterations. Once m Lanczos iterations are performed and the regime is attained where further components of the z array are negligible and the β and γ coefficients display the typical bi-modal behavior of Fig. 19.1, a (much) larger tridiagonal system is solved, where the missing

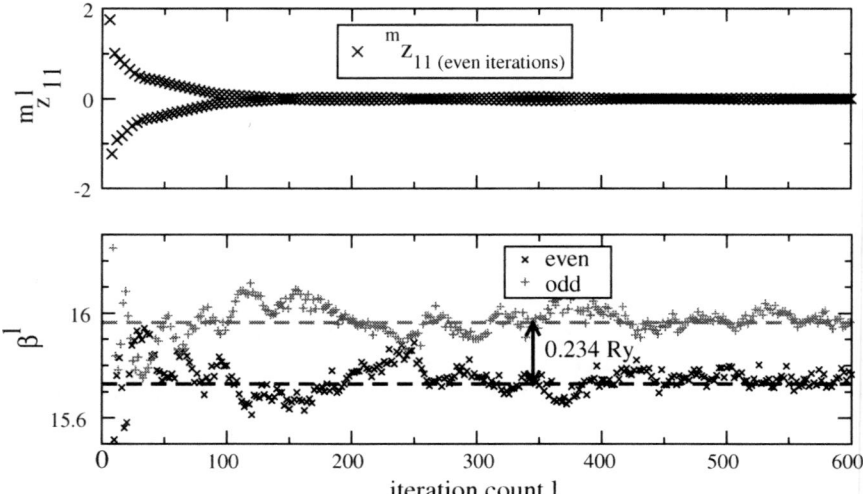

Fig. 19.1 *Upper panel*: values of the components of the ^{m}z array (see Eq. 19.30 and below) as functions of the Lanczos iteration count, as calculated for a fullerene C_{60} molecule, using ultrasoft pseudopotentials and a plane-wave kinetic-energy cutoff of 30 Ry. *Lower panel*: values of the calculated Lanczos β coefficients (see Eq. 19.27) coefficients for the same system

components of z are simply set to zero, whereas the missing values of α and β are set to the average of the values which have been actually calculated. An example of the efficiency of this procedure will be discussed in Sect. 19.5 in the case of an organic dye molecule.

19.4 Optical Sum Rules

Optical susceptibilities satisfy many sum rules, the most fundamental of which is probably the Thomas–Reiche–Kuhn (Thomas 1925; Kuhn 1925; Reiche and Thomas 1925) (or f-sum) rule, which relates the integral of the absorption coefficient of a molecular system to the number of electrons contained in it. As observed in Sect. 19.2, the absorption coefficient is proportional to the product of the frequency times the trace of the molecular polarizability, whose integral reads:

$$f = \sum_{j} \Im \int_{0}^{\infty} d\omega \, \alpha_{jj}(\omega)\omega \tag{19.35a}$$

$$= \frac{3}{2} N_{e}\pi, \tag{19.35b}$$

N_e being the number of electrons. Strictly speaking, the above equation only holds when electrons are subject to a local external potential. A violation of the f-sum rule should be expected when non-local (norm-conserving as well as ultra-soft) pseudopotentials are used (see below).

A remarkable feature of the Liouville-Lanczos approach to TDDFT is that the f-sum rule is satisfied exactly when truncating the Lanczos recursion to any number of iterations. This fact is a slight generalization of a well known result connecting the classical-moment and the Hermitian-Lanczos methods, and stating that the momenta of the local density of states calculated from a Lanczos chain of N steps are exact up to order $2N + 1$ (Nex 1978).

To demonstrate this feature, let us start from Eq. 19.15. Using the spectral resolution of the Liouvillian and the relation $\lim_{\epsilon \to 0} \Im m \ 1/(x + i\epsilon) = -\pi \delta(x)$, one easily sees that:

$$f = \pi \sum_j (r_j, L \cdot y_j),$$ (19.36)

where r_j, y_j, and L are defined in Eqs. 19.16a–c. We now use the tridiagonal representation of the Liouvillian and the bi-orthogonality of the basis resulting from the LBOA, Eqs. 19.26, initiated by $u^1 = v^1 = y_j$, to obtain:

$$
\begin{aligned}
f &= \pi \sum_j \sum_{kl} (r_j, v_j^k)(u_j^l, y_j) T_j^{kl} \\
&= \pi \sum_j (r_j, v_j^2) \beta_j^2.
\end{aligned}
$$ (19.37)

Equation 19.37 already shows that, whatever the value of the f-sum is, it is independent of the number of iterations used to perform the LBOA. In order to complete the demonstration, we notice that, because of the special block structure of the Liouvillian, Eq. 19.16c, and of the u and v arrays, Eqs. 19.16a, b, the matrix element in Eq. 19.37 does not depend on interaction effects (embodied in the $\check{K}$ matrix in Eq. 19.16c) and it is therefore equal to the value it would have for independent electrons, for which the f-sum rule holds.

The validity of the f-sum rule relies on the commutator between the Hamiltonian and the dipole operator to be proportional to the momentum operator, Eq. 19.22, which is only true when the external potential acting on the electrons is local. In the presence of non-local (pseudo-) potentials, corrections to the f-sum rule are to be expected.

Experience shows that these corrections are rather small, though slightly larger for ultra-soft than for norm-conserving pseudo-potentials. For the case of the fullerene C_{60} molecule, our Liouville-Lanczos TDDFPT formalism leads to a spectrum in which the f-sum rule is extremely well satisfied ($\Delta f / f \approx 0.007$) when norm-conserving pseudopotentials as employed. In the case of ultra-soft pseudopotentials the error is one order of magnitude larger ($\Delta f / f \approx 0.078$) (Malcioglu et al. 2011).

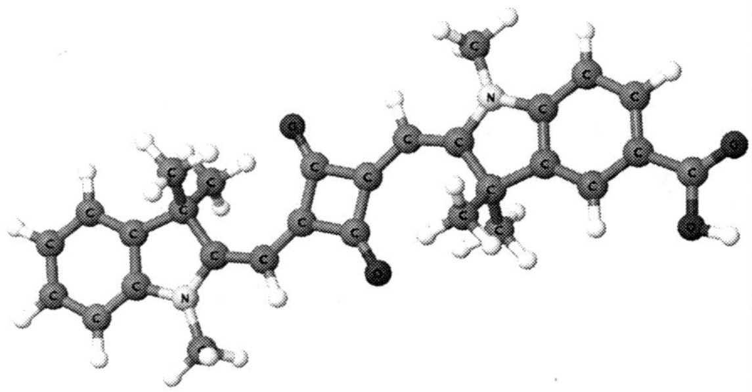

Fig. 19.2 Squaraine
molecule—the dye consists
of 63 atoms and 178
electrons. Carbon atoms are
shown in grey, oxygen in
dark grey, nitrogen in light
gray, and hydrogen in white

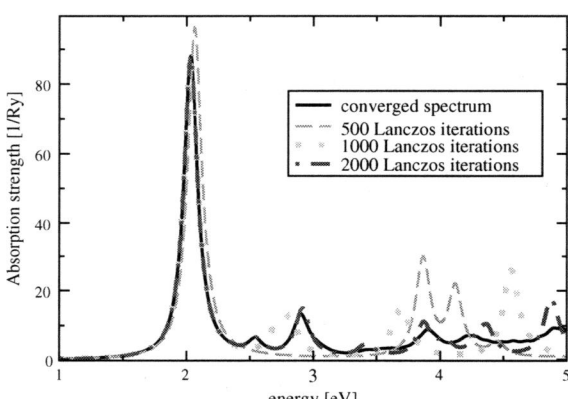

Fig. 19.3 Absorption
spectrum of the squaraine
dye. While the main
absorption peak (2.03 eV) is
converged already with 1000
iterations, reproducing the
features at higher energy
require no less than 2000
Lanczos iterations

19.5 Application to an Organic Dye Molecule

Let us now illustrate the Liouville-Lanczos TDDFPT algorithm by applying it to the
computation of the optical spectrum of a squaraine organic dye molecule. Such dyes
have recently attracted considerable interest in the domain of organic photovoltaics,
due to their easily tunable optical properties, high extinction coefficients, abundant
availability and environmental benefits.

The squaraine molecule considered here is depicted in Fig. 19.2 . The molecule
is simulated in a cubic supercell of size 25 Å, using the PBE functional (Perdew
et al. 1996b) for both the ground-state computation and for the (adiabatic) TDDFT
calculations, while the interaction of the electrons with the ions is described by ultrasoft pseudopotentials (Walker and Gebauer 2007). A plane-wave basis set is used
with a kinetic-energy cutoff of 25 Ry for the KS wavefunctions and of 260 Ry for the
charge density.

Figure 19.3 shows the optical absorption spectrum up to 5 eV, averaged over the
three cartesian directions. The spectrum is characterised by a strong peak at 2.03 eV,

Fig. 19.4 Absorption spectrum of the squaraine dye. After 1700 iterations, the spectrum is well converged only for energies lower than roughly 3.5 eV. The extrapolation of the coefficients allows one to obtain a good spectrum over a much larger energy range with negligible additional computational cost

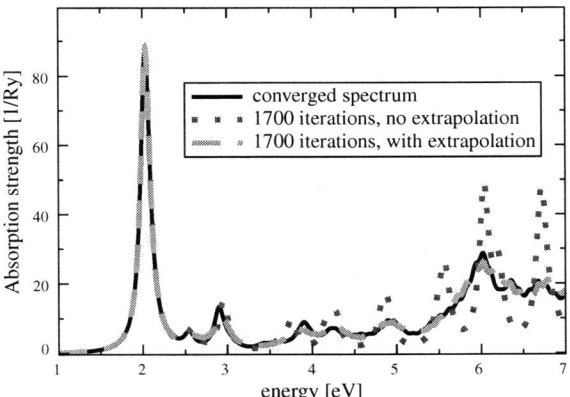

followed by several low intensity peaks at higher energies. In Fig. 19.3 it can be clearly seen that by increasing the number of Lanczos steps the spectrum converges up to higher energies. The main peak is already well represented with 1000 iterations, while the second peak is converged only with 2000 iterations.

As discussed in Sect. 19.3.3, the convergence of the optical spectra can be accelerated by extrapolating the recursion coefficients. In the case of our dye molecule, this behaviour is illustrated in Fig. 19.4. Using for example 1700 Lanczos iterations, the spectrum calculated *without* extrapolation is clearly not converged for energies larger than about 3.5 eV. Using the same 1700 Lanczos steps and extrapolating the remaining coefficients up to 10,000, one obtains a spectrum that is very close to the converged result.

19.6 Conclusions

Lanczos methods provide powerful tools to handle some of the hard numerical problems in the simulation of molecular and solid-state excited-state properties within TDDFT. We believe that the resulting Lanczos-Liouville approach to TDDFT is close to be numerically optimal if detailed information on individual excited states is not required. Of course, no numerical advance can cope with the inadequacy of currently available xc kernels to properly describe excitonic and charge-transfer effects in the excited states. These effects can in principle be successfully addressed within many-body perturbation theory, whose approach to neutral excitations (the Bethe–Salpeter equation) has a structure which is very similar to that of linearized TDDFT. As a matter of fact, a simplified version of the Bethe–Salpeter equation featuring a statically screened exchange kernel has been recently treated with success using a simple extension of the approach presented here for TDDFT (Rocca et al. 2010).

A computer code implementing the Liouville-Lanczos approach to TDDFT described here is freely available within the Quantum Espresso distribution (Giannozzi et al. 2009) and is thoroughly described in Malcioglu et al. (2011).

Chapter 20
The Projector Augmented Wave Method

Lauri Lehtovaara

20.1 Introduction

DFT and TDDFT calculations are computationally intensive, and therefore, many different strategies are employed to reduce the computational burden. As the cost of a (TD)DFT calculation depends heavily on the number of degrees of freedom and active electrons, both should be minimized to speed up calculations while still reproducing the chemical and physical properties of interest. This can be achieved by either using the traditional pseudopotential approach (Pickett 1989; Phillips and Kleinman 1959; Hamann et al. 1979; Troullier and Martins 1991; Vanderbilt 1990) or the more recent projector-augmented wave (PAW) method (Blöchl 1994; Blöchl et al. 2003; Kresse and Joubert 1999).

The external potential in the TD-KS equations includes the electron-nuclear Coulomb attraction, which diverges to negative infinity at the nuclei. Consequently, all-electron KS wavefunctions have rapid oscilations near the nuclei. However, chemistry happens in regions where atomic orbitals of different atoms overlap significantly, i.e., away from the nuclei and close to the middle of bonds. Therefore, chemical properties do not directly depend on the shape of wavefunctions in the close vicinity of nuclei. This allows us to replace rapidly oscillating wavefunctions by something computationally less challenging, i.e., by something smoother, as long as the correct behavior in the bonding region is reproduced. Moreover, deeply bound core states can be "frozen" as they do not extend to the bonding region, i.e., they are chemically inert.

In the pseudopotential approximation, the electron–nuclear and core–valence interactions can be replaced by a single effective potential, called pseudopotential. A pseudopotential is much smoother than the bare Coulomb potential and core

L. Lehtovaara (✉)
Laboratoire de Physique de la Matière Condensée et Nanostructures,
University Claude Bernard Lyon 1, CNRS, 43 boulevard du 11 novembre 1918,
69622 Villeurbanne Cedex, France
e-mail: lauri.lehtovaara@iki.fi

M. A. L. Marques et al. (eds.), *Fundamentals of Time-Dependent Density Functional Theory*, Lecture Notes in Physics 837, DOI: 10.1007/978-3-642-23518-4_20, © Springer-Verlag Berlin Heidelberg 2012

electrons appear only through the pseudopotential. Smoothness of the "pseudized" external potential leads to smooth pseudo wavefunctions, which are easier to discretize and require less degrees of freedom. The pseudopotential approach leads to tremendous savings in computational cost, especially in the case of uniform discretizations (plane-waves and finite-difference real-space grids). Unfortunately, pseudopotentials are nonlocal, and the Hohenberg–Kohn and Runge–Gross theorems hold only for local external potentials.

An alternative to the pseudopotential approach is the projector augmented-wave method. It is a powerful numerical framework that tries to combine the efficiency of pseudopotential based methods with the accuracy of all-electron calculations. The PAW method is in principle an all-electron, full-potential method, i.e., it provides access to the all-electron Kohn–Sham wavefunction and it uses the full KS potential calculated from the all-electron density. In practice, the core electrons are usually frozen (i.e., frozen-core approximation), and a cutoff for the partial-wave expansion is required. Even though PAW is an all-electron method, it is extremely efficient with performance comparable to ultrasoft pseudopotentials and is considerably faster than traditional approaches based on norm-conserving pseudopotentials. Actually, pseudopotentials can be seen as approximation to the PAW method, which to some extent justifies their use within (TD)DFT. The price to pay is that the PAW framework leads to equations that are somewhat more involved, and to added complexity in its numerical implementation.

In this chapter, we give a brief introduction to PAW, and on its use within TDDFT. We start by describing how the PAW wavefunctions are formed from three different components: (i) smooth pseudo wavefunctions, (ii) smooth atomic pseudo wavefunctions, and (iii) atomic all-electron wavefunctions. Then we show how operators in the PAW method are defined. The rest of the chapter briefly reviews how PAW method can be used for DFT and TDDFT.

20.2 The PAW Method

The PAW method divides space into two kind of regions: non-overlapping atomic regions, called augmentation spheres, and an interstitial region. In the interstitial region, Kohn–Sham wavefunctions are expected to be smooth and easily described by an uniform discretization (e.g., an uniform grid or planewaves). The smooth discretization spans also the atomic regions, but in addition each atomic region has spherical augmentation functions called partial-waves. The total wavefunction $\varphi_k(\mathbf{r})$ is represented as a combination of (i) smooth pseudo wavefunction $\tilde{\varphi}_k(\mathbf{r})$, (ii) atomic all-electron wavefunctions, $\xi^a(r_a, \theta_a, \phi_a)$ and (iii) atomic pseudo wavefunctions $\tilde{\xi}^a(r_a, \theta_a, \phi_a)$ (see Fig. 20.1):

$$\varphi_k(\mathbf{r}) = \tilde{\varphi}_k(\mathbf{r}) + \sum_a^{\text{atoms}} \sum_{nlm} c_{k,nlm}^a \left[\xi_{nlm}^a(r_a, \theta_a, \phi_a) - \tilde{\xi}_{nlm}^a(r_a, \theta_a, \phi_a) \right]. \quad (20.1)$$

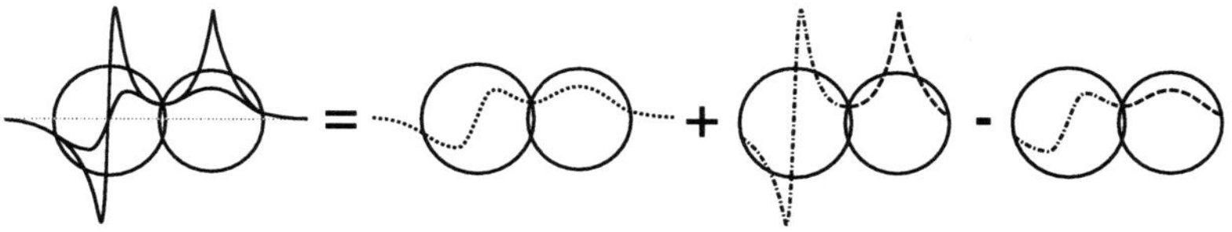

Fig. 20.1 All-electron PAW wavefunction is equal to the pseudo wavefunction plus the atomic all-electron wavefunction minus the atomic pseudo wavefunction

The atomic all-electron and pseudo wavefunctions are written as a linear combination of all-electron partial-waves $\xi^a_{nlm}(r_a, \theta_a, \phi_a)$ and pseudo partial-waves $\tilde{\xi}^a_{nlm}(r_a, \theta_a, \phi_a)$, respectively. Partial waves are radial functions centered at the nuclear positions R_a multiplied by (real) spherical harmonics:

$$f(r - R_a) = f(r_a)Y_{lm}(\theta_a, \phi_a). \tag{20.2}$$

This kind of functions can easily describe sharp oscillations of the all-electron wavefunction near a nucleus.

The purpose of the all-electron partial-waves is to add the all-electron wavefunction inside an augmentation sphere, whereas pseudo partial-waves cancel the smooth pseudo wavefunction $\tilde{\varphi}_k(r)$ inside an augmentation sphere (see Fig. 20.1). The all-electron partial-waves are obtained from an all-electron atomic calculation: they are the part of the atomic all-electron wavefunctions that are inside an augmentation sphere. Each valence state has its corresponding partial-wave but usually a few unbound (or scattering) states are included in the calculation. The unbound states allow more flexibility on the all-electron wavefunction. For example, a hydrogen atom can be described with two s-type partial-waves and one p-type partial-wave. The p-type partial-wave is included to allow the all-electron density to polarize.

Each all-electron partial-wave has its corresponding pseudo partial-wave of the same symmetry $(s, p, d, \dots)$. Its purpose is to cancel the corresponding part of the smooth pseudo wavefunction $\tilde{\varphi}(r)$ inside an augmentation sphere. The pseudo partial-waves are obtained from an atomic calculation with a smooth artificial potential. As the wavefunction and its gradient must be continuous, an all-electron partial-wave and its pseudo partial-wave must have the same value and gradient at the boundary of an augmentation sphere. The pseudo partial-waves must be smooth so that they can be represented on the smooth discretization in addition to the atomic discretization.

The coefficients $c^a_{k,nlm}$ remain to be determined. In the PAW method, this is done by defining smooth functions $\tilde{p}^a_{nlm}(r_a, \theta_a, \phi_a) = \tilde{p}^a_{nlm}(r)$, called projectors, which are localized inside augmentation spheres. The purpose of the projector $\tilde{p}^a_{nlm}(r)$ is to define the coefficient $c^a_{k,nlm}$ from the part of the smooth pseudo wavefunction $\tilde{\varphi}_k(r)$ which is inside the augmentation sphere a. Each partial-wave pair, $\xi^a_{nlm}(r_a, \theta_a, \phi_a)$ and $\tilde{\xi}^a_{nlm}(r_a, \theta_a, \phi_a)$, have their own projector $\tilde{p}^a_{nlm}(r)$. The projectors are defined by requiring them to be orthonormal to the pseudo partial-waves

$$\langle \tilde{p}^a_{nlm}(r_a, \theta_a, \phi_a) | \tilde{\xi}^{a'}_{n'l'm'}(r_a, \theta_a, \phi_a) \rangle = \delta_{a,a'} \delta_{nlm,n'l'm'}, \tag{20.3}$$

and by requiring that the pseudo wavefunction $\tilde{\varphi}_k(\boldsymbol{r})$ is reproduced by the pseudo partial-wave expansion

$$|\tilde{\varphi}_k(\boldsymbol{r})\rangle = \sum_{a,nlm} |\tilde{\xi}^a_{nlm}(r_a, \theta_a, \phi_a)\rangle \langle \tilde{p}^a_{nlm}(r_a, \theta_a, \phi_a) | \tilde{\varphi}_k(\boldsymbol{r}) \rangle \tag{20.4}$$

inside the augmentation sphere. In the limit when $\tilde{\xi}^a_{nlm}(r_a, \theta_a, \phi_a)$ forms a complete set, the sum

$$\sum_{nlm} |\tilde{\xi}^a_{nlm}(r_a, \theta_a, \phi_a)\rangle \langle \tilde{p}^a_{nlm}(r_a, \theta_a, \phi_a)| \tag{20.5}$$

is equal to an identity operator inside the augmentation sphere a, which we denote by 1_a. In practice, this holds only approximately as $\tilde{\xi}^a_{nlm}(r_a, \theta_a, \phi_a)$ does not form a complete set.

We multiply the pseudo wavefunction in Eq. 20.1 by the unity $1 = 1 - \sum_a 1_a + \sum_a 1_a$:

$$\begin{aligned}
\varphi_k(\boldsymbol{r}) &= \left(1 - \sum_a 1_a\right) |\tilde{\varphi}_k(\boldsymbol{r})\rangle \\
&+ \sum_{a,nlm} |\tilde{\xi}^a_{nlm}(r_a, \theta_a, \phi_a)\rangle \langle \tilde{p}^a_{nlm}(r_a, \theta_a, \phi_a) | \tilde{\varphi}_k(\boldsymbol{r}) \rangle \\
&+ \sum_{a,nlm} c^a_{k,nlm} \left[\xi^a_{nlm}(r_a, \theta_a, \phi_a) - \tilde{\xi}^a_{nlm}(r_a, \theta_a, \phi_a) \right], \tag{20.6}
\end{aligned}$$

where we used that $\sum_{nlm} |\tilde{\xi}^a_{nlm}(r_a, \theta_a, \phi_a)\rangle \langle \tilde{p}^a_{nlm}(r_a, \theta_a, \phi_a)| = 1_a$. The terms $(1 - \sum_a 1_a)|\tilde{\varphi}_k(\boldsymbol{r})\rangle$ and $\sum_{a,nlm} c^a_{k,nlm} \xi^a_{nlm}(r_a, \theta_a, \phi_a)$ yield the all-electron wavefunction, therefore the terms including pseudo partial-waves must cancel:

$$\sum_{a,nlm} c^a_{k,nlm} \tilde{\xi}^a_{nlm}(r_a, \theta_a, \phi_a) = \sum_{a,nlm} |\tilde{\xi}^a_{nlm}(r_a, \theta_a, \phi_a)\rangle \langle \tilde{p}^a_{nlm}(r_a, \theta_a, \phi_a) | \tilde{\varphi}_k(\boldsymbol{r}) \rangle, \tag{20.7}$$

and we obtain

$$c^a_{k,nlm} = \langle \tilde{p}^a_{nlm}(r_a, \theta_a, \phi_a) | \tilde{\varphi}_k(\boldsymbol{r}) \rangle. \tag{20.8}$$

Finally, the total wavefunction can be written as a linear operator $\hat{T}$ acting on the pseudo wavefuction

$$\varphi_k(\boldsymbol{r}) = \hat{T} \tilde{\varphi}_k(\boldsymbol{r}), \tag{20.9}$$

where

$$\hat{T} = 1 + \sum_{a,nlm} \left[\xi_{nlm}^a (r_a, \theta_a, \phi_a) - \tilde{\xi}_{nlm}^a (r_a, \theta_a, \phi_a) \right] \langle \tilde{p}_{nlm}^a (r) |. \qquad (20.10)$$

At this point, a few remarks are in place. The augmentation spheres cannot overlap in theory, but in practice, they can slightly overlap without significant changes in results. Moreover, in practice, a small number of partial-waves are enough to satisfy the above equations within required accuracy. For example, for a carbon atom, in addition to the $1s$ (which is frozen), $2s$ and $2p$, only one additional set of s, p, and d type unbound partial-waves are included.

20.3 Operators

PAW pseudo operators can be simply derived from the all-electron operators:

$$\langle \varphi_i | \hat{A} | \varphi_j \rangle = \langle \tilde{\varphi}_i | \hat{T}^\dagger \hat{A} \hat{T} | \tilde{\varphi}_j \rangle = \langle \tilde{\varphi}_i | \tilde{A} | \tilde{\varphi}_j \rangle, \qquad (20.11)$$

where $\hat{A}$ is the all-electron operator and $\tilde{A}$ is the correponding PAW operator. The algebra is lengthy, but straightforward, and yields for $\tilde{A} = \hat{T}^\dagger \hat{A} \hat{T}$:

$$\begin{aligned}
\tilde{A} = \hat{A} &+ \sum_{a,nlm} \left| \tilde{p}_{nlm}^a \right\rangle \left[\left\langle \xi_{nlm}^a \right| - \left\langle \tilde{\xi}_{nlm}^a \right| \right] \hat{A} [(1 - 1_a) + 1_a] \\
&+ \sum_{a,nlm} [(1 - 1_a^\dagger) + 1_a^\dagger] \hat{A} \left[\left| \xi_{nlm}^a \right\rangle - \left| \tilde{\xi}_{nlm}^a \right\rangle \right] \left\langle \tilde{p}_{nlm}^a \right| \\
&+ \sum_{a,nlm} \left| \tilde{p}_{nlm}^a \right\rangle \left[\left\langle \xi_{nlm}^a \right| - \left\langle \tilde{\xi}_{nlm}^a \right| \right] \hat{A} \sum_{a',n'l'm'} \left[\left| \xi_{n'l'm'}^{a'} \right\rangle - \left| \tilde{\xi}_{n'l'm'}^{a'} \right\rangle \right] \left\langle \tilde{p}_{n'l'm'}^{a'} \right|. \quad (20.12)
\end{aligned}$$

If we now expand this expression, the cross-terms of type $\left\langle \xi_{nlm}^a \middle| \hat{A} \middle| \tilde{\xi}_{n'l'm'}^{a'} \right\rangle$ cancel, and we are left with the following expression:

$$\begin{aligned}
\tilde{A} = \hat{A} &+ \sum_{a,nlm} \left| \tilde{p}_{nlm}^a \right\rangle \left[\left\langle \xi_{nlm}^a \middle| \hat{A} \middle| \xi_{nlm}^a \right\rangle - \left\langle \tilde{\xi}_{nlm}^a \middle| \hat{A} \middle| \tilde{\xi}_{nlm}^a \right\rangle \right] \left\langle \tilde{p}_{nlm}^a \right| \\
&+ \sum_{a,nlm} \left| \tilde{p}_{nlm}^a \right\rangle \left[\left\langle \xi_{nlm}^a \right| - \left\langle \tilde{\xi}_{nlm}^a \right| \right] \hat{A} [1 - 1_a] \\
&+ \sum_{a,nlm} [1 - 1_a^\dagger] \hat{A} \left[\left| \xi_{nlm}^a \right\rangle - \left| \tilde{\xi}_{nlm}^a \right\rangle \right] \left\langle \tilde{p}_{nlm}^a \right| \\
&+ \sum_{a \neq a'} \sum_{nlm} \left| \tilde{p}_{nlm}^a \right\rangle \left[\left\langle \xi_{nlm}^a \right| - \left\langle \tilde{\xi}_{nlm}^a \right| \right] \hat{A} \\
&\times \sum_{n'l'm'} \left[\left| \xi_{n'l'm'}^{a'} \right\rangle - \left| \tilde{\xi}_{n'l'm'}^{a'} \right\rangle \right] \left\langle \tilde{p}_{n'l'm'}^{a'} \right|. \quad (20.13)
\end{aligned}$$

For local (e.g., one-particle potential) and semi-local (e.g., kinetic energy) operators, only the first two terms remain as other terms cancel because bra and ket vectors are never nonzero in the same point of space. For example, the density (i.e., operator $\sum_k n_k |r\rangle\langle r|$) reads

$$n(r) = \tilde{n}(r) + \sum_a \left[n^a(r) - \tilde{n}^a(r) \right]. \tag{2.14}$$

For some nonlocal operators, like the Coulomb interaction of the Hartree potential, a special trick can be used. A smooth function, called compensation charge $-\tilde{Z}^a(r, \theta, \phi)$, is added and removed (also known as "an intelligent zero") inside each augmentation sphere. The purpose of the compensation charge is to cancel all multipole moments of the atomic density contributions $n^a(r) - \tilde{n}^a(r)$, and therefore, the potential is zero outside augmentation spheres. The compensation charge is then added to the pseudo charge density to restore the cancelled multipoles. (Note that this is not an approximation.) The Hartree and nuclear-electron attraction potential reads

$$\int d^3r' \frac{n(r') - \sum_a Z^a \delta(r' - R^a)}{|r - r'|} = \int d^3r' \frac{\tilde{n}(r') - \sum_a \tilde{Z}^a(r')}{|r - r'|}$$
$$+ \int d^3r' \frac{1}{|r - r'|} \sum_a \left\{ n^a(r'_a, \theta'_a, \phi'_a) - Z^a \delta(r' - R^a) \right.$$
$$\left. - \left[\tilde{n}^a(r'_a, \theta'_a, \phi'_a) - \tilde{Z}^a(r'_a, \theta'_a, \phi'_a] \right] \right\}, \tag{20.15}$$

where Z^a is the charge of the nucleus a.

20.4 Ground-State Kohn–Sham Equation and Forces

The PAW method causes relatively small changes to an existing pseudopotential code. The ground-state Kohn–Sham eigenvalue problem in the PAW method becomes a generalized eigenvalue problem

$$\tilde{H}\tilde{\varphi}_k = \tilde{\varepsilon}_k \tilde{S}\tilde{\varphi}_k, \tag{20.16}$$

where $\tilde{S} = \hat{T}^\dagger \hat{1} \hat{T}$ is the overlap operator, which arises from the non-orthonormality of the PAW discretization $\langle \tilde{\varphi}_i | \tilde{\varphi}_j \rangle \neq 0$, or actually from the $\tilde{S}$-orthonormality $\langle \tilde{\varphi}_i | \tilde{S} | \tilde{\varphi}_j \rangle = \delta_{ij}$. Also, the forces change as the partial-waves and projectors depend on nuclear positions. The ground-state force becomes

$$F^a = -\frac{\partial E}{\partial R^a} - \sum_k \frac{\partial E}{\partial |\tilde{\varphi}_k\rangle} \frac{d|\tilde{\varphi}_k\rangle}{dR^a} - \sum_k \frac{\partial E}{\partial \langle\tilde{\varphi}_k|} \frac{d\langle\tilde{\varphi}_k|}{dR^a} = -\frac{\partial E}{\partial R^a} + \sum_k n_k \varepsilon_k \langle\tilde{\varphi}_k| \frac{d\tilde{S}}{dR^a} |\tilde{\varphi}_k\rangle. \tag{20.17}$$

20.4.1 Connection to Nonlocal Pseudopotentials

The PAW method can be used to derive the nonlocal pseudopotential scheme in a well justified way (Kresse and Joubert 1999). If we make a Taylor expansion of some density dependent operator $\hat{B}[n]$ (e.g., electronic energy) with respect to the density at the density of a reference atom n_{ref}^a:

$$\hat{B}^a[n^a] = \hat{B}^a[n_{\rm ref}^a] + (n^a - n_{\rm ref}^a) \left.\frac{\partial \hat{B}^a[n^a]}{\partial n^a}\right|_{n^a = n_{\rm ref}^a} + \mathcal{O}\left((n^a - n_{\rm ref}^a)^2\right), \quad (20.18)$$

and we keep only the zeroth and the first order terms, we get the ultrasoft pseudo potential approach (for details, see Kresse and Joubert 1999).

20.5 Time-Dependent DFT

It is quite simple to derive the TD Schrödinger equations within the PAW formalism. We start from the all-electron action principle and apply the PAW transformation operator $\hat{T}$

$$A[\Psi] = \int dt \langle \Psi(t)|i\frac{\partial}{\partial t} - \hat{H}(t)|\Psi(t)\rangle = \int dt \langle \tilde{\Psi}(t)|\hat{T}^\dagger \left[i\frac{\partial}{\partial t} - \hat{H}(t)\right] \hat{T}|\tilde{\Psi}(t)\rangle.$$
$$(20.19)$$

If the operator $\hat{T}$ is time-independent, i.e., nuclei do not move, the equation reduces to

$$A[\Psi] = \int dt \langle \tilde{\Psi}(t)|i\tilde{S}\frac{\partial}{\partial t} - \tilde{H}(t)|\tilde{\Psi}(t)\rangle, \quad (20.20)$$

which corresponds to the following time-dependent Kohn–Sham equation

$$i\tilde{S}\frac{\partial}{\partial t}\tilde{\varphi}(t) = \tilde{H}_{\rm KS}(t)\tilde{\varphi}(t). \quad (20.21)$$

If the nuclei move, for example, when modelling nonadiabatic electron-ion dynamics with the Ehrenfest TDDFT, the PAW transformation operator $\hat{T}$ is time-dependent. The TDKS equations become (Qian et al. 2006)

$$\hat{T}^\dagger i\frac{\partial \hat{T}}{\partial t}\tilde{\varphi}(t) + i\tilde{S}\frac{\partial}{\partial t}\tilde{\varphi}(t) = \tilde{H}(t)\tilde{\varphi}(t) \quad (20.22)$$

that implies

$$i\tilde{S}\frac{\partial}{\partial t}\tilde{\varphi}(t) = [\tilde{H}(t) + \tilde{P}(t)]\tilde{\varphi}(t), \quad (20.23)$$

where

$$\tilde{P}(t) = -i\hat{T}^{\dagger}\frac{\partial \hat{T}}{\partial t} = -i\sum_a \boldsymbol{v}_a(t) \cdot \left(\hat{T}^{\dagger}\nabla_{\boldsymbol{R}_a}\hat{T}\right) \quad (20.24)$$

and $\boldsymbol{v}_a(t)$ is the velocity of ion a. The forces are calculated within the Ehrenfest scheme and taking into account that wavefunctions depend on atomic positions (Di Ventra and Pantelides 2000b):

$$\boldsymbol{F}_a = -i\frac{d}{dt}\left\langle\frac{\partial}{\partial \boldsymbol{R}_a}\right\rangle \quad (20.25)$$

which leads to

$$\boldsymbol{F}_a = \frac{-1}{\langle\Psi|\Psi\rangle}\left\{\left\langle\Psi\left|\frac{\partial H}{\partial \boldsymbol{R}_a}\right|\Psi\right\rangle + \left\langle\frac{\partial\Psi}{\partial \boldsymbol{R}_a}\left|H - i\frac{\partial}{\partial t}\right|\Psi\right\rangle\right. \quad (20.26)$$

$$\left. + \left\langle\Psi\left|H + i\frac{\partial}{\partial t}\left|\frac{\partial\Psi}{\partial \boldsymbol{R}_a}\right\rangle\right\}\right.. \quad (20.27)$$

20.5.1 Time-Propagation

The time-propagation approach for TDDFT has to deal with the overlap matrix $\tilde{S}$. This basically corresponds to replacing the Hamiltonian $\tilde{H}[n](t)$ with $\tilde{S}^{-1}(t)\tilde{H}[n](t)$ in propagation methods. If the augmentation spheres are nonoverlapping, the overlap matrix $\tilde{S}$ is block diagonal, and therefore, also its inverse $\tilde{S}^{-1}$ is block diagonal. Each block of the overlap matrix $\tilde{S}$ can be efficiently inverted via dense matrix algebra. However, in the usual case that the augmentation spheres are slightly overlapping, the inverse of overlap matrix $\tilde{S}^{-1}$ becomes much denser, and its calculation and storage becomes prohibitively expensive. This makes most of the propagation methods which are usually employed in practical applications of TDDFT, unsuitable for PAW. This is particularly true for polynomial expansions of split-operator methods. However, the well-known Crank-Nicholson method

$$\left[\tilde{S} + \frac{i}{2}\tilde{H}[n](t + \Delta t/2)\right]\varphi(t + \Delta t) = \left[\tilde{S} - \frac{i}{2}\tilde{H}[n](t + \Delta t/2)\right]\varphi(t), \quad (20.28)$$

with a predictor-corrector step to handle nonlinearity of the Hamiltonian $\tilde{H}[n](t)$, can be applied without an additional effort (Qian et al. 2006; Walter et al. 2008). The inversion of the overlap matrix $\tilde{S}$ is incorporated in the inversion of the propagator matrix $\tilde{S} + \frac{i}{2}\tilde{H}[n](t + \Delta t/2)$, and therefore does not add further complexities or computational burden to the propagation. Note that the solution of Eq. 20.28 can be performed by efficient sparse linear solvers (e.g., the complex symmetric conjugate gradient) which exhibit good scaling and efficiency in modern computer architectures.

20.5.2 Linear-Response TDDFT

The inclusion of the PAW formalism in linear-response TDDFT is fairly straightforward (Walter et al. 2008), but somewhat cumbersome. The overlap operator $\tilde{S}$ does not appear in the Casida equation (see Chap. 7), as we are working in the Kohn–Sham eigenvector basis and the eigenvectors are $\tilde{S}$-orthonormal. However, the PAW corrections will appear in the Hartree and exchange-correlation kernel K_{Hxc}. If the exchange-correlation functional is local or semilocal and frequency independent, the all-electron xc kernel reads

$$f_{xc}[n](\boldsymbol{r}) = f_{xc}[\tilde{n}](\boldsymbol{r}) + \sum_a [f_{xc}[n^a](\boldsymbol{r}) - f_{xc}[\tilde{n}^a](\boldsymbol{r})]. \tag{20.29}$$

The Hartree contribution K_H to the Hartree and exchange-correlation kernel K_{Hxc} reads

$$K_{ij,pq}^{H} = \int d^3r \int d^3r' n_{ij}^*(\boldsymbol{r}) \frac{n_{pq}(\boldsymbol{r}')}{|\boldsymbol{r} - \boldsymbol{r}'|}, \tag{20.30}$$

where

$$n_{ij}(\boldsymbol{r}) = \varphi_i^*(\boldsymbol{r})\varphi_j(\boldsymbol{r}) \tag{20.31}$$

is the pair density. Its extension to the PAW formalism is slightly more involved, but by using the same trick as with the Hartree potential for the ground-state, we can again simplify it to pseudo and atomic terms:

$$K_{ij,pq}^{H} = \int d^3r \int d^3r' \frac{\left[\tilde{n}_{ij}(\boldsymbol{r}) - \sum_a \tilde{Z}_{ij}^a(\boldsymbol{r})\right]^* \left[\tilde{n}_{pq}(\boldsymbol{r}') - \sum_a \tilde{Z}_{pq}^a(\boldsymbol{r}')\right]}{|\boldsymbol{r} - \boldsymbol{r}'|}$$

$$+ \int d^3r \int d^3r' \sum_a \left\{ \frac{[n_{ij}^a(\boldsymbol{r})]^* n_{pq}^a(\boldsymbol{r}')}{|\boldsymbol{r} - \boldsymbol{r}'|} \right.$$

$$\left. - \frac{\left[\tilde{n}_{ij}^a(\boldsymbol{r}) - \tilde{Z}_{ij}^a(\boldsymbol{r})\right]^* \left[\tilde{n}_{pq}^a(\boldsymbol{r}') - \tilde{Z}_{pq}^a(\boldsymbol{r}')\right]}{|\boldsymbol{r} - \boldsymbol{r}'|} \right\}. \tag{20.32}$$

Again, many terms have cancelled due to fact that all multipole moments of the atomic terms were forced to zero using the compensation charges $\tilde{Z}_{ij}^a(\boldsymbol{r})$.

20.6 Applications

The first TDDFT implementation employing the PAW method appeared only recently (Walter et al. 2008). The time-propagation and Casida approaches were demonstrated for small molecules. Later, the time-propagation approach was applied to large ligand-protected metal nanoparticles (Kacprzak et al. 2009; Enkovaara et al. 2010), the largest of which was $Au_{104}(S-CH_3)_{44}$ with $\sim$1700 active electrons. The PAW method has also been applied to the calculation of optical properties of solids (Gajdos et al. 2006; Ramos et al. 2008).

Chapter 21
Harnessing the Power of Graphic Processing Units

Xavier Andrade and Luigi Genovese

21.1 Introduction

The continuous increment in the power of modern high-performance computing (HPC) platforms has further stimulated the interest of the electronic structure calculations community for more computationally challenging studies. Systems which were intractable only few years ago become now accessible with the advent of modern machines. In the past few years, the possibility of using graphic processing units (GPU) for scientific calculations has raised a lot of interest as alternative to current calculations based on central processing units (CPU). A technology initially developed for home computers has rapidly evolved in the direction of a programmable parallel streaming processor. The features of these devices, in particular the very low price performance ratio, together with the relatively low energy consumption per Flops (floating point operations per second), make them attractive platforms for intensive scientific computations.

X. Andrade (✉)
Department of Chemistry and Chemical Biology, Harvard University,
12 Oxford Street, Cambridge, MA 02138, USA
e-mail: xavier@tddft.org

X. Andrade
ESTF and Departamento de Física de Materiales, University of the Basque Country
UPV/EHU, Av. Tolosa 72, 20018 San Sebastían, Spain

L. Genovese
European Synchtrotron Radiation Facility,
6 rue Horowitz, BP220, 38043 Grenoble, France

L. Genovese
Laboratoire de Simulation Atomistique, Commissariat à l'Énergie
Atomique et aux Energies Alternatives, INAC/SP2M,
17 avenue des Martyrs, 38054 Grenoble, France

M. A. L. Marques et al. (eds.), *Fundamentals of Time-Dependent Density Functional Theory*, Lecture Notes in Physics 837, DOI: 10.1007/978-3-642-23518-4_21,
© Springer-Verlag Berlin Heidelberg 2012

In this chapter we address the usage of GPU architectures in DFT and TDDFT computations. The objective is to give an overview on the problematic of porting a code to a hybrid CPU/GPU platform, in general and in the particular case of the KS picture for modelling electronic systems. We start by briefly discussing GPU architecture and their special characteristics in comparison with CPUs. We then move to the aspects of programming a GPU code. Next we present two applications of GPUs: ground-state DFT calculations in the BIGDFT code (Genovese et al. 2008) and real-time TDDFT propagation in the OCTOPUS code (Marques et al. 2003; Castro et al. 2006). Finally we discuss how GPUs could be used for other types of TDDFT applications and formalisms.

21.2 Basic Concepts in GPU Architectures

In many aspects, GPUs are quite different from CPUs. Understanding the peculiarities of their architecture is essential to plan the GPU port of a program and to write efficient GPU code.

The first particularity of GPUs is that they are co-processors controlled exclusively by a CPU (see Fig. 21.1). GPUs cannot be used alone and must have a CPU associated, thus forming a so-called hybrid architecture. GPUs can only access data that lies in its dedicated memory. Data must be explicitly copied by the programmer between CPU memory and GPU memory. They are normally connected through a PCI Express (PCIe) link, which has a relatively small bandwidth and high latency (see Table. 21.1)

GPUs are massively parallel processors, in a single chip they can include hundreds of floating point execution units,[1] while a multi-core CPU typically has around 32 floating point execution units (considering all cores). As a result, the theoretical floating point throughput of a GPU is around one order of magnitude larger than the one of a CPU (part of the difference is compensated by the CPU higher operating frequency). Both processing units have approximately the same number of transistors and consume an amount of power of the same order, so this difference is mainly explained by the design strategy of each processor type, based on the tasks that each one is targeted to perform. CPUs are designed to run complex programs as fast as possible. GPUs, on the other hand, are designed to run simple programs in parallel over a large amount of data, so they can dedicate most of its transistors to execution units.

In a GPU, execution units are organised in groups that form a multiprocessor. All the execution units in a multiprocessor share a control unit, so they perform the same instruction at the same time.

To exploit the highly parallel nature of the processors, GPU programs use fine-grained threads or tasks. These are intimately different from the CPU threads which are typically used in parallel environments like OpenMP. In a GPU the strategy is to have many more threads than execution units by assigning each thread a very small

---

[1] For marketing reasons these execution units are sometimes called "cores", although they are not comparable to a CPU core.

Fig. 21.1 A GPU associated with a CPU

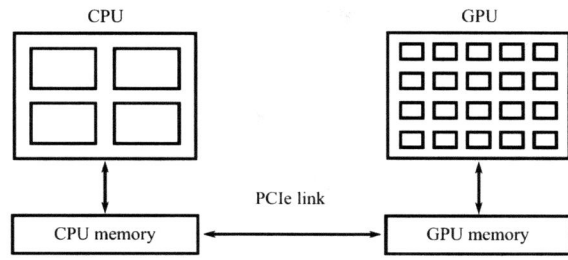

Type	Latency [ns]	Bandwidth [Gb/s]
CPU memory	40	30
GPU memory	300	100
PCIe link	20000	8

Table 21.1 Comparison of typical data transfer capabilities of a GPU, a CPU and the PCI Express (PCIe) link that connects them

amount of the work. As the GPU can switch threads without cost, it can hide the latency of the operations of one thread by processing other threads while waiting.

Memory access is also particular to GPUs. CPUs only have one type of memory and rely on caches to speed up the access to it. GPUs do not always have a cache[2] but instead they have a fast local memory shared by all the execution units in a multi-processor. This local or private memory must be explicitly used by the programmer to store data that needs to be accessed frequently. Main memory access also need to be done carefully. The memory bandwidth is higher than for a CPU but latency is also higher, see Table 21.1. Moreover, to obtain maximum memory transfer rates, the execution units in a multiprocessor must access sequential memory locations.

So we summarise the principal features of GPU computation:

• Due to the high latency of the communication, the programmer should try to limit the data transfer between CPU and GPU as most as possible.
• Calculation workload is parallelised in many little chunks, which perform the same kind of operations on different data.
• Data locality is of great importance to achieve good performance, since different multiprocessors have different local memories.
• Memory access patterns should be as regular and homogeneous as possible.

21.3 GPU Programming

Given their special characteristics, GPUs cannot be programmed directly using traditional serial programming languages, so a special framework is needed. The CUDA programming language of Nvidia was probably the first in the market and it is,

---

[2] Fortunately for programmers, newer GPUs include a small cache. This makes GPU code optimisation much simpler.

nowadays, the most advanced in terms of functionalities and maturity. Nonetheless, being an architecture-specific language, the code is not portable and there is no standard associated to it.

21.3.1 The OpenCL Language

To avoid the problem of having different GPU-programming frameworks specific to each hardware a standard was proposed: the OpenCL specification (Munshi 2009). Its ambition is to open a pathway toward cross-platform parallel computing, of which hybrid CPU/GPU architectures are a first example. Its specifications are similar to the present organisation of the CUDA language, but with some useful generalisations.

It must be pointed out that, while OpenCL code can be executed unchanged in different platforms (a GPU, a CPU, or other OpenCL supported device), this does not necessarily mean that the code optimised for one platform will run efficiently in other platforms.

21.3.2 Evaluation of Benefits: Performance with Complex Codes

The peculiarities of GPU architecture are thus of paramount importance to determine if an application can have benefits from it. A GPU program will be ported conveniently depending of the nature of its operations. An evaluation should be performed to understand the trade-off between rewriting and speed-up. The situation is even more complicated for a complex code with many operations, which may work in a parallel environment. For this case, the evaluation of the benefits of using a GPU-accelerated code must be performed at three different levels.

- Firstly, one has to evaluate the effective speed-up provided by the GPU code with respect to the corresponding CPU routines which perform the same operations. This is the "bare" speed-up.
- At the second level, the "complete" speed-up has to be evaluated; the performances of the whole hybrid CPU/GPU code should be analysed with respect to the pure CPU executions. Clearly, this result depends of the importance of the ported routines in the context of the whole code [i.e., following Amdahl's law (Amdahl 1967)] and the additional cost of CPU/GPU copies.
- For a parallel code, there is still another step which has to be evaluated. This is the behaviour of the hybrid code in a parallel environment. Indeed, for parallel runs the picture is complicated by two things. The first one is that since GPUs can be faster than CPUs, the relative cost of communication increases. The second issue is that we might need to copy data between remote GPUs, actually increasing the communication time.

There are, however, some additional considerations to make. First of all, as with all parallel systems, on a GPU the speed up depends on the size of the studied system. In this context, GPUs might not provide shorter calculation times, but they could be useful to study more complex systems in a similar time [as modelled by Gustafson's law (Gustafson 1988)]. In second place, a direct migration of a CPU code might not be the best strategy to obtain an efficient GPU code. Sometimes restructuring the calculation or even using different algorithms might be necessary to fully exploit the GPU capabilities.

21.4 GPUs for DFT and TDDFT

The advantages offered by GPU programming are of great interest for physics and chemistry calculations. A lot of scientific applications have been recently ported on GPU, including, for example, molecular dynamics (Yang et al. 2007), quantum Monte-Carlo (Anderson et al. 2007), and finite element methods (Göddeke et al. 2007). In the domain of electronic structure calculations, up to now, most efforts have been done in the context of Quantum Chemistry for DFT (Yasuda 2008; Ufimtsev and Martinez 2009) and Moller-Plesset (Vogt et al. 2008; Watson et al. 2010) methods. Given the small size of the basis set, usually the amount of data to be transferred to the GPU is limited and these implementations may benefit of the acceleration of arithmetic operations.

The situation is less developed for the condensed matter physics community, where systematic discretization schemes are used. In this case, the number of degrees of freedom is order of magnitudes larger, and this implies that the amount of data to be treated is large. This poses a problem for GPU computation as transferring data into the GPU memory is more time-consuming, so care should be taken in managing such operations.

In the next sections we illustrate two applications for GPUs in codes that use systematic discretisation strategies. First the implementation of GPU acceleration in the BIGDFT code, where GPUs are used for ground-state DFT calculations based on a wavelet basis set. Next, we discuss the GPU port of the OCTOPUS code to perform real-time TDDFT calculations based on a real-space grid discretisation.

21.5 GPU Implementation in the BIGDFT Code

The BIGDFT code implements DFT calculations based on Daubechies wavelets basis set (Daubechies 1992), it is distributed under GNU–GPL license and integrated in the ABINIT (Gonze et al. 2002) software package. A separate, standalone version of this code (including the hybrid version) is also available (Genovese et al. 2008). Thanks to wavelet properties, this code shows high systematic convergence properties, very good performance and an excellent parallel efficiency.

21.5.1 The Code Structure: Preliminary CPU Investigation

Applying the Hamiltonian operator on the KS wave-function is only one of the operations which are performed in a DFT code. In general, an optimisation iteration of a KS wave-function is organised as follows:

1. Application of the Hamiltonian onto wave-functions,
2. Overlap matrix,
3. Preconditioning,
4. Wave-function update,
5. Orthogonalisation.

Any of these steps is associated to a different set of operations. Steps (2) and (4) are essentially related to linear algebra routines (BLAS calls), whereas step (5) is in general implemented via LAPACK routines. These routines can be accelerated using GPU ported linear algebra libraries. The CUBLAS package developed by NVidia can be easily linked to the code. Steps (1) and (3) are associated to the application of different operators on the wave-functions, and from the viewpoint of GPU porting, they have in general to be recoded.

The details of the BIGDFT code are presented elsewhere (Genovese et al. 2008, 2009). Most of the operators which are associated to the KS Hamiltonian are computationally written as a combination of convolutions with short, separable filters. The lowest level routine which will be ported on GPU is then a set of independent, one dimensional convolutions.

21.5.2 GPU Convolution Routines and CUBLAS Linear Algebra

We have evaluated the performances of the GPU port of the 1D convolutions and their 3D counterpart. For these evaluations, we used a computer with an Intel Xeon Processor X5472 (3.0 GHz) and a NVidia Tesla S1070 card. The CPU version of BIGDFT is deeply optimised with optimal loop unrolling and compiler options. The GPU code is compiled with the Intel Fortran compiler (version 10.1.011). All benchmarks are performed with double precision floating point numbers. With these options the magic filter convolutions run at about 3.4 GFlops.

The GPU versions of the one-dimensional convolutions are about one order of magnitude faster than their CPU counterparts. Since these are the building blocks of the 3D convolutions, this gain in performance is reflected also on the 3D operators.

Also the linear algebra operations can be executed on the card thanks to the CUBLAS routines. We obtain speed-ups between a factor of 20 and 60 for double precision calls to CUBLAS routines for a typical wave-function size of a BIGDFT run as a function of the number of orbitals. These results take into account the amount of time needed for transferring data to and from the card.

From these tests, we can see that both GPU-ported sections are orders of magnitude faster than the corresponding CPU counterpart. We now discuss the performance of the complete code.

21.5.3 Performance Evaluation of Hybrid Code

As a test system, we used the ZnO crystal, which has a wurtzite bulk-like structure. Such system has a relatively high density of valence electrons so that the number of orbitals is rather large even for a moderate number of atoms.

We performed two kinds of tests. The first is related to the behavior of the GPU-accelerated code on a single machine. Results can be found in Fig. 21.2. It can be seen that GPU acceleration contributes to a significant reduction of the overhead of linear algebra operations and convolutions. Both CUDA and the more recent OpenCL implementation of these convolutions are tested. By combining these accelerations with in-node MPI parallelisation, we may achieve a speed-up up to one order of magnitude faster than the mono-core CPU run. The second test compares the behavior of the hybrid code in a multi-node machine. We use the hybrid section of the CCRT Titane machine, with Intel X5570 CPUs and Nvidia Tesla S1070 GPUs. In this test we keep the size of the system fixed and increase the number of MPI processes such as to decrease the number of orbitals per core. We then compare the speed-up of each run with the hybrid code. The parallel efficiency of the code is not particularly affected by the presence of the GPU. For this machine, the time-to-solution speed-up is around three.

21.6 TDDFT on GPUs: Implementation in OCTOPUS

In this section we detail the approach implemented in the OCTOPUS code (Marques et al. 2003, Castro et al. 2006) for TDDFT calculations on GPUs. OCTOPUS is a free software package that implements several TDDFT formalisms. Real-time propagation is the most used method, due to its flexibility. So up to now we have focused our efforts in this formalism.

The implementation is based on the idea of using blocks of KS orbitals. We use the GPU to apply the KS Hamiltonian over these blocks, obtaining important performance gains not only for time propagation but also for ground-state DFT and Sternheimer calculations (Andrade et al. 2007). The scheme is also applied for code optimisation on CPUs with vectorial floating units.

The GPU implementation is based on the OpenCL framework and it has been tested on Nvidia and AMD implementations. However, for the moment the code has only been optimized for Nvidia cards. The CPU vectorial code is designed for x86 CPUs with SSE2 instructions and the IBM Blue Gene/P architecture, using compiler directives. The code is included in the current development version of OCTOPUS and it will be included in the next release.

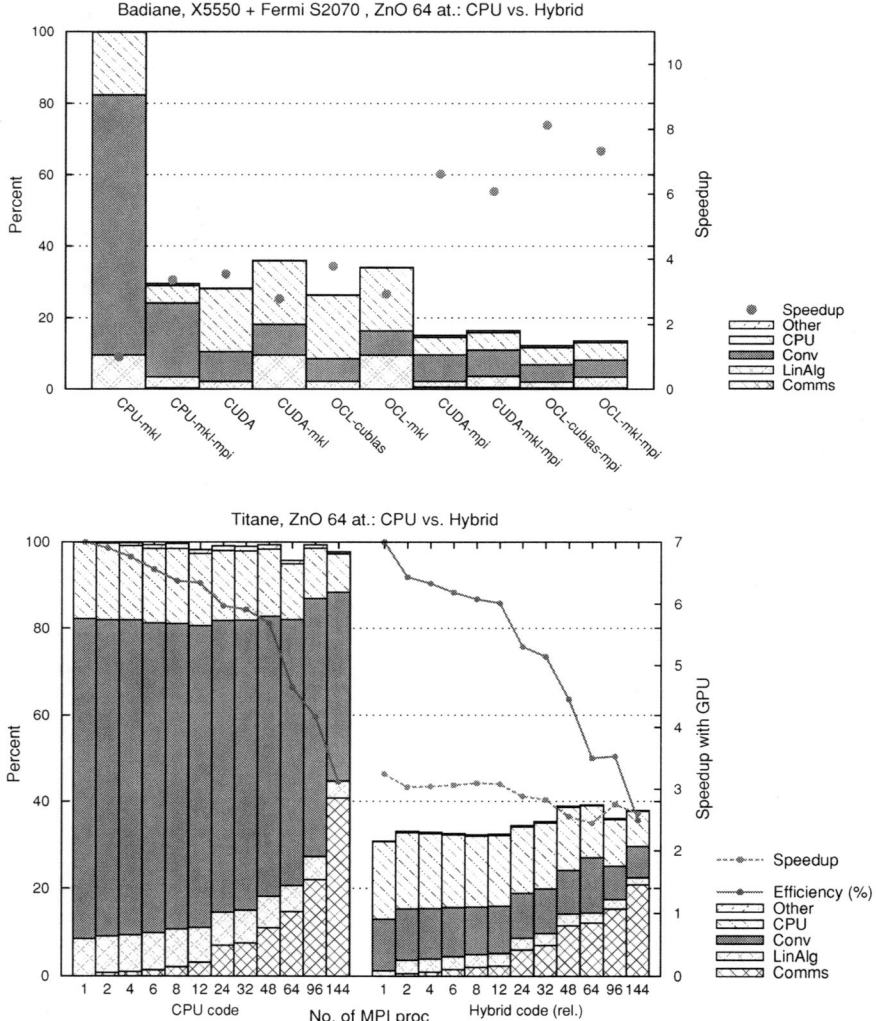

Fig. 21.2 Performances of a run of the BIGDFTcode for a 64 atoms ZnO system. Different acceleration strategies are compared with respect to the time spent for a sequential pure CPU run. Two different platforms were used for this test; in the first run (*top panel*), an Intel X5550 quad-core is associated to a Fermi S2070 card, with CUDA 3.2. Both CUDA and OpenCL (OCL) implementations are tested. In the *bottom panel*, the system have been tested with an increasing number of processors (Intel X5570 2.93 GHz). The scaling efficiency of the calculation is also indicated. In the right side of bottom panel, the same calculation has been done using one Tesla S1070 card per CPU core, for both architectures

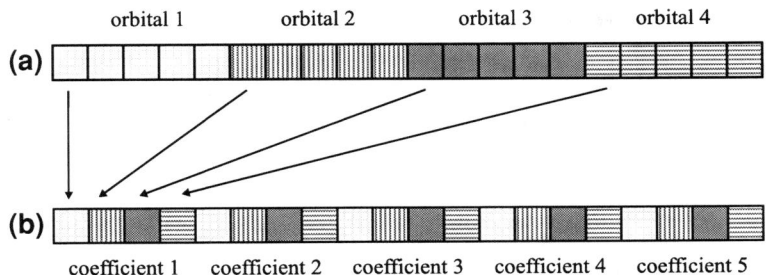

Fig. 21.3 Example of memory layout for a block of four orbitals with five coefficients each: **a** Standard memory layout where each orbital is contiguous in memory. **b** Optimum memory where all the coefficients in a block are contiguous. The arrows indicate the relation of the position of the first coefficient in both schemes

21.6.1 Working with Blocks of Kohn-Sham Orbitals

The key to obtain good performance in a GPU is to apply the same instructions over several streams of independent data. In TDDFT, the operations to be performed to each orbital are exactly the same and fully independent. In this way, we can obtain these independent data-streams by working with a group of KS orbitals, that we call a block. This idea is the central strategy for the GPU porting of OCTOPUS, but the scheme is quite general. It does not depend on the type representation or basis set used, as long as there are operations to be applied simultaneously to several orbitals. It is general also in the sense that it can be a good code optimisation strategy not only for GPUs, but also for CPUs.

For optimal execution, the number of orbitals in the block, or *block size*, needs to be compatible (an exact divisor or multiple) with the number of execution units in each multiprocessor of the GPU (typically 16, 32 or 64 in modern GPUs). So powers of two are a good choice.[3]

Working with blocks of orbitals also allows to optimise memory accesses. However, choosing an appropriate memory layout is crucial. The order used in OCTOPUS, and many other DFT codes, is to store contiguously in memory each orbital, as seen in Fig. 21.3a. This is not optimal since values for the same coefficient of different orbitals are scattered. The optimal memory arrangement is to have all coefficients together, effectively transposing the order of the array, Fig. 21.3b.[4] In this manner the GPU can maximise memory accesses since tasks read or write sequential memory locations. The transposition is done by the GPU after copying to GPU memory and undone before copying back to main memory. The cost of the transposition is negligible in comparison with the cost of the copy.

---

[3] This has the additional advantage that integer multiplications by the block size can be done using the much faster bit shift operations.

[4] Note that it is the block that is transposed with respect to the standard ordering, not the whole set of orbitals.

21.6.2 Application of the Kohn-Sham Hamiltonian

Since for real-time TDDFT the work to be done to each orbital is independent of the other orbitals, we can apply simultaneously the KS Hamiltonian over a block of orbitals. This is not always the case for ground-state DFT. In some algorithms, due to orthogonality constraints the calculation of the ground-state orbitals is sequential. There are, however, eigensolvers based on the idea of direct minimisation (Pulay 1980) that are suitable for this type of parallelism.

In OCTOPUS the orbitals are not always available in GPU memory, they must be copied before the calculation, and the result copied back to main memory. Since these copies are costly they are avoided as much as possible by keeping data in GPU memory if the Hamiltonian is going to be applied again. For example, in real-time TDDFT, for the application of the exponential of the Hamiltonian in the Taylor approximation (Castro et al. 2004a) only the initial orbitals and the resulting ones need to be copied, while the intermediate results are passed in GPU memory.

To fully apply the Hamiltonian using the GPU in a real space code we need to calculate the action of the potential, and the kinetic energy operator. The kinetic part is the most important one and is discussed in the next section. The local part of the potential is in general very simple to apply, since in real space it is only a multiplication of two arrays. The non-local part of the potential, that appears when pseudo-potentials are used, is more interesting. In order to achieve a level of parallelism suitable for the GPU, the projectors corresponding to all atoms must be applied simultaneously.

21.6.3 The Kinetic Energy Operator in Real-Space

OCTOPUS is based on a real-space grid discretisation. The basis for this method is the approximation of the Laplacian in the kinetic energy operator by high-order finite differences (Chelikowsky et al. 1994). This is the most time-consuming part, so it is essential to apply it as efficiently as possible. Memory access is a delicate issue in finite difference operators, since each point is used several times. In OCTOPUS, the order in which grid values are stored in memory is chosen such that points that are neighbours are close. The result is that memory accesses have a good locality and can profit from cache memory.

We can see how critical blocks are to realise the performance potential of the GPU in Fig. 21.4a . It shows the throughput obtained in a GPU and a CPU for the Laplacian for different block sizes. In the GPU, by using blocks we can obtain speed-ups of 5 with respect to the case of working with a single orbital. And, while for a block size of 1 the GPU is only 1.4 times faster than a CPU, for a block size of 32 the GPU is almost 4.5 times faster. Due to the limited cache size, increasing the block size beyond 32 orbitals decreases the performance in both processors.

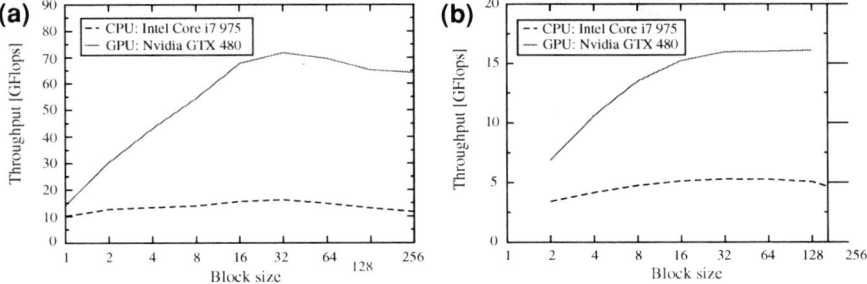

Fig. 21.4 Comparison of the throughput of OCTOPUS for different block sizes. **a** Comparison of the throughput of the Laplacian **a** Execution of a real-time propagation for a C_{60} molecule. GPU: Nvidia GTX 480. CPU: Intel Core i7 975, 3.33 GHz running four threads

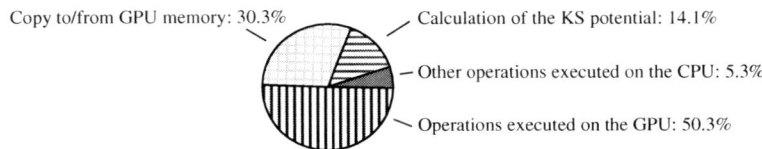

Fig. 21.5 Breakdown of the execution of a real-time propagation of OCTOPUS of a C_{60} molecule. Execution on a Nvidia GTX 480 GPU and a Core i7 975 CPU using four threads

21.6.4 Overall Performance Improvements

In Fig. 21.4b we present the throughput obtained for a real-time propagation with OCTOPUS performed with a GPU and a CPU. For this case the GPU is three times faster than the CPU. This represents a smaller speed-up than the one of the individual components. There are two reasons for this. First, copies between CPU and GPU memory consume a considerable amount of time. Second, the parts of the code that are executed on the CPU limit the speed-up that can be obtained. In Fig. 21.5 we show how the time is spent during the execution of OCTOPUS on a GPU. Roughly half of the time is used executing GPU code, while 30% is spent in GPU/CPU copies. The remaining 20% corresponds to tasks that are performed by the CPU.

21.7 Future Developments in TDDFT

Up to now most of the applications of GPU computing to electronic structure problems have been centred around ground state DFT methods. In many cases these developments can be directly applied to TDDFT formalisms, specially for methods based on the direct representation of the Hamiltonian instead of the spectral representation (in terms of unoccupied orbitals). There are, however, approaches in TDDFT that use different types of operations from ground state DFT, and whose application to GPU computing must still be studied.

In the Casida linear response formalism (Casida 1996), the main operation to be performed is the calculation of Coulomb integrals of the form

$$K_{ijkl} = \int \mathrm{d}^3r \int \mathrm{d}^3r' \frac{\varphi_i^*(\boldsymbol{r})\varphi_j^*(\boldsymbol{r})\varphi_k(\boldsymbol{r})\varphi_l(\boldsymbol{r})}{|\boldsymbol{r} - \boldsymbol{r}'|}. \tag{21.1}$$

When using Gaussian type orbitals, these integrals can be calculated in analytic form. In this context, Ufimtsev and Martinez proposed a method to calculate the analogous integrals that appear in HF using GPUs (Ufimtsev and Martinez 2009). A key point of their approach is to re-order the integrals by the type of orbital, avoiding branches that can degrade performance in GPU code. Up to now, their method is only valid for s and p orbitals. This limits the applicability to TDDFT, since a proper description of the excited states using Gaussian type orbitals requires higher angular momentum components.

In other discretizations, the evaluation of Eq. 22.1 is done by calculating one of the integrals as the solution of a Poisson equation. There exist many Poisson solvers for different boundary conditions [see e.g. (Genovese et al. 2006, 2007)] that can be applied to electronic structure problems. In other areas, some Poisson solvers have already been applied using GPUs (Bolz et al. 2005, Shi et al. 2009, Jeschke et al. 2009, Grossauer and Thoman 2008).

For ground state DFT, a single Poisson solution is required per iteration. While in a scheme like Casida (or HF), several Poisson solutions are required and they can be calculated simultaneously. This gives the possibility of using a scheme of blocks to exploit the parallelism of the GPU.

Another common approach for TDDFT, specially for periodic systems, is to calculate the response functions using perturbation theory. This is usually done in a plane wave basis [as, for example, the yambo code does (Marini et al. 2009)]. The most time consuming operations in this type of calculations are fast Fourier transforms and dense matrix linear algebra. These operations are common and standard high-performance libraries exist or should be available soon for GPU architectures. However, since these calculations require large quantities of memory, it is unlikely that they can fit the whole data-set in the limited memory of a GPU. So, the main challenge for these approaches will come from efficiently managing CPU/GPU copy operations.

21.8 Conclusions

The process of porting a code to a hybrid CPU/GPU architecture requires some effort to learn the GPU programming techniques and to write and optimise the GPU code. This process however, should only affect routines that take care of well defined operations like the application of the Hamiltonian, the solution of the Poisson equation, or orthogonalisation. Once these "building blocks" are ported to a GPU architecture, the different types of calculations that use them do not need to be modified if a proper abstraction layer is used. High performance GPU libraries could be of great help in the porting process.

Currently, one of the main challenges of writing an efficient GPU code is to deal with the problem of copying data between main memory and GPU memory over a relatively slow link. Unlike other previous GPU limitations, like single precision or the lack of cache, it is not clear that this issue will be solved completely by GPU vendors in the near future. While CPU chips will soon include a GPU that shares the same memory, it will most likely correspond to a slow GPU not suitable for HPC. Fast GPUs require higher memory bandwidths than what CPU memory can provide, so in the foreseeable future they will still have their own memory banks. However, we will see increments in the speed of the CPU/GPU link and perhaps other improvements, like direct MPI access to GPU memory.

Programmable GPUs, and the tools to write programs for them, have only been available for a short time. So only a few DFT or TDDFT GPU codes are available, and the application to different types of calculations still needs to be investigated. It is clear, though, that the use of hybrid CPU/GPU architectures is an efficient way to harness high amounts of computing power and that is suitable for TDDFT. For the two applications shown, the calculation time is reduced by around a factor of three with respect to an optimised multi-core CPU calculation. This factor could be increased still, since reported performances are far from the GPU theoretical throughput and there are certainly many more optimisation opportunities left to discover. In fact, larger speed-ups for GPU ports can be found in literature for other applications. However, one should be aware that sometimes reported performance gains are unrealistic[5] and are based on comparisons of optimised GPU code with poorly optimised CPU code.

The lower cost per GFlops for hybrid architectures will surely enhance their diffusion in the near future. For example, in late 2010 already three out to the top four fastest supercomputers are GPU based, including the number one (Stone and Xin 2010). As GPUs are beginning to be integrated in HPC platforms, OpenCL support might become a standard feature of scientific codes, as MPI parallelisation is today. In fact, it is possible that GPUs will set the trend for high performance computing. The combination of highly parallel processors and a standard framework to write code for them, OpenCL, could well be the base for the future of numerically intensive calculations, comprising not only GPUs but also CPUs and accelerator boards. In fact, processor vendors like AMD, IBM or Intel already provide OpenCL implementations for their CPUs. So the key issue behind the GPU "revolution" is the development of in-chip parallel programming languages, that will allow engineers to start putting more execution units in processors of any type, knowing that many applications will be able to profit from them.

---

[5] Unrealistic in the sense that the speed-up is larger than the theoretical GPU/CPU performance ratio.

Part VI
TDDFT Versus Other
Theoretical Techniques

Chapter 22
Dispersion (van der Waals) Forces and TDDFT

John F. Dobson

22.1 Introduction

By "dispersion forces" (Mahanty and Ninham 1976) we mean the attractive part
of the van der Waals (vdW) interaction that cannot be attributed to any permanent
electric multipoles. These ubiquitous forces are typically weaker than ionic and
covalent bonding forces, but are of longer range than the latter, typically decaying
algebraically rather than exponentially with separation. They are important in soft
condensed matter and in rare-gas chemistry, for example. We will work in the electromagnetically non-retarded [non-Casimir (Milton 2001)] limit, which means in
practice that we can treat interacting systems at separations from about a micron
down to full overlap of electronic clouds. We do not aim for a complete review
of vdW phenomena and theories, but will rather concentrate on approaches that
deal with dynamic electron density-density response functions, which in turn can
be calculated by TDDFT methods. In Sect. 22.2 we give simple physical arguments
that motivate the form of the dispersion interaction between small non-overlapping
systems, and in Sect. 22.3 we outline the simplest way to transfer this approach
to larger systems via pairwise summation. Section 22.4 discusses the perturbation
theoretic approach to dispersion forces. In Sect. 22.5 it is pointed out that both the
pairwise additive and low-order perturbative approaches can give poor results for
dispersion forces between highly polarizable, highly anisotropic systems, including
nanostructures of strong current interest. Examples from the literature since the last
edition of this book are included. Section 22.6 introduces the fluctuation dissipation
theorem including a first-principles derivation, and its use along with the adiabatic
connection formula (ACFD) to obtain groundstate xc energies. This Section includes
a detailed proof that ACFD applied to the bare KS response leads to the exact (DFT)

J. F. Dobson (✉)
Micro and Nano Technology Centre and School of BPS,
Griffith University, Nathan QLD 4111, Australia
e-mail: j.dobson@griffith.edu.au

M. A. L. Marques et al. (eds.), *Fundamentals of Time-Dependent Density Functional
Theory*, Lecture Notes in Physics 837, DOI: 10.1007/978-3-642-23518-4_22,
© Springer-Verlag Berlin Heidelberg 2012

exchange energy, correcting errors in the previous edition. Section 22.7 discusses the direct random phase approximation (dRPA) as the simplest fully microscopic implementation of the ACFD approach, suitable for dispersion energy calculations beyond the pair-summation approach. The substantial progress with this approach since the previous edition is discussed. Section 22.8 discusses the latest non-DFT methods of going beyond dRPA. Section 22.9 discusses some general difficulties in using a non-zero kernel f_{XC} go beyond the dRPA in calculating the xc energy, and includes some as yet unpublished results. Section 22.10 discusses the use of direct approximations to the electronic response to obtain nonlocal vdW energies, including the vdWDF.

A new unpublished approach is briefly discussed that goes beyond the pair summation format. Finally the chapter is summarized in Sect. 22.11

22.2 Simple Models of the vdW Interaction Between Small Systems

22.2.1 Coupled-Fluctuation Picture

It is worthwhile to consider first a very simple picture of the vdW interaction between neutral spherical atoms at separation $R \gg b$ where b is an atomic size. [For more detail see e.g. (Dobson et al. 2001) or (Langbein 1974)]. The Hartree field of a neutral spherical atom decays exponentially with distance, and so the Hartree energy cannot explain the algebraic decay of the vdW interaction. However the zero-point motions of the electrons (or thermal motions where significant) can cause a temporary fluctuating dipole moment d_2 to arise on atom #2. The nonretarded Coulomb interaction energy between this dipole, and another dipole of order $\alpha_1 d_2 R^{-3}$ that it induces on atom #1, has a nonzero average value that can be estimated (Dobson et al. 2001) as

$$E = -C_6 R^{-6}, \quad C_6 = K \hbar \omega_0 \alpha_1 \alpha_2. \tag{22.1}$$

Here α_1 and α_2 are the dipolar polarizabilities of the atoms. The "Hamaker constant" C_6 for this geometry contains a dimensionless constant K, not specifiable from the above qualitative argument. The factor R^{-6} can be understood as arising from *two* actions of the dipolar field, each proportional to R^{-3}, showing that this approach relates to *second*-order perturbation theory (PT).

22.2.2 Picture Based on the Static Correlation Hole: Failure of LDA/GGA at Large Separations

The spontaneous dipole d_2 invoked above would be implied *if* we had found an electron at a position r on one side of atom #2. The induced dipolar distortion on atom #1 then represents a very distant part of the correlation hole density $n_c(r'|r)$

(Gunnarsson and Lundqvist 1976) due to discovery of the electron at r. The shape of this hole is entirely determined by the shape of atom #1, and is thus quite unlike the long-ranged part of the xc hole present in a uniform electron gas of density $n(r)$. It is therefore unsurprising that the local density approximation (LDA) misses the long-ranged tail of the vdW interaction. In fact the LDA and the GGAs can only obtain the vdW tail via the distortion of the density of each atom, a distortion that is predicted by these theories to decay exponentially with separation of the two atoms, thus ruling out the correct algebraic decay of the energy. The situation with GGA is less clear when the densities of the interacting fragments overlap. If the principal attractive correlation energy contribution comes from electrons near the overlap region, then treating this region as part of a weakly nonuniform gas might be reasonable. In keeping with this, various different GGAs can give qualitatively reasonable results for vdW systems such as rare-gas dimers. The results are not consistent or reliable, however (Perez-Jorda and Becke 1995; Zhang et al. 1997; Patton and Pederson 1997; Patton et al. 1997) , though surprisingly good results *near the energy minimum* are obtained (Perez-Jorda et al. 1999; Walsh 2005) with Hartree–Fock exchange plus the Wilson-Levy functional (Wilson and Ivanov 1998; Wilson and Levy1990). Some discussion is given in a review (Dobson et al. 2001).

22.2.3 Picture Based on Small Distortions of the Groundstate Density

Instead of considering the energy directly, Feynman (1939) and Allen and Tozer (2002) considered the small separation-dependent changes $\delta n(r, R)$ in the ground-state density $n(r)$ of each fragment, caused by the inter-fragment Coulomb interaction v_{ee12}. The Coulomb field acting at the nucleus of each fragment, created by $\delta n(r, R)$ as source, leads to a force which was identified as the vdW force, in the distant limit. One can then obtain the correct result $F = -\nabla_R(-C_6 R^{-6})$ in the widely-separated limit, in agreement with (22.1). Such a result emerges, for example, if $\delta n(r, R)$ is calculated from a many-electron wavefunction correct to *second* order in v_{ee12}, involving a double summation with two energy denominators. [The *first*-order wavefunction perturbation makes zero contribution to $\delta n(r, R)$.] By contrast, looking at the total *energy* to second order in v_{ee12} one already obtains the dispersion interaction with only a single summation and one energy denominator, a substantially easier task of the same order as obtaining the *first-order* perturbed wavefunction. From here on we restrict attention to approaches based directly on the energy.

22.2.4 Coupled-Plasmon Picture

Another simple way to obtain the R^{-6} interaction is to consider the coupled fluctuating dipoles invoked above as forming a coupled plasmon mode of the two systems (Langbein 1974). One solves coupled equations for the time-dependent density distortions on the two systems, leading to two normal modes (in- and out-of-phase plasmons) of free vibration of the electrons. The R dependence of the sum of the zero-point plasmon energies gives an energy of form $-C_6 R^{-6}$, in qualitative agreement with the coupled-fluctuation approach described above, for the case of two small separated systems [see e.g. (Dobson et al. 2005; Mahanty and Ninham 1976; Langbein 1974)]. A strength of the coupled-plasmon approach is that it is not perturbative, and is equally valid for large or small systems, even for metallic cases where zero energy denominators could render PT suspect. The coupled-plasmon theory is linked to the correlation-hole approach by the fluctuation-dissipation theorem to be discussed starting from Sect. 22.6.2 below

22.3 The Simplest Models for vdW Energetics of Larger Systems

There is a large early literature [see e.g. (Mahanty and Ninham 1976)] calculating forces between macroscopic bodies by adding R^{-6} energy contributions between pairs of atoms, or pairs of volume elements. To describe close contacts or chemical bonds, the small-R divergence has to be "saturated" or cut off and substituted by another form for small R. The well-known "6–12" or Lennard-Jones potential $\phi_{6-12}(R) = -C_6 R^{-6} + C_{12} R^{-12}$ is an example used both in chemical and biological situations, and also quite recently for graphitic structures. Typically the coefficients C_6 and C_{12} are fitted to experimental data, and may differ even between different structures of the same (graphitic) type (Girifalco et al. 2000). Other saturation procedures are currently used (Wu and Yang 2002; Grimme et al. 2010). Recent work has provided a non-empirical way to achieve short-distance saturation (Dion et al. 2004) that appears to be valid at least for small finite systems: see Sect. 22.10.2. We show in Sect. 22.5 below, however, that the $\sum R^{-6}$ tail of fitted Lennard-Jones potentials is not in principle correct for the asymptotic vdW interaction of an important class of condensed matter systems. It is also difficult to see how a more recent fitting scheme (von Lilienfeld et al. 2004) can deal with the severe non-additivity required. These considerations reinforce the need to develop the seamless ACFD approach to be described below in Sect. 22.6

22.4 Formal Perturbation Theory Approach

22.4.1 Casimir–Polder Formula: Second Order Perturbation Theory for Two Finite Nonoverlapping Systems

A more precise result than (22.1) can be obtained by treating the inter-fragment Coulomb interaction $v_{ee12} = e^2/r_{12}$ in second-order PT. This is invalid when the electrons on the two systems overlap, so that electrons in system 1 and system 2 cannot be treated as distinguishable. (A more complex v_{ee12} PT, the symmetry adapted PT [SAPT (Jeziorski et al. 1994)] is available in cases of overlap). Some doubts also occur for infinite systems (see Sect. 22.5). With these caveats in mind, one can write the interaction energy exactly to $O(v_{ee12})^2$ in the following form (Longuet-Higgins 1965; Zaremba and Kohn 1976), frequently termed the Casimir–Polder formula in the chemical literature:

$$E_{12}^{(2)} = -\frac{\hbar}{2\pi} \int d^3r_1 \int d^3r_1' \int d^3r_2 \int d^3r_2' \frac{e^2}{r_{12}} \frac{e^2}{r_{12}'} \times \int_0^\infty du \chi_1(r_1, r_1', iu) \chi_2(r_2, r_2', iu).$$

(22.2)

Here χ_1 and χ_2 are the density-density response functions (see Chap. 4) of the two fragments separately (each treated in the complete absence of the other, but including all interactions inside each fragment, and evaluated at imaginary frequency $\omega = iu$). The arguments r_1 and r_1' are positions inside the first fragment, while r_2 and r_2' are inside the second fragment. There are many equivalent forms (McWeeny 1989; Zaremba and Kohn 1976) of the second-order perturbation energy. The form (22.2) has been chosen for display here because it is useful in establishing connections between different approaches to the dispersion interaction to be discussed below.

The charge-neutrality and constant-potential conditions

$$\int d^3r' \chi(r, r', iu) = \int d^3r \chi(r, r', iu) = 0$$

(22.3)

are automatically satisfied (Dobson et al. 1998; Dobson and Dinte 1996) if one writes the density-density responses as the gradient of nonlocal dynamic polarisability tensors α (Hunt 1983),

$$\chi(r, r', iu) = -e^{-2} \partial_{ri} \partial_{r'j} \alpha_{ij}(r, r', iu).$$

(22.4)

Using integration by parts and defining $t_{ij}(r) = r^{-2}(3r_i r_j - \delta_{ij} |r|^2)$ one then obtains from (22.2)

$$E^{(2)} = -\frac{\hbar}{2\pi} \int d^3r_1 \int d^3r_2 \int d^3r_1' \int d^3r_2' \sum_{ijkl=1}^{3} t_{ik}(r_{12}) t_{jl}(r_{12}')$$

$$\times r_{12}^{-3} r_{12}'^{-3} \int_0^\infty du \alpha_{ij}^{(1)}(r_1, r_1', iu) \alpha_{kl}^{(2)}(r_2, r_2', iu).$$

(22.5)

From (22.5) it is clear that the leading dependence of the dispersion energy $E^{(2)}$ is of $O(R^{-6})$ at large separations (as indicated by the simple arguments in Sect. 22.2 above), but that $E^{(2)}$ depends in general on the orientation of the two systems as embodied in the angular dependence of the polarisability tensors $\alpha^{(1)}$, $\alpha^{(2)}$. By writing $r_{12} = R + x_1 - x_2$, expanding t_{ij} in powers of x_i and x_j', and assuming isotropic dipole polarizabilities $A_{ij} = \delta_{ij} A(iu)$ for each system, (22.5) gives in lowest order the following more familiar formula generalizing (22.1):

$$E^{(2)} = -C_6/R^6, \qquad C_6 = \frac{3\hbar}{\pi} \int du \, A^{(1)}(iu) A^{(2)}(iu). \qquad (22.6)$$

Higher terms in the multipolar expansion in powers of x_1 and x_2 give corrections to (22.5) and (22.6) of order R^{-8}, R^{-10} etc. (Jeziorski et al. 1994; Osinga et al. 1997).

For well-separated finite systems, (22.2) reduces the calculation of accurate vdW interaction parameters to the calculation of sufficiently accurate dynamical density response functions of the isolated components. The simplest case is that of two atoms, but atomic polarizabilities are in fact notoriously difficult to treat via density functional methods, because they involve extreme inhomogeneity of the density. Straightforward ALDA response calculations for atoms, based on LDA KS potentials, lead to dimer C_6 coefficients that are up to 20% too large for rare gases (van Gisbergen et al. 1995) and 50% for the Be dimer. RPA response tends to underestimate C_6. Part of the problems with ALDA come from the need for self-interaction correction (Mahan and Subbaswamy 1990) perhaps both in the groundstate calculation and in the response (Pacheco and Ekardt 1992a, b; Pacheco and Ramalho 1997). Use of the LB94 groundstate KS potential, which has the correct asymptotic behaviour, like SIC theories, leads to improved C_6 and C_8 vdW coefficients (Osinga et al. 1997).

Given the availability of accurate (if cumbersome) correlated quantum chemical approaches for small vdW-bound systems, DFT response approaches to vdW phenomena are probably most needed for larger systems. This situation is familiar from conventional groundstate DFT. The exact response-based formula (22.2) is nevertheless important here because it facilitates the testing of various response-based TDDFT approaches to the vdW force.

Equation 22.2 should not be confused with Moeller-Plesset (MP) PT (Szabo and Ostlund 1989). The latter goes systematically beyond the Hartree–Fock theory by treating the *entire* bare electron-electron interaction v_{ee} as a perturbation. This is in contrast to Eq. 22.2 in which the exact susceptibilities χ_1 and χ_2 contain the *intra*-subsystem interactions v_{ee11}, v_{ee22} in principle to *all* orders: only the *inter*-subsystem interaction v_{ee12} is treated perturbatively. Correspondingly, the separation-dependent part of the second-order MP (MP2) energy is equivalent to (22.2) but with the *bare* (Kohn–Sham or Hartree–Fock) responses χ_{KS1}, χ_{KS2} on the right-hand side. On the other hand, in MP2 and related approaches the overlapped case is not excluded. There have been a number of approaches related to this idea (Engel and Bonetti 2001; Engel et al. 2000; Angyan 2005). Some used perturbation of Eq. 22.19 within the ACFD (Lein et al. 1999) and so could include f_{xc} directly.

22.4.2 vdW and Higher-Order Perturbation Theory

For non-overlapping electronic systems one can go further within PT with respect to the inter-system Coulomb interactions v_{eeij}. In third order one finds an interaction between three separated systems, which cannot be expressed as the pairwise sum of

R^{-6} Hamaker terms such as (22.6). At large separations and spherical systems the leading (dipolar) contribution to this third-order term has the Axilrod-Teller form [see e.g. (Rapcewicz and Ashcroft 1991)] $E^{(3)\text{vdW}} \sim C_9 R_{12}^{-3} R_{23}^{-3} R_{13}^{-3}$, where C_9 contains some angular dependence. There are also corrections to the pair interaction (22.5) and (22.6) from perturbation orders beyond 2 (Jeziorski et al. 1994). A recent discussion from the RPA viewpoint is given in (Lu et al. 2010).

22.4.3 Symmetry-Adapted Perturbation Theory

The perturbative approach is also possible in the regime of electron charge overlap provided that the exchange symmetry of the many-electron wavefunction is taken into account via a projection operator technique. This has been developed to a very high degree in the SAPT of Jeziorski et al. (1994). This approach should probably be considered the state of the art for vdW interactions where it is numerically feasible, namely for molecular pairs of moderate size. Within SAPT the properties of each subsystem can be treated at various levels up to full configuration interaction. Some of the terms can be written in terms of density response functions of the subsystems, and here TDDFT (ususally RPA or ALDA) has been put to good use in reducing the numerical burden of SAPT calculations (Hesselmann et al. 2006; Misquitta et al. 2005). A proper account of this approach would require may pages. The present chapter will concentrate instead on methods suitable for macroscopic and nano systems too large for the SAPT approach.

22.5 Nonuniversality of vdW Asymptotics in Layered and Striated Systems

Unfortunately the finite-order perturbation approaches discussed in the previous section, as well as $\sum R^{-6}$ formulas partly justified by this perturbative approach, are questionable for many large solid-state systems of current technological interest (Dobson et al. 2001, 2005). In highly anisotropic systems, the coefficient of the R^{-6} interaction can be severely mis-estimated by pairwise additive theories (Kim et al. 2006). The most severe cases, in the sense that the net interaction has a non-conventional power law, are those where there is a zero electronic energy gap, leading to zero energy denominators in PT. One example is the case of nanoscopically thin metallic sheets, where by summing the zero point energies of coupled 2D plasmons, one obtains (Bostrom and Sernelius 2000; Dobson et al. 2001, 2005) an interaction $E^{\text{vdW}} \sim -C_{5/2} D^{-5/2}$ where D is the separation of two such layers. By contrast, the $\sum R_{ij}^{-6}$ approach gives $E^{\text{vdW}} \sim -C_4 D^{-4}$, which is necessarily much smaller at large separations. For two parallel metallic nanotubes or nanowires of radius b separated by distance $D \gg 2b$, the $\sum R_{ij}^{-6}$ method gives $E^{\text{vdW}} \sim -C_5 D^{-5}$, whereas the coupled plasmon approach (Chang et al. 1971; Dobson et al. 2001, 2005, 2006) gives $E^{\text{vdW}} \sim -C_a D^{-2} (\ln(D/b))^{-3/2}$, different by almost three powers of D.

This has recently been confirmed by diffusion Monte Carlo calculations (Drummond and Needs 2007). The important case of two graphene planes (zero-gap semiconductors) was shown recently (Dobson et al. 2006) to give $E^{\text{vdW}} \sim -C_3 D^{-3}$ as $D \to \infty$, whereas the $\sum R_{ij}^{-6}$ approach gives $E^{\text{vdW}} \sim -C_4 D^{-4}$, the same result as for 2D insulators (Rydberg et al. 2000). Unusual interactions have been predicted in other geometries also (White and Dobson 2008; Misquitta et al. 2010; Gould et al. 2008; Gould et al. 2009; Dobson et al. 2009). A finite number of higher perturbation terms (see Sect. 22.4.2), added to $\sum R_{ij}^{-6}$, do not lead to these unconventional power laws. For graphite, the D^{-3} term is small near the equilibrium spacing (Lebegue et al. 2010) but the non-additive physics is still important.

In the above cases where $\sum R_{ij}^{-6}$ gives the wrong asymptotic power law, in addition to the zero energy gap there is an incomplete metallic screening because the systems are nanoscopically small in at least one space dimension. Three-dimensional metals (e.g. thick metallic plates) do not exhibit vdW power laws of unusual form (Dzyaloshinskii et al. 1961; Dobson et al. 2001), and this seems to be associated with complete metallic screening, leading to a finite polarizability at small frequency and wavenumber (Dobson et al. 2006).

The above considerations apply to widely separated sub-systems. Recently, evidence has also been given that standard theories [LDA/GGA/fitted Lennard-Jones potentials (Charlier et al. 1994; Girifalco and Hodak 2002)] do not give reliable answers for the energetics of layered metallic or semi-metallic systems near their equilibrium spacing, either. This is despite the fact that the LDA predicts good equilibrium geometries. For theoretical evidence, see (Dobson and Wang 2004), Fig. 4 of (Jung et al. 2004), and (Rydberg et al. 2003; Hasegawa and Nishidate 2004; Tournus et al. 2005; Marini et al. 2006b; Spanu et al. 2009; Lebegue et al. 2010). There is experimental evidence also (Benedict et al. 1998; Zacharia et al. 2004). These are technologically important systems (graphite and derivatives, fullerenes, boron nitride), so these discrepancies are significant.

In principle, therefore, and especially for large solid-state systems, one would like a theory of vdW forces that is not local nor perturbative and is "seamless"—that is, it gives good results at all separations. Below we make the case that appropriate theories can be constructed from the fluctuation-dissipation theorem along with a density-density response function $\chi(\boldsymbol{r}, \boldsymbol{r}', \omega)$ from TDDFT: indeed even the direct random phase approximation (dRPA) version of χ (the case $f_{\text{xc}} = 0$) can sometimes give good results. We first present the necessary theory in some detail.

22.6 Correlation Energies From Response Functions: the Fluctuation-Dissipation Theorem

The approach to be described here has antecedents in theories of the Lifshitz type (Dzyaloshinskii et al. 1961) that included electromagnetic retardation but ultimately approximated the response function of electrons in a local macroscopic fashion.

That theory was mainly applied to bulk systems at macroscopic separation, and represented the response of the electrons through a spatially local but frequency-dependent dielectric function obtained principally from experiment. Here we consider only the electromagnetically non-retarded case but retain the full nonlocal microscopic response functions of the electrons, permitting a "seamless" treatment at any separation where electromagnetic retardation is unimportant, right down to bonding with overlap of the electronic clouds.

22.6.1 Basic Adiabatic Connection Fluctuation-Dissipation Theory

The exact adiabatic connection formula (ACF) for the combined Hartree, exchange and correlation energy in the groundstate of an inhomogeneous many-electron system follows from the Feynman-Hellmann theorem, and is (Langreth and Perdew 1975; Gunnarsson and Lundqvist 1976)

$$E_{\text{Hxc}} = \frac{1}{2} \int_0^1 d\lambda \int d^3r \int d^3r' \frac{e^2}{|r - r'|} n_{2\lambda}(r, r'). \tag{22.7}$$

See also Sect. 5.2.1. A derivation of (22.7) is given in Gunnarsson and Lundqvist (1976) starting from their Eq. 22.28, with our λ denoted "g". In (22.7) $n_{2\lambda}(r, r')$ is the pair density in the reduced-interaction many-electron groundstate $|\Psi_\lambda\rangle$ with electron-electron coulomb potential $\lambda e^2/r_{12}$ The probability of finding an electron in a small volume d^3r near r, and simultaneously another electron in d^3r' near r', is $n_{2\lambda}(r, r')d^3rd^3r'$. The λ integration restores the necessary kinetic correlation energy arising from quantal zero-point motions, and is performed at constant groundstate density $n(r)$. An earlier approach to the vdW energy did not work at constant density (Harris and Griffin 1975).

We now obtain a useful expression for the (spin-dependent) pair distribution, ignoring spin-orbit coupling and assuming that each electron labelled "a" has a definite spin projection $\sigma_a = \pm 1/2$. Introducing an operator $\hat{r}_a$ for the position of the ath electron, and remembering that the second electron found at r' cannot be the same as the first one ($a \neq b$) we have

$$n_{2\lambda}(r\sigma, r'\sigma') = \langle \Psi_\lambda | \sum_{a \neq b} \delta_{\sigma\sigma_a}\delta_{\sigma'\sigma_b}\delta(r - \hat{r}_a)\delta(r' - \hat{r}_b) |\Psi_\lambda\rangle . \tag{22.8}$$

Separating the $a = b$ term, we arrive at

$$n_{2\lambda}(r\sigma, r'\sigma') = \langle \Psi_\lambda | \sum_{\text{all } a,b} \delta_{\sigma\sigma_a}\delta_{\sigma\sigma_b}\delta(r - \hat{r}_a)\delta(r' - \hat{r}_b) |\Psi_\lambda\rangle$$
$$- \delta(r - r')\delta_{\sigma\sigma'} \langle \Psi_\lambda | \sum_a \delta_{\sigma\sigma_a}\delta(r - \hat{r}_a) |\Psi_\lambda\rangle \tag{22.9}$$

which finally leads to

$$n_{2\lambda}(r\sigma, r'\sigma') = \langle \Psi_\lambda | \hat{n}_\sigma(r)\hat{n}_{\sigma'}(r') | \Psi_\lambda \rangle - \delta(r - r')\delta_{\sigma\sigma'}n_\sigma(r). \qquad (22.10)$$

Here we used the electron spin density operator $\hat{n}_\sigma(r) = \sum_a \delta_{\sigma\sigma_a}\delta(r - \hat{r}_a)$, whose expectation value is the electron spin density, $\langle \hat{n}_\sigma(r) \rangle = n_\sigma(r)$. We now also introduce the spin density fluctuation operator

$$\delta\hat{n}_\sigma(r) = \hat{n}_\sigma(r) - n_\sigma(r). \qquad (22.11)$$

(This operator corresponds to spontaneous density fluctuation processes such as that invoked in Sect. 22.2.1, in which a density fluctuation away from the expectation value $n(r)$ produces dipole fields that initiate the vdW energy.)

Then, putting $\hat{n} = n + \delta\hat{n}$ from (22.11) into (22.10) and noting that $\langle \Psi_\lambda | \delta\hat{n}_\sigma(r) | \Psi_\lambda \rangle = 0$, we have from (22.10)

$$n_{2\lambda}(r\sigma, r'\sigma') = \langle \Psi_\lambda | \delta\hat{n}_\sigma(r)\delta n_{\sigma'}(r') | \Psi_\lambda \rangle + n_\sigma(r)n_{\sigma'}(r') - \delta(r - r')\delta_{\sigma\sigma'}n_\sigma(r). \qquad (22.12)$$

We can obtain the usual spin-independent formalism by summing over σ, σ':

$$n_{2\lambda}(r, r') = \langle \lambda | \delta\hat{n}(r)\delta n(r') | \lambda \rangle + n(r)n(r') - \delta(r - r')n(r). \qquad (22.13)$$

When the second term on the right side of (22.13) is put into the ACF (22.7) it simply yields the Hartree energy. The other two terms of (22.13) thus yield the exact xc energy:

$$E_{xc} = \frac{1}{2} \int_0^1 d\lambda \int d^3r \int d^3r' \frac{e^2}{|r - r'|} \left[\langle \Psi_\lambda | \delta\hat{n}(r)\delta n(r') | \Psi_\lambda \rangle - \delta(r - r')n(r) \right]. \qquad (22.14)$$

(The correlation term $\langle \Psi_\lambda | \delta\hat{n}(r)\delta n(r') | \Psi_\lambda \rangle$ represents the fact that density fluctuations at r' can be tied to (correlated with) fluctuations at r, just the kind of process described in Sect. 22.2.1).

The idea now is that the correlations represented by the fluctuation term in (22.14) are *caused* by interactions between the electrons, and the description of this physics is rather subtle: IF a fluctuation occurs, it will cause some interaction energy beyond the Hartree description. It is easier conceptually, and helpful in the construction of approximation schemes, to relate this process to the interaction between the *non*-random density changes caused *when* an small externally-controlled field is applied. Thus we are led to introduce time-dependent density response theory, and we have converted a tricky "IF" scenario into a conceptually simpler "WHEN" scenario. The mathematical tool that justifies this shift in philosophy is the fluctuation-dissipation theorem, described in sufficient generality (i.e. for two unequal operators) in the book by Landau and Lifshitz (1969). For completeness we now give a simple direct derivation of the frequency integrated, zero-temperature form of the theorem needed here.

Suppose that $\hat{A}$ and $\hat{B}$ [e.g., $\delta\hat{n}(r)$ and $\delta\hat{n}(r')$] are hermitian operators with zero groundstate average, $\langle\hat{A}\rangle = \langle\hat{B}\rangle = 0$ where the expectation value is taken in the groundstate of an interacting system with Hamiltonian $\hat{H}$. In the presence of an additional externally applied time-dependent "potential" $\delta v \exp(ut)$ coupling to $\hat{A}$, the Hamiltonian is $\hat{H} - \delta v \exp(ut)\hat{A}$. Calculating the perturbed state $|\Psi(t)\rangle$ by standard first-order time-dependent perturbation methods, one finds that the expectation of property $\hat{B}$ at time t is of form $\langle\Psi(t)|\,\hat{B}\,|\Psi(t)\rangle = \chi_{BA}(iu)\delta v \exp(ut)$, where

$$\chi_{AB}(iu) + \chi_{BA}(iu) = -2\sum_J \frac{E_{IJ}\left(\langle 0|\,\hat{A}\,|J\rangle\,\langle J|\,\hat{B}\,|0\rangle + \langle 0|\,\hat{B}\,|J\rangle\,\langle J|\,\hat{A}\,|0\rangle\right)}{E_{IJ}^2 + \hbar^2 u^2}$$

Here $|J\rangle$ is an eigenstate, and $E_{IJ} = E_I - E_J$ is an eigen-energy difference, of the interacting Hamiltonian $\hat{H}$. The quantity χ_{AB} is the "AB response function", e.g. the electronic density-density response in the case at hand: see also Eq. 50 of Chap. 4. Note that the use of a real external potential $\exp(ut)$ corresponds to the more usual choice $\exp(-i\omega t)$ but with a positive imaginary frequency $\omega = iu$. Using the arctan integral

$$\int_0^\infty du(E_{IJ}^2 + \hbar^2 u^2)^{-1} = \pi/(2\hbar E_{IJ}) \tag{22.15}$$

we have

$$\int_0^\infty du\,[\chi_{AB}(iu) + \chi_{BA}(iu)] = -\frac{\pi}{\hbar}\sum_J \left(\langle 0|\,\hat{A}\,|J\rangle\,\langle J|\,\hat{B}\,|0\rangle + \langle 0|\,\hat{B}\,|J\rangle\,\langle J|\,\hat{A}\,|0\rangle\right)$$

$$= -\frac{\pi}{\hbar}\,\langle 0|\,\hat{A}\hat{B} + \hat{B}\hat{A}\,|0\rangle\,. \tag{22.16}$$

This is a frequency integrated, $T = 0\,K$ form of the very general fluctuation-dissipation theorem (Callen 1951; Landau and Lifshitz 1969) (FDT). The FDT is more usually quoted as a result at finite temperature and for a single frequency lying just above the real axis, where $\mathrm{Im}(\chi_{AA})$ is known to represent energy absorption ("dissipation"). Applying (22.16) to the density fluctuation operators $\hat{A}, \hat{B} = \delta\hat{n}(r), \delta\hat{n}(r')$ from the spin-summed version of (22.11), and noting that the r and r' integrations can be interchanged, in (22.14) we have effectively

$$\langle\Psi_\lambda|\,\delta\hat{n}(r)\delta\hat{n}(r')\,|\Psi_\lambda\rangle = -\frac{\hbar}{\pi}\int_0^\infty du\,\chi_\lambda(r, r', iu) \tag{22.17}$$

where the density-density response χ_λ is just that introduced in Chap. 4, applied to the system with reduced coulomb interaction $\lambda e^2/r_{12}$.

Combining (22.17) with (22.14) we obtain

$$
E_{xc} = \frac{1}{2} \int\limits_0^1 d\lambda \int d^3r \int d^3r' \frac{e^2}{|\boldsymbol{r} - \boldsymbol{r}'|}
$$

$$
\times \left[-\frac{\hbar}{\pi} \int\limits_0^\infty du\, \chi_\lambda(\boldsymbol{r}, \boldsymbol{r}', iu) - \delta(\boldsymbol{r} - \boldsymbol{r}')n(\boldsymbol{r}) \right]. \tag{22.18}
$$

We will refer to the important exact result (22.18) as the "ACFD" xc energy.

In TDDFT the linearized interacting density-density response, given by $\chi_\lambda = \chi_\lambda(\boldsymbol{r}, \boldsymbol{r}', \omega)$, and the KS response $\chi_{KS} = \chi_{\lambda=0}$ are related by the Dyson-like screening equation (Gross 1985)

$$
\chi_\lambda = \chi_{KS} + \chi_{KS} * (\lambda v_{ee} + f_{xc\lambda}) * \chi_\lambda \tag{22.19}
$$

where stars represent convolution in $(\boldsymbol{r}, \boldsymbol{r}')$ space and v_{ee} is the bare electron-electron Coulomb potential (see Chap. 4). In practice, the inputs to (22.19) (namely χ_{KS} and, in almost all existing approximations, f_{xc}) can be computed from the ground-state KS orbitals $\{\varphi_i\}$. In turn the $\{\varphi_i\}$ are directly computable from the groundstate KS potential $v_{KS}(\boldsymbol{r})$. Thus the ACFD energy is often best regarded as a functional $E^{ACFD}[v_{KS}(\boldsymbol{r})]$ of the groundstate KS potential. In principle, one should introduce a high level of self-consistency (the OEP level) by choosing v_{KS} to minimize the nonlocal functional (22.18), with the external potential fixed. In this sense, most detailed calculations to date have been "post-functionals": that is, they have computed $E[v_{KS}^{LDA}]$ rather than $E[v_{KS}^{OEP}]$. In Sect. 22.7.1 below we will argue that this does not significantly affect the predictions of (22.18) for the distant vdW correlation energy, at least in the case $f_{xc} = 0$ (corresponding to the RPA). Thus the energy functionals are not overly sensitive to the v_{KS} used as input. On the other hand, if one uses the Feynman approach described in Sect. 22.2.3 above, in which the vdW force is calculated directly as a force on the nuclei, then the small selfconsistent changes in v_{KS} as a function of distance, calculated beyond the LDA, are crucial (Allen and Tozer 2002).

22.6.2 Exact Exchange: a Strength of the ACFD Approach

In order for a theory to work well for interacting systems with overlap of electronic clouds, it is essential that it accounts adequately for the Pauli exchange energy: indeed the Hartree–Fock theory already describes many covalent bonds quite well. In fact when the independent-electron (Kohn–Sham) response $\chi_{KS} = \chi_{\lambda=0}$ is substituted for χ_λ, (22.18) yields the exact nonlocal exchange energy $E_x^{Exact,DFT}$, defined in the DFT sense. To show this explicitly we assume no spin-orbit coupling so that the groundstate KS eigenfunctions $\varphi_{n\sigma}(r)$ have a definite spin projection $\sigma = +\frac{1}{2}$ or $-\frac{1}{2}$. By applying a one-body potential $dv_{\sigma'}(r')$ to electrons of spin projection σ' and applying time-dependent PT to the occupied ($f_{n\sigma} \neq 0$) orbitals we obtain

$$\chi_{KS}(r\sigma, r'\sigma', iu) = \delta_{\sigma\sigma'} \sum_{nk} (f_{n\sigma} - f_{k\sigma}) \frac{\varphi_{k\sigma}^*(r')\varphi_{n\sigma}(r')\varphi_{n\sigma}^*(r)\varphi_{k\sigma}(r)}{\varepsilon_k - \varepsilon_n - i\hbar u}, \quad (22.20)$$

independent of λ. Then

$$\chi_{KS}(r\sigma, r'\sigma', iu) + \chi_{KS}(r'\sigma', r\sigma, iu)$$
$$= \delta_{\sigma\sigma'} \sum_{nk} (f_{n\sigma} - f_{k\sigma}) \varphi_{k\sigma}^*(r')\varphi_{n\sigma}(r')\varphi_{n\sigma}^*(r)\varphi_{k\sigma}(r) \frac{2\varepsilon_{nk}}{\varepsilon_{nk}^2 + (\hbar u)^2} \quad (22.21)$$

where $\varepsilon_{nk} = \varepsilon_n - \varepsilon_k$ is a KS eigenvalue difference, and the dummy indices n and k were interchanged to combine the two contributions indicated on the left side. Then

$$-\frac{\hbar}{\pi} \int du \left[\chi_{KS}(r\sigma, r'\sigma', iu) + \chi_{KS}(r\sigma, r'\sigma', iu)\right]$$
$$= \delta_{\sigma\sigma'} \sum_{nk} (f_{n\sigma} - f_{k\sigma}) \varphi_{k\sigma}^*(r')\varphi_{n\sigma}(r')\varphi_{n\sigma}^*(r)\varphi_{k\sigma}(r) \text{sign}(\varepsilon_{nk}) \quad (22.22)$$

where $\text{sign}(\varepsilon) = 1$ for $\varepsilon > 0$, -1 for $\varepsilon < 0$, and 0 for $\varepsilon = 0$. Now use the identity

$$f_{n\sigma} - f_{k\sigma} = f_{n\sigma}(1 - f_{k\sigma}) - f_{k\sigma}(1 - f_{n\sigma}) \quad (22.23)$$

which reminds us that, for fermions, transitions giving rise to a response can only occur when one of the orbitals $k\sigma$, $n\sigma$ is occupied and the other is empty. Furthermore we assume that we have an integer number N of fermions *at zero temperature* with

$$f_{k\sigma} = \begin{cases} 1, & \varepsilon_{k\sigma} < \mu \\ 0, & \varepsilon_{k\sigma} > \mu \end{cases}. \quad (22.24)$$

Also if there are any states with $\varepsilon_{k\sigma} = \mu$ we assume that they all have the same occupation, either $f_{k\sigma} = 0$ or $f_{k\sigma} = 1$, so that $f_{k\sigma} - f_{n\sigma} = 0$ if $\varepsilon_{k\sigma} = \varepsilon_{n\sigma}$ in every case and the sum in (22.22) is restricted to $\varepsilon_{n\sigma} \neq \varepsilon_{k\sigma}$. (This may be problematic for degenerate ground states). Then

$$f_{k\sigma}(1 - f_{n\sigma})\text{sign}(\varepsilon_{k\sigma} - \varepsilon_{n\sigma}) = -f_{k\sigma}(1 - f_{n\sigma}) \quad (22.25)$$

since $f_{k\sigma}(1 - f_{n\sigma})$ is nonzero only when $\varepsilon_{k\sigma} < \mu < \varepsilon_{n\sigma}$ so that the sign function is negative. Similarly $f_{n\sigma}(1 - f_{k\sigma})\text{sign}(\varepsilon_{k\sigma} - \varepsilon_{n\sigma}) = f_{n\sigma}(1 - f_{k\sigma})$. Then

$$(f_{n\sigma} - f_{k\sigma}) \text{sign}(\varepsilon_{k\sigma} - \varepsilon_{n\sigma})$$
$$= f_{n\sigma}(1 - f_{k\sigma})\text{sign}(\varepsilon_{k\sigma} - \varepsilon_{n\sigma}) - f_{k\sigma}(1 - f_{n\sigma})\text{sign}(\varepsilon_{k\sigma} - \varepsilon_{n\sigma})$$
$$= f_{n\sigma}(1 - f_{k\sigma}) + f_{k\sigma}(1 - f_{n\sigma}) \quad (22.26)$$
$$= f_{n\sigma} + f_{k\sigma} - 2f_{n\sigma}f_{k\sigma}.$$

Putting (22.26) and (22.22) into the fluctuation dissipation theorem (22.16) we find that the use of χ_{KS} in place of χ yields the xc part of the symmetrized pair distribution as

$$\langle 0| \, \delta\hat{n}_\sigma(\boldsymbol{r})\delta n_{\sigma'}(\boldsymbol{r}') \, |0\rangle + \langle 0| \, \delta\hat{n}_{\sigma'}(\boldsymbol{r}')\delta n_\sigma(\boldsymbol{r}) \, |0\rangle$$
$$= \delta_{\sigma\sigma'} \sum_{nk} (f_{n\sigma} + f_{k\sigma} - 2f_{n\sigma}f_{k\sigma}) \, \varphi_{k\sigma}^*(\boldsymbol{r}')\varphi_{n\sigma}(\boldsymbol{r}')\varphi_{n\sigma}^*(\boldsymbol{r})\varphi_{k\sigma}(\boldsymbol{r}). \quad (22.27)$$

Then from (22.12) and (22.26) the (symmetrized) total pair density is

$$\frac{1}{2}\left[n_2(\boldsymbol{r}\sigma, \boldsymbol{r}'\sigma') + n_2(\boldsymbol{r}'\sigma', \boldsymbol{r}\sigma) \right] = -\delta_{\sigma\sigma'}\delta(\boldsymbol{r} - \boldsymbol{r}')n_\sigma(\boldsymbol{r}) + n_\sigma(\boldsymbol{r})n_{\sigma'}(\boldsymbol{r}')$$
$$+ \frac{1}{2}\delta_{\sigma\sigma'} \sum_{nk} (f_{n\sigma} + f_{k\sigma} - 2f_{n\sigma}f_{k\sigma})$$
$$\times \varphi_{k\sigma}^*(\boldsymbol{r}')\varphi_{n\sigma}(\boldsymbol{r}')\varphi_{n\sigma}^*(\boldsymbol{r})\varphi_{k\sigma}(\boldsymbol{r}). \quad (22.28)$$

By using the identities $\delta(\boldsymbol{r} - \boldsymbol{r}')n_\sigma(r) = \delta(\boldsymbol{r} - \boldsymbol{r}') \sum_n f_{n\sigma}\varphi_{n\sigma}^*(\boldsymbol{r})\varphi_{n\sigma}(\boldsymbol{r}')$ and $\delta(\boldsymbol{r} - \boldsymbol{r}') = \sum_k \varphi_{k\sigma}^*(\boldsymbol{r})\varphi_{k\sigma}(\boldsymbol{r}')$ we arrive at

$$\frac{1}{2}\left[n_2(\boldsymbol{r}\sigma, \boldsymbol{r}'\sigma') + n_2(\boldsymbol{r}'\sigma', \boldsymbol{r}\sigma) \right]$$
$$= \sum_{nk} f_{n\sigma}\varphi_{n\sigma}^*(\boldsymbol{r})\varphi_{n\sigma}(\boldsymbol{r})f_{k\sigma}\varphi_{k\sigma'}^*(\boldsymbol{r}')\varphi_{k\sigma'}(\boldsymbol{r}')$$
$$+ \frac{1}{2}\delta_{\sigma\sigma'} \sum_{nk} [(f_{n\sigma} + f_{k\sigma} - 2f_{n\sigma}f_{k\sigma}) - (f_{n\sigma} + f_{k\sigma})]$$
$$\times \varphi_{k\sigma}^*(\boldsymbol{r}')\varphi_{n\sigma}(\boldsymbol{r}')\varphi_{n\sigma}^*(\boldsymbol{r})\varphi_{k\sigma}(\boldsymbol{r}). \quad (22.29)$$

The $\delta(\boldsymbol{r} - \boldsymbol{r}')n_\sigma(r)$ term was written twice and divided by 2, after interchanging the dummy indices n and k, to obtain the $-(f_{n\sigma} + f_{k\sigma})$ term of (22.29). We finally obtain

$$\frac{1}{2}\left[n_2(\boldsymbol{r}\sigma, \boldsymbol{r}'\sigma') + n_2(\boldsymbol{r}'\sigma', \boldsymbol{r}\sigma) \right]$$
$$= \sum_{nk} f_{n\sigma}f_{k\sigma'}\varphi_{n\sigma}^*(\boldsymbol{r})\varphi_{k\sigma'}(\boldsymbol{r}') \left[\varphi_{k\sigma'}^*(\boldsymbol{r}')\varphi_{n\sigma}(\boldsymbol{r}) - \delta_{\sigma\sigma'}\varphi_{n\sigma}(\boldsymbol{r}')\varphi_{k\sigma}^*(\boldsymbol{r}) \right]. \quad (22.30)$$

Equation (22.30) gives the groundstate Hartree–Fock pair distribution calculated for KS orbitals. Summing over σ and σ' and substituting into (22.7), noting that (22.30) is independent of λ, we recover the electron-electron energy of Hartree–Fock form. That is, using the Kohn–Sham response in the ACFD results in the exact DFT exchange contribution to the electron-electron energy. This means that ACFD correlation energies are suitable for adding to an exact exchange calculation, and indeed correlation and exact Hartree–Fock exchange can be calculated together within the ACFD, if desired.

22.7 The xc Energy in the Direct Random Phase Approximation

The energy in the (d)RPA is defined by (22.18) and (22.19) with f_{xc} set to zero. It can also be obtained diagrammatically as the sum of rings of open bubbles within a Feynman-diagram (Fetter and Walecka 1971) or coupled-cluster (Scuseria et al. 2008) approach. The RPA energy at $T = 0\,K$ is related to the sum of zero-point energies of excitations (Mahanty and Ninham 1976; Furche 2008). This can sometimes lead to simple analytic evaluation of E_c^{RPA} in cases of vdW attraction at large separations, when the relevant excitations are simple plasmons [see e.g. (Mahanty and Ninham 1976; Dobson et al. 2001, 2005)]. There are, however, cases where plasmons do not provide an appropriate approach even for large solid state systems (Dobson 2009b; Dobson et al. 2006).

While analytic solutions based on long-wavelength physics are sometimes available for large subsystem separations as just noted, microscopic numerical evaluation of the RPA energy is required for inhomogeneous systems where subsystems overlap. This is computationally expensive compared to (e.g.) GGAs. It has nevertheless been carried out for jellium slab geometry (Dobson and Wang 1999; Jung et al. 2004) and is now also feasible for molecules of up to 100 atoms (Eshuis et al. 2010; Furche 2002c, 2008). It is also feasible for periodic systems with modest-sized unit cells (Miyake et al. 2002; Marini et al. 2006b; Harl and Kresse 2009; Lebegue et al. 2010; Harl et al. 2010). In particular, dRPA has given a good account of some of the trickiest vdW bonding problems including the interlayer stretching energetics of the laminar vdW-bound systems boron nitride (Marini et al. 2006b) and graphite (Harl and Kresse 2009; Lebegue et al. 2010). It is especially demanding for graphite because of the need for fine sampling of the Brillouin zone, associated with the semi-metallic bandstructure of graphene. It is particularly pleasing that the dRPA works so well for vdW crystals, despite the diseases of dRPA discussed two sections below.

22.7.1 Casimir–Polder Consistency: a Good Feature of the dRPA for vdW Calculations in the Well Separated Limit

The simple physical arguments above show that the *basic* physics of the vdW interaction is present within the ACFD energy with mean-field-like approximations for the response function. However it is not clear a priori why such an approach can produce *quantitative* predictions of vdW energies. It was therefore reassuring to find (Dobson 1994b) that the generalized Casimir–Polder formula (22.2) is exactly true in the widely-separated limit of a seamless dimer ACFFD calculation, when the dRPA approximation is used for all response functions. Any xc approximation for which the dimer calculation reproduces (22.2) will here be called "Casimir–Polder consistent". It has been shown formally that the RPA correlation energy is Casimir–Polder consistent.

The proof (Dobson 1994b) involved a combined dimer system with direct perturbation of the RPA version of the screening Eq. 22.19 with respect to the interaction v_{ee12} between monomers "1" and "2". It was necessary to include (i) terms involving first-order perturbation to the cross response $\chi_{\lambda 12}$, but also (ii) terms involving second order perturbation to the intra-system responses χ_{11} and χ_{22} [sometimes called "spectator" terms (Langreth et al. 2005)]

After an integration with the coulomb potential as per the ACFD, both of these terms yielded contributions of second order in v_{ee12}. The outcome was that the asymptotic vdW interaction from the dimer RPA correlation energy is of form (22.2) but with the individual responses of the isolated systems, naturally, replaced by their RPA versions

$$
E_{12}^{RPA(2)} \sim -\frac{\hbar}{2\pi} \int d^3 r_1 \int d^3 r_1' \int d^3 r_2 \int d^3 r_2' \frac{e^2}{r_{12}} \frac{e^2}{r_{12}'}
$$
$$
\int_0^\infty du\, \chi_1^{RPA}(r_1, r_1', iu)\chi_2^{RPA}(r_2, r_2', iu).
$$

(22.31)

Thus the formula (22.2) is true within the seamless ACFD/RPA formalism for a pair of widely separated systems. That is, *the dRPA is Casimir–Polder consistent*. It is also possible to prove (22.31) by noting that the λ integration in (22.18) can be done analytically for the dRPA case, yielding a logarithmic operator. This is then expanded to second order to obtain (22.31). A recent analysis is given in (Lu et al. 2010). This proof however sheds no light on the spectator terms, and it will be suggested below that these spectator terms, together with the requirement of Casimir–Polder consistency, may be the key to the choice of a suitable approximate nonlocal dynamic xc kernel f_{xc} for vdW correlation energies within the ACFD.

It is also worth noting that the proof of (22.31) in (Dobson 1994b) assumed that the KS (independent-electron) responses $\chi_{1,1}^{KS}$, $\chi_{2,2}^{KS}$ of the subsystems did not vary with separation D, as $D \rightarrow \infty$. This corresponds to the commonest way of implementing the RPA correlation energy as a "post-functional", that is, using χ_{KS} deduced from the KS orbitals obtained from a groundstate LDA or GGA calculation. Of couse, in a fully selfconsistent (OEP) RPA correlation energy calculation, one would vary the groundstate KS potential v_{KS} (and correspondingly χ_{KS}), to minimize the total nonlocal RPA energy. An estimate of the effect of the change in the KS potential in the subsystems can be made as follows. Suppose that we start with the selfconsistent OEP-RPA solution v_{KS}^{OEP} as described above and make a small change δv_{KS} in the KS potential (e.g. toward the LDA post-functional value described above). The change in RPA energy due to this change is of *second* order in δv_{KS} because the energy is minimized by v_{KS}^{OEP}. If we assume that $\delta v_{KS} = O(D^{-p})$ as required to produce the nuclear D^{-7} vdW force via the Feynman–Tozer approach described in Sect. 22.2.3 above, then the (second-order) correction energy is of $O(D^{-2p})$ and so is negligible in the widely-separated limit, compared with the post-functional energy (22.31). Of course the choice of groundstate orbitals is indeed very important at closer separations, as verified e.g. in Toulouse et al. (2010).

22.7.2 Problems with the dRPA

A serious deficiency of dRPA, apparent already in the uniform electron gas, is that it greatly over-estimates the correlation hole at short range. Fortunately this often largely cancels out when one forms energy differences between iso-electronic configurations, for example dissociation energy curves or surface energies (Kurth and Perdew 1999). Nevertheless for accurate work, even where differences are involved, one should correct this, either via introduction of a suitable f_{xc} or by other means.

Another disadvantage of the dRPA response function is that it allows a spurious dynamic self-interaction of a localized orbital, leading to a corresponding spurious self-correlation contribution to the ACFD energy [see e.g. (Dobson et al. 2005; Gruneis et al. 2009; Paier et al. 2010)]

In vdW applications, possibly even more important than correcting the dynamic self-interaction is choosing a set of starting orbitals φ_n that come from a self-interaction-free approximation to the groundstate KS potential. Because of orbital self-interaction, LDA orbitals φ_n^{LDA} of highly localized systems tend to be too diffuse and too polarizable, leading to an over-estimate of the RPA response and hence also of the vdW interaction. For the dissociation energy curve $E(D)$ of small molecules where groundstate SIC is very important, the starting orbitals have indeed been found to have a very strong effect: see for example the dRPA $D(E)$ curves for He$_2$ with Hartree–Fock and PBE orbitals in Fig. 2 of (Toulouse et al. 2010). In principle, for any given functional $E_{xc}[\{\varphi\}]$, one should construct the $\{\varphi_n\}$ from the 1-body potential v_{eff} such that the total energy is minimized (optimized effective potential (OEP) approach). This has mostly not been used to date for functionals of the RPA class [but see (Kotani 1998)]. For solids, the use of v_{KS} and $\{\varphi_n\}$ from OEP seems to have a less dramatic effect on dRPA energies than it does for small molecules (Harl et al. 2010). Most of the realistic RPA calculations for periodic systems have started from GGA orbitals that have reasonably realistic diffusenesss because of the reasonable SIC properties of the chosen GGA. Furthermore, for the delicate but technologically significant case of the layered systems BN and graphite, the π_z orbitals involved in the layer-layer vdW interaction are already quite diffuse so that groundstate self-interaction effects are relatively unimportant, which may partially account for the success of the recent ACFD/RPA calculations.

Nevertheless there are certainly cases where use of the dRPA correlation energy is quite disastrous, such as the dissociation energy curves $E(D)$ of diatomic molecules in the regime of bond-breaking atomic separations $D \approx D_0$. (Toulouse et al. 2010). This seems to be due partly to inadequate SIC behavior of the dRPA when the degree of localization is changing with D. It was also found that $E(D)$ is very strongly dependent on the starting orbitals in these cases.

22.8 Beyond dRPA: Non-TDDFT Methods

The self-interaction error of dRPA can be removed via the "RPAx", also known as "full RPA" [ACFD with time dependent Hartree–Fock response: see e.g. (Toulouse et al. 2009)], though this carries a computational cost and has its own diseases. The short-ranged failures of the dRPA correlation hole can be partly remedied by the range separation approach in which different treatments are given to the short-ranged and long-ranged components of the bare electron-electron potential v_{ee} (Toulouse et al. 2009). Recently the second order screened exchange (SOSEX) approach has been favored (Paier et al. 2010) for self-interaction correction of the dRPA. This addresses both the orbital self-interaction problem and, to some extent, the short range inadequacies of dRPA. It appears to provide a systematic improvement to the dRPA results for lattice spacings and binding energies of crystals (Gruneis et al. 2009), and can make very substantial improvements to energetics of small molecules (Paier et al. 2010).

The dRPA can be related (Scuseria et al. 2008) to the "doubles" contribution in the coupled cluster expansion of quantum chemistry. For molecules the state of the art is typically the CCSD(T) theory in which some of the triples terms of the cluster expansion are also included. At present this approach is computationally out of reach for large molecules and for most solids.

For small systems (Sorella et al. 2007), or large systems with especially simple geometry (Drummond and Needs 2007; Spanu et al. 2009), the vdW energetics can be calculated via the electron diffusion Monte Carlo approach, resulting in benchmark numbers, provided that noise and size convergence issues of diffusion Monte Carlo can be controlled.

The self-interaction error of dRPA can also be addressed in principle by including a sufficiently realistic f_{xc} in the TDDFT screening Eq. 22.19.

22.9 Beyond the dRPA: ACFD with a Nonzero xc Kernel

22.9.1 The Case of Two Small Distant Systems in the ACFD with a Nonzero xc Kernel

The situation is not entirely simple when one uses a TDDFT response with a nonzero kernel f_{xc} within the ACFD energy (22.18), rather than the RPA response. Suppose that one assumes (i) that the dynamic kernel $f_{xc}(r_1, r_2, \omega)$ is zero when r_1 and r_2 lie in different fragments, to second order in v_{ee12} and (ii) that the presence of the other fragment does not affect the value of $f_{xc}(r_1, r'_1, \omega)$ inside a given fragment, again to second order in v_{ee12}—i.e., we neglect "spectator" effects in f_{xc}. (These assumptions are satisfied within the ALDA provided one uses a "post-functional" approach in which the changes in density and χ_{KS} in each fragment, beyond the isolated case, are ignored.) Then in fact one can show that the seamless

ACFD approach does *not* yield the perturbative result (22.2) in the well-separated limit, there being an additional energy term involving $\partial f_{xc}/\partial \lambda$. When used with simple local approximations for f_{xc} the additional term is non-vanishing. That is, ACFD is *not* Casimir–Polder consistent when used with a non-zero *local* kernel f_{xc}.

The solution to this dilemma may be that, when r and r' are both in system 1, $f_{xc}[n](r, r')$ contains a "spectator" contribution that depends on the response function of the distant system 2. This dependence amounts to a highly nonlocal dependence of f_{xc} on the groundstate density $n(r'')$. It arises from nonlocal screening of internal v_{ee} lines of system 1 via polarization of system 2. This type of term must be included in a seamless beyond-dRPA ACFD dimer calculation in order for it to recover (22.2)—i.e., to obtain Casimir–Polder consistency. This could be a useful guide to construction of approximate f_{xc} functionals for use in ACFD energy calculations for vdW-bonded systems.

22.9.2 Beyond the dRPA in the ACFD: Energy-Optimized f_{xc} Kernels

The standard non-memory approximation for f_{xc} has been the ALDA (Zangwill and Soven 1980a; Gross and Kohn 1985) given by

$$f_{xc}^{ALDA}(r, r', \omega) = \delta(r - r')\frac{d^2[n\epsilon_{xc}^{hom}(n)]}{dn^2}. \tag{22.32}$$

It is "optimized" for describing low-frequency, long-wavelength excitations in near-homogeneous systems, and is therefore quite unsuitable for the calculation of groundstate xc energies from Eqs. 22.18 and 22.19, because these formulae effectively sample all of frequency and wavenumber space. For example, in a uniform electron gas, $f_{xc}^{ALDA}(r, r', \omega)$ leads to a very poor evaluation of the correlation energy when substituted into (22.18) and (22.19), even exhibiting the wrong sign at low densities (i.e., at high Wigner-Seitz radius r_s) (Dobson and Wang 2000).

The simplest way to remedy this is to find a local frequency-independent kernel $f_{xc}^{en.opt}(\lambda, n)$ that *does* give the correct uniform-gas E_{xc} when substituted into (22.18) and (22.19). It turns out (Dobson and Wang 2000) that this requirement, applied for every homogeneous density n, together with the scaling rule

$$f_{xc}(\lambda, n) = \lambda^{-1}\hbar^4 e^{-2} m^{-2} F(\lambda^{-1}r_s), \qquad \frac{4\pi}{3}r_s^3 a_B^3 = n^{-1} \tag{22.33}$$

uniquely determines the dimensionless energy-optimized kernel $f_{xc}^{en.opt}(r)$. The optimized kernel has to produce a magnitude of corrrelation energy lying between the (too large) RPA value corresponding to $f_{xc} = 0$ and the (too small) value corresponding to the ALDA kernel. Unsurprisingly, then, $f_{xc}(r_s)$ lies between zero and f_{xc}^{ALDA}. A parametrized form accurate for metallic densities is (Dobson and Wang 2000)

$$f_{xc}^{en.opt}(r_s) = -0.5004(4)r_s^2 + 4.5365(3) \times 10^{-3}r_s^3 - 3.366(0) \times 10^{-5}r_s^4. \quad (22.34)$$

The theory is of course constrained to give accurate correlation energies for the uniform gas. Its performance for a *nonuniform* gas has been tried in the context of layered jellium (Dobson and Wang 2000) where it had little effect on the energy *differences* required for a vdW energy calculation, compared with a pure RPA calculation. Other than this, there has been little direct testing of this kernel in other systems.

The purely local (q-independent) character of the $f_{xc}^{en.opt}$ kernel results in a divergent on-top hole (Dobson and Wang 2000, Furche and van Voorhis 2005), though this is integrable in a correlation energy calculation. It can nevertheless cause numerical difficulties in the solution of the screening Eq. 22.19, when a q-space algorithm is used. For this reason, and to include some of the physical aspects of nonlocality, a static but spatially nonlocal energy-optimized kernel was tried by (Jung et al. 2004). They used the λ-scaling law

$$f_{xc\lambda}(r_1, r_2; \omega) = \lambda^{-1}f_{xc}^{(0)}(\lambda^{-3}\tilde{n}, \lambda r_{12}) \quad (22.35)$$

where $\tilde{n}(r_1, r_2)$ is an effective density taken in their applications to be $[n(r_1) + n(r_2)]/2$ or $\sqrt{n(r_1)n(r_2)}$, both leading to similar results. The form (22.35) (and indeed the scaling in the Dobson-Wang scheme) is consistent with the frequency-independent limit of the general scaling law derived in (Lein et al. 2000b). The spatial nonlocality was assumed by Jung et al. (2004) to take a Hubbard-like form fitted to the correlation energy of the 3D electron gas [see Fig. 2 of (Jung et al. 2004)]. With the nonlocal kernel of Jung et al the energy-versus separation curve of two high-density jellium slabs was found to be very similar to the pure RPA result. This energy-optimized kernel has also been tested (Garcia-Gonzalez et al. 2007) for the correlation energy of jellium spheres, where it gave a systematic improvement over dRPA.

It seems clear from Sect. 22.7.1 above that the local and quasi-local energy-optimized kernels will not be Casimir–Polder consistent, so that for vdW situations they may require supplementation with an additional term that is a very nonlocal functional of the groundstate density, when used in a seamless vdW calculation.

22.9.3 Beyond the RPA in the ACFD: More Realistic Uniform-Gas Based f_{xc} Kernels

Over recent decades more realistic q- and/or ω-dependent kernels have been obtained for the uniform electron gas (Richardson and Ashcroft 1994; Corradini et al. 1998) that rather accurately (Lein et al. 2000b) give ε_c for the 3D electron-gas, when substituted into the ACFDT formulae (22.18) and (22.19). The static kernel of Corradini et al. (1998) has been used with the ACFD under the appropriate λ scaling formula (22.35) for layered nonuniform jellium systems (Pitarke and Perdew 2003;

Jung et al. 2004). The surface energy results based on a single jellium slab were not very different from the pure RPA nor from the LDA results. The vdW results for two jellium layers were not very different from the energy-optimized schemes described above, nor from the pure RPA calculation. There were some differences from the LDA, and in one case (Jung et al. 2004) all the microscopic schemes including the Corradini version did give about a 50% increase in binding energy of the slabs, compared with an LDA energy calculation. The equilibrium separation D_0 of the slabs was quite similar in all of the f_{xc}-corrected schemes and the LDA, but was slightly smaller in the pure RPA energy calculation.

It should also be possible to graft the more sophisticated frequency-dependent Richardson-Ashcroft uniform-gas kernel (Richardson and Ashcroft 1994) onto an ACFD energy calculation for nonuniform systems, but that has not yet been done to the author's knowledge.

Another consideration not yet explored in the vdW context is the use of tensor memory kernels such as that used in the Vignale-Kohn-Ullrich current-current response formula (Vignale and Kohn 1996; Vignale et al. 1997; Conti and Vignale 1999). These kernels are based on the near-homogeneous gas with sinusoidal density variation, and have been proposed (van Faassen et al. 2002) for the description of the polarization response in 1D systems. This method obtains rather good polymer properties (see Chap. 24), and so should be a good candidate for seamless vdW energetics of these systems.

Once again, any schemes based on the uniform gas may require supplementation via a term with a highly nonlocal density dependence, if Casimir–Polder consistency is to be achieved.

22.9.4 xc Kernels not Based on the Uniform Electron Gas

All of the xc kernels discussed above have been based on the uniform or near-uniform electron gas, and grafted onto an ACFD energy calculation for a nonuniform system, in a modified local-density manner. Kernels have also been developed that are based specifically on the KS orbitals of the nonuniform system. Perhaps the simplest of these is the kernel of Petersilka, Gossman and Gross (PGG) (Petersilka et al. 1996, 1998) which is a static, spatially nonlocal exchange-only kernel. It has been tested (Fuchs and Gonze 2002) on the energetics of the Be-Be dimer, where it performed somewhat similarly to the RPA for the binding energy, and better than the RPA for the bond length. It did not perform as well as the energy optimized kernel (Jung et al. 2004) for the correlation energy of jellium spheres (Garcia-Gonzalez et al. 2007). The PGG kernel is self-interaction free, which suggests that for small systems it might out-perform the RPA when used with an appropriate SIC groundstate calculation.

The nonlocal kernel of Reining et al. (2002) is designed for finite-freqency response of semiconductors, and may be relevant to vdW interactions between these systems.

A further possibility, similar in spirit to a system-specific f_{xc}, is the use of the inhomogeneous Singwi-Tosi-Land-Sjolander (ISTLS) formalism (Dobson et al. 2002; Constantin et al. 2008) to generate χ_λ. The ISTLS formalism has been shown (Dobson 2009a) to be equivalent to an adiabatic spatially nonlocal xc kernel within time dependent *current* DFT. This scheme is free from orbital self-interaction. To date no vdW calculations appear to have been performed in this way, however.

22.10 Density-Based Approximations for the Response Functions in ACFD vdW Theory

The previous sections have explored corrections to the non-local dRPA correlation energy via approximations to the xc kernel f_{xc} that sample the groundstate density $n(r)$, leading to numerics that are at least as heavy as the dRPA itself. In order to reduce the computational load one can also consider approximating the KS response χ_{KS} or other responses χ directly in terms of $n(r)$. This removes the need for numerically intensive computations of χ_{KS} based on KS orbitals, as in dRPA and its extensions.

22.10.1 Density-Based Approximations for the Non-overlapping Regime

Consider first the result (22.5), in which the long-ranged effects of v_{ee12} in (22.19) have already been taken into account via PT. The remaining response functions $\alpha_1(r_1, r_1', iu)$ and $\alpha_2(r_2, r_2', iu)$ can be approximated in a local fashion based on the local fragment density, without losing the vdW tail. This leads to an approximation to the second-order vdW energy of two small systems by a functional of their individual groundstate electron densities $n_1(r_1)$, $n_2(r_2)$:

$$E^{(2)} \approx -\frac{3\hbar e}{2(4\pi)^{3/2} m^{1/2}} \int\limits_1 d^3 r_1 \int\limits_2 d^3 r_2 \frac{1}{r_{12}^6} \frac{\sqrt{n_1 n_2}}{\sqrt{n_1} + \sqrt{n_2}} \qquad (22.36)$$

Equation 22.36 is a highly nonlocal functional of the groundstate electron density n. It was first obtained (Andersson et al. 1996), by considering some asymptotics, starting from a semi-empirical plasmon-motivated form due to Rapcewicz and Ashcroft (1991). Independently, (22.36) was derived straightforwardly (Dobson and Dinte 1996) by a constrained local approximation for the response functions in (22.5). In practice, (22.36) gives answers sometimes greatly too large because of the contributions from the tails of the atomic densities, and it requires a cutoff in the spatial integrations, based on gradients of n and described in detail in Andersson et al. (1996). The answers are sensitive to this cutoff, but do provide results for C_6 that are mostly quite good for distant atom-atom interactions, and err in the worst cases by about a factor of two. Given the relatively poor performance of the RPA and ALDA for atomic polarizabilities via (22.6), this kind of error does not look so bad. However, one must also note the very simple and rather successful formulas in the chemistry literature (Atkins and Friedman 1997), involving the electron affinity I and based loosely on the arguments given in Sect. 22.2.1 above.

A related approach to the vdW interaction between atoms in molecules have been given by Sato and Nakai (2009) .These approaches also involve a short-ranged cutoff function to allow for a seamless calculation. Tkatchenko and Scheffler (2009) have given a somewhat different approach based on the compression of atoms in molecules, in which the volume compression of an atom within a molecule is used to reduce the corresponding bare-atom polarizability.

22.10.2 "Seamless" Density-Based vdW Approximations Valid into the Overlapped Regime

Within the ACFD, the essential nonlocality of the vdW energy arises from the long range of the electron-electron interaction v_{ee} in the Dyson-like screening Eq. 22.19. This fact is the key to obtaining density-based approximations that do not suppress the tail of the vdW interaction. The approach of Dobson and Wang (1999) to the general case, including overlap, was to evaluate the ACFD energy by approximating the "bare" (Kohn–Sham) response χ_{KS} in the screening Eq. 22.19, using the groundstate density as input. The screening Eq. 22.19 is still solved numerically with retention of the long-ranged character of the electron-electron interaction v_{ee}. The quantity approximated from uniform-gas data is the polarizability $\alpha = (\epsilon - 1)/4\pi$ [somewhat as in the macroscopic Lifshitz approach (Dzyaloshinskii et al. 1961)]: this choice conserves charge (Dobson and Dinte 1996), and it is evaluated at a mean density chosen to ensure that the approximate χ_{KS} does not introduce unphysical flow of electrons from one subsystem to the other, in the non-overlapping limit. This approach was tested successfully (Dobson and Wang 1999; Dobson and Wang 2000) for the vdW interaction of a pair of jellium slabs. Here comparison could be made with an accurate numerical solution of the full ACFD equations at all separations from the overlapping contact situation out to the asymptotic non-overlapping vdW regime. The correct vdW-RPA interaction was obtained at large separations where the usual LDA/GGA produces no interaction at all, and the RPA energies near the equilibrium separation were reproduced somewhat better than in the LDA. The tests were subsequently extended (Dobson and Wang 2000) to the inclusion of a local f_{xc} which turned out to make a negligible difference in the cases studied. This type of approach has not yet been tried in other geometries because it is numerically intensive. A related theory by Gould and Dobson is more tractable: see the next Section.

The "vdW Density Functional" (vdWDF) (Dion et al. 2004; Langreth et al. 2005) was derived starting from the ACFD. While it goes beyond the dRPA in principle, in practice it embodies four distinct approximations/assumptions that render it hard to judge its accuracy *a priori*. It has the form of a sum of pairwise contributions from space points r_A and r_B:

$$E \approx \int d^3 r_A \int d^3 r_B g(n_A(r_A), \nabla n_A(r_A); n_B(r_B), \nabla n_B(r_B); r_A, r_B), \quad (22.37)$$

where n is the groundstate density and $g \propto |r_A - r_B|^{-6}$ as $|r_A - r_B| \to \infty$. The pairwise form of (22.37) results because the TDDFT coulomb screening Eq. 22.19 has effectively been solved pertubatively via a second-order expansion of an operator logarithm. This PT is not a direct perturbation in the bare Coulomb potential, so the theory performs much better for large systems than the well-known MP2 second-order theory, while nevertheless retaining a pairwise form. A good feature of this approach is that it treats overlapped cases while also providing a vdW energy "tail" of classic R^{-6} form at large distances, while remaining much more tractable computationally than full RPA E_c calculations. When used with a specific semi-local exchange energy approximation, the vdWDF is found to give very sensible energetics for a wide range of vdW bound crystals and nanostructures [see e.g. (Dion et al. 2004; Langreth et al. 2005; Berland and Hyldgaard 2010)]. The binding energies are quite good but equilibrium lattice constants are typically over-estimated by $O(10\%)$ and the elastic constants are not reliable [see e.g. (Marini et al. 2006b; Lebegue et al. 2010)]. Vydrov and van Voorhis (2009b, 2010a, 2010b) have started from the vdWDF derivation and the form (22.37) but have further optimized various aspects, using empirical fits where appropriate (see details in Chap. 23). As a result they obtain significant improvement over the original vdWDF for the interaction between pairs of molecules. The original authors have also recently fine-tuned their vdWDF functional (Lee et al. 2010). The present author's opinion is that, for large nanostructures, further improvement beyond these developments may require a non-perturbative solution of the TDDFT screening equation so that the restriction to the pairwise form (22.37) is relaxed. Various situations [e.g. (Kim et al. 2006; Dobson et al. 2006)] require such a non-pairwise-additive approach. A very recent step in this direction by Gould and Dobson uses continuum mechanics (Gao et al. 2010) (see also Chap. 25) based on the groundstate stress tensor, coupled with full solution of the Dyson-like RPA screening equation within the ACFD. This gives a highly non-pairwise vdW energy functional. It is computationally more tractable than the related functional of (Dobson and Wang 1999), but performs well for low-dimensional metals, in contrast to most simplified approaches to ACFD.

22.11 Summary

The SAPT theory is probably the state of the art for vdW energy calculations of molecular pairs of moderate size at all separations down to overlap, but is not feasible for macroscopic systems. There the prediction of vdW energetics is still a controversial area for highly anisotropic "soft" systems, especially those with small or zero energy gaps. For such systems, a conventional sum of pairwise atom-atom contributions of R^{-6} form is sometimes inadequate to describe the vdW physics. Even for such "difficult" systems the ACFD approach, based on various approximations for the density-density response χ or related quantities, appears to be very promising. For fully macroscopic systems at macroscopic separations, the Lifshitz theory is available. Its electromagnetically non-retarded form can be derived from the ACFD by approximating the dielectric function locally using data for the bulk system, and otherwise taking an RPA-like approach. For overlapped systems, density-based approximation of the response in ACFD has yielded the "vdWDF." It is computationally convenient and yields sensible energies for a wide range of vdW-interacting nanosystems. Lattice spacings and elastic constants are less satisfactory in vdWDF, however, and a more accurate approach is desirable. In particular one might wish to avoid approximating the solution of the Dyson-like Coulomb screening equation of TDDFT, as it leads to a pairwise summation form of the vdW interaction. A preliminary version of such an approach has now been given.

The direct RPA (ACFD with $f_{xc} = 0$), applied without further approximation, leads to very demanding numerical correlation energy calculations that yield vdW terms, but these have now been carried out for systems of up to 100 atoms, and for crystals with modest-sized unit cells. While it has some serious shortcomings for the binding curves of small molecules, the dRPA based on GGA orbitals has given very good results for the lattice spacing, elastic constants and binding energies of vdW-bound periodic systems, mostly in agreement with experiment where available. This success is probably due to the relative unimportance of dynamic self-interaction in the diffuse outer p orbitals that give rise to the vdW interactions in the cases studied. By contrast, the lack of self-interaction correction is disastrous in the dRPA dissociation curves of small molecules at intermediate bond lengths.

Various microscopic approaches such as SOSEX can be used to improve the accuracy of dRPA, but are computationally demanding. In principle the remaining deficiencies of the dRPA can alternatively be remedied by inclusion of a sufficiently realistic xc kernel f_{xc} in the Dyson-like screening equation of TDDFT, prior to application of the FDT. f_{xc} will have to be quite nonlocal, however, in order to effect dynamic self interaction correction in the case of localized orbitals. It may also have to be a highly nonlocal function of groundstate density $n(\mathbf{r})$ in order to achieve Casimir–Polder consistency. Some of the burden of these developments may be removed via range-separation approaches. Much work needs to be done, but this task should be possible starting from the relatively good success of the dRPA version of the ACFDF theory for large systems.

For a more computationally tractable approach, one will also want to derive an approximation that avoids construction of RPA-type response functions directly from excited orbitals or their equivalent.

Chapter 23
Nonlocal Van Der Waals Density Functionals Based on Local Response Models

Oleg A. Vydrov and Troy Van Voorhis

23.1 Introduction

As described in Chap. 22, dispersion interactions, also known as van der Waals interactions, arise from long-range correlated fluctuations of the electron charge density. The dispersion energy is thus a nonlocal component of the correlation energy. In practical applications of Kohn–Sham DFT, the correlation energy is usually approximated as a local or a semilocal density functional (Fiolhais et al. 2003). (Semi)local functionals cannot in principle include the proper physics of long-range dispersion interactions. Empirical dispersion corrections are quite popular and reasonably successful (Grimme et al. 2010), but they typically entail a departure from pure DFT into the realm of classical force fields, and hence they fall outside the scope of this book. The rigorous description of long-range van der Waals interactions requires fully nonlocal treatment of correlation (Riley et al. 2010). Unfortunately, the rigor usually comes at the cost of an explicit and cumbersome dependence on both occupied and virtual orbitals. Within the framework of Kohn–Sham DFT, substantial advances have been made in the development of orbital-dependent nonlocal correlation functionals, exemplified by the random phase approximation (RPA) and other closely related methods, usually derived via the adiabatic connection/fluctuation-dissipation theorem, as described in Chap. 22, the RPA and its variants have been successfully applied to weakly interacting systems (Eshuis and Furche 2011; Schimka et al. 2010; Zhu et al. 2010; Nguyen and Galli 2010; Janesko et al. 2009). The practical usefulness of these methods is limited by their

O. A. Vydrov · T. Van Voorhis (✉)
Department of Chemistry, Massachusetts Institute of Technology,
Cambridge, MA 02139, USA
e-mail: tvan@mit.edu

O. A. Vydrov
e-mail: vydrov@mit.edu

M. A. L. Marques et al. (eds.), *Fundamentals of Time-Dependent Density Functional Theory*, Lecture Notes in Physics 837, DOI: 10.1007/978-3-642-23518-4_23, © Springer-Verlag Berlin Heidelberg 2012

high computational cost and by the lack of self-consistency in current implementations.

An elegant compromise between rigor and computational tractability has been
achieved in the recently introduced class of nonlocal correlation functionals that
treat the entire range of dispersion interactions in a general and seamless fashion, yet
include no explicit orbital dependence and use only the electron density as input
(Dion et al. 2004; Lee et al. 2010; Vydrov and Van Voorhis 2009a, b; 2010c).
These functionals are relatively computationally inexpensive and lend themselves to
efficient self-consistent implementations. This group of methods can be considered
as an extension of the asymptotic theories of Andersson et al. (1996), Dobson and
Dinte (1996). Since the long-range non-overlapping regime is the only limit where
dispersion interactions are unambiguously and uniquely defined, we will begin our
discussion by considering this asymptotic limit.

23.2 Long-Range Asymptote of Dispersion Interaction

23.2.1 Local Polarizability Formalism

The second-order dispersion interaction energy $E^{(2)}$ between two finite non-
overlapping systems was derived in Chap. 22 and expressed in terms of nonlocal
dynamic dipole polarizability tensors in (22.5). A tractable model for $E^{(2)}$ can be
obtained by adopting a local isotropic response approximation:

$$\alpha_{ij}(\boldsymbol{r}, \boldsymbol{r}', iu) = \delta_{ij}\alpha(\boldsymbol{r}, iu)\delta(\boldsymbol{r} - \boldsymbol{r}'). \tag{23.1}$$

Under this approximation, (22.5) transforms into

$$E^{(2)} = -\frac{3\hbar}{\pi} \int_0^\infty du \int_A d^3r \int_B d^3r' \frac{\alpha(\boldsymbol{r}, iu)\alpha(\boldsymbol{r}', iu)}{|\boldsymbol{r} - \boldsymbol{r}'|^6}, \tag{23.2}$$

where A and B define the domains of the non-overlapping subsystems, and $\boldsymbol{r} \in A$
while $\boldsymbol{r}' \in B$. The local polarizability density $\alpha(\boldsymbol{r}, iu)$ is connected to the experimentally measurable average dynamic polarizability $\bar{\alpha}(iu)$ via

$$\bar{\alpha}(iu) = \int d^3r \alpha(\boldsymbol{r}, iu). \tag{23.3}$$

The f-sum rule requires that, in the $u \to \infty$ limit, $\bar{\alpha}(iu) \to Ne^2/mu^2$, where N is
the number of electrons in the system. This constraint is usually imposed by writing
$\alpha(\boldsymbol{r}, iu)$ in the following way:

$$\alpha(\boldsymbol{r}, iu) = \frac{e^2}{m} \frac{n(\boldsymbol{r})}{\omega_0^2(\boldsymbol{r}) + u^2}. \tag{23.4}$$

Since $\int d^3 r n(r) = N$, the f-sum rule is recovered by (23.4). Plugging (23.4) into (23.2) and integrating over u, we arrive at

$$E^{(2)} = -\frac{3\hbar e^4}{2m^2} \int_A d^3 r \int_B d^3 r' \frac{n(r)n(r')}{\omega_0(r)\omega_0(r')\left[\omega_0(r) + \omega_0(r')\right]|r - r'|^6}. \quad (23.5)$$

When the distance between species A and B is large compared to the size of these systems, $|r - r'|^{-6}$ in (23.5) can be taken out of the integral as R^{-6}, leading to the $-C_6^{AB} R^{-6}$ form. Hence, (23.5) can be used for computing asymptotic van der Waals C_6 coefficients.

23.2.2 Practical Local Polarizability Models

Various local polarizability models differ substantially in how ω_0 in (23.4) is defined. In the Andersson, Langreth, and Lundqvist (ALL) model (Andersson et al. 1996; Dobson and Dinte 1996), $\omega_0(r)$ is taken to be equal to the local plasma frequency: $\omega_0(r) = \omega_p(r) = \sqrt{4\pi n(r)e^2/m}$. Thus the local polarizability at r is determined entirely by the local electron density $n(r)$ and modeled by the long-wavelength dielectric response of the uniform electron gas (UEG) of density $n = n(r)$. The ALL energy expression was given in (22.36). Nesbet argued (Nesbet 1997) that for describing polarizability of a free atom it is more appropriate to use $\omega_0(r) = \omega_p(r)/\sqrt{3}$, which makes the theory consistent with the classical Clausius-Mossotti formula. Nesbet's suggestion is corroborated by the fact that $\omega_0 = \omega_p/\sqrt{3}$ gives the correct polarizability for an isolated sphere of uniform electron density ("jellium sphere"), whereas the ALL formula gives an incorrect result for this model system. Regardless of the proportionality coefficient, using $\omega_0^2(r) \propto \omega_p^2(r) \propto n(r)$ in (23.4) leads to the inadequate treatment of static ($u = 0$) polarizability of an atom or a molecule: in such a model, $\alpha(r, 0)$ is an r-independent and density-independent constant and hence $\bar{\alpha}(0)$ is determined entirely by the (arbitrary) choice of the integration limits in (23.3). A density-based prescription for a sharp integration cutoff has been proposed (Rapcewicz and Ashcroft 1991; Andersson et al. 1996), but it is not entirely satisfactory from either the numerical or formal point of view. Such an integration cutoff discards density tail regions, reducing N in violation of the f-sum rule.

The need for an explicit integration cutoff is obviated altogether if $\omega_0(r)$ is defined in such a way that $n/\omega_0^2 \rightarrow 0$ in the density tails. In the more recent theories described below, ω_0 is constructed to satisfy this condition. Nonlocal van der Waals density functionals (vdW-DFs) of Langreth and coworkers (Dion et al. 2004; Lee et al. 2010) and their variants (Vydrov and Van Voorhis 2009a) are essentially based on the local polarizability approximation. In the long-range asymptotic limit, all of these functionals reduce to the form of (23.5) with different choices of $\omega_0(r)$, as summarized in Table 23.1.

Table 23.1 Definitions of $\omega_0(r)$ in several methods

Model	Definition of ω_0	MAPE (%)
vdW-DF-04	$\frac{9\hbar}{8\pi m}\left[k_F\left(1+\mu s^2\right)-\frac{4\pi}{3e^2}\varepsilon_c^{LDA}\right]^2$ with $\mu=0.09434$	18.5
vdW-DF-10	same as in vdW-DF-04, but with $\mu=0.20963$	60.9
vdW-DF-09	$\frac{\hbar}{3m}k_F^2\left(1+\mu s^2\right)^2$ with $\mu=0.22$	10.4
VV09/10	$\sqrt{\frac{\omega_p^2}{3}+C\frac{\hbar^2}{m^2}\left|\frac{\nabla n}{n}\right|^4}$ with $C=0.0089$	10.7

The last column gives the mean absolute percentage error (MAPE) for a set of 34 C_6 coefficients of closed-shell species assembled in Table II of Vydrov and Van Voorhis (2010b). C_6 were computed using (23.5) with the corresponding ω_0. Computational details are given in Vydrov and Van Voorhis (2010b)

The nonlocal correlation in vdW-DFs is more conveniently expressed not in terms of ω_0, but in terms of another quantity q_0, which can be written as $q_0(r) = k_F(r)F(r)$, where $k_F(r) = [3\pi^2 n(r)]^{1/3}$ is the local Fermi wavevector and F is an enhancement factor. In vdW-DF-04 (Dion et al. 2004) as well as in its later re-parameterization vdW-DF-10 [termed vdW-DF2 in Lee et al. (2010)], q_0 is related to ω_0 via

$$\omega_0(r) = \frac{9\hbar}{8\pi m}q_0^2(r), \qquad (23.6)$$

whereas in vdW-DF-09 (Vydrov and Van Voorhis 2009a) this relation is slightly different:

$$\omega_0(r) = \frac{\hbar}{3m}q_0^2(r). \qquad (23.7)$$

In practice, the enhancement factor $F = q_0/k_F$ is required to circumvent the afore-mentioned problem with static polarizability. In the $u = 0$ limit, the vdW-DF local polarizability model behaves as

$$\alpha(r,0) \propto \frac{n(r)}{\omega_0^2(r)} \propto \frac{n(r)}{k_F^4(r)F^4(r)} \propto \frac{1}{n^{1/3}(r)F^4(r)}.$$

Using $F = 1$ would lead to $\alpha(r,0)$ that is exponentially divergent (as $n^{-1/3}$) in the density tails. It is essential to use an F such that $n^{1/3}F^4 \to \infty$ in the density tails. In all three versions of vdW-DF, this is accomplished by including into F a term proportional to s^2, where $s = |\nabla n|/(2k_F n)$ is a dimensionless density gradient. In vdW-DF-09, a particularly simple choice was made: $F = 1+\mu s^2$. Polarizabilities resulting from (23.4) and C_6 coefficients computed via (23.5) are very sensitive to the value of μ—the coefficient of s^2. Table 23.1 summarizes the mean errors of several methods for a C_6 benchmark set. The definition of $\omega_0(r)$ in vdW-DF-10 is the same as in vdW-DF-04 apart from the difference in the μ value, yet these two functionals yield very different C_6 coefficients. Among the methods included in Table 23.1, vdW-DF-10 is by far the worst performer: it underestimates C_6 by a factor of 2.6 on average. In vdW-DF-09, μ was fitted to yield accurate C_6 coefficients (Vydrov and Van Voorhis 2009a). As a result, vdW-DF-09 gives a respectably low mean error of about 10% for a diverse C_6 test set. Due to its good performance, the vdW-DF-09 local polarizability model was employed by Sato and Nakai as a basis for their pairwise atom-atom dispersion correction (Sato and Nakai 2009).

All three versions of vdW-DF fail to give the proper C_6 coefficients for jellium spheres, yielding qualitatively incorrect dependence of C_6 on the electron density. Vydrov and Van Voorhis (VV) proposed (Vydrov and Van Voorhis 2009b; 2010b) to define $\omega_0(r)$ in such a way that in the uniform density limit $\omega_0 = \omega_p/\sqrt{3}$, giving the correct result for jellium spheres. VV also argued that the local response in nonuniform systems can be made more realistic by introducing a "local band gap". At the elementary level, the metallic UEG response function can be transformed into that of a semiconductor with a band gap $\hbar\omega_g$ by subtracting ω_g^2 from ω^2, where ω is the perturbing field frequency. We are interested in imaginary frequencies $\omega = iu$, and hence u^2 should be replaced by $(\omega_g^2 + u^2)$ in (23.4). In the denominator of (23.4), ω_g^2 can be absorbed into ω_0^2, resulting in

$$\omega_0^2(r) = \frac{\omega_p^2(r)}{3} + \omega_g^2(r). \tag{23.8}$$

A suitable model for $\omega_g(r)$ can be deduced by examining the typical shape of the electron density $n(r)$. In atoms, $n(r)$ can be approximated as piecewise exponential. In the density tails, the exact behavior (Levy et al. 1984) is known:

$$n(r) \sim \exp(-\alpha|r|), \quad \text{with} \quad \alpha = \frac{2}{\hbar}\sqrt{2mI}, \tag{23.9}$$

where I is the ionization potential. Generalizing the result of (23.9), we can define a "local ionization potential" as

$$I(r) = \frac{\hbar^2}{8m}\left|\frac{\nabla n(r)}{n(r)}\right|^2. \tag{23.10}$$

Taking $\hbar\omega_g(r) \propto I(r)$, we write

$$\omega_g^2(r) = C\frac{\hbar^2}{m^2}\left|\frac{\nabla n(r)}{n(r)}\right|^4, \tag{23.11}$$

where C is a parameter that can be adjusted such that (23.5) with $\omega_0(r)$ of (23.8) produces accurate van der Waals C_6 coefficients for atoms and molecules (Vydrov and Van Voorhis 2009b, 2010b). The optimal value of $C = 0.0089$ was obtained by fitting to a training set [however, as discussed in Vydrov and Van Voorhis (2010c), the fitted value of C can vary slightly, depending on the source of the input densities]. In the uniform density limit, ω_g of (23.11) vanishes, and hence ω_0 of (23.8) reduces to $\omega_p/\sqrt{3}$.

We have to mention that there exist a number of simple and accurate methods for computing asymptotic dispersion coefficients of atoms and molecules using the electron density and/or occupied orbitals as input (Becke and Johnson 2007; Tkatchenko and Scheffler 2009). However, a much more challenging task is to devise a theory that would not be limited to the asymptotic regime of non-overlapping subsystems, but would treat the entire range of dispersion interactions in a seamless fashion. The nonlocal van der Waals density functionals described in the next section achieve this aim without a significant increase in computational complexity.

23.3 General and Seamless Nonlocal Van Der Waals Density Functionals

23.3.1 Functional Form

In the growing family of van der Waals density functionals (Dion et al. 2004; Lee et al. 2010; Vydrov and Van Voorhis 2009a, b; 2010c), the correlation energy is divided into two contributions,

$$E_c = E_c^0 + E_c^{nl}, \qquad (23.12)$$

with the larger fraction, E_c^0, approximated by a (semi)local functional. The quantity E_c^0 is typically chosen to give the exact UEG correlation energy in the uniform density limit. The second piece in (23.12), E_c^{nl}, is a fully nonlocal functional that includes long-range dispersion interactions. E_c^{nl} is designed to vanish for a uniform electron density, such that double-counting in (23.12) is avoided at least in the UEG limit.

Approximations to E_c^{nl}, described below, are written in the form

$$E_c^{nl} = \frac{\hbar}{2} \int d^3r \int d^3r' n(r) \Phi(r, r') n(r'). \qquad (23.13)$$

The correlation kernel Φ is symmetric, $\Phi(r, r') = \Phi(r', r)$, and depends only on $|r - r'|$ and charge densities and density gradients at r and r'. Φ is designed in such a way that in the $|r - r'| \to \infty$ limit,

$$\Phi \to -\frac{3e^4}{2m^2\omega_0(r)\omega_0(r')[\omega_0(r) + \omega_0(r')]|r - r'|^6}, \qquad (23.14)$$

so that E_c^{nl} tends to the asymptotic form of (23.5), with various choices of ω_0 summarized in Table 23.1. The extra factor of $1/2$ in (23.13) as compared to (23.5) arises because these formulas compute different things: (23.5) computes the interaction energy between subsystems A and B, hence $r \in A$ and $r' \in B$; whereas (23.13) gives the nonlocal correlation energy, which includes inter- and intramolecular contributions, hence both r and r' integrals in (23.13) are over the entire space. The E_c^{nl}

functional of (23.13) has a very general and seamless form that requires neither splitting the system into interacting fragments nor any kind of atomic partitioning.

23.3.2 vdW-DF-04 and its Variants

In the vdW-DF-04 theory (Dion et al. 2004), an expression for $\Phi(r, r')$ was obtained using a second-order perturbation expansion and a number of other approximations (see Sect. 22.10.2). The power series expansion used in the derivation of vdW-DFs was recently analyzed in detail in Lu et al. (2010). The construction of vdW-DFs hinges on a single function $q_0(r)$, which is paramount to the theory. As shown in Sect. 23.2.2, $\omega_0 \propto q_0^2$, and hence q_0 controls the long-range asymptote of Φ via (23.14). Furthermore, q_0 also defines the length scale for "damping" of Φ at shorter range, as elaborated below. In vdW-DF-04 (Dion et al. 2004), this function is defined as

$$q_0 = k_F\left(1 + \mu s^2\right) - \frac{4\pi}{3e^2}\varepsilon_c^{LDA}, \qquad (23.15)$$

where $\mu = 0.09434$ and ε_c^{LDA} is the LDA correlation energy density per electron. In the recent re-parameterization, denoted vdW-DF-10 here but called vdW-DF2 in (Lee et al. 2010), the same form for q_0 is used, but with $\mu = 0.20963$. Note that Dion et al. (2004) and Lee et al. (2010) use a notation that is somewhat different from ours, with $Z_{ab} = -9\mu$. In vdW-DF-09 (Vydrov and Van Voorhis 2009a), a simpler form for q_0 was adopted:

$$q_0 = k_F(1 + \mu s^2), \qquad (23.16)$$

where $\mu = 0.22$ was adjusted to give accurate C_6 coefficients, as discussed in Sect. 23.2.2.

It is convenient to introduce two new variables that are sufficient to represent Φ in vdW-DFs:

$$D = \frac{q_0(r) + q_0(r')}{2}\left|r - r'\right| \quad \text{and} \quad \delta = \frac{\left|q_0(r) - q_0(r')\right|}{q_0(r) + q_0(r')}.$$

The expression for Φ cannot be obtained in a closed form, but it can be computed numerically for a given pair of D and δ. In practical applications, it is most efficient to precompute Φ for a set of D and δ and compile a look-up table. The functional vdW-DF-10 uses exactly the same $\Phi(D, \delta)$ as in vdW-DF-04 [see the plot of Φ in the Erratum to Dion et al. (2004)]. In vdW-DF-09, the shape of $\Phi(D, \delta)$ is slightly different (Vydrov and Van Voorhis 2009a). In these theories, the kernel Φ is negative (attractive) at large and intermediate scaled distances D. For very large D values, Φ reduces to the asymptotic form of (23.14). For small D values, Φ is positive, i.e. repulsive. In the uniform density limit, $\delta = 0$ and Φ has the useful property that

$$4\pi \int_0^\infty D^2 \Phi(D, 0)\, \mathrm{d}D = 0,$$

making E_c^{nl} vanish. In nonuniform systems, E_c^{nl} is strictly nonnegative, meaning that the positive short-range contribution outweighs the negative long-range part.

The earliest seamless functional of the form of (23.13), vdW-DF-04, gives reasonable asymptotic C_6 coefficients, as shown in Table 23.1. However, vdW-DF-04 strongly overestimates the magnitude of dispersion attraction near equilibrium inter-monomer distances (Vydrov and Van Voorhis 2010a). By increasing the value of μ in vdW-DF-10, the magnitude of dispersion interaction was attenuated at all distances. This proved beneficial for equilibrium binding energies, but it led to severe underestimation of C_6 coefficients in vdW-DF-10. These results suggest that a more flexible form of Φ may be needed in order to accurately describe dispersion interactions for the whole range of distances.

Several features of vdW-DFs pose inconveniences for numerical implementation: as mentioned above, Φ is not expressible in a closed analytic form and has to be numerically tabulated; in addition, Φ diverges to $+\infty$ for $|\boldsymbol{r} - \boldsymbol{r}'| \to 0$. Below we describe the progress made in our group towards designing improved models for Φ, that are more flexible and easily implementable.

23.3.3 VV09 and VV10

In the VV09 model (Vydrov and Van Voorhis 2009b), the correlation kernel is written as

$$\Phi = \frac{3e^4 \mathcal{D}(\boldsymbol{r}, \boldsymbol{r}')}{2m^2 \omega_0(\boldsymbol{r})\omega_0(\boldsymbol{r}')\left[\omega_0(\boldsymbol{r}) + \omega_0(\boldsymbol{r}')\right] |\boldsymbol{r} - \boldsymbol{r}'|^6}, \tag{23.17}$$

with ω_0 given by (23.8). In the above equation, $\mathcal{D}$ serves as a "damping function". A closed-form expression for $\mathcal{D}$ was constructed in Vydrov and Van Voorhis (2009b) in such a way as to ensure that the VV09 formalism has the following desirable features

- In the $|\boldsymbol{r} - \boldsymbol{r}'| \to \infty$ limit, $\mathcal{D} \to -1$, recovering the asymptotic form of (23.14). With ω_0 of (23.8), this method gives accurate C_6 coefficients, as shown in Table 23.1.
- In the $|\boldsymbol{r} - \boldsymbol{r}'| \to 0$ limit, Φ goes to a finite value.
- E_c^{nl} vanishes in the uniform density limit.
- E_c^{nl} has a realistic second-order gradient expansion in the slowly varying density limit.

We do not give the detailed expression for $\mathcal{D}(\boldsymbol{r}, \boldsymbol{r}')$ here, because VV09 has been superseded by a simpler yet more accurate model—VV10, described below.

The main idea in the design of the VV10 model (Vydrov and Van Voorhis 2010c) was to choose an analytic functional form that is as simple and intuitive as possible. VV10 has the same long-range behavior as its precursor VV09, but the damping mechanism of dispersion interactions at short range is greatly simplified. This simplification not only makes the VV10 model more efficient and computationally tractable, but it also leads to improved performance.

In VV10 (Vydrov and Van Voorhis 2010c), the nonlocal correlation kernel in written as

$$\Phi = -\frac{3e^4}{2m^2 g g'(g + g')} \tag{23.18}$$

with

$$g = \omega_0(\mathbf{r}) \left|\mathbf{r} - \mathbf{r}'\right|^2 + \kappa(\mathbf{r}) \tag{23.19}$$

and similarly

$$g' = \omega_0(\mathbf{r}') \left|\mathbf{r} - \mathbf{r}'\right|^2 + \kappa(\mathbf{r}'), \tag{23.20}$$

where ω_0 is defined in the same way as in VV09—see (23.8) and Table 23.1. In Eqs. 23.19 and 23.20, a new quantity was introduced:

$$\kappa(\mathbf{r}) = b\frac{v_F^2(\mathbf{r})}{\omega_p(\mathbf{r})} = 3b\frac{\omega_p(\mathbf{r})}{k_s^2(\mathbf{r})}, \tag{23.21}$$

where $v_F = \hbar k_F/m$ is the local Fermi velocity, $k_s = \sqrt{3}\omega_p/v_F$ is the local Thomas–Fermi screening wave vector, and b is an adjustable parameter that controls the short-range damping of the asymptotic form of (23.14).

In the uniform density limit, (23.18) reduces to

$$\Phi^{uni} = -\frac{3e^4}{4m^2}\left[\frac{\omega_p}{\sqrt{3}}r^2 + b\frac{v_F^2}{\omega_p}\right]^{-3} \tag{23.22}$$

(with $r = |\mathbf{r} - \mathbf{r}'|$) and (23.13) gives the following energy density per electron:

$$\varepsilon_c^{uni} = 2\pi\hbar n\int_0^\infty r^2\Phi^{uni}\mathrm{d}r = -\frac{3\pi^2\hbar e^4 n}{32m^2 v_F^3}\left[\frac{3}{b^2}\right]^{3/4}$$

$$= -\frac{e^2}{32a_0}\left[\frac{3}{b^2}\right]^{3/4} = -\beta, \tag{23.23}$$

where $a_0 = \hbar^2/me^2$ is the Bohr radius and β is a density-independent constant. It is instructive to rewrite (23.22) in a different form:

$$\Phi^{\text{uni}} = -\frac{9\sqrt{3}e^4}{4m^2\omega_{\text{p}}^3 r^6} \left[1 + \frac{3\sqrt{3}b}{(k_s r)^2} \right]^{-3}. \tag{23.24}$$

The above equation shows that the r^{-6} asymptote is damped at short range on the length scale given by $k_s r$, which is the scaled distance relevant for correlation energy (Perdew and Wang 1992b).

The definition of the VV10 density functional is finalized by writing

$$E_c^{\text{VV}10} = E_c^{\text{nl}} + \beta N = \int d^3 r\, n(r) \left[\beta + \frac{\hbar}{2} \int d^3 r'\, n(r') \Phi(r, r') \right], \tag{23.25}$$

where β is determined by (23.23). By construction, $E_c^{\text{VV}10}$ of (23.25) vanishes in the uniform density limit. An essential aspect of the VV10 formalism is the additional flexibility introduced with the help of an adjustable parameter b which controls the short-range behavior of the nonlocal correlation energy. When $E_c^{\text{VV}10}$ of (23.25) is added as a correction to an existing xc functional, b is adjusted to attain a balanced merging of interaction energy contributions at short and intermediate ranges (Vydrov and Van Voorhis 2010c).

23.3.4 Implementation

One of the most attractive features of the orbital-independent functionals of the type of (23.13) is that they enable self-consistent treatment of dispersion interactions (Thonhauser et al. 2007; Gulans et al. 2009; Román-Pérez 2009; Vydrov et al. 2008; Vydrov and Van Voorhis 2010a). By contrast, practical applications of RPA methods or force-field-like dispersion corrections are typically performed in a post-self-consistent fashion, using densities and orbitals produced by a semilocal or a hybrid functional. Self-consistency may not be important for computing binding energy curves of weakly interacting complexes (Thonhauser et al. 2007), but for a number of useful applications, such as gradient-based geometry optimizations, a self-consistent implementation is required.

In the first self-consistent implementation of vdW-DF-04 (Thonhauser et al. 2007), the nonlocal correlation potential $v_c^{\text{nl}}(r) = \delta E_c^{\text{nl}}/\delta n(r)$ was evaluated on a numerical real-space grid. Soon thereafter, we reported an implementation of vdW-DF-04 within a Gaussian basis set code (Vydrov et al. 2008). In our methodology, the electron density is expressed in terms of atom-centered basis functions $\{\chi_\mu\}$ as

$$n(r) = \sum_{\mu\nu} P_{\mu\nu} \chi_\mu(r) \chi_\nu(r), \tag{23.26}$$

where $P_{\mu\nu}$ are the density matrix elements. The self-consistent treatment is simplified by the fact that we do not need to compute $v_c^{\text{nl}}(r)$ explicitly, but only need its matrix elements—the derivatives of E_c^{nl} with respect to $P_{\mu\nu}$:

$$\frac{dE_c^{nl}}{dP_{\mu\nu}} = \int d^3r \, \chi_\mu(r) \frac{\delta E_c^{nl}}{\delta n(r)} \chi_\nu(r) = \left\langle \mu | v_c^{nl} | \nu \right\rangle. \tag{23.27}$$

These matrix elements can be straightforwardly computed and used in the standard formalism developed for semilocal functionals (Johnson et al. 1993). We utilized the same approach in our self-consistent implementations of VV09 and VV10 (Vydrov and Van Voorhis 2010a, c).

The double integration over the space variables in (23.13) is in practice evaluated as a double sum over a numerical grid. The nonlocal correlation kernels in vdW-DFs diverge logarithmically for $|r - r'| \to 0$. This divergence causes difficulties in practical implementations. Omitting the singular $r = r'$ terms in the double sum leads to numerical instabilities, such as errors in the gradients with respect to nuclear displacements. It also causes substantial grid superposition errors in binding energies, if atom-centered quadrature grids are used. Implementational tricks have been devised to deal with the singularity in the vdW-DF-04 kernel (Román-Pérez and Soler 2009). In VV09 and VV10, the source of the problem is eliminated altogether, since Φ is finite for $|r - r'| = 0$.

Even if implemented in the most straightforward fashion, functionals of the form of (23.13) scale as N^2, meaning that when the system size is doubled, the computational cost quadruples. This scaling is rather modest as compared to correlated wavefunction methods, for which the typical scaling is N^5 or higher. It has been recently shown that with specially tailored numerical algorithms (Gulans et al. 2009; Román-Pérez and Soler 2009) the scaling of the vdW-DF-04 nonlocal functional can be further reduced below N^2. Unlike correlated wavefunction techniques that require high-quality basis sets, nonlocal functionals of the type of (23.13) are rather undemanding in terms of the basis set size. The correlation kernel in VV10 has a particularly featureless and smooth analytic form with finite $\Phi(r, r)$. Consequently, VV10 is quite insensitive to the fineness of the numerical grid, so that even rather coarse grids can be used for evaluating E_c^{VV10}.

Self-consistent treatment of the energy functional is a prerequisite for computing forces on nuclei. Within an atom-centered basis set implementation, the gradients of the energy with respect to nuclear displacements contain not only the usual Hellmann–Feynman terms, but also the so-called "Pulay forces" arising due to the fact that the atom-centered basis functions as well as numerical quadrature grid points move together with the nuclei. All these terms can be straightforwardly computed (Vydrov and Van Voorhis 2008a; Vydrov et al. 2008). Since the energy gradients for the nonlocal van der Waals functionals are readily available, structural optimizations can be performed routinely and efficiently.

23.4 Dispersionless Correlation and Exchange Components

The term E_c^{nl} of (23.13) accounts for a rather small part of correlation energy. In practice, E_c^{nl} is added as a correction to an existing xc functional. The E_c^0 component in (23.12) is typically represented by the LDA correlation functional in the parameterization of Perdew and Wang (1992a) (used with vdW-DF-04, vdW-DF-10, and VV09), or by a semilocal correlation functional, such as the generalized gradient approximation of Perdew, Burke, and Ernzerhof (PBE) (Perdew et al. 1996b) used in VV10.

A very nontrivial problem is the proper choice of the exchange model to be used alongside the correlation functional of (23.12). For molecular complexes bound predominantly by van der Waals interactions, Hartree–Fock theory provides adequate representation of the repulsive wall ('Pauli repulsion'). However, Hartree–Fock exchange is unsuitable for pairing with LDA or GGA correlation functionals (used for E_c^0)—such combinations are known to perform poorly for many properties.

The so-called "long-range correction" (LRC) scheme preserves the Hartree–Fock's proper treatment of Pauli repulsion in van der Waals complexes, but also describes covalent bonds well. In the LRC method, the Coulomb operator $1/r$ is split into the long-range part $\mathrm{erf}(\omega r)/r$, treated by Hartree–Fock, and the short-range counterpart $\mathrm{erfc}(\omega r)/r$, treated by a semilocal exchange functional. The LRC method is quite versatile: with the proper adjustment of the range-separation parameter ω, LRC exchange can be successfully used not only with LDA or GGA correlation [see e.g. Vydrov and Scuseria (2006)], but also with fully nonlocal correlation functionals. LRC exchange has been incorporated into a number of DFT methodologies that aim at accurate description of van der Waals complexes (Kamiya et al. 2002; Sato et al. 2007; Sato and Nakai 2009; Janesko et al. 2009; Zhu et al. 2010).

A more computationally affordable option is to employ a semilocal exchange functional of GGA type. Unfortunately, most GGA approximations cannot properly describe the Pauli repulsion in weakly interacting systems. Many GGA exchange functionals even yield substantial spurious attraction in van der Waals complexes at short range. When dispersion interactions are treated by a nonlocal correlation functional E_c^{nl}, double-counting should be avoided, hence the exchange functional must not give any "pseudo-dispersion" binding. Comparative assessments of various semilocal exchange approximations (Lacks and Gordon 1993; Kannemann and Becke 2009; Murray et al. 2009) showed that the PW86 exchange functional (Perdew and Wang 1986) excels in the treatment of repulsive components of van der Waals potentials. A refitted version of PW86 was recently proposed (Murray et al. 2009). We denote this 'refitted PW86' as PW86R here.

vdW-DF-04 is known to be incompatible with accurate exchange functionals: combinations of vdW-DF-04 with Hartree–Fock, LRC, or PW86R exchange produce considerable overbinding of van der Waals complexes (Vydrov et al. 2008; Vydrov and Van Voorhis 2010a). More reasonable binding energies are obtained if vdW-DF-04 is paired with revPBE exchange (Zhang and Yang 1998), but this combination tends to give too long equilibrium intermonomer distances (Langreth et al. 2009). As shown in Klimeš et al. (2010), Cooper (2010), the performance can be substantially improved by tailoring an exchange functional specially fitted to be used alongside vdW-DF-04. The re-parameterized functional vdW-DF-10 works rather well with PW86R exchange (Lee et al. 2010). The VV10 methodology can be adapted for a particular exchange functional by fine-tuning the parameter b. For

Table 23.2 Performance of several xc functionals for the binding energies of the S22 test set

E_x	E_c^0	E_c^{nl}	ME	MAE	MAPE
revPBE	LDA	vdW-DF-04	−1.40	1.43	21.0
PW86R	LDA	vdW-DF-04	1.04	1.04	31.9
PW86R	LDA	VV09	−1.17	1.19	19.9
PW86R	LDA	vdW-DF-10	−0.92	0.94	14.7
PW86R	PBE	VV10[a]	0.16	0.31	4.4
LC-ωPBE[b]	PBE	VV10[c]	0.09	0.21	4.6

ME denotes mean signed error in kcal/mol, MAE—mean absolute error in kcal/mol, and MAPE—mean absolute percentage error in %. Fixed molecular geometries from Jurečka et al. (2006) are used. Reference binding energies are from Podeszwa et al. (2010). Computational details can be found in Vydrov and Van Voorhis (2010a, 2010c)
[a] Using $C = 0.0093$ and $b = 5.9$. The βN term is included
[b] Long-range-corrected PBE with $\omega = 0.45\ a_0^{-1}$
[c] Using $C = 0.0089$ and $b = 6.3$. The βN term is included

instance, using $b = 5.9$ renders VV10 correlation compatible with PW86R exchange (Vydrov and Van Voorhis 2010c).

23.5 Benchmark Tests on Binding Energies

To give an idea about the accuracy of nonlocal van der Waals functionals for equilibrium binding energies, in Table 23.2 we summarize the error statistics for several xc approximations that include an E_c^{nl} model in combination with some typical exchange (E_x) and semilocal correlation (E_c^0) functionals. Table 23.2 reports the errors for the S22 benchmark test set (Jurečka et al. 2006; Podeszwa et al. 2010) that includes 22 molecular duplexes representing noncovalent interactions typical in biological molecules, including hydrogen-bonded, dispersion-dominated, and mixed duplexes. We define binding energies to be positive, hence a negative mean error (ME) in Table 23.2 indicates an underbinding trend, while a positive ME means overbinding. Table 23.2 clearly shows that the nonlocal functionals of the new generation, vdW-DF-10 and VV10, are substantially more accurate than older models. Note that the results reported in (Lee et al. 2010) for the same S22 set are somewhat different from ours, because intermonomer separations were optimized in Lee et al. (2010), whereas the results in Table 23.2 were obtained at fixed geometries from Jurečka et al. (2006).

Applications of vdW-DF-04 to a broad range of molecular complexes and materials were recently reviewed in Langreth et al. (2009). Binding energy curves for a number of weakly-interacting molecular complexes, computed with the newest nonlocal functionals vdW-DF-10 and VV10 can be found in Lee et al. (2010), Vydrov and Van Voorhis (2010c).

23.6 Known Limitations and Avenues for Improvement

The nonlocal correlation formalism described in Sect. 23.3 has a number of limitations and shortcomings. Further improvements and refinements of the methodology are certainly possible and desirable.

E_c^{nl} of the form of (23.13) accounts for two-body contributions to the dispersion energy, but neglects nonadditive many-body contributions, such as the Axilrod-Teller-type three-body terms. These neglected contributions can be significant in some cases, as argued in Lu et al. (2010). It is possible to derive an expression for the Axilrod-Teller three-body correction within the local polarizability approximation. However, such a term would scale as N^3 and would significantly increase computational cost.

For non-overlapping fragments, E_c^{nl} of (23.13) reduces to the second-order energy $E^{(2)}$ of (23.5). Such a theory is inherently inadequate for describing interactions between spatially extended metallic (or semimetallic) fragments of reduced dimensionality at very large distances, e.g. between parallel metallic sheets or wires (Dobson et al. 2006) (see Sect. 22.5).

The accuracy of the formalism at short and intermediate range depends not only on the quality of the E_c^{nl} model, but also on the choice of E_x and E_c^0 functionals. The question of the proper choice of the exchange component has been actively discussed (Murray et al. 2009; Klimeš et al. 2010; Cooper 2010). At the same time, very little attention has been devoted to the choice of correlation functional used for E_c^0 in (23.12). The terms E_c^{LDA} and E_c^{PBE} contribute sizably and rather unpredictably to binding energies of van der Waals complexes near equilibrium intermonomer separations. At present, it is not clear whether (and to what degree) these contributions are valid or spurious.

Finally, (23.12) gives one particular prescription for splitting the correlation energy into two parts. Other ways of dividing the correlation functional into (semi)local and nonlocal contributions can be envisioned. Alternatively, the whole of correlation energy may be treated by a nonlocal functional.

Notwithstanding their imperfections, modern nonlocal van der Waals density functionals are already sufficiently accurate to provide useful predictions for a broad variety of weakly interacting systems (Langreth et al. 2009; Lee et al. 2010; Vydrov and Van Voorhis 2010c).

Chapter 24
Time-Dependent Current Density Functional Theory

Giovanni Vignale

24.1 Introduction

The nonlocality of the exchange-correlation (xc) potential, i.e., the fact that the xc potential at a certain position depends on the global distribution of the particles in space, is the curse of density functional theory. It is mainly because of this fact that, even after years of intensive studies, the exact form of the xc potential as a functional of the density remains unknown. Nevertheless, it is true that many accurate and useful results can be obtained from the use of an approximation—the local density approximation (LDA)—which ignores the problem altogether. Apparently, the nonlocal dependence of the Kohn–Sham orbitals on the density is sufficient in many cases to give about the right quantum chemistry. Furthermore, a number of successful strategies have been designed to go beyond the LDA when needed: in one such approach (the generalized gradient approximation—GGA) one goes beyond the LDA by including the dependence of the xc potential on the *gradient* of the local density; in another approach, one expresses the xc potential as a functional of the Kohn–Sham orbitals, and, finally, in the "meta-GGA" approach one fights the problem by including additional local variables, such as the kinetic energy density.

In this chapter we are going to see that the nonlocality problem affects in a more severe form the time-dependent density functional theory. This complication arises as a consequence of *memory*. The xc potential at a time t (now) depends on the density at earlier times t'. But at these earlier times a small volume element of the system, which is now located at r, was located at a different position r'. Retardation in time thus implies nonlocality in space. We will see that, when retardation is taken into account, the local density approximation breaks down *even in the limit of slowly varying density*. We will refer to this feature of the time-dependent theory as *ultranonlocality*,

G. Vignale (✉)
Department of Physics, University of Missouri-Columbia,
Columbia, MO 65211, USA
e-mail: vignaleg@missouri.edu

M. A. L. Marques et al. (eds.), *Fundamentals of Time-Dependent Density Functional Theory*, Lecture Notes in Physics 837, DOI: 10.1007/978-3-642-23518-4_24,
© Springer-Verlag Berlin Heidelberg 2012

to distinguish it from the ordinary nonlocality, which becomes harmless in the limit of slowly varying density. Furthermore, we will see that "ultranonlocality" is not related to the presence of a long-range interaction between the particles, but, more in depth, implies that the particle density is not a well-chosen variable (although, in principle a legitimate one) for the description of effects that involve retardation in time. It is also evident that the inclusion of additional variables might "cure" the ultranonlocality: for example, by looking at the current density of an infinitesimal volume element of the fluid at position r at a certain time we might be able to estimate its position at an earlier time, and thus arrive at a local or quasi-local expression for the retarded xc potential. These general ideas will be explored in some detail in the following sections. We will see that the the introduction of the current density as a basic variable does indeed cure the ultranonlocality in the linear response regime. However, in the general nonlinear case, a consistent cure of the problem requires the introduction of the *deformation tensor*, which is non-locally related to the current density. The ensuing "deformation functional theory" will be the topic of the next chapter.

24.2 First Hints of Ultranonlocality: the Harmonic Potential Theorem

Historically the first hint of ultranonlocality in TDDFT came from the work of John Dobson (1994a) on the collective dynamics of electrons in parabolic quantum wells. Under the action of a uniform time-dependent electric field the electronic density in the quantum well oscillates back and forth without changing its shape, i.e., one has $n(r, t) = n_{GS}(r - r(t))$ where $n_{GS}(r)$ is the ground-state density and $r(t)$ is the position of the center of mass of the electrons. The latter moves exactly as a single classical particle of mass m and charge $-e$ under the action of the external electric field: this is the content of the "harmonic potential theorem" (HPT). It is easy to see that the exact TDDFT satisfies the HPT for, according to the exact condition (5.40), the xc potential created by the oscillating density $n_{GS}(r - r(t))$ is given by $v_{xc}(r, t) = v_{xc\,GS}(r - r(t))$, where $v_{xc\,GS}(r)$ is the xc potential in the ground-state. In an accelerated frame of reference that moves together with the center of mass of the system, the external electric field is cancelled by the inertial force, while the xc potential has exactly the form that is needed to preserve the ground-state density distribution.

Dobson observed that a naive application of the local density approximation, including a local but retarded xc potential (Gross and Kohn 1985), leads to results that are in conflict with the HPT. For example, one finds a density-dependent shift in the frequency of the oscillatory motion of the center of mass, and this motion becomes "damped". The reason for this difficulty is that the local density approximation is unable to distinguish between a situation in which the density variation is due to local compression/rarefaction of the electron liquid (as in the case of a long-wavelength plasmon) and the present one in which this variation is due to a global translation of a system, without compression or rarefaction. The "obvious" choice of (Gross and Kohn 1985) amounts to choosing the first option in both cases: this introduces spurious dissipation in a situation in which there is no dissipation whatsoever.

From a mathematical point of view the link between ultranonlocality and the HPT can be seen as follows (Vignale and Kohn 1998). First of all, we have just seen that the satisfaction of the HPT in exact TDDFT is a direct consequence of the identity (5.40) (Vignale 1995a). That identity is intimately related to the zero-force theorem, Eq. 5.42 which in turn implies that the exact xc kernel of any system must satisfy the equation

$$\int d^3r' f_{xc}(\mathbf{r}, \mathbf{r}', \omega)\nabla n_{GS}(\mathbf{r}') = \nabla v_{xc\,GS}(\mathbf{r}) \qquad (24.1)$$

where $n_{GS}(\mathbf{r})$ and $v_{xc\,GS}(\mathbf{r})$ are the density and the xc potential in the ground-state. Notice that the quantity on the right hand side of this equation is frequency-independent, implying that the integral over $\mathbf{r}'$ on the left hand side must somehow "wash out" the frequency dependence of the integrand. Now assume that f_{xc} has a finite range in the sense that the integral

$$\int d^3r' f_{xc}(\mathbf{r}, \mathbf{r}', \omega) \qquad (24.2)$$

is finite. Indeed, this condition is satisfied by the xc kernel of a strictly homogeneous electron liquid, since it is known that the Fourier transform $f_{xc}(\mathbf{k}, \omega)$ of the homogeneous xc kernel has finite limit for $\mathbf{k} \to 0$. Suppose now that $n_{GS}(\mathbf{r})$ is very slowly varying on the scale of the range of $f_{xc}(\mathbf{r}, \mathbf{r}', \omega)$. Then we can pull $\nabla n_{GS}(\mathbf{r}')$ out of the integral of Eq. 24.1 and get

$$\nabla n_{GS}(\mathbf{r}) \int d^3r' f_{xc}(\mathbf{r}, \mathbf{r}', \omega) = \nabla v_{xc\,GS}(\mathbf{r}). \qquad (24.3)$$

In the limit that the density approaches uniformity, the integral on the left hand of this expression ought to converge (if it converges at all) to the $\mathbf{k} \to 0$ limit of the homogeneous electron gas kernel $f_{xc}(\mathbf{k}, \omega)$, which, as we have just stated, is a function of frequency. Since the right hand side of the expression is still frequency-independent we have arrived at a contradiction, which proves the fallacy of the initial assumption, namely, the existence of the integral (24.2) in a weakly inhomogeneous system. Indeed, the divergence of the integral (24.2) is the mathematical signature of the ultranonlocality problem. Notice that, unlike ordinary nonlocality, this problem is present in systems that are arbitrarily close to a homogeneous electron liquid.

24.3 TDDFT and Hydrodynamics

In hindsight one can easily understand why the description of many-body forces as gradients of a density-dependent potential becomes inadequate as soon as one attempts to go beyond the adiabatic approximation. In the classical theory of fluid motion, hydrodynamics, the adiabatic approximation amounts to assuming that the fluid is, at every point in space, in a state of quasi equilibrium characterized by the instantaneous values of density, velocity, and temperature. Then the local conservation laws of particle number and momentum lead to the equations

$$\frac{\partial n(\mathbf{r}, t)}{\partial t} = -\nabla \cdot \mathbf{j}(\mathbf{r}, t), \qquad \text{continuity equation} \qquad (24.4a)$$

$$m\left(\frac{\partial}{\partial t} + \mathbf{v} \cdot \nabla\right) \mathbf{j}(\mathbf{r}, t) = \mathbf{F}(\mathbf{r}, t) - \nabla p(\mathbf{r}, t), \qquad \text{Euler's equation} \qquad (24.4b)$$

where $\mathbf{F}(\mathbf{r}, t)$ is the external volume force density, $p(\mathbf{r}, t)$ is the pressure, related to the local density and temperature by the equation of state, and $\mathbf{v}(\mathbf{r}, t)$—the velocity field—is defined as the ratio of the current density to the particle density, i.e., $\mathbf{v}(\mathbf{r}, t) \equiv \frac{\mathbf{j}(\mathbf{r}, t)}{n(\mathbf{r}, t)}$. It should be noted that already at this level the equations of motion for the density and the velocity are coupled. However, the force term $-\nabla p(\mathbf{r}, t)$ can be expressed as an instantaneous function of the density: from this point of view the theory is analogous to TDDFT in the adiabatic local density approximation. Unfortunately, Euler's equation suffers from a major limitation, namely the viscosity of the fluid is not taken into account. To include viscosity, one must go beyond the adiabatic approximation. To do this, one must recognize that the local state of the fluid deviates from quasi-equilibrium by an amount that is proportional to the gradients of the velocity fields. When this deviation is taken into account, Euler's equation acquires an extra term and becomes the classical Navier-Stokes equation (Landau and Lifshitz 1987)

$$m\left(\frac{\partial}{\partial t} + \mathbf{v} \cdot \nabla\right) \mathbf{j}(\mathbf{r}, t) = \mathbf{F}(\mathbf{r}, t) - \nabla p(\mathbf{r}, t) + \nabla \cdot \overset{\leftrightarrow}{\sigma}'(\mathbf{r}, t). \qquad (24.5)$$

The viscous stress tensor σ'_{ij} is given by

$$\sigma'_{ij}(\mathbf{r}, t) = \eta\left(\frac{\partial v_i}{\partial r_j} + \frac{\partial v_j}{\partial r_i} - \frac{2}{3}\nabla \cdot \mathbf{v}\delta_{ij}\right) + \zeta\nabla \cdot \mathbf{v}\delta_{ij}, \qquad (24.6)$$

where the coefficients η and ζ are, respectively, the shear viscosity and the bulk viscosity of the fluid. Its divergence,

$$[\nabla \cdot \overset{\leftrightarrow}{\sigma}']_i = \sum_j \frac{\partial \sigma'_{ij}}{\partial r_j} = \eta\nabla^2 v_i + \left(\zeta - \frac{2}{3}\eta\right)\nabla_i(\nabla \cdot \mathbf{v}) \qquad (24.7)$$

is the viscous force exerted on the volume element by the surrounding medium. *This force cannot be expressed as a local functional of the density.* One might think to express the current density (and hence the velocity) in terms of the density by inverting the continuity equation (24.4a). But this cannot be done, because Eq. 24.4a determines only the longitudinal component of the current density, $\mathbf{j}_L$, as opposed

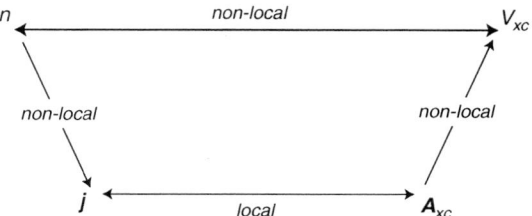

Fig. 24.1 Diagram showing the relation between current density functional theory and density functional theory. The ultranonlocal relation between n and v_{xc} is transformed into a local relation between j and A_{xc} by means of two non-local transformations from n to j and from v_{xc} to A_{xc}

to the full current density. Furthermore, $j_L(r)$ is a highly nonlocal functional of the density, as can be seen from the explicit solution of Eq. 24.4a:

$$j_L(r, t) = \int d^3 r' \frac{\partial n(r', t)}{\partial t} \nabla_r \frac{1}{4\pi |r - r'|}. \tag{24.8}$$

This is in a sense the crux of the ultranolocality problem in TDDFT and the reason why, just as in classical hydrodynamics, it is better to describe exchange-correlation effects in terms of the current density, rather than in terms of the density. The basic idea is that the ultra-nonlocal relation between density and exchange-correlation potential can be replaced, in the linear response regime, by a local relation between current density and exchange-correlation vector potential, as shown schematically in Fig. 24.1. The analogy with hydrodynamics also shows clearly why the current-density description is unavoidable if one wants to go beyond the adiabatic approximation by including dissipative effects, such as the viscosity of the electron fluid while retaining a spatially-local dependence on the density. Viscous forces are naturally expressed in terms of the gradients of the velocity field, and do not have a local expression in terms of the density.

The path in front of us is now clear. In the next three sections we will develop a time-dependent current-density functional theory (TDCDFT) in which the basic variable is the current density and the ordinary xc potential is replaced by an xc vector potential. We will see that in this theory the exchange-correlation force density has the form of the contact force density of hydrodynamics, i.e., it is the divergence of a stress tensor, and that this stress tensor can be safely approximated as a local functional of the current density.

This theory will enable us to calculate not only the density but also the current density from the single-particle orbitals of an effective Kohn–Sham theory. In order to accomplish this we will have to generalize Eq. 24.6 by endowing the viscosity constants η and ζ with both real and imaginary parts (the latter representing the dynamical bulk and shear moduli of the liquid) and making them functions of the frequency as well as the local particle density. In the next chapter it will be shown how to bypass the Kohn–Sham equation, by constructing a closed equation of motion for the current density in terms of a stress tensor, whose functional dependence on the current is, in principle, obtainable from the solution of a universal many-body problem.

Our discussion will be mostly restricted to the linear response regime. By this we mean that the time dependent density has the form

$$n(\mathbf{r}, t) = n_{\text{GS}}(\mathbf{r}) + n_1(\mathbf{r})e^{-i\omega t} + \text{c.c.} \tag{24.9}$$

where $n_{\text{GS}}(\mathbf{r})$ is the ground-state (equilibrium) density, and $n_1(\mathbf{r}) \ll n_{\text{GS}}(\mathbf{r})$. It will also be assumed that the frequency is high in the sense that $\omega \gg qv_F$ and $\omega \gg kv_F$ where q^{-1} is the characteristic length scale for density variations in the ground-state, $\mathbf{k}$ is the wave vector of the external field, and v_F is the local Fermi velocity.

24.4 Current Density Functional Theory

In TDCDFT we consider a broader class of Hamiltonians than those considered in the original Runge–Gross formulation, namely Hamiltonians of the form

$$H = \sum_i \left\{ \frac{1}{2m} \left[\left(\mathbf{p}_i + \frac{e}{c} \mathbf{A}_i \right)^2 \right] + v_{\text{ext},i} \right\} + v_{ee} \tag{24.10}$$

where $\mathbf{A}_i$ is the external vector potential evaluated at the position $\mathbf{r}_i$ of the ith particle, $v_{\text{ext},i}$ is the scalar potential at the position of the ith particle, and v_{ee} represents the electron-electron interaction. The reason why this is a proper generalization of the Runge–Gross (RG) Hamiltonian is that every scalar potential $v_{\text{ext}}(\mathbf{r})$ can be represented as a *longitudinal* vector potential $\mathbf{A}_v(\mathbf{r})$ by choosing the latter as the solution of the equation

$$\frac{e}{c} \frac{\partial \mathbf{A}_v(\mathbf{r}, t)}{\partial t} = \nabla v_{\text{ext}}(\mathbf{r}, t). \tag{24.11}$$

Of course, *transverse* vector potential represents different physics (magnetic fields).

It can be easily proved that for Hamiltonians of the form (24.10) the time-dependent current density, together with the initial state, uniquely determine the scalar and the vector potential, up to a gauge transformation that leaves the initial state unchanged. A first proof of this generalized RG theorem was provided by Ghosh and Dhara (1988), and later I found a simpler proof (Vignale 2004). Therefore, following the usual arguments, one hopes to be able to construct, uniquely, a Kohn–Sham Hamiltonian, $\hat{H}_{\text{KS}}$, that produces the correct current of the many-body system (see Sect. 4.4.4). This Hamiltonian will have the form

$$H_{\text{KS}} = \sum_i \left\{ \frac{1}{2m} \left[\left(\mathbf{p}_i + \frac{e}{c} \mathbf{A}_{\text{KS},i} \right)^2 \right] + v_{\text{KS},i} \right\} \tag{24.12}$$

and notice that the effective vector potential $\mathbf{A}_{\text{KS}}$ will have in general longitudinal and transverse components even though the original external vector potential $\mathbf{A}$

was purely longitudinal. This equation (unlike the Kohn–Sham equation of ordinary TDDFT) determines in principle the whole current — not just the longitudinal component of it. The particle density is, of course, an immediate by-product of the longitudinal current.

So far goes the formalism. Now in order to find a concrete expression for $A_{KS} = A + A_{xc}$ we resort to linear response theory; namely we assume that we are close enough to equilibrium that A_{xc} can be approximated as a linear functional of the current, with coefficients that depend on the equilibrium density. In other words we assume that A_{xc} has the form

$$A_{xc}(r, \omega) = \int d^3 r' \overleftrightarrow{f}_{xc}(r, r', \omega) \cdot j(r', \omega) \qquad (24.13)$$

where the tensor kernel $\overleftrightarrow{f}_{xc}$ is a generalization of the scalar xc kernel of TDDFT. We will discuss its structure in the next section. It must be borne in mind, however, that after doing the linear response approximation on A_{xc}, we lose control on the terms proportional to A_{xc}^2, which arise from the expansion of the kinetic energy operator in the Kohn–Sham equation.

24.5 The xc Vector Potential for the Homogeneous Electron Liquid

Let us first consider the tensor exchange-correlation kernel $\overleftrightarrow{f}_{xc}(r, r', \omega)$ in a homogeneous electron liquid of density n. Translational invariance makes $\overleftrightarrow{f}_{xc}(r, r', \omega)$ a function of $r - r'$ and we will therefore focus on its Fourier transform

$$\overleftrightarrow{f}_{xc}(k, \omega) = \int d^3 r \, \overleftrightarrow{f}_{xc}(r, \omega) e^{ik \cdot r}. \qquad (24.14)$$

Furthermore, we make use of rotational invariance to express the full kernel in terms of just two independent scalar functions of $k = |k|$—the longitudinal component $f_{xc\,L}(k, \omega)$ and the transverse component $f_{xc\,T}(k, \omega)$—in the following manner:

$$[\overleftrightarrow{f}_{xc}(k, \omega)]_{ij} = \left[f_{xc\,L}(k, \omega) \hat{k}_i \hat{k}_j + f_{xc\,T}(k, \omega)(\delta_{ij} - \hat{k}_i \hat{k}_j) \right] \frac{ck^2}{e\omega^2}, \qquad (24.15)$$

where $\hat{k}$ is the unit vector in the direction of k. The factor $ck^2/e\omega^2$ is introduced here as a matter of convenience, in order to make $f_{xc\,L}(k, \omega)$ coincide with the xc kernel of the ordinary density functional theory.

Given the kernels $f_{xc\,L}$ and $f_{xc\,T}$ it is easy to construct the linear response of the homogeneous electron liquid to an external vector potential $A(k, \omega)$. As discussed in the previous section, this is just the response of the non-interacting electron gas (at the same density n) to the effective field $A(k, \omega) + A_{xc}(k, \omega)$. (Once again, the

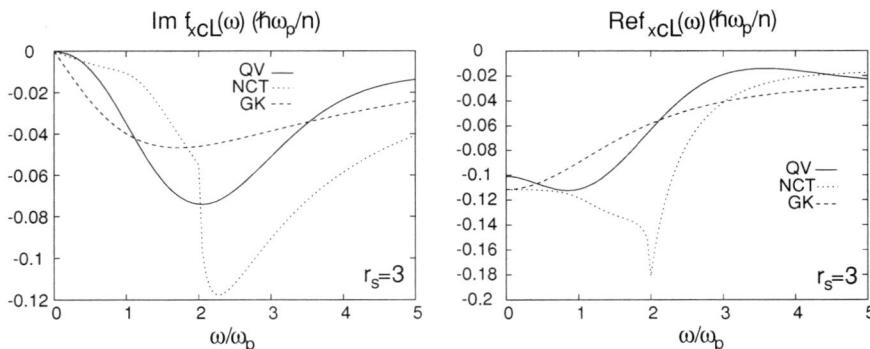

Fig. 24.2 The imaginary and the real parts of $f_{xc\,L}(\omega)$ (in units of $\hbar\omega_p/n$) in a homogeneous electron liquid at $r_s = 3$. The *short-dashed line* (NCT) is the result of a mode-coupling calculation (Conti et al. 1997). The *long-dashed line* (GK) (Gross and Kohn 1985) and the *solid line* (QV) (Qian and Vignale 2002) are interpolation formulas based on exact limiting forms

"external" field A is assumed to include the Hartree potential $A_H(k, \omega) = 4\pi e^2 \hat{k}/\omega^2$ in three dimensions.)

The connection between the xc kernels and the linear response functions of the electron liquid is the basis of the microscopic calculation of $f_{xc\,L}$ and $f_{xc\,T}$ (Conti et al. 1997; Qian and Vignale 2002, 2003). These calculations are slightly too technical to be described here, but the following features should be noted:

(i) Both $f_{xc\,L}(k, \omega)$ and $f_{xc\,T}(k, \omega)$ tend to finite limits, denoted by $f_{xc\,L}(\omega)$ and $f_{xc\,T}(\omega)$, when $k \to 0$ at finite ω (this is a consequence of translational invariance, as it implies that the electron liquid accelerates as a whole in response to a uniform electric field)

(ii) The $k = 0$ kernels $f_{xc\,L}(\omega)$ and $f_{xc\,T}(\omega)$ have both real and imaginary parts. The real parts have finite limiting values at $\omega = 0$ and $\omega = \infty$ and may have either sign; the imaginary parts are negative at all frequencies and tend to zero linearly for $\omega \to 0$ and as $\omega^{-d/2}$, where d is the dimension, for $\omega \to \infty$: the coefficients of these asymptotic behaviors are known analytically.

Representative plots of the longitudinal kernel $f_{xc\,L}(\omega)$ and of the transverse kernel $f_{xc\,T}(\omega)$ vs ω are shown in Figs. 24.2 and 24.3.

Let us now return to the full xc vector potential. Combining the longitudinal and transverse components, and making use of the existence of the $k \to 0$ limit of $f_{xc\,L(T)}(k, \omega)$ we see that up to order k^2 the xc vector potential can be written as

$$\frac{e}{c}A_{xc}(k, \omega) = [f_{xc\,L}(\omega)\hat{k} \cdot j - f_{xc\,T}(\omega)\hat{k} \times (\hat{k} \times j)]\frac{k^2}{\omega^2}. \qquad (24.16)$$

From this we want to separate the adiabatic LDA contribution. Recall that in adiabatic LDA the xc potential is just the xc component of the chemical potential μ_{xc} evaluated at the intantaneous local density $n = n_{GS} + n_1 e^{i(k\cdot r - \omega t)} + \text{c.c.}$ Thus, in the linear

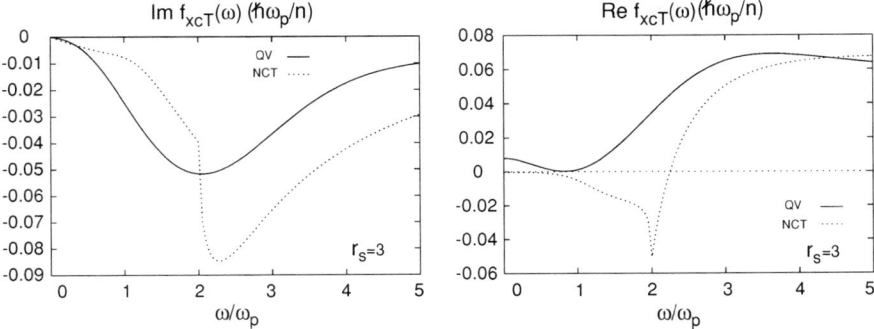

Fig. 24.3 Same as the previous figure for the imaginary and real parts of $f_{xc\,T}(\omega)$ at $r_s = 3$

response approximation we have

$$v_{xc}^{ALDA}(k, \omega) = \mu'_{xc}(n_{GS})n(k, \omega) \tag{24.17}$$

where the prime denotes a derivative with respect to n. Finally, let us use the continuity equation $n(k, \omega) = k \cdot j/\omega$ to express $n(k, \omega)$ in terms of the current and recast v_{xc}^{ALDA} as a longitudinal vector potential according to the formula

$$\frac{e}{c}A_{xc}^{ALDA}(k, \omega) = \frac{kv_{xc}^{ALDA}(k, \omega)}{\omega} \tag{24.18}$$

then we arrive at

$$\frac{e}{c}A_{xc}(k, \omega) = \frac{e}{c}A_{xc}^{ALDA}(k, \omega)$$
$$+ \left\{ [f_{xc\,L}(\omega) - \mu'_{xc}]\hat{k} \cdot j - f_{xc\,T}(\omega)\hat{k} \times (\hat{k} \times j) \right\} \frac{k^2}{\omega^2}. \tag{24.19}$$

We are now very close to the promised hydrodynamic form. All that remains to be done is to Fourier-transform the expression for A_{xc} back to real space, keeping in mind that under this transformation ik becomes the ∇ operator. It is also convenient to focus on the force exerted by the vector potential on the volume element rather than the vector potential itself: this is given by $F_{xc}(k, \omega) = -neE_{xc}(k, \omega) = -i\omega n\frac{e}{c}A_{xc}(k, \omega)$ (This is strictly speaking only the electric force. The magnetic Lorentz force is neglected, being of higher order in both j and k). Thus, after some straightforward algebra we arrive at the following expression for the force density:

$$F_{xc}(r, \omega) = F_{xc}^{ALDA}(r, \omega) - \nabla \cdot \overleftrightarrow{\sigma}_{xc}(r, t) \tag{24.20}$$

where

$$[\overleftrightarrow{\sigma}_{xc}(r, t)]_{ij} = \tilde{\eta}\left(\frac{\partial v_i}{\partial r_j} + \frac{\partial v_j}{\partial r_i} - \frac{2}{d}\nabla \cdot v\delta_{ij}\right) + \tilde{\zeta}\nabla \cdot v\delta_{ij}. \tag{24.21}$$

Here $\tilde{\eta}$, $\tilde{\zeta}$ are *generalized viscosities* which depend on density and frequency and are related to the $k \rightarrow 0$ limit of the xc kernel in the following manner:

$$\tilde{\eta} = -\frac{n^2}{i\omega} f_{xc\ T}(\omega), \tag{24.22a}$$

$$\tilde{\zeta} = -\frac{n^2}{i\omega} \left[f_{xc\ L}(\omega) - \frac{2(d-1)}{d} f_{xc\ T}(\omega) - \mu'_{xc} \right]. \tag{24.22b}$$

Notice that, at variance with the original hydrodynamic viscosities of Eq. 24.6, the generalized viscosities have both a real and an imaginary part:

$$\tilde{\eta}(\omega) = \eta(\omega) - \frac{S_{xc}(\omega)}{i\omega} \tag{24.23a}$$

$$\tilde{\zeta}(\omega) = \zeta(\omega) - \frac{B_{xc}^{dyn}(\omega)}{i\omega}, \tag{24.23b}$$

where η, ζ, S_{xc}, and B_{xc}^{dyn} are all real quantities. Clearly $\eta(\omega)$ and $\zeta(\omega)$ describe the physical viscosity of the liquid. On the other hand, $S_{xc}(\omega)$ and $B_{xc}^{dyn}(\omega)$ describe the *elasticity of the electron liquid* (Conti and Vignale 1999): they are identified as the dynamical shear modulus and the dynamical bulk modulus, respectively (notice that there is no static shear modulus in a liquid, hence the superscript "dyn" is not needed for S_{xc}). These elastic constants are absent in hydrodynamics because hydrodynamics deal with a collisional regime $\omega\tau \ll 1$ where frequent collisions between the particles bring about local equilibrium (i.e., a spherical distribution of the velocities) in a time that is short compared to the period of the oscillations. But at finite frequency this condition may not be satisfied: the Fermi surface becomes distorted (i.e., non-spherical) and there is a energy cost to pay for such a change in the form of the velocity distribution. The elastic constants are precisely the stiffnesses associated with this energy cost.

Based on the above discussion one could expect that S_{xc} and B_{xc}^{dyn} vanish in the $\omega \rightarrow 0$ limit. This expectation is borne out for the dynamical bulk modulus, but, surprising not for the shear modulus. The point is that in the regime we are considering the system remains dynamical down to the lowest frequency - the inverse relaxation time vanishing as T^2 when the temperature, T, tends to zero. Indeed, since the real part of $f_{xc\ T}(\omega)$ tends to a finite limit for $\omega \rightarrow 0$, while the imaginary part of this quantity tends to zero in the same limit, it turns out that the shear modulus is the dominant contribution to the xc field in the $\omega \rightarrow 0$ limit. It is precisely this term that makes all the difference between current-DFT and ordinary adiabatic LDA in the calculation of the polarizability of long polymer chains. In any case the lesson to be learned is that the $\omega \rightarrow 0$ limit of the time-dependent current-DFT is not the same as the adiabatic ALDA provided the limit is taken in such a way that local equilibrium is not reached.

24.6 The xc Vector Potential for the Inhomogeneous Electron Liquid

The main result of the previous section, Eq. 24.20, is written so that it can immediately be turned into a local density approximation for the xc electric field of an
inhomogeneous electron liquid through the replacement $n \rightarrow n_{GS}(r)$, where $n_{GS}(r)$
is the ground-state density of the inhomogeneous liquid. Of course, the xc kernels
must also be evaluated at the local density. The correctness of this procedure is
confirmed by a careful study of the structure of the tensorial exchange-correlation
kernel in a weakly inhomogeneous electron liquid. This study was carried out by
Vignale and Kohn (VK) (1996) and is reviewed in (Vignale and Kohn 1998). In
1996 VK considered an electron liquid modulated by a charge-density wave of small
amplitude γ and small wave vector q. The wave vector k of the external field and q
were assumed to be small not only in comparison to the Fermi wave vector k_F but
also in comparison to ω/v_F (v_F being the Fermi velocity). The second condition
ensures that the phase velocity of the density disturbance is higher than the Fermi
velocity, so that no form of static screening can occur. Under these assumptions, all
the components of the tensorial kernel $\overleftrightarrow{f}_{xc}$ could be calculated, up to first order in
the amplitude of the charge density wave, and to second order in the wave vectors
k and q. The calculations were greatly facilitated by a set of sum rules which are
mathematically equivalent to the zero-force and zero-torque requirements discussed
earlier in the book. The result of the analysis was a rather complicated, but regular
expression for the various components of f_{xc} in the limit of small k and q. Finally,
this expression could be rearranged [Vignale e al. 1997, Ullrich and Vignale 2002b]
in the elegant form

$$\frac{e}{c} A_{xc}(r, \omega) = \frac{e}{c} A_{xc}^{ALDA}(r, \omega) - \frac{1}{i\omega n_{GS}(r)} \nabla \cdot \overleftrightarrow{\sigma}_{xc}(r, t), \qquad (24.24)$$

which is of course equivalent to Eq. 24.20.

Recently, gradient corrections to the xc kernels of the homogeneous electron gas
were worked out by Tao et al. (2007). Also, a more accurate expression for the
homogeneous f_{xc} in the high-frequency (anti-adiabatic) limit was worked out by
Nazarov et al. (2010) in terms of the best available Monte Carlo calculations of the
static structure factor of the electron liquid. We refer the reader to the original papers
for the details of these rather technical calculations.

Because the occurrence of two spatial derivatives of the velocity field in the post-
ALDA term is dictated by general principles, Eq. 24.20, with the xc stress tensor
given by Eq. 24.21, is expected to remain valid even for large values of the velocity,
i.e., in the nonlinear regime, provided v and n are slowly varying. The argument
goes as follows. Suppose we tried to extend Eq. 24.20 into the nonlinear regime by
including terms of order v^2. Because the stress tensor must depend on first derivatives
of v, such corrections would have to go as $(\nabla v)^2$. But then the force density, given
by the derivative of the stress tensor, would have to involve at least three derivatives.

Thus, for sufficiently small spatial variation of the density and velocity fields, the nonlinear terms can be neglected.

Since the ALDA is an intrinsically nonlinear approximation, Vignale, Ullrich, and Conti (1997) proposed that Eq. 24.20, written in the time domain, could provide an appropriate description of both linear and nonlinear response properties. A nonlinear, retarded expression for $v_{xc,1}$ was also proposed by Dobson et al. (1997). The two approximations coincide in "one-dimensional systems" (i.e., when one has a unidirectional current density field that depends only on one coordinate), but differ in the general case.

24.7 Irreversibility in TDCDFT

One of the attractive features of TDCDFT is that it allows us to study the time evolution of isolated many-body systems, which evolve from non equilibrium initial conditions under the action of a time-independent Hamiltonian. The initial state might be, for example, the ground-state of the many-body Hamiltonian in the presence of an external potential v_{ext}. At the initial time $t = 0$ the external potential is turned off and we look at the subsequent evolution of the system. Strictly speaking, a finite isolated system will never fully relax to its ground-state, for it has no way to get rid of the extra energy. However, if the system is sufficiently large, the extra energy will be distributed over an enormous number of microscopic degrees of freedom, while individual collective variables, e.g. the particle density, relax to their ground-state value. This physical picture implies that the density of the Kohn–Sham reference system evolves irreversibly toward the ground-state value, whereas the wave function of the Kohn–Sham system undergoes a unitary time-evolution, which shows no trace of irreversibility. Evidently, this nearly paradoxical behavior can only occur if the Kohn–Sham Hamiltonian includes a vector potential or some other effective potential that produces the analogue of a classical viscous force: otherwise, the density will keep oscillating back and forth forever around the ground-state solution. To see that viscosity indeed drives the density to its equilibrium value, consider the quantity

$$E(t) = \langle \Phi_{KS}(t)|\hat{H}_{KS}(t)|\Phi_{KS}(t)\rangle - \int d^3r\, v_{Hxc}(r, t)n(r, t) + E_{Hxc}[n(t)]. \quad (24.25)$$

The first term in this expression is the expectation value of the Kohn–Sham Hamiltonian, H_{KS}, in the Kohn–Sham wave function, $|\Phi_{KS}\rangle$. From this we subtract the part of the energy that pertains to the scalar xc and Hartree potentials to obtain the kinetic energy functional, and finally we add the ground-state Hartree and xc energy *functional*. The final expression is the energy the system would have at time t, were it to remain in the instantaneous ground-state. Now, making use of $\delta \hat{H}_{KS}/\delta A_{xc}(r) = \hat{j}(r)$ it is straightforward to show that

$$\frac{dE}{dt} = \int d^3r\, j(r, t) \cdot \frac{dA_{xc}(r, t)}{dt}. \quad (24.26)$$

Equation (24.26) immediately allows us to identify dE/dt as the work done on the system by the self-consistent xc vector potential. Substituting our expression (24.24) for $\mathbf{A}_{xc}$ in Eq. 24.26 we get

$$\frac{dE}{dt} = -2 \int d^3r \, \eta \text{Tr} \left\{ \omega - \frac{1}{3}\text{Tr}\{\omega\} \right\}^2 - \int d^3r \, \zeta \, (\text{Tr}\{\omega\})^2, \qquad (24.27)$$

where, $\omega_{ij} = (\nabla_i v_j + \nabla_j v_i)/2$. The negativity of $dE(t)/dt$ follows from the fact that the viscosity constants η and ζ are positive. Therefore $E(t)$ is a monotonically decreasing function of time. Furthermore, it is possible to show that $E(t)$ is bounded from below by the ground-state energy of the corresponding system of bosons (D'Agosta and Vignale 2006). Therefore $dE(t)/dt$ must vanish in the limit $t \to \infty$ and this can only happen if the velocity field, and hence the current density vanishes everywhere. Finally, the vanishing of the current density implies, via the continuity equation, the constancy of the density.

We conclude that the solution of the Kohn–Sham equation of TDCDFT has a time-independent density in the limit of $t \to \infty$. Such a time-independent density can arise only if the Kohn–Sham system settles into one of its own eigenstates. While we cannot rule out theoretically the possibility that this could still be an excited eigenstate, it is natural to expect that the Kohn–Sham equation will settle in its own ground-state, and thus lead to the ground-state density.

Chapter 25
Time-Dependent Deformation Functional Theory

Ilya V. Tokatly

25.1 Introduction

Interpretation of most experimental results in condensed matter physics does not require the knowledge of all microscopic details of complicated many-body dynamics. Normally it is sufficient to know only a few reduced collective variables, such as the density $n(r, t)$ and/or the current $j(r, t)$. Indeed, typically the behavior of a system is controlled and probed by applying external electro-magnetic fields, while the density and the current are the observables conjugated to the scalar $v_{ext}(r, t)$ and the vector $A(r, t)$ potentials, respectively. TDDFT is perfectly adjusted to this observation as the RG theorem (see Chap. 4) implies the possibility to exactly trace out irrelevant microscopic degrees of freedom, and to formulate a closed theory that operates with only one observable of interest—the density $n(r, t)$. Similarly TDCDFT (see Chap. 24) is a reduced theory which describes the behavior of the density n and the current j, formally ignoring the rest of microscopic dynamics. Obviously TDCDFT provides us with a more general theoretical setup as it gives a direct access to the current and allows to describe systems driven by transverse external fields. In the previous chapter we have seen that in the linear response regime switching to TDCDFT also solves the problem of ultranonlocality of nonadiabatic TDDFT. However beyond the linear regime a more radical modification of the theory is required to cure the ultranonlocality problem. The physical reason for the ultranonlocality is a convection in the nonequilibrium electron fluid—the motion of infinitesimal volume elements which in general retain memory on their trajectories. The theory, which allows us to exactly treat the convective nonlocality is discussed in this chapter.

I. V. Tokatly (✉)
Departamento de Física de Mate-riales, Universidad del País Vasco UPV/EHU,
Av. Tolosa 72, 20018 San Sebastian, Spain
e-mail: ilya_tokatly@ehu.es

I. V. Tokatly
IKERBASQUE, Basque Foundation for Science, 48011 Bilbao, Spain

M. A. L. Marques et al. (eds.), *Fundamentals of Time-Dependent Density Functional Theory*, Lecture Notes in Physics 837, DOI: 10.1007/978-3-642-23518-4_25,
© Springer-Verlag Berlin Heidelberg 2012

25.2 Hydrodynamic Formulation of TDCDFT

Let us consider a system of N identical particles in the presence of time dependent external scalar $v_{\text{ext}}(r, t)$ and vector $A(r, t)$ potentials. The corresponding many-body wave function $\Psi(r_1, \ldots, r_N, t)$ is a solution to the time-dependent Schrödinger equation

$$i\partial_t \Psi(r_1, \ldots, r_N, t) = \hat{H}\Psi(r_1, \ldots, r_N, t) \tag{25.1}$$

with the following Hamiltonian

$$H = \sum_{j=1}^{N} \left[\frac{1}{2m} [i\nabla_{r_j} + A(r_j, t)]^2 + v_{\text{ext}}(r_j, t) \right] + \frac{1}{2} \sum_{j \neq k} v_{\text{ee}}(|r_j - r_k|) \tag{25.2}$$

where $v_{\text{ee}}(|r - r'|)$ is the interaction potential. For a given initial condition,

$$\Psi(r_1, \ldots, r_N, 0) = \Psi_0(r_1, \ldots, r_N), \tag{25.3}$$

the dynamics of the system is completely specified by Eq. 25.1.

Usually the experimentally measurable response of the system to external probes can be described in terms of reduced "collective" variables: the density of particles $n(r, t)$, and the density of current $j(r, t)$

$$n(r, t) = \rho(r, r, t), \tag{25.4a}$$

$$j(r, t) = \frac{i}{2m} \lim_{r' \to r} (\nabla_r - \nabla_{r'})\rho(r, r', t) - \frac{n}{m}A(r, t), \tag{25.4b}$$

where $\rho(r, r', t)$ is the one particle reduced density matrix

The main idea of TDCDFT is to reduce, at the formally exact level, the problem of calculation of the density and the current to solving a closed system of equations which involve only $n(r, t)$ and $j(r, t)$. Below we describe a few equivalent ways to formulate such a closed theory.

25.2.1 Local Conservation Laws and TDCDFT Hydrodynamics in Eulerian Formulation

Using the microscopic definitions of Eqs. 25.4a and b, and the Schrödinger equation (25.1) one can derive the following hydrodynamic equations of motion for the density and the current

$$\partial_t n + \partial_{r^\mu} j_\mu = 0, \tag{25.5a}$$

$$m\partial_t j_\mu - [j \times B]_\mu - nE_\mu + \partial_{x^\nu} \Pi_{\mu\nu} = 0 \tag{25.5b}$$

where $E(r, t)$ and $B(r, t)$ are electric and magnetic fields generated by the external time-dependent scalar and vector potentials

$$E(r, t) = -\partial_t A(r, t) - \nabla v_{\text{ext}}(r, t), \tag{25.6a}$$

$$B(r, t) = \nabla \times A(r, t). \tag{25.6b}$$

Equation 25.5a is the usual continuity equation, i.e., a local conservation law of the number of particles. The equation of motion for the current, Eq. 25.5b, corresponds to a local momentum conservation law (or a local force balance equation): the time derivative of the current equals to a sum of the external and internal forces. Importantly, the local internal force (the last term in Eq. 25.5b) has a form of a divergence of a second rank tensor, which implies a vanishing net internal force, in agreement with Newton's third law. The momentum flow tensor $\Pi_{\mu\nu}$ entering Eq. 25.5b contains a kinetic and an interaction contributions (Puff and Gillis 1968; Tokatly 2005a), $\Pi_{\mu\nu} = \Pi_{\mu\nu}^{\text{kin}} + V_{\mu\nu}^{\text{ee}}$, which are expressed in terms of the many-body wave function as follows

$$\Pi_{\mu\nu}^{\text{kin}}(r) = \frac{1}{2m} \left[\lim_{r' \to r} (\hat{P}_\mu^* \hat{P}_\nu' + \hat{P}_\nu^* \hat{P}_\mu') \rho(r, r') - \frac{\delta_{\mu\nu}}{2} \nabla^2 n(r) \right], \tag{25.7a}$$

$$V_{\mu\nu}^{\text{ee}}(r) = -\frac{1}{2} \int d^3 r' \frac{r'^\mu r'^\nu}{r'} \frac{\partial v_{\text{ext}}(r')}{\partial r'} \int_0^1 d\lambda G_2(r + \lambda r', r - (1 - \lambda)r') \tag{25.7b}$$

where $\hat{P}_\mu = -i\partial_{r^\mu} - A_\mu(r, t)$ is the kinematic momentum operator, and $G_2(r, r', t)$ is a two particle reduced density matrix

$$G_2(r, r', t) = N(N-1) \int d^3 r_3 \cdots \int d^3 r_N \Psi^*(r, r', r_3, \ldots, r_N, t) \Psi(r, r', r_3, \ldots, r_N, t). \tag{25.8}$$

We note that the kinetic part $\Pi_{\mu\nu}^{\text{kin}}$ of momentum flow tensor is closely related to the momentum-stress tensor $T_{\mu\nu}$ introduced in Chap. 9, Eq. 9.10a, while the object defined by Eq. 9.10b is a divergence of the interaction part $V_{\mu\nu}^{\text{ee}}$ of the momentum flow tensor.

The existence of a closed theory for calculation of the density and the current follows from the mapping theorem of TDCDFT (Ghosh and Dhara 1988; Vignale 2004) (see also Sect. 24.4), which states that the external potentials are unique (modulo a gauge transformation) functionals of the initial state and the current density, $v_{\text{ext}} = v_{\text{ext}}[\Psi_0, j]$ and $A = A[\Psi_0, j]$. This implies that the many-body wave function $\Psi(t)$ and, therefore, any physical observable is also a functional of Ψ_0 and j. In particular, inserting the functionals $A[\Psi_0, j]$ and $\Psi = \Psi[\Psi_0, j]$ into the definitions of Eqs. 25.7a and b we get the exact momentum flow tensor as a unique functional of the initial wave function and the current $\Pi_{\mu\nu} = \Pi_{\mu\nu}[\Psi_0, j]$. This functional is universal in a sense that it does not explicitly contain the external potentials, but is uniquely recovered from a given current and an initial state (in the following for the sake of brevity we omit Ψ_0 in the arguments of the functionals).

Substituting the functional $\Pi_{\mu\nu}[j]$ into Eq. 25.5b we obtain a closed system of equations for $n(r, t)$ and $j(r, t)$. Hence from the system of Eqs. 25.5a, b we can in principle determine the dynamics of the density of particles and the density of current avoiding, at least formally, the explicit solution of the full many-body problem. The closed system (25.5a, b) can be viewed as an exact quantum hydrodynamics. TDCDFT in this form is analogous to the formulation of the static DFT in a form of the direct Hohenberg–Kohn variational principle (Hohenberg and Kohn 1964).

A connection of TDCDFT hydrodynamics to the standard mechanics of fluids (Landau and Lifshitz 1987) can be made more obvious if we switch the basic variable from the current j to the velocity field $v = j/n$. It is also useful to extract from the full momentum flow tensor $\Pi_{\mu\nu}$ its exactly known part—the flow of momentum due to convective motion of the fluid, $mnv_\mu v_\nu$,

$$\Pi_{\mu\nu} = mnv_\mu v_\nu + P_{\mu\nu} \tag{25.9}$$

where $P_{\mu\nu}$ is the stress tensor which is responsible for a local internal force related to a relative motion of particles "inside" n small moving fluid element (see Sect. 25.3 for a more detailed discussion). Using the representation (25.9) and expressing all currents in terms of the velocity field we transform equations of motion (25.5a, b) to the standard "Navier–Stokes" form

$$(\partial_t + v \cdot \nabla)n + n\partial_{x^\mu}v_\mu = 0, \tag{25.10a}$$

$$m(\partial_t + v \cdot \nabla)v_\mu - [v \times B]_\mu - E_\mu + \frac{1}{n}\partial_{x^\nu}P_{\mu\nu}[v] = 0. \tag{25.10b}$$

Since the map $(n, j) \mapsto (n, v)$ is one-to-one we are allowed to replace the functional dependence of the stress tensor on the current by the functional dependence on the velocity field, $P_{\mu\nu}[j] \mapsto P_{\mu\nu}[v]$.

25.2.2 Kohn–Sham Construction in TDCDFT

Most practical applications of any DFT rely on the KS construction (Kohn and Sham 1965). In the time-dependent setting it can be introduced as follows (Tokatly and Pankratov 2003; Tokatly 2005b). Let us consider a fictitious system of N noninteracting particles in the presence of an electromagnetic field generated by the external 4-potential (v_{ext}, A), and by some selfconsistent vector A^{xc} and scalar $v_{\text{Hxc}} = v_{\text{H}} + v_{\text{xc}}$ potentials, where v_{H} is the Hartree potential. The dynamics of this system is described by N one-particle Schrödinger equations for KS orbitals $\phi_j(r, t)$, $j = 1 \ldots N$

$$i\partial_t\phi_j = \frac{1}{2m}(i\partial_r + A + A^{\text{xc}})^2\phi_j + (v_{\text{ext}} + v_{\text{Hxc}})\phi_j. \tag{25.11}$$

Obviously, the density n_{KS} and the velocity v_{KS} of the KS system satisfy the continuity equation (25.10a) and the force balance equation of the form of Eq. 25.10b, but with the stress tensor $P_{\mu\nu}$, and the external Lorentz force being replaced, respectively, by the kinetic stress tensor $T^{KS}_{\mu\nu}$ of the noninteracting KS particles, and by the Lorentz force corresponding to the total effective 4-potential. From the requirement that the KS density and current reproduce the density and the current in the real interacting system we get the following equation connecting the xc potential to the stress tensor functional

$$\partial_t A^{xc}_\mu - [v \times (\nabla \times A^{xc})]_\mu + \partial_{r^\mu} v_{Hxc} = -\frac{1}{n} \partial_{r^\nu} P^{xc}_{\mu\nu}[v], \qquad (25.12)$$

where $P^{xc}_{\mu\nu}[v] = P_{\mu\nu}[v] - T^{KS}_{\mu\nu}[v]$ is the difference of the stress tensors in the interacting and noninteracting KS systems. Equation 25.12 is the most general definition of the xc potentials. For a given stress tensor functional (the right hand side) it defines the xc 4-potential (v_{xc}, A^{xc}) up to a gauge transformation.

It is important to note that the KS construction is only an auxiliary device for solving the general collective variable theory in a form of closed equations of motion, Eqs. 25.5a, b (or, equivalently, Eqs. 25.10a, b for the basic variables: the current/velocity and the density. This is similar to the static DFT where the KS construction is merely a useful mathematical trick for transforming the fundamental Hohenberg–Kohn variational principle to a system of differential equations for one-particle orbitals.

25.2.3 TDCDFT Hydrodynamics in the Lagrangian Form

In general TDCDFT is a closed formalism to describe a convective motion of a many-body system driven by external fields. Usually the convective motion is characterized by the density of particles $n(r, t)$ and the density of current $j(r, t)$ or a velocity field $v(r, t)$. An alternative way to completely characterize the convective motion is commonly referred to as a Lagrangian description. Let us consider the system as a collection of infinitesimal fluid elements (the so-called "materials"). Every element can be uniquely labeled by a continuous variable ξ—its position at the initial time $t = 0$. The Lagrangian description can be viewed as tracking the motion of those infinitesimal elements of the fluid. In other words, the convective motion is characterized by a (continuous) set of trajectories $r(\xi, t)$, where the argument ξ indicates the starting point of the trajectory (the unique label of the element). One can show that the map $v(r, t) \mapsto r(\xi, t)$ is unique and invertible.

For a given velocity $v(r, t)$ the Lagrangian trajectory $r(\xi, t)$ is solution to the following Cauchy problem

$$\partial_t r(\xi, t) = v(r(\xi, t), t), \qquad r(\xi, 0) = \xi. \qquad (25.13)$$

Hence from the velocity field we uniquely construct a set of Lagrangian trajectories. Every trajectory $r(\xi, t)$ is specified by its initial point ξ, which means that, given the initial position ξ of a fluid element, we can always find its coordinate $r = r(\xi, t)$ at any instant t, and, tracing the trajectory in the reverse order, from a given position r at time t one uniquely recovers the initial point of the trajectory, $\boldsymbol{\xi} = \boldsymbol{\xi}(r, t)$. In a more formal language, the map $\boldsymbol{\xi} \mapsto r : r = r(\boldsymbol{\xi}, t)$ is unique and invertible. This fact allows us to recover the Eulerian variables, i.e., the velocity field and the density, from given Lagrangian trajectories

$$
v(r, t) = \left[\frac{\partial r(\boldsymbol{\xi}, t)}{\partial t} \right]_{\boldsymbol{\xi} = \boldsymbol{\xi}(r,t)}, \quad n(r, t) = \left[\frac{n_0(\boldsymbol{\xi})}{\sqrt{g(\boldsymbol{\xi}, t)}} \right]_{\boldsymbol{\xi} = \boldsymbol{\xi}(r,t)}, \quad (25.14)
$$

where $\boldsymbol{\xi}(r, t)$ is the inverse of $r(\boldsymbol{\xi}, t)$, $n_0(r)$ is the initial density, and $\sqrt{g(\boldsymbol{\xi}, t)} = J(\boldsymbol{\xi}, t) = \det(\partial r^\mu / \partial \xi^\nu)$ is the Jacobian of the transformation of coordinates $r \to \boldsymbol{\xi}$. The first equality in (25.14) is a consequence of Eq. 25.13, while the equation for $n(r, t)$ can be easily checked by a direct substitution into the continuity equation (25.10a). Hence the function $r(\boldsymbol{\xi}, t)$ indeed completely characterizes the convective motion of a system.

The basic equation in the Lagrangian description of collective dynamics is the equation of motion for a fluid element. This equation can be derived from the equation of motion for Eulerian velocity $v(r, t)$, Eq. 25.10b, by making a transformation of coordinates $r \to \boldsymbol{\xi}$, i.e., by considering the initial points $\boldsymbol{\xi}$ of Lagrangian trajectories as independent spatial coordinates. Under this transformation the convective derivative, $\partial_t + v \cdot \nabla$ becomes simply ∂_t so that the first term in Eq. 25.10b transforms to $m\ddot{r}(\boldsymbol{\xi}, t)$, while the divergence of the stress tensor in the last term in Eq. 25.10b becomes a covariant divergence in the space with metrics $g_{\mu\nu}(\boldsymbol{\xi}, t)$ induced by the transformation from r- to $\boldsymbol{\xi}$-coordinates. Hence after the transformation of coordinates we arrive at the following equation of motion for a fluid element

$$
m\ddot{r}^\mu - E_\mu(r, t) - [\dot{r} \times B(r, t)]_\mu + \frac{\sqrt{g}}{n_0} \frac{\partial \xi^\alpha}{\partial r^\mu} \hat{D}_\nu \widetilde{P}_\alpha^\nu[r(\boldsymbol{\xi}, t)] = 0, \quad (25.15)
$$

where $\widetilde{P}_{\mu\nu}(\boldsymbol{\xi}, t)$ is the original stress tensor $P_{\alpha\beta}(r, t)$ transformed to the new coordinates according to the standard rules (Dubrovin et al. 1984)

$$
\widetilde{P}_{\mu\nu}(\boldsymbol{\xi}, t) = \frac{\partial r^\alpha}{\partial \xi^\mu} \frac{\partial r^\beta}{\partial \xi^\nu} P_{\alpha\beta}(r(\boldsymbol{\xi}, t), t). \quad (25.16)
$$

The $\hat{D}$ operator in Eq. 25.15 stands for the covariant divergence

$$
\hat{D}_\nu \widetilde{P}_\mu^\nu = \frac{1}{\sqrt{g}} \partial_{\xi^\nu} \sqrt{g} \widetilde{P}_\mu^\nu - \frac{1}{2} \widetilde{P}^{\alpha\beta} \partial_{\xi^\mu} g_{\alpha\beta}, \quad (25.17)
$$

and the metric tensor in the $\boldsymbol{\xi}$-space of "initial positions" is defined as follows

$$
g_{\mu\nu}(\boldsymbol{\xi}, t) = \frac{\partial r^\alpha}{\partial \xi^\mu} \frac{\partial r^\alpha}{\partial \xi^\mu}; \quad [g_{\mu\nu}]^{-1} = g^{\mu\nu} = \frac{\partial \xi^\mu}{\partial r^\alpha} \frac{\partial \xi^\nu}{\partial r^\alpha}. \quad (25.18)
$$

The equation of motion (25.15) has to be solved with the initial conditions $r(\xi, 0) = \xi$ and $\dot{r}(\xi, 0) = v_0(\xi)$, where $v_0(r)$ is the initial velocity distribution calculated from the initial many-body wave function of Eq. 25.3.

The first three terms in Eq. 25.15 correspond to a classical Newton equation for a particle moving in the external electromagnetic field, while the last, the stress term, takes care of all complicated quantum and many-body effects. Because of the uniqueness and invertibility of the map $v(r, t) \mapsto r(\xi, t)$ the transformed stress tensor can be considered as a unique functional of the Lagrangian trajectories, $\widetilde{P}_{\mu\nu} = \widetilde{P}_{\mu\nu}[r(\xi, t)]$. Hence Eq. 25.15 is a closed equation of motion that, at the formally exact level, completely determines the collective dynamics of the system. This is the basic equation of TDCDFT in the Lagrangian form.

On first sight the representation of TDCDFT hydrodynamics in the Lagrangian form of Eq. 25.15 does not bring anything fundamentally new. This is indeed true if one follows a route outlined in this section: starting from the traditional formulation of the TDCDFT mapping theorem and via the Eulerian equation of motion for the density and the current. However, in the next section we will see that using the ideas of the Lagrangian description one can reformulate the whole theory in a constructive way that also ends up with the equation of motion (25.15), but provides us with a clear procedure for calculating the basic stress tensor functional. As an additional benefit we get a complete solution of the ultranonlocality problem.

25.3 Time-Dependent Deformation Functional Theory

25.3.1 Many-Body Theory in a Co-moving Reference Frame

TDCDFT is a reduced theory aimed at describing only the convective motion of the system without a detailed knowledge of the full dynamics of all microscopic degrees of freedom. Therefore it is natural to separate the "convective" degrees of freedom at the very beginning, i.e. at the level of the full many-body theory. The Lagrangian description is perfectly suited for this purpose. Since in the Lagrangian formalism the convective dynamics is characterized by the motion of fluid elements, it can be easily separated from the microscopic dynamics of quantum particles by transforming the many-body theory to a local noninertial reference frame moving along the Lagrangian trajectories.

At the formal level one proceeds as follows. Consider a reference frame defined by some (unspecified for the moment) velocity field $v(r, t)$. By solving the Cauchy problem of Eq. 25.13 with the above velocity in the right hand side we get the local trajectories $r(\xi, t)$ of the frame. The transformation of the theory to the new reference frame corresponds to a transformation of coordinates $r_j \to \xi_j$, with $r(\xi, t)$ being the transformation function, i.e. $r_j = r(\xi_j, t)$, in the many-body Schrödinger equation (25.1). We emphasize that the same function $r(\xi, t)$ is used to transform the coordinates of each particle in the many-body system, which means that after the transformation each particle is seen from a local frame moving along the trajectory $r(\xi, t)$. It is convenient to define the many-body wave function $\widetilde{\Psi}(\xi_1, \ldots, \xi_N, t)$ in the new frame as follows (Tokatly 2007)

$$\widetilde{\Psi}(\xi_1, \ldots, \xi_N, t) = \prod_{j=1}^{N} g^{\frac{1}{4}}(\xi_j, t) e^{-iS_{cl}(\xi_j, t)} \Psi(r(\xi_1, t), \ldots, r(\xi_N, t), t), \quad (25.19)$$

where $S_{cl}(\xi, t)$ is the action of a particle moving along the trajectory $r(\xi, t)$

$$S_{cl}(\xi, t) = \int_0^t dt' \left[\frac{m}{2} [\dot{r}(t')]^2 + \dot{r}(t') \cdot A(r(t'), t') - v_{ext}(r(t'), t') \right]. \quad (25.20)$$

Equation 25.19 is a generalization of the transformation to a homogeneously acceler-ated frame, which is used, for example, in the proofs of a harmonic potential theorem (Dobson 1994a; Vignale 1995a). The exponential prefactor accounts for the phase acquired due to the motion of the frame, while the factor $\prod_{j=1}^{N} g^{\frac{1}{4}}(\xi_j, t)$ ensures the standard normalization of the wave function $\langle \widetilde{\Psi} | \widetilde{\Psi} \rangle = 1$ under a non-volume-preserving transformation of coordinates.

Performing a transformation of coordinates, $r_j \rightarrow \xi_j$, in Eq. 25.1, and using the definition (25.19) we obtain the many-body Schrödinger equation in the frame moving with some velocity $v(r, t)$

$$i\partial_t \widetilde{\Psi}(\xi_1, \ldots, \xi_N, t) = \widetilde{H}[g_{ij}, \mathcal{A}] \widetilde{\Psi}(\xi_1, \ldots, \xi_N, t). \quad (25.21)$$

The Hamiltonian in the new frame takes the form

$$\widetilde{H}[g_{ij}, \mathcal{A}] = \sum_{j=1}^{N} g_j^{-\frac{1}{4}} \hat{K}_{j,\mu} \frac{\sqrt{g_j} g_j^{\mu\nu}}{2m} \hat{K}_{j,\nu} g_j^{-\frac{1}{4}} + \frac{1}{2} \sum_{k \neq j} v_{ee}(l_{\xi_k \xi_j}) \quad (25.22)$$

where $\hat{K}_{j,\mu} = -i\partial_{\xi_j^\mu} - \mathcal{A}_\mu(\xi_j, t)$, $g_j^{\mu\nu} = g^{\mu\nu}(\xi_j, t)$, and $l_{\xi_k \xi_j}$ is the distance between jth and kth particles in the moving frame (the length of geodesic connecting points ξ_j and ξ_k in the space with metric $g_{\mu\nu}$ (Dubrovin et al. 1984). An effective vector potential $\mathcal{A}(\xi, t)$ in the Hamiltonian is defined as

$$\mathcal{A}_\mu = \frac{\partial x^\nu}{\partial \xi^\mu} \dot{x}^\nu + \frac{\partial x^\nu}{\partial \xi^\mu} A_\nu(r, t) - \partial_{\xi^\mu} S_{cl}(\xi, t). \quad (25.23)$$

Physically $\mathcal{A}(\xi, t)$ describes a combined action of the external and inertial forces in a local noninertial frame. Since at $t = 0$ the moving frame coincides with the laboratory one (this follows from the condition $r(\xi, 0) = \xi$) the initial condition for the transformed Schrödinger equation (25.21) is unchanged:

$$\widetilde{\Psi}(\xi_1, \ldots, \xi_N, 0) = \Psi_0(\xi_1, \ldots, \xi_N). \quad (25.24)$$

Equations 25.15–25.24 completely determine the dynamics of the quantum N-particle system in an arbitrary local noninertial frame.

Since our aim is to reformulate the theory in a particular frame moving with a physical flow (this frame is called co-moving, or Lagrangian) we need to impose an additional local condition to specify the required frame. By definition, in the co-moving frame the current density is zero everywhere and at all times, while the density of particles stays stationary and equal to the initial density $n_0(\boldsymbol{\xi})$. Hence it is natural to fix the frame by the requirement of zero transformed current density $\widetilde{j}(\boldsymbol{\xi}, t) = 0$. Explicitly this condition reads

$$n_0(\boldsymbol{\xi})\boldsymbol{\mathcal{A}}(\boldsymbol{\xi}, t) = \frac{i}{2} \lim_{\boldsymbol{\xi}' \to \boldsymbol{\xi}} (\nabla_{\boldsymbol{\xi}} - \nabla_{\boldsymbol{\xi}'})\widetilde{\rho}(\boldsymbol{\xi}, \boldsymbol{\xi}', t) \tag{25.25}$$

where $\widetilde{\rho}(\boldsymbol{\xi}, \boldsymbol{\xi}', t)$ is the one particle reduced density matrix calculated from the transformed wave function

$$\rho(\boldsymbol{\xi}, \boldsymbol{\xi}', t) = N \int d^3\xi_2 \cdots \int d^3\xi_N \Psi^*(\boldsymbol{\xi}, \boldsymbol{\xi}_2, \ldots, \boldsymbol{\xi}_N, t)\Psi(\boldsymbol{\xi}', \boldsymbol{\xi}_2, \ldots, \boldsymbol{\xi}_N, t). \tag{25.26}$$

The frame-fixing condition (25.25) simply states that in the co-moving frame a "paramagnetic" current [the right hand side of (25.25)] is exactly cancelled by the "diamagnetic" contribution [the left hand side of (25.25)].

The Schrödinger equation (25.21), the definition of the effective vector potential (25.23), and the zero current condition (25.25) constitute a closed system of equations that determine the dynamics of the many-body system in the co-moving reference frame. In principle one can eliminate the effective vector potential by substituting $\boldsymbol{\mathcal{A}}(\boldsymbol{\xi}, t)$ from Eq. 25.25 into Eqs. 25.23 and 25.21. The result is a system of two first order (in time) differential equations for two functions: (i) the trajectory $r(\boldsymbol{\xi}, t)$ which describes the convective motion on the system, and (ii) the transformed wave function $\widetilde{\Psi}(\boldsymbol{\xi}_1, \ldots, \boldsymbol{\xi}_N, t)$ describing the rest of the microscopic degrees of freedom in the frame moving with the convective flow. The equations have to be solved with the initial conditions Eq. 25.24 for $\widetilde{\Psi}(t)$ and $r(\boldsymbol{\xi}, 0) = \boldsymbol{\xi}$ for the Lagrangian trajectory.

25.3.2 Emergence of TDDefFT: A Universal Many-Body Problem

Let us discuss possible procedures for solving the system of Eqs. 25.21, 25.23 and 25.25, but first we rewrite it in a more physical and clear form.

Because Eq. 25.23 contains the classical action $S_{cl}(\boldsymbol{\xi}, t)$ of Eq. 25.20, it is nonlocal in time. This nonlocality can be removed by differentiating Eq. 25.23 with respect to t. The time derivative of Eq. 25.23 takes the form

$$m\ddot{r}^\mu = E_\mu(r, t) + [\dot{r} \times B(r, t)]_\mu + \frac{\partial\xi^\nu}{\partial r^\mu}\partial_t\mathcal{A}_\nu, \tag{25.27}$$

which is exactly the classical Newtonian equation for a particle moving in the external electromagnetic field and, in addition, in the "electric" field generated by the effective vector potential $\mathcal{A}$.

Comparing Eq. 25.27 with Eq. 25.15 we observe that these two equations become identical if the time derivative of the effective vector potential is equal to the covariant divergence of the stress tensor. To see that this is indeed the case we consider the local force balance equation that follows from the many-body Schrödinger equation (25.21) in the local noninertial frame. Apparently the force balance equation should be of the general form of Eq. 25.5b, but with the usual divergence of the momentum flow tensor replaced by a covariant divergence of the stress tensor in the $\boldsymbol{\xi}$-space. Namely,

$$\partial_t \widetilde{j}_\mu - \widetilde{j}^\nu (\partial_{\xi^\nu} \mathcal{A}_\mu - \partial_{\xi^\mu} \mathcal{A}_\nu) + \widetilde{n} \partial_t \mathcal{A}_\mu + \sqrt{g} D_\nu \widetilde{P}^\nu_\mu = 0. \tag{25.28}$$

where the stress tensor in the space with metric $g_{\mu\nu}$ can be determined from the following universal formula (Tokatly 2005a, 2007; Rogers and Rappe 2002)

$$\widetilde{P}_{\mu\nu}(\boldsymbol{\xi}, t) = \frac{2}{\sqrt{g}} \left\langle \widetilde{\Psi} \left| \frac{\delta \widetilde{H}[g_{\alpha\beta}, \mathcal{A}]}{\delta g^{\mu\nu}(\boldsymbol{\xi}, t)} \right| \widetilde{\Psi} \right\rangle \equiv \langle \widetilde{\Psi} | \hat{\widetilde{P}}_{\mu\nu}[g_{\alpha\beta}, \mathcal{A}] | \widetilde{\Psi} \rangle \tag{25.29}$$

with the Hamiltonian defined by Eq. 25.22. An explicit form of the stress tensor operator $\hat{\widetilde{P}}_{\mu\nu}$ entering Eq. 25.29 can be found, for example, in (Tokatly 2005a). The important point is that $\hat{\widetilde{P}}_{\mu\nu}[g_{\alpha\beta}, \mathcal{A}]$ is an explicitly known and local in time functional of the metric tensor and the effective vector potential. Since in the co-moving frame the transformed current density $\widetilde{j}$ is zero, only the last two terms survive in Eq. 25.28. Therefore in our frame of interest the force balance equation reduces to the following identity

$$\partial_t \mathcal{A}_\mu = -\frac{\sqrt{g}}{n_0} D_\nu \widetilde{P}^\nu_\mu. \tag{25.30}$$

Inserting this identity into Eq. 25.27 we exactly recover the basic equation of hydrodynamics in the Lagrangian description, Eq. 25.15. The important progress is that the stress tensor in this equation is now defined entirely in terms of the variables entering the many-body problem in the co-moving frame. Hence we have transformed Eq. 25.23 to a clear physical form by reducing it to the equation of motion for the fluid elements.

Using Eqs. 25.27 and 25.30 we can write down the following final system of equations describing the dynamics of the full many-body system

$$i\partial_t \widetilde{\Psi}(\boldsymbol{\xi}_1, \ldots, \boldsymbol{\xi}_N, t) = \widetilde{H}[g_{\mu\nu}, \mathcal{A}] \widetilde{\Psi}(\boldsymbol{\xi}_1, \ldots, \boldsymbol{\xi}_N, t) \tag{25.31a}$$

$$\mathcal{A}(\boldsymbol{\xi}, t) = \frac{i}{2n_0(\boldsymbol{\xi})} \lim_{\boldsymbol{\xi}' \to \boldsymbol{\xi}} (\nabla_\xi - \nabla_{\xi'}) \widetilde{\rho}(\boldsymbol{\xi}, \boldsymbol{\xi}', t) \tag{25.31b}$$

$$m\ddot{r}^\mu = E_\mu(\boldsymbol{r}, t) + [\dot{\boldsymbol{r}} \times \boldsymbol{B}(\boldsymbol{r}, t)]_\mu - \frac{\sqrt{g}}{n_0} \frac{\partial \xi^\alpha}{\partial r^\mu} D_\nu \widetilde{P}^\nu_\alpha, \tag{25.31c}$$

where the Hamiltonian $\widetilde{H}[g_{\mu\nu}, \mathcal{A}]$, the reduced density matrix $\widetilde{\rho}[\widetilde{\Psi}](\xi, \xi', t)$, and the stress tensor $\widetilde{P}_{\mu\nu}[g_{\alpha\beta}, \mathcal{A}, \Psi]$ are defined after Eqs. 25.22, 25.26, and 25.29, respectively. The metric tensor $g_{\mu\nu}(\xi, t)$ entering Eqs. 25.31a and 25.31c is related to the Lagrangian trajectory via Eq. 25.18.

The system of Eqs. 25.31a, b, c is equivalent to the original many-body Schrö-dinger equation (25.1). Everything we did to derive Eqs. (25.31a, b, c) from Eq. 25.1 was an identical change of variables aimed at separating the "convective" and the "relative" motions of quantum particles. However after this identical transformation the structure of the many-body theory becomes quite remarkable. Now the physical external fields enter only the equation of motion for the fluid elements, Eq. 25.31c, while the many-body dynamics, which is governed by Eqs. 25.31a and b, depends only on the fundamental geometric characteristic of the co-moving frame: the metric tensor $g_{\mu\nu}(\xi, t)$. Equations 25.31a and b describe *universal* dynamics of N particles driven by a given time-dependent metric and constrained by the requirement of zero current density. By solving the universal many-body problem of Eqs. 25.31a and b for a given metric of the form 25.18 we get the wave function $\widetilde{\Psi}$ and the effective vector potential $\mathcal{A}$ as universal functionals of the metric tensor: $\widetilde{\Psi} = \widetilde{\Psi}[g_{\mu\nu}]$ and $\mathcal{A} = \mathcal{A}[g_{\mu\nu}]$. Substitution of these functionals into Eq. 25.29 gives the universal stress tensor functional $\widetilde{P}_{\mu\nu}[g_{\alpha\beta}]$. Thus Eq. 25.31c becomes a closed equation of motion for fluid elements, which determines the Lagrangian trajectories of the system. As all basic quantities are functionals of the metric tensor, which is exactly the Green's deformation tensor of the classical elasticity theory (Masson 1964), we call this approach the time-dependent deformation functional theory (TDDefFT).

The solution of Eq. 25.31c with a known functional $\widetilde{P}_{\mu\nu}[g_{\alpha\beta}](\xi, t)$ gives the description of the convective motion in terms of the Lagrangian picture. Alternatively we can transform $\widetilde{P}_{\mu\nu}[g_{\alpha\beta}](\xi, t)$ to the laboratory frame to recover the tensor $P_{\mu\nu}[v](r, t)$,

$$P_{\mu\nu}[v](r, t) = \frac{\partial \xi^\alpha}{\partial r^\mu} \frac{\partial \xi^\beta}{\partial r^\nu} \widetilde{P}_{\alpha\beta}[g_{\alpha\beta}(\xi(r, t), t)](\xi(r, t), t), \qquad (25.32)$$

which can be used either in the hydrodynamic formulation of Eqs. 25.5a, b or to calculate the xc potentials for the KS formulation of TDCDFT, Eqs. 25.11, 25.12. Finally, since in the laboratory frame the function $\xi(r, t)$ (the initial point of the trajectory that arrives to r at the time t) can be found from the equation

$$[\partial_t + v(r, t) \cdot \nabla]\xi(r, t) = 0, \quad \xi(r, 0) = r \qquad (25.33)$$

the stress tensor determined by Eq. 25.32 is indeed a universal functional of the Eulerian velocity $v(r, t)$.

Therefore we recovered the full formal structure of the traditional TDCDFT, but at the fundamentally new level of understanding. The main point is that in TDDefFT formalism a closed theory of convective motion appears from a regular and conceptually clean procedure: it is simply a natural and regular step in solving the many-body problem in the co-moving frame. Now we clearly understand where the universal functionals entering the theory come from and why they are universal.

25.4 Approximate Functionals from TDDefFT

The formalism of TDDefFT provides us with the following general recipe to construct xc potentials entering the time-dependent KS equations (25.11). The exact xc 4-potential $(v_{\text{xc}}, \boldsymbol{A}^{\text{xc}})$ is determined by Eq. 25.12, where the xc stress tensor $P_{\mu\nu}^{\text{xc}}[v](\boldsymbol{r}, t)$ in the laboratory frame is related to the xc stress tensor $\widetilde{P}_{\alpha\beta}^{\text{xc}}[g_{\mu\nu}](\boldsymbol{\xi}, t)$ in the Lagrangian frame as follows

$$P_{\mu\nu}^{\text{xc}}[v](\boldsymbol{r}, t) = \frac{\partial \xi^\alpha}{\partial r^\mu} \frac{\partial \xi^\beta}{\partial r^\nu} \widetilde{P}_{\alpha\beta}^{\text{xc}}[g_{\alpha\beta}(\boldsymbol{\xi}(\boldsymbol{r}, t), t)](\boldsymbol{\xi}(\boldsymbol{r}, t), t). \tag{25.34}$$

Finally, the stress tensor $\widetilde{P}_{\alpha\beta}^{\text{xc}}[g_{\mu\nu}](\boldsymbol{\xi}, t)$ is calculated from the solution of the universal many-body problem (25.31a, b) for the interacting and noninteracting systems:

$$\widetilde{P}_{\mu\nu}^{\text{xc}}[g_{\mu\nu}](\boldsymbol{\xi}, t) = \frac{2}{\sqrt{g}} \left\langle \widetilde{\Psi} \left| \frac{\delta \widetilde{H}[g_{\alpha\beta}, \boldsymbol{A}]}{\delta g^{\mu\nu}(\boldsymbol{\xi}, t)} \right| \widetilde{\Psi} \right\rangle - \frac{2}{\sqrt{g}} \left\langle \widetilde{\Phi} \left| \frac{\delta \widetilde{T}[g_{\alpha\beta}, \boldsymbol{A}_{\text{KS}}]}{\delta g^{\mu\nu}(\boldsymbol{\xi}, t)} \right| \widetilde{\Phi} \right\rangle \tag{25.35}$$

Here $\widetilde{H}[g_{\alpha\beta}, \boldsymbol{A}]$ is defined by Eq. 25.22, $\widetilde{T}[g_{\alpha\beta}, \boldsymbol{A}_{\text{KS}}]$ is the Hamiltonian for noninteracting system in the co-moving frame (the kinetic energy operator given by the first term in (25.22)), $\boldsymbol{A}$ and $\boldsymbol{A}_{\text{KS}}$ are the effective vector potentials entering the interacting and the noninteracting universal problems, and $\widetilde{\Psi}(t)$ and $\widetilde{\Phi}(t)$ are the solutions of the interacting and the noninteracting universal many-body problems, respectively.

Clearly, the exact solution of the universal problem (25.31a, b) is at least as difficult as the full solution of the original many-body Schrödinger equation. However, various approximate solutions to Eqs. 25.31a and b can be used to construct approximate universal xc functionals that satisfy a number of exact constraints and become exact in certain limiting cases. The simplest and historically the first approximation is a time-dependent local deformation approximation (TDLDefA), which has been introduced in (Tokatly 2005b) and further analyzed in (Ullrich and Tokatly 2006a). TDLDefA is a consistent time-dependent analog of LDA in the ground state DFT. Formally this approximation is obtained by inserting into Eq. 25.35 the solutions of the universal problem for a *homogeneous* electron gas driven by a *homogeneous* time-dependent metric tensor $g_{\mu\nu}(t)$. Since a homogeneous deformation does not induce any current, the effective vector potential vanishes, $\boldsymbol{A} = 0$, and the wave function becomes a local, but in general retarded functional of $g_{\mu\nu}(t)$. Therefore in TDLDefA the stress tensor in the definition of the xc potential (25.12) is also a local functional of the deformation tensor. Importantly, this approximation, being absolutely local in terms of $g_{\mu\nu}(t)$, is free of the ultranonlocality problem . The reason is that in TDLDefA the convective nonlocality is treated exactly via the dependence on the Lagrangian coordinate $\boldsymbol{\xi}(\boldsymbol{r}, t)$, which is obtained from the exact equation (25.33).

In the limit of small deformations the wave functions of the homogeneously deformed uniform electron gas can be calculated using the linear response theory. The corresponding linearized TDLDefA (Tokatly 2005b) exactly recovers the VK approximation introduced in the previous chapter (see also [Vignale and Kohn 1996,

Vignale et al. 1997, Vignale and Kohn 1998]). A connection to the VK approximation can be easily understood from the exact Eq. 25.33. By solving Eq. 25.33 to the linear order in velocity gradients we find

$$\boldsymbol{\xi}(\boldsymbol{r}, t) \approx \boldsymbol{r} - \int_0^t dt' \boldsymbol{v}(\boldsymbol{r}, t'), \tag{25.36a}$$

$$g_{\mu\nu}(\boldsymbol{r}, t) \approx \delta_{\mu\nu} + \int_0^t dt' [\partial_{r^\mu} v_\nu(\boldsymbol{r}, t') + \partial_{r^\nu} v_\mu(\boldsymbol{r}, t')]. \tag{25.36b}$$

Hence in this limit the Lagrangian coordinates and the deformation tensor become local functionals of the velocity, which explains the ability of TDCDFT to cure the ultranonlocality in the linear regime. However, in general, a consistent local description is only possible in terms of TDLDefA, that can be viewed as a nonlinear extension of VK approximation. At this point it is worth noting that the regime of small deformations does not necessarily assume that the velocity/current itself is small. The linearized form of the deformation tensor, Eq. 25.36b, and the corresponding linearization of the stress tensor only require a smallness of the velocity gradients. A bright example is provided by the harmonic potential theorem-type of motion when the velocity $\boldsymbol{v}(t)$ is constant in space, and thus the system remains strictly undeformed $g_{\mu\nu}(\boldsymbol{\xi}, t) = \delta_{\mu\nu}$, while the absolute value of $\boldsymbol{v}(t)$ can be arbitrary large. Therefore, there are situations when the linearized TDLDefA is valid well beyond the limits of the formal linear response theory. Such a "semilinear" extension of VK approximation has been proposed on phenomenological grounds by Vignale, Ulrich and Conti (VUC) (Vignale et al. 1997). Later (Ullrich and Tokatly 2006) the xc functional constructed by VUC has been formally derived as a small deformation limit of TDLDefA.

Unfortunately, beyond the regime of small deformations a full solution of the time-dependent universal problem, even for a homogeneous system, becomes very difficult. Therefore some other simplifying assumptions are required. One of such assumptions, which is valid in the case of very fast variations of the deformation tensor, leads to a so called elastic approximation in TDDefFT. In the short-time limit we can set $\mathcal{A} = \mathcal{A}_{KS} = 0$, $\widetilde{\Psi}(t) = \Psi_{GS}$, and $\widetilde{\Phi}(t) = \Phi_{GS}$ in Eq. 25.35, where Ψ_{GS} and Φ_{GS} are the ground state wave functions for the interacting and the noninteracting systems (we assume that the evolution starts from the ground state). Since the wave functions Ψ_{GS} and Φ_{GS} do not depend on the metric, the Eq. 25.35 for the xc stress tensor simplifies as follows

$$\widetilde{P}_{\mu\nu}^{xc}[g_{\mu\nu}](\boldsymbol{\xi}, t) = \frac{2}{\sqrt{g}} \frac{\delta \widetilde{E}_{xc}[g_{\alpha\beta}]}{\delta g^{\mu\nu}(\boldsymbol{\xi}, t)}, \tag{25.37}$$

where $\widetilde{E}_{xc}[g_{\alpha\beta}] = \langle \Psi_{GS} | \widetilde{H}[g_{\alpha\beta}] | \Psi_{GS} \rangle - \langle \Phi_{GS} | \widetilde{T}[g_{\alpha\beta}] | \Phi_{GS} \rangle$ is the xc energy of *instantaneously* deformed electron gas. Formally Eq. 25.37 is similar to the stress-deformation relation in the classical elasticity theory with $\widetilde{E}_{xc}[g_{\alpha\beta}]$ being the elastic deformation energy. The elastic approximation becomes exact in the limit of very fast variations of the deformation tensor, which corresponds to the *antiadiabatic* regime of electron dynamics . By linearizing the general elastic xc potential one obtains the exact antiadiabatic xc kernel $f_{xc}^{\infty}(r, r')$, which fully accounts for all spatial nonlocalities (Nazarov et al. 2010). The practical implementation of the elastic approximation requires, as an input, a knowledge of the ground state one particle density matrix and the pair correlation functions of the inhomogeneous system of interest. Using instead the density matrix and the pair corelation function of the homogeneous electron gas we obtain the elastic TDLDefA (Tokatly 2005b, Ullrich and Tokatly 2006). In the latter case, all we need is the usual xc energy $e_{xc}(n)$ of the homogeneous electron gas. Therefore the elastic TDLDefA requires the same input as ALDA. However, both formally and physically, these two approximations are essentially different. ALDA is a purely ad hoc construction which is hard to justify formally. In fact, a partial success of ALDA still remains one of the most puzzling features of TDDFT. In contrast to this, the formal limits of validity for TDLDefA are well defined. It becomes essentially exact in the limit $v_F T / L \ll 1$, where T and L are the characteristic time and length scales of the dynamical process, and v_F is a local Fermi velocity (for a discussion of this condition in the context VK approximation see Chap. 24).

Up to now we discussed the construction of xc potentials entering the KS equations. However the above approximations are also applicable to the hydrodynamic formulation of TDCDFT. Foe example, the explicit elastic stress tensor

$$\widetilde{P}_{\mu\nu}^{el}[g_{\mu\nu}](\xi, t) = \frac{2}{\sqrt{g}} \frac{\delta}{\delta g^{\mu\nu}(\xi, t)} \langle \Psi_{GS} | \widetilde{H}[g_{\alpha\beta}] | \Psi_{GS} \rangle, \qquad (25.38)$$

can be used directly in the Lagrangian hydrodynamics equation (25.15), or, after the transformation (25.32), in the "Navier–Stokes" equation (25.10b). As a result we obtain a closed quantum continuum mechanics. Explicit equations of the linearized quantum continuum mechanics were derived in (Tao et al. 2009; Gao et al. 2010). A great advantage of this theory is that the complexity of the equations does not depend on the number of particles, which makes this theory promising for applications to large systems.

Therefore the formalism of TDDefFT allows us to regularly derive a variety of approximations. However up to now only the elastic TDLDefA and the linearized quantum continuum mechanics have been tested for model systems. Implementation of the above approximations into the electronic structure codes and analysis of their performance remain important problems for the nearest future.

Chapter 26
Time-Dependent Reduced Density Matrix Functional Theory

Klaas J. H. Giesbertz, Oleg V. Gritsenko and Evert Jan Baerends

26.1 Introduction

In this chapter we will give an introduction into one-body reduced density matrix functional theory (RDMFT). This is a rather new method to deal with the quantum many-body problem. Especially the development of a time-dependent version, TDRDMFT, is very recent. Therefore, there are many open questions and the formalism has not crystalized yet into a standard form such as in (TD)DFT. Although RDMFT has similarities with DFT, there are many more differences. This chapter is too short for a full introduction into the wondrous world of RDMFT, but we hope to give an idea what (TD)RDMFT might bring.

Although TDDFT has proven to be often a useful method to calculate response properties and excitation energies, there are still a number of situations where TDDFT with its current approximations fails badly. One failure which received much attention is charge transfer excitations (Dreuw et al. 2003). Using the standard local approximations for the xc kernel like the ALDA, these excitations are generally predicted too low and they lack the typical Coulombic $1/R$ behavior for increasing donor-acceptor distance (see Sect. 4.8.2). The failure of the charge transfer excitations can easily be attributed to the the fact that only the local density is used in the approximations to the

K. J. H. Giesbertz
Department of Physics, Nanoscience Center, University of Jyväskylä,
40014 Jyväskylä, Finland
e-mail: klaas.giesbertz@jyu.fi

O. V. Gritsenko · E. J. Baerends
Section Theoretical Chemistry, VU University, De Boelelaan 1083,
1081 HV Amsterdam, The Netherlands
e-mail: ov.gritsenko@few.vu.nl

E. J. Baerends (✉) · O. V. Gritsenko
Pohang University of Science and Technology,
San 31, Hyojadong, Namgu, 790-784 Pohang, Republic of Korea
e-mail: baerends@few.vu.nl

M. A. L. Marques et al. (eds.), *Fundamentals of Time-Dependent Density Functional Theory*, Lecture Notes in Physics 837, DOI: 10.1007/978-3-642-23518-4_26,
© Springer-Verlag Berlin Heidelberg 2012

functional. The charge transfer excitation is essentially non-local, since an electron is excited from one region of the system to an other region. Therefore, two regions are important and the approximation will need to depend non-locally on the density to handle these excitations. Indeed, TDHF, which has a kernel connecting densities at two different points in space, already performs much better for these excitations (Dreuw et al. 2003). This led to a new hybrid functional which uses 100% exact exchange at long distances (Yanai et al. 2004). Similar, it comes as no surprise that the EXX functional also gives an improved description of charge transfer excitations (Hesselmann et al. 2009).

Another important class of systems for which adiabatic TDDFT fails badly are strongly correlated systems . Typical examples are Mott insulators and dissociating molecules. These systems have in common that the description of the ground-state is already problematic for DFT. The wave-function has an essential multi-reference character (important contributions of multiple determinants), so the single determinant wave-function of the Kohn-Sham (KS) system, even though it may represent the electron density perfectly, is very poor as a wave-function. Of course, the xc functional should make up for the bad description by the KS wave-function, but our approximations are by no means up to this task. It is therefore not surprising that also TDDFT with the current approximations fails badly for these systems.

An example is provided with the comparison of the results of a (TD)DFT calculation to exact potential energy surfaces from a full configuration interaction calculation in Fig. 26.1. The excited potential energy surfaces for (TD)DFT are obtained by adding the excitation energy from the TDDFT calculation to the ground-state DFT energy. Around equilibrium distance (1.4 bohr) DFT does a good job for the ground state. The $1^1 \Sigma_g^+$ surface closely resembles the curve from the exact calculation. However, at elongated distances ($\gtrsim 3.5$ bohr) the DFT (LDA or GGA) surfaces start to deviate significantly from the exact potential energy surfaces.

The performance of (TD)DFT for the excited states is much poorer. Only at very short distances (1–1.5 bohr), the general trend of the excited potential energy surfaces is allright, although the quantitative agreement (error of ca. 2.5 eV) is not particularly good. However, at longer distances both excited surfaces predicted by (TD)DFT have little resemblance with the exact surfaces. The $1^1 \Sigma_u^+$ excitation energy actually goes to zero in the dissociation limit, so it approaches the ground-state energy surface from above. Therefore, this (TD)DFT calculation predicts that the H_2 molecule would immediately dissociate after an excitation to the $1^1 \Sigma_u^+$, whereas the molecule should remain bound according to the exact $1^1 \Sigma_u^+$ surface. Although the excitation energy to the $2^1 \Sigma_g^+$ surface at least does not collapse to zero, its performance cannot be considered to be much better. It has wrong behavior at short distance, being not only too low but also missing the double minimum structure of the exact potential energy surface. At longer distance it turns to higher energy than the exact surface and at still larger distance it lacks the $1/R$ behavior. For more examples we refer to (Giesbertz et al. 2009).

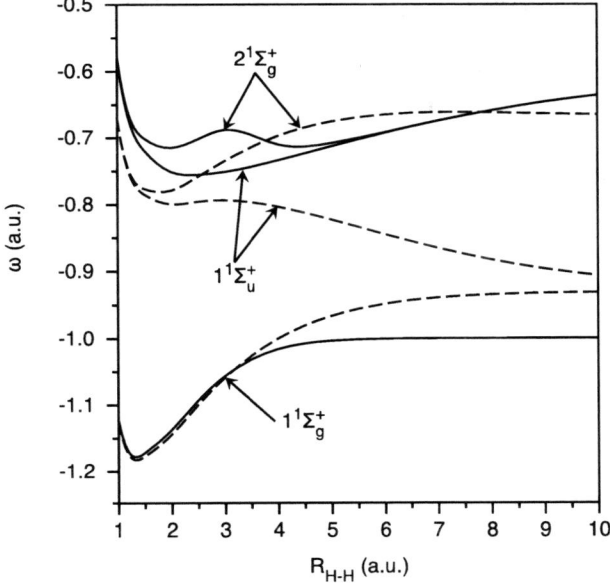

Fig. 26.1 The $1^1 \Sigma_g^+$ (ground-state), $1^1 \Sigma_u^+$ and $2^1 \Sigma_g^+$ potential energy surfaces of the hydrogen molecule (H_2) calculated in a cartesian aug-cc-pVTZ basis. The *full* lines show the results from full configuration interaction and the *dashed* lines show the (TD)DFT results using the Becke 88 + Perdew 86 functional and the adiabatic approximation

26.2 The One-Body Reduced Density Matrix

The problem of DFT is that it is very hard to describe the changing physics of electron correlation along the bond breaking coordinate of a molecule with just the electron density. The density is a rather structureless function of r, except for the cusps at the nuclei. The problem is exemplified by the kinetic energy, which is one example of a property that is very hard to obtain from an accurate density functional. It has been very important for DFT to use the KS system to have some physically reasonable approximation to the kinetic energy. The KS independent particle system reproduces the density with only a single Slater determinant. However, strong correlation means that multiple determinants become important in the wave-function. This means that the KS system is not a physically useful starting point for property evaluations. The approximate functionals and kernels will have to make up for the deficiencies in the KS determinant as a wave-function. This places a too heavy burden on the existing, rather poor, approximations to the true xc functional.

There are two different ways to deal with this problem. The first one is to try to be very smart and invent some super-duper functional of the one-electron density which can actually deal with strong correlations. The second approach is to make life easier by not using only the density, but by using the additional information that is included in orbitals, for instance in the KS orbitals. This is the idea behind the optimized effective potential methods (OEP) (see Chap. 6). This is still within the density functional theory context, since the KS orbitals are implicit functionals of the ground state density. Stepping beyond DFT proper, one can contemplate employing the complete interacting one-body reduced density matrix (1RDM), or equivalently the natural orbitals (NO) and NO occupation numbers. The feasibility of a 1RDM functional theory has been provided by the extension of the Hohenberg-Kohn proof for the unique relation between the density and the ground state wave-function to such a one-to-one relation between 1RDM and ground state wave-function by Gilbert (1975). The 1RDM is defined as

$$\gamma(x, x', t) \equiv \langle \Psi | \hat{\psi}_H^\dagger(x', t) \hat{\psi}_H(x, t) | \Psi \rangle, \qquad (26.1)$$

where we used a combined space-spin coordinate, $x = (r, \sigma)$. The 1RDM has the advantage over the density that *any* one-body property can be calculated from it. For example the kinetic energy can be calculated directly from the 1RDM. In DFT it is necessary to represent part of the kinetic energy in the exchange-correlation functional, together with all the other exchange-correlation effects. This is no longer needed with RDMFT. Another important feature comes from the spectral representation of the 1RDM

$$\gamma(x, x', t) = \sum_k n_k(t) \phi_k(x, t) \phi_k^*(x', t). \qquad (26.2)$$

The eigenfunctions, $\phi_k(x, t)$, are called natural (spin) orbitals and the eigenvalues, $n_k(t)$, the (natural) occupation numbers (Löwdin 1955). In accordance with their name, the occupation numbers have to be positive and in the case of fermions cannot be larger than one. The occupation numbers sum to the total number of electrons, N.

For spin-compensated systems ($M = 0$ or $N_\uparrow = N_\downarrow$) the $\uparrow\uparrow$-block and the $\downarrow\downarrow$ block are necessarily equal, so often the spin-integrated 1RDM is used

$$\gamma(r, r', t) \equiv \sum_\sigma \gamma(r\sigma, r'\sigma, t). \qquad (26.3)$$

The spin-integrated 1RDM can also be diagonalized and has the same properties as the spin-dependent 1RDM. The only difference is that the NOs can contain two electrons (up and down) instead of only one. This is reflected in the upper-bound of the occupation numbers as $0 \leq n_k \leq 2$.

The fractional occupation numbers are important indicators of strong correlation. Let us study the ground-state of H_2 in more detail as an example. Expanding the wave-function in NOs, only two determinants are important to give a good description of H_2 dissociation. Since the singlet wave-function is already anti-symmetric (with respect to permutation of electron coordinates) in the spin part, the spatial part is symmetric

$$\Psi(r_1, r_2) = c_g \sigma_g(r_1) \sigma_g(r_2) + c_u \sigma_u(r_1) \sigma_u(r_2). \qquad (26.4)$$

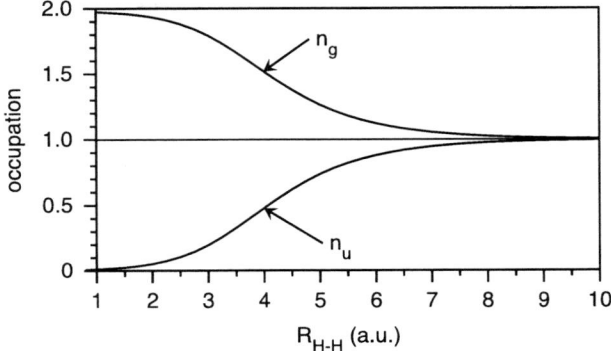

Fig. 26.2 The occupation numbers of the $1\sigma_g$ and $1\sigma_u$ NO of the ground-state $(1\,^1\Sigma_g^+)$ of the hydrogen molecule (H_2).

The occupation numbers of the 1RDM are simply related to the coefficients in the wave-function expansion as $n_k = 2|c_k|^2$, so the occupation numbers are clear indicators which determinant is important. The occupation numbers for the ground state of the H_2 calculation are shown in Fig. 26.2. The wave-function (26.4) with only the doubly occupied $|(\sigma_g)^2|$ and $|(\sigma_u)^2|$ determinants is well justified, since $n_g + n_u \approx 2$. At short distance (around equilibrium) only the $|(\sigma_g)^2|$ determinant is important since $n_g \approx 2$. However, at elongated bond-lengths, the $|(\sigma_u)^2|$ determinant also becomes important ($n_g \approx n_u \approx 1$ at 10 bohr). Since the KS system has only one determinant, it is a poor wave-function. This should not be detrimental for application of DFT if one would know the exact exchange-correlation functional. However, with the approximate functionals that are generally used even the gross details of the changing electron correlation, which are immediately apparent from the behavior of the NO occupation numbers, are not captured. In adiabatic TDDFT, where not only the first functional derivative of the exchange-correlation energy (the KS potential) but also the second derivative (the exchange-correlation kernel) are important, the problems are amplified, and understandably TDDFT fails in practice for cases of more intricate electron correlation (Fig. 26.1). It appears that a theory that uses the much more extensive information available in the orbitals and in the occupation numbers, i.e. a theory that uses the 1RDM, might be an avenue to much higher accuracy. For ground state problems this has proven to be the case (Rohr et al. 2008; Piris et al. 2010b). In this chapter we investigate to what extent the same benefits can be obtained with a time-dependent extension of RDMFT for the treatment of response properties.

26.3 1RDM Functionals

Instead of using the 1RDM directly to approximate functionals, it is more common, as noted above, to use the NOs and occupation numbers to build approximations.

An example of a functional for which we do know both forms is the Hartree-Fock (HF) functional. The HF electron-electron interaction energy E^{HF} can be written as an explicit 1RDM functional

$$E^{HF}[\gamma] \equiv \frac{1}{2} \int d^4x \int d^4x' \left[\gamma(x,x)\gamma(x',x') - |\gamma(x,x')|^2 \right] v_{ee}(x,x') . \quad (26.5)$$

By inserting the spectral expansion for the 1RDM (26.2), one finds an equivalent definition in terms of the NOs and occupation numbers

$$E^{HF}[\{\phi, n\}] \equiv \frac{1}{2} \sum_{rs} n_r n_s (w_{rssr} - w_{rsrs}), \quad (26.6)$$

where the two-electron integrals are defined as

$$w_{klba} \equiv \int d^4x_1 \int d^4x_2 \, \phi_k^*(x_1)\phi_l^*(x_2) v_{ee}(x_1, x_2)\phi_b(x_2)\phi_a(x_1). \quad (26.7)$$

The only other functional for which an explicit 1RDM expression is known is the Müller functional (1984), which is identical to the functional of Buijse (1991); Buijse and Baerends (2002). The latter was inspired by electron correlation in two-electron systems (see below). All the other functionals are only known in the NO representation. It is perfectly fine to define the functional in terms of the NOs and occupation numbers and have no explicit 1RDM form. We only have to keep in mind that a 1RDM functional cannot depend on the phases of the NOs, since these are arbitrary, the NOs being the eigenfunctions of a Hermitian kernel, the 1RDM.

The two-electron system takes a special role in RDMFT, since the wave-function can be reconstructed almost completely from the 1RDM in a trivial manner. This relation follows from the decomposition of the wave-function in NOs, first shown by Löwdin and Shull(1956). This is actually a special case of the expansion of the wave-function in eigenfunctions of the pRDM and $(N - p)$RDM derived by Carlson and Keller (1961). In this section we will only show the simplest case: the singlet two-body system.

Since the spin part of the singlet two-body wave-function is anti-symmetric, the spatial part is symmetric under permutation. Therefore the expansion of the spatial part of the wave-function in a basis of products of two one-electron functions (from a complete one-electron basis) will involve a symmetric matrix of expansion coefficients, which can be diagonalized. So the wave-function can be written in the spectral form

$$\Psi(r_1, r_2, t) = \sum_k c_k(t)\psi_k(r_1, t)\psi_k(r_2, t) , \quad (26.8)$$

where the orbitals ψ_k are orthonormal and the squares of the coefficients sum to one for the wave-function to be normalized. The corresponding spin-integrated 1RDM can be calculated as

$$\gamma(\boldsymbol{r}, \boldsymbol{r}', t) = 2 \int d^3 r_2 \Psi(\boldsymbol{r}, \boldsymbol{r}_2, t) \Psi^*(\boldsymbol{r}', \boldsymbol{r}_2, t) = 2 \sum_k |c_k(t)|^2 \psi_k(\boldsymbol{r}, t) \psi_k^*(\boldsymbol{r}', t),$$

$$(26.9)$$

since the orbitals ψ_k are orthonormal. We find that the eigenfunctions of the wave-function are the NOs and that the eigenvalues of the wave-function are related to the occupation numbers as $n_k(t) = 2|c_k(t)|^2$. Therefore, given the NOs, ϕ_k, and occupation numbers, $n_k(t)$, of a 1RDM corresponding to a singlet two-body system, the corresponding wave-function can be expressed as

$$\Psi(\boldsymbol{r}_1, \boldsymbol{r}_2, t) = \sum_k f_k(t) \sqrt{\frac{n_k(t)}{2}} \phi_k(\boldsymbol{r}_1, t) \phi_k(\boldsymbol{r}_2, t), \qquad (26.10)$$

where $f_k(t)$ are phase-factors. Now the constrained search formulation (Levy 1979) of the exact ground state functional for the interaction part of the energy will be

$$E[\gamma] = \min_{\Psi \to \gamma} \langle \Psi | \hat{V}_{ee} | \Psi \rangle = \frac{1}{2} \min_{\{f\}} \sum_{r,s} f_r f_s^* \sqrt{n_r n_s} w_{rrss}. \qquad (26.11)$$

In practice, this minimization over the phase-factors is rarely carried out. Usually only the two highest occupied NOs give a significant contribution and $f_1 = +1$, $f_2 = -1$ (this sign choice can be shown to lead to a correlated electron movement). In most known cases (e.g. the He iso-electronic series, and H_2 at equilibrium) the higher f_k are also negative. It has only been shown that the dissociation of H_2 requires, at very elongated bond lengths, an alternating sign pattern (Cioslowski and Pernal 2006). Since all the occupation numbers go to zero in that case, except for the first σ_g and σ_u NO, keeping even in that case a negative sign for $k > 2$ hardly affects the energy. Fixing the phases according to the above prescription, one can define a functional as

$$E^{PILS}[\{\phi, n\}] = \frac{1}{2} \sum_{r,s} f_r f_s^* \sqrt{n_r n_s} w_{rrss}. \qquad (26.12)$$

This functional is not a proper 1RDM functional, since it depends on the phases of the NOs. For example, multiplying an NO ϕ_s by i gives an additional minus sign in the two-body energy expression. Therefore, this functional is named the phase including Löwdin-Shull (PILS) functional, since it corresponds to the energy expression obtained with the Löwdin-Shull expansion of the two-electron wave-function.

It is possible to convert this functional into a proper DMFT functional, independent of the NO phases, by simply swapping the stars between one r and one s NO in the integral. The two-electron integral then becomes the normal exchange integral and we obtain a functional we call the density matrix Löwdin-Shull (DMLS) functional

$$E^{DMLS} = \frac{1}{2} \sum_{r,s} f_r f_s^* \sqrt{n_r n_s} w_{rsrs}. \qquad (26.13)$$

Note that, for the ground-state, changing the complex conjugation in the definition usually does not matter, since one then varies only over real NOs. However, since we have to work with complex NOs in time-dependent systems, we can expect the difference between these functionals to show up.

26.4 The Equation of Motion

The basic equation for TDRDMFT will be the equation of motion of the 1RDM. Its derivation is rather straightforward by using the equations of motion for the creation and annihilation operators (Fetter and Walecka 1971)

$$
\begin{aligned}
i\partial_t \gamma(\boldsymbol{x}_1, \boldsymbol{x}'_1, t) &= i\langle\Psi| \left(\partial_t \hat{\psi}^\dagger_{\mathrm{H}}(\boldsymbol{x}'_1, t)\right) \hat{\psi}_{\mathrm{H}}(\boldsymbol{x}_1, t) + \hat{\psi}^\dagger_{\mathrm{H}}(\boldsymbol{x}'_1, t)\partial_t \hat{\psi}_{\mathrm{H}}(\boldsymbol{x}_1, t)|\Psi\rangle \\
&= \left[\hat{h}(\boldsymbol{x}_1, t) - \hat{h}(\boldsymbol{x}'_1, t)\right]\gamma(\boldsymbol{x}_1, \boldsymbol{x}'_1, t) \\
&\quad + \int d^4x_2 \left[v_{\mathrm{ee}}(\boldsymbol{x}_1, \boldsymbol{x}_2) - v_{\mathrm{ee}}(\boldsymbol{x}'_1, \boldsymbol{x}_2)\right] \Gamma(\boldsymbol{x}_1\boldsymbol{x}_2, \boldsymbol{x}_2\boldsymbol{x}'_1, t), \quad (26.14)
\end{aligned}
$$

where $\hat{h}(t) \equiv -\frac{1}{2}\nabla^2 + \hat{v}_{\mathrm{ext}}(t)$ and we introduced the 2RDM

$$
\Gamma(\boldsymbol{x}_1\boldsymbol{x}_2, \boldsymbol{x}'_2\boldsymbol{x}'_1, t) \equiv \langle\Psi|\hat{\psi}^\dagger_{\mathrm{H}}(\boldsymbol{x}'_1, t)\hat{\psi}^\dagger_{\mathrm{H}}(\boldsymbol{x}'_2, t)\hat{\psi}_{\mathrm{H}}(\boldsymbol{x}_2, t)\hat{\psi}_{\mathrm{H}}(\boldsymbol{x}_1, t)|\Psi\rangle. \quad (26.15)
$$

The equation of motion of the 1RDM depends on the 2RDM, so we should also propagate the 2RDM to solve the equation exactly. However, it is not hard to convince oneself that the equation of motion of the 2RDM depends on the 3RDM and the equation for the 3RDM depends on the 4RDM, etc. till we reach the complete N RDM. This chain of equations for the RDMs is known as the Bogoliubov-Born-Green-Kirkwood-Yvon hierarchy. The only way to use this hierarchy is to break the chain at some suitable level. In the case of TDRDMFT we will use that the 2RDM is a functional of the 1RDM, $\Gamma[\gamma]$, under certain conditions.

Often, the field operators are expanded in a basis, $\{\chi\}$ (which in practice has to be finite) as

$$
\hat{\psi}_{\mathrm{H}}(\boldsymbol{x}, t) = \sum_r \hat{c}_r(t)\chi_r(\boldsymbol{x}). \quad (26.16)
$$

The 1RDM and 2RDM in the basis $\{\chi\}$ are now defined by explicit matrices

$$
\gamma_{ka}(t) = \langle\Psi|\hat{c}^\dagger_a(t)\hat{c}_k(t)|\Psi\rangle, \quad (26.17a)
$$

$$
\Gamma_{klba}(t) = \langle\Psi|\hat{c}^\dagger_a(t)\hat{c}^\dagger_b(t)\hat{c}_l(t)\hat{c}_k(t)|\Psi\rangle \quad (26.17b)
$$

We can now also write the equation of motion for the 1RDM in terms of these matrices by multiplying (26.14) by $\chi^*_k(\boldsymbol{x})$ and $\chi_l(\boldsymbol{x}')$ and integrating over $\boldsymbol{x}$ and $\boldsymbol{x}'$ to obtain

$$i\partial_t \gamma_{kl}(t) = \sum_r \left[h_{kr}(t)\gamma_{rl}(t) - \gamma_{kr}(t)h_{rl}(t) \right] + \left[\widetilde{W}_{kl}^{\dagger}[\gamma](t) - \widetilde{W}_{kl}^{\dagger}[\gamma](t) \right], \quad (26.18)$$

where we used $h_{kl}(t) = \langle \chi_k | \hat{h}(t) | \chi_l \rangle$ and for the contraction of the two-electron integrals with the 2RDM introduced the definition

$$\widetilde{W}_{kl}[\gamma](t) \equiv \sum_{rst} \Gamma_{krst}[\gamma](t) w_{tsrl}. \quad (26.19)$$

Instead of using the stationary basis $\{\chi\}$, we can also use the time-dependent NO basis. The basis functions themselves will be time-dependent and the 1RDM in the NO basis will be simply diagonal, $\gamma_{kl}(t) = n_k(t)\delta_{kl}$. Although not particularly useful for practical calculations, the NO representation will be useful to explain some properties of the adiabatic approximation. The equation of motion for the 1RDM in NO basis, $\{\phi(t)\}$, can be written as

$$i \left\{ \dot{n}_k(t)\delta_{kl} + [n_l(t) - n_k(t)] \langle \phi_k(t)|\dot{\phi}_l(t)\rangle \right\}$$
$$= [n_l(t) - n_k(t)] h_{kl}(t) + \left[\widetilde{W}_{kl}^{\dagger}(t) - \widetilde{W}_{kl}^{\dagger}(t) \right]. \quad (26.20)$$

26.5 Response Equations

The central equations in TDRDMFT are, as in most applications of TDDFT, the linear response equations. They can be used to calculate first order dynamic properties such as the polarizability, but also the excitation energies can be obtained as the response function diverges at these energies (Fetter and Walecka 1971). One starts with the time-independent (stationary) situation, where the NOs are solutions of a time-independent Hamiltonian, $\hat{h}$. The starting 1RDM, $\gamma^{(0)}$, is stationary, so $\partial_t \gamma^{(0)} = 0$. Although not necessary, we assume the stationary 1RDM to be represented in the basis of the initial NOs, so it is diagonal, $\gamma_{kl}^{(0)} = n_k \delta_{kl}$. Now we apply an infinitesimal time-dependent potential, $\delta v_{ext}(t)$, so the one-body Hamiltonian becomes $\hat{h}(t) = \hat{h} + \delta v_{ext}(t)$. Due to the change in the potential, also the 1RDM will change, $\gamma(t) = \gamma^{(0)} + \delta \gamma(t) + \delta^{(2)}\gamma(t) + \cdots$. To calculate the linear response of the 1RDM, $\delta\gamma(t)$, we simply use the equation of motion of the 1RDM (26.18) and retain only the first order terms

$$i\delta\dot{\gamma}_{kl}(t) = (n_l - n_k)\delta v_{kl}(t) + \sum_r \left[h_{kr}\delta\gamma_{rl}(t) - \delta\gamma_{kr}(t)h_{rl} \right] + \delta \left[\widetilde{W}_{kl}^{\dagger}(t) - \widetilde{W}_{kl}^{\dagger}(t) \right].$$
$$(26.21)$$

To get an equation in $\delta\gamma(t)$ only, we define the coupling matrix

$$K_{kl,ba}[\gamma^{(0)}](t, t') \equiv \left. \frac{\delta \left[\widetilde{W}_{kl}^{\dagger}(t) - \widetilde{W}_{kl}(t) \right]}{\delta\gamma_{ab}(t')} \right|_{\gamma=\gamma^{(0)}}. \quad (26.22)$$

The coupling matrix is the TDRDMFT analogue of the Hartree-exchange-correlation kernel, f_{Hxc}, in TDDFT (see Sect. 4.5). The linear response equations can now be expressed in terms of $\delta v_{ext}(t)$ and $\delta \gamma(t)$

$$i\delta \dot{\gamma}_{kl}(t) = (n_l - n_k)\delta v_{kl}(t) + \sum_r \left[h_{kr}\delta \gamma_{rl}(t) - \delta \gamma_{kr}(t)h_{rl} \right]$$

$$+ \sum_{ab} \int_{-\infty}^{\infty} \mathrm{d}t' K_{kl,ba}[\gamma^{(0)}](t - t')\delta \gamma_{ab}(t). \tag{26.23}$$

Note that we wrote that the coupling matrix only depends on the time difference. Because we use a stationary state as a reference, the response function depends only on $t - t'$, so also the coupling matrix only depends on $t - t'$. We need this result, since later on we will take the Fourier transform which facilitates the calculation of frequency-dependent properties and excitation energies. Before taking the Fourier transform, we will first introduce the adiabatic approximation, since this will eliminate some parts in the linear response equations. The adiabatic approximation is made in the same way as in TDDFT (see Sect. 4.7.1). The functionals are approximated to be local in time and further approximated by their ground-state counterpart. In particular, the time-dependence of the coupling matrix becomes a simple delta-function

$$K_{kl,ba}[\gamma^{(0)}](t - t') \approx K^{GS}_{kl,ba}[\gamma^{(0)}]\delta(t - t'). \tag{26.24}$$

The linear response equations in the adiabatic approximation can be expressed as

$$i\delta \dot{\gamma}_{kl}(t) = (n_l - n_k)\delta v_{kl}(t) + \sum_{ab} A_{kl,ba}\delta \gamma_{ab}(t), \tag{26.25}$$

where we used the definition

$$A_{kl,ba} \equiv h_{ka}\delta_{bl} - \delta_{ka}h_{bl} + K^{GS}_{kl,ba}. \tag{26.26}$$

Since the coupling matrix is now of such a simple form, it becomes useful to expand the response equations using explicitly the real and imaginary parts of the matrix elements of the response 1RDM and the perturbation. The perturbed 1RDM has to remain Hermitian, so we have $\Re e\, \delta \gamma_{kl} = \Re e\, \delta \gamma_{lk}$ and $\Im m\, \delta \gamma_{kl} = -\Im m\, \delta \gamma_{lk}$. Using these symmetries, we can write

$$i\partial_t \Re e\, \delta \gamma_{kl}(t) = (n_l - n_k)i\Im m\, \delta v_{kl}(t) + \sum_{a>b} A^-_{kl,ba}i\Im m\, \delta \gamma_{ab}(t), \tag{26.27a}$$

$$i\partial_t i\Im m\, \delta \gamma_{kl}(t) = (n_l - n_k)\Re e\, \delta v_{kl}(t) + \sum_{a>b} A^+_{kl,ba}\Re e\, \delta \gamma_{ab}(t)$$

$$+ \sum_a A_{kl,aa}\delta n_a(t), \tag{26.27b}$$

where we used $A_{kl,ba}^{\pm} \equiv A_{kl,ba} \pm A_{kl,ab}$. To obtain the frequency-dependent linear response equations, we simply have to take the Fourier transform. Since we have simple linear equations, we can arrange them in a nice matrix form

$$
\begin{pmatrix} \omega \mathbf{1}_M & A_{MM}^- & \mathbf{0} \\ A_{MM}^+ & \omega \mathbf{1}_M & A_{Mm} \\ \mathbf{0} & A_{mM}^- & \omega \mathbf{1}_m \end{pmatrix} \begin{pmatrix} \delta\gamma^R(\omega) \\ i\delta\gamma^I(\omega) \\ \delta n(\omega) \end{pmatrix} = \begin{pmatrix} N i\delta v^I(\omega) \\ N\delta v^R(\omega) \\ 0 \end{pmatrix},
\tag{26.28}
$$

where we used $f^R(\omega) \equiv \mathcal{F}[\Re e\ f](\omega)$ and $f^I(\omega) \equiv \mathcal{F}[\Im m\ f](\omega)$ as a short-hand notation for the Fourier transform of the real and imaginary parts of the function $f(t)$ respectively and $N_{kl,ba} \equiv (n_l - n_k)\delta_{ka}\delta_{bl}$. The big matrix on the l.h.s. is an $(M, M, m) \times (M, M, m)$ matrix, where m denotes the size of the basis set and $M = m(m - 1)/2$ is the number of unique off-diagonal elements. Thus the first two entries in the vectors only denote the unique off-diagonal elements and $\delta n(\omega)$ the diagonal of the perturbation in the 1RDM. As an indication of which submatrices of the matrices A and $A^{\pm}$ are meant, we used subscripts m and M (diagonal and off-diagonal part respectively).

From the structure of the adiabatic frequency-dependent response equations (26.28), one might already suspect a problem for adiabatic TDRDMFT. The response should be symmetric for positive and negative frequencies, since it should not matter if we go forward or backward in time. In particular, the excitation energies are found at $+\omega$ and $-\omega$ with eigenvectors of the form real-part pm imaginary-part (Giesbertz 2010). However, in the adiabatic response equations (26.28) we do not have an even number of roots, there are more real elements than imaginary elements. In fact one can show that the adiabatic response equations result in m zero excitations. As discussed in Giesbertz et al. (2009) these m zero excitations are directly related to the m δn response elements, which describe double excitations. Mathematically this prevents a problem with the odd number of roots of (26.28), which could lead to lack of symmetry between + and - ω solutions, since zero is the only number equal to minus itself. But physically it means that the present formalism does not work for the problem of the double excitations mentioned in the beginning. We might expect TDRDMFT to do better, since the possibility to describe double excitations is explicitly built in with the presence of the δn response elements. We refer to (Giesbertz et al. 2010b) for an extension of the TD1RDM method in order to deal with this problem.

26.6 Excitations of H_2

The problem of adiabatic TDRDMFT with the double excitations signaled in the previous section, is rather fundamental. It can be shown that any pure 1RDM functional in the adiabatic approximation will suffer from the problem that there is no response in the occupation numbers (Pernal et al. 2007b; Appel 2007; Giesbertz et al. 2009, 2010). This problem is intimately connected with the the fact that a genuine

1RDM functional does not depend on the phases of the NOs. It can be remedied by introducing functionals that do depend on the phases of the NOs. In this chapter we will not deal with this extension of the theory, but we will restrict ourselves to a demonstration of how important the step of using *phase-including* NO in the functionals is. This can be done using the two functionals for the two-electron system we have defined earlier. The DMLS functional of (26.13) is a genuine 1RDM functional. It is perfect for the ground state energy, and only applies the complex conjugation to the NOs in a different way than the Löwdin-Shull energy expression does. The PILS functional of (26.12) is not a 1RDM functional but is an example of a functional with the additional phase degrees of freedom included. Upon optimization of the phases, this functional reproduces the total energy of the two-electron system, i.e. it is the exact phase-including NO functional in this case.

In Fig. 26.3 we show the results of a linear response calculation for the excitation energies. The starting NOs were simply obtained by performing a full configuration interaction calculation and diagonalizing the 1RDM. For numerical stability reasons we have not included the full set of $\delta\gamma_{kl}^{R/I}$ matrix elements, but we have restricted the vector of response matrix elements $\delta\gamma_{kl}$ to $k \leq 22$, and no restriction on l. Only the first three diagonal elements $\delta\gamma_{kk} = \delta n_k$ have been taken into account. Loosely speaking, we include all "excitations" from the first 22 NOs (with the highest occupation numbers) to all NOs, which is still the majority of the excitations. One might have expected the DMLS functional to give very good results, or even the best, since it is a proper 1RDM functional. However, nothing is farther from the truth. The DMLS excitation energies make a real "spaghetti" of potential energy surfaces in Fig. 26.3. On the other hand, the PILS functional gives practically the same results as the exact calculation. It can actually be shown analytically that the PILS results in this case should be equal to the exact ones (Giesbertz 2010). The slight discrepancy in the fifth excitation is due to the limited excitation space used for the TDRDMFT calculation. In particular we note that the problem of TDDFT that the first $^1\Sigma_u^+$ excitation goes to zero with increasing bond length has disappeared.

26.7 Further Reading

This chapter on (TD)RDMFT is too short to give an exhaustive introduction into the theory. For the theoretical foundations of ground-state RDMFT apart from the references in the text (Löwdin 1955; Carlson and Keller 1961; Gilbert 1975; Levy 1979) also the work of Coleman (1963) (N-representability) is relevant. For the calculation of ionization energies, consult (Morrell et al. 1975; Pernal and Cioslowski 2005; Leiva and Piris 2006), for optimization of the ground-state consider (Cohen and Baerends 2002; Staroverov and Scuseria 2002; Kollmar and Hes 2003; Pernal 2005; Cancès and Pernal 2008; Requist and Pankratov 2008; Piris and Ugalde 2009; Giesbertz and Baerends 2010). The first RDMFT functional was formulated by Müller (1984). It has also been derived from attempts to describe both

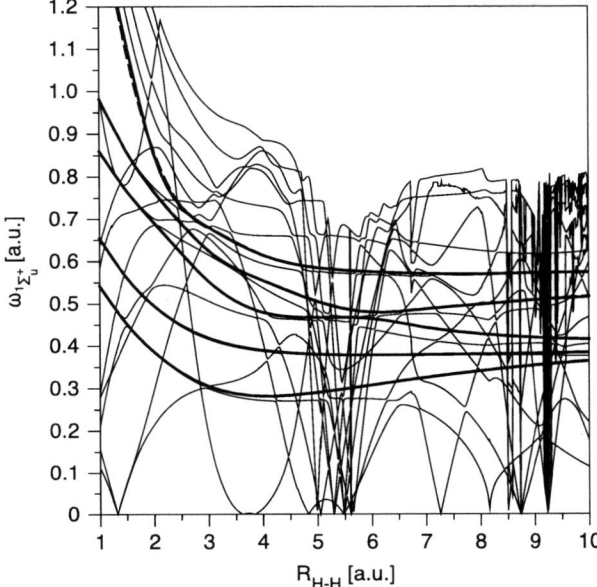

Fig. 26.3 Results for the calculation of the $^1\Sigma_u^+$ excitation energies of H_2 at varying bond distance in a cartesian aug-cc-pVTZ basis. The *thin* lines show the first 14 excitations using the DMLS functional. The exact results are shown by *thick, straight* lines. PILS results (*thick, dashed* lines) are barely visible, since they almost exactly coincide with the exact results

the dynamic and the nondynamic correlation with an orbital based modeling of the exchange-correlation hole (Buijse 1991; Buijse and Baerends 2002). The list of functionals is too long to mention them all, but consider (Goedecker and Umrigar 1998; Lopéz-Sandoval and Pastor 2000; Csányi and Arias 2000; Kollmar and Hes 2003; Gritsenko et al. 2005; Sharma et al. 2008; Rohr et al. 2008; Piris et al. 2010a) for some interesting ones. For more detailed information TDRDMFT consider (Pernal et al. 2007; Appel 2007; Giesbertz et al. 2008, 2009, 2010; Giesbertz 2010).

References

Adragna, G., del Sole, R., Marini, A.: Ab initio calculation of the exchange-correlation kernel in extended systems. Phys. Rev. B **68**, 165108-1–165108-5 (2003)

Adler, S.L.: Quantum theory of the dielectric constant in real solids. Phys. Rev. **126**, 413–420 (1962)

Agostini, P., Fabre, F., Mainfray, G., Petite, G., Rahman, N.K.: Free-free transitions following six-photon Ionization of xenon atoms. Phys. Rev. Lett. **42**, 1127–1130 (1979)

Alavi, A., Kohanoff, J., Parrinello, M., Frenkel, D.: Ab initio molecular dynamics with excited electrons. Phys. Rev. Lett. **73**, 2599–2602 (1994)

Alavi, A., Parrinello, M., Frenkel, D.: Ab initio calculation of the sound velocity of dense hydrogen: Implications for models of Jupiter. Science **269**, 1252–1254 (1995)

Albrecht, S., Reining, L., Sole, R.Del, Onida, G.: Ab initio calculation of excitonic effects in the optical spectra of semiconductors. Phys. Rev. Lett. **80**, 4510–4513 (1998)

Allen, M.P., Tildesley, D.J.: Computer Simulation of Liquids. Oxford University Press, USA (1989)

Allen, M.J., Tozer, D.J.: Helium dimer dispersion forces and correlation potentials in density functional theory. J. Chem. Phys. **117**, 11113–11120 (2002)

Almbladh C.-O., Hedin L.: In: Koch, E.E. (ed.) Handbook on Synchrotron Radiation, vol 1B, chapter 8, North-Holland, Amsterdam (1983)

Almbladh, C.-O.: Photoemission beyond the sudden approximation. J. Phys. Conf. Ser. **35**, 127–144 (2006)

Alnaser, A.S., Voss, S., Tong, X.-M., Maharjan, C.M., Ranitovic, P., Ulrich, B., Osipov, T., Shan, B., Chang, Z., Cocke, C.L.: Effects of molecular structure on ion disintegration patterns in ionization of O_2 and N_2 by short laser pulses. Phys. Rev. Lett. **93**, 113003-1–113003-4 (2004a)

Alnaser, A.S., Tong, X.M., Osipov, T., Voss, S., Maharjan, C.M., Ranitovic, P., Ulrich, B., Shan, B., Chang, Z., Lin, C.D., Cocke, C.L.: Routes to control of H_2 Coulomb explosion in few-cycle laser pulses. Phys. Rev. Lett. **93**, 183202-1–183202-4 (2004b)

Alonso, J.L., Andrade, X., Echenique, P., Falceto, F., Prada-Gracia, D., Rubio, A.: Efficient formalism for large-scale ab initio molecular dynamics based on time-dependent density functional theory. Phys. Rev. Lett. **101**, 096403-1–096403-4 (2008)

Alonso, J.L., Castro, A., Echenique, P., Polo, V., Rubio, A., Zueco, D.: Ab initio molecular dynamics on the electronic Botzmann equilibrium distribution. New J. Phys. **12**, 083064-1–083064-10 (2010)

Amdahl, G.: Validity of the single processor approach to achieving large-scale computing capabilities. AFIPS Conf. Proc. **30**, 483–485 (1967)

M. A. L. Marques et al. (eds.), *Fundamentals of Time-Dependent Density Functional Theory*, Lecture Notes in Physics 837, DOI: 10.1007/978-3-642-23518-4,
© Springer-Verlag Berlin Heidelberg 2012

Ament L.J.P., van Veenendaal M., Devereaux T.P., Hill J.P., van den Brink J.: Resonant inelastic X-ray scattering studies of elementary excitations. Rev. Mod. Phys. **83**, 705–767 (2011)

Anderson, A.G., Goddard III, W.A., Schröder, P.: Quantum Monte Carlo on graphical processing units. Comput. Phys. Commun **177**, 298–306 (2007)

Andersson, Y., Langreth, D.C., Lundqvist, B.I.: van der Waals interactions in density- functional theory. Phys. Rev. Lett **76**, 102–105 (1996)

Ando, T.: Inter-subband optical-absorption in space-charge layers on semiconductor surfaces. Z. Phys. B **26**, 263–272 (1977a)

Ando, T.: Inter-subband optical transitions in a surface space-charge layer. Solid. State. Commun. **21**, 133–136 (1977b)

Andrade, X., Botti, S., Marques, M.A.L., Rubio, A.: Time-dependent density functional theory scheme for efficient calculations of dynamic (hyper)polarizabilities. J. Chem. Phys. **126**, 184106-1–184106-8 (2007)

Andrade, X., Castro, A., Zueco, D., Alonso, J.L., Echenique, P., Falceto, F., Rubio, A.: Modified Ehrennfest formalism for efficient large-scale ab initio molecular dynamics. J. Chem. Theor. Comp. **5**, 728–742 (2009)

Anglada, J.M., Bofill, J.M.: Reduced-restricted-Quasi-Newton-Raphson method for locating and optimizing energy crossing points between two potential energy surfaces. J. Comput. Chem. **18**, 992–1003 (1997)

Angyan, J.G., Savin, A.: van der Waals forces in density functional theory: Perturbational long-range electron interaction corrections. Phys. Rev. A **72**, 012510-1–012510-9 (2005)

Anisimov, V.I., Zaanen, J., Andersen, O.K.: Band theory and Mott insulators: Hubbard U instead of Stoner I. Phys. Rev. B **44**, 943–954 (1991)

Appel, H., Gross, E.K.U., Burke, K.: Excitations in time-dependent density- functional theory. Phys. Rev. Lett. **90**, 043005-1–043005-4 (2003)

Appel H.: Time-dependent quantum many-body systems: Linear response, Electronic transport and reduced density matrices, Ph.D. Thesis, Freie Universität, Berlin, (2007)

Appel, H., Di Ventra, M.: Stochastic quantum molecular dynamics. Phys. Rev. B **80**, 212303-1–212303-4 (2009)

Appel, H., Gross, E.K.U.: Time-dependent natural orbitals and occupation numbers. Europhys. Lett. **92**, 23001-1–23001-5 (2010)

Appel H., Di Ventra M.: Stochastic quantum molecular dynamics for finite and extended systems. Chem. Phys. (2011, in press). http://dx.doi.org/10.1016/ j.chemphys.2011.05.001

Aquino, A.J.A., Plasser, F., Barbatti, M., Lischka, H.: Ultrafast excited-state proton transfer processes: Energy surfaces and on-the-fly dynamics simulations. Croatica Chem. Acta **82**, 105–114 (2009)

Aryasetiawan, F., Gunnarsson, O., Knupfer, M., Fink, J.: Local-field effects in NiO and Ni. Phys. Rev. B **50**, 7311–7321 (1994)

Atchity, G.J., Xantheas, S.S., Ruedenberg, K.: Potential energy surfaces near intersections. J. Chem. Phys. **95**, 1862–1876 (1991)

Atkins, P.W., Friedman, R.S.: Molecular Quantum Mechanics, 3rd edn. Oxford University Press, New York (1997)

Autschbach, J.: Charge-transfer excitations and time-dependent density functional theory: Problems and some proposed solutions. Chem. Phys. Chem. **10**, 1757–1760 (2009)

Awasthi, M., Vanne, Y.V., Saenz, A., Castro, A., Decleva, P.: Single-active-electron approximation for describing molecules in ultrashort laser pulses and its application to molecular hydrogen. Phys. Rev. A **77**, 063403-1–063403-17 (2008)

Baer, R.: Non-adiabatic couplings by time-dependent density functional theory. Chem. Phys. Lett. **364**, 75–79 (2002)

Baer, R., Neuhauser, D., Ždánská, P.R., Moiseyev, N.: Ionization and high-order harmonic generation in aligned benzene by a short intense circularly polarized laser pulse. Phys. Rev. A **68**, 043406-1–043406-8 (2003)

Baer, R., Siam, N.: Real-time study of the adiabatic energy loss in an atomic collision with a metal cluster. J. Chem. Phys. **121**, 6341–6346 (2004)

Baer, R.: Prevalence of the adiabatic exchange-correlation potential approximation in time-dependent density functional theory. J. Mol. Struct. (THEOCHEM) **914**, 19–21 (2009)

Baer, R.: Ground-state degeneracies leave recognizable topological scars in the electronic density. Phys. Rev. Lett. **104**, 073001-1–073001-4 (2010)

Baer, R., Livshits, E., Salzner, U.: Tuned range-separated hybrids in density functional theory. Annu. Rev. Phys. Chem. **61**, 85–109 (2010)

Bailey, P.B., Everitt, W.N., Zettl, A.: The SLEIGN2 Sturm-Liouville Code. ACM Trans. Math. Software **21**, 143–192 (2001)

Balasubramanian, M., Johnson, C.S., Cross, J.O., Seidler, G.T., Fister, T.T., Stern, E.A., Hamner, C., Mariager, S.O.: Fine structure and chemical shifts in nonresonant inelastic x-ray scattering from Li-intercalated graphite. Appl. Phys. Lett. **91**, 031904-1–031904-3 (2001)

Baldereschi, A., Tosatti, E.: Mean-value point and dielectric properties of semiconductors and insulators. Phys. Rev. B **17**, 4710–4717 (1978)

Bandrauk, A.D., Shen, H.: Exponential split operator methods for solving coupled time-dependent Schrödinger equations. J. Chem. Phys. **99**, 1185–1193 (1993)

Bandrauk, A.D.: Molecules in Laser Fields. M. Dekker, New York (1994a)

Bandrauk, A.D.: Molecular multiphoton transitions Computational spectroscopy for perturbative and non-perturbative regimes. Int. Rev. Phys. Chem. **13**, 123–162 (1994b)

Bandrauk, A.D., Shen, H.: Exponential operator methods for coupled timedependent nonlinear Schroedinger equations. J. Phys. A: Math. Gen. **27**, 7147–7155 (1994)

Bandrauk, A.D., Yu, H.: High-order harmonic generation at long range in intense laser pulses. J. Phys. B: At. Mol. Opt. Phys. **31**, 4243–4255 (1998)

Bandrauk, A.D., Yu, H.: High-order harmonic generation by one- and two-electron molecular ions with intense laser pulses. Phys. Rev. A **59**, 539–548 (1999)

Bandrauk, A.D., Chelkowski, S.: Dynamic imaging of nuclear wave functions with ultrashort UV laser pulses. Phys. Rev. Lett. **87**, 273004-1–273004-4 (2001)

Bandrauk, A.D., Kono, H.: Molecules in intense laser fields: Nonlinear multiphoton spectroscopy and near-femtosecond to sub-femtosecond (attosecond) dynamics. In: Lin, S.H., Villaeys, A.A., Fujimura, Y. (eds.) Ad-vances in MultiPhoton Processes, Spectroscopy, vol. 15, pp. 147–214. World Scientific, Singapore (2003)

Bandrauk, A.D., Chelkowski, S., Kawata, I.: Molecular above-threshold-ionization spectra: The effect of moving nuclei. Phys. Rev. A **67**, 013407-1–013407-13 (2003)

Bandrauk, A.D., Lu, H.Z.: Electron correlation and double ionization of a 1D H_2 in an intense laser field. J. Mod. Opt. **53**, 35–44 (2006)

Barbatti, M., Granucci, G., Persico, M., Ruckenbauer, M., Vazdar, M., Eckert-Maksi, M., Lischka, H.: The on-the-fly surface-hopping program system Newton-X: Application to ab initio simulation of the nonadiabatic photodynamics of benchmark systems. J Photochem. Photobiol. A: Chem. **190**, 228–240 (2007)

Barbatti, M., Pitner, J., Pederzoli, M., Werner, U., Mitrić, R., Bonačić-Koutecký, V., Lischka, H.: Non-adiabatic dynamics of pyrrole: Dependence of deactivation mechanisms on the excitation energy. Chem. Phys. **375**, 26–34 (2010)

Baroni, S., Resta, R.: Ab initio calculation of the macroscopic dielectric constant in silicon. Phys. Rev. B **33**, 7017–7021 (1986a)

Baroni, S., Resta, R.: Ab initio calculation of the low-frequency Raman cross section in silicon. Phys. Rev. B **33**, 5969–5971 (1986b)

Baroni, S., Giannozzi, P., Testa, A.: Elastic constants of crystals from linear- response theory. Phys. Rev. Lett. **59**, 2662–2665 (1987a)

Baroni, S., Giannozzi, P., Testa, A.: Green's-function approach to linear response in solids. Phys. Rev. Lett. **58**, 1861–1864 (1987b)

Baroni, S., de Gironcoli, S., Dal Corso, A., Giannozzi, P.: Phonons and related crystal properties from density-functional perturbation theory. Rev. Mod. Phys. **73**, 515–562 (2001)

Baroni, S., Gebauer, R., Malcioğlu, O.B., Saad, Y., Umari, P., Xian, J.: Harnessing molecular excited states with Lanczos chains, J. Phys.: Condens. Matter **22**, 074204-1–074204-8 (2010)

Bartolotti, L.J.: Time-dependent extension of the Hohenberg-Kohn-Levy energy-density functional. Phys. Rev. A **24**, 1661–1667 (1981)

Bartolotti, L.J.: Time-dependent Kohn-Sham density-functional theory. Phys. Rev. A **26**, 2243–2244 (1982)

Bartolotti, L.J.: Variation-perturbation theory within a time-dependent Kohn Sham formalism: An application to the determination of multipole polarizabilities, spectral sums, and dispersion coefficients. J. Chem. Phys. **80**, 5687–5695 (1984)

Bartolotti, L.J.: Velocity form of the Kohn-Sham frequency-dependent polarizability equations. Phys. Rev. A **36**, 4492–4493 (1987)

Bassani, F., Altarelli, M.: Interaction of radiation with condensed matter. In: Koch, E.-E. (ed.) Handbook of Synchrotron Radiation, vol. 1(a), pp. 465–597. Amsterdam, North Holland (1983)

Bastida, A., Cruz, C., Zúñiga, J., Requena, A., Miguel, B.: The Ehrenfest method with quantum corrections to simulate the relaxation of molecules in solution: Equilibrium and dynamics. J. Chem. Phys. **126**, 014503–014511 (2007)

Batani, D., Joachain C. J., Martellucci S., Chester A. J. (eds.): Atoms, solids, and plasmas in super-intense laser fields. Kluwer Academic, New York (2000)

Bauer, D.: Two-dimensional, two-electron model atom in a laser pulse: Exact treatment, single-active-electron analysis, time-dependent density-functional theory, classical calculations, and nonsequential ionization. Phys. Rev. A **56**, 3028–3039 (1997)

Bauer, D., Ceccherini, F.: Electron correlation versus stabilization: A two-electron model in an intense laser pulse. Phys. Rev. A **60**, 2301–2307 (1999)

Bauer, D., Ceccherini, F.: Time-dependent density functional theory applied to nonsequential multiple ionization of Ne at 800 nm. Opt. Express **8**, 377–382 (2001)

Bauernschmitt, R., Ahlrichs, R.: Treatment of electronic excitations within the adiabatic approximation of time dependent density functional theory. Chem. Phys. Lett **256**, 454–464 (1996a)

Bauernschmitt, R., Ahlrichs, R.: Stability analysis for solutions of the closed shell Kohn-Sham equation. J. Chem. Phys. **104**, 9047–9052 (1996b)

Bauernschmitt, R., Häser, M., Treutler, O., Ahlrichs, R.: Calculation of excitation energies within time-dependent density functional theory using auxiliary basis set expansions. Chem. Phys. Lett. **264**, 573–578 (1997)

Bearpark, M.J., Robb, M.A., Schlegel, H.B.: Direct method for the location of the lowest energy point on a potential surface crossing. Chem. Phys. Lett. **223**, 269–274 (1994)

Becke, A.D.: Density-functional exchange-energy approximation with correct asymptotic behavior. Phys. Rev. A **38**, 3098–3100 (1988a)

Becke, A.D.: A multicenter numerical integration scheme for polyatomic molecules. J. Chem. Phys. **88**, 2547–2553 (1988b)

Becke, A.D., Edgecombe, K.E.: A simple measure of electron localization in atomic and molecular systems. J. Chem. Phys. **92**, 5397–5403 (1990)

Becke, A.D.: A new mixing of Hartree-Fock and local density-functional theories. J. Chem. Phys. **98**, 1372–1377 (1993a)

Becke, A.D.: Density-functional thermochemistry. III. The role of exact exchange. J. Chem. Phys. **98**, 5648–5652 (1993b)

Becke, A.D., Johnson, E.R.: Exchange-hole dipole moment and the dispersion interaction revisited. J. Chem. Phys. **127**, 154108-1–154108-6 (2007)

Becker, W., Grasbon, F., Kopold, R., Milošević, D.B., Paulus, G.G., Walther, H.: Abovethreshold ionization: From classical features to quantum effects. Adv. Atom. Mol. Opt. Phys. **48**, 35–98 (2002)

Becker, A., Faisal, F.H.M.: Intense-field many-body S-matrix theory. J. Phys. B: At. Mol. Opt. Phys. **38**, R1–R56 (2005)

Benedict, L.X., Chopra, N.G., Cohen, M.L., Zettl, A. , Louie, S.G., Crespi, V.H.: Microspcopic determination of the interlayer binding energy in graphite. Chem. Phys. Lett., **286**, 490–496 (1998)

Berglund, C.N., Spicer, W.E.: Photoemission studies of copper and silver: Theory. Phys. Rev. **136**, A1030–A1044 (1964)

Berland, K., Hyldgaard, P.: Structure and binding in crystals of cagelike molecules: Hexamine and platonic hydrocarbons. J. Chem. Phys. **132**, 134705-1–13470510 (2010)

Bernard, W., Callen, H.B.: Irreversible thermodynamics of nonlinear processes and noise in driven systems. Rev. Mod. Phys. **31**, 1017–1044 (1959)

Bernasconi, L., Sprik, M., Hutter, J.: Time dependent density functional theory study of charge-transfer and intramolecular electronic excitations in acetone-water systems. J. Chem. Phys. **119**, 12417–12431 (2003)

Berova, N., Nakanishi, K. Woody R.W. (eds.): Circular Dichroism: Principles and Applications. VCH, New York (1994)

Bertoni, G., Verbeeck, J.: Accuracy and precision in model based EELS quantification. Ultramicroscopy **108**, 782–790 (2008)

Bethe, H.A., Salpeter, E.E.: Quantum Mechanics of One and Two-Electron Atoms. Springer, Berlin (1957)

Billeter, S.R., Curioni, A.: Calculation of nonadiabatic couplings in density-functional theory. J. Chem. Phys. **122**, 034105-1–034105-7 (2005)

Bishop, D.M.: Molecular vibrational and rotational motion in static and dynamic electric fields. Rev. Mod. Phys. **62**, 343–374 (1990)

Bloch, F.: Bremsvermögen von Atomen mit mehreren electronen. Z. Phys. **81**, 363–376 (1933)

Blöchl, P.E.: Projector augmented-wave method. Phys. Rev. B **50**, 17953–17979 (1994)

Blöchl, P.E., Först, C.J., Schimpl, J.: Projector augmented wave method: Ab-initio molecular dynamics with full wave functions. Bull. Mater. Sci. **26**, 33–41 (2003)

Blume, M.: Magnetic scattering of x rays. J. Appl. Phys. **57**, 3615–3618 (1985)

Bolivar, A.O.: Quantization of non-Hamiltonian physical systems. Phys. Rev. A **58**, 4330–4335 (1998)

Bolz, J., Farmer, I., Grinspun, E., Schröder, P.: Sparse matrix solvers on the GPU: Conjugate gradients and multigrid. In: Fujii, J. (ed.) ACM SIGGRAPH 2005 Courses, pp. 171–178. ACM, New York (2005)

Born, M., Huang, K.: Dynamical Theory of Crystal Lattices. Oxford University Press, Oxford (1954)

Bornemann, F.A., Nettesheim, P., Schütte, C.: Quantum-classical molecular dynamics as an approximation to full quantum dynamics. J. Chem. Phys. **105**, 1074–1085 (1996)

Bostrom, M., Sernelius, B.E.: Fractional van der Waals interaction between thin metallic films. Phys. Rev. B **61**, 2204–2210 (2000)

Bothma, J.P., Gilmore, J.B., McKenzie, R.H.: The role of quantum effects in proton transfer reactions in enzymes: Quantum tunneling in a noisy environment. New. J. Phys. **055002-1**, 27 (2010)

Botti, S., Fourreau, A., Nguyen, F., Renault, Y.-O., Sottile, F., Reining, L.: Energy dependence of the exchange-correlation kernel of time-dependent density functional theory: A simple model for solids. Phys. Rev. B **72**, 125203-1–125203-9 (2005)

Botti, S., Schindlmayr, A., Del Sole, R., Reining, L.: Time-dependent density-functional theory for extended systems. Rep. Prog. Phys. **70**, 357–407 (2007)

Botti, S., Castro, A., Andrade, X., Rubio, A., Marques, M.A.L.: Cluster-surface and cluster-cluster interactions: Ab initio calculations and modeling of asymptotic van der Waals forces. Phys. Rev. B **78**, 035333-1–035333-10 (2008)

Botti, S., Castro, A., Lathiotakis, N.N., Andrade, X., Marques, M.A.L.: Optical and magnetic properties of boron fullerenes. Phys. Chem. Chem. Phys. **11**, 4523–4527 (2009)

Brabec, T. (ed.): Strong Field Laser Physics. Springer, Berlin (2008)

Brabec, T., Krausz, F.: Intense few-cycle laser fields: Frontiers of nonlinear optics. Rev. Mod. Phys. **72**, 545–591 (2000)

Bradley, J.A., Seidler, G.T., Cooper, G., Vos, M., Hitchcock, A.P., Sorini, A.P., Schlimmer, C., Nagle, K.P.: Comparative study of the valence electronic excitations of N_2 by inelastic X-ray and electron scattering. Phys. Rev. Lett. **105**, 053202-1–053202-4 (2010)

Brancato, G., Rega, N., Barone, V.: Accurate density functional calculations of near-edge X-ray and optical absorption spectra of liquid water using nonperiodic boundary conditions: The role of self-interaction and long-range effects. Phys. Rev. Lett. **100**, 107401-1–107401-4 (2008)

Brandbyge, M., Mozos, J.-L., Ordejón, P., Taylor, J., Stokbro, K.: Density-functional method for nonequilibrium electron transport. Phys. Rev. B **65**, 165401-1–165401-17 (2002)

Branschädel, A., Schneider, G., Schmitteckert, P.: Conductance of inhomogeneous systems: Real-time dynamics. Ann. Phys. (Berlin) **522**, 657–678 (2010)

Breidbach, J., Cederbaum, L.S.: Universal attosecond response to the removal of an electron. Phys. Rev. Lett. **94**, 033901-1–033901-4 (2005)

Breuer, H.-P., Huber, W., Petruccione, F.: Stochastic wave-function method versus density matrix: A numerical comparison. Comp. Phys. Comm. **104**, 46–58 (1997)

Breuer, H.-P., Petruccione, F.: The Theory of Open Quantum Systems. Oxford University Press, Oxford (2002)

Brif, C., Chakrabarti, R., Rabitz, H.: Control of quantum phenomena: Past, present and future. New J. Phys. **12**, 075008-1–075008-68 (2010)

Bruneval, F., Sottile, F., Olevano, V., Del Sole, R., Reining, L.: Many-body perturbation theory using the density-functional concept: Beyond the GW approximation. Phys. Rev. Lett. **94**, 186402-1–186402-4 (2005)

Bruneval, F., Sottile, F., Olevano, V., Reining, L.: Beyond time-dependent exact exchange: The need for long-range correlation. J. Chem. Phys **124**, 144113-1–144113-9 (2006)

Buijse, M., 1991, Electron Correlation, Ph.D. Thesis, VU University, Amsterdam

Buijse, M., Baerends, E.J.: An approximate exchange-correlation hole density as a functional of the natural orbitals. Mol. Phys. **100**, 401–421 (2002)

Bullet, D.W., Haydock, R., Heine, V., Kelly, M.: Solid state physics, vol. 35. In: Ehrenreich, H., Seitz, F., Turnbull D. (eds.). Academic, New York (1980)

Bunker, P.R., Jensen, P.: Molecular Symmetry and Spectroscopy. NRC Research Press, Ottawa (1998)

Burke, P.G., Francken, P., Joachain, C.J.: R-matrix-floquet theory of multiphoton processes. Europhys. Lett. **13**, 617–622 (1990)

Burke, P.G., Francken, P., Joachain, C.J.: R-matrix-floquet theory of multiphoton processes. J. Phys. B: At. Mol. Opt. Phys. **24**, 761–790 (1991)

Burke, K., Perdew, J.P., Ernzerhof, M.: Why semi-local functionals work: Accuracy of the on-top pair density and importance of system averaging. J. Chem. Phys. **109**, 3760–3771 (1998)

Burke, K., Gross, E.K.U.: A guided tour of time-dependent density-functional theory. In: Joubert, D. (ed.) Density Functionals: Theory and Applications, pp. 116–146. Springer, Berlin (1998)

Burke, K., Werschnik, J., Gross, E.K.U.: Time-dependent density functional theory: Past, present, and future. J. Chem. Phys. **123**, 062206-1–062206-9 (2005a)

Burke, K., Car, R.: Electron transport with dissipation: A quantum kinetic approach. Int. J. Quant. Chem. **101**, 564–571 (2005)

Burke, K., Car, R., Gebauer, R.: Density-functional theory of the electrical conductivity of molecular devices. Phys. Rev. Lett. **94**, 146803-1–146803-4 (2005b)

Burnett, K., Reed, V.C., Knight, P.L.: Atoms in ultra-intense laser fields. J. Phys. B At. Mol. Opt. Phys. **26**, 561–598 (1993)

Burnus, T., Marques, M.A. L., Gross, E.K.U.: Time-dependent electron localization function. Phys. Rev. A **71** 010501-1–010501-4(R) (2005)

Bushong, N., Sai, N., Di Ventra, M.: Approach to steady-state transport in nanoscale conductors. Nano. Lett. **5**, 2569–2572 (2005)

Bussi, G., Donadio, D., Parrinello, M.: Canonical sampling through velocity rescaling. J. Chem. Phys. **126**, 014101-1–014101-7 (2007)

Butriy, O., Ebadi, H., de Boeij, P.L., van Leeuwen, R., Gross, E.K.U.: Multicomponent density-functional theory for time-dependent systems. Phys. Rev. **76**, 052514-1–052514-17 (2007)

Caillat, J., Zanghellini, J., Kitzler, M., Koch, O., Kreuzer, W., Scrinzi, A.: Correlated multielectron systems in strong laser fields—an mctdhf approach. Phys. Rev. A **71**, 012712-1–012712-13 (2005)

Caillat, J., Maquet, A., Haessler, S., Fabre, B., Ruchon, T., Salières, P., Mairesse, Y., Ta, R.: Attosecond resolved electron release in two-color near-threshold photoionization of N2. Phys. Rev. Lett. **106**, 093002-1–093002-4 (2011)

Callen, H.B., Welton, T.A.: Irreversibility and generalized noise. Phys. Rev. **83**, 34–40 (1951)

Calvayrac, F., Reinhard, P.-G., Suraud, E.: Coulomb explosion of an Na13 cluster in a diabatic electron-ion dynamical picture. J. Phys. B At. Mol. Opt. Phys. **31**, 5023–5030 (1998)

Cancès, E., Pernal, K.: Projected gradient algorithms for Hartree–Fock and density matrix functional theory calculations. J. Chem. Phys. **128**, 134108-1–134108-8 (2008)

Capitani, J.F., Nalewajski, R.F., Parr, R.G.: Non-Born-Oppenheimer density functional theory of molecular systems. J. Chem. Phys. **76**, 568–573 (2000)

Car, R., Parrinello, M.: Unified approach for molecular dynamics and density—functional theory. Phys. Rev. Lett. **55**, 2471–2474 (1985)

Cardona, M., Pollak, F.H.: Energy-Band structure of germanium and silicon: The k.p Method. Phys. Rev. **142**, 530–543 (1966)

Caricato, M., Mennucci, B., Tomasi, J., Ingrosso, F., Cammi, R., Corni, S., Scalmani, G.: Formation and relaxation of excited states in solution: A new time dependent polarizable continuum model based on time dependent density functional theory. J. Chem. Phys. **124**, 124520-1–124520-13 (2006)

Caricato, M., Trucks, G.W., Frisch, M.J., Wiberg, K.B.: Electronic transition energies: A study of the performance of a large range of single reference density functional and wave function methods on valence and Rydberg states compared to experiment. J. Chem. Theory Comput. **6**, 370–383 (2010)

Caricato, M., Trucks, G.W., Frisch, M.J., Wiberg, K.B.: Oscillator strength: How does TDDFT compare to EOM-CCSD? J. Chem. Theory Comput. **7**, 456–466 (2011)

Carlson, B.C., Keller, J.M.: Eigenvalues of density matrices. Phys. Rev. **121**, 659–661 (1961)

Carmichael, H.J., Singh, S., Vyas, R., Rice, P.R.: Photoelectron waiting times and atomic state reduction in resonance fluorescence. Phys. Rev. A **39**, 1200–1218 (1989)

Carrier, P., Wentzcovitch, R., Tsuchiya, J.: First-principles prediction of crystal structures at high temperatures using the quasiharmonic approximation. Phys. Rev. B **76**, 064116-1–064116-5 (2007)

Casida, M.E.: Time-dependent density functional response theory for molecules. In: Chong, D.E. (ed.) Recent Advances in Density Functional Methods. Recent Advances in Computational Chemistry, vol. 1, pp. 155–192. World Scientific, Singapore (1995)

Casida, M.E.: Time-dependent density functional response theory of molecular systems: Theory, computational methods, and functionals. In: Seminario, J.M. (ed.) Recent Developments and Application of Modern Density Functional Theory, pp. 391–439. Elsevier, Amsterdam (1996)

Casida, M.E., Casida, K.C., Salahub, D.R.: Excited-state potential energy curves from time-dependent density-functional theory: A cross-section of formaldehyde's $^1A_1$ manifold. Int. J. Quant. Chem. **70**, 933–941 (1998a)

Casida, M.E., Jamorski, C., Casida, K.C., Salahub, D.R.: Molecular excitation energies to high-lying bound states from time-dependent density-functional response theory: Characterization and correction of the time-dependent local density approximation ionization threshold. J. Chem. Phys. **108**, 4439–4449 (1998b)

Casida, M.E., Gutierrez, F., Guan, J., Gadea, F.-X., Salahub, D.R., Daudey, J.-P.: Charge-transfer correction for improved time-dependent local density approximation excited-state potential energy curves: Analysis within the two-level model with illustration for H2 and LiH. J. Chem. Phys. **113**, 7062–7071 (2000)

Casida, M.E.: Jacob's Ladder for time-dependent density-functional theory: Some rungs on the way to photochemical heaven. In: Hoffmann, M.R., Dyall, K.G. (eds.) Accurate Description of Low-Lying Molecular Excited States and Potential Energy Surfaces of ACS Symposium Series, vol. 828, pp. 199–220. ACS Press, Washington, D.C (2002)

Casida, M.E., Wesolowski, T.A.: Generalization of the Kohn–Sham equations with constrained electron density formalism and its time-dependent response theory formulation. Int. J. Quantum Chem. **96**, 577–588 (2004)

Casida, M.E.: Propagator corrections to adiabatic time-dependent density-functional theory linear response theory. J. Chem. Phys. **122**, 054111-1–054111-9 (2005)

Castro, A., Marques, M.A.L., Rubio, A.: Propagators for the time-dependent Kohn–Sham equations. J. Chem. Phys. **121**, 3425–3433 (2004a)

Castro, A., Marques, M.A.L., Alonso, J.A., Bertsch, G.F., Rubio, A.: Excited states dynamics in time-dependent density functional theory. Eur. Phys. J. D **28**, 211–218 (2004b)

Castro, A., Appel, H., Oliveira, M., Rozzi, C.A., Andrade, X., Lorenzen, F., Marques, M.A.L., Gross, E.K.U., Rubio, A.: Octopus: A tool for the application of time-dependent density functional theory. Phys. Stat. Sol. B **243**, 2465–2488 (2006)

Castro, A., Burnus, T., Marques, M.A.L., Gross, E.K.U.: Time-dependent electron localization function: A tool to visualize and analyze ultrafast processes. In: Kühn, O., Wöste, L. (eds.) Analysis and Control of Ultrafast Photoinduced Reactions, Springer Series in Chemical Physics, vol. 87, pp. 553–574. Springer, Berlin (2007)

Castro, A., Räsänen, E., Rubio, A., Gross, E.K.U.: Femtosecond laser pulse shaping for enhanced ionization. Europhys. Lett. **87**, 5300-1–5300-6 (2009)

Cave, R.J., Zhang, F., Maitra, N.T., Burke, K.: A dressed TDDFT treatment of the 21Ag states of butadiene and hexatriene. Chem. Phys. Lett. **389**, 39–42 (2004)

Cederbaum, L.: Born-Oppenheimer approximation and beyond. In: Domcke, W., Yarkony, D.R., Noppel, H.K. (eds.) Conical Intersections: Electronic Structure, Dynamics and Spectroscopy, pp. 3–40. World Scientific, Singapore (2004)

Chai, J.-D., Head-Gordon, M.: Systematic optimization of long-range corrected hybrid density functionals. J. Chem. Phys. **128**, 084106-1–084106-15 (2008)

Chakraborty, A., Pak, M.V., Hammes-Schiffer, S.: Development of electron-proton density functionals for multicomponent density functional theory. Phys. Rev. Lett. **101**, 153001-1–153001-4 (2009)

Chang, D.B., Cooper, H.L., Drummond, J.E., Young, A.C.: van der Waals attraction between two conducting chains. Phys. Lett. A **37**, 311–312 (1971)

Charlier, J.-C., Gonze, X., Michenaud, J.-P.: Graphite interplanar bonding: Electronic delocalization and van der Waals interaction. Europhys. Lett. **28**, 403–408 (1994)

Chayes, J.T., Chayes, L., Ruskai, M.B.: Density functional approach to quantum lattice systems. J. Stat. Phys. **38**, 497–518 (1985)

Chelikowsky, J.R., Troullier, N., Saad, Y.: Finite-difference-pseudopotential method: Electronic structure calculations without a basis. Phys. Rev. Lett. **72**, 1240–1243 (1994)

Chelkowski, S., Bandrauk, A.D.: Two-step Coulomb explosions of diatoms in intense laser fields. J. Phys. B At. Mol. Opt. Phys. **28**, L723–L731 (1995)

Cheng, C.-L., Evans, J.S., Van Voorhis, T.: Simulating molecular conductance using real-time density functional theory. Phys. Rev. B **74**, 155112-1–155112-11 (2006)

Chernyak, V., Mukamel, S.: Density-matrix representation of nonadiabatic couplings in time-dependent density functional (TDDFT) theories. J. Chem. Phys. **112**, 3572–3579 (2000)

Chernyak, V., Schulz, M.F., Mukamel, S., Tretiak, S., Tsiper, E.V.: Krylov-space algorithms for time-dependent Hartree–Fock and density functional computations. J. Chem. Phys. **113**, 36–43 (2000)

Chiba, M., Tsuneda, T., Hirao, K.: Excited state geometry optimizations by analytical energy gradient of long-range corrected time-dependent density functional theory. J. Chem. Phys. **124**, 144106-1–144106-11 (2006)

Chin, S.L., Golovinski, P.A.: High harmonic generation in the multiphoton regime: Correlation with polarizability. J. Phys. B At. Mol. Opt. Phys. **28**, 55–63 (1995)

Choquet-Bruhat, Y., DeWitt-Morette, C., Dillard-Bleick, M.: Analysis, Manifolds and Physics, Part I: Basics, Section IIC. North-Holland, Amsterdam (1991)

Chu, X., Chu, S.-I.: Self-interaction-free time-dependent density-functional theory for molecular processes in strong fields: High-order harmonic generation of H2 in intense laser fields. Phys. Rev. A **63**, 023411-1–023411-10 (2001a)

Chu, X., Chu, S.-I.: Time-dependent density-functional theory for molecular processes in strong fields: Study of multiphoton processes and dynamical response of individual valence electrons of N2 in intense laser fields. Phys. Rev. A **64**, 063404-1–063404-9 (2001b)

Chu, S.-I., Telnov, D.A.: Beyond the Floquet theorem: Generalized Floquet formalisms and quasienergy methods for atomic and molecular multiphoton processes in intense laser fields. Phys. Rep. **390**, 1–131 (2004)

Chu, X., and S.-I Chu: Role of the electronic structure and multielectron responses in ionization mechanisms of diatomic molecules in intense short-pulse lasers: An all-electron ab initio study, Phys. Rev. A **70**, 061402-1–061402-4(R) (2004)

Chu, S.-I.: Recent development of self-interaction-free time-dependent density-functional theory for nonperturbative treatment of atomic and molecular multiphoton processes in intense laser fields. J. Chem. Phys. **123**, 062207-1–062207-16 (2005)

Ciminelli, C., Granucci, G., Persico, M.: The photoisomerization mechanism of azobenzene: A semiclassical simulation of nonadiabatic dynamics. Chem. Eur. J. **10**, 2327–2341 (2004)

Cinal, M., Holas, A.: Noniterative accurate algorithm for the exact exchange potential of density-functional theory. Phys. Rev. A **76**, 042510-1–042510-4 (2007)

Cini, M.: Time-dependent approach to electron transport through junctions: General theory and simple applications. Phys. Rev. B **22**, 5887–5899 (1980)

Ciofini, I., Adamo, C.: Accurate evaluation of valence and low-lying Rydberg states with standard time-dependent density functional theory. J. Phys. Chem. A **111**, 5549–5556 (2007)

Cioslowski, J., Pernal, K.: Unoccupied natural orbitals in two-electron Coulombic systems. Chem. Phys. Lett. **430**, 1–3 (2006)

Čížek, J., Paldus, J.: Stability conditions of the Hartree–Fock equations for atomic and molecular systems. Application to the Pi-electron model of cyclic polyenes. J. Chem. Phys. **47**, 3976–3985 (1967)

Cococcioni, M., de Gironcoli, S.: Linear response approach to the calculation of the effective interaction parameters in the LDA+U method. Phys. Rev. B **71**, 035105-1–035105-16 (2005)

Cohen-Tannoudji, C., Dupont-Roc, J., Grynberg, G.: Atom-Photon Interactions. Wiley-VCH, Weinheim (2004)

Cohen, A.J., Baerends, E.J.: Variational density matrix functional calculations for the corrected Hartree and corrected Hartree–Fock functionals. Chem. Phys. Lett. **364**, 409–419 (2002)

Coleman, A.J.: Structure of fermion density matrices. Rev. Mod. Phys. **35**, 668–687 (1963)

Constantin, L.A., Pitarke, J.M., Dobson, J.F., Garcia-Lekue, A., Perdew, J.P.: Highlevel correlated approach to the jellium surface energy, without uniform-electron-gas input. Phys. Rev. Lett. **100**, 036401-1–036401-4 (2008)

Conti, S., Nifosì, R., Tosi, M.P.: The exchange-correlation potential for current-density functional theory of frequency-dependent linear response. J. Phys. Condens. Matter **9**, L475–L482 (1997)

Conti, S., Vignale, G.: Elasticity of an electron liquid. Phys. Rev. B **60**, 7966–7980 (1999)

Cooper, M.J., Mijnarends, P.E., Shiotani, N., Sakai, N., Bansil A. (eds.): X-ray Comp- ton scattering. Oxford University Press, Oxford (2004)

Cooper, V.R.: van der Waals density functional: An appropriate exchange functional. Phys. Rev. B **81**, 161104-1–161104-4 (2010)

Cordova, F., Joubert Doriol, L., Ipatov, A., Casida, M.E.: Troubleshooting time-dependent density-functional theory for photochemical applications: Oxirane. J. Chem. Phys. **127**, 164111-1–164111-18 (2007)

Corkum, P.B.: Plasma perspective on strong field multiphoton ionization. Phys. Rev. Lett. **71**, 1994–1997 (1993)

Corkum, P.B., Ellert, C., Mehendale, M., Dietrich, P., Hankin, S., Aseyev, S., Rayner, D., Villeneuve, D.: Molecular science with strong laser fields. Faraday Discuss. **113**, 47–59 (1999)

Cormier, E., Lambropoulos, P.: Above-threshold ionization spectrum of hydrogen using B-spline functions. J. Phys. B At. Mol. Opt. Phys. **30**, 77–91 (1997)

Cornagia, C.: Molecular rescattering signature in above-threshold ionization. Phys. Rev. A **78**, 041401-1–041401-4(R) (2008)

Cornagia, C.: Electron ion elastic scattering in molecules probed by laser-induced ionization. J. Phys. B **42**, 161002-1–161002-5 (2009)

Cornagia, C.: Enhancements of rescattered electron yields in above-threshold ionization of molecules. Phys. Rev. A **82**, 053410-1–053410-4 (2010)

Corradini, M., Del Sole, R., Onida, G., Pallummo, M.: Analytical expressions for the local-field factor g(q) and the exchange-correlation kernel kxc(r) of the homogeneuos electron gas. Phys. Rev. B **57**, 14569–14571 (1998)

Craig, C.F., Duncan, W.R., Prezhdo, O.V.: Trajectory surface hopping in the time-dependent Kohn–Sham approach for electron-nucelar dynamics. Phys. Rev. Lett. **95**, 163001-1–163001-4 (2005)

Csányi, G., Arias, T.A.: Tensor product expansions for correlation in quantum many-body systems. Phys. Rev. B **61**, 7348–7352 (2000)

Cuniberti, C., Fagas, G., Richter, K. (eds.): Introducing Molecular Electronics. Lecture Notes in Physics, vol. 680. Springer, Berlin (2005)

D'Agosta, R., Vignale, G.: Non-V-representability of currents in time-dependent many particle systems. Phys. Rev. B **71**, 245103-1–245103-6 (2005)

D'Agosta, R., Vignale, G.: Relaxation in time-dependent current-density-functional theory. Phys. Rev. Lett. **96**, 016405-1–016405-4 (2006)

D'Agosta, R., Di Ventra, M.: Stochastic time-dependent current-density-functional theory: A functional theory of open quantum systems. Phys. Rev. B. **78**, 165105-1–165105-16 (2008)

Dahlen, N.E., van Leeuwen, R.: Double ionization of a two-electron system in the time-dependent extended Hartree–Fock approximation. Phys. Rev. A **64**, 023405-1–023405-7 (2001)

Dahlen, N.E.: Effect of electron correlation on the two-particle dynamics of a helium atom in a strong laser pulse. Int. J. Mod. Phys. B **16**, 415–452 (2002)

Dalibard, J., Castin, Y., Mølmer, K.: Wave-function approach to dissipative processes in quantum optics. Phys. Rev. Lett. **68**, 580–583 (1992)

Dal Corso, A., Mauri, F.: Wannier and Bloch orbital computation of the nonlinear susceptibility. Phys. Rev. B **50**, 5756–5759 (1994)

Dal Corso, A., Mauri, F., Rubio, A.: Density-functional theory of the nonlinear optical susceptibility: Application to cubic semiconductors. Phys. Rev. B **53**, 15638–15642 (1996)

Damascelli, A., Hussain, Z., Shen, Z.-X.: Angle-resolved photoemission studies of the cuprate superconductors. Rev. Mod. Phys. **75**, 473–541 (2003)

d'Amico, I., Ullrich, C.A.: Dissipation through spin Coulomb drag in electronic spin transport and optical excitations. Phys. Rev. B. **74**, 121303-1–121303-4 (2006)

d'Amico, I., Ullrich, C.A.: Coulomb interactions and spin transport in semiconductors: The spin Coulomb drag effect. Phys. Status Solidi B **247**, 235–247 (2010)

Dancoff, S.M.: Non-adiabatic meson theory of nuclear forces. Phys. Rev. **78**, 382–385 (1950)

Daubechies, I.: Ten Lectures on Wavelets. SIAM, Philadelphia (1992)

Daul, C.: Density-functional theory applied to excited-states of coordination compounds. Int. J. Quant. Chem. **52**, 867–877 (1994)

Davidson, E.R., Feller, D.: Basis set selection for molecular calculations. Chem. Rev. **86**, 681–696 (1987)

de Boeij, P.L., Kootstra, F., Berger, J.A., van Leeuwen, R., Snijders, J.G.: Current density functional theory for optical spectra: A polarization functional. J. Chem. Phys. **115**, 1995–1999 (2001)

de Boer, M.P., Hoogenraad, J.H., Vrijen, R.B., Noordam, L.D., Muller, H.G.: Indications of high-intensity adiabatic stabilization in neon. Phys. Rev. Lett. **71**, 3263–3266 (1993)

de Boer, M.P., Hoogenraad, J.H., Vrijen, R.B., Constantinescu, R.C., Noordam, L.D., Muller, H.G.: Adiabatic stabilization against photoionization: An experimental study. Phys. Rev. A **50**, 4085–4098 (1994)

de Gironcoli, S., Baroni, S., Resta, R.: Piezoelectric properties of III–V semiconductors from first-principles linear-response theory. Phys. Rev. Lett. **62**, 2853–2856 (1989)

de Gironcoli, S.: Lattice dynamics of metals from density-functional perturbation theory. Phys. Rev. B **51**, 6773–6776 (1995)

De Wijn, A.S., Lein, M., Kümmel, S.: Strong-field ionization in time-dependent density functional theory. Eur. Phys. Lett. **84**, 43001-1–43001-6 (2008)

Deb, B.M., Ghosh, S.K.: Schrödinger fluid dynamics of many-electron systems in a time-dependent density-functional framework. J. Chem. Phys. **77**, 342–348 (1982)

Debernardi, A., Baroni, S.: Third-order density-functional perturbation theory: A practical implementation with applications to anharmonic couplings in Si. Solid State Commun. **91**, 813–816 (1994)

Della Sala, F., Görling, A.: The asymptotic region of the Kohn–Sham exchange potential in molecules. J. Chem. Phys. **116**, 5374–5388 (2002a)

Della Sala, F., Görling, A.: Asymptotic Behavior of the Kohn–Sham Exchange Potential. Phys. Rev. Lett. **89**, 033003-1–033003-4 (2002b)

Delone, N.B., Krainov, V.P.: Multiphoton Processes in Atoms. Springer, Berlin (2000)

Del Sole, R., Fiorino, E.: Macroscopic dielectric tensor at crystal surfaces. Phys. Rev. B **29**, 4631–4645 (1984)

Del Sole, R., Girlanda, R.: Optical properties of semiconductors within the independent-quasiparticle approximation. Phys. Rev. B **48**, 11789–11795 (1993)

Devereaux, T.P., Hackl, R.: Inelastic light scattering from correlated electrons. Rev. Mod. Phys. **79**, 175–233 (2007)

Dhar, A., Sen, D.: Nonequilibrium Green's function formalism and the problem of bound states. Phys. Rev. B **73**, 085119-1–085119-14 (2006)

Dhara, A.K., Ghosh, S.K.: Density-functional theory for time-dependent systems. Phys. Rev. A **35**, 442–444 (1987)

Diau, E.W.-G., Kötting, C., Zewail, A.H.: Femtochemistry of norrish type-I reactions: I. Experimental and theoretical studies of acetone and related ketones on the S_1 surface. Chem. Phys. Chem. **2**, 273–293 (2001a)

Diau, E.W.-G., Kötting, C., Zewail, A.H.: Femtochemistry of Norrish Type-I Reactions: II. The Anomalous Predissociation Dynamics of Cyclobutanone on the S_1 Surface. Chem. Phys. Chem **2**, 294–309 (2001b)

Diau, E.W.-G., Kötting, C., Sølling, T.I., Zewail, A.H.: Femtochemistry of norrish type-I reactions: III. Highly excited ketones—Theoretical. Chem. Phys. Chem **3**, 57–78 (2002)

Diau, E.W.-G., Zewail, A.H.: Femtochemistry of trans-azomethane: A combined experimental and theoretical study. Chem. Phys. Chem **4**, 445–456 (2003)

DiChiara, A.D., Ghebregziabher, I., Waesche, J.M., Stanev, T., Ekanayake, N., Barclay, L.R., Wells, S.J., Watts, A., Videtto, M., Mancuso, C.A., Walker, B.C.: Photoionization by an ultraintense laser field: Response of atomic xenon. Phys. Rev. A **81**, 043417-1–043417-6 (2010)

Dierksen, M., Grimme, S.: The Vibronic structure of electronic absorption spectra of large molecules: A time-dependent density functional study on the influence of "Exact" Hartree–Fock exchange. J. Phys. Chem. A **108**, 10225–10237 (2004)

Dion, M., Rydberg, H., Schröder, E., Langreth, D.C., Lundqvist, B.I.: van der Waals Density Functional for General Geometries. Phys. Rev. Lett. **92**, 246401-1–246401-4 (2004)

Dion, M., Rydberg, H., Schröder, E., Langreth, D.C., Lundqvist, B.I.: van der Waals Density Functional for General Geometries. Phys. Rev. Lett. **95**, 109902(E) (2005)

Dion, M., Burke, K.: Coordinate scaling in time-dependent current density functional theory. Phys. Rev. A **72**, 020502-1–020502-4 (2005)

Diosi, L.: Stochastic pure state representation for open quantum systems. Phys. Lett. A. **114**, 451–454 (1986)

Diosi, L.: Quantum stochastic processes as models for state vector reduction. J. Phys. A **21**, 2885–2898 (1988)

Ditmire, T., Zweiback, J., Yanovsky, V.P., Cowan, T.E., Hays, G., Wharton, K.B.: Nuclear fusion from explosions of femtosecond laser-heated deuterium clusters. Nature (London) **398**, 489–492 (1999)

Di Ventra, M., Pantelides, S.T., Lang, N.D.: First-principles calculation of transport properties of a molecular device. Phys. Rev. Lett. **84**, 979–982 (2000)

Di Ventra, M., Pantelides, S.T.: Hellmann–Feynman theorem and the definition offorces in quantum time-dependent and transport problems. Phys. Rev. B **61**, 16207–16212 (2000)

Di Ventra, M., D'Agosta, R.: Stochastic time-dependent current-density-functional theory. Phys. Rev. Lett. **98**, 226403-1–226403-4 (2007)

Dobson, J.F.: Harmonic-Potential Theorem: Implications for approximate many-body theories. Phys. Rev. Lett. **73**, 2244–2247 (1994a)

Dobson, J.F.: Quasi-local approximation for a van der Waals energy functional. In: Das D.P. (ed.) Topics in Condensed Matter Physics, chapter 7. Nova, New York. Reproduced in arXiv:0311371 (1994b)

Dobson, J.F., Dinte, B.P.: Constraint satisfaction in local and gradient susceptibility approximations: Application to a van der Waals density functional. Phys. Rev. Lett. **76**, 1780–1783 (1996)

Dobson, J.F., Bünner, M., Gross, E.K.U.: Time-dependent density functional theory beyond linear response: An exchange-correlation potential with memory. Phys. Rev. Lett. **79**, 1905–1908 (1997)

Dobson, J.F., Dinte, B., Wang, J.: van der Waals Functionals via Local Approximations for Susceptibilities. Plenum, New York (1998)

Dobson, J.F., Wang, J.: Successful test of a seamless van der Waals density functional. Phys. Rev. Lett. **82**, 2123–2126 (1999)

Dobson, J.F., Wang, J.: Energy-optimized,local exchange-correlation kernel for the electron gas, with application to van der Waals forces. Phys. Rev. B **62**, 10038–10045 (2000)

Dobson, J.F., McLennan, K., Rubio, A., Wang, J., Gould, T., Le, H.M., Dinte, B.P.: Prediction of dispersion forces: Is there a problem? Australian J. Chem. **54**, 513–527 (2001)

Dobson, J.F., Wang, J., Gould T.: Correlation energies of inhomogeneous manyelectron systems. Phys. Rev. B **66**, 081108-1–081108-4(R) (2002)

Dobson, J.F., Wang, J.: Testing the local density approximation for energy-versus separation curves of jellium slab pairs. Phys. Rev. B **69**, 235104-1–235104-9 (2004)

Dobson, J.F., Wang, J., Dinte, B.P., McLennan, K., Le, H.M.: Soft cohesive forces. Int. J. Quantum Chem. **101**, 579–598 (2005)

Dobson, J.F., White, A., Rubio, A.: Asymptotics of the dispersion interaction: Analytic benchmarks for van der Waals energy functionals. Phys. Rev. Lett. **96**, 073201-1–073201-4 (2006)

Dobson, J.F., Gould, T., Klich, I.: Dispersion interaction between crossed conducting wires. Phys. Rev. A **80**, 012506-1–012506-5 (2009)

Dobson, J.F.: Inhomogeneous STLS theory and TDCDFT. Phys. Chem. Chem. Phys. **11**, 4528–4534 (2009a)

Dobson, J.F.: Validity comparison between asymptotic dispersion energy formalisms for nanomaterials. J. Comput. Theor. Nanosci. **6**, 960–971 (2009b)

Doltsinis, N.L., Marx, D.: First principles molecular dynamics involving excited states and nonadiabatic transitions. J. Theo. Comput. Chem. **1**, 319–349 (2002)

Doltsinis, N.L., Kosov, D.S.: Plane wave/pseudopotential implementation of excited state gradients in density functional linear response theory: A new route via implicit differentiation. J. Chem. Phys. **122**, 144101-1–144101-7 (2005)

Domcke, W., Stock, G.: Theory of ultrafast nonadiabatic excited-state processes and their spectroscopic detection in real time. Adv. Chem. Phys. **100**, 1–169 (1997)

Domcke, D., Yarkony, D.R., Köppel H. (eds.): Conical Intersections, Electronic Structure, Dynamics & Spectroscopy. World Scientific, Singapore (2004)

Dorner, R., Mergel, V., Jagutzki, O., Spielberger, L., Ullrich, J., Moshammer, R., Schmidt-Bocking, H.: Cold target recoil Ion momentum spectroscopy: A momentum microscope to view atomic collision dynamics. Phys. Rep. **330**, 95–192 (2000)

Dörner, R., Weber, T., Weckenbrock, M., Staudte, A., Hattass, M., Moshammer, R., Ullrich, J., Schmidt-Böcking, H.: Multiple ionization in strong laser fields. Adv. Atom. Mol. Opt. Phys. **48**, 1–34 (2002)

Dreizler, R.M., Gross, E.K.U.: Density Functional Theory: An Approach to the Quantum Many-Body Problem. Springer, Berlin (1990)

Drescher, M., Hentschel, M., Kienberger, R., Tempea, G., Spielmann, C., Reider, G.A., Corkum, P.B., Krausz, F.: X-ray pulses approaching the attosecond frontier. Science **291**, 1923–1927 (2001)

Drescher, M., Hentschel, M., Kienberger, R., Uiberacker, M., Yakovlev, V., Scrinzi, A., Westerwalbesloh, T., Kleineberg, U., Heinzmann, U., Krausz, F.: Time-resolved atomic inner-shell spectroscopy. Nature (London) **419**, 803–807 (2002)

Dreuw, A., Weisman, J.L., Head-Gordon, M.: Long-range charge-transfer excited states in time-dependent density functional theory require non-local exchange. J. Chem. Phys. **119**, 2943–2946 (2003)

Dreuw, A., Head-Gordon, M.: Failure of time-dependent density functional theory for long-range charge-transfer excited states: The zincbacteriochlorin-bacteriochlorin and bacteriochloro-phyll-spheroidene complexes. J. Am. Chem. Soc. **126**, 4007–4016 (2004)

Dreuw, A., Head-Gordon, M.: Single-reference ab Initio methods for the calculation of excited states of large molecules. Chem. Rev. **105**, 4009–4037 (2005)

Drummond, N.D., Needs, R.J.: van der Waals interactions betweem thin metallic wires and layers. Phys. Rev. Lett. **99**, 166401-1–166401-4 (2007)

Dubrovin, B.A., Fomenko, A.T., Novikov, S.P.: Modern Geometry–Methods and Applications, vol. 1. Springer, New York (1984)

Dum, R., Zoller, P., Ritsch, H.: Monte Carlo simulation of the atomic master equation for spontaneous emission. Phys. Rev. A **45**, 4879–4887 (1992a)

Dum, R., Parkins, A.S., Zoller, P., Gardiner, C.W.: Monte Carlo simulation of master equations in quantum optics for vacuum, thermal, and squeezed reservoirs. Phys. Rev. A **46**, 4382–4396 (1992b)

Dundas, D., Rost, J.M.: Molecular effects in the ionization of N_2, O_2, and F_2 by intense laser fields. Phys. Rev. A **71**, 013421-1–013421-8 (2005)

Dunlap, B.I., Connolly, J.W.D., Sabin, J.R.: On some approximations in applications of Xá theory. J. Chem. Phys. **71**, 3396–3402 (1979)

Dunning Jr., T.H.: Gaussian basis sets for use in correlated molecular calculations. I. The atoms boron through neon and hydrogen. J. Chem. Phys. **90**, 1007–1023 (1989)

Dzyaloshinskii, I.E., Lifshitz, E.M., Pitaevskii, L.P.: The general theory of van der Waals forces. Adv. Phys. **10**, 165–209 (1961)

Eberly, J.H., Grobe, R., Law, C.K., Su, Q.: Numerical experiments in strong and super-strong fields. In: Gavrila, M. (ed.) Atoms in Intense Laser Fields, pp. 301–334. Academic Press, Boston (1992)

Eberly, J.H., Kulander, K.C.: Atomic stabilization by super-intense lasers. Science **262**, 1229–1233 (1993)

Echenique, P., Alonso, J.L.: A mathematical and computational review of Hartree-Fock SCF methods in quantum chemistry. Mol. Phys. **105**, 3057–3098 (2007)

Eckart, C.: Some studies concerning rotating axes and polyatomic molecules. Phys. Rev. **47**, 552–558 (1935)

Ehrenfest, P.: Bemerkung über die agenäherte Gültigkeit der klassichen Mechanik innerhalb der Quantenmechanik. Z. Physik **45**, 455–457 (1927)

Ehrenreich, H.: The Optical Properties of Solids: Proceedings of the International School of Physics "Enrico Fermi", In: Tauc, J. (ed.), pp. 106–154. Academic Press, New York (1966)

Eichkorn, K., Treutler, O., Öhm, H., Häser, M., Ahlrichs, R.: Auxiliary basis sets to approximate Coulomb potentials. Chem. Phys. Lett. **242**, 652–660 (1995)

Eichkorn, K., Weigend, F., Treutler, O., Ahlrichs, R.: Auxiliary basis sets for main row atoms and transition metals and their use to approximate Coulomb potentials. Theor. Chem. Acc. **97**, 119–124 (1997)

Elliott, P., Furche, F., Burke, K.: Excited states from time-dependent density functional theory. In: Lipkowitz, K.B., Cundari, T.R. (eds.) Reviews in Computational Chemistry, pp. 91–165. Wiley, Hoboken (2009)

Elliott, P., Goldson, S., Canahui, C., Maitra, N.T.: Perspectives on double-excitations in TDDFT. Chem. Phys. (in press). http://dx.doi.org/10.1016/j.chemphys.2011.03.020

Engel, E., Höck, A., Dreizler, R.M.: van der Waals bonds in density functional theory. Phys. Rev. A **61**, 032502-1-032502-5 (2000)

Engel, E., Bonetti, F.: Implicit density functionals for the exchange-correlation energy: Description of van der Waals bonds. Int. J. Mod. Phys. B **15**, 1703–1713 (2001)

Enkovaara, J., Rostgaard, C., Mortensen, J.J., Chen, J., Dulak, M., Ferrighi, L., Gavnholt, J., Glinsvad, C., Haikola, V., Hansen, H.A., Kristoffersen, H.H., Kuisma, M., Larsen, A.H., Lehtovaara, L., Ljungberg, M., Lopez-Acevedo, O., Moses, P.G., Ojanen, J., Olsen, T., Petzold, V., Romero, N.A., Stausholm-Møller, J., Strange, M., Tritsaris, G.A., Vanin, M., Walter, M., Hammer, B., Häkkinen, H., Madsen, G.K.H., Nieminen, R.M., Nørskov, J.K., Puska, M., Rantala, T.T., Schiøtz, J., Thygesen, K.S., Jacobsen, K.W.: Electronic structure calculations with GPAW: A real-space implementation of the projector augmented-wave method. J. Phys. Condens. Matter **22**, 253202–253224 (2010)

Eppink, A.T.J.B., Parker, D.H.: Velocity map imaging of ions and electrons using electrostatic lenses: Application in photoelectron and photofragment ion imaging of molecular oxygen. Rev. Sci. Instrum. **68**, 3477–3484 (1997)

Eremina, E., Liu, X., Rottke, H., Sandner, W., Schätzel, M.G., Dreischuh, A., Paulus, G.G., Walther, H., Moshammer, R., Ullrich, J.: Influence of molecular structure on double ionization of N_2 and O_2 by high intensity ultrashort laser pulses. Phys. Rev. Lett. **92**, 173001-1-173001-4 (2004)

Erhard, S., Gross, E.K.U.: High harmonic generation in hydrogen and helium atoms subject to one- and two-color laser pulses. In: Lambropoulos, P., Walther, H. (eds.) Multiphoton Processes 1996, pp. 37–46. IOP Publishing, Bristol (1997)

Errea, L.F., Méndez, L., Riera, A., Yáñez, M., Hanssen, J., Harel, C., Salin, A.: The LiH^{2+} quasimolecule: A comparison between the configuration interaction and the OEDM approaches. J. Phys. **46**, 709–718 (1985)

Eschrig, H.: The Fundamentals of Density Functional Theory, 2nd edn. Teubner, Leipzig (2003)

Eschrig, H.: T > 0 ensemble-state density functional theory via Legendre transform. Phys. Rev. B **82**, 205120-1-205120-9 (2010)

Eshuis, H., Yarkony, J., Furche, F.: Fast computation of molecular random phase approximation correlation energies using resolution of the identity and imaginary frequency integration. J. Chem. Phys. **132**, 234114-1-234114-9 (2010)

Eshuis, H., Furche, F.: A parameter-free density functional that works for noncovalent interactions. J. Phys. Chem. Lett. **2**, 983–989 (2011)

Evans, J.S., Cheng, C.L., Van Voorhis, T.: Spin-charge separation in molecular wire conductance simulations. Phys. Rev. B **78**, 165108-1-165108-11 (2008)

Evans, J.S., Van Voorhis, T.: Dynamic current suppression and gate voltage response in metal-molecule-metal junctions. Nano. Lett. **9**, 2671–2675 (2009)

Evans, J.S., Vydrov, O.A., Van Voorhis, T.: Exchange and correlation in molecular wire conductance: Nonlocality is the key. J. Chem. Phys. **131**, 034106-1–034106-10 (2009)

Evans, L.C.: Partial Differential Equations. American Mathematical Society, Rhode Island (2010)

Evers, F., Weigend, F., Koentopp, M.: Conductance of molecular wires and transport calculations based on density-functional theory. Phys. Rev. B **69**, 235411-1–235411-9 (2004)

Fano, U.: Sullo spettro di assorbimento dei gas nobili presso il limite dello spettro d'arco. Nuovo Cimento **12**, 154–161 (1935)

Fano, U.: Effects of configuration interaction on intensities and phase shifts. Phys. Rev. **124**, 1866–1878 (1961)

Fano, U., Cooper, J.W.: Spectral distribution of atomic oscillator strengths. Rev. Mod. Phys. **40**, 441–507 (1968)

Ferray, M., L'Huillier, A., Li, X.F., Lompré, L.A., Mainfray, G., Manus, C.: Multipleharmonic conversion of 1064 nm radiation in rare gases. J. Phys. B At. Mol. Opt. Phys. **21**, L31–L35 (1988)

Feshbach, H.: Unified theory of nuclear reactions. Ann. Phys. **5**, 357–390 (1958)

Fetter, A.L., Walecka, J.D.: Quantum Theory of Many-Particle Systems. McGraw-Hill, New York (1971)

Feynman, R.P.: Forces in molecules. Phys. Rev. **56**, 340–343 (1939)

Filatov, M., Shaik, S.: Application of spin-restricted open-shell Kohn–Sham method to atomic and molecular multiplet states. J. Chem. Phys. **110**, 116–125 (1999)

Filippi, C., Umrigar, C.J., Gonze, X.: Excitation energies from density functional perturbation theory. J. Chem. Phys. **107**, 9994–10002 (1997)

Fiolhais, C., Nogueira, F., Marques M.A.L. (eds.): A Primer in Density Functional Theory. Springer, Berlin (2003)

Fittinghoff, D.N., Bolton, P.R., Chang, B., Kulander, K.C.: Observation of nonsequential double ionization of helium with optical tunneling. Phys. Rev. Lett. **69**, 2642–2645 (1992)

Francl, M.M., Pietro, W.J., Hehre, W.J., Binkley, J.S., Gordon, M.S., DeFrees, D.J., Pople, J.A.: Self-consistent molecular orbital methods. XXIII. A polarization-type basis set for second-row elements. J. Chem. Phys. **77**, 3654–3665 (1982)

Frank, I., Hutter, J., Marx, D., Parrinello, M.: Molecular dynamics in low-spin excited states. J. Chem. Phys. **108**, 4060–4069 (1998)

Freeman, R.R., Bucksbaum, P.H.: Investigations of above-threshold ionization using subpicosecond laser pulses. J. Phys. B At. Mol. Opt. Phys. **24**, 325–347 (1991)

Friedrich, H.: Theoretical Atomic Physics. Springer, Berlin (2006)

Friedrichs, J., Frank, I.: Mechanism of electrocyclic ring opening of diphenyl oxirane: 40 years after Woodward and Hoffmann. Chem. Eur. J. **15**, 10825–10829 (2009)

Fritsche, L., Yuan, J.: Alternative approach to the optimized effective potential method. Phys. Rev. A. **57**, 3425–3432 (1998)

Frydel, D., Terilla, W., Burke, K.: Adiabatic connection from accurate wave-function calculations. J. Chem. Phys. **112**, 5292–5297 (2000)

Fowe, E.P., Bandrauk, A.D.: Nonlinear time-dependent density functional theory studies of the ionization of CO_2 by ultrashort intense laser pulses. Can. J. Chem. **87**, 1081–1089 (2009)

Fowe, E.P., Bandrauk, A.D.: Nonlinear time-dependent density-functional-theory study of ionization and harmonic generation in CO_2 by ultrashort intense laser pulses: Orientational effects. Phys. Rev. A **81**, 023411-1–023411-8 (2010a)

Fowe, E.P., Bandrauk, A.D.: Nonlinear time-dependent density functional investigation and visualization of ionizations in CO_2–Effects of laser intensities and molecular orientations. Can. J. Chem. **89**, 1186–1194 (2010b)

Fuchs, M., Gonze, X.: Accurate density functionals: Approaches using the adiabaticconnection fluctuation-dissipation theorem. Phys. Rev. B **65**, 235109-1–235109-4 (2002)

Fuchs, M., Niquet, Y.-M., Gonze, X., Burke, K.: Describing static correlation in bond dissociation by Kohn–Sham density functional theory. J. Chem. Phys. **122**, 094116-1–09411613 (2005)

Fujiwara, H.: Spectroscopic Ellipsometry—Principles and Applications. Wiley, New York (2007)

Furche, F.: On the density matrix based approach to time-dependent density functional response theory. J. Chem. Phys. **114**, 5982–5992 (2001)

Furche, F.: Molecular tests of the random phase approximation to the exchange-correlation energy functional. Phys. Rev. B **64**, 195120-1–195120-8 (2002)

Furche, F., Ahlrichs, R.: Adiabatic time-dependent density functional methods for excited state properties. J. Chem. Phys. **117**, 7433–7447 (2002a)

Furche, F., Ahlrichs, R.: Absolute configuration of D_2-symmetric fullerene C_{84}. J. Am. Chem. Soc. **124**, 3804–3805 (2002b)

Furche, F., Ahlrichs, R.: Adiabatic time-dependent density functional methods for excited state properties. J. Chem. Phys. **121**, 12772 (E) (2004)

Furche, F., van Voorhis, T.: Fluctuation-dissipation theorem density-functional theory. J. Chem. Phys. **122**, 164106-1–164106-10 (2005)

Furche, F., Rappoport, D.: Density functional methods for excited states: Equilibrium structure and excited spectra. In: Olivucci, M. (ed.) Computational Photochemistry, pp. 93–128. Elsevier, Amsterdam (2005)

Furche, F., Burke, K.: Time-dependent density functional theory in quantum chemistry. In: Spellmeyer, D.C. (ed.) Annual Reports in Computational Chemistry, vol. 1, ch. 2, pp. 19–30. Elsevier, Amsterdam (2005)

Furche, P.: Developing the random phase approximation into a practical post-Kohn–Sham correlation model. J. Chem. Phys. **129**, 114105-1–114105-8 (2008)

Gajdos, M., Hummer, K., Kresse, G., Furthmller, J., Bechstedt, F.: Linear optical properties in the PAW methodology. Phys. Rev. B **73**, 045112-1–045112-9 (2006)

Gao, X., Tao, J., Vignale, G., Tokatly, I.V.: Continuum mechanics for quantum many-body systems: Linear response regime. Phys. Rev. B **81**, 195106-1–195106-22 (2010)

García de Abajo, F.J.: Optical excitations in electron microscopy. Rev. Mod. Phys. **82**, 209–275 (2010)

Garcia-Gonzalez, P., Fernandez, J.J., Marini, A., Rubio, A.: Advanced correlation functionals: Application to bulk materials and localized systems. J. Phys. Chem. A **111**, 12458–12465 (2007)

Gardiner, C.W.: Stochastic Methods: A Handbook for the Natural and Social Sciences, Springer Series in Synergetics, vol. 13, 2nd edn, Springer, New York (1985)

Gardiner, C.W., Parkins, A.S., Zoller, P.: Wave-function quantum stochastic differential equations and quantum-jump simulation methods. Phys. Rev. A **46**, 4363–4381 (1992)

Gaspard, P., Nagaoka, M.: Slippage of initial conditions for the Redfield master equation. J. Chem. Phys. **111**, 5668–5675 (1999a)

Gaspard, P., Nagaoka, M.: Non-Markovian stochastic Schrödinger equation. J. Chem. Phys. **111**, 5676–5690 (1999b)

Gatti, M., Olevano, V., Reining, L., Tokatly, I.V.: Transforming nonlocality into a frequency dependence: A shortcut to spectroscopy. Phys. Rev. Lett. **99**, 057401-1–057401-4 (2007a)

Gatti, M., Bruneval, F., Olevano, V., Reining, L.: Understanding correlations in vanadium dioxide from first principles. Phys. Rev. Lett. **99**, 266402-1–266402-4 (2007b)

Gaudoin, R., Burke, K.: Lack of Hohenberg–Kohn theorem for excited states. Phys. Rev. Lett. **93**, 173001-1–173001-4 (2004)

Gavnholt, J., Rubio, A., Olsen, T., Thygesen, K.S., Schiøtz, J.: Hot-electron-assisted femtochemistry at surfaces: A time-dependent density functional theory approach. Phys. Rev. B **79**, 195405-1–195405-10 (2009)

Gavrila, M. (ed.): Atoms in Intense Laser Fields. Academic Press, Boston (1992)

Gavrila, M.: Atomic stabilization in superintense laser fields. J. Phys. B At. Mol. Opt. Phys. **35**, R147–R193 (2002)

Gebauer, R., Car, R.: Current in open quantum systems. Phys. Rev. Lett. **93**, 160404-1–160404-4 (2004a)

Gebauer, R., Car, R.: Kinetic theory of quantum transport at the nanoscale. Phys. Rev. B **70**, 125324-1–125324-5 (2004b)

Gebauer, R., Car, R.: Electron transport with dissipation: A quantum kinetic approach. Int. J. Quantum Chem. **101**, 564–571 (2005)

Gebauer, R., Piccinin, S., Car, R.: Quantum collision current in electronic circuits. Chem. Phys. Chem. **6**, 1727–1730 (2005)

Genovese, L., Deutsch, T., Neelov, A., Goedecker, S., Beylkin, G.: Efficient solution of Poisson's equation with free boundary conditions. J. Chem. Phys. **125**, 074105-1–074105-5 (2006)

Genovese, L., Deutsch, T., Goedecker, S.: Efficient and accurate three-dimensional Poisson solver for surface problems. J. Chem. Phys. **127**, 054704-1–054704-6 (2007)

Genovese, L., Neelov, A., Goedecker, S., Deutsch, T., Ghasemi, S.A., Willand, A., Caliste, D., Zilberberg, O., Rayson, M., Bergman, A., Schneider, R.: Daubechies wavelets as a basis set for density functional pseudopotential calculations. J. Chem. Phys. **129**, 014109-1–014109-14 (2008)

Genovese, L., Ospici, M., Deutsch, T., Méhaut, J.-F., Neelov, A., Goedecker, S.: Density functional theory calculation on many-cores hybrid central processing unit-graphic processing unit architectures. J. Chem. Phys. **131**, 034103-1–034103-8 (2009)

Geppert, D., Seyfarth, L., de Vivie-Riedle, R.: Laser control schemes for molecular switches. Appl. Phys. B **79**, 987–992 (2004)

Gerber, R.B., Buch, V., Ratner, M.A.: Time-dependent self-consistent field approximation for intramolecular energy transfer. I. Formulation and application to dissociation of van der Waals molecules. J. Chem.Phys. **77**, 3022–3030 (1982)

Ghosez, P., Michenaud, J.-P., Gonze, X.: Dynamical atomic charges: The case of ABO3 compounds. Phys. Rev. B **58**, 6224–6240 (1998)

Ghosh, S.K., Deb, B.M.: Schrödinger fluid dynamics of many electron systems in a time-dependent density-functional framework. J. Chem. Phys. **77**, 342–348 (1982)

Ghosh, S.K., Deb, B.M.: A density-functional calculation of dynamic dipole polarizabilities of noble gas atoms. Theor. Chim. Acta **62**, 209–217 (1983a)

Ghosh, S.K., Deb, B.M.: A simple density-functional calculation of frequency-dependent multipole polarizabilities of noble gas atoms. J. Mol. Struct. Theochem. **103**, 163–176 (1983b)

Ghosh, S.K., Dhara, A.K.: Density-functional theory of many-electron systems subjected to time-dependent electric and magnetic fields. Phys. Rev. A **38**, 1149–1158 (1988)

Giannozzi, P., Baroni, S., Bonini, S., Calandra, M., Car, R., Cavazzoni, C., Ceresoli, D., Chiarotti, G.L., Cococcioni, M., Dabo, I., Dal Corso, A., de Gironcoli, S., Fabris, S., Fratesi, G., Gebauer, R., Gerstmann, U., Gougoussis, C., Kokalj, A., Lazzeri, M., Martin-Samos, L., Marzari, N., Mauri, F., Mazzarello, R., Paolini, S., Pasquarello, A., Paulatto, A., Sbraccia, C., Scandolo, S., Sclauzero, G., Seitsonen, A.P., Smogunov, A., Umari, P., Wentzcovitch, R.M.: QUANTUM ESPRESSO: A modular and open-source software project for quantum simulations of materials. J. Phys. Condens. Mat. **21**, 395502-1–395502-19 (2009)

Gidopoulos, N.: Kohn–Sham equations for multicomponent systems: The exchange and correlation energy functional. Phys. Rev. B **57**, 2146–2152 (1998)

Giesbertz, K.J.H.: Time-dependent one-body reduced density matrix functional theory; adiabatic approximations and beyond. Ph.D. thesis, VU University, Amsterdam (2010)

Giesbertz, K.J.H., Baerends, E.J.: Aufbau derived from a unified treatment of occupation numbers in Hartree–Fock, Kohn–Sham, and natural orbital theories with the Karush–Kuhn–Tucker conditions for the inequality constraints $n_i \le 1$ and $n_i \ge 0$. J. Chem. Phys. **132**, 194108-1–194108-7 (2010)

Giesbertz, K.J.H., Gritsenko, O.V., Baerends, E.J.: Charge transfer, double and bond-breaking excitations with time-dependent density matrix functional theory. Phys. Rev. Lett. **101**, 033004-1–033004-4 (2008)

Giesbertz, K.J.H., Baerends, E.J., Gritsenko, O.V.: Excitation energies with time-dependent density matrix functional theory: Singlet two-electron systems. J. Chem. Phys. **130**, 114104-1–114104-16 (2009)

Giesbertz, K.J.H., Gritsenko, O.V., Baerends, E.J.: The adiabatic approximation in time-dependent density matrix functional theory: Response properties from dynamics of phase-including natural orbitals. J. Chem. Phys. **133**, 174119-1–174119-13 (2010)

Gilbert, T.L.: Hohenberg–Kohn theorem for nonlocal external potentials. Phys. Rev. B **12**, 2111–2120 (1975)

Girifalco, L.A., Hodak, M.: van der Waals binding energies in graphitic structures. Phys. Rev. B **65**, 125404-1–125404-5 (2002)

Girifalco, L.A., Hodak, M., Lee, R.S.: Carbon nanotubes, buckyballs, ropes, and a universal graphitic potential. Phys. Rev. B **62**, 13104–13110 (2000)

Giuliani, G.F., Vignale, G.: Quantum Theory of the Electron Liquid. Cambridge University Press, Cambridge (2005)

Giustino, F., Yates, J.R., Souza, I., Cohen, M.L., Louie, S.G.: Electron–phonon interaction via electronic and lattice Wannier functions: Superconductivity in boron-doped diamond reexamined. Phys. Rev. Lett. **98**, 047005-1–047005-4 (2007)

Giustino, F., Cohen, M.L., Louie, S.G.: Small phonon contribution to the photoemission kink in the copper oxide superconductors. Nature **452**, 975–978 (2008)

Giustino, F., Cohen, M.L., Louie, S.G.: GW method with the self-consistent Sternheimer equation. Phys. Rev. B **81**, 115105-1–115105-17 (2010)

Göddeke, D., Strzodka, R., Mohd-Yusof, J., McCormick, P., Buijssen, S.H.M., Grajewski, M., Turek, S.: Exploring weak scalability for FEM calculations on a GPU-enhanced cluster. Parallel Comput **33**, 685–699 (2007)

Goedecker, S., Umrigar, C.: Natural orbital functional for the many-electron problem. Phys. Rev. Lett. **81**, 866–869 (1998)

Goerigk, L., Moellmann, J., Grimme, S.: Computation of accurate excitation energies for large organic molecules with double-hybrid density functionals. Phys. Chem. Chem. Phys. **11**, 4611–4620 (2009)

Gomer, E., Noyes, W.A.: Photochemical studies. XLII. Ethylene oxide. J. Am. Chem. Soc. **72**, 101–108 (1950)

Gonze, X.: Perturbation expansion of variational principles at arbitrary order. Phys. Rev. A **52**, 1086–1095 (1995)

Gonze, X.: First-principles responses of solids to atomic displacements and homogeneous electric fields: Implementation of a conjugate-gradient algorithm. Phys. Rev. B **55**, 10337–10354 (1997)

Gonze, X., Lee, C.: Dynamical matrices, Born effective charges, dielectric permittivity tensors, and interatomic force constants from density-functional perturbation theory. Phys. Rev. B **55**, 10355–10368 (1997)

Gonze, X., Scheffler, M.: Exchange and correlation kernels at the resonance frequency: Implications for excitation energies in density-functional theory. Phys. Rev. Lett. **82**, 4416–4419 (1999)

Gonze, X., Vigneron, J.-P.: Density-functional approach to nonlinear-response coefficients of solids. Phys. Rev. B **39**, 13120–13128 (1989)

Gonze, X., Ghosez, P., Godby, R.W.: Density-polarization functional theory of the response of a periodic insulating Solid to an electric field. Phys. Rev. Lett. **74**, 4035–4038 (1995)

Gonze, X., Beuken, J.M., Caracas, R., Detraux, F., Fuchs, M., Rignanese, G.M., Sindic, L., Verstraete, M., Zerah, G., Jollet, F., Torrent, M., Roy, A., Mikami, M., Ghosez, P., Raty, J.Y., Allan, D.C.: First-principles computation of material properties: The ABINIT software project. Comput. Mater. Sci. **25**, 478–492 (2002)

Gorini, V., Kossakowski, A., Sudarshan, E.C.G.: Completely positive dynamical semigroups of n-level systems. J. Math. Phys. **7**, 821–825 (1976)

Görling, A.: Time-dependent Kohn–Sham formalism. Phys. Rev. A. **55**, 2630–2639 (1997)

Görling, A.: Exact exchange-correlation kernel for dynamic response properties and excitation energies in density-functional theory. Phys. Rev. A **57**, 3433–3436 (1998a)

Görling, A.: Exact exchange kernel for time-dependent density-functional theory. Int. J. Quant. Chem. **69**, 265–277 (1998b)

Görling, A.: Density-functional theory beyond the Hohenberg–Kohn theorem. Phys. Rev. A **59**, 3359–3374 (1999)

Görling, A., Levy, M.: Correlation-energy functional and its high-density limit obtained from a coupling-constant perturbation expansion. Phys. Rev. B **47**, 13105–13113 (1993)

Görling, A., Levy, M.: Hybrid schemes combining the Hartree–Fock method and density-functional theory: Underlying formalism and properties of correlation functionals. J. Chem. Phys. **106**, 2675–2680 (1997)

Görling, A., Ipatov, A., Götz, W., Hesselmann, A.: Density-functional theory with orbital-dependent functionals: Exact-exchange Kohn–Sham and density-functional response methods. Z. Phys. Chem. **224**, 325–342 (2010)

Gould, T., Simpkins, K., Dobson, J.F.: A theoretical and semiemprical correction to the long-range dispersion power law of stretched graphite. Phys. Rev. B **77**, 165134-1–165134-5 (2008)

Gould, T., Gray, E.M., Dobson, J.F.: van der Waals dispersion power laws for cleavage, exfoliation and stretching in multi-scale, layered systems. Phys. Rev. B **79**, 113402-1–113402-4 (2009)

Goulielmakis, E., Loh, Z.-H., Wirth, A., Santra, R., Rohringer, N., Yakovlev, V.S., Zherebtsov, S., Pfeifer, T., Azzeer, A.M., Kling, M.F., Leone, S.R., Krausz, F.: Real-time observation of valence electron motion. Nature **466**, 739–743 (2010)

Grabo, T., Petersilka, M., Gross, E.K.U.: Molecular excitation energies from time-dependent density functional theory. J. Mol. Struct. Theochem. **501–502**, 353–367 (2000)

Grasbon, F., Paulus, G.G., Walther, H., Villoresi, P., Sansone, G., Stagira, S., Nisoli, M., De Silvestri, S.: Above-threshold ionization at the few-cycle limit. Phys. Rev. Lett. **91**, 173003-1–173003-4 (2003)

Griffel, D.H.: Applied Functional Analysis. Ellis Horwood, Chichester (1985)

Grimme, S.: Density functional calculations with configuration interaction for the excited states of molecules. Chem. Phys. Lett. **259**, 128–137 (1996)

Grimme, S.: Calculation of the electronic spectra of large molecules. In: Lipkowitz, K.B., Cundari, T.R. (eds.) Reviews in Computational Chemistry, vol. 20, Chap. 3, pp. 153–218. VCH Publishers, New York (2004)

Grimme, S.: Semiempirical hybrid density functional with perturbative second-order correlation. J. Chem. Phys. **124**, 034108-1–034108-16 (2006)

Grimme, S., Parac, M.: Substantial errors from time-dependent density functional theory for the calculation of excited states of large π systems. ChemPhysChem **4**, 292–295 (2003)

Grimme, S., Neese, F.: Double-hybrid density functional theory for excited electronic states of molecules. J. Chem. Phys. **127**, 154116-1–154116-18 (2007)

Grimme, S., Waletzke, M.: A combination of Kohn–Sham density functional theory and multi-reference configuration interaction methods. J. Chem. Phys. **111**, 5645–5655 (1999)

Grimme, S., Antony, J., Ehrlich, S., Krieg, H.: A consistent and accurate ab initio parametrization of density functional dispersion correction (DFT-D) for the 94 elements H-Pu. J. Chem. Phys. **132**, 154104-1–154104-19 (2010)

Gritsenko, O.V., Baerends, E.J.: Effect of molecular dissociation on the exchange-correlation Kohn–Sham potential. Phys. Rev. A **54**, 1957–1972 (1996)

Gritsenko, O.V., Baerends, E.J.: Asymptotic correction of the exchange-correlation kernel of time-dependent density functional theory for long-range charge-transfer excitations. J. Chem. Phys. **121**, 655–660 (2004)

Gritsenko, O.V., Baerends, E.J.: Double excitation effect in non-adiabatic time-dependent density functional theory with an analytic construction of the exchange-correlation kernel in the common energy denominator approximation. Phys. Chem. Chem. Phys. **11**, 4640–4646 (2009)

Gritsenko, O.V., van Gisbergen, S.J.A., Schipper, P.R.T., Baerends, E.J.: Origin of the field-counteracting term of the Kohn–Sham exchange-correlation potential of molecular chains in an electric field. Phys. Rev. A **62**, 012507-1–012507-10 (2000)

Gritsenko, O.V., Pernal, K., Baerends, E.J.: An improved density matrix functional by physically motivated repulsive corrections. J. Chem. Phys. **122**, 204102-1–204102-13 (2005)

Gross, E.K.U., Kohn, W.: Local density-functional theory of frequency-dependent linear response. Phys. Rev. Lett. **55**, 2850–2852 (1985)

Gross, E.K.U., Kohn, W.: Local density-functional theory of frequency-dependent linear response. Phys. Rev. Lett. **57**, 923(E) (1986)

Gross, E.K.U., Kohn, W.: Time-dependent density functional theory. Adv. Quant. Chem. **21**, 255–291 (1990)

Gross, E.K.U., Mearns, D., Oliveira, L.N.: Zeros of the frequency-dependent linear density response. Phys. Rev. Lett. **61**, 1518 (1988)

Gross, E.K.U., Dobson, J., Petersilka, M.: Density functional theory of time-dependent phenomena. In: Nalewajski, R.F. (ed.) Density Functional Theory II, Topics in Current Chemistry, vol. 181, pp. 81–172. Springer, Berlin (1996)

Grossauer, H., Thoman, P.: GPU-based multigrid: Real-time performance in high resolution nonlinear image processing. In: Gasteratos, A., Vincze, M., Tsotsos, J. (eds.) Computer Vision Systems, Lecture Notes in Computer Science, vol. 5008, pp. 141–150. Springer, Berlin (2008)

Grosso, G., Parravicini, G.P.: Solid State Physics. Academic Press, London (2000)

Grumbach, M.P., Hohl, D., Martin, R.M., Car, R.: Ab initio molecular dynamics with a finite-temperature density functional. J. Phys. Condens. Mat. **6**, 1999–2014 (1994)

Gruneis, A., Marsman, M., Harl, J., Schimka, L., Kresse, G.: Making the random phase approximation to electronic correlation accurate. J. Chem. Phys. **131**, 154115-1–154115-5 (2009)

Grüning, M., Gritsenko, O.V., van Gisbergen, S.J.A., Baerends, E.J.: On the required shape corrections to the local density and generalized gradient approximations to the Kohn–Sham potentials for molecular response calculations of (hyper)polarizabilities and excitation energies. J. Chem. Phys. **116**, 9591–9601 (2002)

Grüning, M., Marini, A., Rubio, A.: Density functionals from many-body perturbation theory: The band gap for semiconductors and insulators. J. Chem. Phys. **124**, 154108–154117 (2006)

Guan, J., Casida, M.E., Salahub, D.R.: Time-dependent density-functional theory investigation of excitation spectra of open-shell molecules. J. Mol. Struct. Theochem. **527**, 229–244 (2000)

Gulans, A., Puska, M.J., Nieminen, R.M.: Linear-scaling self-consistent implementation of the van der Waals density functional. Phys. Rev. B **79**, 201105-1–201105-4 (2009)

Gunnarsson, O., Lundqvist, B.I.: Exchange and correlation in atoms, molecules and solids by the spin-density functional method. Phys. Rev. B **13**, 4274–4298 (1976)

Guo, C., Gibson, G.N.: Ellipticity effects on single and double ionization of diatomic molecules in strong laser fields. Phys. Rev. A **63**, 40701-1–40701-4 (2001)

Gustafson, J.L.: Reevaluating Amdahl's law. Commun. ACM **31**, 532–533 (1988)

Guzzo, M., Lani, G., Sottile, F., Romaniello, P., Gatti, M., Rehr J.J., Sirotti, F., Reining, L.: Satellites in Bulk Silicon: Beyond the GW Approximation (2011, in preparation)

Haas, F.: The damped Pinney equation and its applications to dissipative quantum mechanics. Phys. Scripta **81**, 025004-1–025004-7 (2010)

Habenicht, B.F., Craig, C.F., Prezhdo, O.V.: Time-domain Ab initio simulation of electron and hole relaxation dynamics in a single-walled semiconducting carbon nanotube. Phys. Rev. Lett. **96**, 187401-1–187401-4 (2006)

Haessler, S., Caillat, J., Boutu, W., Giovannetti-Teixeira, C., Ruchon, T., Auguste, T., Diveki, Z., Breger, P., Maquet, A., Carré, B., Ta, R., Salières, P.: Attosecond imaging of molecular electronic wavepackets. Nat. Phys. **6**, 200–206 (2010)

Hamann, D.R., Schlüter, M., Chiang, C.: Norm-conserving pseudopotentials. Phys. Rev. Lett. **43**, 1494–1497 (1979)

Hamann, D.R., Wu, X., Rabe, K.M., Vanderbilt, D.: Metric tensor formulation of strain in density-functional perturbation theory. Phys. Rev. B **71**, 035117-1–035117-13 (2005)

Hammes-Schiffer, S., Tully, J.C.: Proton transfer in solution: Molecular dynamics with quantum transitions. J. Chem. Phys. **101**, 4657–4667 (1994)

Handy, N.C., Schaefer III, H.F.: On the evaluation of analytic energy derivatives for correlated wave functions. J. Chem. Phys. **81**, 5031–5033 (1984)

Hanke, W.: Dielectric theory of elementary excitations in crystals. Adv. Phys. **27**, 287–341 (1978)

Hariharan, P.C., Pople, J.A.: Influence of polarization functions on molecular-orbital hydrogenation energies. Theor. Chim. Acta **28**, 213–222 (1973)

Harl, J., Kresse, G.: Accurate bulk properties from approximate many-body techniques. Phys. Rev. Lett. **103**, 056401-1–056401-4 (2009)

Harl, J., Schimka, L., Kresse, G.: Assessing the quality of the random phase approximation for lattice constants and atomization energies of solids. Phys. Rev. B **81**, 115126-1–115126-18 (2010)

Harriman, J.E.: Orthonormal orbitals for the representation of an arbitrary density. Phys. Rev. A **24**, 680–682 (1981)

Harris, J., Griffin, A.: Correlation energy and van der Waals interaction of coupled metal films. Phys. Rev. B **11**, 3669–3677 (1975)

Harumiya, K., Kono, H., Fujimura, Y., Kawata, I., Bandrauk, A.D.: Intense laserfield ionization of H_2 enhanced by two-electron dynamics. Phys. Rev. A **66**, 043403-1–043403-14 (2002)

Hasegawa, M., Nishidate, K.: Semiempirical approach to the energetics of interlayer binding in graphite. Phys. Rev. B **70**, 205431–205437 (2004)

Hatcher, R., Beck, M., Tackett, A., Pantelides, S.T.: Dynamical effects in the interaction of ion beams with solids. Phys. Rev. Lett. **100**, 103201-1–103201-4 (2008)

Hedin, L.: New method for calculating the one-particle Green's function with application to the electron-gas problem. Phys. Rev. **139**, A796–A823 (1965)

Hedin, L.: On correlation effects in electron spectroscopies and the GW approximation. J. Phys. Condens. Mat. **11**, R489–R528 (1999)

Hedin, L., Lundqvist, S.: Effects of electron–electron and electron–phonon interactions on the one-electron states of solids. Sol. Stat. Phys. **23**, 1–181 (1969)

Hedin, L., Michiels, J., Inglesfield, J.: Transition from the adiabatic to the sudden limit in core-electron photoemission. Phys. Rev. B **58**, 15565–15582 (1998)

Heidrich-Meisner, F., Feiguin, A.E., Dagotto, E.: Real-time simulations of nonequilibrium transport in the single-impurity Anderson model. Phys. Rev. B **79**, 235336-1–235336-6 (2009)

Helbig, N., Tokatly, I.V., Rubio, A.: Exact Kohn–Sham potential of strongly correlated finite systems. J. Chem. Phys. **131**, 224105-1–224105-8 (2009)

Helgaker, T., Jørgensen, P.: Configuration–interaction energy derivatives in a fully variational formulation. Theor. Chim. Acta **75**, 111–127 (1989)

Helgaker, T., Jørgensen, P., Olsen, J.: Molecular Electronic-Structure Theory. Wiley, Chichester (2000)

Hellgren, M., von Barth, U.: Exact-exchange approximation within time-dependent density-functional theory: Frequency-dependence and photoabsorption spectra of atoms. J. Chem. Phys. **131**, 044110-1–044110-13 (2009)

Henkel, N., Keim, M., Lüdde, H.J., Kirchner, T.: Density functional theory investigation of antiproton-helium collisions. Phys. Rev. A. **80**, 032704-1–032704-7 (2009)

Hentschel, M., Kienberger, R., Spielmann, C., Reider, G.A., Milosevic, N., Brabec, T., Corkum, P., Heinzmann, U., Drescher, M., Krausz, F.: Attosecond metrology. Nature **414**, 509–513 (2001)

Hertel, I.V., Radloff, W.: Ultrafast dynamics in isolated molecules and molecular clusters. Rep. Prog. Phys. **69**, 1897–2003 (2006)

Hesselmann, A., Görling, A.: Efficient exact-exchange time-dependent densityfunctional theory methods and their relation to time-dependent Hartree–Fock. J. Chem. Phys. **134**, 034120-1–034120-17 (2011)

Hesselmann, A., Jansen, G., Schutz, M.J.: Interaction energy contributions of H-bonded and stacked structures of the at and GC DNA base pairs from the combined density functional theory and intermolecular perturbation theory approach. J. Am. Chem. Soc. **128**, 11730–11731 (2006)

Hesselmann, A., Ipatov, A., Görling, A.: Charge-transfer excitation energies with a time-dependent density-functional method suitable for orbital-dependent exchange-correlation kernels. Phys. Rev. A **80**, 012507-1–012507-6 (2009)

Hessler, P., Park, J., Burke, K.: Several theorems in time-dependent density functional theory. Phys. Rev. Lett. **82**, 378–381 (1999)

Hessler, P., Park, J., Burke, K.: Several theorems in time-dependent density functional theory. Phys. Rev. Lett. **83**, 5184(E) (1999)

Hessler, P., Maitra, N.T., Burke, K.: Correlation in time-dependent density-functional theory. J. Chem. Phys. **117**, 72–81 (2002)

Higuet, J., Ruf, H., Thiré, N., Cireasa, R., Constant, E., Cormier, E., Descamps, D., Mével, E., Petit, S., Pons, B., Mairesse, Y., Fabre, B.: High-order harmonic spectroscopy of the Cooper minimum in argon: Experimental and theoretical study. Phys. Rev. A **83**, 053401-1–053401-12 (2011)

Hippert, F., Geissler, E., Hodeau, J.L., Leliévre-Berna, E., Regnard J.-R. (eds.): Neutron and X-ray Spectroscopy. Springer, Dordrecht (2006)

Hirai, H., Sugino, O.: A time-dependent density-functional approach to nonadiabatic electron-nucleus dynamics: Formulation and photochemical application. Phys. Chem. Chem. Phys. **11**, 4570–4578 (2009)

Hirata, S., Head-Gordon, M.: Time-dependent density functional theory within the Tamm–Dancoff approximation. Chem. Phys. Lett. **314**, 291–299 (1999)

Hirschfelder, J.O., Byers Brown, W., Epstein, S.T.: Recent developments in perturbation theory. In: Löwdin, P.-O. (ed.) Advances in Quantum Chemistry, vol. 1, pp. 255–374. Academic Press, New York (1964)

Hitchcock, A.P.: Inner shell excitation spectroscopy of molecules using inelastic electron scattering. J. El. Spectr. Rel. Phen. **112**, 9–29 (2000)

Hohenberg, P., Kohn, W.: Inhomogeneous electron gas. Phys. Rev. **136**, B864–B871 (1964)

Holas, A., Balawender, R.: Maitra–Burke example of initial-state dependence in time-dependent density-functional theory. Phys. Rev. A **65**, 034502-1–034502-4 (2002)

Horiba, K., Taguchi, M., Chainani, A., Takata, Y., Ikenaga, E., Miwa, D., Nishino, Y., Tamasaku, K., Awaji, M., Takeuchi, A., Yabashi, M., Namatame, H., Taniguchi, M., Kumigashira, H., Oshima, M., Lippmaa, M., Kawasaki, M., Koinuma, H., Kobayashi, K., Ishikawa, T., Shin, S.: Nature of the well screened state in hard X-Ray Mn $2p$ core-level photoemission measurements of $La_{1-x}Sr_xMnO_3$ Films. Phys. Rev. Lett. **93**, 236401-1–236401-4 (2004)

Horowitz, C.M., Proetto, C.R., Rigamonti, S.: Kohn–Sham exchange potential for a metallic surface. Phys. Rev. Lett. **97**, 026802-1–026802-4 (2006)

Hsu, C., Hirata, S., Head-Gordon, M.: Excitation energies from time-dependent density functional theory for linear polyene oligomers: Butadiene to decapentaene. J. Phys. Chem. A **105**, 451–458 (2001)

Hu, C., Sugino, O.: Average excitation energies from time-dependent density-functional theory. J. Chem. Phys. **126**, 074112-1–074112-10 (2007)

Hu, C., Hirai, H., Sugino, O.: Nonadiabatic couplings from time-dependent density functional theory: Formulation in the Casida formalism and practical scheme within modified linear response. J. Chem. Phys. **127**, 064103-1–064103-9 (2007)

Hu, C., Hirai, H., Sugino, O.: Nonadiabatic couplings from time-dependent density functional theory. II. Successes and challenges of the pseudopotential approximation. J. Chem. Phys. **128**, 154111-1–154111-9 (2008)

Hu, C., Sugino, O., Tateyama, Y.: All-electron calculation of nonadiabatic couplings from time-dependent density functional theory: Probing with the Hartree–Fock exact exchange. J. Chem. Phys. **131**, 114101-1–114101-8 (2009)

Hübener, H., Luppi, E., Véniard, V.: Ab initio calculation of second harmonic generation in solids. Phys. Stat. Sol. (b) **247**, 1984–1991 (2010)

Hüfner, S.: Photoelectron Spectroscopy: Principles and Applications. Springer, Berlin (2003)

Hüfner, S. (ed.): Very High Resolution Photoelectron Spectroscopy. Springer, Heidelberg (2007)

Huismans, Y., Rouzée, A., Gijsbertsen, A., Jungmann, J., Smolkowska, A., Logman, P., Lépine, F., Cauchy, C., Zamith, S., Marchenko, T., Bakker, J., Berden, G., Redlich, B., van der Meer, L., Muller, H.G., Vermin, W., Schafer, K.J., Smirnova, O., Ivanov, M.Y., Spanner, M., Bauer, D., Prophuzhenko, S., Vrakking, M.J.J.: Time-resolved holography with photoelectrons. Science **331**, 61–64 (2010)

Huix-Rotllant, M., Natarajan, B., Ipatov, A., Wawire, C.M., Casida, M.E.: Assessment of noncollinear spin-flip Tamm–Dancoff approximation time-dependent density-functional theory for the photochemical ring-opening of oxirane. Phys. Chem. Chem. Phys. **12**, 12811–12825 (2010)

Huix-Rotllant, M., Ipatov, A., Rubio, A., Casida, M.E.: Assessment of dressed time-dependent density-functional theory for the low-lying valence states of 28 organic chromophores. Chem. Phys. (2011, in press). Available online at http://dx.doi.org/10.1016/j.chemphys.2011.03.019

Hunt, K.L.C.: Nonlocal polarizability densities and van der Waals interactions. J. Chem. Phys. **78**, 6149–6155 (1983)

Huotari, S., Soininen, J.A., Vankó, G., Monaco, G., Olevano, V.: Screening in $YBa_2Cu_3O_7$-δ at large wave vectors. Phys. Rev. B **82**, 064514-1–064514-6 (2010a)

Huotari, S., Soininen, J.A., Pylkkänen, T., Hämäläinen, K., Issolah, A., Titov, A., McMinis, J., Kim, J., Ceperley, D.M., Holzmann, M., Olevano, V.: Momentum distribution and renormalization factor in sodium and the electron gas. Phys. Rev. Lett **105**, 086403-1–086403-4 (2010b)

Hutter, J.: Excited state nuclear forces from the Tamm–Dancoff approximation to time-dependent density functional theory within the plane wave basis set framework. J. Chem. Phys. **118**, 3928–3934 (2003)

Hybertsen, M.S., Louie, S.G.: Electron correlation in semiconductors and insulators: Band gaps and quasiparticle energies. Phys. Rev. B **34**, 5390–5413 (1986)

Hybertsen, M.S., Louie, S.G.: Ab initio static dielectric matrices from the densityfunctional approach. I. Formulation and application to semiconductors and insulators. Phys. Rev. B **35**, 5585–5601 (1987)

Iikura, H., Tsuneda, T., Yanai, T., Hirao, K.: A long-range correction scheme for generalized-gradient-approximation exchange functionals. J. Chem. Phys. **115**, 3540–3544 (2001)

Inokuti, M.: Inelastic collisions of fast charged particles with atoms and molecules—the Bethe theory revisited. Rev. Mod. Phys. **43**, 297–347 (1971)

Ipatov, A., Heßelmann, A., Görling, A.: Molecular excitation spectra by TDDFT with the nonadiabatic exact exchange kernel. Int. J. Quantum Chem. **110**, 2202–2220 (2010)

Ismail-Beigi, S., Chang, E.K., Louie, S.G.: Coupling of nonlocal potentials to electromagnetic fields. Phys. Rev. Lett. **87**, 087402-1–087402-4 (2001)

Itatani, J., Levesque, J., Zeidler, D., Niikura, H., Pépin, H., Kieffer, J.C., Corkum, P.B., Villeneuve, D.M.: Tomographic imaging of molecular orbitals. Nature (London) **432**, 867–871 (2004)

Ivanov, M.Yu., Smirnova, O., Spanner, M.: Anatomy of strong field ionization. J. Mod. Opt. **52**, 165–184 (2005)

Iyengar, S.S., Sumner, I., Jakowski, J.: Hydrogen tunneling in an enzyme active site: A quantum wavepacket dynamical perspective. J. Phys. Chem. B **112**, 7601–7613 (2008)

Izzo, R., Klessinger, M.: Optimization of conical intersections using the semiempirical MNDOC-CI method with analytic gradients. J. Comp. Chem. **21**, 52–62 (2000)

Jackiw, R., Kerman, A.: Time-dependent variation principle and the effective action. Phys. Lett. A **71**, 158–162 (1979)

Jackson, J.D.: Classical electrodynamics. John Wiley, New York (1962)

Jacquemin, D., Perpète, E.A., Scalmani, G., Frisch, M.J., Kobayashi, R., Adamo, C.: Assessment of the efficiency of long-range corrected functionals for some properties of large compounds. J. Chem. Phys. **126**, 144105-1–144105-12 (2007a)

Jacquemin, D., Perpète, E.A., Vydrov, O.A., Scuseria, G.E., Adamo, C.: Assessment of long-range corrected functionals performance for $n \to \pi^*$ transitions in organic dyes. J. Chem. Phys. **127**, 094102-1–094102-6 (2007b)

Jacquemin, D., Wathelet, V., Perpète, E.A., Adamo, C.: Extensive TD-DFT benchmark: Singlet-excited states of organic molecules. J. Chem. Theory Comput. **5**, 2420–2435 (2009)

Jacquemin, D., Perpète, E.A., Ciofini, I., Adamo, C., Valero, R., Zhao, Y., Truhlar, D.G.: On the performances of the M06 family of density functionals for electronic excitation energies. J. Chem. Theory Comput. **6**, 2071–2085 (2010a)

Jacquemin, D., Perpète, E.A., Ciofini, I., Adamo, C.: Assessment of Functionals for TD-DFT Calculations of Singlet–Triplet Transitions. J. Chem. Theory Comput. **6**, 1532–1537 (2010b)

Jamorski, C., Casida, M.E., Salahub, D.R.: Dynamic polarizabilities and excitation spectra from a molecular implementation of time-dependent density-functional response theory: N2 as a case study. J. Chem. Phys. **104**, 5134–5147 (1996)

Janesko, B.G., Henderson, T.M., Scuseria, G.E.: Long-range-corrected hybrid density functionals including random phase approximation correlation: Application to noncovalent interactions. J. Chem. Phys. **131**, 034110-1–034110-9 (2009)

Jeschke, S., Cline, D., Wonka, P.: A GPU Laplacian solver for diffusion curves and Poisson image editing. ACM Trans. Graph. **28**, 116-1–116-8 (2009)

Jeziorski, B., Moszynski, R., Szalewicz, K.: Perturbation theory approach to intermolecular potential energy surfaces of van der Waals complexes. Chem. Rev. **94**, 1887–1930 (1994)

Joachain, C.J., Norr, M.D., Kylstra, N.J.: High intensity laser–atom physics. Adv. Atom. Mol. Opt. Phys. **42**, 225–286 (2000)

Johnson, B.G., Gill, P.M.W., Pople, J.A.: The performance of a family of density functional methods. J. Chem. Phys. **98**, 5612–5626 (1993)

Joyce, S.A., Yates, J.R., Pickard, C.J., Mauri, F.: A first principles theory of nuclear magnetic resonance J-coupling in solid-state systems. J. Chem. Phys. **127**, 204107-1–204107-9 (2007)

Jung, J., Garcia-Gonzalez, P., Dobson, J.F., Godby, R.W.: Effects beyond the randomphase approximation in calculating the interaction between metal films. Phys. Rev. B **70**, 205107-1–205107-11 (2004)

Jurečka, P., Sponer, J., Černý, J., Hobza, P.: Benchmark database of accurate (MP2 and CCSD(T) complete basis set limit) interaction energies of small model complexes, DNA base pairs, and amino acid pairs. Phys. Chem. Chem. Phys. **8**, 1985–1993 (2006)

Kacprzak, K.A., Lehtovaara, L., Akola, J., Lopez-Acevedo, O., Häkkinen, H.: A density functional investigation of thiolate-protected bimetal $PdAu_{24}(SR)_{18}^z$ clusters: Doping the superatom complex. Phys. Chem. Chem. Phys. **11**, 7123–7129 (2009)

Kaduk, B., Van Voorhis, T.: Conical intersections using constrained density functional theory—configuration interaction. J. Chem. Phys. **133**, 061102-1–061102-4 (2010)

Kamiya, M., Tsuneda, T., Hirao, K.: A density functional study of van der Waals interactions. J. Chem. Phys. **117**, 6010–6015 (2002)

Kannemann, F.O., Becke, A.D.: van der Waals interactions in density-functional theory: Rare-gas diatomics. J. Chem. Theory Comput. **5**, 719–727 (2009)

Kapral, R., Ciccotti, G.: Mixed quantum-classical dynamics. J. Chem. Phys. **110**, 8919–8929 (1999)

Kapral, R.: Progress in the theory of mixed quantum-classical dynamics. Annu. Rev. Phys. Chem. **57**, 129–157 (2006)

Karolewski, A., Stein, T., Baer, R., Kümmel, S.: Tailoring the optical gap in lightharvesting molecules. J. Chem. Phys. **134**, 151101–151104 (2011)

Kato, T., Kono, H.: Time-dependent multiconfiguration theory for electronic dynamics of molecules in an intense laser field. Chem. Phys. Lett. **392**, 533–540 (2004)

Kawashita, Y., Nakatsukasa, T., Yabana, K.: Time-dependent density-functional theory simulation for electronon dynamics in molecules under intense laser pulses. J. Phys. Condens. Matter **21**, 064222-1–064222-6 (2009)

Kawata, I., Kono, H., Fujimura, Y., Bandrauk, A.D.: Intense-laser-field-enhanced ionization of two-electron molecules: Role of ionic states as doorway states. Phys. Rev. A **62**, 031401-1–031401-4 (2000)

Kawata, I., Kono, H., Bandrauk, A.D.: Mechanism of enhanced ionization of linear H_+^3 in intense laser fields. Phys. Rev. A **64**, 043411-1–043411-15 (2001)

Keal, T.W., Koslowski, A., Thiel, W.: Comparison of algorithms for conical intersection optimisation using semiempirical methods. Theor. Chem. Acc. **118**, 837–844 (2007)

Keldysh, L.V.: Diagram technique for nonequilibrium processes. Sov. Phys. JETP **20**, 1018–1026 (1965)

Kelkensberg, F., Siu, W., Pérez-Torres, J.F., Morales, F., Gademan, G., Rouzée, A., Johnson, P., Lucchini, M., Calegari, F., Martín F., Vrakking, M.J.J.: Attosecond time-resolved electron dynamics in the hydrogen molecule. Phys. Rev. Lett. **107**, 043002-1–043002-4 (2011).

Kendall, R.A., Dunning Jr., T.H., Harrison, R.J.: Electron affinities of the first-row atoms revisited. Systematic basis sets and wave functions. Chem. Phys. **96**, 6796–6806 (1992)

Khosravi, E., Kurth, S., Stefanucci, G., Gross, E.K.U.: The role of bound states in time-dependent quantum transport. Appl. Phys. A **93**, 355–364 (2008)

Khosravi, E., Stefanucci, G., Kurth, S., Gross, E.K.U.: Bound states in time-dependent quantum transport: Oscillations and memory effects in current and density. Phys. Chem. Chem. Phys. **11**, 4535–4538 (2009)

Kienberger, R., Hentschel, M., Uiberacker, M., Spielmann, C., Kitzler, M., Scrinzi, A., Wieland, M., Westerwalbesloh, T., Kleineberg, U., Heinzmann, U., Drescher, M., Krausz, F.: Steering attosecond electron wave packets with light. Science **297**, 1144–1148 (2002)

Kienberger, R., Goulielmakis, E., Uiberacker, M., Baltuska, A., Yakovlev, V., Bammer, F., Scrinzi, A., Westerwalbesloh, T., Kleineberg, U., Heinzmann, U., Drescher, M., Krausz, F.: Atomic transient recorder. Nature (London) **427**, 817–821 (2004)

Kim, Y.-H., Görling, A.: Excitonic optical spectrum of emiconductors obtained by time-dependent density-functional theory with the exact-exchange kernel. Phys. Rev. Lett. **89**, 096402–096404 (2002)

Kim, H.Y., Sofo, J.O., Velegol, D., Cole, M.W., Lucas, A.A.: van der Waals forces between nanoclusters: Importance of many-body effects. J. Chem. Phys. **124**, 074504-1–074504-4 (2006)

Kitzler, M., Zanghellini, J., Jungreuthmayer, C., Smits, M., Scrinzi, A., Brabec, T.: Ionization dynamics of extended multielectron systems. Phys. Rev. A **70**, 041401-1–041401-4(R) (2004)

Klimeš, J., Bowler, D.R., Michaelides, A.: Chemical accuracy for the van der Waals density functional. J. Phys. Condens. Matter **22**, 022201-1–022201-5 (2010)

Kloeden, P.E., Platen, E.: Numerical Solution of Stochastic Differential Equations. Springer, Berlin (1992)

Klünder, K., Dahlström, J.M., Gisselbrecht, M., Fordell, T., Swoboda, M., Guénot, D., Johnsson, P., Caillat, J., Mauritsson, J., Maquet, A., Ta, R., L'Huillier, A.: Probing single- photon ionization on the attosecond time scale. Phys. Rev. Lett. **106**, 143002-1–143002-4 (2011)

Knorr, W., Godby, R.W.: Investigating exact density-functional theory of a model semiconductor. Phys. Rev. Lett. **68**, 639–641 (1992)

Koentopp, M., Burke, K., Evers, F.: Zero-bias molecular electronics: Exchangecorrelation corrections to Landauers formula. Phys. Rev. B. **73**, 121403-1–4(R) (2006)

Koentopp, M., Chang, C., Burke, K., Car, R.: Density functional calculations of nanoscale conductance. J. Phys.: Condens. Matter **20**, 083203-1–083203-21 (2008)

Koga, N., Morokuma, K.: Determination of the lowest energy point on the crossing seam between two potential surfaces using the energy gradient. Chem. Phys. Lett. **119**, 371–374 (1985)

Kohl, H., Dreizler, R.M.: Time-dependent density functional theory: Conceptual and practical aspects. Phys. Rev. Lett. **56**, 1993–1995 (1986)

Kohn, W., Sham, L.J.: Self-consistent equations including exchange and correlation effects. Phys. Rev. **140**, A1133–A1138 (1965)

Kohn, W.: v-Representability and density functional theory. Phys. Rev. Lett. **51**, 1596–1598 (1983)

Kollmar, C., Hes, B.A.: A new approach to matrix functional theory. J. Chem. Phys. **119**, 4655–4661 (2003)

Kondo, K., Tamida, T., Nabekawa, Y., Watanabe, S.: High-order harmonic generation and ionization using ultrashort KrF and Ti:sapphire lasers. Phys. Rev. A **49**, 3881–3889 (1994)

Kono, A., Hattori, S.: Accurate oscillator strengths for neutral helium. Phys. Rev. A **29**, 2981–2988 (1984)

Koopmans, T.: Über die Zuordnung vonWellenfunktionen und Eigenwerten zu den Einzelnen Elektronen Eines Atoms. Physica **1**, 104–113 (1934)

Körzdörfer, T., Kümmel, S., Mundt, M.: Self-interaction correction and the optimized effective potential. J. Chem. Phys. **129**, 014110-1–014110-12 (2008)

Körzdörfer, T., Kümmel, S.: Single-particle and quasiparticle interpretation of Kohn-Sham and generalized Kohn-Sham eigenvalues for hybrid functionals. Phys. Rev. B **82**, 155206-1–155206-9 (2010)

Kostin, M.D.: On the Schrödinger-Langevin equation. J. Chem. Phys. **57**, 3589–3591 (1972)

Kostin, M.D.: Friction and dissipative phenomena in quantum mechanics. J. Stat. Phys. **12**, 145–151 (1975)

Kotani, T.: An optimized-effective-potential method for solids with exact exchange and random-phase approximation correlation. J. Phys. Condens. Matt. **10**, 9241–9261 (1998)

Kozak, C.R., Kistler, K.A., Lu, Z., Matsika, S.: Excited-state energies and electronic couplings of DNA base dimers. J. Phys. Chem. B **114**, 1674–1683 (2010)

Kramers, H.A.: La diffusion de la lumiere par les atomes. Atti Cong. Intern. Fisica, (Transactions of Volta Centenary Congress) Como **2**, 545–557 (1927)

Krause, J.L., Schafer, K.J., Kulander, K.C.: Optical harmonic generation in atomic and molecular hydrogen. Chem. Phys. Lett. **178**, 573–578 (1991)

Krause, P., Klamroth, T., Saalfrank, P.: Time-dependent configuration-interaction calculations of laser-pulse-driven many-electron dynamics: Controlled dipole switching in lithium cyanide. J. Chem. Phys. **123**, 074105-1–074105-7 (2005)

Krause, P., Klamroth, T., Saalfrank, P.: Molecular response properties from explicitly time-dependent configuration interaction methods. J. Chem. Phys. **127**, 034107-1–034107-10 (2007)

Krausz, F., Ivanov, M.Yu.: Attosecond physics. Rev. Mod. Phys. **81**, 163– 234 (2009)

Kreibich, T., Kurth, S., Grabo, T., Gross, E.K.U.: Asymptotic properties of the optimized effective potential. Adv. Quantum. Chem. **33**, 31–49 (1999)

Kreibich, T.: Multicomponent density-functional theory for molecules in strong laser fields. Ph.D. Thesis, Universität Würzburg, Shaker-Verlag (2000)

Kreibich, T., Gross, E.K.U.: Multicomponent density-functional theory for electrons and nuclei. Phys. Rev. Lett. **86**, 2984–2987 (2001)

Kreibich, T., Lein, M., Engel, V., Gross, E.K.U.: Even-harmonic generation due to Beyond-Born-Oppenheimer dynamics. Phys. Rev. Lett. **87**, 103901-1–103901-4 (2001)

Kreibich, T., Gidopoulos, N.I., van Leeuwen, R., Gross, E.K.U.: Towards timedependent density-functional theory for molecules in strong laser pulses. Prog. Theor. Chem. Phys. **14**, 69–78 (2003) (Springer, Berlin)

Kreibich, T., van Leeuwen, R., Gross, E.K.U.: Time-dependent variational approach to molecules in strong laser fields. Chem. Phys. **304**, 183–202 (2004)

Kreibich, T., van Leeuwen, R., Gross, E.K.U.: Multicomponent density-functional theory for electrons and nuclei. Phys. Rev. A **78**, 022501-1–022501-22 (2008)

Kresse, G., Joubert, D.: From ultrasoft pseudopotentials to the projector augmentedwave method. Phys. Rev. B **59**, 1758–1775 (1999)

Krieger, J.B., Li, Y., Iafrate, G.J.: Systematic approximations to the optimized effective potential: Application to orbital-density-functional theory. Phys. Rev. A **46**, 5453–5458 (1992a)

Krieger, J.B., Li, Y., Iafrate, G.J.: Construction and application of an accurate local spin-polarized Kohn-Sham potential with integer discontinuity: Exchange-only theory. Phys. Rev. A **45**, 101–126 (1992b)

Krikunova, M., Maltezopoulos, T., Wessels, P., Schlie, M., Azima, A., Wieland, M., Drescher, M.: Ultrafast photofragmentation dynamics of molecular iodine driven with timed XUV and near-infrared light pulses. J. Chem. Phys. **134**, 024313-1–024313-7 (2011)

Krishna, V.: Time-dependent density-functional theory for nonadiabatic electronic dynamics. Phys. Rev. Lett. **102**, 053002-1–053002-4 (2009)

Kronig, R., de, L.: On the theory of the dispersion of X-rays. J. Opt. Soc. Am. **12**, 547–557 (1926)

Krueger, A.J., Maitra, N.T.: Autoionizing resonances in time-dependent density functional theory. Phys. Chem. Chem. Phys. **11**, 4655–4663 (2009)

Kruit, P., Kimman, J., Muller, H.G., van der Wiel, M.J.: Electron spectra from multiphoton ionization of xenon at 1064, 532, and 355 nm. Phys. Rev. A **28**, 248–255 (1983)

Kubo, R.: Statistical-mechanical theory of irreversible processes. I. General theory and simple applications to magnetic and conduction problems. J. Phys. Soc. Jpn. **12**, 570–586 (1957)

Kuhn, W.: Über die Gesamtstärke der von einem Zustande ausgehenden Absorptionslinien. Z. Phys. **33**, 408–412 (1925)

Kühne, T.D., Krack, M., Mohamed, F.R., Parrinello, M.: Efficient and accurate Car- Parrinello-like approach to Born-Oppenheimer molecular dynamics. Phys. Rev. Lett. **98**, 066401-1–066401-4 (2007)

Kulander, K.C.: Multiphoton ionization of hydrogen: A time-dependent theory. Phys. Rev. A **35**, 445–447 (1987)

Kulander, K.C.: Time-dependent theory of multiphoton ionization of xenon. Phys. Rev. A **38**, 778–787 (1988)

Kulander, K.C., Schafer, K.J., Krause, J.L.: Dynamic stabilization of hydrogen in an intense, high-frequency, pulsed laser field. Phys. Rev. Lett. **66**, 2601–2604 (1991a)

Kulander, K.C., Schafer, K.J., Krause, J.L.: Single active electron calculations of multiphoton processes in krypton. Int. J. Quant. Chem. Symp. **25**, 415–429 (1991b)

Kulander, K.C., Schafer, K.J., Krause, J.L.: Time-dependent studies of multiphoton processes. In: Gavrila, M. (ed.) Atoms in Intense Laser Fields, pp. 247–300. Academic Press, Boston (1992)

Kulander, K.C., Schafer, K.J., Krause, J.L.: Dynamics of short-pulse excitation, ionization and harmonic conversion. In: Piraux, B., L'Huillier, A., Rzazewski, K. (eds.) Super-Intense Laser-Atom Physics. NATO ASI Series B316, pp. 95–110 (Plenum Press, New York) (1993)

Kümmel, S., Perdew, J.P.: Two avenues to self-interaction correction within Kohn-Sham theory: Unitary invariance is the shortcut. Mol. Phys. **101**, 1363–1368 (2003a)

Kümmel, S., Perdew, J.P.: Optimized effective potential made simple: Orbital functionals, orbital shifts, and the exact Kohn-Sham exchange potential. Phys. Rev. B **68**, 035103-1–035103-15 (2003b)

Kümmel, S., Kronik, L., Perdew, J.P.: Electrical response of molecular chains from density functional theory. Phys. Rev. Lett. **93**, 213002-1–213002-4 (2004)

Kümmel, S., Kronik, L.: Orbital-dependent density functionals: Theory and applications. Rev. Mod. Phys. **80**, 3–60 (2008)

Kunert, T., Schmidt, R.: Excitation and fragmentation mechanisms in ion-fullerene collisions. Phys. Rev. Lett. **86**, 5258–5261 (2001)

Kurth, S., Perdew, J.P.: Density-functional correction of rpa correlation, with results for jellium surface energies. Phys. Rev. B **59**, 10461–10468 (1999)

Kurth, S., Stefanucci, G., Almbladh, C.-O., Rubio, A., Gross, E.K.U.: Time-dependent quantum transport: A practical scheme using density functional theory. Phys. Rev. B **72**, 035308-1–035308-13 (2005)

Kurth, S., Stefanucci, G., Khosravi, E., Verdozzi, C., Gross, E.K.U.: Dynamical Coulomb blockade and the derivative discontinuity of time-dependent density functional theory. Phys. Rev. Lett. **104**, 236801-1–236801-4 (2010)

Kurth, S., Stefanucci, G.: Time-dependent bond-current functional theory for lattice Hamiltonians: Fundamental theorem and application to electron transport", Chem. Phys. (2011, in press). Available online at http://dx.doi.org/10.1016/j.chemphys.2011.01.016

Kurzweil, Y., Baer, R.: Time-dependent exchange-correlation current density functionals with memory. J. Chem. Phys. **121**, 8731–8741 (2004)

Laarmann, T., Shchatsinin, I., Singh, P., Zhavoronkov, N., Gerhards, M., Schulz, C.P., Hertel, I.V.: Coherent control of bond breaking in amino acid complexes with tailored femtosecond pulses. J. Chem. Phys. **127**, 201101-1–201101-4 (2007)

Lacks, D.J., Gordon, R.G.: Pair interactions of rare-gas atoms as a test of exchange energy-density functionals in regions of large density gradients. Phys. Rev. A **47**, 4681–4690 (1993)

Lam, P.K., Cohen, M.L.: Ab initio calculation of the static structural properties of Al. Phys. Rev. B **24**, 4224–4229 (1981)

Lambropoulos, P., Maragakis, P., Zhang, J.: Two-electron atoms in strong fields. Phys. Rep. **305**, 203–293 (1998)

Lammert, P.E.: Well-behaved coarse-grained model of density-functional theory. Phys. Rev. A **82**, 012109-1–012109-14 (2010)

Landau, L.D.: On the theory of transfer of energy at collisions II. Phys. Z. Sowjetunion **2**, 46–51 (1932)

Landau, L.D., Lifshitz, E.: Statistical Physics. Addison-Wesley, Reading (1969)

Landau, L.D., Lifshitz, E.M.: Mechanics of Fluids, vol. 6 of Course of Theoretical Physics, 2nd edn. Pergamon, New York (1987)

Lang, N.D.: Resistance of atomic wires. Phys. Rev B **52**, 5335–5342 (1995)

Langbein, D.: Theory of van der Waals Attraction, vol. 72 of Springer Tracts in Modern Physics. Springer, Berlin (1974)

Langhoff, P.W., Epstein, S.T., Karplus, M.: Aspects of time-dependent perturbation theory. Rev. Mod. Phys. **44**, 602–644 (1972)

Langreth, D.C., Perdew, J.P.: The exchange-correlation energy of a metallic surface. Sol. State Commun. **17**, 1425–1429 (1975)

Langreth, D.C., Perdew, J.P.: Theory of nonuniform electronic systems. I. Analysis of the gradient approximation and a generalization that works. Phys. Rev. B **21**, 5469–5493 (1980)

Langreth, D., Dion, M., Rydberg, H., Schröder, E., Hyldegaard, P., Lundqvist, B.I.: van der Waals density functional theory with applications. Int. J. Quantum Chem. **101**, 599–610 (2005)

Langreth, D.C., Lundqvist, B.I., Chakarova-Käck, S.D., Cooper, V.R., Dion, M., Hyldgaard, P., Kelkkanen, A., Kleis, J., Kong, L., Li, S., Moses, P.G., Murray, E., Puzder, A., Rydberg, H., Schröder, E., Thonhauser, T.: A density functional for sparse matter. J. Phys. Condens. Matter **21**, 084203-1–084203-15 (2009)

Lappas, D., van Leeuwen, R.: Electron correlation effects in the double ionization of He. J. Phys. B At. Mol. Opt. Phys. **31**, L249–L256 (1998)

Larochelle, S., Talebpour, A., Chin, S.L.: Non-sequential multiple ionization of rare gas atoms in a Ti:sapphire laser field. J. Phys. B At. Mol. Opt. Phys. **31**, 1201–1214 (1998)

Lautenschlager, P., Garriga, M., Viña, L., Cardona, M.: Temperature dependence of the dielectric function and interband critical points in silicon. Phys. Rev. B **36**, 4821–4830 (1987)

Lazzeri, M., Mauri, F.: First-principles calculation of vibrational raman spectra in large systems: Signature of small rings in crystalline SiO_2. Phys. Rev. Lett. **90**, 036401-1–036401-4 (2003)

Lazzeri, M., Mauri, F.: Nonadiabatic Kohn anomaly in a doped graphene monolayer. Phys. Rev. Lett. **97**, 266407-1–266407-4 (2006)

Lebegue, S., Harl, J., Gould, T., Angyan, J.G., Kresse, G., Dobson, J.F.: Cohesive properties and asymptotics of the dispersion interaction in graphite by the random phase approximation. Phys. Rev. Lett. **105**, 196401-1–196401-4 (2010)

Lee, C., Yang, W., Parr, R.G.: Development of the Colle-Salvetti correlation-energy formula into a functional of the electron density. Phys. Rev. B **37**, 785–789 (1988)

Lee, K., Murray, É.D., Kong, L., Lundqvist, B.I., Langreth, D.C.: Higher-accuracy van der Waals density functional. Phys. Rev. B **82**, 081101-1–081101-4 (2010)

Légaré, F., Litvinyuk, I.V., Dooley, P.W., Quéré, F., Bandrauk, A.D., Villeneuve, D.M., Corkum, P.B.: Time-resolved double ionization with few cycle laser pulses. Phys. Rev. Lett. **91**, 093002-1–093002-4 (2003)

Legrand, C., Suraud, E., Reinhard, P.-G.: Comparison of self-interaction-corrections for metal clusters. J. Phys. B At. Mol. Opt. Phys. **35**, 1115–1123 (2002)

Lein, M., Dobson, J.F., Gross, E.K.U.: Towards the description of van der Waals interactions in density functional theory. J. Comp. Chem. **20**, 12–22 (1999)

Lein, M., Gross, E.K.U., Engel, V.: Intense-field double ionization of Helium: Identifying the mechanism. Phys. Rev. Lett. **85**, 4707–4710 (2000a)

Lein, M., Gross, E.K.U., Perdew, J.P.: Electron correlation nergies from scaled exchange-correlation kernels: Importance of spatial versus temporal nonlocality. Phys. Rev. B **61**, 13431–13437 (2000b)

Lein, M., Kümmel, S.: Exact time-dependent exchange-correlation potential for strong-field electron dynamics. Phys. Rev. Lett. **94**, 143003-1–143003-4 (2005)

Leiva, P., Piris, M.: Calculation of vertical ionization potentials with the Piris natural orbital functional. J. Mol. Struct. THEOCHEM **770**, 45–49 (2006)

Levine, Z.H., Allan, D.C.: Linear optical response in silicon and germanium including self-energy effects. Phys. Rev. Lett. **63**, 1719–1722 (1989)

Levine, I.G.: Quantum Chemistry. Prentice-Hall, New Jersey (2000)

Levine, B.G., Ko, C., Quenneville, J., Martínez, T.J.: Conical intersections and double excitations in time-dependent density functional theory. Mol. Phys. **104**, 1039–1051 (2006)

Levine, B.G., Coe, J.D., Martínez, T.J.: Optimizing conical intersections without derivative coupling vectors: Application to mutistate multireference second-order perturbation theory (MS-CASPT2). J. Phys.Chem. B **112**, 405–413 (2008)

Levy, M.: Universal variational functionals of electron densities, first order density matrices, and natural-spinorbitals and solutions of the v-representability problem. Proc. Natl. Acad. Sci. USA **76**, 6062–6065 (1979)

Levy, M., Perdew, J.P., Sahni, V.: Exact differential equation for the density and ionization energy of a many-particle system. Phys. Rev. A **30**, 2745–2748 (1984)

Levy, M., Perdew, J.P.: Hellmann.Feynman, virial, and scaling requisites for the exact universal density functionals. Shape of the correlation potential and diamagnetic susceptibility for atoms. Phys. Rev. A **32**, 2010–2021 (1985)

Levy, M., Nagy, Á.: Variational density-functional theory for an individual excited state. Phys. Rev. Lett. **83**, 4361–4364 (1999)

Lewenstein, M., Balcou, P., Yu Ivanov, M., L'Huillier, A., Corkum, P.: Theory of highharmonic generation by low-frequency laser fields. Phys. Rev. A. **49**, 2117–2132 (1994)

L'Huillier, A., Lompré, L.A., Mainfray, G., Manus, C.: Multiply charged ions induced by multiphoton absorption in rare gases at 0.53 μm. Phys. Rev. A **27**, 2503–2512 (1983)

L'Huillier, A., Lompré, L.A., Mainfray, G., Manus, C.: High-order harmonic generation in rare gases. In: Gavrila, M. (ed.) Atoms in Intense Laser Fields, pp. 139–206. Academic Press, Boston (1992)

L'Huillier, A., Balcou, P.: High-order harmonic generation in rare gases with a 1–ps 1053–nm laser. Phys. Rev. Lett. **70**, 774–777 (1993)

L'Huillier, A.: Atoms in strong laser fields. Europhys. News **33**, 205–207 (2002)

Li, T., Tong, P.: Hohenberg-Kohn theorem for time-dependent ensembles. Phys. Rev. A **31**, 1950–1951 (1985)

Li, T., Li, Y.: Kohn-Sham equation for time-dependent ensembles. Phys. Rev. A. **31**, 3970–3971 (1985)

Li, T., Tong, P.: Time-dependent density-functional theory for multicomponent systems. Phys. Rev. A **34**, 529–532 (1986)

Li, X.F., L'Huillier, A., Ferray, M., Lompré, L.A., Mainfray, G.: Multiple-harmonic generation in rare gases at high laser intensity. Phys. Rev. A **39**, 5751–5761 (1989)

Liang, Y., Augst, S., Chin, S.L., Beaudoin, Y., Chaker, M.: High harmonic generation in atomic and diatomic molecular gases using intense picosecond laser pulses–a comparison. J. Phys. B At. Mol. Opt. Phys. **27**, 5119–5130 (1994)

Lieb, E.H., Oxford, S.: An improved lower bound on the indirect Coulomb energy. Int. J. Quant. Chem. **19**, 427–439 (1981)

Lieb, E.H.: Density functionals for coulomb systems. Int. J. Quant. Chem. **24**, 243–277 (1983)

Lima, N.A., Oliveira, L.N., Capelle, K.: Density-functional study of the mott gap in the Hubbard model. Europhys. Lett. **60**, 601–607 (2002)

Lima, N.A., Silva, M.F., Oliveira, L.N., Capelle, K.: Density functionals not based on the electron gas: Local-density approximation for a luttinger liquid. Phys. Rev. Lett. **90**, 146402-1–146402-4 (2003)

Lindblad, G.: On the generators of quantum dynamical semigroups. Commun. Math. Phys. **48**, 119–130 (1976)

Lindner, F., Schätzel, M.G., Walther, H., Baltuška, A., Goulielmakis, E., Krausz, F., Milošević, D.B., Bauer, D., Becker, W., Paulus, G.G.: Attosecond double-slit experiment. Phys. Rev. Lett. **95**, 040401-1–040401-4 (2005)

Littlejohn, R.G., Reinsch, M.: Gauge fields in the separation of rotations and internal motions in the n-body problem. Rev. Mod. Phys. **69**, 213–276 (1997)

Liu, K.L., Vosko, S.H.: A time-dependent spin density functional theory for the dynamical spin susceptibility. Can. J. Phys. **67**, 1015–1021 (1989)

Liu, X., Rottke, H., Eremina, E., Sandner, W., Goulielmakis, E., Keeffe, K.O., Lezius, M., Krausz, F., Lindner, F., Schätzel, M.G., Paulus, G.G., Walther, H.: Nonsequential double ionization at the single-optical-cycle limit. Phys. Rev. Lett. **93**, 263001-1–263001-4 (2004)

Liu, J., Guo, Z., Sun, J., Liang, W.: Theoretical studies on electronic spectroscopy and dynamics with the real-time time-dependent density functional theory. Front. Chem. China **5**, 11–28 (2010)

Löwdin, P.O.: On the non-orthogonality problem connected with the use of atomic wave functions in the theory of molecules and crystals. J. Chem. Phys. **18**, 365–375 (1950)

Longuet-Higgins, H.C.: Spiers memorial lecture. Intermolecular forces. Discuss. Faraday Soc. **40**, 7–18 (1965)

Lopéz-Sandoval, R., Pastor, G.M.: Density-matrix functional theory of the Hubbard model: An exact numerical study. Phys. Rev. B **61**, 1764–1772 (2000)

López-López, S., Nest, M.: Analysis of the continuous-configuration time-dependent self-consistent field method applied to system-bath dynamics. J. Chem. Phys. **132**, 104103-1–104103-7 (2010)

Louck, J.D.: Derivation of the molecular vibration-rotation hamiltonian from the Schrödinger equation for the molecular model. J. Mol. Spectr. **61**, 107–137 (1976)

Lovesey, S.W., Collins, S.P.: X-Ray Scattering and Absorption in Magnetic Materials. Oxford University Press, Oxford (1996)

Löwdin, P.-O.: Quantum theory of many-particle systems. I. Physical interpertations by means of density matrices, natural spin-orbitals, and convergence problems in the method of configurational interaction. Phys. Rev. **97**, 1474–1489 (1955)

Löwdin, P.-O., Shull, H.: Natural orbitals in the quantum theory of two-electron systems. Phys. Rev. **101**, 1730–1739 (1956)

Löwdin, P.-O., Mukherjee, P.K.: Some comments on the time-dependent variation principle. Chem. Phys. Lett. **14**, 1–7 (1972)

Lu, D., Nguyen, H.-V., Galli, G.: Power series expansion of the random phase approximation correlation energy: The role of the third- and higher-order contributions. J. Chem. Phys. **133**, 154110-1–154110-11 (2010)

Luth, H.: Solid Surfaces, Interfaces and Thin Films. Springer, Berlin (2001)

Macklin, J.J., Kmetec, J.D., Gordon III, C.L.: High-order harmonic generation using intense femtosecond pulses. Phys. Rev. Lett. **70**, 766–769 (1993)

Maeda, S., Ohna, K., Morokuma, K.: Automated global mapping of minimal energy points on seams of crossing by the anharmonic downward distortion following method: A case study of H_2CO. J. Phys. Chem. A **113**, 1704–1710 (2009)

Maeda, S., Ohno, K., Morokuma, K.: Updated branching plane for finding conical intersections without coupling derivative vectors. J. Chem. Theo. Comput. **6**, 1538–1545 (2010)

Magyar, R.J., Tretiak, S.: Dependence of spurious charge-transfer excited states on orbital exchange in TDDFT: Large molecules and clusters. J. Chem. Theory Comput. **3**, 976–987 (2007)

Mahan, G.D.: Theory of photoemission in simple metals. Phys. Rev. B **2**, 4334–4350 (1970)

Mahan, G.D., Subbaswamy, K.R.: Local Density Theory of Polarizability. Plenum, New York (1990)

Mahanty, J., Ninham, B.W.: Dispersion Forces. Academic Press, London (1976)

Mainfray, G., Manus, C.: Multiphoton ionization of atoms. Rep. Prog. Phys. **54**, 1333–1372 (1991)

Maitra, N.T., Burke, K.: Demonstration of initial-state dependence in time-dependent density-functional theory. Phys. Rev. A **63**, 042501-1–042501-7 (2001a)

Maitra, N.T., Burke, K.: Demonstration of initial-state dependence in time-dependent density-functional theory. Phys. Rev. A **64**, 039901(E) (2001b)

Maitra, N.T., Burke, K.: On the floquet formulation of time-dependent density functional theory. Chem. Phys. Lett. **359**, 237–240 (2002)

Maitra, N.T., Burke, K., Woodward, C.: Memory in time-dependent density functional theory. Phys. Rev. Lett. **89**, 023002-1–023002-4 (2002a)

Maitra, N.T., Burke, K., Appel, H., Gross, E.K.U., van Leeuwen, R.: Ten topical questions in time-dependent density-functional theory. In: Sen, K.D. (ed.) Reviews in Modern Quantum Chemistry, A Celebration of the Contributions of Robert Parr, pp. 1186–1225. World Scientific, Singapore (2002b)

Maitra, T.N., Souza, I., Burke, K.: Current-density functional theory of the response of solids. Phys. Rev. B **68**, 045109-1–045109-5 (2003)

Maitra, N.T., Zhang, F., Cave, R., Burke, K.: Double excitations within time-dependent density functional theory linear response. J. Chem. Phys. **120**, 5932–5937 (2004)

Maitra, N.T.: Memory formulas for perturbations in time-dependent density functional theory. Int. J. Quant. Chem. **102**, 573–581 (2005a)

Maitra, N.T.: Undoing static correlation: Long-range charge transfer in time-dependent density-functional theory. J. Chem. Phys. **122**, 234104-1–234104-6 (2005b)

Maitra, N.T., Tempel, D.G.: Long-range excitations in time-dependent density functional theory. J. Chem. Phys. **125**, 184111-1–184111-6 (2006a)

Maitra, N.T.: On correlated electron-nuclear dynamics using time-dependent density functional theory. J. Chem. Phys. **125**, 014110-1–014110-4 (2006b)

Maitra, N.T., van Leeuwen, R., Burke, K.: Comment on critique of the foundations of time-dependent density functional theory. Phys. Rev. A. **78**, 056501-1–056501-4 (2008)

Maitra, N.T., Todorov, T.N., Woodward, C., Burke, K.: Density-potential mapping in time-dependent density functional theory. Phys. Rev. A. **81**, 042525-1–042525-7 (2010)

Makri, N., Miller, W.H.: Time-dependent self-consistent field (TDSCF) approximation for a reaction coordinate coupled to a harmonic bath: Single and multiple configuration treatments. J. Chem. Phys. **87**, 5781–5787 (1987)

Malcioglu, O.B., Gebauer, R., Rocca, D., Baroni, S.: TurboTDDFT— A code for the simulation of molecular spectra using the Liouville-Lanczos approach to time-dependent density-functional perturbation theory. Comp. Phys. Comm. **182**, 1744–1754 (2011)

Manaa, M.R., Yarkony, D.R.: On the intersection of two potential energy surfaces of the same symmetry. Systematic characterization using a lagrange multiplier constrained procedure. J. Chem. Phys. **99**, 5251–5256 (1993)

Marangos, J.P.: Molecules in a strong laser field. In: Batani, D., Joachain, C.J., Martellucci, S. (eds.) Atoms and Plasmas in Super- Intense Laser Fields, SIF Conference Proceedings, **88**, 213–243. Societ'a Italiana di Fisica, Bologna (2004)

Marini, A., Del Sole, R., Rubio, A.: Bound excitons in time-dependent density- functional theory: Optical and energy-loss spectra. Phys. Rev. Lett. **91**, 256402–256404 (2003)

Marini, A., Del Sole, R., Rubio, A.: Optical properties of solids and nanostructures from a many-body fxc kernel. In: Marques, M.A.L., Ullrich, C.A., Nogueira, F., Rubio, A., Burke, K., Gross, E.K.U. (eds.) Time-Dependent Density Functional Theory, pp. 301–316. Springer, Berlin Heidelberg (2006a)

Marini, A., Garcia-Gonzalez, P., Rubio, A.: First-principle description of correlation effects in layered materials. Phys. Rev. Lett. **96**, 136404-1–136404-4 (2006b)

Marini, A., Hogan, C., Grüning, M., Varsano, D.: Yambo: An ab initio tool for excited state calculations. Comp. Phys. Comm. **180**, 1392–1403 (2009)

Markevitch, A.N., Smith, S.M., Romanov, D.A., Schlegel, D.A., Yu Ivanov, D.A., Levis, R. J.: Nonadiabatic dynamics of polyatomic molecules and ions in strong laser fields. Phys. Rev. A **68**, 011402-1(R)–011402-1(R) (2003)

Markevitch, A.N., Romanov, D.A., Smith, D.A., Schlegel, D.A., Yu Ivanov, M., Levis, R.J.: Sequential nonadiabatic excitation of large molecules and ions driven by strong laser fields. phys. Rev. A **69**, 013401-1–013401-13 (2004)

Marques, M.A.L., Castro, A., Rubio, A.: Assessment of exchange-correlation functionals for the calculation of dynamical properties of small clusters in time-dependent density functional theory. J. Chem. Phys. **115**, 3006–3014 (2001)

Marques, M.A.L., Castro, A., Bertsch, G.F., Rubio, A.: Octopus: A first-principles tool for excited electron-ion dynamics. Comp. Phys. Comm. **151**, 60–78 (2003)

Marques, M.A. L., Ullrich, C.A., Nogueira, F., Rubio, A., Burke, K., Gross, E.K.U. (eds.): Time-Dependent Density Functional Theory. Lecture Notes in Physics, vol. 706, Springer, Berlin (2006a)

Marques, M.A.L., Rubio, A.: Time versus frequency space techniques. In: Marques, M.A.L., Ullrich, C.A., Nogueira, F., Rubio, A., Burke, K., Gross, E.K.U. (eds.) Time-Dependent Density Functional Theory, pp. 227–240. Springer, Heidelberg (2006)

Marques, M.A.L., Castro, A., Malloci, G., Mulas, G., Botti, S.: Efficient calculation of van der Waals dispersion coefficients with time-dependent density functional theory in real time: Application to polycyclic aromatic hydrocarbons. J. Chem. Phys. **127**, 014107-1–014107-6 (2007)

Martin, P.C., Schwinger, J.: Theory of many-particle systems. I. Phys. Rev. **115**, 1342–1373 (1959)

Martin, J.D.D., Hepburn, J.W.: Electric field induced dissociation of molecules in Rydberg-like highly vibrationally excited ion-pair states. Phys. Rev. Lett. **79**, 3154–3157 (1997)

Martinazzo, R., Nest, M., Saalfrank, P., Tantardini, G.F.: A local coherent-state approximation to system-bath quantum dynamics. J. Chem. Phys **125**, 194102-1–194102-16 (2006)

Marx, D., Hutter, J.: Ab Initio Molecular Dynamics: Theory and Implementation. In: Grotendorstin, J. (ed.) Modern Methods and Algorithms of Quantum Chemistry. NIC Series, vol. 1, pp. 301–449. John von Neumann Institute for Computing, Jülich (2000)

Marzari, N., Vanderbilt, D., Payne, M.C.: Ensemble density-functional theory for Ab initio molecular dynamics of metals and finite-temperature insulators. Phys. Rev. Lett. **79**, 1337–1340 (1997)

Masson, W.P. (ed.): Physical Acoustic, vol. I, Academic Press, New York (1964)

Mauri, F., Car, R., Tosatti, E.: Canonical statistical averages of coupled quantum—classical systems. Europhys. Lett. **24**, 431–436 (1993)

Mauri, F., Louie, S.G.: Magnetic susceptibility of insulators from first principles. Phys. Rev. Lett. **76**, 4246–4249 (1996)

Mauri, F., Pfrommer, B.G., Louie, S.G.: Ab initio theory of NMR chemical shifts in solids and liquids. Phys. Rev. Lett. **77**, 5300–5303 (1996)

Maxwell-Garnett, J.C.: Colours in metal glasses and in metallic films, philos. Trans. R. Soc. London **A203**, 385–420 (1904)

May, V., Kühne, O.: Charge and Energy Transfer Dynamics in Molecular Systems. Wiley-VCH, New York (2004)

Mazur, G., Włodarczyk, R.: Application of the dressed time-dependent density- functional theory for the excited states of linear polyenes. J. Comput. Chem. **30**, 811–817 (2009)

Mazur, G., Marcin, M., Włodarczyk, R., Aoki, Y.: Dressed TDDFT study of low- lying electronic excited states in selected linear polyenes and diphenylopolyenes. Int. J. Quant. Chem. **111**, 810–825 (2011a)

Mazur, G., Makowski, M., Wlodarcyk, R., Aoki, Y.: Dressed TDDFT study of low-lying electronic excited states in selected linear polyenes and dipheylopolyenes. Int. J. Quant. Chem. **111**, 819–825 (2011b)

McLachlan, A.D., Ball, M.A.: Time-dependent Hartree-Fock theory for molecules. Rev. Mod. Phys. **36**, 844–855 (1964)

McPherson, A., Gibson, G., Jara, H., Johann, U., Luk, T.S., McIntyre, I.A., Boyer, K., Rhodes, C.K.: Studies of multiphoton production of vacuum-ultraviolet radiation in the rare gases. J. Opt. Soc. Am. B **4**, 595–601 (1987)

McWeeny, R.: Methods of Molecular Quantum Mechanics, 2nd edn. Academic Press, London (1989)

Mearns, D., Kohn, W.: Frequency-dependent v-representability in density-functional theory. Phys. Rev. A **35**, 4796–4799 (1987)

Meier, C., Tannor, D.J.: Non-markovian evolution of the density operator in the presence of strong laser fields. J. Chem. Phys. **111**, 3365–3377 (1999)

Mennucci, B., Cappelli, C., Guido, C.A., Cammi, R., Tomasi, J.: Structures and properties of electronically excited chromophores in solution from the polarizable continuum model coupled to the time-dependent density functional theory. J. Phys. Chem. A **113**, 3009–3020 (2009)

Mermin, N.D.: Thermal properties of the inhomogeneous electron gas. Phys. Rev. **137**, A1441–A1443 (1965)

Meyer, H.D., Manthe, U., Cederbaum, L.S.: The multi-configurational time-dependent Hartree approach. Chem. Phys. Lett. **165**, 73–78 (1990)

Meyer, H.: The molecular hamiltonian. Annu. Rev. Phys. Chem. **53**, 141–172 (2002)

Michl, J., Bonačić-Koutecký, V.: Electronic Aspects of Organic Photochemistry. Wiley, New York (1990)

Milton, K.A.: The Casimir Effect: Physical Manifestations of Zero-Point Energy. World Scientific, Singapore (2001)

Minezawa, N., Gordon, M.S.: Optimizing conical intersections by spin-flip density functional theory: Application to ethylene. Phys. Chem. A **113**, 12749–12753 (2009)

Mirić, R., Werner, U., Bonačić-Koutecký, V.: Nonadiabatic dynamics and simulation of time resolved photoelectron spectra within time-dependent density functional theory: Ultrafast photoswitching in benzylideneaniline. J. Chem. Phys. **129**, 164118–164119 (2008)

Misquitta, A.J., Podeszwa, R., Jeziorski, B., Szalewicz, K.: Intermolecular potentials based on symmetry-adapted perturbation theory with dispersion energies from timedependent density-functional calculations. J. Chem. Phys. **123**, 214103-1–214103-14 (2005)

Misquitta, A.J., Spencer, J., Stone, A.J., Alavi, A.: Dispersion interactions between semiconducting wires. Phys. Rev. B **82**, 075312-1–075312-7 (2010)

Mitrić, R., Werner, U., Wohlgemuth, M., Seifert, G., Bonačić-Koutecký, V.: Nonadiabatic dynamics within time-dependent density functional tight binding method. J. Phys. Chem. A **113**, 12700–12705 (2009)

Miura, M., Aoki, Y., Champagne, B.: Assessment of time-dependent density functional schemes for computing the oscillator strengths of benzene, phenol, aniline, and fluorobenzene. J. Chem. Phys. **127**, 084103-1–084103-16 (2007)

Miyake, T., Aryasetiawan, F., Kotani, T., van Schilfgaarde, M., Usuda, M., Terakura, K.: Total energy of solids: An exchange and random-phase approximation correlation study. Phys. Rev. B **66**, 245103-1–245103-4 (2002)

Miyazaki, K., Sakai, H.: High-order harmonic generation in rare gases with intense subpicosecond dye laser pulses. J. Phys. B At. Mol. Opt. Phys. **25**, L83–L89 (1992)

Mohammed, A., Agren, H., Norman, P.: Time-dependent density functional theory for resonant properties: Resonance enhanced Raman scattering from the complex electricdipole polarizability. Phys. Chem. Chem. Phys. **11**, 4539–4548 (2009)

Moroni, S., Ceperley, D.M., Senatore, G.: Static response and local field factor of the electron gas. Phys. Rev. Lett. **75**, 689–692 (1995)

Morrell, M.M., Parr, R.G., Levy, M.: Calculation of ionization potentials from density matrices and natural functions, and the long-range behavior of natural orbitals and electron density. J. Chem. Phys. **62**, 549–554 (1975)

Moshammer, R., Feuerstein, B., Schmitt, W., Dorn, A., Schröter, C.D., Ullrich, J., Rottke, H., Trump, C., Wittmann, M., Korn, G., Hoffmann, K., Sandner, W.: Momentum distributions of Ne^{n+} ions created by an intense ultrashort laser pulse. Phys. Rev. Lett. **84**, 447–450 (2000)

Mühlbacher, L., Rabani, E.: Real-time path integral approach to nonequilibrium many-body quantum systems. Phys. Rev. Lett. **100**, 176403-1–176403-4 (2008)

Mukamel, S.: Generalized time-dependent density-functional-theory response functions for spontaneous density fluctuations and response: Resolving the causality paradox in real time. Phys. Rev. A **71**, 024503-1–024503-4 (2005)

Müller, A.M.K.: Explicit approximate relation between reduced two- and one-particle density matrices. Phys. Lett. **105A**, 446–452 (1984)

Mulliken, R.S.: Intensities of electronic transitions in molecular spectra II. Charge- transfer spectra. J. Chem. Phys. **7**, 20–34 (1939)

Mulser, P., Bauer, D.: High Power Laser-Matter Interaction. Springer, Berlin (2010)

Mundt, M., Kümmel, S.: Derivative discontinuities in time-dependent density- functional theory. Phys. Rev. Lett. **95**, 203004-1–203004-4 (2005)

Mundt, M., Kümmel, S.: Optimized effective potential in real time: Problems and prospects in time-dependent density-functional theory. Phys. Rev. A **74**, 022511-1–022511-7 (2006)

Mundt, M., Kümmel, S.: Photoelectron spectra of anionic sodium clusters from timedependent density-functional theory in real time. Phys. Rev. B **76**, 035413-1–035413-8 (2007)

Mundt, M., Kümmel, S., van Leeuwen, R., Reinhard, P.-G.: Violation of the zero-force theorem in the time-dependent Krieger-Li-Iafrate approximation. Phys. Rev. A **75**, 050501-1(R)–050501-4(R) (2007)

Munshi, A. (ed.): The Open CL Specification. Khronos group, Philadelphia (2009)

Murray, E.D., Lee, K., Langreth, D.C.: Investigation of exchange energy density functional accuracy for interacting molecules. J. Chem. Theory Comput. **5**, 2754–2762 (2009)

Myöhänen, P., Stan, A., Stefanucci, G., van Leeuwen, R.: A many-body approach to quantum transport dynamics: Initial correlations and memory effects. Europhys. Lett. **84**, 67001-1–67001-5 (2008)

Myöhänen, P., Stan, A., Stefanucci, G., van Leeuwen, R.: Kadanoff-Baym approach to quantum transport through interacting nanoscale systems: From the transient to the steady-state regime. Phys. Rev. B **80**, 115107-1–115107-16 (2009)

Nakajima, S.: On quantum theory of transport phenomena—steady diffusion. Rep. Prog. Phys. **20**, 948–959 (1958)

Nazarov V.U., Pitarke J.M., Kim C.S., Takada Y.: Time-dependent densityfunctional theory for the stopping power of an interacting electron gas for slow ions. Phys. Rev. B **71**, 121106-1–121106-4(R) (2005)

Nazarov, V.U., Pitarke, J.M., Takada, Y., Vignale, G., Chang, Y.-C.: Including nonlocality in the exchange-correlation kernel from time-dependent current density functional theory: Application to the stopping power of electron liquids. Phys. Rev. B **76**, 205103-1–205103-6 (2007)

Nazarov, V.U., Tokatly, I.V., Pittalis, S., Vignale, G.: Antiadiabatic limit of the exchange-correlation kernels of an inhomogeneous electron gas. Phys. Rev. B **81**, 245101-1–245101-9 (2010)

Nesbet, R.K.: Local response model of the generalized polarization potential. Phys. Rev. A **56**, 2778–2783 (1997)

Nest, M., Klamroth, T., Saalfrank, P.: The multiconfiguration time-dependent Hartree-Fock method for quantum chemical calculations. J. Chem. Phys. **122**, 124102-1–124102-7 (2005)

Neugebauer, J., Louwerse, M.J., Baerends, E.J., Wesolowski, T.A.: The merits of the frozen-density embedding scheme to model solvatochromic shifts. J. Chem. Phys. **122**, 094115-1–0941151-3 (2005)

Neuhauser, D., Lopata, K.: Quantum drude friction for time-dependent density functional theory. J. Chem. Phys. **129**, 134106-1–1341061-3 (2008)

Nex, C.M.M.: Estimation of integrals with respect to a density of states. J. Phys. A **11**, 653–663 (1978)

Ng, T.K., Singwi, K.S.: Time-dependent density functional theory in the linear-response regime. Phys. Rev. Lett. **59**, 2627–2630 (1987)

Ng, T.K.: Transport properties and a current-functional theory in the linear-response regime. Phys. Rev. Lett. **62**, 2417–2420 (1989)

Nguyen, H.S., Bandrauk, A.D., Ullrich, C.A.: Asymmetry of above-threshold ionization of metal clusters in two-color laser fields: A time-dependent density-functional study. Phys. Rev. A **69**, 063415-1–063415-8 (2004)

Nguyen, N.A., Bandrauk, A.D.: Electron correlation of one-dimensional H_2 in intense laser fields: Time-dependent extended Hartree-Fock and time-dependent densityfunctional-theory approaches. Phys. Rev. A **73**, 032708-1–032708-7 (2006)

Nguyen, H.-V., Galli, G.: A first-principles study of weakly bound molecules using exact exchange and the random phase approximation. J. Chem. Phys. **132**, 044109-1–044109-8 (2010)

Nguyen, K.A., Day, P.N., Pachter, R.: Analytical energy gradients of Coulomb-attenuated time-dependent density functional methods for excited states. Int. J. Quantum Chem. **110**, 2247–2255 (2010)

Nielsen, O.H., Martin, R.M.: Quantum-mechanical theory of stress and force. Phys. Rev. B **32**, 3780–3791 (1985)

Nielsen, S., Kapral, R., Ciccotti, G.: Statistical mechanics of quantum-classical systems. J. Chem. Phys. **115**, 5805–5815 (2001)

Niquet, Y.M., Gonze, X.: Band-gap energy in the random-phase approximation to density-functional theory. Phys. Rev. B **70**, 245115-1–245115-12 (2004)

Nitzan, A.: Chemical Dynamics in Condensed Phases. Oxford University Press, Oxford (2006)

Nobusada, K., Yabana, K.: High-order harmonic generation from silver clusters: Laser-frequency dependence and the screening effect of d electrons. Phys. Rev A **70**, 043411-1–043411-7 (2004)

Nordholm, S., Rice, S.A.: A quantum ergodic theory approach to unimolecular fragmentation. J. Chem. Phys. **62**, 157–168 (1975)

Nosé, S.: A molecular dynamics method for simulations in the canonical ensemble. Mol. Phys. **52**, 255–268 (1984)

Nosé, S.: Constant temperature molecular dynamics. Prog. Theor. Phys. Suppl. **103**, 1–46 (1991)

Ohtsuki, Y., Turinici, G., Rabitz, H.: Generalized monotonically convergent algorithms for solving quantum optimal control problems. J. Chem. Phys. **120**, 5509–5517 (2004)

Olevano, V., Palummo, M., Onida, G., Del Sole, R.: Exchange and correlation effects beyond the LDA on the dielectric function of silicon. Phys. Rev. B **60**, 14224–14233 (1999)

Oliveira, M.J.T., Castro, A., Marques, M.A.L., Rubio, A.: On the use of Neumann's principle for the calculation of the polarizability tensor of nanostructures. J. Nanosci. Nanotechnol. **8**, 3392–3398 (2008)

Onida, G., Reining, L., Rubio, A.: Electronic excitations: Density-functional versus many-body Green's-function approaches. Rev. Mod. Phys. **74**, 601–659 (2002)

Osinga, V.P., van Gisbergen, S.J.A., Snijders, J.G., Baerends, E.J.: Density functional results for isotropic multipole polarizabilities and c6, c7 and c8 van der Waals dispersion coefficients for molecules. J. Chem. Phys. **106**, 5091–5101 (1997)

Ozawa, L.: Cathodoluminescence: Theory and Applications. VHC, New York (1990)

Pacheco, J.M., Ekardt, W.: A new formulation of the dynamical response of manyelectron systems and the photoelectron cross-section of small metal clusters. Z. Phys. D **24**, 65–69 (1992a)

Pacheco, J.M., Ekardt, W.: Response of finite many-electron systems beyond the time-dependent LDA: Application to small metal clusters. Annalen der Physik **1**, 254–269 (1992b)

Pacheco, J.M., Ramalho, J.P.P.: First-principles determination of the dispersion interaction between fullerenes and their intermolecular potential. Phys. Rev. Lett. **79**, 3873–3876 (1997)

Paier, J., Janesko, B.G., Henderson, T.M., Scuseria, G.E., Grüneis, A., Kresse, G.: Hybrid functionals including random phase approximation correlation and second-order screened exchange. J. Chem. Phys. **132**, 94103-1–94103-10 (2010)

Palao, J., Kosoff, R.: Quantum computing by an optimal control algorithm for unitary transformations. Phys. Rev. Lett. **89**, 188301–188304 (2002)

Panaccione, G., Altarelli, M., Fondaraco, A., Georges, A., Huotari, S., Lacovig, P., Lichenstein, A., Metcalf, P., Monaco, G., Offi, F., Paolasini, L., Poteryaev, A., Tjernberg, O., Sacchi, M.: Coherent peaks and minimal probing depth in photoemission spectroscopy of Mutt-Hubbard systems. Phys. Rev. Lett. **97**, 116401-1–116401-4 (2006)

Papalazarou, E., Gatti, M., Marsi, M., Brouet, V., Iori, F., Reining, L., Annese, E., Vobornik, I., Offi, F., Fondacaro, A., Huotari, S., Lacovig, P., Tjernberg, O., Brookes, N.B., Sacchi, M., Metcalf, P., Panaccione, G.: Valence-band electronic structure of V_2O_3: Identification of V and O bands. Phys. Rev. B **80**, 155115-1–155115-6 (2009)

Parac, M., Grimme, S.: Comparison of multireference møller–plesset theory and time-dependent methods for the calculation of vertical excitation energies of molecules. J. Chem. Phys. **106**, 6844–6850 (2002)

Parac, M., Grimme, S.: A TDDFT study of the lowest excitation energies of polycyclic aromatic hydrocarbons. Chem. Phys. **292**, 11–21 (2003)

Parandekar, P.V., Tully, J.C.: Mixed quantum-classical equilibrium. J. Chem. Phys. **122**, 094102-1–094102-6 (2005)

Parandekar, P.V., Tully, J.C.: Detailed balance in Ehrenfest mixed quantum-classical dynamics. J. Chem. Theor. Comput. **2**, 217–228 (2006)

Parker, J., Taylor, K.T., Clark, C.W., Blodgett-Ford, S.: Intense-field multiphoton ionization of a two-electron atom. J. Phys. B At. Mol. Opt. Phys. **29**, L33–L42 (1996)

Parker, J.S., Smyth, E.S., Taylor, K.T.: Intense-field multiphoton ionization of helium. J. Phys. B At. Mol. Opt. Phys. **31**, L571–L578 (1998)

Pasquarello, A., Car, R.: Dynamical charge tensors and infrared spectrum of Amorphous SiO_2. Phys. Rev. Lett. **79**, 1766–1769 (1997)

Pasquarello, A., Resta, R.: Dynamical monopoles and dipoles in a condensed molecular system: The case of liquid water. Phys. Rev. B **68**, 174302-1–174302-10 (2003)

Patton, D.C., Pederson, M.R.: A theoretical study of rare-gas diatomic molecules within the generalized-radient approximation to density-functional theory. Phys. Rev. A **56**, R2495–R2498 (1997)

Patton, D.C., Porezag, D.V., Pederson, M.R.: Simplified generalized-gradient approximation and anharmonicity: Benchmark calculations on molecules. Phys. Rev. B **55**, 7454–7459 (1997)

Paul, P.M., Toma, E.S., Breger, P., Mullot, G., Augé, F., Balcou, P., Muller, H.G., Agostini, P.: Observation of a train of Attosecond pulses from high harmonic generation. Science **292**, 1689–1692 (2001)

Paulus, G.G., Nicklich, W., Xu, H., Lambropoulos, P., Walther, H.: Plateau in above threshold ionization spectra. Phys. Rev. Lett. **72**, 2851–2854 (1994)

Paulus, G.G., Grasbon, F., Walther, H., Kopold, R., Becker, W.: Channel-closinginduced resonances in the above-threshold ionization plateau. Phys. Rev. A **64**, 021401-1–021401-4(R) (2001)

Peach, M.J.G., Cohen, A.J., Tozer, D.J.: Influence of Coulomb-attenuation on exchange–correlation functional quality. Phys. Chem. Chem. Phys. **8**, 4543–4549 (2006)

Peach, M.J.G., Benfield, P., Helgaker, T., Tozer, D.J.: Excitation energies in density functional theory: An evaluation and a diagnostic test. J. Chem. Phys. **128**, 044118-1–044118-8 (2008)

Peirce, P.A., Dahleh, M.A., Rabitz, H.: "Optimal control of quantum-mechanical systems: Existence, numerical approximation, and applications. Phys. Rev. A **37**, 4950–4964 (1988)

Pendry, J.B.: Theory of photoemission. Surf. Sci. **57**, 679–705 (1976)

Peng, D., Zou, W., Liu, W.: Time-dependent quasirelativistic density-functional theory based on the zeroth-order regular approximation. J. Chem. Phys. **123**, 144101-1–144101-13 (2005)

Penz, M., Ruggenthaler, M.: Domains of time-dependent density-potential mappings. J. Phys. A: Math. Theor. **44**, 335208 (2011)

Perdew, J.P., Zunger, A.: Self-interaction correction to density-functional approximations for many-electron systems. Phys. Rev. B **23**, 5048–5079 (1981)

Perdew, J.P., Parr, R.G., Levy, M., Balduz Jr, J.L.: Density-functional theory for fractional particle number: Derivative discontinuities of the energy. Phys. Rev. Lett. **49**, 1691–1694 (1982)

Perdew, J.P.: What do the Kohn-Sham orbitals mean? How do atoms dissociate? In: Dreizler, R.M., da Providencia, J. (eds.) Density Functional Methods in Physics, p. 265. Plenum, New York (1985)

Perdew, J.P.: Density-functional approximation for the correlation energy of the inhomogenous electron gas. Phs. Rev. B **33**, 8822–8824 (1986)

Perdew, J.P.: Size-consistency, self-interaction correction, and derivative discontinuity in density functional theory. Adv. Quantum Chem. **21**, 113–134 (1990)

Perdew, J.P., Wang, Y.: Accurate and simple analytic representation of the electrongas correlation energy. Phys. Rev. B **45**, 13244–13249 (1992a)

Perdew, J.P., Wang, Y.: Pair-distribution function and its coupling-constant average for the spin-polarized electron gas. Phys. Rev. B **46**, 12947–12954 (1992b)

Perdew, J.P., Wang, Y.: Pair-distribution function and its coupling-constant average for the spin-polarized electron gas. Phys. Rev. B **56**, 7018(E) (1997)

Perdew, J.P., Burke, K., Wang, Y.: Generalized gradient approximation for the exchange-correlation hole of a many-electron system. Phys. Rev. B **54**, 16533–16539 (1996a)

Perdew, J.P., Burke, K., Wang, Y.: Generalized gradient approximation for the exchange-correlation hole of a many-electron system. Phys. Rev. B **57**, 14999(E) (1998)

Perdew, J.P., Burke, K., Ernzerhof, M.: Generalized gradient approximation made simple. Phys. Rev. Lett. **77**, 3865–3868 (1996b)

Perdew, J.P., Burke, K., Ernzerhof, M.: Generalized gradient approximation made simple. Phys. Rev. Lett. **78**, 1396(E) (1997)

Perdew, J.P., Ernzerhof, M., Burke, K.: Rationale for mixing exact exchange with density functional approximations. J. Chem. Phys. **105**, 9982–9985 (1996c)

Perdew, J.P., Kurth, S.: Density functionals for non-relativistic Coulomb systems in the new century. In: Primer, A., Fiolhais, C., Nogueira, F., Marques, M.A.L. (eds.) A Primer in Density Functional Theory, pp. 1–55. Springer, Berlin (2003)

Perdew, J.P., Ruzsinszky, A., Tao, J., Staroverov, V.N., Scuseria, G.E., Csonka, G.I.: Prescription for the design and selection of density functional approximations: More constraint satisfaction and fewer fits. J. Chem. Phys. **123**, 062201-1–062201-9 (2005)

Perez-Jorda, J.M., Becke, A.D.: A density functional study of van der Waals forces: Rare gas diatomics. Chem. Phys. Lett. **233**, 134–137 (1995)

Perez-Jorda, J.M., San-Fabian, E., Perez-Jimenez, A.J.: Density-functional study of van der Waals forces on rare-gas diatomics: Hartree Fock exchange. J. Chem. Phys. **110**, 1916–1920 (1999)

Pernal, K.: Effective potential for natural spin orbitals. Phys. Rev. Lett. **23**, 233002-1–233002-4 (2005)

Pernal, K., Cioslowski, J.: Ionization potentials from the extended Koopmans' theorem applied to density matrix functional theory. Chem. Phys. Lett. **412**, 71–75 (2005)

Pernal, K., Gritsenko, O., Baerends, E.J.: Time-dependent density matrix functional theory. Phys. Rev. A **75**, 012506-1–012506-8 (2007a)

Pernal, K., Giesbertz, K.J.H., Gritsenko, O., Baerends, E.J.: Adiabatic approximation of time dependent density matrix functional theory. J. Chem. Phys **1237**, 214101-1–214101-11 (2007b)

Perry, M.D., Szöke, A., Landen, O.L., Campbell, E.M.: Nonresonant multiphoton ionization of noble gases: Theory and experiment. Phys. Rev. Lett. **60**, 1270–1273 (1988)

Peskin, U., Steinberg, M.: A temperature-dependent Schrödinger equation based on a time-dependent self consistent field approximation. J. Chem. Phys **109**, 704–710 (1998)

Petersilka, M., Gossmann, U.J., Gross, E.K.U.: Excitation energies from time-dependentdensity-functional theory. Phys. Rev. Lett. **76**, 1212–1215 (1996)

Petersilka, M., Gross, E.K.U.: Spin-multiplet energies from time-dependent densityfunctional theory. Int. J. Quant. Chem. Symp. **30**, 1393–1407 (1996)

Petersilka, M., Gossman, U.J., Gross, E.K.U.: Time-dependent optimized effective potential in the linear response regime. In: Dobson, J.F., Vignale, G., Das, M. (eds.) Electronic Density Functional Theory: Recent Progress, New Directions, New York, Plenum pp. 177–197. (1998)

Petersilka, M., Gross, E.K.U.: Strong-field double ionization of helium: A density- functional perspective. Laser Phys. **9**, 105–114 (1999)

Petersilka, M., Gross, E.K.U., Burke, K.: Excitation energies from time-dependent density-functional theory using exact and approximate potentials. Int. J. Quant.Chem. **80**, 534–554 (2000)

Peterson, R.L.: Formal theory of nonlinear response. Rev. Mod. Phys. **39**, 69–77 (1967)

Peuckert, V.: A new approximation method for electron systems. J. Phys. C **11**, 4945–4956 (1978)

Phillips, J.C., Kleinman, L.: New method for calculating wave functions in crystals and molecules. Phys. Rev. **116**, 287–294 (1959)

Pickard, C.J., Mauri, F.: All-electron magnetic response with pseudopotentials: NMR chemical shifts. Phys. Rev. B **63**, 245101-1–245101-13 (2001)

Pickard, C.J., Mauri, F.: First-principles theory of the EPR g tensor in solids: Defects in quartz. Phys. Rev. Lett. **88**, 086403-1–086403-4 (2002)

Pickard, C.J., Mauri, F.: Nonlocal pseudopotentials and magnetic fields. Phys. Rev. Lett. **91**, 196401-1–196401-4 (2003)

Pickett, W.E.: Pseudopotential methods in condensed matter applications. Comput. Phys. Rep. **9**, 115–197 (1989)

Pindzola, M.S., Robicheaux, R., Gavras, P.: Double multiphoton ionization of a model atom. Phys. Rev. A **55**, 1307–1313 (1997)

Piraux, B., L'Huillier, A., Rzazewski K. (eds.): Super-Intense Laser-Atom Physics, NATOASI Series, vol. B316. Plenum Press, New York (1993)

Piris, M., Ugalde, J.M.: Iterative diagonalization for orbital optimization in natural orbital functional theory. J. Comput. Chem. **30**, 2078–2086 (2009)

Piris, M., Matxain, J.M., Lopez, X., Ugalde, J.M.: Accurate description of atoms and molecules by natural orbital functional theory. J. Chem. Phys. **132**, 031103-1–031103-4 (2010a)

Piris, M., Matxain, J.M., Lopez, X., Ugalde, J.M.: The role of positivity N-representability conditions in natural orbital functional theory. J. Chem. Phys. **133**, 111101-1–111101-4 (2010b)

Pitarke, J.M., Perdew, J.P.: Metal surface energy: Persistent cancellation of shortrange correlation effects beyond the random-phase approximation. Phys. Rev. B **67**, 045101-1–045101-5 (2003)

Platzman, P.M., Wolff, P.A.: Waves and Interactions in Solid State Plasmas. Academic Press, New York (1973)

Plenio, M.B., Knight, P.L.: The quantum-jump approach to dissipative dynamics in quantum optics. Rev. Mod. Phys. **70**, 101–144 (1998)

Plötner, J., Tozer, D.J., Dreuw, A.: Dependence of excited state potential energy surfaces on the spatial overlap of the Kohn–Sham orbitals and the amount of nonlocal Hartree–Fock exchange in time-dependent density functional theory. J. Chem. Theory Comput. **6**, 2315–2324 (2010)

Podeszwa, R., Patkowski, K., Szalewicz, K.: Improved interaction energy benchmarks for dimers of biological relevance. Phys. Chem. Chem. Phys. **12**, 5974–5979 (2010)

Pohl, A., Reinhard, P.-G., Suraud, E.: Towards single-particle spectroscopy of small metal clusters. Phys. Rev. Lett. **84**, 5090–5093 (2000)

Polli, D., Altoè, P., Weingart, O., Spillane, K.M., Manzoni, C., Brida, D., Tomasello, G., Orlandi, G., Kurkura, P., Mathies, R.A., Garavellli, M., Cerullo, G.: Conical intersection dynamics in the primary photoisomerization event in vision. Nature (London) **467**, 440–443 (2010)

Pont, M., Walet, N.R., Gavrila, M., McCurdy, C.W.: Dichotomy of the hydrogen atom in superintense, high-frequency laser fields. Phys. Rev. Lett. **61**, 939–942 (1998)

Pont, M., Walet, N.R., Gavrila, M.: Radiative distortion of the hydrogen atom in superintense, high-frequency fields of linear polarization. Phys. Rev. A **41**, 477–494 (1990)

Pople, J.A., Johnson, B.G., Gill, P.M.W.: Kohn–Sham density-functional theory within a finite basis set. Chem. Phys. Lett. **199**, 557–560 (1992)

Posthumus, J.H.: The dynamics of small molecules in intense laser fields. Rep. Prog. Phys. **67**, 623–665 (2004)

Prodan, E.: Raising the temperature on density functional theory. Physics **3**, 99 (2010)

Protopapas, M., Keitel, C.H., Knight, P.L.: Atomic physics with super-high intensity lasers. Rep. Prog. Phys. **60**, 389–486 (1997)

Puff, R.D., Gillis, N.S.: Fluctuations and transport properties of many-particle systems. Ann. Phys. (Amsterdam, Neth.) **46**, 364–397 (1968)

Pulay, P.: Convergence acceleration of iterative sequences. The case of scf iteration. Chem. Phys. Lett. **73**, 393–398 (1980)

Pulay, P.: Analytical derivative methods in quantum chemistry. In: Lawley, K.P. (ed.) Advances in Chemical Physics, vol. 69, pp. 241–286. Wiley, Chichester (1987)

Putrino, A., Parrinello, M.: Anharmonic Raman spectra in high-pressure ice from Ab initio simulations. Phys. Rev. Lett. **88**, 176401-1–176401-4 (2002)

Qian, Z., Vignale, G.: Dynamical exchange-correlation potentials for an electron liquid. Phys. Rev. B **65**, 235121-1–235121-12 (2002)

Qian, Z., Vignale, G.: Dynamical exchange-correlation potentials for the electron liquid in the spin channel. Phys. Rev. B **68**, 195113-1–195113-14 (2003)

Qian, X., Li, J., Lin, X., Yip, S.: Time-dependent density functional theory with ultrasoft pseudopotentials: Real-time electron propagation across a molecular junction. Phys. Rev. B **73**, 035408-1–035408-11 (2006)

Quong, A.A., Eguilez, A.G.: First-principles evaluation of dynamical response and plasmon dispersion in metals. Phys. Rev. Lett. **70**, 3955–3958 (1993)

Rabitz, H.: Focus on quantum control. New J. Phys. **11**, 105030-1–105030-5 (2009)

Ragazos, I.N., Robb, M.A., Bernardi, F., Olivucci, M.: Optimization and characterization of the lowest energy point on a conical intersection using an MC-SCF lagrangian. Chem. Phys. Lett. **197**, 217–223 (1992)

Rajagopal, A.K.: Time-dependent variational principle and the effective action in densityfunctional theory and Berry's phase. Phys. Rev. A **54**, 3916–3922 (1996)

Rajam, A.K., Hessler, P., Gaun, C., Maitra, N.T.: Phase-space explorations in time-dependent density functional theory. J. Mol. Struct. THEOCHEM **914**, 30–37 (2009)

Ramos, L.E., Paier, J., Kresse, G., Bechstedt, F.: Optical spectra of Si nanocrystallites: Bethe–Salpeter approach versus time-dependent density-functional theory. Phys. Rev. B **78**, 195423-1–195423-9 (2008)

Rapcewicz, K., Ashcroft, N.W.: Fluctuation attraction in condensed matter: A nonlocal functional approach. Phys. Rev. B **44**, 4032–4035 (1991)

Rappoport, D., Furche, F.: Photoinduced intramolecular charge transfer in 4-(dimethyl)amino-benzonitrile—A theoretical perspective. J. Am. Chem. Soc. **126**, 1277–1284 (2004)

Rappoport, D., Furche, F.: Analytical time-dependent density functional derivative methods within the RI-J approximation an approach to excited states of large molecules. J. Chem. Phys. **122**, 064105-1–064105-8 (2005)

Rappoport, D., Crawford, N.R.M., Furche, F., Burke, K.: Which functional should I choose? In: Solomon, E.I., King, R.B., Scott, R.A. (eds.) Computational Inorganic and Bioinorganic Chemistry, pp. 159–172. Wiley, Chichester (2009)

Rappoport, D., Furche, F.: Property-optimized Gaussian basis sets for molecular response calculations. J. Chem. Phys. **133**, 134105-1–134105-11 (2010)

Rassolov, V.A., Pople, J.A., Ratner, M.A., Windus, T.L.: 6–31G basis set for atoms K through Zn. J. Chem. Phys. **109**, 1223–1229 (1998)

Rau, A.R.P., Fano, U.: Transition matrix elements for large momentum or energy transfer. Phys. Rev. **162**, 68–70 (1967)

Reading, J.F., Ford, A.L.: The forced impulse method applied to the double ionization of helium by collision with high-energy protons, antiprotons, and alpha particles. J. Phys. B **20**, 3747–3769 (1987)

Rehr, J.J., Albers, R.C.: Theoretical approaches to x-ray absorption fine structure. Rev. Mod. Phys. **72**, 621–654 (2000)

Reiche, F., Thomas, W.: Über die Zahl der Dispersionselektronen die einem stationn Zustand zugeordnet sind. Z. Phys. **34**, 510–525 (1925)

Reinhard, P.-G., Suraud, E.: On electron dynamics in violent cluster excitations. J. Cluster Sci. **10**, 239–270 (1999)

Reinhard, P.-G., Suraud, E.: Introduction to Cluster Dynamics. Wiley, New York (2003)

Reining, L., Olevano, V., Rubio, A., Onida, G.: Excitonic effects in solids described by time-dependent density-functional theory. Phys. Rev. Lett. **88**, 066404-1–066404-4 (2002)

Remetter, T., Jonhsson, P., Varjù, K., Mauritsson, J., Ni, Y.F., Lépine, F., Kling, M., Khan, J., Lopez-Martens, R., Schafer, K.J., Vrakking, M.J.J., Huillier, A.L.: Attosecond electron wave packet interferometry. Nat. Phys. **2**, 323–326 (2006)

Requist, R., Pankratov, O.: Generalized Kohn–Sham system in one-matrix functional theory. Phys. Rev. B **77**, 235121-1–235121-16 (2008)

Resta, R.: Macroscopic polarization in crystalline dielectrics: The geometric phase approach. Rev. Mod. Phys. **66**, 899–915 (1994)

Rettig, W.: Charge separation in excited states of decoupled systems-TICT compounds and implications regarding the development of new laser dyes and the primary processes of vision and photosynthesis. Angew. Chem. Int. Ed. **25**, 971–988 (1986)

Rice, S.A., Zhao, M.: Optical Control of Molecular Dynamics. Wiley, New York (2000)

Richardson, C.F., Ashcroft, N.W.: Dynamical local field factors and effective interaction in the 3d electron fluid. Phys. Rev. B **50**, 8170–8181 (1994)

Richter, M., Amusia, M.Y., Bobashev, S.V., Feigl, T., Juranić, P.N., Martins, M., Sorokin, A.A., Tiedke, K.: Extreme ultraviolet laser excites atomic giant resonance. Phys. Rev. Lett. **102**, 163002-1–163002-4 (2009)

Riess, J., Münch, W.: The theorem of Hohenberg and Kohn for subdomains of a quantum system. Theor. Chim. Acta **58**, 295–300 (1981)

Riley, K.E., Pitoňák, M., Jurečka, P., Hobza, P.: Stabilization and structure calculations for noncovalent interactions in extended molecular systems based on wave function and density functional theories. Chem. Rev. **110**, 5023–5063 (2010)

Rinkevicius, Z., Vahtras, O., Ågren, H.: Spin-flip time dependent density functional theory applied to excited states with single, double, or mixed electron excitation character. J. Chem. Phys. **133**, 114104-1–114104-12 (2010)

Rocca, D., Gebauer, R., Saad, Y., Baroni, S.: Turbo charging time-dependent densityfunctional theory with Lanczos chains. J. Chem. Phys. **128**, 154105-1–154105-14 (2008)

Rocca, D., Lu, D.Y., Galli, G.: Ab initio calculations of optical absorption spectra: Solution of the Bethe Salpeter equation within density matrix perturbation theory. J. Chem. Phys. **133**, 164109-1–164109-10 (2010)

Roessel, D.M., Walker, W.C.: Optical constants of magnesium oxide and lithium fluoride in the far ultraviolet. J. Opt. Soc. Am. **57**, 835–836 (1967)

Rogers, C.L., Rappe, A.M.: Geometric formulation of quantum stress fields. Phys. Rev. B **65**, 224117-1–224117-8 (2002)

Rohlfing, M., Louie, S.G.: Electron-hole excitations and optical spectra from first principles. Phys. Rev. B **62**, 4927–4944 (2000)

Rohr, D., Pernal, K., Gritsenko, O.V., Baerends, E.J.: A density matrix functional with occupation number driven treatment of dynamical and nondynamical correlation. J. Chem. Phys. **129**, 164105-1–164105-11 (2008)

Rohrdanz, M.A., Herbert, J.M.: Simultaneous benchmarking of ground-and excitedstate properties with long-range-corrected density functional theory. J. Chem. Phys. **129**, 034107-1–034107-9 (2008)

Rohrdanz, M.A., Martins, K.M., Herbert, J.M.: A long-range-corrected density functional that performs well for both ground-state properties and time-dependent density functional theory excitation energies, including charge-transfer excited states. J. Chem. Phys. **130**, 054112-1–054112-8 (2009)

Rohringer, N., Peter, S., Burgdörfer, J.: Calculating state-to-state transition probabilities with time-dependent density-functional theory. Phys. Rev. A **74**, 042512-1–042512-7 (2006)

Román-Pérez, G., Soler, J.M.: Efficient implementation of a van der Waals density functional: Application to double-wall carbon nanotubes. Phys. Rev. Lett. **103**, 096102-1–096102-4 (2009)

Roman, E., Yates, J.R., Veithen, M., Vanderbilt, D., Souza, I.: Ab initio study of the nonlinear optics of III–V semiconductors in the terahertz regime. Phys. Rev. B **74**, 245204-1–245204-9 (2006)

Romaniello, P., de Boeij, P.L.: Relativistic two-component formulation of timedependent current-density functional theory: Application to the linear response of solids. J. Chem. Phys **127**, 174111-1–174111-12 (2007)

Romaniello, P., Sangalli, D., Berger, J.A., Sottile, F., Molinari, L.G., Reining, L., Onida, G.: Double excitations in finite systems. J. Chem. Phys. **130**, 044108-1–044108-11 (2009)

Rosa, A., Ricciardi, G., Gritsenko, O., Baerends, E.J.: Excitation energies of metal complexes with time-dependent density functional theory. In: Kaltsoyannis, N., McGrady, J. E. (eds.) Structure and Bonding, vol. 112, ch. 2, pp. 3–17. Springer, Heidelberg (2004)

Rouzée, A., Siu, W., Huismans, Y., Lépine, F., Marchenko, T., Dusterer, S., Tavella, F., Stojanovic, N., Azima, A., Treusch, R., Kling, M., Vrakking, M.: Field-free molecular alignment probed by the free electron laser in Hamburg (FLASH). J. Phys. B **42**, 134017-1–134017-9 (2009)

Rudenko, A., Zrost, K., Feuerstein, B., de Jesus, V.L.B., Schröter, C.D., Moshammer, R., Ullrich, J.: Correlated multielectron dynamics in ultrafast laser pulse interactions with atoms. Phys. Rev. Lett. **93**, 253001-1–253001-4 (2004)

Ruggenthaler, M., Penz, M., Bauer, D.: On the existence of effective potentials in timedependent density functional theory. J. Phys. A Math. Theor. **42**, 425207-1–425207-11 (2009)

Ruggenthaler, M., Penz, M., Bauer, D.: General Runge–Gross-type theorem for dipole laser–matter interaction. Phys. Rev. A **81**, 062108-1–062108-4 (2010)

Ruggenthaler, M., Mackenroth, F., Bauer, D.: Kohn–Sham approach to Quantum Electrodynamics. arXiv:1011.4162v1 (2011a)

Ruggenthaler, M., van Leeuwen, R.: Global fixed point proof of time-dependent density-functional theory. Europhys. Lett. **95**, 13001-1–6 (2011b)

Runge, E., Gross, E.K.U.: Density-functional theory for time-dependent systems. Phys. Rev. Lett. **52**, 997–1000 (1984)

Runge, E., Gross, E.K.U., Heinonen, O.: Many-Particle Theory. Adam Hilger, Bristol (1991)

Rydberg, H., Lundqvist, B.I., Langreth, D.C., Dion, M.: Tractable nonlocal correlation densities for flat surfaces and slabs. Phys. Rev. B. **62**, 6997–7006 (2000)

Rydberg, H., Jacobson, N., Hyldgaard, P., Simak, S., Lundqvist, B.I., Langreth, D.C.: Hard numbers on soft matter. Surf. Sci. **532**, 606–610 (2003)

Saad, Y.: Iterative Methods for Sparse Linear Systems, 2nd edition (SIAM, Philadelphia) (2003).

Saalmann, U., Schmidt R.: Non-adiabatic quantum molecular dynamics: Basic formalism and case study. Z. Phys. D **38**, 153–163 (1996)

Saalmann, U., Schmidt, R.: Excitation and relaxation in atom-cluster collisions. Phys. Rev. Lett. **80**, 3213–3216 (1998)

Sagmeister, S., Ambrosch-Draxl, C.: Time-dependent density functional theory versus Bethe–Salpeter equation: An all-electron study. Phys. Chem. Chem. Phys. **11**, 4451–4457 (2009)

Sagvolden, E., Furche, F., Köhn, A.: Förster energy transfer and Davydov splittings in time-dependent density functional theory: Lessons from 2-pyridone dimer. J. Chem. Theory Comput. **5**, 873–880 (2009)

Sai, N., Bushong, N., Hatcher, R., Di Ventra, M.: Microscopic current in nanoscale junctions. Phys. Rev. B **75**, 115410-1–115410-8 (2007)

Saitta, A.M., Lazzeri, M., Calandra, M., Mauri, F.: Giant nonadiabatic effects in layer metals: Raman spectra of intercalated graphite explained. Phys. Rev. Lett. **100**, 226401-1–226401-4 (2008)

Sakai, H., Larsen, J.J., Wendt-Larsen, I., Olesen, J., Corkum, P.B., Stapelfeldt, H.: Nonsequential double ionization of D_2 molecules with intense 20-fs pulses. Phys. Rev. A **67**, 063404-1–063404-4 (2003)

Sakko, A., Rubio, A., Hakala, M., Hämäläinen, K.: Time-dependent density functional approach for the calculation of inelastic x-ray scattering spectra of molecules. J. Chem. Phys **133**, 174111-1–174111-6 (2010)

Salières, P., L'Huillier, A., Antoine, P., Lewenstein, M.: Study of the spatial and temporal coherence of high order harmonies. Adv. Atom. Mol. Opt. Phys. **41**, 83–142 (1999)

Salpeter, E.E., Bethe, H.A.: A relativistic equation for bound-state problems. Phys. Rev. **84**, 1232–1242 (1951)

Samson, J.A.R., He, Z.X., Yin, L., Haddad, A.: Precision measurements of the absolute photoionization cross sections of He. J. Phys. B **27**, 887–898 (1994)

Sansone, G., Kelkensberg, F., Pérez-Torres, J. F., Morales, F., Kling, M. F., Siu, W.O., Ghafur, Johnsson, P., Swoboda, M., Benedetti, E., Ferrari, F., Lépine, F., Sanz-Vicario, J. L., Zherebtsov, S., Znakovskaya, I., L'Huillier, A., Yu Ivanov, M., Nisoli, M., Martín, F., Vrakking, M. J. J.: Electron localization following attosecond molecular photoionization. Nature (London) **465**, 763–767 (2010)

Sarukura, N., Hata, K., Adachi, T., Nodomi, R., Watanabe, M., Watanabe, S.: Coherent soft-x-ray generation by the harmonics of an ultrahigh-power KrF laser. Phys. Rev. A **43**, 1669–1672 (1991)

Sato, T., Tsuneda, T., Hirao, K.: Long-range corrected density functional study on weakly bound systems: Balanced descriptions of various types of molecular interactions. J. Chem. Phys. **126**, 234114-1–234114-12 (2007)

Sato, T., Nakai, H.: Density functional method including weak interactions: Dispersion coefficients based on the local response approximation. J. Chem. Phys. **131**, 224104-1–224104-12 (2009)

Savin, A.: On degeneracy, near-degeneracy and density functional theory. In: Seminario, J.M. (ed.) Recent Developments and Applications of Modern Density Functional Theory, pp. 327–357. Elsevier, Amsterdam (1996)

Savin, A., Umrigar, C.J., Gonze, X.: Relationship of Kohn-Sham eigenvalues to excitation energies. Chem. Phys. Lett. **288**, 391–395 (1998)

Savrasov, S.Y.: Linear response calculations of spin fluctuations. Phys. Rev. Lett. **81**, 2570–2573 (1998)

Sawatzky, G.A., Allen, J.W.: Magnitude and origin of the band gap in NiO. Phys. Rev. Lett. **53**, 2339–2342 (1984)

Scalmani, G., Frisch, M.J., Mennucci, B., Tomasi, J., Cammi, R., Barone, V.: Geometries and properties of excited states in the gas phase and in solution: Theory and application of a time-dependent density functional theory polarizable ontinuum model. J. Chem. Phys. **124**, 094107-1–094107-15 (2006)

Schäfer, A., Horn, H., Ahlrichs, R.: Fully optimized contracted Gaussian basis sets for atoms Li to Kr. J. Chem. Phys. **97**, 2571–2577 (1992)

Schäfer, A., Huber, C., Ahlrichs, R.: Fully optimized contracted Gaussian basis sets of triple zeta valence quality for atoms Li to Kr. J. Chem. Phys. **100**, 5829–5835 (1994)

Schaich, W.L., Ashcroft, N.W.: Model calculations in the theory of photoemission. Phys. Rev. B **3**, 2452–2465 (1970)

Schattke, W., Van Hove, M.A.: Solid-State Photoemission and Related Methods. Wiley, Weinheim (2003)

Schimka, L., Harl, J., Stroppa, A., Grüneis, A., Marsman, M., Mittendorfer, F., Kresse, G.: Accurate surface and adsorption energies from many-body perturbation theory. Nature Mater. **9**, 741–744 (2010)

Schinke, R.: Photodissociation Dynamics. Cambridge University Press, Cambridge (1993)

Schmidt, J.R., Parandekar, P.V., Tully, J.C.: Mixed quantum-classical equilibrium: Surface hopping. J. Chem. Phys. **129**, 044104-1–044104-6 (2008)

Schnürer, M., Spielmann, C., Wobrauschek, P., Streli, C., Burnett, N.H., Kan, C., Ferencz, K., Koppitsch, R., Cheng, Z., Brabec, T., Krausz, F.: Coherent 0.5-keV X-Ray emission from helium driven by a sub-10-fs laser. Phys. Rev. Lett. **80**, 3236–3239 (1998)

Schreiber, M., Silva-Junior, M.R., Sauer, S.P.A., Thiel, W.: Benchmarks for electronically excited states: CASPT2, CC2, CCSD, and CC3. J. Chem. Phys. **128**, 134110-1–134110-25 (2008)

Schülke, W.: Electron Dynamics by Inelastic X-Ray Scattering. Oxford University Press, Oxford (2007)

Schultze, M., Fieß, M., Karpowicz, N., Gagnon, J., Korbman, M., Hofstetter, M., Neppl, S., Cavalieri, A.L., Komninos, Y., Mercouris, Th., Nicolaides, C.A., Pazourek, R., Nagele, S., Feist, J., Burgdörfer, J., Azzeer, A.M., Ernstorfer, R., Kienberger, R., Kleineberg, U., Goulielmakis, E., Krausz, F., Yakovlev, V.S.: Delay in photoemission. Science **328**, 1658–1662 (2010)

Schumacher, B., Westmoreland, M.: Quantum Processes, Systems, and Information. Cambridge University Press, Cambridge (2010)

Scopigno, T., Ruocco, G., Sette, F.: Microscopic dynamics in liquid metals: The experimental point of view. Rev. Mod. Phys. **77**, 881–933 (2005)

Scrinzi, A., Ivanov, M.Yu., Kienberger, R., Villeneuve, D.M.: Attosecond physics. J. Phys. B: At. Mol. Opt. Phys. **39**, R1–R37 (2006)

Scuseria, G.E., Henderson, T.M., Sorensen, D.C.: The ground state correlation energy of the random phase approximation from a ring coupled cluster doubles approach. J. Chem. Phys. **129**, 231101-1–231101-4 (2008)

Seideman, T., Ivanov, M.Yu., Corkum, P.B.: Role of electron localization in intense- field molecular ionization. Phys. Rev. Lett. **75**, 2819–2822 (1995)

Seidl, A., Görling, A., Vogl, P., Majewski, J.A., Levy, M.: Generalized Kohn-Sham schemes and the band-gap problem. Phys. Rev. B **53**, 3764–3774 (1996)

Seidl, M., Perdew, J.P., Kurth, S.: Simulation of all-order density-functional perturbation theory, using the second order and the strong-correlation limit. Phys. Rev. Lett. **84**, 5070–5073 (2000)

Senatore, G., Subbaswamy, K.R.: Nonlinear response of closed-shell atoms in the density-functional formalism. Phys. Rev. A **35**, 2440–2447 (1987)

Send, R., Furche, F.: First-order nonadiabatic couplings from time-dependent hybrid density functional response theory: Consistent formalism, implementation, and performance. J. Chem. Phys. **132**, 044107-1–044107-12 (2010)

Send, R., ValssonO, Filippi C.: Electronic excitations of simple cyanine dyes: Reconciling density functional and wave function methods. J. Chem. Theory Comput. **7**, 444–455 (2011)

Serban, I., Werschnik, J., Gross, E.K.U.: Optimal control of time-dependent targets. Phys. Rev. A **71**, 053810-1–053810-9 (2005)

Seth, M., Mazur, G., Ziegler, T.: Time-dependent density functional theory gradients in the Amsterdam density functional package: Geometry optimizations of spin-flip excitations. Theor. Chim. Acta **129**, 331–342 (2010)

Shao, Y., Head-Gordon, M., Krylov, A.I.: The spin-flip approach within timedependent density functional theory: Theory and applications to diradicals. J. Chem. Phys. **118**, 4807–4818 (2003)

Shapiro, M., Brumer, P.: Principles of the Quantum Control of Molecular Processes. Wiley, New York (2003)

Sharma, S., Dewhurst, J.K., Lathiotakis, N.N., Gross, E.K.U.: Reduced density matrix functional for many-electron systems. Phys. Rev. B **78**, 201103-1–201103-4 (2008)

Shen, Y.R.: The Principles of Nonlinear Optics. John Wiley and Sons, New York (1984)

Sherwood, P.: Hybrid quantum mechanics/molecular mechanics approaches. In: Grotendorst, J. (ed.) Modern Methods and Algorithms of Quantum Chemistry, pp. 285–305. John von Neumann Institute for Computing, Forschungszentrum Jülich (2000)

Shi, J., Cai, Y., Hou, W., Ma, L., Tan, S. X.-D., Ho, P. Wang, X.: GPU friendly fast Poisson solver for structured power grid network analysis. In: Proceedings of the 46th Annual Design Automation Conference (DAC '09), ACM, New York, 178–183 (2009)

Shigeta, Y., Hirao, K., Hirata, S.: Exact-exchange time-dependent density-functional theory with the frequency-dependent kernel. Phys. Rev. A. **73**, 010502-1–4(R) (2006)

Shiner, A.D., Schmidt, B.E., Trallero-Herrero, C., Wörner, H.J., Patchkovskii, S., Corkum, P.B., Kieffer, J.-C., Légaré, F., Villeneuve, D.M.: Probing collective multi-electron dynamics in xenon with high-harmonic spectroscopy. Nat. Phys. **7**, 464–467 (2011)

Shinohara, Y., Yabana, K., Kawashita, Y., Iwata, J.-I., Otobe, T., Bertsch, G.F.: Coherent phonon generation in time-dependent density functional theory. Phys. Rev. B **82**, 155110-1–155110-10 (2010)

Sicilia, F., Blancafort, L., Bearpark, M.J., Robb, M.A.: New algorithms for optimizing and linking conical intersection points. J. Chem. Theor. Comput **4**, 257–266 (2008)

Silva-Junior, M.R., Schreiber, M., Sauer, S.P.A., Thiel, W.: Benchmarks for electronically excited states: Time-dependent density functional theory and density functional theory based multireference configuration interaction. J. Chem. Phys. **129**, 104103-1–104103-14 (2008)

Silvestrelli, P.L., Bernasconi, M., Parrinello, M.: Ab initio infrared spectrum of liquid water. Chem. Phys. Lett. **277**, 478–482 (1997)

Slipchenko, L.V., Krylov, A.I.: Electronic structure of the trimethylenemethane diradical in is ground and electronically excited states: Bonding, equilibrium geometries, and vibrational frequencies. J. Chem. Phys. **118**, 6874–6883 (2003)

Sølling, T.I., Diau, E.W.-G., Kötting, C., De Feyter, S., Zewail, A.H.: Femtochemistry of Norrish Type-I reactions: IV. Highly excited ketones–experimental. Chem. Phys. Chem. **3**, 79–97 (2002)

Sorella, S., Casula, M., Rocca, D.: Weak binding between two aromatic rings: Feeling the van der Waals attraction by quantum monte carlo methods. J. Chem. Phys. **127**, 014105-1–014105-12 (2007)

Sottile, F., Olevano, V., Reining, L.: Parameter-free calculation of response functions in time-dependent density-functional theory. Phys. Rev. Lett. **91**, 056402-1–056402-4 (2003)

Sottile, F., Bruneval, F., Marinopoulos, A.G., Dash, L.K., Botti, S., Olevano, V., Vast, N., Rubio, A., Reining, L.: TDDFT from molecules to solids: The role of long-rangeinteractions. Int. J. Quant. Chem. **102**, 684–701 (2005)

Sottile, F., Marsili, M., Olevano, V., Reining, L.: Efficient ab initio calculations of bound and continuum excitons in the absorption spectra of semiconductors and insulators. Phys. Rev. B **76**, 161103-1–4(R) (2007)

Souza, I., Íñiguez, J., Vanderbilt, D.: First-principles approach to insulators in finite electric fields. Phys. Rev. Lett. **89**, 117602-1–117602-4 (2002)

Spanu, L., Sorella, S., Galli, G.: Nature and strength of interlayer binding in graphite. Phys. Rev. Lett. **103**, 196401-1–196401-4 (2009)

Spielmann, C., Burnett, N.H., Sartania, S., Koppitsch, R., Schnürer, M., Kan, C., Lenzner, M., Wobrauschek, P., Krausz, F.: Generation of coherent X-rays in the water window using 5-femtosecond laser pulses. Science **278**, 661–664 (1997)

Stapelfeldt, H., Seideman, T.: Colloquium: Aligning molecules with strong laser pulses. Rev. Mod. Phys. **75**, 543-1–54315 (2003)

Staroverov, V.N., Scuseria, G.E.: Optimization of density matrix functionals by the Hartree–Fock–Bogoliubov method. J. Chem. Phys. **117**, 11107–11112 (2002)

Staroverov, V.N., Scuseria, G. E., Tao, J., Perdew, J.: Tests of a ladder of density functionals for bulk solids and surfaces. Phys. Rev. B. **69**, 075102-1–9 (2004)

Staroverov, V.N., Scuseria, G. E., Tao, J., Perdew, J.: Tests of a ladder of density functionals for bulk solids and surfaces. Phys. Rev. B **78**, 239907(E) (2008)

Stefanucci, G., Almbladh, C.-O.: Time-dependent quantum transport: An exact formulation based on TDDFT. Europhys. Lett. **67**, 14–20 (2004a)

Stefanucci, G., Almbladh, C.-O.: Time-dependent partition-free approach in resonant tunneling systems. Phys. Rev. B **69**, 195318-1–195318-17 (2004b)

Stefanucci, G., Kurth, S., Gross, E.K.U., Rubio, A.: Time dependent transport phenomena. In: Seminario, J.M. (ed.) Molecular and Nano Electronics: Analysis, Design and Simulation, vol. 17, pp. 247–284. Elsevier, Amsterdam (2007)

Stefanucci, G.: Bound states in ab initio a proaches to quantum transport: A timedependent formulation. Phys. Rev. B **75**, 195115-1–195115-11 (2007)

Stefanucci, G., Kurth, S., Rubio, A., Gross, E.K.U.: Time-dependent approach to electron pumping in open quantum systems. Phys. Rev. B **77**, 075339-1–075339-14 (2008a)

Stefanucci, G., Perfetto, E., Cini, M.: Ultrafast manipulation of electron spins in a double quantum dot device: A real-time numerical and analytical study. Phys. Rev. B **78**, 075425-1–075425-14 (2008b)

Stefanucci, G., Perfetto, E., Cini, M.: Time-dependent quantum transport with superconducting leads: A discrete-basis Kohn-Sham formulation and propagation scheme. Phys. Rev. B **81**, 115446-1–115446-18 (2010)

Stein, T., Kronik, L., Baer, R.: Reliable prediction of charge transfer excitations in molecular complexes using time-dependent density functional theory. J. Am. Chem. Soc. **131**, 2818–2820 (2009a)

Stein, T., Kronik, L., Baer, R.: Prediction of charge-transfer excitations in coumarinbased dyes using a range-separated functional tuned from first principles. J. Chem. Phys. **131**, 244119-1–244119-5 (2009b)

Stener, M., Decleva, P., Görling, A.: The role of exchange and correlation in timedependent density-functional theory for photoionization. J. Chem. Phys. **114**, 7816–7830 (2001)

Stener, M., Toffoli, D., Fronzoni, G., Decleva, P.: Recent advances in molecular photoionization by density functional theory based approaches. Theor. Chem. Acc. **117**, 943–956 (2007)

Sternemann, C., Huotari, S., Vankó, G., Volmer, M., Monaco, G., Gusarov, A., Lustfeld, H., Sturm, K., Schülke, W.: Correlation-induced double-plasmon excitation in simple metals sudied by inelastic X-Ray scattering. Phys. Rev. Lett. **95**, 157401–157404 (2005)

Sternheimer, R.M.: Electronic polarizabilities of ions from the Hartree–Fock wave functions. Phys. Rev. **96**, 951–968 (1954)

Stöger-Pollach, M.: Optical properties and bandgaps from low loss EELS: Pitfalls and solutions. Micron **39**, 1092–1110 (2008)

Stöhr, J.: NEXAFS Spectroscopy. Springer, Berlin (1992)

Stoll, H., Savin, A.: Density functionals for correlation energies of atoms and molecules. In: Dreizler, R.M., da Providência, J. (eds.) Density Functional Methods in Physics, pp. 177–207. Plenum Press, New York (1985)

Stolow, A.: Femtosecond time-resolved photoelectron spectroscopy of polyatomic molecules. Annu. Rev. Phys. Chem. **54**, 89–119 (2003)

Stone, R., Xin, H.: Supercomputer leaves competition–and users–in the dust. Science **330**, 746–747 (2010)

Stratmann, R.E., Scuseria, G.E., Frisch, M.J.: An efficient implementation of timedependent density-functional theory for the calculation of excitation energies of large molecules. J. Chem. Phys. **109**, 8218–8224 (1998)

Strinati, G.: Application of the Green's functions methods to the study of the optical properties of semiconductors. Riv. Nuovo Cimento Soc. Ital. Fis. **11**, 1–86 (1988)

Stubner, R., Tokatly, I.V., Pankratov, O.: Excitonic effects in time-dependent density-functional theory: An analytically solvable model. Phys. Rev. B **70**, 245119-1–245119-12 (2004)

Su, Q., Eberly, J.H.: Model atom for multiphoton physics. Phys. Rev. A **44**, 5997–6008 (1991)

Suarez, A., Silbey, R., Oppenheim, I.: Memory effects in the relaxation of quantum open systems. J. Chem. Phys. **97**, 5101–5107 (1992)

Sulpizi, M., Carloni, P., Hutter, J., Röthlisberger, U.: A hybrid TDDFT/MM investigation of the optical properties of aminocoumarins in water and acetonitrile solution. Phys. Chem. Chem. Phys. **5**, 4798–4805 (2003)

Sulpizi, M., Röhrig, U.F., Hutter, J., Öthlisberger, U.: Optical properties of molecules in solution via hybrid TDDFT/MM simulations. Int. J. Quantum Chem. **101**, 671–682 (2005)

Sutcliffe, B.: The decoupling of electronic and nuclear motions in the isolated molecule Schrödinger Hamiltonian. Adv. Chem. Phys. **114**, 1–121 (2000)

Suzuki, T., Minemoto, S., Kanai, T., Sakai, H.: Optimal control of multiphoton ionization processes in aligned I2 molecules with time-dependent polarization pulses. Phys. Rev. Lett. **92**, 133005-1–133005-4 (2004)

Szabo, A., Ostlund, N.S.: Modern Quantum Chemistry: Introduction to Advanced Electronic Structure Theory. Mc Graw Hill, New York (1989)

Takimoto, Y., Vila, F.D., Rehr, J.J.: Real-time time-dependent density functional theory approach for frequency-dependent nonlinear optical response in photonic molecules. J. Chem. Phys. **127**, 154114-1–154114-10 (2007)

Tamm, I.: Relativistic interaction of elementary particles. J. Phys. **9** , 449 (1945)

Tang, X., Rudolph, H., Lambropoulos, P.: Nonperturbative time-dependent theory of helium in a strong laser field. Phys. Rev. A **44**, R6994–R6997 (1991)

Tao, J., Perdew, J.P., Staroverov, V.N., Scuseria, G.E.: Climbing the density functional ladder: Nonempirical meta-generalized gradient approximation designed for molecules and solids. Phys. Rev. Lett. **91**, 146401-1–146401-4 (2003)

Tao, J., Vignale, G., Tokatly, I.V.: Time-dependent density functional theory: Derivation of gradient-corrected dynamical exchange-correlational potentials. Phys. Rev. B **76**, 195126-1–195126-13 (2007)

Tao, J., Gao, X., Vignale, G., Tokatly, I.V.: Linear continuum mechanics for quantum many-body systems. Phys. Rev. Lett. **103**, 086401-1–086401-4 (2009)

Tapavicza, E., Tavernelli, I., Röthlisberger, U.: Trajectory surface hopping within linear response time-dependent density-functional theory. Phys. Rev. Lett. **98**, 023001-1–023001-4 (2007)

Tapavicza, E., I. Tavernelli, U. Röthlisberger, C. Filippi, Casida, M. E.: Mixed timedependent density-functional theory/classical trajectory surface hopping study of oxirane photochemistry. J. Phys. Chem. **129**, 124108-1–124108-1-19 (2008)

Taut, M.: Two electrons in an external oscillator potential: Particular analytic solutions of a coulomb correlation problem. Phys. Rev. A **48**, 3561–3566 (1993)

Tavernelli, I.: Electronic density response of liquid water using time-dependent density functional theory. Phys. Rev. B **73**, 094204-1–094204-7 (2006)

Tavernelli, I., Tapavicza, E., Röthlisberger, U.: Nonadiabatic coupling vectors within linear response time-dependent density functional theory. J. Chem. Phys. **130**, 124107-1–124107-10 (2009a)

Tavernelli, I., Curchod, B.F.E., Röthlisberger, U.: On nonadiabatic coupling vectors in time-dependent density functional theory. J. Chem. Phys. **131**, 196101-1–196101-2 (2009b)

Tavernelli, I., Tapavicza, E., Röthlisberger, U.: Non-adiabatic dynamics usingtimedependent density functional theory: Assessing the coupling strengths. J. Mol. Struct. THEOCHEM **914**, 22–29 (2009c)

Tavernelli, I., Curchod, B.F.E., Laktionov, A., Röthlisberger, U.: Nonadiabatic coupling vectors for excited states within time-dependent density functional theory in the Tamm–Dancoff approximation and beyond. J. Chem. Phys. **133**, 194104-1–194104-10 (2010a)

Tavernelli, I., Curchod, B.F.E., Röthlisberger, U.: Mixed quantum-classical dynamics with time-dependent external fields: A time-dependent density-functional-theory approach. Phys. Rev. A **81**, 052508-1–052508-13 (2010b)

Tawada, T., Tsuneda, T., Yanagisawa, S., Yanai, T., Hirao, K.: A long-range-corrected time-dependent density functional theory. J. Chem. Phys. **120**, 8425–8433 (2004)

Tempel, D.G., Martínez, T.J., Maitra, N.T.: Revisiting Molecular Dissociation in Density Functional Theory: A simple model. J. Chem. Theory Comput. **5**, 770–780 (2009)

Tempel, D.G., Olivares-Amaya, R., Watson, M.A., Aspuru-Guzik, A.: Time- dependent density functional theory of open quantum systems in the linear-response regime. J. Chem. Phys. **134**, 074116-1–074116-18 (2011a)

Tempel, D.G., Aspuru-Guzik, A.: Relaxation and dephasing in open quantum systems time-dependent density functional theory: Properties of exact functionals from an exactly-solvable model system. Chem. Phys. (2011b, in press) . Available online at http://dx.doi.org/10.1016/j.chemphys.2011.03.014

Tersigni, S.H., Gaspard, P., Rice, S.A.: On using shaped light pulses to control the selectivity of product formation in a chemical reaction: An application to a multiple level system. J. Chem. Phys. **93**, 1670–1680 (1990)

Tesch, C.M., Kurtz, L., de Vivie-Riedle, R.: Applying optimal control theory for elements of quantum computation in molecular systems. Chem. Phys. Lett. **343**, 633–641 (2001)

Theilhaber, J.: Ab initio simulations of sodium using time-dependent density-functional theory. Phys. Rev. B **46**, 12990–13003 (1992)

Theophilou, A.K.: The energy density functional formalism for excited states. J. Phys. C. Solid State **12**, 5419–5430 (1979)

Thiele, M., Gross, E.K.U., Kümmel, S.: Adiabatic approximation in nonperturbative time-dependent density-functional theory. Phys. Rev. Lett. **100**, 153004-1–153004-4 (2008)

Thiele, M., Kümmel, S.: Photoabsorption spectra from adiabatically exact timedependent density-functional theory in real time. Phys. Chem. Chem. Phys. **11**, 4631–4639 (2009)

Thole, B.T., Carra, P., Sette, F., van der Laan, G.: X-Ray circular dichroism as a probe of orbital magnetization. Phys. Rev. Lett. **68**, 1943–1946 (1992)

Thomas, W., 1925, "Über die Zahl der Dispersionselektronen, die einem stationn Zustande zugeordnet sind. (Vorläufige Mitteilung)", Naturwissenschaften, vol. 13, p. 627

Thonhauser, T., Cooper, V.R., Li, S., Puzder, A., Hyldgaard, P., Langreth, D.C.: van der Waals density functional: Self-consistent potential and the nature of the van der Waals bond. Phys. Rev. B **76**, 125112-1–125112-11 (2007)

Thouless, D.J.: Stability conditions and nuclear rotations in the Hartree-Fock theory. Nucl. Phys. **21**, 225–232 (1960)

Thouless, D.J.: The quantum mechanics of many-body systems, 2nd edn. Academic Press, New York (1972)

Tiago, M.L., Chelikowsky, J.R.: First-principles GW-BSE excitations in organic molecules. Solid State Commun. **136**, 333–337 (2005)

Tkatchenko, A., Scheffler, M.: Accurate molecular van der Waals interactions from ground-state electron density and free-atom reference data. Phys. Rev. Lett. **102**, 073005-1–073005-4 (2009)

Tobik, J., Dal Corso, A.: Electric fields with ultrasoft pseudo-potentials: Applications to benzene and anthracene. J. Chem. Phys. **120**, 9934–9941 (2004)

Toher, C., Filippetti, A., Sanvito, S., Burke, K.: Self-interaction errors in density functional calculations of electronic transport. Phys. Rev. Lett. **95**, 146402-1–146402-4 (2005)

Tokatly, I.V., Pankratov, O.: Local exchange-correlation vector potential with memory in time-dependent density functional theory: The generalized hydrodynamics approach. Phys. Rev. B **67**, 201103-1–201103-4(R) (2003)

Tokatly, I.V.: Quantum many-body dynamics in a lagrangian frame: I. Equations of motion and conservation laws. Phys. Rev. B **71**, 165104-1–165104-13 (2005a)

Tokatly, I.V.: Quantum many-body dynamics in a Lagrangian frame: II. Geometric formulation of time-dependent density functional theory. Phys. Rev. B **71**, 165105-1–165105-17 (2005b)

Tokatly, I.V.: Time-dependent deformation functional theory. Phys. Rev. B **75**, 125105-1–125105-15 (2007)

Tokatly, I.V.: Time-dependent current density functional theory via time-dependent deformation functional theory: A constrained search formulation in the time domain. Phys. Chem. Chem. Phys. **11**, 4621–4630 (2009)

Tokatly, I.V.: A unified approach to the density-potential mapping in a family of time-dependent density functional theories. Chem. Phys. (2011a, in press). Available online at http://dx.doi.org/10.1016/j.chemphys.2011.04.005

Tokatly, I.V.: Time-dependent current density functional theory on a lattice. Phys. Rev. B **83**, 035127-1–035127-8 (2011b)

Tomasi, J., Mennucci, B., Cammi, R.: Quantum mechanical continuum solvation models. Chem. Rev. **105**, 2999–3093 (2005)

Tompkins, H.G., McGahan, W.A.: Spectroscopic Ellipsometry and Reflectometry: A User's Guide. Wiley, New York (1999)

Tong, X.M., Chu, S.-I.: Theoretical study of multiple high-order harmonic generation by intense ultrashort pulsed laser fields: A new generalized pseudospectral time-dependent method. Chem. Phys. **217**, 119–130 (1997)

Tong, X.M., Chu, S.-I.: Time-dependent density-functional theory for strong-field multiphoton processes: Application to the study of the role of dynamical electron correlation in multiple high-order harmonic generation. Phys. Rev. A **57**, 452–461 (1998)

Toulouse, J., Colonna, F., Savin, A.: Short-range exchange and correlation energy density functional: Beyond the local-density approximation. J. Chem. Phys. **122**, 014110-1–014110-10 (2005)

Toulouse, J., Gerber, I.C., Jansen, G., Savin, A., Angyan, J.G.: Adiabatic-connection fluctuation-dissipation density-functional theory based on range separation. Phys. Rev. Lett. **102**, 096404-1–096404-4 (2009)

Toulouse, J., Zhu, W., Angyan, J.G., Savin, A.: Range-separated density-functional theory with the random-phase approximation: Detailed formalism and illustrative applications. Phys. Rev. A **82**, 032502-1–032502-5 (2010)

Tournus, F., Charlier, J.-C., Mélinon, P.: Mutual orientation of two C60 molecules: An ab initio study. J. Chem. Phys. **122**, 094315-1–094315-9 (2005)

Tozer, D.J., Handy, N.C.: Improving virtual Kohn-Sham orbitals and eigenvalues: Application to excitation energies and static polarizabilities. J. Chem. Phys. **109**, 10180–10189 (1998)

Tozer, D.J., Amos, R.D., Handy, N.C., Roos, B.O., Serrano-Andres, L.:Does density functional theory contribute to the understanding of excited states of unsaturated organic compounds? Mol. Phys. **97**, 859–868 (1999)

Tozer, D.J., Handy, N.C.: On the determination of excitation energies using density functional theory. Phys. Chem. Chem. Phys. **2**, 2117–2121 (2000)

Tozer, D.J.: Relationship between long-range charge-transfer excitation energy error and integer discontinuity in Kohn-Sham theory. J. Chem. Phys. **119**, 12697–12699 (2003)

Trallero-Herrero, C., Schmidt, B.E., Shiner, A.D., Lassonde, P., Bisson, É., Kieffer, J.-C., Corkum, P.B., Villeneuve, D.M., Légaré, F.: High harmonic generation in ethylene with infrared pulses. Chem. Phys. **366**, 33–36 (2009)

Treutler, O., Ahlrichs, R.: Efficient molecular numerical integration schemes. J. Chem. Phys. **102**, 346–354 (1995)

Troullier, N., Martins, J.L.: Efficient pseudopotentials for plane-wave calculations. Phys. Rev. B **43**, 1993–2006 (1991)

Tuckerman, M.E.: Statistical Mechanics: Theory and Molecular Simulation. Oxford University Press, Oxford (2010)

Tully, J.C., Preston, R.K.: Trajectory Surface Hopping Approach to Nonadiabatic Molecular Collisions: The reaction of H_+ with D_2. J. Chem. Phys. **55**, 562–572 (1971)

Tully, J.C.: Molecular dynamics with electronic transitions. J. Chem. Phys. **93**, 1061–1071 (1990)

Tully, J.C.: Nonadiabatic Dynamics. In: Thompson, D.L. (ed.) Modern Methods in Multidimen- sional Dynam- ics Computations in Chemistry, pp. 34–72. World Scientific, Singapore (1998a)

Tully, J.C.: Mixed quantum-classical dynamics: Mean-field and surface-hopping. In: Berne, B.J., Ciccoti, G., Coker, G. (eds.) Classical and Quantum Dynamics in Condensed Matter Simulations, pp. 489–515. World Scientific, Singapore (1998b)

Turchi, P., Ducastelle, F., Treglia, G.: Band gaps and asymptotic behaviour of continued fraction coefficients. J. Phys. C **15**, 2891–2924 (1982)

Ufimtsev, I.S., Martinez, T.J.: Quantum chemistry on graphical pocessing units. 2. Direct self- consistent-field implementation. J. Chem. Theory Comput. **5**, 1004–1015 (2009)

Ullrich, C.A., Gossmann, U.J., Gross, E.K.U.: Time-dependent optimized effective potential. Phys. Rev. Lett. **74**, 872–875 (1995a)

Ullrich, C.A., Gossmann, U.J., Gross, E.K.U.: Density-functional approach to atoms in strong laser pulses. Ber. Bunsenges. Phys. Chem. **99**, 488–497 (1995b)

Ullrich, C.A., Erhard, S., Gross, E.K.U.: Density-functional approach to atoms in srong laser pulses. In: Muller, H.G., Fedorov, M.V. (eds.) Super Intense Laser Atom Physics IV, pp. 267– 284. Dordrecht, Kluwer (1996). Nato ASI 3/13

Ullrich, C.A., Gross, E.K.U.: Many-electron atoms in strong femtosecond laser pulses: A density- functional study. Comments At. Mol. Phys. **33**, 211–236 (1997)

Ullrich, C.A.: Time-dependent Kohn-Sham approach to multiple ionization. J. Mol. Struct. THEOCHEM **501–502**, 315–325 (2000a)

Ullrich, C.A., Reinhard, P.-G., Suraud, E.: Simplified implementation of selfinteraction correction in sodium clusters. Phys. Rev. A **62**, 053202-1–053202-9 (2000b)

Ullrich, C.A., Vignale, G.: Theory of the linewidth of intersubband plasmons in quantum wells. Phys. Rev. Lett. **87**, 037402-1–037402-4 (2001)

Ullrich, C.A., Kohn, W.: Degeneracy in density functional theory: Topology in the v and n spaces. Phys. Rev. Lett. **89**, 156401-1–156401-4 (2002)

Ullrich, C.A., Vignale, G.: Time-dependent current-density-functional theory for the linear response of weakly disordered systems. Phys. Rev. B **65**, 245102-1–245102-19 (2002)

Ullrich, C.A., Burke, K.: Excitation energies from time-dependent density-functional theory beyond the adiabatic approximation. J. Chem. Phys. **121**, 28–35 (2004)

Ullrich, C.A., Tokatly, I.V.: Nonadiabatic electron dynamics in time-dependent density-functional theory. Phys. Rev. B **73**, 235102-1–235102-15 (2006)

Ullrich, C.A.: Time-dependent density-functional theory beyond the adiabatic approximation: Insights from a two-electron model system. J. Chem. Phys. **125**, 234108-1–234108-10 (2006)

Ullrich, C.A., Bandrauk, A.D.: Atoms and molecules in strong laser fields. In: Marques, M.A.L., Ullrich, C.A., Nogueira, F., Rubio, A. (eds.) Time-Dependent Density Functional Theory, pp. 357–375. Springer, Berlin Heidelberg (2006)

Umari, P., Pasquarello, A., Dal Corso, A.: Raman scattering intensities in α-quartz: A first-principles investigation. Phys. Rev. B **63**, 094305-1–094305-9 (2001)

Umrigar, C.J., Gonze, X.: Accurate exchange-correlation potentials and total-energy components for the helium isoelectronic series. Phys. Rev. A **50**, 3827–3837 (1994)

Vahtras, O., Rinkevicius, Z.: General excitations in time-dependent density functional theory. J. Chem. Phys. **126**, 114101-1–114101-11 (2007)

van Caillie, C., Amos, R.E.: Geometric derivatives of Excitation Energies using SCF and DFT. Chem. Phys. Lett. **308**, 249–255 (1999)

van Caillie, C., Amos, R.D.: Geometric derivatives of density functional theory excitation energies using gradient-corrected functionals. Chem. Phys. Lett. **317**, 159–164 (2000)

van Druten, N.J., Constantinescu, R., Schins, J.M., Nieuwenhuize, H., Muller, H.G.: Adiabatic stabilization: Observation of the surviving population. Phys. Rev. A **55**, 622–629 (1997)

van Faassen, M., de Boeij, P.L., van Leeuwen, R., Berger, J.A., Snijders, J.G.: Ultranonlocality in time-dependent current-density-functional theory: Application to conjugated polymers. Phys. Rev. Lett. **88**, 186401-1–186401-4 (2002)

van Faassen, M., de Boeij, P.L., Kootstra, F., Berger, J.A., van Leeuwen, R., Snijders, J.G.: Application of time-dependent current-density-functional theory to nonlocal exchange-correlation effects in polymers. J. Chem. Phys. **118**, 1044–1053 (2003)

van Faassen, M., de Boeij, P.L.: Excitation energies of π-conjugated oligomers within time-dependent current-density-functional theory. J. Chem. Phys. **121**, 10707–10714 (2004)

van Faassen, M., Burke, K.: The quantum defect: The true measure of time-dependent density-functional results for atoms. J. Chem. Phys. **124**, 094102-1–094102-9 (2006a)

van Faassen, M., Burke, K.: A new challenge for time-dependent density-functional theory. Chem. Phys. Lett. **431**, 410–414 (2006b)

van Faassen, M., Wasserman, A., Engel, E., Zhang, F., Burke, K.: Time-dependent density functional calculation of e-H scattering. Phys. Rev. Lett. **99**, 043005-1–043005-4 (2007)

van Faassen, M., Burke, K.: Time-dependent density functional theory of high excitations: To infinity, and beyond. Phys. Chem. Chem. Phys. **11**, 4437–4450 (2009)

van Gisbergen, S.J.A., Snijders, J.G., Baerends, E.J.: A density functional theory study of frequency-dependent polarizablities and van der Waals dispersion coefficients for polyatomic molecules. J. Chem. Phys. **103**, 9347–9354 (1995)

van Gisbergen, S.J.A., Schipper, P.R.T., Gritsenko, O.V., Baerends, E.J., Snijders, J.G., Champagne, B., Kirtman, B.: Electric field dependence of the echange-correlation potential in molecular chains. Phys. Rev. Lett. **83**, 694–697 (1999)

van Kampen, N.G.: Stochastic Processes in Physics and Chemistry, 2nd edn. North- Holland, Amsterdam (1992)

van Kampen, N.G.: Stochastic Processes in Physics and Chemistry, 3rd edn. North Holland, Amsterdam (2007)

van Leeuwen, R., Baerends, E.J.: Exchange-correlation potential with correct asymptotic behavior. Phys. Rev. A **49**, 2421–2431 (1994)

van Leeuwen, R.: Causality and symmetry in time-dependent density-functional theory. Phys. Rev. Lett. **80**, 1280–1283 (1998)

van Leeuwen, R.: Mapping from densities to potentials in time-dependent density-functional theory. Phys. Rev. Lett. **82**, 3863–3866 (1999) van Leeuwen, R.: Key concepts in time-dependent density-functional theory. Int. J. Mod. Phys. B **15**, 1969–2024 (2001)

van Leeuwen, R.: Density functional approach to the many-body problem: Key concepts and exact functionals. Adv. Quant. Chem. **43**, 25–94 (2003)

van Leeuwen, R.: First-principles approach to the electron-phonon interaction. Phys. Rev. B. **69**, 115110-1–115110-20; ibid 199901(E) (2004)

van Lenthe, E., Baerends, E.J.: Optimized slater-type basis sets for the elements 1–118. J. Comput. Chem. **24**, 1142–1156 (2003)

Vanderbilt, D.: Soft self-consistent pseudopotentials in a generalized eigenvalue formalism. Phys. Rev. B **41**, 7892–7895 (1990)

Vanderbilt, D., Resta, R.: Quantum electrostatics of insulators: Polarization, wannier functions, and electric fields. In: Louie, S.G., Cohen, M.L. (eds.) Conceptual Foundations of Materials, pp. 139–164. Elsevier, Amsterdam (2006)

Varsano, D., Espinosa-Leal, L.A., Andrade, X., Marques, M.A.L., di Felice, R., Rubio, A.: Towards a gauge invariant method for molecular chiroptical properties in TDDFT. Phys. Chem. Chem. Phys. **11**, 4481–4489 (2009)

Veithen, M., Gonze, X., Ghosez, P.: Nonlinear optical susceptibilities, Raman efficiencies, and electro-optic tensors from first-principles density functional perturbation theory. Phys. Rev. B **71**, 125107-1–125107-14 (2005)

Véniard, V., Ta, R., Maquet, A.: Atomic clusters submitted to an intense short laser pulse: Adensity-functional approach. Phys. Rev. A **65**, 013202-1–013202-7 (2002)

Véniard, V., Ta, R., Maquet, A.: Photoionization of atoms using time-dependent density functional theory. Laser Phys. **13**, 465–474 (2003)

Verdozzi, C.: Time-Dependent Density-Functional Theory and Strongly Correlated Systems: Insight from Numerical Studies. Phys. Rev. Lett. **101**, 166401-1–166401-4 (2008)

Vignale, G.: Center of mass and relative motion in time dependent density functional theory. Phys. Rev. Lett. **74**, 3233–3236 (1995a)

Vignale, G.: Sum rule for the linear density response of a driven electronic system. Phys. Lett. A **209**, 206–210 (1995b)

Vignale, G., Kohn, W.: Current-dependent exchange-correlation potential for dynamical linear response theory. Phys. Rev. Lett. **77**, 2037–2040 (1996)

Vignale, G., Ullrich, C.A., Conti, S.: Time-dependent density functional theory beyond the adiabatic local density approximation. Phys. Rev. Lett. **79**, 4878–4881 (1997)

Vignale, G., Kohn, W.: Current-density functional theory of linear response to time-dependent electromagnetic fields. In: Dobson, J.F., Vignale, G., Das, M.P. (eds.) Electronic Density Functional Theory: Re-cent Progress and New Directions, pp. 199–216. Plenum Press, New York (1998)

Vignale, G.: Mapping from current densities to vector potentials in time-dependent current density functional theory. Phys. Rev. B **70**, 201102-1–201102-4(R) (2004)

Vignale, G.: Real-time resolution of the causality paradox of time-dependent density-functional theory. Phys. Rev. A **77**, 062511-1–062511-9 (2008)

Vignale, G., Di Ventra, M.: Incompleteness of the Landauer formula for electronic transport. Phys. Rev. B **79**, 014201-1–014201-10 (2009)

Vila, F.D., Strubbe, D.A., Takimoto, Y., Andrade, X., Rubio, A., Louie, S.G., Rehr, J.J.: Basis set effects on the hyperpolarizability of CHCl3: Gaussian-type orbitals, numerical basis sets and real-space grids. J. Chem. Phys. **133**, 034111-1–034111-10 (2010)

Villars, F.M.H., Cooper, G.: Unified theory of nuclear rotations. Ann. Phys. **56**, 224–258 (1970)

Villeneuve, D., Ivanov, M.Yu., Corkum, P.B.: Enhanced ionization of diatomic molecules in strong laser fields: A classical model. Phys. Rev. A **54** , 736–741 (1996)

Vogt, L., Olivares-Amaya, R., Kermes, S., Shao, Y., Amador-Bedolla, C., Aspuru-Guzik, A.: Accelerating resolution-of-the-identity second-order Moller-Plesset quantum chemistry calculations with graphical processing units. J. Phys. Chem. A **112**, 2049–2057 (2008) von Barth, U.: Basic density-functional theory — an overview. Physica Scripta **T109**, 9–39 (2004)

von Barth, U., Dahlen, N.E., van Leeuwen, R., Stefanucci, G.: Conserving approximations in time-dependent density functional theory. Phys. Rev. B **72**, 235109-1–235109-10 (2005)

von Friesen, M.P., Verdozzi, C., Almbladh, C.-O.: Kadanoff-Baym dynamics of hubbard clusters: Performance of many-body schemes, correlation-induced damping and multiple steady and quasi-steady states. Phys. Rev. B **82**, 155108-1–155108-19 (2010)

von Lilienfeld, O.A., Tavernelli, I., Rothlisberger, U., Sebastiani, D.: Optimization of effective atom centered potentials for London dispersion forces in density functional theory. Phys. Rev. Lett. **93**, 153004-1–153004-4 (2004)

Vosko, S.H., Wilk, L., Nusair, M.: Accurate spin-dependent electron-liquid correlation energies for local spin density calculations: A critical analysis. Can. J. Phys. **58**, 1200–1211 (1980)

Vydrov, O.A., Scuseria, G.E.: Assessment of a long-range corrected hybrid functional. J. Chem. Phys. **125**, 234109-1–234109-9 (2006)

Vydrov, O.A., Heyd, J., Krukau, A.V., Scuseria, G.E.: Importance of short-range versus long-range Hartree–Fock exchange for the performance of hybrid density functionals. J. Chem. Phys. **125**, 074106-1–074106-9 (2006)

Vydrov, O.A., Wu, Q., Van Voorhis, T.: Self-consistent implementation of a nonlocal van der Waals density functional with a Gaussian basis set. J. Chem. Phys. **129**, 014106-1–014106-8 (2008)

Vydrov, O.A., Van Voorhis, T.: Improving the accuracy of the nonlocal van der Waals density functional with minimal empiricism. J. Chem. Phys. **130**, 104105-1–104105-7 (2009a)

Vydrov, O.A., Van Voorhis, T.: Nonlocal van der Waals density functional made simple. Phys. Rev. Lett. **103**, 063004-1–063004-4 (2009b)

Vydrov, O.A., Van Voorhis, T.: Implementation and assessment of a simple nonlocal van der Waals density functional. J. Chem. Phys. **132**, 164113-1–164113-6 (2010a)

Vydrov, O.A., Van Voorhis, T.: Dispersion interactions from a local polarizability model. Phys. Rev. A **81**, 062708-1–062708-4 (2010b)

Vydrov, O.A., Van Voorhis, T.: Nonlocal van der Waals density functional: The simpler the better. J. Chem. Phys. **133**, 244103-1–244103-9 (2010c)

Wacker, O.-J., Kümmel, R., Gross, E.K.U.: Time-dependent density-functional theory for superconductors. Phys. Rev. Lett. **73**, 2914–2918 (1994)

Wahlström, C.-G., Larsson, J., Persson, A., Starczewski, T., Svanberg, S., Salières, P., Balcou, P., L'Huillier, A.: High-order harmonic generation in rare gases with an intense short-pulse laser. Phys. Rev. A **48**, 4709–4720 (1993)

Walker, B., Sheehy, B., DiMauro, L.F., Agostini, P., Schafer, K.J., Kulander, K.C.: Precision measurement of strong field double ionization of helium. Phys. Rev. Lett. **73**, 1227–1230 (1994)

Walker, B., Saitta, A.M., Gebauer, R., Baroni, S.: Efficient approach to time-dependent density-functional perturbation theory for optical spectroscopy. Phys. Rev. Lett. **96**, 113001-1–113001-4 (2006)

Walker, B., Gebauer, R.: Ultrasoft pseudopotentials in time-dependent density-functional theory. J. Chem. Phys. **127**, 164106-1–164106-9 (2007)

Wallace, D.C.: Thermodynamics of Crystals. Wiley, New York and London (1972)

Walsh, T.R.: Exact exchange and Wilson & Levy correlation: A pragmatic device for studying complex weakly-bonded systems. Phys. Chem. Chem. Phys. **7**, 443–451 (2005)

Walter, M., Häkkinen, H., Lehtovaara, L., Puska, M., Enkovaara, J., Rostgaard, C., Mortensen, J.J.: Time-dependent density-functional theory in the projector augmentedwave method. J. Chem. Phys. **128**, 244101-1–244101-10 (2008)

Wang, F., Ziegler, T.: Time-dependent density functional theory based on a noncollinear formulation of the exchange-correlation potential. J. Chem. Phys. **121**, 12191–12196 (2004)

Wang, F., Ziegler, T.: The performance of time-dependent density functional theory based on a noncollinear exchange-correlation potential in the calculation of excitation energies. J. Chem. Phys. **122**, 074109-1–074109-9 (2005)

Wang, F., Ziegler, T., van Lenthe, E., van Gisbergen, S., Baerends, E.J.: The calculation of excitation energies based on the relativistic two-component zeroth-order regular approximation and time-dependent density-functional with full use of symmetry. J. Chem. Phys. **122**, 204103-1–204103-12 (2005)

Wang, X., Spataru, C.D., Hybertsen, M.S., Millis, A.J.: Electronic correlation in nanoscale junctions: Comparison of the GW approximation to a numerically exact solution of the single-impurity Anderson model. Phys. Rev. B **77**, 045119-1–045119-10 (2008)

Wasserman, A., Maitra, N.T., Burke, K.: Accurate Rydberg Excitations from the Local Density Approximation. Phys. Rev. Lett. **91**, 263001-1–263001-4 (2003)

Wasserman, A.: Scattering states from time-dependent density functional theory, Ph.D. thesis, Rutgers University (2005a)

Wasserman, A., Maitra, N.T., Burke, K.: Continuum states from time-dependent density functional theory. J. Chem. Phys. **122**, 144103-1–144103-5 (2005)

Wasserman, A., Burke, K.: Rydberg transition frequencies from the local density approximation. Phys. Rev. Lett. **95**, 163006-1–163006-4 (2005)

Watson, M., Olivares-Amaya, R., Edgar, R.G., Aspuru-Guzik, A.: Accelerating correlated quantum chemistry calculations using graphical processing units. Comput. Sci. Eng. **12**, 40–51 (2010)

Weber, T., Weckenbrock, M., Staudte, A., Spielberger, L., Jagutzki, O., Mergel, V., Afaneh, F., Urbasch, G., Vollmer, M., Giessen, H., Dörner, R.: Recoil-ion momentum distributions for single and double ionization of helium in strong laser fields. Phys. Rev. Lett. **84**, 443–446 (2000)

Wehrum, R.P., Hermeking, H.: On the response of arbitrary finite order and its relation to imaginary-time. J. Phys. C **7**, L107–L110 (1974)

Weigend, F., Furche, F., Ahlrichs, R.: Gaussian basis sets of quadruple zeta quality for atoms H–Kr. J. Chem. Phys. **119**, 12753–12762 (2003)

Weigend, F., Ahlrichs, R.: Balanced basis sets of split valence, triple zeta valence and quadruple zeta valence quality for H to Rn: Design and assessment of accuracy. Phys. Chem. Chem. Phys. **7**, 3297–3305 (2005)

Weigend, F.: Accurate Coulomb-fitting basis sets for H to Rn. Phys. Chem. Chem. Phys. **8**, 1057–1065 (2006)

Weiner, A.M.: Femtosecond pulse shaping using spatial light modulators. Rev. Sci. Inst. **71**, 1929–1960 (2000)

Weiss, H., Ahlrichs, R., Häser, M.: A direct algorithm for self-consistent-field linear response theory and application to C_{60}: Excitation energies, oscillator strengths, and frequency-dependent polarizabilities. J. Chem. Phys. **99**, 1262–1270 (1993)

Weiss, U.: Quantum Dissipative Systems. World Scientific, Singapore (2007)

Weissker, H.-C., Serrano, J., Huotari, S., Bruneval, F., Sottile, F., Monaco, G., Krisch, M., Olevano, V., Reining, L.: Signatures of short-range many-body effects in the dielectric function of silicon for finite momentum transfers. Phys. Rev. Lett. **97**, 237602-1–237602-4 (2006)

Wentzel, G.: Eine verallgemeinerung der quantenbedingungen fr die zwecke der wellenmechanik. Z. Phys. **38**, 518–529 (1926)

Werner, U., Mitrić, R., Suzuki, T., Bonačić-Koutecký, V.: Nonadiabatic dynamics within the time dependent density functional theory: Ultrafast photodynamics in pyrazine. Chem. Phys. **349**, 319–324 (2008)

Werner, P., Oka, T., Eckstein, M., Millis, A.J.: Weak-coupling quantum Monte Carlo calculations on the Keldysh contour: Theory and application to the current-voltage characteristics of the Anderson model. Phys. Rev. B **81**, 035108-1–035108-11 (2010)

Wernet, Ph., Odelius, M., Godehusen, K., Gaudin, J., Schwarzkopf, O., Eberhardt, W.: Real-time evolution of the valence electronic structure in a dissociating molecule. Phys. Rev. Lett. **103**, 013001-1–013001-4 (2009)

Werschnik, J.: Quantum Optimal Control Theory: Filter Techniques, Time Dependent Targets, and Time Dependent Density Functional Theory. Cuvillier Verlag, Göttingen (2006)

Werschnik, J., Gross, E.K.U.: Quantum optimal control theory. J. Phys. B At. Mol. Opt. Phys. **40**, R175–R211 (2007)

Wesolowski, T.A., Warshel, A.: Frozen density-functional approach for ab-initio calculations of solvated molecules. J. Phys. Chem. **97**, 8050–8053 (1993)

White, A., Dobson, J.F.: Enhanced van der Waals interaction between quasi-one dimensional conducting collinear structures. Phys. Rev. B **77**, 075436-1–075436-9 (2008)

Wiberg, K.B., de Oliveira, A.E., Trucks, G.: A comparison of the electronic Transition energies for ethene, isobutene, formaldehyde, and acetone calculated using RPA, TDDFT, and EOM-CCSD. Effect of basis sets. J. Phys. Chem. A **106**, 4192–4199 (2002)

Wickenhauser, M., Burgdörfer, J., Krausz, F., Drescher, M.: Time resolved Fano resonances. Phys. Rev. Lett. **94**, 023002-1–023002-4 (2005)

Wigner, E.: On the quantum correction for thermodynamic equilibrium. Phys. Rev. **40**, 749–759 (1932)

Wijewardane, H.O., Ullrich, C.A.: Real-time electron dynamics with exact- exchange time-dependent density-functional theory. Phys. Rev. Lett. **100**, 056404-1–056404-4 (2008)

Wilken, F., Bauer, D.: Adiabatic approximation of the correlation function in the density-functional treatment of ionization processes. Phys. Rev. Lett. **97**, 203001-1–203001-4 (2006)

Wilken, F., Bauer, D.: Momentum distributions in time-dependent density-functional theory: Product-phase approximation for nonsequential double ionization in strong laser fields. Phys. Rev. A **76**, 023409-1–023409-8 (2007)

Willetts, A., Rice, J.E., Burland, D.M., Shelton, D.P.: Problems in the comparison of theoretical and experimental hyperpolarizabilities. J. Chem. Phys. **97**, 7590–7599 (1992)

Wilson, L.C., Levy, M.: Nonlocal Wigner-like correlation-energy density functional through coordinate scaling. Phys. Rev. B **41**, 12930–12932 (1990)

Wilson, L.C., Ivanov, S.: A correlation-energy functional for addition to the Hartree-Fock energy. In: Dobson, J.F., Vignale, G., Das, M.P. (eds.) Electronic Density Functional Theory: Recent Progress and New Directions, p. 133. Plenum, New York (1998)

Wilson, A.K., Woon, D.E., Peterson, K.A., Dunning Jr., T.H.: Gaussian basis sets for use in correlated molecular calculations. IX. The atoms gallium through krypton. J. Chem. Phys. **110**, 7667–7676 (1999)

Wiser, N.: Dielectric constant with local field effects included. Phys. Rev. **129**, 62–69 (1963)

Wittig, C.: The Landau-Zener formula. J. Phys. Chem. B **109**, 8428–8430 (2005)

Wloka, J.: Funktionalanalysis und Anwendungen. de Gruyter, Berlin (1971)

Wong, B.M., Piacenza, M., Della Sala, F.: Absorption and fluorescence properties of oligothiophene biomarkers from long-range-corrected time-dependent density functional theory. Phys. Chem. Chem. Phys. **11**, 4498–4508 (2009)

Wong, B.M., Hsieh, T.H.: Optoelectronic and excitonic properties of oligoacenes: Substantial improvements from range-separated time-dependent density functional theory. J. Chem. Theory Comput. **6**, 3704–3712 (2010)

Woon, D.E., Dunning Jr., T.H.: Gaussian basis sets for use in correlated molecular calculations. III. The atoms aluminium through argon. J. Chem. Phys. **98**, 1358–1371 (1993)

Woon, D.E., Dunning Jr., T.H.: Gaussian basis sets for use in correlated molecular calculations. IV. Calculation of static electrical response properties. J. Chem. Phys. **100**, 2975–2988 (1994)

Wörner, H.J., Bertrand, J.B., Corkum, P.B., Villeneuve, D.M.: High-harmonic homodyne detection of the ultrafast dissociation of Br_2 molecules. Phys. Rev. Lett. **105**, 103002-1–103002-4 (2010)

Wu, Q., Yang, W.: Empirical correction to density functional theory for van der Waals interactions. J. Chem. Phys. **116**, 515–524 (2002)

Wu, Q., Cheng, C.L., Van Voorhis, T.: Configuration interaction based on constrained density functional theory: A multireference method. J. Chem. Phys. **127**, 164119-1–164119-9 (2007)

Xu, B.-X., Rajagopal, A.K.: Current-density-functional theory for time-dependent systems. Phys. Rev. A **31**, 2682–2684 (1985)

Yabana, K., Bertsch, G.F.: Time-dependent local-density approximation in real time. Phys. Rev. B **54**, 4484–4487 (1996)

Yabana, K., Tazawa, T., Abe, Y., Bozek, P.: Time-dependent mean-field description for multiple electron transfer in slow ion-cluster collisions. Phys. Rev. A **57**, R3165–R3168 (1998)

Yabana, K., Bertsch, G.F.: Application of the time-dependent local density approximation to optical activity. Phys. Rev. A **60**, 1271–1279 (1999)

Yamakawa, K., Akahane, Y., Fukuda, Y., Aoyama, M., Inoue, N., Ueda, H., Utsumi, T.: Many-electron dynamics of a Xe atom in strong and superstrong laser fields. Phys. Rev. Lett. **92**, 123001-1–123001-4 (2004)

Yanai, T., Tew, D.P., Handy, N.C.: A new hybrid exchange-correlation functional using the coulomb-attenuating method (CAM-B3LYP). Chem. Phys. Lett. **393**, 1–3 (2004)

Yang, W.: Dynamic linear response of many-electron systems: An integral formulation of density-functional theory. Phys. Rev. A **38**, 5512–5519 (1988)

Yang, J., Wang, Y., Chen, Y.: GPU accelerated molecular dynamics simulation of thermal conductivities. J. Comput. Phys. **221**, 799–804 (2007)

Yang, Z., van Faassen, M., Burke, K.: Must Kohn–Sham oscillator strengths be accurate at threshold? J. Chem. Phys. **131**, 114308-1–114308-7 (2009)

Yarkony, D.R.: Diabolical conical intersections. Rev. Mod. Phys. **68**, 985–1013 (1996)

Yarkony, D.R.: Conical intersections: The new conventional wisdom. J. Phys. Chem. A **105**, 6277–6293 (2001)

Yasuda, K.: Accelerating density functional calculations with graphics processing unit. J. Chem. Theory Comput. **4**, 1230–1236 (2008)

Yu, H., Bandrauk, A.D.: Three-dimensional Cartesian finite element method for the time dependent Schrödinger equation of molecules in laser fields. J. Chem. Phys. **102**, 1257–1265 (1995)

Yu, P.Y., Cardona, M.: Fundamentals of Semiconductors, 2nd edn. Springer, Berlin (1999)

Yudin, G. L., Ivanov, M. Yu.: Physics of correlated double ionization of atoms in intense laser fields: Quasistatic tunneling limit. Phys. Rev. A 63:033404-1–14 (2001)

Yuen-Zhou, J., Rodríguez-Rosario, C., Aspuru-Guzik, A.: Time-dependent current-density functional theory for generalized open quantum systems. Phys. Chem. Chem. Phys. **11**, 4509–4522 (2009)

Yuen-Zhou, J., Tempel, D.G., Rodríguez-Rosario, C., Aspuru-Guzik, A.: Time-dependent density-functional theory for open quantum systems with unitary propagation. Phys. Rev. Lett. **104**, 043001-1–043001-4 (2010)

Zacharia, R., Ulbricht, H., Hertel, T.: Interlayer cohesive energy of graphite from thermal desorption of polyaromatic hydrocarbons. Phys. Rev. B **69**, 155406-1–155406-7 (2004)

Zanghellini, J., Kitzler, M., Fabian, C., Brabec, T., Scrinzi, A.: A MCTDHF approach to multi-electron dynamics in laser fields. Laser Phys. **13**, 1064–1068 (2003)

Zanghellini, J., Kitzler, M., Brabec, T., Scrinzi, A.: Testing the multi-configuration time-dependent Hartree-Fock method. J. Phys. B At. Mol. Opt. Phys. **37**, 763–773 (2004)

Zanghellini, J., Kitzler, M., Zhang, Z., Brabec, T.: Multi-electron dynamics in strong laser fields. J. Mod. Opt. **52**, 479–488 (2005)

Zangwill, A., Soven, P.: Density-functional approach to local-field effects in finite systems: Photoabsorption in the rare gases. Phys. Rev. A **21**, 1561–1572 (1980a)

Zangwill, A., Soven, P.: Resonant photoemission in Barium and Cerium. Phys. Rev. Lett. **45**, 204–207 (1980b)

Zangwill, A., Soven, P.: Resonant two-electron excitation in copper. Phys. Rev. B **24**, 4121–4127 (1981)

Zangwill, A.: Physics at Surfaces. Cambridge University Press, Cambridge (1988)

Zaremba, E., Kohn, W.: van der Waals interaction between an atom and a solid surface. Phys. Rev. B **13**, 2270–2285 (1976)

Zener, C.: Non-adiabatic crossing of energy levels. Proc. R. Soc. London A **137**, 696–702 (1932)

Zhang, Y., Pan, W., Yang, W.: Describing the van der Waals interaction in diatomic molecules with generalized gradient approximations: The role of the exchange functional. J. Chem. Phys. **107**, 7921–7925 (1997)

Zhang, Y., Yang, W.: Comment on generalized gradient approximation made simple. Phys. Rev. Lett. **80**, 890 (1998)

Zhang, F., Burke, K.: Adiabatic connection for near degenerate excited states. Phys. Rev. A **69**, 052510-1–052510-6 (2004)

Zhao, Y., Truhlar, D.G.: Density functional for spectroscopy: No long-range selfinteraction error,good performance for Rydberg and charge-transfer states, and better performance on average than B3LYP for ground states. J. Phys. Chem. A. **110**, 13126–13130 (2006)

Zheng, X., Wang, F., Yam, C.Y., Mo, Y., Chen, G.-H.: Time-dependent density functional theory for open systems. Phys. Rev. B **75**, 195127-1–195127-16 (2007)

Zheng, X., Chen, G., Mo, Y., Koo, S., Tian, H., Yam, C., Yan, Y.: Time-dependent density functional theory for quantum transport. J. Chem. Phys. **133**, 114101-1–114101-11 (2010)

Zhou, J., Peatross, J., Murnane, M.M., Kapteyn, H.C.: Enhanced high-harmonic generation using 25 fs laser pulses. Phys. Rev. Lett. **76**, 752–755 (1996)

Zhu, W., Botina, J., Rabitz, H.: Rapidly convergent iteration methods for quantum optimal control of population. J. Chem. Phys. **108**, 1953–1963 (1998)

Zhu, C., Jasper, A.W., Truhlar, D.G.: Non-born-Oppenheimer Liouville-von Neumann dynamics. Evolution of a subsystem controlled by linear and population-driven decay of mixing with decoherent and coherent sswitching. J. Chem. Theor. Comp. **1**, 527–540 (2005)

Zhu, W., Toulouse, J., Savin, A., Ángyán, J.G.: Range-separated density-functional theory with random phase approximation applied to noncovalent intermolecular interactions. J. Chem. Phys. **132**, 244108-1–244108-9 (2010)

Ziegler, T., Rauk, A., Baerends, E.J.: Calculation of multiplet energies by Hartree-Fock-Slater method. Theor. Chim. Acta **43**, 261–271 (1977)

Zumbach, G., Maschke, K.: New approach to the calculation of density functionals. Phys. Rev. A **28**, 544–554 (1983)

Zumbach, G., Maschke, K.: New approach to the calculation of density functionals. Phys. Rev. A **29**, 1585–1587(E) (1984)

Zuo, T., Bandrauk, A.D.: Charge-resonance-enhanced ionization of diatomic molecular ions by intense lasers. Phys. Rev. A **52**, R2511–R2514 (1995)

Zuo, T., Bandrauk, A.D., Corkum, P.B.: Laser-induced electron diffraction: A new tool for probing ultrafast molecular dynamics. Chem. Phys. Lett. **259**, 313–320 (1996)

Zuo, T., Bandrauk, A. D.: Phase control of molecular ionization: H_2^+ and H_3^{2+} in intense two-color laser fields. Phys. Rev. A **54**, 3254–3260 (1996)

Zwanzig, R.: Ensemble method in the theory of irreversibility. J. Chem. Phys. **33**, 1338–1342 (1960)

Zwanzig, R.: Nonequilibrium Statistical Mechanics. Oxford University Press, Oxford (2001)

Index

M. A. L. Marques et al. (eds.), *Fundamentals of Time-Dependent Density Functional Theory*, Lecture Notes in Physics 837, DOI: 10.1007/978-3-642-23518-4,
© Springer-Verlag Berlin Heidelberg 2012